PLEASURES OF THE BRAIN

PLEASURES OF THE BRAIN

Edited by

Morten L. Kringelbach

and

Kent C. Berridge

OXFORD
UNIVERSITY PRESS

2010

OXFORD
UNIVERSITY PRESS

Oxford University Press, Inc., publishes works that further
Oxford University's objective of excellence
in research, scholarship, and education.

Oxford New York
Auckland Cape Town Dar es Salaam Hong Kong Karachi
Kuala Lumpur Madrid Melbourne Mexico City Nairobi
New Delhi Shanghai Taipei Toronto

With offices in
Argentina Austria Brazil Chile Czech Republic France Greece
Guatemala Hungary Italy Japan Poland Portugal Singapore
South Korea Switzerland Thailand Turkey Ukraine Vietnam

Published by Oxford University Press, Inc.
198 Madison Avenue, New York, New York 10016

www.oup.com

Library of Congress Cataloging-in-Publication Data
Pleasures of the brain / edited by Morten L. Kringelbach and Kent C.
Berridge.
p. cm.
Includes bibliographical references and index.
ISBN 978-0-19-533102-8
1. Pleasure. 2. Reward (Psychology) I. Kringelbach, Morten L.
II. Berridge, Kent C.
BF515.P55 2009
152.4'2—dc22
2009004010

9 8 7 6 5 4 3

Printed in the United States of America
on acid-free paper

Contents

Part II. Human Pleasures

Part III. Clinical Applications

Contributors

J. Wayne Aldridge
Department of Psychology
University of Michigan

Tipu Z. Aziz
Department of Neurosurgery
University of Oxford

Bernard Balleine
Department of Psychology
University of California, Los Angeles

Carlos Beyer
Department of Psychology
Rutgers University

Kent C. Berridge
Department of Psychology
University of Michigan

Kathryn A. Burke
Program in Neuroscience
University of Maryland, Baltimore

Michel Cabanac
Faculty of Medicine
Laval University

Anthony Dickinson
Department of Experimental Psychology
University of Cambridge

Theresa Franz
Department of Anatomy and Neurobiology
University of Maryland, Baltimore

Nico H. Frijda
Department of Social and Behavioral Sciences
Amsterdam University

Janniko R. Georgiadis
University Medical Center Groningen
University of Groningen

Jay A. Gottfried
Department of Neurology
Northwestern University

Alexander L. Green
Department of Neurosurgery
University of Oxford

Barry R. Komisaruk
Department of Psychology
Rutgers University

Rudie Kortekaas
University Medical Center Groningen
University of Groningen

Morten L. Kringelbach
Department of Psychiatry
University of Oxford
Center of Functionally Integrative Neuroscience
Aarhus University

Siri Leknes
Department of Psychology
University of Oslo and
Institute of Neuroscience and Physiology
University of Gothenburg

Marco Leyton
Department of Psychiatry
McGill University

Stephen V. Mahler
Department of Psychology
University of Michigan

Iris B. Mauss
Department of Psychology
University of Denver

Danielle Miller
Baltimore Polytechnic Institute

Susana Peciña
Department of Psychology
University of Michigan

Erlick A. C. Pereira
Department of Neurosurgery
University of Oxford

Predrag Petrovic
Karolinska Institute

Michael L. Platt
Department of Neurobiology
Duke University

Kristin J. Rudenga
The John B. Pierce Laboratory
Interdepartmental Neuroscience Program, Yale University School of Medicine

Geoffrey Schoenbaum
Department of Anatomy and Neurobiology
University of Maryland Baltimore County, Baltimore

Jonathan W. Schooler
Department of Psychology
University of California, Santa Barbara

Stephen V. Shepherd
Neuroscience Institute
Duke University

Peter Shizgal
Department of Psychology
Concordia University

Martin Skov
University of Copenhagen

Dana M. Small
The John B. Pierce Laboratory
Department of Psychiatry, Yale University School of Medicine
Interdepartmental Neuroscience Program, Yale University School of Medicine
Department of Psychology, Yale University

Kyle S. Smith
Department of Psychology
University of Michigan

Irene Tracey
Department of Clinical Neurology
University of Oxford

Maria G. Veldhuizen
The John B. Pierce Laboratory
Department of Psychiatry, Yale University School of Medicine

Peter Vuust
Center of Functionally Integrative Neuroscience
Aarhus University

Karli K. Watson
Department of Neurobiology
Duke University

Beverly Whipple
Department of Psychology
Rutgers University

PLEASURES OF THE BRAIN

Introduction: The Many Faces of Pleasure

MORTEN L. KRINGELBACH AND KENT C. BERRIDGE

The American writer John Steinbeck wrote of "the tragic miracle of consciousness" and how our "species is not set, has not jelled, but is still in a state of becoming" (Steinbeck and Ricketts, 1941). He wrote about how consciousness offers us pleasures, desires, and the freedom of choice, but how this freedom is always accompanied by the certainty of the end. The negative side of this sentiment was emphasized by the French philosopher Jean-Paul Sartre who memorably wrote that "hell is other people" (Sartre, 1947).

Life may ultimately meet a tragic end, but the pleasure along the way is what makes it worthwhile. Pleasure is central to our sense of well-being. The very survival of every large-brained creature as an individual and the evolutionary survival of each species have depended on the pleasures afforded by its hedonic neural systems. We are rewarded by food, sex, and many other sensory and abstract incentives, and as members of a very social species, we also take great pleasure in the company of other people.

A better understanding of the pleasures of the brain might thus offer us fundamental insights into our own nature, into how brains work in daily life, and even into better ways to enhance our quality of life. Pleasures are of many sorts and occur in many different brains. The purpose of this book is set them together in one place, and as far as possible come to an understanding of how diverse pleasures arise from neural systems. While some of this pleasure is clearly consciously experienced, there are also nonconscious components, as convincingly shown by the some of the chapters in this book.

We were encouraged to begin this attempt to capture pleasure in a scientific net by the enormous progress in recent years of affective neuroscience as an important and exciting discipline (LeDoux, 1996; Panksepp, 1999). Through the studies of animals as well as humans, many important insights have been made regarding the brain mechanisms of pleasure, and related motivation and emotion.

It has become increasingly apparent that pleasure and reward are at the heart of affective neuroscience and the psychology of well-being (Berridge, 2003; Berridge and Kringelbach, 2008; Kahneman, 1999; Kringelbach, 2005; Leknes and Tracey, 2008). Pleasure is essential to a normal healthy life. The loss of pleasure, anhedonia, is a common theme in many mental illnesses such as depression, schizophrenia, and addiction, and any progress in understanding the functional neuroanatomy of pleasure thus holds the promise of better treatments.

At the same time, pleasure has sometimes been seen in psychology and neuroscience as perhaps a bit too subjective to be studied scientifically. But pleasure exists as a natural phenomenon, and we believe that what exists can be studied scientifically. While it is certainly true that pleasure is linked with our most subjective states of consciousness, at the same time, it is equally true that pleasure is a multifaceted psychological phenomenon with many constituent nonconscious components. A large part of the failure to make progress in understanding the psychological and neural properties of pleasure may have simply been the reluctance of the scientific community to devote attention and effort to the task. This book is a beginning to redress this omission.

A multifaceted view of pleasure (and of emotion in general) can be helpful in studying pleasure in people and certainly in other animals—and crucially without having to determine whether consciousness is present in these animals (Kringelbach, 2004). As shown in this book, many highly successful experimental paradigms have been developed, which have subsequently given us new insights in the nature and mechanisms of pleasure.

In this book, we have asked many experts to present the state-of-the-art of their neuroscientific research into pleasure and reward. Ground-breaking developments have occurred on several fronts, and recently, there has been a convergence of interesting new data on pleasure coming from many disparate fields. The time seems ripe to present these important findings in a single volume. We hope this book will come to serve both as a starting point and as a reference volume to graduate students and scientists who are fresh to the world as well as to scientists coming from other related and unrelated fields.

The Chapters of this Book

The many faces of pleasure and reward raise interesting questions. We believe that it can be a strength rather than a weakness to have disagreements about certain fundamental concepts, as is the case with many emerging fields, in order to eventually develop the best concepts. To reflect the many different views, we have therefore opened the book with a special section designed extract, distill, and contrast alternative views on fundamentals. We invited the authors of the book to provide us with their answers to a number of common "fundamental questions" regarding the role of pleasure in the brain. It was optional for the authors, and some contributed to the section while others did not.

Contributing authors were encouraged to provide answers to only the questions they felt most passionate about. In other words, the "fundamental questions" section is an opportunity to see at a glance what various authors think are the bedrock conceptual foundations and guiding principles for their scientific studies of pleasure. We hope that this question section will be of great interest to readers on its own.

The rest of the chapters of the book are divided into three sections: animal, human, and clinical applications. This organization is merely for convenience; many issues span the sections and alternative groupings could easily be imagined.

Animal Pleasures

In the opening chapter of the first section, Smith, Mahler, Peciña, and Berridge offer a overview of some affective neuroscience research on finding hedonic hotspots in the rodent brain. The authors show how activity in cubic-millimeters of certain brain areas such as the nucleus accumbens and ventral pallidum can be manipulated to change the generation of pleasure 'liking'. They also discuss some aspects of the distinction 'liking' and 'wanting' of the same pleasure and show how dopamine is clearly more linked to the latter rather than the former.

In the next chapter, Burke, Miller, and Schoenbaum investigate the role of specialized corticolimbic circuits linked to pleasure in rats. These corticolimbic circuits connect together limbic forebrain structures to mediate conditioned reinforcement. They focus especially on three important brain regions: the basolateral amygdala, the orbitofrontal cortex, and the nucleus accumbens, and use devaluation paradigms and selective lesions to study how these three brain regions interact in a coherent circuit.

The chapter by Aldridge and Berridge focuses in more detail on the nature of the neural coding for pleasure in the hedonic hotspot of the ventral pallidum. Interestingly, neurons here code the hedonic impact of a pleasant taste and lesions to this brain region can abolish 'liking' reactions completely. The authors propose that neuronal events in this brain region may play a central role in applying the pleasure gloss to stimuli that makes them rewarding.

Dickinson and Balleine offer an overview and a hedonic interface between pleasure and cognition in their chapter. They show how the function of hedonic and affective experience may be to act as a goal interface between cognitive and motivational systems, interface that is required because these systems use incommensurate psychologies embedded in their somewhat separable neural systems. Aspects of this theory have, in their own words, many "similarities to of the Freudian process of *cathexis*," and reveal a remarkable subtlety in the psychology of pleasure that is shared by humans and other animals.

The animal section concludes with a chapter by Watson, Shepherd, and Platt who investigate the neuroethology of pleasure in nonhuman primates. In particular, the authors show how neuroeconomics and neuroethology can come together to inform the research in pleasure and reward.

Human Pleasures

The second section begins with Frijda's thoughtful chapter on the nature and function of pleasure in daily human life. The chapter is a thorough investigation of the psychology of pleasure. The following chapter by Cabanac provides an overview of the physiological and philosophical investigations of pleasure by the chief originator of the scientific study of "alliesthesia" that has played such an important role in studies of the affective neuroscience of pleasure. Cabanac takes an evolutionary approach to pleasure and discusses links among primary sensory and social pleasures linked to survival and procreation.

The sensory pleasures of food, taste, and smell and their brain bases are the subject of the next two chapters in the human section. The chapter by Gottfried provides an authoritative overview of the human olfactory system. In particular, the author presents recent neuroimaging data on olfaction, which have confirmed that smells are intimately linked to hedonics, pleasure, and emotion. Similarly, the following chapter by Veldhuizen, Rudenga, and Small provides an overview of the human taste system, and in particular describes important new neuroimaging data that help reveal human brain bases of flavor pleasures and show the close links between taste and smell in food hedonics.

Sexual pleasures are also a prominent sensory hedonic experience that has been shaped by evolutionary selection pressures on brain systems, and the book has two chapters devoted to our current understanding of this all-too often taboo subject. In one chapter, Komisaruk, Whipple, and Beyer investigate how sex is good for our health and describe studies in particular of the neural systems and neurotransmitters involved in sexual excitement and in orgasm. Next, Georgiadis and Kortekaas review in their chapter an array of important functional neuroimaging studies to bring together what is known about brain mechanisms of sexual pleasure in people and describe neuropsychological and pharmacoendocrinological anomalies that affect human sexuality.

Both food sensory pleasure and sexual pleasure are compared and linked to research on the social pleasures in the chapter on fundamental pleasure systems by Kringelbach, which proposes a general theory for the mechanisms and functional neuroanatomy of pleasure. Kringelbach gives a special analysis of the role of orbitofrontal cortex in human hedonic reactions, a prefrontal region in cortex that has sometimes been viewed as the apex of pleasure processes in the brain.

Dopamine has long been a favorite topic for neuroscientists interested in pleasure and reward, but it has only recently become possible to link pharmacological and neuroimaging studies together in penetrating experimental designs that reveal whether dopamine actually produces pleasure in humans. Leyton's work leads the way in these efforts, and his chapter links the animal research on dopamine with his exciting new neuroimaging research in people to show how dopamine is implicated in the regulation of mood and motivational states in humans but perhaps not pleasure per se. In particular, the chapter shows how dopamine strongly influences sustained interest and approach, weakly influences positive emotions, yet elegantly shows that dopamine affects human pleasure ratings only tenuously, if at all.

Higher pleasures such as monetary, artistic, musical, altruistic, and transcendent pleasures can perhaps be studied only in people, and recent neuroimaging studies have made some headway in exploring these important human pleasures. The chapter by Vuust and Kringelbach explores the pleasures evoked by music. It traces what is known about brain activity patterns during musical enjoyment and shows how much remains to be discovered about this powerful, and perhaps unique human, positive reward. In a related analysis of human art, the emerging field of neuroesthetics is described in the chapter by Skov. The author in a sense links artistic and social pleasures in a proposal that creating art always involve a desire to affect some hedonic impact in an observer.

Finally, many of the strands of what humans know about their own pleasure are pulled together in the chapter by Schooler and Mauss. The authors describe the psychological research on the experience and meta-awareness of pleasure. They show how many of our most pleasurable experiences occur with little meta-awareness of the fact that we are experiencing pleasure and how conscious attention to pleasure can distort or even destroy the underlying hedonic process.

Clinical Applications

The final section consists of three chapters describing how our current knowledge of pleasure can come to impact on our understanding and treatment of pain. The clinical chapter by Petrovic describes neuroimaging studies of how placebo modulates pain and relates these findings to the underlying processes of pain relief, i.e., pleasure, in the human brain.

Next, Green, Pereira, and Aziz explore the important topic of pleasure electrodes and brain stimulation therapy in human patients, describing growing evidence of how deep brain stimulation can give pain relief to patients with severe chronic pain from, for example, phantom limbs.

The final chapter by Leknes and Tracey provides a conceptual and empirical overview of pleasure in mind and brain. They revisit the important questions raised by the English philosopher Jeremy Bentham about whether pleasure and pain are in fact the "masters of mankind," and link those questions to many of the new scientific developments described by earlier chapters.

The Future of Pleasure in Affective Neuroscience

We hope that the reader will come to enjoy the richness of the chapters in this book. A book on pleasure ought to give some. We hope readers might obtain at least the pleasure of seeing progress in understanding of how hedonic psychological processes are instantiated in brain mechanisms and of a sense that scientific perspectives are gaining a better handle on the slippery topic of pleasure. The contributors here are all leaders in their fields of hedonic psychology and the affective neuroscience of pleasure. They each provide important pieces to the puzzle, which constitutes our current knowledge of the nature of the many faces of pleasure as embedded in our biological brains.

Neuroscientists, psychologists, and related investigators have come a long way in this exploration though our current state of knowledge could equally well be described as a state of only slightly mitigated ignorance. Ignorance is, we all agree, not bliss when it comes to pleasure and brain, and we hope that a better understanding of the functional neuroscience underlying hedonic impact will ultimately come to help more people who live currently without pleasure in their lives. At the very least, we hope that the challenges and opportunities of this exciting scientific adventure will attract many other neuroscientists and lead to further progress in the affective neuroscience of pleasure and insight into the very core of what makes us humans.

References

Berridge, K. C. (2003) Pleasures of the brain. *Brain and Cognition* 52, 106–128.

Berridge, K. C. and Kringelbach, M. L. (2008) Affective neuroscience of pleasure: Reward in humans and animals. *Psychopharmacology*, 199, 457–480.

Kahneman, D. (1999) Objective happiness. In: *Foundations of Hedonic Psychology: Scientific perspectives on Enjoyment and Suffering*, pp. 3–25. Eds. D. Kahneman, E. Diener, and N. Schwartz. Sage: New York.

Kringelbach, M. L. (2004) Emotion. In: *The Oxford Companion to the Mind,* 2nd ed, pp. 287–290. Ed. R. L. Gregory. Oxford University Press: Oxford.

Kringelbach, M. L. (2005) The orbitofrontal cortex: Linking reward to hedonic experience. *Nat. Rev. Neurosci.* 6, 691–702.

LeDoux, J. E. (1996) *The Emotional Brain*. Simon and Schuster: New York.

Leknes, S. and Tracey, I. (2008) A common neurobiology for pain and pleasure. *Nat. Rev. Neurosci.* 9, 314–320.

Panksepp, J. (1999) *Affective Neuroscience*. Oxford University Press: Oxford.

Sartre, J.-P. (1947) *Huis-clos* [*No Exit*]. Editions Gallimard: Paris.

Steinbeck, J. and Ricketts, E. F. (1941) *Sea of Cortez: A Leisurely Journal of Travel and Research*. Viking: New York.

Fundamental Pleasure Questions

Basic Pleasures

1. Is pleasure necessarily a conscious feeling? Or can hedonic reactions ever be unconscious?

Berridge: Surprisingly, hedonic reactions can be unconscious, even though a conscious feeling of enjoyment is central to traditional definitions of pleasure. For example, unconscious 'liking' reactions can occur in people without any subjective awareness at all of the reaction at the moment it is caused (by a subliminal happy face), yet go on to influence later consumption behavior and evaluative ratings of a valence-laden ingestive target (e.g., Winkielman et al., 2005), presumably by directly activating brain limbic systems (Morris et al., 2001). It seems fair to say that there is an unconscious pleasure when a brain generates a positive hedonic 'liking' reaction of which the introspecting mind remains unaware.

My answer does not mean that all instances of behavioral positive reinforcement must entail pleasure, regardless of pleasure reports (Rolls, 2005). There are other routes to behavioral reinforcement besides pleasure, conscious or unconscious (e.g., pure 'wanting' without any 'liking' at all; procedural habits, etc.). But independent evidence for unconscious 'liking' reactions, even if rare, must force us to expand our definition of pleasure.

Cabanac: Yes, which implies the answer to the question that follows is ... no.

Aldridge: I assume that pleasure requires consciousness. Hedonic reactions might not require consciousness (e.g., reflexive taste reactions). Hedonic reactions, which an observer might interpret as indicating a pleasure, may or may not be "actual pleasure" in the subject.

Frijda: Pleasure, if not defined as conscious feeling, is not necessarily conscious. That is, pleasure as a feeling is based on pleasure processes that by themselves are nonconscious and can remain so. Felt pleasure is but one of the outputs of those processes.

Leknes: It is certainly possible to exclude unconscious feelings from one's definition of hedonic reactions such as pleasure. In my opinion, such a definition would miss most of the processing underlying conscious hedonic feelings. Who has never felt their attention drawn away from the task at hand due to a feeling of discomfort, which, upon introspection, has been mounting over time without one's awareness? If this is the case for unpleasant sensations, I can see no reason why pleasant feelings should be different in this respect.

Dickinson: Yes, pleasure is necessarily a conscious experience because this experience grounds the attribution of incentive value to objects, people, and events. However, behavioral responses that accompany pleasure may well be mediated through unconscious processes.

Shizgal: Yes and yes. In common parlance, pleasure refers to a component of conscious experience.

In this view, pleasure is a conscious feeling by definition, and an unconscious pleasure is an oxymoron. The experience of pleasure depends on higher levels of the hedonic apparatus. In contrast, processing at lower levels may operate in the absence of awareness.

An analogy to visual processing serves to illustrate the distinction between the more limited meaning of the first part of the question and the broader meaning of the second one. Information flows from the retina through the multilevel thalamo-cortical division of the visual system. The crucial work performed by cells at lower levels of the pathway, in the retina and visual thalamus, appears to be beyond the ken of the conscious processor. For example, such cells fail to show the lightness constancy that allows our conscious perception of surface reflectance to remain so remarkably stable under varying ambient illumination. In contrast, the responses of cells in the primary visual cortex do show lightness constancy and are thus correlated with visual experience (Shimojo et al., 2001).

The conscious processor is typically described as serial in nature, severely bandwidth limited, and slow. In order to allow a huge number of computations to be performed in parallel by the nervous system, most must occur below the waterline of awareness, and only certain signals are capable of gaining access to consciousness. We are incapable of bringing a retinal image into consciousness, and we should be grateful for this inability—the two-dimensional retinal image is highly ambiguous and contains high spatial resolution only in a small portion of the central field. Extensive lower-level processing is required to organize edges, surfaces, and the results of numerous eye movements into what ultimately emerge as stable, conscious percepts of objects arrayed in a three-dimensional visual world. If the conscious processor had to worry about the details of these crucial lower-level processes, it would be overwhelmed. The bandwidth limitation of the conscious processor is also evident in the hedonic domain. We close our eyes when experiencing intense pleasure and may do so as well when making demanding hedonic evaluations such as determining the relative merits of different wines. However, when distraction undermines the experience of pleasure, hedonic processing at lower levels continues unabated. No matter how engaging the dinner conversation or how breathtaking our companion, we don't tend to eat distasteful items on our plate. Habitual users will continue to work for injections of weak doses of an addictive drug even when unable to accurately report the presence of the drug in the injected solution (Lamb et al., 1991). Thus, hedonic signals can be divided into a class that cannot enter into awareness and a second class that can; whether or not a signal in the latter class succeeds in entering consciousness depends on its fate in the competition for attentional and working-memory resources.

Gottfried: Yes, pleasure is necessarily a conscious feeling, if that is how one wishes to define pleasure. Such a definition would seemingly limit pleasure to the rarefied society of humankind. Certainly the scientific challenges of determining whether a nonhuman animal has feelings, or is conscious of them, have not yet been overcome.

On the other hand, no. Pleasure does not have to be a conscious feeling, if one considers it more simply as a hedonic reaction to particular sensory inputs without reference to consciousness or feelings. Importantly, by this standard, hedonic reactions can be measured. Being measurable they have been shown to influence behavior at an unconscious level. Putative pheromones are one example in which hedonic reactions occur outside of conscious awareness. These chemosensory signals can operate at subthreshold concentrations and have been shown to influence human behavior, mood, and perhaps even mating selection. In the visual domain, studies of affective blindsight and unconscious emotional learning also indicate that the affective content of unseen pictures and faces alter physiological and neural indices of hedonic processing. I favor this more inclusive definition of "pleasure" as it embraces human animals and nonhuman animals alike.

Kringelbach: Pleasure can be defined as the conscious experience of reward but it is questionable whether such a narrow definition is meaningful or useful. Much of our brain activity is not available for conscious introspection and neuroscientific evidence from humans and other animals has made it clear that nonconscious brain activity is essential for controlling our behavior. Some of this nonconscious brain activity is related to hedonic processing and may lead to hedonic reactions, where we are not conscious of their origin but where we are nevertheless happy to confabulate about the causes.

In a similar way to how it is has proven useful to divide emotion into the nonconscious and conscious subcomponents of emotions and feelings, it might be more useful and meaningful to divide pleasure into both nonconscious and conscious subcomponents of evaluative hedonic processing. Such a definition would hold that while pleasure plays a central role for emotions and conscious feelings, it is not itself a conscious feeling.

2. Is pleasure simply a sensation, like sweetness? Or is the hedonic impact of sweetness and other sensory pleasures somehow added to the pure sensation signal?

Berridge: Pleasure is more than the sensation that causes it. Pleasure is an additional niceness gloss painted upon the sensation (as Frijda puts it). Pleasure always must be actively generated by brain hedonic circuits to transform a mere sensation such as sweetness into something nice.

Aldridge: Given my answer to question 1 (above), I would say pleasure is more than a sensation. Hedonic reactions may be responses to simple sensations and may look to an observer like pleasure, but they are only reactions. In this view, "pleasure" requires a human to report it.

Frijda: Pleasure is not a sensation. It is a "pleasantness gloss" added to whatever is pleasant. Pleasure is always pleasantness of something. When the feeling is focused on, it disappears (it is "evanescent").

Leknes: It is likely that the pleasantness of chocolate is related to our perception of its sweetness, fattiness, etc. Eating chocolate is not pleasurable to a sated subject, however, although it is probably safe to assume that the sensory properties remain unchanged (Small et al., 2001). A simple model would propose that pleasure arises from a weighted combination of the sensory signals and of signals about homeostatic state (i.e., how useful the stimulus is for the organism).

Dickinson: No, pleasure is not a sensation, but an affective experience that accompanies but is usually also integrated with sensation in experience.

Shizgal: No and no. The purpose of sensory systems is to provide facts about the world. These systems are engineered to function as objectively as possible. Thus, the lightness constancy mechanism to which I referred in my answer to question 1 does a remarkable job of accurately reporting the reflectances of surfaces, regardless of whether they are in full sunlight or deep shadow. Similarly, the color-constancy mechanism largely compensates for the spectral changes in the illuminant over the course of the day, preventing a forager from confusing an unripe fruit viewed at dawn with a ripe one viewed at noon.

Hedonic systems provide a subjective commentary on the information provided to them by sensory systems. Both a warm stimulus encountered when one is hypothermic and a cool stimulus encountered when hyperthermic are experienced as pleasant. As Michel Cabanac has argued, their subjective meaning is similar—they are both good for us and are sought out because they help return a crucial physiological variable to its regulated value. Nonetheless, we do not confuse the sensations arising from the two stimuli. We perceive them as objectively different even if their subjective hedonic values are the same.

Mixing the objective and subjective signals could prove harmful. For example, if judgments about the sugar and fat content of prey items depended on the hedonic experience that accompanies their consumption, a forager could make errors in trading off amount, procurement costs, and quality, thus failing to maximize net energy intake. Thus, accurate sensory assessment is crucial to determining the relative value of prey items. However, hedonic signals, which depend on the physiological and ecological state of the forager, could provide information about absolute value and thus adjust key decision variables such as risk appetite.

Kringelbach: Pleasure does not fit most common definitions of sensations, as pointed out by Ryle (1954). Instead, pleasure would appear to be part of the subsequent valuation of sensory stimuli needed in decision making, including most importantly the hedonic valence.

The pleasure or hedonic impact of sweetness will elicit what has been termed "acceptance wriggles" by Frijda (see this volume), which adds the hedonic gloss to the sensation, which we experience as conscious pleasure. These pleasure-elicited behaviors are also present in other animals including rodents who will lick their lips to sweet foods as convincingly described by Berridge (see this volume) and can be taken as an objective measure of the pleasure elicited. While human infants initially exhibit similar kinds licking of their lips for sweet foods, these stereotyped behaviors disappear after a while. Humans still, however, exhibit much pleasure behavior from the carefree smiles and laughter of pleasant social interactions to the deep groans of sensory and sexual pleasure. Most people would instinctly feel that our pleasure would somehow not be quite the same without these pleasure-elicited behaviors and the case could be made that it would in fact not be pleasure but "false" pleasure. Consciously engaging the pleasure-elicited behaviors even without conscious or nonconscious elicited pleasure may start a positive feedback loop, which recruits hedonic processes, as in the experiments of Strack et al. (1988) where affective responses became stronger when participants were required to hold a pen in their mouth in ways typically associated with smiling without requiring them to pose in a smiling face—and significantly less strong when not engaging these smiling muscles.

3. Is human pleasure similar to or different from that of other animals?

Berridge: The answer is yes, both. Human pleasure is unique in the sense that unmatched human cognitive capacities transform our mental representation of pleasant events into accompanying elaborate thoughts. Human cognition adds richness and alters the attention we pay to pleasures, elaborates our plans to get them, vastly expands the range of events that can trigger pleasure to include cognitive and cultural sources (art, music, social rewards, etc.), and provides new top-down regulatory ways to amplify or dampen a pleasure or displeasure.

But as an affective neuroscientist trying to find out how brains generate basic sensory pleasures, my answer is: human pleasure is essentially the same as other animals (at least other mammals, and possibly beyond). Humans and animals share the same limbic brain circuits and likely have the same hedonic hotspots to generate pleasure. Those hedonic limbic circuits operate as far as we know by the same neurochemical signals and circuit rules in humans and nonhumans alike.

Cabanac: May I answer that question with another question? Is YOUR pleasure similar to or different from MY pleasure? Or with a similar question: is female orgasm similar to male orgasm? Yet we know that pleasure fulfils the same function in animals and humans: optimization of behavioral decisions.

Aldridge: It is not possible to determine if pleasure is the same in animals and humans. Hedonic reactions may appear to an animal observer to be similar. Neuroscientists may be able to demonstrate that hedonic reactions involve homologous brain systems. Scanning or pharmacological experiments may also demonstrate similarities between neurochemically defined brain systems of animals and humans. Neither of these would prove that pleasure is the same in animals and humans.

Frijda: Human pleasure is similar to and different from that of other animals, like a glass is both half full and half empty. The reason is simple. See my answer to question 2: what pleasure is about is different between humans and other animals because animals do not know that they are feeling pleasure, but functionally (e.g., in evoking acceptance wriggles). I assume they are the same.

Komisaruk: I think pleasure is not unique to humans. Therefore, I think that pleasure does not require language ability. When my sons come home, I would say that their dogs greet them happily. If and when they scold their dogs, their dogs do not look happy. A dog wagging its tail looks to me like a happy dog—that is, a dog that is feeling pleasure (I sense my animal behavioral, anti-anthropomorphizing colleagues gritting their teeth). Similarly, when my cat purred, she looked to me as if she were content, that is, feeling pleasure. She never purred if she appeared to be disturbed in any way. It would perhaps be useful to see whether purring and tail-wagging could be used as valid indicators of pleasure—pharmacologically speaking. Questions such as these, while difficult to answer, are far more manageable than the question of which and how neurons produce any bit of awareness. However, still more difficult is the question of which and how neurons produce a bit of the feeling of pleasure.

Dickinson: Depends on what animals—ape, rat, or cockroach? My view would be that any cognitive animal (i.e., one capable of true goal-directed action) experiences states of pleasure that are similar to our own.

Kringelbach: Pleasure serves a central role in fulfilling the evolutionary imperative of survival and procreation. This means that for all animals the sensory pleasures linked to food intake is likely to be a basic pleasure. Similarly, the social interactions with other members of the same species, which could potentially lead to the propagation of genes, have probably been selected for, which means that social pleasures must also be basic. Also progeny may elicit social pleasure as in the very important social bond between parents and infants. In social species such as most mammals, it might be that social interactions are at least as pleasurable as the sensory pleasures related to food intake.

Careful neuroscientific experimentation in humans and other animals have shown that evolution appears to have preserved many brain circuits between species. Some of these brain networks must be involved in pleasure and pleasure-elicited behaviors. Thus it is likely that human pleasure will share many features with other animals and particular those closest related such as other mammals and primates. Yet, it may well be that human conscious experience of pleasure is different not only in degree but also in kind from other animals. Activities combining sensory and social pleasures such as those involved in a dinner party could have a synergistic effect on the higher-order pleasures experienced in humans, which might be hard to find in other animals.

4. Is pleasure simply the experience of getting what you want? Are liking and wanting simply two words for the same pleasure process? Or can pleasure liking or pleasure wanting exist without the other?

Berridge: Getting what you want is different from liking what you got. Getting what you want is not always pleasurable. And even when it is, its pleasure is quite different from the wanting and getting. Taking pleasure in what you get requires the additional 'liking' gloss, a distinctive and hedonic brain process of its own. If that hedonic process is lacking, then getting what you want will produce no true pleasure.

Aldridge: *First question:* I don't know. *Second question:* The Berridge scheme divides 'liking' and 'wanting' into two separate psychological processes. In that scheme, it doesn't make sense to call them the same thing and there is good evidence that 'liking' and 'wanting' can be manipulated independently. It is yet to be determined how 'liking' and 'wanting' map onto pleasure. In my view, pleasure is a process involving all brain systems processing reward information (cortical and subcortical) so 'liking' and 'wanting' would be combined. *Third question:* I don't know what "pleasure liking" or "pleasure wanting" are or how they differ from 'liking' and 'wanting'.

Frijda: Pleasure does not consist in getting what you want but, more generally, meeting what befits you (since pleasure signals "well-functioning"), which includes getting what you want, but also many other things (like perhaps getting what you want, or meeting what you might want, or what allows you to do what you can do). And pleasure exists without any wanting, such as walking in the sunshine when you are twenty and healthy and reasonably well-fed (the same for when you are eighty).

Dickinson: It depends upon what you mean by 'wanting'. In the nontechnical sense (i.e., not in the Berridge–Robinson sense), the pleasure or liking induced by an experience brings about a wanting for that experience through the process of incentive learning.

Kringelbach: Many theories of desire have taken pleasure to simply be the fulfilment of desire. Spinoza wrote that "pleasure is the transition of a man from a less to a greater perfection," where perfection is the completeness of which an individual has realized her desires. Schroeder (2004) has argued against such standard accounts of desire, since getting what you might desire does not always lead to pleasure, and he has instead proposed a theory, which links intrinsic desire directly with the reward systems of the brain. Berridge (see this volume) has convincingly argued that the hedonic impact, 'liking', and the incentive salience, 'wanting' are partly dissociable in terms of their underlying neural circuitry and pathways. In terms of neurotransmitters, it has been shown that dopamine is more related to the 'wanting' or the desire, while opioids are more related to the 'liking' or the pleasure. Malignant desires such as addiction can then be conceptualized as 'wanting' without 'liking' as argued by Robinson and Berridge (1993). Similarly 'liking' without 'wanting' would be akin to what has been described by some world religions as bliss or "true" happiness. Whether such a state truly exists has not yet been demonstrated but the aforementioned conceptualization may offer the scientific tools to test it.

5. Can pleasure be measured by objective physiological or behavioral techniques? (e.g., facial reaction or EMG, pupil dilation, GSR, neuronal firing, neurotransmitter release, neuroimaging)

Berridge: Yes, at least, basic or core 'liking' reactions to pleasure can be measured by objective neural or behavioral techniques. Conscious liking, admittedly, is more difficult to objectively measure (though even here, properly constructed rating scales can provide replicable and meaningful measures of subjective pleasure).

The measurement glass is more than half full. Psychologists and neuroscientists can use objective hedonic measures of core 'liking' reactions to discover which neural systems generate the brain's basic pleasure gloss. Eventually they may be able to recognize reliable electrophysiological–neuroimaging brain signatures of 'liking'. Scientists can also explore psychological features of the core pleasure process, including the relation of hedonic 'liking' to motivational 'wanting'. And they can compare basic 'liking' reactions to subjective rating measures of conscious pleasure, perhaps uncovering commonalities and differences in the underlying mechanisms.

Cabanac: Two words from the question may be answered separately: "Reliably," yes. Our body reacts to pleasure and these physiological responses such as hypertension, tachycardia, fever, etc. can be reliably recorded.

"Measure," definitely no. The word measure implies quantifying parametrically a mental event that takes place at the same time of a physiological response. The latter can be parametrically measured. The former can be quantified, but not parametrically. There always remains a doubt about a mental event report by any participant, even when the experimenter is self-testing.

Aldridge: Neural activity, neuroimages, and other physiological responses correlated with hedonic reactions can be measured. If humans report pleasure when they are scanned or being measured, then one could say that the scans or physiological responses are correlates of pleasure. It is likely, however, that the same regions of the brain may be active or the same physiological responses might occur in other contexts apart from reported pleasure. Based on my assumption that pleasure requires consciousness, physiological correlates of pleasure can only be measured in humans who report pleasure. Hedonic reactions can be measured at other times, but these may or may not be correlated with "pleasure."

Frijda: Can pleasure be measured objectively? I do not know whether all pleasure can, when the criterion is subjective report. But I suppose one can get fairly close by behavioral techniques: remaining longer with a stimulus or event than necessary for identification or preparation of escape.

Petrovic: Certainly pleasure cannot be measured. We know for fact that physiological responses correlating with the report of pleasure can be measured, including various muscular reactions, sweating, activation of certain regions in the brain and involvement of specific neurotransmitter systems. We can only study the mirror image of pleasure. However, our problem is that none of these responses are involved in just pleasure, thus the specificity is low. So in a way studying pleasure systems is a complex task relying on putting together a large amount of bits of a puzzle and trying to see the big picture.

Dickinson: No. Only indirectly.

Kringelbach: The pleasure-elicited behaviors can be measured in animals and include stereotyped behaviors such as facial expressions, pupil dilations, and orgasms. These behavioral changes must correspond to physiological changes in brain activity such as the temporal unfolding of neural activity and neurotransmitter release linked to specific brain regions, which then presumably can be used as objective measurements. In order to establish the relevant physiological changes, causal interventions are needed such as those carried out by Berridge and colleagues in the nucleus accumbens and ventral pallidum, where they shown that microinjections of opioids can change the hedonic gloss on subsequent pleasure-elicited behaviors.

We have used the causal technique of deep brain stimulation (DBS) in humans to show pain relief when stimulating the periacqueductal gray. At the same time, we have used magnetoencephalopathy (MEG) to measure the whole brain activity associated with this intervention (Kringelbach et al., 2007a,b). This is a promising technique for studying pleasure in the human brain where different brain targets can be switched on and off and the effects measured on the whole-brain activity and on pleasure-elicited behaviors, which can be compared to subjective conscious reports.

6. Are pleasure and pain on a continuum?

Berridge: Controversy persists on how positive affect relates to negative. The brain often seems to produce affective responses as if it generated pleasure and pain (or displeasure) along a single continuum. For example, increases in positive 'liking' expressions typically are accompanied by decreases in 'disliking' expressions for the same target and vice versa. Reciprocity between pleasure and pain has led many psychologists to posit a single continuum for affect.

And yet, teasing bits of evidence from psychology and neuroscience continue to support a contrary argument that pleasure and pain–displeasure may have separable mechanisms. Pleasure and displeasure may be capable of being produced independently and perhaps even sometimes simultaneously by the same target. If so, two separate dimensions would seem in order. In short, the evidence remains a bit contradictory, and our field still needs a more conclusive proof.

Aldridge: Pain systems activate brain regions not usually included in those thought to be processing hedonic reactions and/or reward. Thus, it seems unlikely that pleasure and pain are on a continuum.

Frijda: Like the half full, half empty glasses. They are on a continuum in some regard (objects can be placed on a continuum with reasonable confidence, or on some preference continuum), but they also and always differ in some regards, like the discontinuity or categorical jump between credit and debt. And they are not on a continuum in the sense that both can exist simultaneously, as in mixed feelings, hedonic uncertainty, and nostalgia.

Leknes: In everyday speech (and in the writings of philosophers (Bentham, 1907)), pain and pleasure often represent opposite sides of a hedonic continuum, where pains describe unpleasant and unwanted feelings as varied as boredom, pain in a medical sense, or embarrassment. The scientific literature usually refers to these pains and pleasures as punishments and rewards. In general, pleasurable feelings are usually rewarding and pain is usually a punishment. There are some notable exceptions to this heuristic, such as pleasurable pain in sexual masochism, and also interesting mixtures like the bittersweet quality of unrequited love or the guilty pleasure of eating the last piece of pie.

Petrovic: Some studies indicate that there is a continuum at least in the involvement of specific neurosystems. We know that activation of the opioid neurosystem will induce a sensation of pleasure but also suppress pain. It has also been shown that induction of sadness will suppress the opioid system. It seems that several neurosystems work antagonistically in this way, for example, activation of the cholecyctokinin (CCK) system induces anxiety (and even panic attacks in larger doses), and moreover this system will make pain to be perceived as more intense and unpleasant. Also, if the opioid system is inhibited, the CCK systems will be more active and vice versa. In this way, these systems seem to work together in a continuum stretching from pleasure and suppression of unpleasantness to anxiety and increased unpleasantness.

Green: Pleasure and pain can certainly be regarded as two opposite extremes. On the one hand, pleasure is associated with a feeling of well-being as opposed to the feeling of misery or doom associated with pain. However, the subjective feeling of pain has tangible benefits for the survival of the organism. For example, a limb that feels pain will withdraw from a hot stimulus. On the other hand, what are the tangible benefits of pleasure to an organism's survival? Is pleasure simply the conscious awareness of a higher being's state of safety or a recognition that direct actions do not need to be taken to aid survival?

Pleasure actually appears much more complex than "the opposite of pain." If it is simply "the opposite," how do we explain the fact that some people derive pleasure from pain? One extreme example may be masochistic sexual experiences. However, a more subtle example involves the experience of pain that will eventually lead to a benefit. For example, training for a marathon can be very painful and difficult, but the individual will derive pleasure from the satisfaction that they are becoming physically stronger and knowing that they will be able to undertake the race. Does this "no pain–no gain" phenomenon disprove the continuum hypothesis or is it that we are prepared to put up with pain in order to defer a greater pleasure?

Dickinson: No. They are on orthogonal continua but usually with a mutual inhibitory interrelationship.

Kringelbach: Pleasure and pain are closely linked with each other but opinions differ over whether they are opposites or different kinds. As with most controversies, the answer depends primarily on focus and definition. Pain is not exactly the same as the lack of pleasure and does not necessarily solely correspond to displeasure. While pleasure is mostly stable, pain is more unstable and calls out for change. A stimulus will rarely make animal both approach and avoid it at the same time, but it is nevertheless clear that at least humans can feel both pleasure and displeasure as part of mixed feeling states. One example of such a mixed feeling is the Portuguese word *saudade*, which is akin to nostalgia but not fully translatable as such. Both words describe bittersweet emotions that are linked to painful memories from pleasures past, which at the same time are also pleasant memories. In addition, the word *saudade* also includes future expectations by evoking the pleasant and painful feelings of longing for pleasures past, which might return in a distant future.

Reward and punishment are intimately connected to pleasure and pain. Some scientific evidence would seem to indicate that there are different pathways involved in reward and punishment. At the same time, there is also evidence that reward and punishment make use of shared pathways. Depending on which levels of the brain processing one is focusing on, the answer could be one of opposition or difference of kind and most likely a combination of the two—but more evidence is needed.

7. Does pleasure have an evolutionary function?

Berridge: Yes, pleasure has an evolutionary function—probably more than one. Brain evolution cannot afford to wastefully dispense the massive amounts of neural machinery that process pleasure on major psychological processes that have no fitness benefit. Pleasure and displeasure reactions are so prominent in our own lives and in the behavior of other animals, and the underlying limbic neural mechanisms for generating affective reactions so well developed in the brains of both, that we are forced to conclude the capacity for pleasure reaction is an evolutionary trait that was selected and conserved. It is difficult to imagine an evolutionary scenario that would have led to such prominent and similar limbic brains in so many species if pleasure were not adaptive.

How could pleasure have had evolutionary functions? Basic core pleasure reactions have always had objective consequences for an individual's behavior, physiology, and eventual gene fitness. In a sense, hedonic reactions have been too important to survival for hedonia to be exclusively subjective. And subjective pleasure itself, in creatures that have it, carries an additional function: providing a declarative goal to guide flexible cognitive systems that operate at least partly in conscious modes (see Dickinson and Balleine, this volume). Brains have had to actually do many things based on hedonic impact, and the doing of those things has given evolutionary functions to pleasure.

Cabanac: Any answer to that question belongs to the realm of belief, because it is not possible to "prove" anything regarding evolutionary usefulness. Yet, I believe that the answer is *yes*. The emergence of pleasure in the Amniotes gave them such an efficacy that this property remained and most likely contributed to the evolution from reptiles to birds and mammals.

Aldridge: It seems likely. Pleasure focuses behavior toward evolutionary useful ends, for example, eating, drinking, sex.

Frijda: Pleasure has the evolutionary function of signaling functioning well of any function that impacts overall function monitoring (either in consciousness or state of well-being).

Petrovic: The conscious part of pleasure must have a similar function as other conscious phenomena. It has been suggested that consciousness may be a way of selection of the very most important information processes going on in the brain, and that this "hyper-attention" has a direct evolutionary benefit. Possibly, the same idea may be suggested for conscious experience of pleasure. Pleasure may "simply" represent an extreme form of motivation and learning of what is good in our surroundings to drive complex behavior in the future.

Dickinson: Yes—that of allowing the control of behavior by cognitive process by supplying these processes with their goal values.

Kringelbach: As mentioned above, pleasure is likely to play a central role for the central evolutionary principles of survival and procreation of the species. The function of the basic sensory and social pleasures could be to help optimize our decisions such that survival and procreation remain possible. This is demonstrated by those individuals temporarily without pleasure which is common in depression and mental illness. The suicides involved in these afflictions would seem to indicate that without pleasure even survival and procreation become meaningless.

Brain Pleasures

8. What brain substrates actually cause pleasure?

Berridge: The brain is surprisingly frugal in its number of neural substrates able to directly cause pleasure. Pleasure causation implies that the substrate is either a sufficient cause to increase hedonic impact, or a necessary cause that must remain intact for normal hedonic impact. The causation question is especially knotty because several putative brain pleasure substrates have turned out to probably not cause pleasure after all (for example, mesolimbic dopamine systems and many so-called "pleasure electrodes").

Only a few subcortical brain substrates so far have compelling positive evidence for pleasure causation. For example, hedonic hotspots in nucleus accumbens, ventral pallidum, and brainstem have been found where opioid or related neurochemical activation causes increases in natural 'liking' reactions to sweet pleasure. Conversely, damage in some hotspots may disrupt normal pleasure reactions. But not many other sites can be listed yet for which necessary or sufficient criteria are met by strong evidence. Other limbic sites, and especially cortical sites, need a closer look regarding pleasure causation.

Aldridge: I don't expect that we will find that a single brain region "causes" pleasure. Rather, I expect that distributed patterns of activity across sets of brain regions may "represent" a pleasure state. When that representation is engaged, a subject may report pleasure. It seems likely that many sites including cortical and subcortical regions are activated during pleasure. One might find that particular patterns of activated sites are correlated with reports of pleasure or with observations of hedonic reactions. If a stimulus triggers activation in these same sites in a human, it is reasonable to predict that it would be reported as pleasurable. Further, depending on the flow of activation through brain circuits, experimentally stimulating one brain site directly may lead to activation in an entire set of sites; however, stimulation in this one site should not be viewed as causal. The stimulus would just be triggering the representation.

Petrovic: I believe that complex networks of regions are involved in processing what we experience as pleasure. I think that it is possible to dissociate specific subcomponents of pleasure. If we again study where the opioid system (highly associated to the experience of pleasure) is located in the brain, it is spread out over many different, but specific, regions from the brainstem and the nucleus accumbens to the anterior insula and the anterior cingulate cortex. Possibly, nucleus accumbens is relevant for the motor response in pleasure such as smiling while the insula may be involved perceiving secondly derived bodily feelings when we experience pleasure and the anterior cingulate cortex may be involved in the interaction between pleasure and cognition.

Schoenbaum: Pleasure seems to be an extraordinarily subjective and complex emotion. Presumably, pleasure emerges from signaling across multiple brain areas (VTA, amygdala, accumbens, ventral pallidum, hypothalamus, etc.) that are intimately involved in processing information about biological rewards. Humans and almost certainly animals are able to recognize a particular neural state in these circuits with the attainment of biological goals/rewards. We would speculate that this recognition, perhaps occurring in cortical regions (prefrontal?), would be what we'd call pleasure. Because it can be recognized, that neural state can then be mapped on to higher constructs or more abstract goals, so that it can be evoked by them. The fact that these constructs/goals are a step (or more) removed from the biological goal/reward triggering the original state may explain why pleasure derived from attaining these secondary goals may be variable, less intense, more abstract, and different in subtle ways from pleasure derived directly from meeting biological needs. Thus pleasure, whether it is derived from a primary reward or secondary reward, may be processed in both several regions of the brain, both cortical and subcortical.

Leknes: Here, I will restrict my comment to substrates of human pleasure. It is notoriously difficult to experimentally induce pleasure in an MRI scanner environment, and usually fMRI studies of pleasure rely on an experimentally induced homeostatic imbalance such as hunger, thirst, or, in my own work, a pain state. The good news is that it is easy to measure pleasure in these studies since subjects can give subjective reports on pleasure rating scales. To my knowledge, not a single area implicated in pleasure in the human literature has failed to be implicated in aversive processing as well. Examples are the amygdala (Becerra et al., 2001; Gottfried et al., 2003; Paton et al., 2006) and the nucleus accumbens (Menon and Levitin, 2005; Zubieta et al., 2005).

Komisaruk: Pleasure for me is like what Associate Justice of the Supreme Court Potter Stewart once said about pornography: "I could never succeed in intelligibly defining [it]...but I know it when I see it." The question of what brain systems produce pleasure raises the nasty question of which neurons produce consciousness and how they do it. With brain imaging, we see particular brain regions activated during orgasm, which is pleasurable. The nucleus accumbens and the hypothalamic paraventricular nucleus become particularly activated at orgasm. This indicates that the neurons that respond to dopamine and those that secrete oxytocin are both activated during this intensely pleasurable experience. However, we have (yet) no way of knowing whether the activation of these two groups of neurons themselves is what produces the feeling of orgasmic pleasure, or whether it is activity that they relay to other neurons that creates the feeling of pleasure. If it is to other neurons, then which ones, and even so, how does *their* activity produce the feeling of pleasure? That, of course, raises the question of how *any* neuron activity produces *any* feeling or cognitive experience, and the different qualities thereof, such as pleasure, pain, red, cold, sweet, or melody.

Dickinson: No idea but suspect that there is a major cortical involvement—insula?

Kringelbach: Berridge and colleagues (see this volume) have convincingly shown that in rodents subcortical regions such as the nucleus accumbens and ventral pallidum have hedonic hotspots where the activity modulates the pleasure-elicited behaviors related to food intake. They have also shown that dopamine is mostly related to 'wanting' and opioids are most likely linked to 'liking.' There is also some evidence that direct stimulation of the PAG in humans can elicit pain relief, which is reported as pleasurable (Kringelbach et al., 2007a,b), presumably linked to the engagement of the opioid system but not exclusively (see Green and Aziz, this volume).

These subcortical structures interact with cortical structures such as the orbitofrontal cortex (OFC), the insula, and anterior cingulate cortex (ACC; both anterior and posterior parts). The directionality of this causation has not been demonstrated but it is known that in mammals the structures of the basal ganglia are mainly on the output side of the OFC (see Schoenbaum, this volume). Using MEG, it has been demonstrated that the pain relief obtained from direct stimulation of the periacqueductal gray (PAG) in humans will elicit activity in the mid-anterior OFC (Kringelbach et al., 2007a,b). Other human neuroimaging experiments have shown that this part of the OFC is the most likely candidate for the subjective hedonic experience of pleasure (Kringelbach, 2005). It is currently not known whether this brain region causes pleasure or whether it is the point of integration between nonconscious and conscious hedonic processing.

9. Do the same brain substrates mediate conscious pleasure and trigger basic behavioral–physiological hedonic reactions? Or is conscious pleasure mediated separately?

Berridge: Conscious pleasure must be mediated separately from basic or core 'liking' reactions. Behavioral–physiological hedonic reactions can sometimes occur unconsciously even in normal people, thus separating conscious and basic forms of hedonic reaction. Independent phenomena must have separable causes, and so only two conclusions are possible about the relevant brain substrates. One is that diverging anatomical brain circuits must mediate subjective conscious pleasure versus objective core 'liking' reactions. The other is that, at the very least, if the same neural substrates mediate both conscious pleasure and unconscious pleasure reaction, then conscious (subjective plus basic) and unconscious (basic only) hedonic reactions must correspond to different modes of activation for that substrate.

Aldridge: Behavioral–physiological hedonic reactions are responses to sensations. I would not define hedonic reactions as pleasure. Given my answer above (question 8), I expect that activation in circuits related to hedonic reactions could be a subset of circuits activated during conscious pleasure. Basic behavioral–physiological hedonic reactions are not pleasure on their own, although they may occur during pleasurable activation and may even trigger patterns of activation in more widespread areas.

Small: There is very strong evidence that the conscious pleasure associated with eating is encoded in the OFC but not the amygdala. Neuroimaging studies in humans, in which perceived pleasantness can be ascertained with rating scales, consistently demonstrate strong positive correlations between perceived pleasantness ratings of taste (O'Doherty et al., 2001; Small et al., 2003), smell (Anderson et al., 2003), flavor (de Araujo et al., 2003), and food reward (Kringelbach et al., 2003; Small et al., 2001) and activation of OFC. This is true whether pleasantness is derived from variation in stimulus attributes or internal state.

Additionally, in a recent study from our laboratory, we asked subjects to evaluate several dimensions of sweet, sour, salty, and tasteless solutions (Bender et al., 2005). Activation of the caudolateral OFC was selectively associated with evaluation of stimulus pleasantness, and this region was preferentially connected to earlier taste relays when a taste compared to a tasteless solution was experienced. This suggests that the caudolateral OFC organizes retrieval of sensory information from earlier taste relays in the service of computing perceived pleasantness. Neural responses in the amygdala do not correlate with perceived pleasantness of taste, flavor, and food reward (Anderson et al., 2003; de Araujo et al., 2003; Small et al., 2003), nor are they sensitive to alliesthesia (Kringelbach et al., 2003; Small et al., 2001)—the reduction in food pleasantness associated with satiety (Cabanac, 1971). Instead, the amygdala responds to food cues (LaBar et al., 2001; O'Doherty et al., 2002) and its response is sensitive to devaluation (Gottfried et al., 2003), indicating that it encodes the incentive value of food cues and that it is sensitive to changes in the incentive value related to internal state.

However, despite its critical role in encoding predictive food reward, there is preliminary evidence from the laboratory of Marci Pelchat that it does not mediate the conscious perception of desire or food craving (Pelchat et al., 2004). Pelchat and colleagues examined neural response to food cues that did or did not elicit subjective cravings. They found that although the amygdala responded to food cues, it was the insula and dorsal striatum that respond during time periods in which subjects reported experiencing cue induce cravings. Together, these findings suggest that neural representation of conscious pleasure experienced during eating and conscious desire experienced during food anticipation is at least partially segregated from the encoding of the predictive value of food cues.

A related issue is whether emotion and affect consist of explicit hedonic feelings, such as perceived pleasure, as well as "implicit affect," and whether these two components of emotion are represented by separable neural systems. Berridge and Robinson (2003) have argued that implicit affective reactions can exist objectively without necessarily being experienced subjectively. For example, subliminally perceived happy and angry faces produced opposite effects on the value of a beverage despite no change in reported feelings (Winkielman et al., 2005). In a landmark study, Morris et al. (1998) showed that the amygdala distinguished sensory stimuli solely based upon whether they had been previously associated with a subliminal happy or angry face. In our recent study, referred to above, we found that the amygdala preferentially communicated with the primary gustatory cortex during passive perception of taste compared to active evaluation (when a judgment about a stimulus feature was required). This contrasts with preferential connectivity between primary gustatory regions and OFC during the conscious evaluation of pleasantness. Thus, although these findings are a long way from providing proof of concept, given the role of the amygdala in encoding subliminally presented faces, they at least hint at the existence of separable systems for explicit and implicit emotion.

Schoenbaum: We would speculate that subcortical areas intimately involved in reward processing are the substrate or detector of situations in which pleasure is possible. In other words, these regions must signal by their activity pattern that critical needs have been met. However the actual experienced emotion of pleasure is derived from cortical recognition of this state. For this reason, we can have pleasure imposed upon us to some extent by external circumstances (such as winning the lottery), but we almost always have substantial control over it (i.e., we can ruin it). Moreover, we can anticipate pleasure, which may involve cortical areas invoking a pleasure-like state in these downstream regions (either in reality or virtually in their own local synapses that retrieve information from these areas). Finally it may also be possible for cortical areas to selectively influence different parts of these circuits, perhaps due to there own anatomical specificity, leading to different forms of pleasure (e.g., satisfaction vs. joy).

Dickinson: My own view is that conscious pleasure result from a re-entrant transformation of basic behavioral–physiological hedonic reactions?

Shizgal: The signals that give rise to behavioral–physiological hedonic reactions may also trigger an accompanying experience of pleasure. However, awareness of these signals depends on whether they have gained access to working memory. Thus, the brain substrates appear to be organized hierarchically. Conscious experience arises from the higher levels of the system.

According to Baars' global workspace theory, the cognitive architecture consists of a multitude of specialized modules that can work independently and locally. Signals must gain access to consciousness in order to be broadcast simultaneously to many different modules and thus to coordinate their activity. When we sail on a brisk day, local modules adjust our posture and exposure to sun and wind without requiring the intervention of consciousness. However, when thermal discomfort crosses a threshold, we become aware of our predicament, a state than entails broadcast of the information. This enables the recruitment of the multiple modules required to formulate and execute a plan to go below and fetch a windbreaker.

Kringelbach: Given the demonstrations of nonconscious hedonic processing, it would seem likely that there is a separation and perhaps part overlap of brain mechanisms and substrates. In terms of correlation, it would appear from various neuroimaging experiments that the mid-anterior OFC correlates with conscious subjective pleasure reports as shown in an experiments involving "selective satiety" (Kringelbach et al., 2003). This evidence has recently been corroborated by causal evidence when combining the causal intervention of deep brain stimulation with MEG, which showed that pain relief (which was reported as more pleasant) through stimulation in the PAG elicited brain activity in this region. It is unlikely that the mid-anterior region of the OFC is the only node in what is likely to be an extended network of cortical and subcortical regions mediating conscious pleasure, which is also likely to include the cingulate cortex and the insular cortex (e.g., Craig, 2003).

10. Is there common currency for all sensory pleasures (food, sex, drugs, etc)? Or are different sensory pleasures mediated by different neural circuits?

Berridge: Brain hedonic mechanisms probably overlap heavily, at least for sensory pleasures. This is only a guess; admittedly these are still early days regarding evidence. But from what we know so far, many of the same cortical and subcortical substrates participate in pleasures as diverse as food, drugs, sex, parental, romantic and social interaction, money, music, and various cultural rewards. Of course, individual pleasures might also have their own pockets of unique neural substrate within the brain. Yet even if sweet-unique, sex-unique, or other pleasure-unique pockets exist, the general rule for mediation of sensory pleasures seems likely to be brain overlap and a neural common currency.

Cabanac: The term "common currency" implies that we are dealing with a mental mechanism that allows to compare, sort, and rank the various motivations present at a given time, in order to satisfy the most urgent. Such an emergence into cognition does not necessarily mean that the nervous substrate is common to all motivations. Especially, positive and negative hedonic impacts may result from the activation of different nervous substrates.

Aldridge: I predict there will be separate circuits for food, sex, drugs and, rock 'n' roll; however, I also predict that there will be extensive overlap between these different circuits. It may be that some circuits such as ventral basal ganglia or cortical regions are activated in all pleasure responses.

Frijda: I do not see that common currency and involving common neural circuitry are the same. As to common currency: I think it is an open question to what extent pleasures are substitutable. Pleasures in part are pleasures that contribute to higher order pleasure (also known as sense of well-being); but they can also give contour to absence of other pleasures; since all pleasures are pleasures of/about something.

Leknes: Surely there can be a common currency without the neural circuitry of different pleasures overlapping completely? People make decisions about gains and losses even when these are in different modalities; just think about guilty pleasures, when the addition of a mere touch of extra guilt can cause the pleasure to vanish altogether.

Dickinson: My theoretical prejudice is to answer yes to this question.

Shizgal: Yes and yes. As Michel Cabanac has argued, sensory pleasures are often tied to the capacity of a stimulus to redress a physiological imbalance. Multiple physiological variables are regulated and different sets of physical resources in the world must be procured to keep each of these variables within the required range. For example, the macronutrients required to maintain the short- and long-term energy store are of little use in maintaining hydromineral balance, and no amount of salt or water will provide an energy source for metabolism. Thus, local currencies are required to evaluate the energy and hydromineral content of prey items, and each may be reflected in hedonic signals. Given that both types of resources are found in the same prey, a mechanism is required to translate the local currencies into a more global (common) one. This is what microeconomists call a substitution problem. Salt and fat are considered complements in microeconomic parlance because they satisfy different needs; one cannot substitute for the other. Carbohydrates and fats are partial substitutes; both are energy sources, but the former is better suited to replenishing the short-term store and the latter, the long-term store. In order to obtain an optimal combination of resources that are not perfect substitutes, a nonlinear combinatorial rule is required. This is a fundamental problem that has long been neglected by students of hedonic processing. Once again, the answer to the question depends on the level of processing under consideration. At the early levels, local currencies are employed. At a higher level, the local currencies are converted into a common one. This argument can be generalized readily to a broader class of objectives and sources of hedonic signals.

Kringelbach: From a computational point of view, it would seem to make sense to have a common currency, which could be used for the comparisons of sensory stimuli needed for decision-making. Kahneman et al. (2003) proposed a distinction between "experience utility" and "decision utility," where the "experience utility" is the degree of like or dislike of the choices or the hedonic value involved—and as such a measure of pleasure. In contrast, the "decision utility" relates to whether the object of choice is wanted or unwanted and this concept thus shares features with desire. These decision-making processes are related to the present, while the memories and expectations of these are called "remembered utility" and "predictive utility." These processes can be thought of as beliefs about the wants and likes involved in the past and future decisions.

The neuroscientific data are currently inconclusive about the possible nature of such a common currency. I would include social pleasure as a basic pleasure at the same level as sensory pleasures, and my hunch is that the basic pleasures use partly overlapping neural circuits on which the higher-order pleasures are parasitic. I have proposed a model where some these functions are served by the OFC in humans (Kringelbach, 2005). The OFC is one of the most polymodal regions of the brain. Sensory information from all the senses is received and combined in multimodal integration in the posterior parts of the orbitofrontal cortex. The reward value of the stimuli is assigned in more anterior parts of the orbitofrontal cortex, from where it can be used to influence subsequent behavior (in lateral parts of the anterior orbitofrontal cortex with connections to the anterior cingulate cortex), stored for monitoring/prediction/learning (in medial parts of the anterior orbitofrontal cortex) and made available for subjective hedonic experience (in mid-anterior orbitofrontal cortex). The reward value and the subjective hedonic experience can be modulated by hunger and other internal states. Human neuroimaging experiments have shown that affective sensory and social stimuli affect the activity in various regions of the OFC in similar ways to higher order stimuli such as monetary and esthetic stimuli.

11. Do brain substrates for basic sensory pleasures also participate in mediating higher social, esthetic, or intellectual pleasures?

Berridge: Yes, I think many of the pleasure mechanisms activated in the brain by basic sensory pleasures also participate in at least some higher human pleasures. This reflects the brain's conservation and common currency of neural circuitry for hedonic reaction. However, human higher pleasures also undoubtedly have their own complicated and unique brain signatures and certainly unique routes to activation. Higher cognitive mechanisms of induction are quite different from direct sensory pleasures. It is even conceivable that some few higher pleasures might turn out to be entirely separate from sensory pleasures, involving no overlap at all. But in the end, my bet is on substantial overlap for virtually all pleasures.

Aldridge: If there can be esthetic or intellectual assessments of food, sex, drugs and, rock 'n' roll, and

I believe there are these kinds of appreciation, then I expect that brain representations of these esthetic pleasures would invoke activity in the same pleasure circuits. Perhaps one could make the same kind of argument for social pleasures.

Frijda: I do not know whether brain substrates for "basic sensory pleasures" also participate in "higher" pleasures. There are too many presuppositions in this question. Sensory pleasures seem to me no more basic than the pleasures of behaving without impediment or social pleasures.

Kringelbach: As stated above, it would seem likely that the basic sensory pleasures form building blocks for higher-order such as esthetic and intellectual pleasures. Note also that I regard social pleasure as a basic and necessary pleasure in the mammalian brain. By including the social pleasures in the basic building blocks, it becomes possible to see how higher-order pleasures such as schadenfreude or killjoy can be extracted from the higher-dimensional space of basic sensory, sexual and social pleasures.

12. What are the relative roles in pleasure of subcortical limbic structures versus cortex?

Berridge: Perhaps it is a blow to our cerebral self-image, but the subcortical limbic brain probably contains the chief generating circuitry for many of our most intense pleasures. So far the most effective sensory pleasure generators, at least, have been found by experiments that manipulate subcortical brain structures, such as subcortical nucleus accumbens and connected limbic subcortical sites. For example, activating opioid or related neurochemical signals in those sites is sufficient to directly cause increases in hedonic 'liking' reactions to a sweet pleasure. Likewise, only subcortical lesions (e.g., ventral pallidum) appear to eliminate normal 'liking' reactions to sweetness and to replacing them with negative 'disliking' reactions that are usually associated with bitter or other nasty tastes.

Similar evidence about the causation of pleasure does not yet exist for any region of cortex as far as I know. Even many "anhedonia" patients with cortical lesions may still retain most basic pleasures, despite deficits in how they act on their emotions. However, impressive neuroimaging and electrophysiological activation studies have shown that orbitofrontal and related cortex limbic regions do clearly code pleasure (described by Kringelbach, Small, Schoenbaum, and other authors in this book). And the cortex is undoubtedly a controller of subcortical structures, so that, like a domino falling earlier in the chain, downward causation may give cortex a once-removed role in triggering pleasure via activation of subcortical hedonic circuits. Finally, nearly everyone agrees that the cortex is important to conscious pleasure feelings and to cognitive representations of pleasant events. But it might be truer to characterize the cortex role in subjective feelings as causing the consciousness of an underlying pleasure reaction, rather than causing the basic pleasure reaction itself.

Aldridge: I expect that cortical and subcortical structures cooperate and interact extensively. The anatomy suggests that cortical inputs might "enable" or "gate" activity in subcortical circuits, which can in turn drive activity in cortical circuits. By the patterns of cortical gating, subcortical circuits might have differential levels of access to inform the cortex. In a way then, cortical circuits can control their own inputs. I predict that pleasure can't exist without both cortical and subcortical circuits. Hedonic reactions may not proceed without activity in both.

Gottfried: In some ways, this question captures the basic distinction between emotion and feeling. If pleasure is taken to reflect a biologically meaningful emotional state, then subcortical limbic structures may play the major role. However, if pleasure is taken to reflect subjective positive feeling, then the cortex would have a more prominent role. In all likelihood, the answer is that there is a role for both systems. A neurological syndrome known as "pseudobulbar affect" or "pathological laughing and crying" sheds some light on the topic. This condition was noted by Darwin as long ago as 1872 and characterized in detail by the eminent neurologist Kinnear Wilson in 1924 (for review and discussion, see Schiffer and Pope, 2005). Patients with pseudobulbar affect exhibit spontaneous, intense emotional outbursts, typically laughing or crying, which are usually incongruent to their mood and inappropriate to the immediate situation. This disorder is often observed with bilateral hemispheric lesions of the frontal cortex or internal capsule, and it is thought that the interruption of descending (inhibitory) motor information onto subcortical brainstem structures causes a release (disinhibition) of motor programs underlying emotional expression. Thus, the clinical and pathological features of pseudobulbar affect suggest a heuristically useful dichotomy between emotional control (cortical) and emotional output (subcortical).

Kringelbach: The data suggests that the ancient evolutionary developed brain structures can override our cortical structures. Yet, it is also clear that the cortex, and especially the OFC and ACC regions can also drive subcortical structures. More empirical evidence for the interactions is needed.

Higher Pleasures

13. What is the relation of pleasure to cognition?

Berridge: Pleasure is essentially affective, whereas cognition is not. Cognition and affect are mutually intertwined but never wholly identical. They trigger and modulate each other, but remain distinguishable at least in principle.

Cabanac: Mental objects of cognition possess four dimensions: 1. quality (nature), 2. intensity (magnitude), 3. duration (time), and 4. hedonicity (pleasure/displeasure). Dimensions 1–3 cannot be nil, but dimension 4 can. In that case, hedonicity is indifference.

Aldridge: Pleasure requires cognition. Hedonic reactions don't require cognition. I am assuming that cognition means consciousness.

Frijda: What is the relation of pleasure to cognition? It is like asking what is the relation of one person to another. But if the question means: "is there pleasure without any cognition" the answer depends on the meaning of "cognition" and if cognition means information processing the answer is: no pleasure without information processing because all pleasure is of/about something, including assessment of one's current overall state of functioning.

Shizgal: The machinery of cognition is required in order to produce the experience of pleasure, and pleasure may result from various cognitive activities, such as problem solving. I will address the first of these statements, which relates to my answers to other questions.

In Baars' global workspace theory, we become conscious of a signal, such as a sensation, only once it has gained entry to working memory. Attention plays a crucial role as a gatekeeper to this evanescent, capacity-limited, mnemonic store. Thus, thinking hard about something else should negate the experience of pleasure. Once in working memory, a signal can be accessed by executive processes involved in goal selection and is broadcast to the numerous special-purpose components of the cognitive apparatus that operate outside of awareness. This is essential to the formulation and execution of plans for maximizing, prolonging, and re-initiating pleasurable experiences. These objectives can be attained, to a more limited degree, by lower-level modules (c.f., the example of the sailor described in my answer to question 10), but the crucial element of a stable, long-term plan would be missing, without the intervention of executive processes.

Kringelbach: Some psychologists have tended to see cognition as separate from pleasure, emotion, and motivation. Yet it is difficult to see how cognition could proceed without these processes. Pleasure clearly influences cognition. Take the example of the human dorsolateral prefrontal cortex, which is the structure that many psychologists would point to as the main brain region involved in cognition and higher-order cognitive concepts like working memory and selection for action. It turns out that this brain region also has valenced representations of taste, which could aid higher cognitive processes in guiding complex motivational and emotional behavior (Kringelbach et al., 2004).

Similarly, neurophysiological recordings in a reward preference task have demonstrated that neurons in the dorsolateral prefrontal cortex encode both the reward amount and the monkeys' forthcoming response, while neurons in the orbitofrontal cortex more often encode the reward amount alone (Wallis and Miller, 2003). It would seem high time to integrate pleasure, motivation, and emotion into the cognitive neurosciences. As an example, Dickinson and Balleine (in this volume) have argued that subjective pleasure may allow animals to have declarative goals with can come to guide flexible cognition.

14. What is the relation of pleasure to social cognition?

Berridge: Social cognition is a distinctive trigger, though I think its pleasure shares brain circuitry with nonsocial pleasures.

Cabanac: The relation is the same as with other cognitive objects. Hedonicity indicates what is (or what was in the evolutionary past of our species) useful. With sensations, pleasure indicates physiological usefulness; with social cognition, pleasure indicates social usefulness. In the case of aggressiveness, passive behaviors and highly aggressive behaviors arouse displeasure. But medium-intensity aggression can be agreeable (to the aggressor).

Aldridge: I don't know.

Kringelbach: Pleasure is central to social interaction, which in its simplest form is not a higher pleasure but a basic pleasure, as argued above. Our liking of infant faces is an example of such a basic social pleasure. Darwin pointed out that in order for infants to survive and to perpetuate the human species, adults need to respond and care for their young and Lorenz proposed that it is the specific structure of the infant face that serves to elicit these parental responses. Using MEG, we have recently found a key difference in the early brain activity of normal, nonparental adults to infant faces compared to adult faces (Kringelbach et al., 2008). Only infant faces elicited early activity at around 130 milliseconds in the medial OFC, which has previously been shown to reflect the reward value of a wide variety of stimuli, where the brain activity was correlating with their reported pleasantness. Higher-order social cognition such as theory of mind arises later in primate development but it is likely to build on combinations of the basic pleasures.

15. What is the relation between language and pleasure?

Berridge: The way we talk about pleasure is perhaps why the conscious feeling traditionally has been its defining feature. But dictionary definitions are never the last word on the true nature of any psychological process.

Cabanac: As with any other mental experience, pleasure indicates what to decide. We have evidence that participants selected the grammatical formulas that gave them pleasure and avoided those that arose displeasure. Thus, grammatical optimization is achieved through the maximization of pleasure.

Aldridge: I don't know. Perhaps language is important for representations of some kinds of pleasure such as intellectual or esthetic pleasure.

Frijda: There is pleasure without language. Ask my cat (she can't talk but can purr).

Kringelbach: Pleasure is possible without language as argued above. Human language and our subsequent linguistic reports of subjective experience may, however, come to change our pleasure. The evidence also suggests that we have limited conscious access to nonconscious processing and that at least some subjective linguistic reports are post hoc and confabulatory (e.g., Johansson et al., 2005).

16. How do sensory pleasures relate to higher positive affects generated by social-cognitive (social pleasures, money) or esthetic (art, music) or moral (altruistic or transcendent loves)?

Berridge: Again, I think that overlap exists. Even unique human cultural pleasures may be pleasurable precisely because they act as new psychological keys in the same old brain hedonic locks that generate sensory pleasure. Of course, massive differences also exist between sensory pleasure and some higher pleasures, and a few higher affects might turn out to entirely different from sensory pleasures. Still, overlap is the rule for many.

Cabanac: These pleasures permit to rank the behavioral responses in terms of which to satisfy first. The hedonic dimension of consciousness is what triggers decisions in that realm of activity also. Pleasure is the common currency that ranks the urgencies. The most pleasant (or least unpleasant) is always ranked first. If sensory pleasure is more intense than esthetics, then physiology will be satisfied first. If altruistic pleasure is more intense than money, then moral behavior will be accomplished first.

Aldridge: One may like the feel of money, the sound of music, the sight of paintings. These sensations might trigger pleasure representations in the brain.

Frijda: Interesting question. I do not know.

Gottfried: What sets these higher positive affects apart from the "lower" sensory pleasures is that by and large they represent "civilized" pleasures unique to humans. Frequently these higher-order pleasures are abstractions of biologically salient stimuli or affective states. After the manner of learning theorists, with their models of S-S (stimulus–stimulus) and S-R (stimulus–response) learning, one could reasonably think of these distinctly human pleasures as a form of "I-S learning," whereby a positive stimulus (S) is effectively linked to an idea (I). These ideas could be concrete or abstract, and might take the form of symbols, signs, multisensory perceptual events, mental states, concepts, or thoughts, such as pounds sterling, or sonatas, or sapphires, or love. In this framework, any organism with the capacity for abstractive learning (what one might call I-S learning) needs to be able to satisfy certain criteria. First, it needs to be able to have a central nervous system, for the high level of integrative processing necessary for I-S learning could not be accomplished without a brain. This basic stipulation would disqualify many animals (and perhaps some humans). Second, it needs to be able to store and retain (neural) representations of ideas.

Third, it needs to be able to form associations between idea representations and pleasurable sensory (or affective) representations. Based on the distinctive evolutionary features of the human brain, I would speculate that the prefrontal cortex is critical for the development and realization of these civilized pleasures. As a final note, insofar as the experience of pleasure has biological (survival) value, it follows that higher positive affects should be behaviorally beneficial. Some of the earliest examples of abstract pleasure are found in the cave paintings at Lascaux, circa 15,000 B.C. Here, the cavemen's depiction of bison and woolly mammoths probably had less to do with esthetic contemplation, and more to do with finding ways of overcoming their innate fears of these large beasts, in order to improve their chances on the hunt.

Kringelbach: Higher-order pleasures are likely to be higher-dimensional combinations of the basic sensory and social pleasures and as such may re-use some of the same brain mechanisms. The inclusion of social pleasures in the basic pleasures makes this into a higher-dimensional space of which it becomes easier to form even apparently maladaptive pleasures such as schadenfreude and killjoy.

17. In what ways are pleasure and happiness linked?

Berridge: Happiness cannot be reduced to pleasure alone. But the attainment of happiness must surely include the ready capacity for pleasure reactions.

Cabanac: There is a fundamental misunderstanding with the word "happiness". Because hedonicity is the common currency that allows motivations to "talk" to one another, the mechanisms must be homologous to all motivations. Thus happiness (general) must follow the same rules as comfort (physiology). In physiology, comfort is the absence of hedonic experience. Thus, comfort is indifference and can be stable. On the other hand, pleasure indicates that a stimulus is useful, and maximizing pleasure optimizes behavior. But, as soon as we have maximized pleasure, we thus reduce the physiological need and usefulness disappears. Thus, pleasure is always transient, while comfort is stable and can be permanent.

The general case follows the same rules as sensation and physiology. The equivalent of pleasure is joy; the equivalent of comfort is happiness. Thus joy is hedonically positive but transient and happiness is indifferent and stable.

Aldridge: I don't know. Maybe they are the same thing.

Frijda: There is no happiness without pleasure; there is much pleasure without happiness. Pleasure is a core evaluative process; happiness is an emotion or long-term evaluation.

Kringelbach: Pleasure is but a fleeting moment in the state which is happiness. It is possible that "true" happiness or bliss might be a state of 'liking' without 'wanting', which with the current available neuroscientific evidence is becoming a testable hypothesis.

References

Anderson, A. K., Christoff, K., Stappen, I., Panitz, D., Ghahremani, D. G., Glover, G., Gabrieli, J. D. E. and Sobel, N. (2003) Dissociated neural representations of intensity and valence in human olfaction. *Nat. Neurosci.* **6,** 196–202.

Becerra, L., Breiter, H. C., Wise, R., Gonzalez, R. G. and Borsook, D. (2001) Reward circuitry activation by noxious thermal stimuli. *Neuron* **32,** 927–946.

Bender, G., Mak, Y. E. and Small, D. M. (2005) The interaction between evaluative and passive response to taste in the human brain. *Chemical Senses* **30,** A188.

Bentham, J. (1789) *An Introduction to the Principles of Morals and Legislation.* Library of Economics and Liberty. Retrieved September 16, 2005 from http://www.econlib.org/library/Bentham/bnthPML1.html.

Berridge, K. C. and Robinson, T. E. (2003) Parsing reward. *Trends Neurosci.* **26,** 507–513.

Cabanac, M. (1971) Physiological role of pleasure. *Science* **173,** 1103–1107.

Craig, A. D. (2003) A new view of pain as a homeostatic emotion. *Trends Neurosci.* **26,** 303–307.

de Araujo, I. E., Rolls, E. T., Kringelbach, M. L., McGlone, F. and Phillips, N. (2003) Taste–olfactory convergence, and the representation of the pleasantness of flavour, in the human brain. *Eur. J. Neurosci.* **18,** 2059–2068.

Gottfried, J. A., O'Doherty, J. and Dolan, R. J. (2003) Encoding predictive reward value in human amygdala and orbitofrontal cortex. *Science* **301,** 1104–1107.

Johansson, P., Hall, L., Sikstrom, S. and Olsson, A. (2005) Failure to detect mismatches between intention and outcome in a simple decision task. *Science* **310,** 116–119.

Kahneman, D., Diener, E. and Schwarz, M. (2003) *Well-Being: The Foundations of Hedonic Psychology.* Russell Sage Foundation: New York.

Kringelbach, M. L. (2004) Food for thought: Hedonic experience beyond homeostasis in the human brain. *Neuroscience* **126,** 807–819.

Kringelbach, M. L. (2005) The human orbitofrontal cortex: Linking reward to hedonic experience. *Nat. Rev. Neurosci.* **6,** 691–702.

Kringelbach, M. L., O'Doherty, J., Rolls, E. T. and Andrews, C. (2003) Activation of the human orbitofrontal cortex to a liquid food stimulus is correlated with its subjective pleasantness. *Cereb. Cortex* **13,** 1064–1071.

Kringelbach, M. L., Jenkinson, N., Green, A. L., Owen, S. L., Hansen, P. C., Cornelissen, P. L., Holliday, I. E., Stein, J. and Aziz, T. Z. (2007a) Deep brain stimulation for chronic pain investigated with magnetoencephalography. *Neuroreport* **18,** 223–228.

Kringelbach, M. L., Jenkinson, N., Owen, S. L. and Aziz, T. Z. (2007b) Translational principles of deep brain stimulation. *Nat. Rev. Neurosci.* **8,** 623–635.

Kringelbach, M. L., Lehtonen, A., Squire, S., Harvey, A. G., Craske, M. G., Holliday, I. E., Green, A. L., Aziz, T. Z., Hansen, P. C., Cornelissen, P. L. and Stein, A. (2008) A specific and rapid neural signature for parental instinct. *PLoS ONE* **3,** e1664. doi:1610.1371/journal.pone.0001664.

LaBar, K. S., Gitelman, D. R., Parrish, T. B., Kim, Y. H., Nobre, A. C. and Mesulam, M. M. (2001) Hunger selectively modulates corticolimbic activation to food stimuli in humans. *Behav. Neurosci.* **115,** 493–500.

Lamb, R. J., Preston, K. L., Schindler, C. W., Meisch, R. A., Davis, F., Katz, J. L., Henningfield, J. E. and Goldberg, S. R.

(1991) The reinforcing and subjective effects of morphine in post-addicts: a dose–response study. *J. Pharmacol. Exp. Ther.* **259,** 1165–1173.

Menon, V. and Levitin, D. J. (2005) The rewards of music listening: Response and physiological connectivity of the mesolimbic system. *NeuroImage* **28,** 175–184.

Morris, J. S., Ohman, A. and Dolan, R. J. (1998) Conscious and unconscious emotional learning in the human amygdala.[comment]. *Nature.* **393,** 467–470.

Morris, J. S., DeGelder, B., Weiskrantz, L. and Dolan, R. J. (2001) Differential extrageniculostriate and amygdala responses to presentation of emotional faces in a cortically blind field. *Brain* **124,** 1241–1252.

O'Doherty, J., Rolls, E. T., Francis, S., Bowtell, R. and McGlone, F. (2001) Representation of pleasant and aversive taste in the human brain. *J. Neurophysiol.* **85,** 1315–1321.

O'Doherty, J., Deichmann, R., Critchley, H. D. and Dolan, R. J. (2002) Neural responses during anticipation of a primary taste reward. *Neuron* **33,** 815–826.

Paton, J. J., Belova, M. A., Morrison, S. E. and Salzman, C. D. (2006) The primate amygdala represents the positive and negative value of visual stimuli during learning. **439,** 865–870.

Pelchat, M. L., Johnson, A., Chan, R., Valdez, J. and Ragland, J. D. (2004) Images of desire: Food-craving activation during fMRI. *NeuroImage* **23,** 1486–1493.

Robinson, T. E. and Berridge, K. C. (1993) The neural basis of drug craving: an incentive-sensitization theory of addiction. *Brain Res. Brain Res. Rev.* **18,** 247–291.

Rolls, E. (2005) *Emotions Explained.* Oxford University Press: Oxford.

Rolls, E. T., Kringelbach, M. L. and de Araujo, I. E. (2003) Different representations of pleasant and unpleasant odours in the human brain. *Eur. J. Neurosci.* **18,** 695–703.

Ryle, G. (1954) Pleasure. *Proc. Aristotelian Soc.* **28,** 135–146.

Schiffer, R. and Pope, L. E. (2005) Review of pseudobulbar affect including a novel and potential therapy. *J. Neuropsychiatr. Clin. Neurosci.* **17,** 447–454.

Schroeder, T. (2004) *Three Faces of Desire.* Oxford University Press: Oxford.

Shimojo, S., Paradiso, M. and Fujita, I. (2001) What visual perception tells us about mind and brain. *Proc. Natl. Acad. Sci. USA* **98,** 12340–12341.

Small, D. M., Zatorre, R. J., Dagher, A., Evans, A. C. and Jones-Gotman, M. (2001) Changes in brain activity related to eating chocolate: From pleasure to aversion. *Brain* **124,** 1720–1733.

Small, D. M., Gregory, M. D., Mak, Y. E., Gitelman, D., Mesulam, M. M. and Parrish, T. (2003) Dissociation of neural representation of intensity and affective valuation in human gustation. *Neuron* **39,** 701–711.

Strack, F., Martin, L. L. and Stepper, S. (1988) Inhibiting and facilitating conditions of the human smile: a nonobtrusive test of the facial feedback hypothesis. *J. Pers. Social Psychol.* **54,** 768–777.

Wallis, J. D. and Miller, E. K. (2003) Neuronal activity in primate dorsolateral and orbital prefrontal cortex during performance of a reward preference task. *Eur. J. Neurosci.* **18,** 2069–2081.

Winkielman, P., Berridge, K. C. and Wilbarger, J. L. (2005) Unconscious affective reactions to masked happy versus angry faces influence consumption behavior and judgments of value. *Pers. Soc. Psychol. Bull.* **31,** 121–135.

Zubieta, J. K., Bueller, J. A., Jackson, L. R., Scott, D. J., Xu, Y., Koeppe, R. A., Nichols, T. E. and Stohler, C. S. (2005) Placebo effects mediated by endogenous opioid activity on mu-opioid receptors. *J. Neurosci.* **25,** 7754–7762.

PART I

ANIMAL PLEASURES

1

Hedonic Hotspots: Generating Sensory Pleasure in the Brain

KYLE S. SMITH, STEPHEN V. MAHLER, SUSANA PECIÑA,
AND KENT C. BERRIDGE

A vital question concerning sensory pleasure is how brain mechanisms cause stimuli to become pleasurable and liked. Pleasure is not an intrinsic feature of any stimulus, but instead reflects an affective evaluation added to the stimulus by the brain. That is, as Frijda expresses it (Frijda, 2006, Chapter 6, this book), a pleasure gloss or hedonic value must be actively "painted" on sweet or other sensations to make them pleasant. Brain mechanisms of pleasure, whatever they are, must take a mere sensory signal and transform it into a hedonic and 'liked' reward.

Finding the brain mechanisms responsible for painting a pleasure gloss is a major challenge for affective neuroscience (Barrett and Wager, 2006; Berridge, 2003b; Damasio, 1999; Davidson and Irwin, 1999; Kringelbach, 2005, Chapter 12, this book; LeDoux, 1996; Panksepp, 1991; Peciña et al., 2006). Fortunately, progress on finding hedonic generators in the brain is being made. In this chapter, we focus specifically on the neuroanatomical hedonic hotspots in the brain where neurochemical signals actually contribute causally to the generation of pleasure.

We define a hedonic hotspot as a brain site where pleasure mechanisms are sufficiently concentrated together in one anatomical locus to cause pleasure enhancement when neurally activated (while recognizing that a hotspot's contribution to pleasure enhancement depends also on its participation in larger brain circuits). A hotspot might also be a site where natural pleasures are reduced below normal levels by neural suppression or damage. However, being able to enhance or stimulate pleasure is slightly different from being needed for normal pleasure, and so the contributions of "sufficient cause" (enhancement) and "necessary cause" (normal) need to be assessed separately. Finally, these and many other brain substrates may code the occurrence a pleasurable event by their neural activation. But, for many substrates, the neural activation may be neither necessary nor sufficient to produce pleasure (and presumably, those substrates instead transfer pleasure information to cause other functions that guide decisions) (Dickinson and Balleine, Chapter 4, this book; Kringelbach, Chapter 12, this book). A particular brain substrate may have all three of these hedonic roles (code, sufficient cause, and necessary cause), but alternatively, it may code without causing, or it may act as a sufficient cause but not a necessary cause (Table 1.1). For example, some sites in nucleus accumbens may enhance pleasure but may not be needed for all normal pleasures. It is an important task for affective neuroscience to assign pleasure causation to the proper brain substrates. The goal of this chapter is to identify some of the hedonic hotspots in the brain most able to cause enhancements of pleasure.

To identify a neural substrate that causes pleasure, it is often helpful to manipulate the brain and observe whether this manipulation causes a change in hedonic reactions to a sensory stimulus. Using experimental techniques to manipulate neurochemicals in focused brain locations, we have recently mapped several hedonic hotspots that contribute in causal ways to

Types of Roles in Pleasure

"Sufficient Cause" Example Nucleus Accumbens 	**Neural stimulation is sufficient to cause increase in pleasure** Caveat: Causation is distributed beyond stimulated substrate *The stimulated substrate doesn't contain all causation itself, but rather interacts with other distributed components of a larger brain circuit to cause pleasure. Condition of other brain substrates and external events may modulate impact of neural activation.*
"Necessary Cause" Example Ventral Pallidum 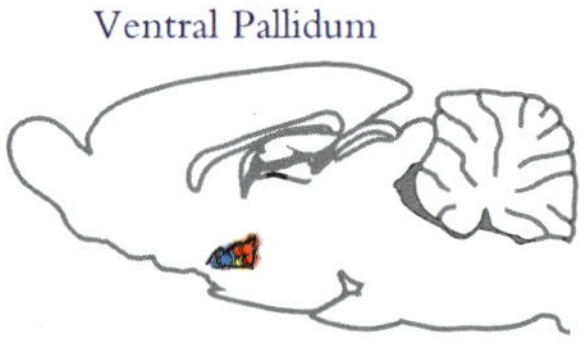	**Neural blockade/lesion produces loss of pleasure** Caveat: Deficit may not always be mirror image of normal function *Loss of pleasure after a lesion may mean that the substrate was the pleasure generator, but alternatively could mean that its function was to facilitate pleasure generation in other structures that still remain (e.g. removal of a transistor may make a radio sequeal, but the transistor's funtion was not merely a 'squeal suppressor').*
"Code" Example Orbitofrontal Cortex 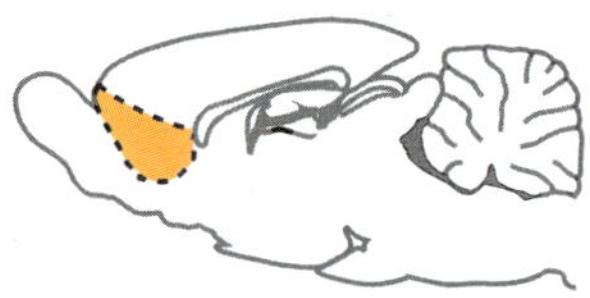	**Neural activation during pleasure** Caveat: Code may or may not be cause *Some neural activations may cause the pleasure they code. Other neural activations may be instead a consequence of a hedonic reaction generated elsewhere in the brain, rather than cause the hedonic reaction themselves (and presumably help cause some other psychological function).*

Table 1.1 Types of Pleasure Mediation: Sufficient Cause, Necessary Cause, and Code. The phrase "brain structure X mediates pleasure" has three different possible meanings, which may or may not coincide, though they are often meant together. It is useful to distinguish between cause and code, and even to distinguish among different ways of causing (caveats apply to each shorthand distinction). Examples are neither exhaustive nor exclusive (e.g., ventral pallidum also codes pleasure, and orbitofrontal cortex may turn out to cause pleasure); see text for discussion.

pleasure. These hotspots are scattered across locations that span almost the entire brain and are embedded in a larger pleasure circuit in the brain that operates as a whole to increase hedonic experience.

How Can We Measure Hedonic 'Liking' in Animals?

Traditional studies of pleasure 'liking' have focused on human adult subjects who can describe their feelings (Cabanac, 1971). But how can pleasure be measured in nonverbal animals like rats, in which most research on neurobiological causes must be conducted? The premise that underlies our affective neuroscience research on hedonics is that 'liking' is a basic evaluative reaction of the brain, with objective neural and behavioral indicators that can be quantified by appropriate methods in animals and humans alike.

These objective indicators include emotional facial expressions (Berridge, 2000; Darwin, 1872; Ekman, 1999). Many animals including humans, primates, and rats exhibit homologous, hedonic, and aversive facial reactions to pleasant and unpleasant tastes. For example, a human infant, even on its first day of life, will rhythmically lick its lips when a drop of sugar water is placed on its tongue (Steiner, 1973). By contrast, a bitter taste like quinine elicits characteristic aversive reactions including gaping of the mouth (Steiner, 1973). Like humans, nonhuman primates and rodents display homologous 'liking' and 'disliking' reactions to sweet and bitter tastes (Steiner et al., 2001). Rats, for example, display the same rhythmic tongue protrusions as human infants, as well as paw-licking and related movements, when presented with a sugary solution in the mouth (Grill and Berridge, 1985; Grill and Norgren, 1978a) (Figure 1.1). Similarly, in response to a bitter taste, rats emit the same aversive gaping reactions that human infants show, along with headshakes and frantic wiping of the mouth (Figure 1.1).

Importantly, these animal affective reactions fluctuate in similar ways to human subjective pleasure when relevant circumstances change (Berridge, 2000). For example, just as food is more pleasant to us when hungry, sweet tastes elicit more 'liking' reactions when rats are hungry than when they are full (Berridge, 2000; Cabanac, 1971). Such homeostatically induced changes in sensory pleasure have been called "alliesthesia" (Cabanac, 1971, Chapter 7, this book; Leknes and Tracey, Chapter 19, this book). Similarly, the intense taste of salt at concentrations higher than seawater is not pleasant to either people or rats, and normal rats accordingly respond to this taste with gapes and other aversive reactions. However, if one physiologically depletes a rat of sodium, thus eliciting a state of "salt appetite," affective reactions to this very same salty taste suddenly flip from negative to positive and hedonic tongue protrusions are observed instead of aversive gapes (Berridge et al., 1984; Schulkin, 1991; Tindell et al., 2006). Thus 'liking' facial reactions to tastes reflect not simply the sensory properties of the stimulus, but rather a hedonic evaluation of it that incorporates physiological needs. 'Liking' reactions also incorporate psychological influences on hedonic

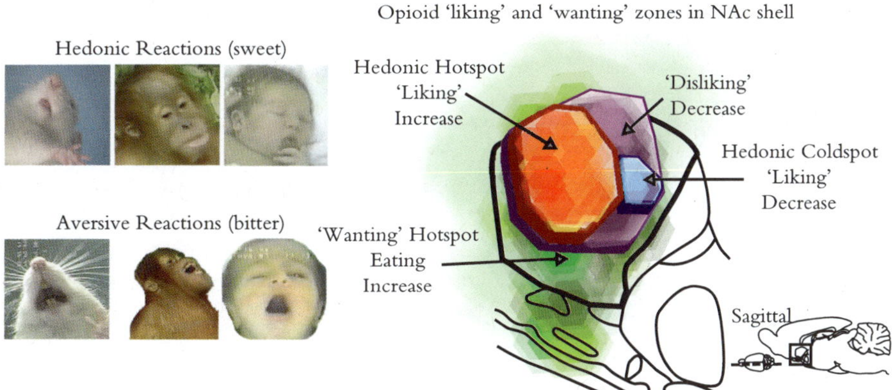

Figure 1.1 Taste 'liking' reactions and contrast map of nucleus accumbens hotspots. Positive 'liking' reactions to pleasant sweet tastes shared by human newborn, young orangutan, and adult rat (tongue protrusion; left top), and aversive 'disliking' reactions to unpleasant bitter tastes (gape; left bottom). Affective facial expressions like these provide an objective index of 'liking' and 'disliking' reactions to the hedonic impact of tastes. Opioid hotspots and coldspots for hedonic 'liking.' 'disliking,' and motivational 'wanting' are mapped and stacked within the nucleus accumbens (medial shell region shown in sagittal view; right). Virtually the entire medial shell stimulates 'wanting' for reward (e.g., increased food intake) in response to opioid stimulation (green hexagons represent individual microinjection Fos plumes) and so do other nearby structures including the core of nucleus accumbens as well as parts of the ventral neostriatum above the accumbens, and the olfactory tubercle beneath the accumbens. The much smaller hedonic hotspot for 'liking,' where opioid stimulation actually increases positive hedonic reactions to sucrose taste (red), is contained within the anterior and dorsal quarter of shell. 'Liking' reactions to sucrose are reduced by opioid stimulation in a small posterior hedonic coldspot (though still stimulating 'wanting'; blue), whereas an intermediate region that contains both hotspot and coldspot mediates opioid suppression of aversive 'disliking' for bitter quinine (purple). The hotspot zone map is modified from Peciña and Berridge (2005).

impact, such as preference learning, as well as many neural factors (Berridge, 2000). For these reasons, we and other neuroscientists have been able to use such affective reactions in rodents as an index of core 'liking' or hedonic impact. By measuring how 'liked' or 'disliked' a particular taste is, we can then experimentally manipulate 'liking' and 'disliking' to reveal the neural mechanisms responsible for adding pleasure to gustatory sensation.

Neuroscience Tools for Identifying Hedonic Hotspots: Microinjections and Fos Plumes

One way to reveal brain substrates that cause 'liking' is to activate a mechanism in a neuroanatomically and neurochemically focused fashion in order to observe increased hedonic reactions characteristic of pleasure impact. A useful technique is microinjection of a tiny droplet of drug directly into the brain, because it can painlessly activate a neural system in a fashion that is highly specific both neuroanatomically and neurochemically (it is painless because it is made via the intracranial cannulae that were previously implanted under anesthesia). If a microinjection boosts the pleasure processes, then the brain site must contain a neurochemically coded mechanism sufficient to cause amplification of hedonic impact.

For precision mapping of hedonic mechanisms, however, it is not enough to know where a drug has been microinjected in the brain. Drugs diffuse from the site of injection, which makes pinpointing functional 'liking' effects imprecise unless one knows exactly how far the impact spreads. To help pinpoint 'liking' mechanisms, we have developed a microinjection "Fos plume" mapping tool based on the ability of many drugs to activate local protein production in neurons they impact (Mahler et al., 2007; Peciña and Berridge, 2000; 2005; Peciña et al., 2006; Smith and Berridge, 2005). A drug that activates neurotransmitter receptors on a neuron can trigger intracellular second messengers and cascades of molecular signals to the cell nucleus to quickly make proteins that influence the neuron's function.

As a step to altering neuronal function, several reward-related drugs trigger transcription of the c-*fos* gene on neuronal chromosomes into RNA and its translation into the Fos protein. Thus, looking to see which neurons produce more Fos protein is a useful way to see which neurons are most activated by a drug microinjection. The functional spread of Fos activation can be seen as a plume of darkly stained neurons around a drug microinjection site when a slice of brain tissue is chemically processed in a way that stains Fos-containing neurons. The size of this Fos plume reveals the location in the brain a drug microinjection has acted when it elevates the 'liking' reactions to sweet taste. By assigning the behavioral 'liking' enhancements that we observe to the particular microinjection Fos plumes that cause them, we can map objective and precise plots of hedonic hotspots in the brain (Mahler et al., 2007; Peciña and Berridge, 2000; 2005; Peciña et al., 2006; Smith and Berridge, 2005).

Hedonic Hotspots

It turns out that several brain limbic structures have tiny pleasure-generating sites tucked within them—hedonic hotspots. These hotspots combine together to form a distributed causal circuit to add pleasure to sensory experience.

Opioid Hedonic Hotspot in the Nucleus Accumbens

We have identified an opioid hedonic hotspot of approximately 1 cubic millimeter volume within the medial shell of the nucleus accumbens in rats. If the hotspot size is proportional to the whole brain size, then in humans, the hotspot will be roughly 750 times bigger in volume than in the rat, or between 0.7 and 1 cm^3 (Figures 1.1 and 1.2). The nucleus accumbens has long been linked to affective processes (Balleine, 2005; Berridge, 2003b; Cardinal et al., 2002; Kelley et al., 2002; Knutson and Cooper, 2005; Leyton, Chapter 13, this book; Panksepp, 1991; Petrovic, Chapter 17, this book; Roitman et al., 2005; Taha and Fields, 2005; Veldhuizen et al., Chapter 9, this book; Yamamoto, 2006). However, the existence of localized hotspots within nucleus accumbens specialized to amplify hedonic impact was not previously known. At the most, distinctions have been made between the large shell and core subregions of nucleus accumbens, but the shell has generally been presumed to act as a whole in generating hedonic impact.

The discovery of a specialized hotspot shows that reality is more complex. The specialized opioid hedonic hotspot constitutes only one-third of the medial portion of shell (Figures 1.1 and 1.2). This hotspot is only about one-fifth of the whole shell (medial and lateral parts combined) and only one-seventh of the entire

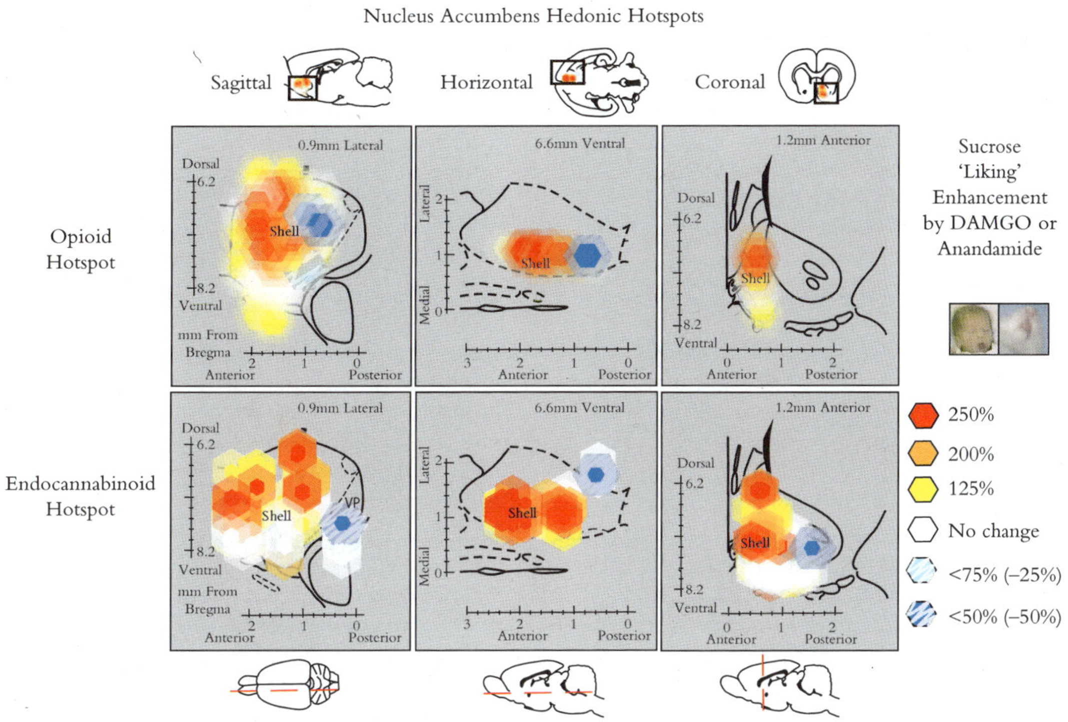

Figure 1.2 Opioid and endocannabinoid hedonic hotspots in the nucleus accumbens. In these Fos plume maps, symbol color denotes the intensity of sucrose 'liking' amplification by opioid microinjection (yellow-to-red = increase above normal) and symbol size denotes estimated drug functional spread based on Fos plumes. Top row depicts the medial shell containing the opioid hedonic hotspot in sagittal, horizontal, and coronal views (modified from Peciña and Berridge, 2005). The hedonic hotspot for 'liking' is revealed in the anterior shell, whereas a smaller coldspot for 'liking' suppression is revealed behind it (blue = decrease below normal). Bottom row depicts the endocannabinoid hedonic hotspot that overlaps the same location (modified from Mahler et al., 2007). The endocannabinoid hotspot covers the entire opioid hotspot and may be larger, but anatomically both accumbens hedonic hotspots are roughly similar.

nucleus accumbens (shell and core). Thus, this opioid hotspot represents a focused area that is specialized for hedonic causation within the nucleus accumbens.

The critical finding is that within the 1 mm^3 hedonic hotspot, microinjection of an opioid drug (DAMGO) that stimulates mu opioid receptors—mu receptors are one of several subtypes of receptors for opioid neurotransmitters—elevates hedonic 'liking' reactions to a sucrose taste by up to quadruple the usual number (Peciña and Berridge, 2005). This accumbens hedonic hotspot is located in the dorsal part in the anterior half of the medial shell. In terms of anatomical landmarks, the cubic millimeter hotspot is just anterior to the posterior edge of the islands of Calleja but posterior to the rear edge of the dorsal tenia tecta and the lateral septum and anterior to the level of the anterior commissure.

An equivalent nucleus accumbens hotspot might well be expected to play hedonic roles in humans, perhaps mediating the intensely rewarding effects of opiate drugs themselves (e.g., heroin), as well as mediating the pleasure of such natural rewards as the taste of sugar—and it seems noteworthy that food pleasantness in humans is modulated by systemic administration of opioid drugs (Yeomans and Gray, 2002).

At all other parts of the nucleus accumbens tested so far in rats, microinjection of the same opioid drug fails to increase hedonic 'liking' reactions to sweetness. These other areas include the posterior or ventral subregions of medial shell and as far as we know the core as well. In fact, DAMGO microinjections in a small cold spot in the posterior half of the medial shell actually appear to suppress 'liking' reactions to sucrose below vehicle-control levels (Figure 1.1).

In contrast to the tight localization of 'liking' mechanisms in the hotspot, motivational 'wanting' mechanisms appear to be widely distributed throughout almost all of the medial and lateral shell and probably also extend to cover the core of nucleus accumbens and ventral neostriatum (Figure 1.1). For example, DAMGO microinjection robustly stimulates increases in eating behavior and food intake at all of those accumbens sites (Bakshi and Kelley, 1993; Kelley et al., 2002; Peciña and Berridge, 2005; Zhang and Kelley, 2000). Thus, for opioid mechanisms of reward, the nucleus accumbens hotspot generates both 'liking' and 'wanting' for sweet tastes, whereas other areas of the nucleus accumbens can generate only 'wanting' (Figure 1.1).

Of course, a drug microinjection is an unnatural stimulus, and brains ordinarily would not experience such intense or localized chemical stimulation. Still, brains do experience many naturally induced increases in normal opioid neurotransmitter release, which might have different effects in different locations. Microinjection maps essentially use an artificial manipulation to reveal brain mechanisms that paint a pleasure gloss onto sensation in ordinary life. We have focused here on enhancing the 'liking' of sweetness, because that is what we are most able to test. A number of questions remain open, such as whether the same mechanisms paint pleasure onto other sensations too, or whether pleasure would be generated even in the absence of any sensory stimulus. Current evidence suggests the nucleus accumbens participates in many rewards for people and animals, including sex, music, drugs, social rewards, humor, winning money, and so on (Carelli and Wightman, 2004; Gottfried, Chapter 8, this book; Insel and Fernald, 2004; Kalivas and Volkow, 2005; Knutson and Cooper, 2005; Komisaruk and Whipple, 2005; Komisaruk et al., Chapter 10, this book; Leknes and Tracey, Chapter 19, this book; Leyton, Chapter 13, this book; Menon and Levitin, 2005; Mobbs et al., 2003; Robbins and Everitt, 1996; Robinson and Berridge, 1993; Skov, Chapter 16, this book; Wang and Aragona, 2004). Still, more research is needed on hotspot roles in such diverse pleasures. For now, we can only say that, if the brain is organized parsimoniously and uses a "common neural currency" to mediate multiple kinds of pleasures, the answer to questions about other pleasures may well turn out to be "yes."

Endocannabinoid Hedonic Hotspot in the Nucleus Accumbens

Opioid signals are not the only neurochemical signals in nucleus accumbens that cause increases in pleasure. Endocannabinoids are another type of natural brain messengers and are chemically similar to plant cannabinoids such as Δ^9-THC, a chief psychoactive ingredient in marijuana. An example is anandamide, a brain endocannabinoid named after the word for *bliss* in Sanskrit (an endocannabinoid is a natural brain messenger that is chemically similar to a cannabinoid drug). Cannabinoid drugs have appetite-enhancing effects and increase intake of palatable food and sucrose solution in rats and humans (Hart et al., 2002; Kirkham, 2005).

Endocannabinoid and opioid receptors sometimes coexist on the same neurons in the accumbens shell and have been found nearly side by side on the same spine of the same dendrite on neurons in striatum (Pickel et al., 2004; Schoffelmeer et al., 2006). Thus, both opioid and cannabinoid receptors may exist in many of the same synapses within the hedonic hotspot and beyond (Rios et al., 2006; Schoffelmeer et al., 2006). The two signals might also interact in function. For example, opioid blockers (e.g., naloxone) have been shown to prevent many cannabinoid drug effects (including food intake enhancements) and vice versa (Tanda et al., 1997; Williams and Kirkham, 2002).

Anandamide signals in the nucleus accumbens participate in generating sensory pleasure similar to opioid signals. We have identified an endocannabinoid hedonic hotspot in the nucleus accumbens for enhancing sweetness 'liking,' which seems to completely cover the opioid hotspot already described (and possibly extend beyond it) (Mahler et al., 2007) (Figure 1.2). Microinjections of anandamide directly into this 1.6 mm^3 hotspot, located in the dorsal portion of the medial nucleus accumbens shell, doubled hedonic 'liking' reactions to sucrose above normal levels (Figure 1.2).

The endocannabinoid hotspot for 'liking' may be slightly larger than the opioid hotspot although differences in the experiments that mapped them make it difficult to compare sizes directly. In any case, in the same endocannabinoid hotspot, anandamide also doubled the amount of food eaten and the time spent engaged in eating behavior. These results indicate that anandamide signals, like mu opioid signals in its overlapping hotspot in medial shell of accumbens, enhance both hedonic 'liking' of tasty rewards and 'wanting' to consume those rewards.

The enhancement of affective 'liking' reactions by anandamide appears specific to positive 'liking' and not negative 'disliking.' In contrast to its amplification of positive hedonic reactions to sucrose, anandamide failed to change affective reactivity to a bitter taste of quinine. Selective amplification of sweet 'liking' may possibly suggest a hedonic explanation of why the "marijuana munchies" are often directed toward especially palatable foods, as well as reveal an endogenous brain mechanism for generating the pleasure gloss for natural sensations.

Opioid Hedonic Hotspot in the Posterior Ventral Pallidum

One of the major output structures for nucleus accumbens reward signals is the ventral pallidum, a forebrain structure located just posterior to the nucleus accumbens near the bottom of the brain (Heimer and Wilson, 1975). The ventral pallidum is a limbic "final common pathway." It receives projections from a host of reward-related brain areas in addition to the nucleus accumbens, such as amygdala, orbitofrontal cortex, anterior cingulate cortex and infralimbic cortex, lateral hypothalamus, ventral tegmental area, and parabrachial nucleus. In turn, the ventral pallidum projects reciprocally to many of them, including the nucleus accumbens, and projects upward to the forebrain's mediodorsal nucleus of the thalamus to form a limbic-cortico-limbic loop, connecting to the limbic prefrontal cortex and back down to the accumbens and ventral pallidum (Aldridge and Berridge, Chapter 3, this book; Grove, 1988a,b; Haber et al., 1985; Kalivas and Nakamura, 1999; Mogenson and Yang, 1991 ; Zahm, 2000). Thus, anatomically, the ventral pallidum is in a key position to mediate pleasure signals in the brain.

In fact, it does. In mapping sites where microinjections cause 'liking' enhancement, we have found that the ventral pallidum contains its own opioid hedonic hotspot. The ventral pallidum's hedonic hotspot is an approximately 0.80 mm^3 area in its posterior end where mu opioid stimulation magnifies hedonic 'liking' (Smith and Berridge, 2005) (Figure 1.3). This hotspot is slightly smaller than the 1 mm^3 nucleus accumbens opioid 'liking' hotspot, but it is roughly equal in the proportion of the structure that it fills. The ventral pallidum is only about two-thirds the size of the accumbens medial shell, so both hotspots fill approximately one-third to one-half of their containing structure.

The hedonic features of the ventral pallidum hotspot are similar to those of the nucleus accumbens. In the posterior hotspot, microinjections of the mu opioid agonist DAMGO roughly double the number of hedonic 'liking' reactions to a sucrose taste compared to control-vehicle microinjections (Smith and Berridge, 2005) (Figure 1.3). Opioid receptor activation in the ventral pallidum hedonic hotspot also stimulates food 'wanting' (eating behavior) as well as 'liking' (Shimura et al., 2006; Smith and Berridge,

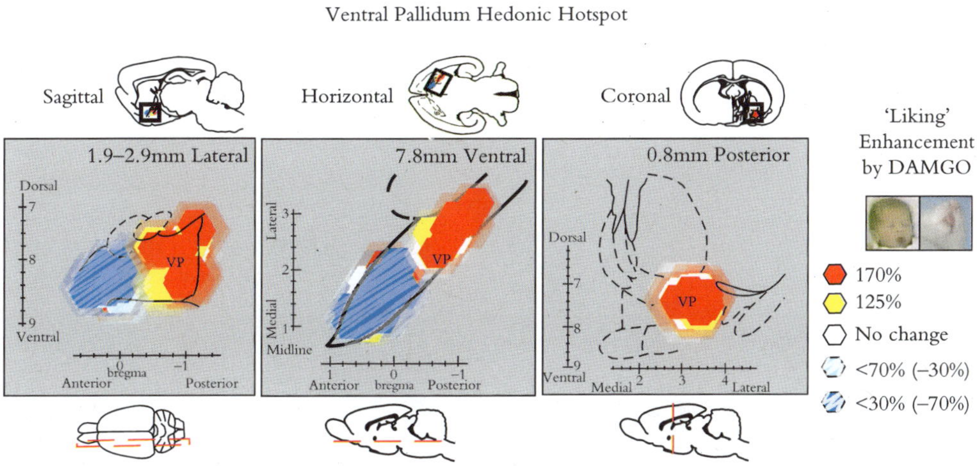

Figure 1.3 Opioid hedonic hotspot in the ventral pallidum. The ventral pallidum hedonic hotspot is contained in the posterior one-third of ventral pallidum, represented in three planes by red and yellow shading (modified from Smith and Berridge, 2005). The Fos plume map shows the intensity of 'liking' amplification caused by opioid microinjections (DAMGO), similar to Figure 1.2. Both 'liking' and 'wanting' are increased simultaneously by opioid stimulation in the hedonic hotspot, whereas both are suppressed together by microinjections in an anterior coldspot (blue area).

2005). In contrast to these positive effects on 'liking' and 'wanting,' a negative suppression of 'liking' and 'wanting' is produced if the same DAMGO microinjections are made in a more anterior coldspot of the ventral pallidum (Figure 1.3). Recently, exciting evidence has emerged that humans might share the same ventral pallidum hotspot and coldspot for food pleasure. Calder and colleagues found that the posterior hotspot of ventral pallidum was activated in people who looked at appetizing pictures of foods like chocolate cake, whereas their anterior coldspot was activated when looking at disgusting pictures of rotten food (Calder et al., 2007).

The ventral pallidum hotspot uses multiple neurochemical signals to generate motivational 'wanting,' but not all generate hedonic 'liking' as well. For example, microinjections of a drug (bicuculline) that blocks $GABA_A$ signals from accumbens causes increases in 'wanting' just as opioid stimulation does, and so makes rats robustly eat more food (Shimura et al., 2006; Smith and Berridge, 2005; Stratford et al., 1999). The GABA-related 'wanting' site extends everywhere in the ventral pallidum (roughly two cubic millimeters), not just the posterior third, and so is much larger than the opioid hotspot. But GABA-related 'wanting' never causes an increase in hedonic 'liking' reactions to sugar taste, not even in the posterior hotspot, even though the GABA motivational enhancement of food 'wanting' is as powerful as the opioid enhancement (Smith and Berridge, 2005). Instead, bicuculline-stimulated eating for ventral pallidum always appears as pure 'wanting' without 'liking.'

Why should blocking GABA receptors in ventral pallidum ever cause increases in 'wanting'? One explanation favored by some neuroscientists is that GABA ordinarily is itself inhibitory (suppressing activity in neurons that receive it) and is released by neurons projecting from the nucleus accumbens to cause inhibition of ventral pallidum neuronal activity. Some nucleus accumbens neurons inhibit firing during a reward or incentive cue, and direct *neural inhibition* of some accumbens neurons (e.g., by microinjection of a GABA agonist that inhibits neurons) causes *psychological excitation* of 'wanting' and 'liking' reward functions (Berridge, 2007a; Day and Carelli, 2007; Kelley et al., 2005; Reynolds and Berridge, 2002). It is possible that accumbens inhibition would shut off the release of GABA in ventral pallidum, and thus free the ventral pallidum neurons to become more active. Our GABA-blocking microinjection would similarly free neurons and might mimic this particular aspect of incentive motivation.

Is the Caudal Ventral Pallidum Hotspot Also Necessary for 'liking'?

Amplification of 'liking' demonstrates that opioid signals in the hedonic hotspot in ventral pallidum are a sufficient cause to increase hedonic impact of a sensory pleasure. Other evidence from brain lesions suggests that this same hotspot may also be a necessary cause for normal hedonic reactions to sweet rewards (perhaps consistent with its special role as a final common pathway for reward).

It has long been known that aversive 'disliking' reactions (e.g., gapes) to normally palatable tastes can accompany the aphagia (failure to eat) caused by very large electrolytic or excitotoxic lesions of lateral hypothalamus, at least if the lesions extend far enough anteriorly and laterally to penetrate the caudal ventral pallidum (Anand and Brobeck, 1951; Berridge, 1996; Schallert and Whishaw, 1978; Stellar et al., 1979; Teitelbaum and Epstein, 1962; Teitelbaum and Stellar, 1954).

An early lesion mapping study by Casey Cromwell in our laboratory aimed to better define the site responsible for lesion-increased aversion and found that the only lesions that caused aversion to sucrose taste were those that damaged the ventral pallidum hotspot region, whereas lesions restricted to the lateral hypothalamus did not cause aversion (even if hypothalamic lesions caused aphagia or failure to eat as much as pallidal lesions) (Cromwell and Berridge, 1993). Hedonic reactions to a normally 'liked' sucrose taste were completely abolished after ventral pallidal lesions that included the hedonic hotspot and replaced by aversive reactions, which are normally evoked by 'disliked' tastes such as quinine (Cromwell and Berridge, 1993).

Such observations suggest that the same hedonic hotspot in ventral pallidum may contain neural substrates that are both a sufficient cause for pleasure (able to amplify above normal) and a necessary cause (needed for normal pleasure), a hypothesis that studies may test in the future. So far, the ventral pallidum is the only brain site known to be a necessary cause for normal pleasure.

Intriguingly, in a recently reported human case, bilateral partial lesions to the ventral pallidum (overlapping with external and internal globus) due to a drug overdose left the patient with "a depressed mood" and "anhedonia" (Miller et al., 2006). The patient was a drug addict prior to the lesion, but over the ensuing year "reported the disappearance of all drug cravings and remained abstinent from all recreational drugs other than an occasional glass of wine with dinner" and "reported that he no longer experienced pleasure from drinking alcohol" (p. 786). Contrary to our earlier description of sensory 'disliking' and aphagia in animals with complete lesions of ventral pallidum, the patient also gained 20 lb in weight over the year. However, the extent of bilateral neuron death in ventral pallidum is not known for this patient, nor is the precise location of his damage compared with the hedonic hotspot that we have identified in the rat ventral pallidum. At the moment, it simply seems striking that ventral pallidum lesions in both humans and other animals appear to induce distortions of hedonic impact and to change the consumption of rewards.

Neurons in the ventral pallidum hotspot of rats code the hedonic impact of taste pleasures in their activation patterns, as well as cause them by affecting psychological–behavioral hedonic reactions. For example, in collaboration with the laboratory of J. Wayne Aldridge at the University of Michigan, we have found that neuronal firing rates in the hedonic hotspot of ventral pallidum code the degree of 'liking' for sweet and salty tastes (Aldridge and Berridge, Chapter 3, this book; Tindell et al., 2004, 2006). Neurons in the ventral pallidum hotspot fire in a faster burst when a rat tastes a sugar or salt that it 'likes' than when it tastes something it 'dislikes.' Normally, the neurons fire very little to a 'disliked' taste such as an intensely salty taste that is three times saltier than seawater. But, when rats are put into a physiological state of "salt appetite" by hormone injections that deplete their bodies of salt, the same intense salty taste suddenly becomes positive and 'liked.' Simultaneously, the neurons in the ventral pallidum hotspot may suddenly fire at least as fast to the intense salt taste as they do to sugar (Aldridge and Berridge, Chapter 3, this book; Tindell et al., 2004, 2006).

Interaction between Accumbens and Pallidum Opioids

How do isolated hotspots in the nucleus accumbens and ventral pallidum combine into integrated brain hedonic circuits? Observations in our laboratory indicate that nucleus accumbens and ventral pallidum hotspots exchange information in both directions to form a single integrated circuit that acts to amplify the hedonic impact of a sensory reward (Smith and Berridge, 2007) (Figure 1.4).

'Liking' amplification by an opioid microinjection in the accumbens hotspot can be blocked if naloxone (an opioid-blocking drug) is simultaneously microinjected in the ventral pallidum hotspot (Smith and Berridge, 2007). The same microinjection of naloxone in ventral pallidum feeds back to inhibit the nucleus accumbens, where it reduces the size of the Fos plume caused by a DAMGO microinjection in the accumbens hotspot. That naloxone-induced suppression of accumbens neurons seems to reflect a suppression of the entire 'liking' circuit, and so no pleasure enhancement can be produced.

Yet, despite this suppression of 'liking' mechanisms, DAMGO microinjection in the accumbens hotspot still generates an increase in food 'wanting' that is as great as if naloxone had not been given into the ventral pallidum at all. This persistent eating enhancement may be due to alternate outgoing opioid-dependent pathways, allowing accumbens 'wanting' signals to bypass the ventral pallidum. Accumbens projections to the lateral hypothalamus provide one potential alternate 'wanting' circuit that might circumvent blockade of the ventral pallidum (Kelley et al., 2005; Smith and Berridge, 2007; Will et al., 2003).

Hedonic Hotspots: Pleasure Valuation Rather Than Motor Expression

Does a hotspot enhancement reflect a true magnification in pleasure 'liking' rather than merely its motor expression? It is an important question, and the answer becomes quite complex. Still, several lines of clear evidence indicate an enhancement of true hedonic 'liking.'

The enhancement caused by hedonic hotspot activation does not fit any motor category: first, the enhancement is not of a single movement of the sort often produced by focused stimulation of a motor structure (because a signature hedonic configuration of several coordinated reactions is enhanced, not just one reaction); second, it does not directly activate the configuration as a fixed motor pattern (because the motor reactions are not generated by the microinjection in the absence of a palatable taste stimulus, indicating the hotspot did not simply turn on a "hedonic orofacial movement generator"; finally, it does not increase all movements as a general motor activation (because aversive reactions or other reactions are actually decreased and because hedonic enhancement occurs at different drugs/doses from locomotor movement enhancements).

In addition, supporting evidence that hotspot neurons are truly hedonic comes from electrophysiology demonstrations that firing of neurons in the ventral pallidum hotspot tracks the hedonic value of a taste

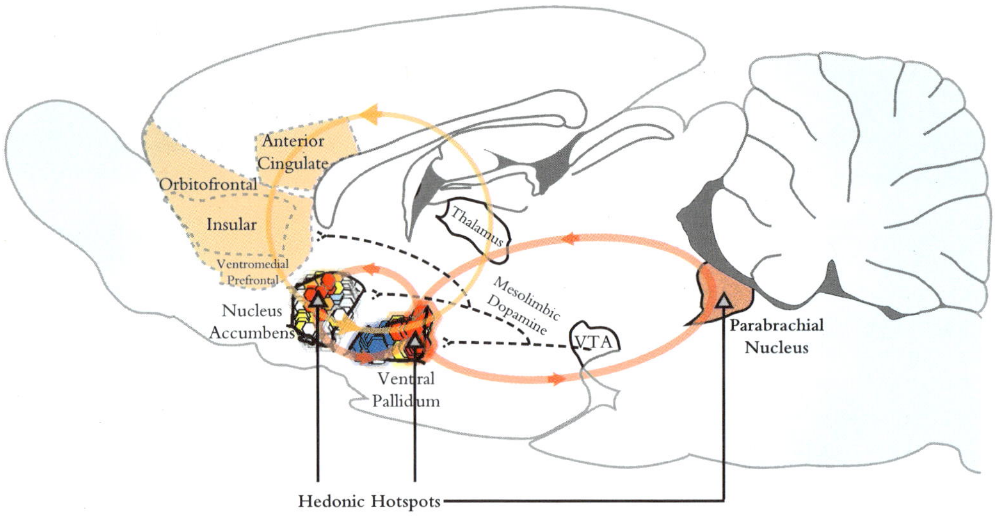

Figure 1.4 Hedonic hotspots and hedonic circuits of the brain. Opioid hedonic hotspots are shown in nucleus accumbens, ventral pallidum, and brainstem parabrachial nucleus. Neurochemical signals in each hedonic hotspot can cause amplification of core 'liking' reactions to sweetness. Hedonic circuits connect hotspots (red) into integrated loops for causation of 'liking' (orange and red loops). Additional forebrain loops relay 'liking' signals to limbic regions of prefrontal cortex and back to hotspots, perhaps for translation of core 'liking' into conscious feelings of pleasure and cognitive representations (dotted, orange cortex). Dashed, black subcortical lines show mesolimbic dopamine projections, which we suggest fail to cause 'liking' after all.

and is not tightly associated to any motor details of reaction movements (Aldridge and Berridge, Chapter 3, this book; Tindell et al., 2006). Such considerations lead us to believe that hotspot maps, based on behavioral 'liking' reaction studies, truly show the location of brain substrates for hedonic valuation of pleasure rather than simply generators of 'liking' movements.

Levels of Pleasure in the Brain

Sensory pleasure does not arise from any one hedonic hotspot, of course. Rather, as indicated by the 'liking' circuit between accumbens–pallidum hotspots already described, pleasure results from their connection together into larger hedonic brain circuits that operate as a whole. These integrated circuits stretch across the brain from forebrain to brainstem, forming a hedonic generating system for natural pleasure.

Brainstem Hedonic Roles?

The notion that the brainstem plays any role in sensory pleasure might come as a surprise to anyone used to thinking of brainstem areas solely in terms of reflexive functions. Yet, for over a century, the brainstem has been recognized to participate in the generation of basic affective reactions, as well as other psychological functions. John Hughlings Jackson, an innovative 19th century neurologist, proposed that brainstem function provided an essential first level in a neural hierarchy of "re-re-representation" of affective and other functions. According to this principle, low levels of the brain (the brainstem) generate a basic and concrete representation of events or functions, sufficient just for basic affective reactions and behavioral responses.

Examples of basic brainstem 'liking' function date back over a century ago when Goltz showed that after surgical removal of its forebrain, a dog would still reject a piece of meat soaked in bitter quinine (Goltz, 1892). Miller and Sherrington subsequently showed that decerebrated cats (surgically transected above the hindbrain) responded with ingestive "elaborate movements of the tongue" to meat-flavored water but with "retching and reflexes of rejection" to quinine (p. 167) (Miller and Sherrington, 1915). In the 1970s, Grill and Norgren showed that chronic decerebrate rats, with only a hindbrain and midbrain intact, still emitted normal positive tongue protrusions and lip-licking reactions to sucrose taste, but emitted aversive gapes and other rejection reactions to quinine taste (Grill and Norgren, 1978b). In humans, Steiner (1973) showed that anencephalic infants (born without the forebrain due to a congenital malformation but with a normal brainstem) similarly emitted normal tongue protrusions to sucrose taste, but aversive gapes and headshakes in response to bitter tastes (and cried as normal infants do). Thus, the brainstem examined in isolation seems capable of generating elemental forms of affective 'liking' or 'disliking' reactions.

In normal animals with intact brains, however, the brainstem does not react in isolation but rather is wired into a larger brain hierarchy of affect generation involving forebrain structures, including the hedonic hotspots in ventral pallidum and nucleus accumbens described earlier. Forebrain levels in a Jacksonian brain hierarchy re-represent and re-re-represent the signals that have been initially represented in brainstem, taking control of lower functions and adding new abstract features (Hughlings Jackson, 1958). By a hierarchical account, a complete affective (or other) function requires the entire system. The full affective function cannot be provided by the brainstem alone in the absence of cortex. But conversely, if 'liking' is truly organized as a Jacksonian brain hierarchy, then full-blown affective function cannot be generated by the cortex alone in the absence of brainstem.

The concept of neural hierarchy and multiple brain levels for affect generation is still present in contemporary thought on emotion, and the brainstem is still posited to make key contributions (Berridge, 2003a; Damasio, 1999; LeDoux, 1996; Panksepp, 1991). For example, Damasio has suggested that the parabrachial nucleus in the pons of the brainstem participates in generating what he calls a "protoself," a coherent representation of the momentary state of the body used by higher brain levels to generate conscious feelings. A consequence is that brainstem lesions that disrupt generation of protoself functions may uniquely cause coma and loss of conscious awareness (Damasio, 1999).

Regarding 'liking' reactions, a surprising feature of the affective brain hierarchy is that ascending levels can be differentially balanced between positive and negative reactions (Grill and Berridge, 1985). As a consequence, less brain can sometimes actually be better affectively balanced than more brain. For example, animals with an isolated brainstem (decerebrates) generate balanced 'liking' and 'disliking' reactions to tastes: positive to sweet but negative to bitter (Grill and Norgren, 1978b). But, adding one more brain layer called the diencephalon or lower forebrain (thalamus, pineal and hypothalamus) actually unbalances the hierarchy in a negative direction toward complete 'disliking.'

For example, a surgical preparation that creates this brainstem-plus-lower-forebrain has sometimes been called a "thalamic" animal, involving ablation of everything above the thalamus. It lacks not only neocortex, but also the subcortical upper forebrain, including ventral pallidum, nucleus accumbens, amygdala, hippocampus, and neostriatum (all these structures together with neocortex belong to the brain level called the telencephalon). A thalamic rat or cat shows only aversive quinine-like rejection reactions even to a sweet taste and lacks any positive hedonic response to normally pleasant stimuli (Bard, 1934; Grill and Norgren, 1978b). The thalamic animal's unbalanced affective negativity suggests that the diencephalon contains circuitry, which pushes brainstem reactions into 'disliking' unless opposed by signals from forebrain structures further above.

What structure above the thalamus adds enough positive affect to flip affective reactions back to 'liking' balance again? That could be answered by adding back forebrain structures "one-by-one", or more practically, taking one or several structures above the thalamus away from normal animals, to find out which one is needed for normal 'liking.' One might have thought that the answer would be the neocortex. However, it turns out that affective balance can be restored by merely adding the subneocortical parts of the upper forebrain or telencephalon. Adding the cortex itself beyond that may add little more to basic 'liking' reactions. This is shown by observations that "decorticate rats," which have had the neocortex completely removed but still have all their subneocortical forebrain structures, upper as well as lower, show completely normal positive 'liking' reactions to sweet tastes and 'disliking' to bitter tastes (and can even learn complex tasks to get rewards) (Bard, 1934; Grill and Norgren, 1978b; Wirsig and Grill, 1982).

Of the subcortical upper forebrain structures needed for normal pleasure, we suggest that the ventral pallidum might be particularly important; perhaps especially its positive hedonic hotspot due to its necessary and sufficient causal roles in generating 'liking' reactions to pleasure (Cromwell and Berridge, 1993; Smith and Berridge, 2005). This would mean that the addition of all of the forebrain to the brainstem, *except* the ventral pallidum, would actually unbalance affect in a negative direction as much as a total "thalamic ablation" of everything above the thalamus.

Essentially, this anatomical configuration is what a brain with only ventral pallidum lesions has. Mapping the exact forebrain sites responsible for anhedonia, or loss of normal 'liking,' is an interesting goal for future exploration. It is interesting to note that normal levels of 'liking' are relatively robust in the face of damage to widespread brain areas. Hedonic robustness may reflect the evolutionary importance of pleasure reactions, as well as the neural re-re-representation of 'liking' function at several levels. Robustness of normal pleasure reactions also contrasts to the relative fragility of 'liking' *enhancement* above normal, which requires unanimous "opioid consent" by multiple forebrain hotspots simultaneously as described earlier (Smith and Berridge, 2007).

In summary, the brainstem has not lost hedonic functions when higher brain areas are present, but rather has been incorporated into a larger neural hierarchy of pleasure controlled by forebrain circuits. Hierarchy means that brainstem has lost its autonomy, so that the forebrain adds new hedonic functions and overrides the preexisting ones (Gallistel, 1980). The hotspots we described in nucleus accumbens and ventral pallidum are examples of forebrain 'liking' mechanisms that can override brainstem functions to enhance sensory pleasure.

Benzodiazepine/GABA Hedonic Substrate in the Parabrachial Nucleus

A concrete residue of basic hedonic function in the brainstem is the existence of a hedonic hotspot in the pons of rats. The brainstem hedonic hotspot appears to be located near the parabrachial nucleus of the pons and uses a benzodiazepine/GABA signal to augment hedonics (Peciña and Berridge, 1996; Soderpalm and Berridge, 2000b) (Figure 1.4). Benzodiazepine drugs are probably most famous for their antianxiety and tranquilizing effects. However, benzodiazepines also stimulate appetite via separate brain mechanisms and were originally suggested by Cooper in the 1980s to augment the hedonic impact of food rewards (Cooper, 1980; Cooper and Estall, 1985).

Subsequent studies identified the brainstem, particularly its parabrachial nucleus area in the pons, as the chief site where benzodiazepines appear to act to enhance taste palatability and appetite. Microinjections of a benzodiazepine drug into the rat brainstem as a whole or directly into the parabrachial nucleus causes a doubling of the number of 'liking' reactions to sugar (Peciña and Berridge, 1996; Soderpalm and Berridge, 2000b) (Figure 1.4). Similar parabrachial microinjections also make rats 'want' to eat more food (Higgs and Cooper, 1996).

The existence of the hedonic hotspot in the parabrachial nucleus of the pons in the brainstem may explain why a brainstem microinjection of a benzodiazepine causes higher increases in 'liking' reactions than forebrain microinjections of the same drug: the forebrain has no known hotspot for benzodiazepine amplification of hedonic impact (Berridge and Peciña, 1995; Peciña and Berridge, 1996; Soderpalm and Berridge, 2000a). It may also explain why even decerebrate rats, which have only a brainstem (hindbrain and midbrain), still show an elevation in positive reactions to sucrose taste if given a systemic injection of benzodiazepine drug to activate their remaining brainstem GABA signals (Berridge, 1988).

The parabrachial nucleus is a relay nucleus where ascending taste sensation signals are processed after leaving the hindbrain nucleus of the solitary tract in the rodent brain (Norgren, 1995; Spector, 2000). In human and other primates, a few studies have indicated that the ascending taste pathway may bypass the parabrachial nucleus on its way to forebrain targets (Beckstead et al., 1980; Pritchard et al., 2000). Until more is known, it is difficult to be sure about whether primate brains really lack a parabrachial taste relay. However, even if the parabrachial nucleus is not part of the direct taste pathway, it is still possible that the human parabrachial nucleus contributes indirectly to taste 'liking.' That is because the parabrachial nucleus also receives indirect descending projections from limbic forebrain sites, which are able to modulate taste sensation (Lundy and Norgren, 2004). Indeed, in humans, taste deficits can occur with pontine lesions near the parabrachial nucleus and taste intensity discrimination recruits parabrachial activity (Landis et al., 2006; Small et al., 2003).

A retained hedonic role would also be compatible with the suggestion that the parabrachial nucleus mediates emotional representations of body states in humans (Damasio, 1999). Thus, although differences may exist between rats and people in ascending taste circuits, it is possible that the parabrachial nucleus might still contribute as a hedonic hotspot in humans too.

In addition, some evidence from rat experiments indicates that the parabrachial GABA signal may require opioid signals, perhaps in the forebrain, to amplify 'liking' reactions. Pretreatment with an injection of the opioid antagonist naloxone can block the typical 200% elevation of sucrose 'liking' reactions that is usually caused by an injection of a benzodiazepine drug (Richardson et al., 2005). A possible neural explanation for naloxone blocking is if the parabrachial nucleus activates endogenous opioid signals in hedonic hotspots, perhaps in the nucleus accumbens and ventral pallidum, as the next step in the neural circuit for enhancing 'liking.' This is consistent with the notion that a distributed brain circuit connects together hotspots in brainstem and forebrain and functions as an integrated whole to amplify sensory pleasure (Figure 1.4).

Hedonics at the Top End of the Brain: Pleasure-Causing Substrates in the Neocortex?

We have described so far how taste pleasure can arise from a number of hedonic hotspots in brainstem and subcortical forebrain. What about at the very top of the brain? Does the neocortex contain hotspots of its own capable of elevating hedonic reward?

In favor of the possibility, impressive neuroimaging studies have demonstrated that sites in prefrontal and related limbic regions of neocortex *code* positive affect and the hedonic impact of many pleasures (Bechara et al., 2000; Burke et al., Chapter 2, this book; Davidson and Irwin, 1999; Kringelbach, Chapter 12, this book; O'Doherty, 2004; Veldhuizen et al., Chapter 9, this book). Most prominent among cortical sites activated by pleasure may be the orbitofrontal region of prefrontal cortex (Knutson et al., 2001; Kringelbach, 2005; Rolls, 2000; Small, 2006). In humans, the orbitofrontal cortex, particularly its medial region, is activated by pleasant tastes and odors, pleasant touch sensations, and other pleasant stimuli (de Araujo et al., 2003; Francis et al., 1999; O'Doherty, 2004; Rolls et al., 2003b; Small et al., 2003). Orbitofrontal cortex activity in rats, monkeys, and humans also tracks changes in pleasure of a constant food stimulus or the alliesthetic reductions in hedonic impact caused by eating foods to satiety (Burke et al., Chapter 2, this book; Faurion et al., 1998; Hollerman et al., 2000; Kringelbach, Chapter 12, this book; Kringelbach et al., 2003; O'Doherty, 2004; Rolls et al., 1989; Simon et al., 2006; Small et al., 2001). For example, the taste of chocolate activates the orbitofrontal cortex in hungry people who like chocolate, but activation declines after subjects eat chocolate to satiety (Veldhuizen et al., Chapter 9, this book; Small et al., 2001). More complex human pleasures, such as pleasurable music or winning money, have also been reported to activate orbitofrontal cortex and other sites (Blood and Zatorre, 2001).

Other cortical regions that might possibly play a role in causing pleasure include anterior cingulate cortex and insular cortex. Cingulate cortex has been observed to be activated by a number of hedonic stimuli, including sexual arousal, taste and olfactory rewards, pleasant music, and rewarding drug stimulation (Breiter et al., 1997; Brown et al., 2004; de Araujo et al., 2003; Firestone et al., 1996; Gottfried, Chapter 8, this book; Komisaruk et al., Chapter 10, this book; McCoy et al., 2003; Platt et al., Chapter 5, this book; Rauch et al., 1999; Veldhuizen et al., Chapter 9, this book). The insular cortex has been suggested to contain an anterior gustatory site and a posterior hedonic site (Kringelbach et al., 2003; Yaxley et al., 1988). Insular cortex is activated by pleasant tastes or odors in hungry humans and rats, and satiety causes a decline in activation to the same stimuli (de Araujo et al., 2006; Kringelbach et al., 2003; O'Doherty et al., 2000; Small et al., 2001; Small et al., 2003). Insular cortex has been suggested to perhaps be especially important for mediating learned likes for initially aversive stimuli, such as the taste of cigarette smoke (Naqvi et al., 2007), and also for learned dislikes such as nausea-induced taste aversions or pictures of rotten foods (Gutierrez et al., 1999).

However, it remains an open question to what extent any of these cortical areas actually *cause* basic hedonic 'liking' reactions to pleasant events beyond coding pleasure for cognitive or other functions (including hedonic consciousness, discussed later). As yet, little direct evidence exists to know if activity in a cortical area is ever sufficient to generate increases in hedonic impact, or necessary for normal hedonic impact, in the same sense as in hedonic hotspots of subcortical brain structures. Alternatively, cortical hedonic coding may not actually cause basic pleasure, but rather re-represent subcortical pleasure reactions for other functions, such as cognitive representations or even conscious awareness. Cognition and consciousness are crucial causal functions too, of course, but distinct from the generation of a basic 'liking' reaction. Thus, the issue of whether specific cortical areas actually cause pleasure 'liking' reactions awaits future evidence (Kringelbach, Chapter 12, this book).

Subcortical Hedonic Systems: Conscious or Unconscious?

A related fascinating question concerns how in the brain the consciousness of pleasure arises. Do subcortical hedonic hotspots or generating circuits ever directly cause a conscious pleasure feeling, in addition to causing core 'liking' reactions? Or is the subjective awareness of pleasure something that must be added by cortex re-representations? Terminologically, it is easy to distinguish between objective and subjective senses of pleasure. We have always used the term 'liking' (in quotes) to mean objective hedonic reactions, whether or not accompanied by subjective feelings (which might not even exist in decerebrates, anencephalics, and similar cases). A 'liking' reaction is held to be a core component of normal hedonic feelings, but can sometimes occur by itself without those conscious feelings. By contrast, we use the word liking (without quotes) to mean its normal sense of a conscious experience of pleasure. This use helps to distinguish between conscious and unconscious forms of pleasure and to highlight the possibility of unconscious pleasure in basic 'liking' reactions.

Beyond mere words, there is also reason to consider conscious pleasure and unconscious pleasure both as real psychological processes with distinct brain mechanisms (Frijda, Chapter 6, this book; Schooler and Mauss, Chapter 14, this book). Although the idea of an unconscious pleasure is counterintuitive to many people, evidence is accumulating that unconscious pleasure processes may exist, often tucked within normal conscious experiences of pleasure and sometimes even on its own.

For example, Winkielman et al. (2005) recently demonstrated that normal human adults can have unconscious 'liking' and 'disliking' reactions that fail to reach conscious awareness. Participants were subliminally shown happy facial expressions (or neutral or angry expressions), followed by a masking stimulus in a task designed to wipe out any subjective feelings produced by the subliminal expressions, using a modification of subliminal emotional priming procedures (Monahan et al., 2000; Winkielman et al., 1997). Participants then rated their own hedonic and arousal feelings and also sampled and rated a novel fruit beverage. No changes in ratings of conscious hedonic/arousal feelings were produced by subliminal exposure to emotional expressions (and participants reported afterwards that they had not seen any emotional expression and were unable to pick the one they saw out of a lineup). Yet, thirsty participants who had subliminally seen a happy subliminal expression poured and drank twice as much of the beverage as those who had seen angry expressions and gave up to four times higher ratings of value to the beverage (Winkielman et al., 2005). These results indicated that under appropriately masked conditions, ordinary people could have core 'liking' and 'disliking' reactions to emotional expressions that were completely unfelt at the moment they were caused, yet were strong enough to go on to markedly shift consumption behavior and reactions to a valence-laden stimulus.

Several human clinical cases also seem consistent with the notion of unconscious pleasure under certain conditions. For example, human drug addicts have been reported to self-administer drugs like cocaine even at doses too low to produce subjective effects or autonomic responses (Fischman and Foltin, 1992; Hart et al., 2001). Similarly, after gustatory cortex damage, a patient has been described to display a clear preference for a sweet beverage over a salty one, yet was unable to tell the two tastes apart in a subjective sensory test and subjectively rated them as equally pleasant (Adolphs et al., 2005).

Some evidence suggests that subliminal stimuli might trigger core 'liking' reactions in the brains of normal people by activating subcortical hedonic hotspots in the absence of conscious awareness. Such studies use neuroimaging measures to show that subliminal presentation of positive hedonic stimuli, too brief to be consciously seen, can still activate limbic brain structures such as ventral pallidum or amygdala? For example, the ventral pallidum is reported to be activated by subliminal presentation of pictures of happy faces (Whalen et al., 1998) and by subliminal presentation of money cues that signal that a large reward is about to be earned (Pessiglione et al., 2007). Such examples suggest that subjective awareness may not always have access to underlying core affective reactions in subcortical brain structures, which might conceivably mediate behavioral manifestations of unconscious 'liking' (Winkielman et al., 2005).

We presume that conscious feelings of liking always incorporate these core 'liking' reactions, but also involve an additional neural and psychological stage that elaborates the core affective reaction into conscious awareness. A traditional and relatively simple brain-based explanation might be that activation of subcortical hedonic hotspots could generate a core 'liking' signal that is not itself directly accessible to consciousness and that higher brain systems such as cortex might use coded 'liking' signals as an input to generate conscious pleasure experience (liking, without quotes).

Rethinking Old Pleasure Sources: Electrodes and Dopamine

False Pleasure Electrodes?

In contrast to the pleasure substrates described above, some brain substrates once thought to cause pleasure may be turning out not to do so. Perhaps the most famous candidate for a brain substrate that generates pleasure were so-called pleasure electrodes, which used brain electrical stimulation of the subcortical limbic forebrain to reinforce self-administration behavior such as pressing a lever or pushing a button (Delgado, 1969; Green et al., Chapter 18, this book; Heath, 1972; Kringelbach, Chapter 12, this book; Kringelbach et al., 2007; Olds and Milner, 1954; Sem-Jacobsen, 1976). Pleasure electrodes were typically aimed at the septum or lateral hypothalamus though a number of the sites fell within what neuroanatomists would now call the nucleus accumbens, and most electrodes likely activated mesolimbic dopamine systems (Heath, 1972; Hernandez et al., 2006; Olds, 1961; Olds and Milner, 1954) (Figure 1.5). Some patients stimulated these "pleasure electrodes" thousands of times in a single 3-h session (Heath, 1972; Sem-Jacobsen, 1976; Valenstein, 1974). Many textbooks cite these cases as examples of intense brain-induced pleasure. But, despite such dramatic self-administration, it is questionable whether many of those electrodes ever actually caused pleasure (Berridge, 2003b; Peciña et al., 2006). If one reads the transcripts of verbal responses closely, it is not clear that the patients experienced intense pleasure per se during stimulation.

For example, "B-19," a young man with chronic electrodes implanted by Heath and colleagues in the 1960s, voraciously self-stimulated his electrode located in septum and nucleus accumbens (Figure 1.5) and protested when the stimulation button was taken away (Heath, 1996). Still, B-19 was never reported to utter exclamations of delight or to say that the electrodes caused pleasure thrills. Instead, B19 reported that stimulating his electrode evoked desire to stimulate again, as well as a strong desire to engage in sexual activities. Another Heath patient said his electrode "made him feel as if he were building up to sexual orgasm" but left him "unable to achieve the orgastic end point," an outcome, which often "was frustrating and produced a "nervous feeling" that seems nearly opposite to pleasure (Heath, 1964). Although stimulation caused patients to become strongly sexually aroused, or to want to eat or drink or pursue other incentives, it never produced feelings like sexual orgasm, and it did not serve as a substitute for sexual acts or other reward consumption.

In more recent years, brain electrodes programmed to spontaneously stimulate reward structures have been implanted in a number of patients with Parkinson's disease in attempts to alleviate problems with movement and low mood (Kringelbach et al., 2007). One interesting motivational effect of such brain stimulation has been to make people and objects in the environment sometimes be perceived as more attractive. For example, one patient developed "fondness" of other people in the clinic and "was in love with two neurologists, and tried to embrace and kiss people" (Herzog et al., 2003). Compulsive pursuits of objects, stimuli, or activities may also sometimes result. For example, the same patient above also "engaged in unrestrained buying of clothing." Urges to engage in activities, such as the desire to visit particular tourist sites or to take up again former hobbies that had lapsed, have been reported (Schlaepfer et al., 2008), as has the development of compulsive gambling, compulsive sex, or stealing (Houeto et al., 2002; Mandat et al., 2006).

Why would anyone press a self-stimulation button thousands of times for electrode stimulation if it is not intensely pleasant? Or why engage in compulsive and intense levels of motivated behavior during brain stimulation if the electrode does not make those acts more pleasurable? One possible alternative to pleasure is that the electrode causes incentive salience to be attributed to events associated with the stimulation—such as the button stimulus and the act of pressing it. That might cause people to 'want' to press again even if they did not especially 'like' it. This incentive salience explanation was originally suggested by observations that stimulation of a rewarding electrode also makes rats 'want' to eat more food—but does so without ever causing them to 'like' the food more (Berridge and Valenstein, 1991). For the rats, the presence of food had been repeatedly paired with the electrode stimulation. For humans in self-stimulation situations, the button and pressing it are the events most closely paired with stimulation, and therefore likely to be the target of greatest 'wanting.' For the patient who suddenly perceives the whole world as motivationally brighter, other people as more desirable, and certain pursuits as compulsively attractive, the act of button pressing should be even more attractive, especially after pressing it several times. In such cases the button itself could become the greatest motivational magnet. A person therefore could intensely come to 'want' to press the button again even if the electrode never caused a hedonic pleasure or 'liking.'

We hasten to say that our claim that most "pleasure electrodes" failed to generate true 'liking' is not to say that none ever did. A few electrode cases sound more plausibly like true pleasure. For example, chronic electrode stimulation of the subthalamic nucleus in the forebrain basal ganglia was described as "morphine-like" or similar to "sexual climax" (Morgan et al., 2006), which might be a candidate for true pleasure (though even here it is still open as to whether the stimulation was truly hedonic in a morphine–euphoric sense or rather a mere sensation of visceral relaxation, and whether it was the hedonic feeling of climax or merely sexual sensations, either of which could be caused by deep forebrain stimulation).

In future research, it would be useful to ask questions that more specifically assess the pleasure of electrode stimulation. Is the stimulation nice? How nice and compared to what? If the electrode makes a person want to eat or drink or engage in sex, then does the stimulation make those targets any more liked when they are consumed? Affirmative answers to such questions should be found if electrode stimulation acts to activate a true hedonic hotspot (Kringelbach et al., 2007).

Dopamine: Not a Pleasure Transmitter?

Another false pleasure-causing substrate may be brain dopamine, especially the mesolimbic system that projects from midbrain to nucleus accumbens (which was likely to have been stimulated, directly or indirectly, by many of the electrodes described above) (Figure 1.5). Dopamine has been famous as a so-called pleasure neurotransmitter for over 30 years (Hoebel et al., 1999; Shizgal, 1999; Wise and Bozarth, 1985). One reason dopamine was thought to mediate pleasure is that dopamine neurons are turned on by pleasurable

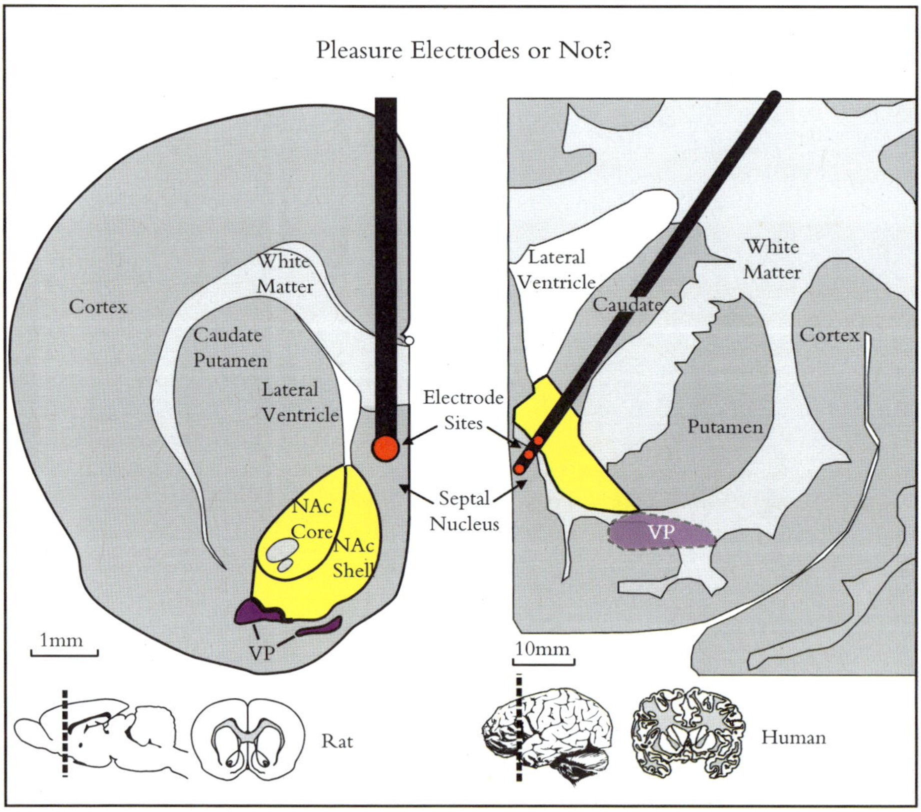

Figure 1.5 Pleasure electrodes or not? Examples of famous so-called pleasure electrode placements in rat (from Olds, 1961) and in human (patient B-19, a young man, from Heath, 1972). Thick black lines show the electrodes (insulated except at tips; red dots indicate their stimulating tips) near the nucleus accumbens. The nearby ventral pallidum is also shown, though it is mostly posterior to the depicted coronal section. Both the rat and the human pressed for electrode stimulation up to thousands of times, but we suggest both electrodes might have produced merely a pure form of 'wanting' (incentive salience) rather than actual 'liking' (true hedonic pleasure).

stimuli ranging from foods, sex, and drugs to social and cognitive rewards (Ahn and Phillips, 1999; Aragona et al., 2006; Becker et al., 2001; Fiorino et al., 1997; Robinson et al., 2005; Schultz, 1998; Wise, 1998). Further, if dopamine was blocked in animals, all rewards seemed to lose rewarding properties in certain instrumental paradigms, becoming no longer 'wanted' in a way that led many neuroscientists to conclude the rewards were no longer 'liked' (Hoebel et al., 1999; Shizgal, 1999; Wise and Bozarth, 1985).

But dopamine is probably not a pleasure neurotransmitter, even if it causes some other component of reward (which we have suggested is incentive salience 'wanting') (Berridge, 2007b; Robinson and Berridge, 2003). Dopamine is not needed to cause normal pleasure of food or drugs of abuse. For example, even massive destruction of ascending dopamine projections does not impair affective 'liking' reactions elicited by a sweet taste (Berridge and Robinson, 1998; Berridge et al., 1989). Similarly, complete gene-based elimination of dopamine has been suggested to not impair 'liking' in dopamine-deficient mutant mice (Robinson et al., 2005). Nor does dopamine blockade by neuroleptic drugs reduce 'liking' facial reactions of rats to sweetness (Peciña et al., 1997). In humans, the perceived pleasantness of chocolate milk is not reduced by the loss of brain dopamine neurons in Parkinson's disease (Sienkiewicz-Jarosz et al., 2005). Similarly, human subjective ratings of the pleasantness of amphetamine, cocaine, or cigarettes have been reported to persist unsuppressed by dopamine-blocking drugs or dietary-induced dopamine depletion, even when those treatments do suppress wanting for more of the same drug (Brauer et al., 2001; Brauer and de Wit, 1997; Leyton, Chapter 13, this book; Leyton et al., 2005).

Elevation of dopamine is not a sufficient cause for pleasure any more than a reduction of dopamine impairs pleasure as a necessary cause (Leyton, Chapter 13, this book). Elevation of dopamine neurotransmission in mutant mice by a gene that raises released dopamine levels to more than one-and-a-half times above normal does not enhance their hedonic 'liking' reactions to sweetness, even though the mutant mice appear to 'want' food rewards more (working harder, faster, and longer to obtain sweet rewards, and resisting distractions more) (Cagniard et al., 2006; Peciña et al., 2003). Similarly, raising dopamine levels by administering amphetamine, either systemically or directly into the nucleus accumbens, also completely fails to increase hedonic 'liking' reactions even when 'wanting' of the same reward is increased (Tindell et al., 2005; Wyvell and Berridge, 2000). Also, in humans, dopamine increases caused by amphetamine or l-Dopa are reported to correlate well with subjective ratings of 'wanting' to take more drug, but not with ratings of liking for the same drug (e.g., "Do you like the effects you are feeling right now"?) (Evans et al., 2006; Leyton et al., 2002).

Overall in both animals and humans, dopamine now appears neither necessary for generating normal pleasure nor sufficient for enhancing pleasure (Leknes and Tracey, Chapter 19, this book; Leyton, Chapter 13, this book).

Questions for Future Research

Many questions remain for future research on pleasure generation in the brain. We end simply by highlighting a few outstanding ones.

Are there additional hedonic hotspots in the brain? Beyond the hedonic hotspots described here, it seems likely that other brain sites may participate in causal generation of 'liking' reactions. Chief among them might be the orbitofrontal cortex in the prefrontal lobe, which is perhaps the most promising candidate for pleasure generation among all cortical structures. Activation of the orbitofrontal cortex appears to show the best cortical correlation with pleasure in humans (Kringelbach, 2005; Kringelbach, Chapter 12, this book; Small et al., 2001; Veldhuizen et al., Chapter 9, this book) and other animals (Burke et al., Chapter 2, this book; Rolls, 2000; Rolls et al., 1989; Schoenbaum and Roesch, 2005). Most intriguingly, orbitofrontal cortex has been suggested to segregate positive and negative affective valence into separate areas, coding positive rewards by medial activation and negative aversion by more lateral activation (Kringelbach, 2005; Rolls et al., 2003a; Small et al., 2001).

If orbitofrontal cortex acts to cause basic affective reactions, then local stimulation of it might increase positive or negative affective reactions, respectively, or focused lesions might disrupt particular affective reactions. It would be of great interest to find a cortical region that exerts clear causal influence on a core 'liking' reaction. Other cortical candidates for causal hedonic hotspots might include the insular cortex or anterior cingulate cortex. Other subcortical candidates also remain to be examined more thoroughly: these include the lateral shell of nucleus accumbens (only the medial shell has been thoroughly mapped so far) and perhaps the core of the nucleus accumbens, amygdala nuclei, and the related "extended amygdala" and other limbic structures that are closely wired to hotspots in the nucleus accumbens and ventral pallidum. Discovery of new hedonic hotspots will be useful to extend neuroscientific understanding of the unique brain circuit that is able to generate amplification of pleasure.

A related wonderful opportunity to assess whether specific brain sites in humans can actually cause pleasure is offered by the new crop of deep brain stimulation procedures that have recently begun to be reapplied to pathological conditions such as Parkinson's disease and depression (Kringelbach et al., 2007). We have argued that some of the classic cases of brain stimulation–induced pleasure may be equivocal at best, but future studies may be more successful at demonstrating true pleasure electrodes (Green et al., Chapter 18, this book; Kringelbach, Chapter 12, this book). Finding stimulation sites that do support pleasure in this way would be a significant step forward in mapping human hedonic hotspots. It would be useful to have clear evidence that electrode stimulation was truly 'liked' in a genuinely hedonic sense, more than merely 'wanted.'

Beyond finding more hotspots in the brain, it will also be important in the future to better understand how hotspots work normally to generate pleasure in the brain. This issue touches on an essential question: what does it mean to have a brain-based explanation of pleasure? Something more is needed than merely pointing to neurochemical activation in a crucial brain hotspot. We need a better explanation of why hotspot activation causes pleasure. Tracing that path of neural–psychological causation will require considerably more information on hedonic brain mechanisms and perhaps major conceptual developments as well.

Conclusions

The question of how pleasure is caused in the brain is fundamentally one of how hedonic value gets added to a mere sensation. Modern experimental tools, such as microinjection Fos plume mapping techniques combined with behavioral 'liking' reaction measures, have revealed an interactive network of hedonic hotspots in the nucleus accumbens, ventral pallidum, and other brain structures (Figure 1.4). These hedonic hotspots appear to be the crucial brain mechanisms that actively paint a gloss of pleasure onto sensations such as sweetness and cause them to be 'liked.'

Acknowledgments

We thank Morten L. Kringelbach, J. Wayne Aldridge, and Sarah W. Newman for helpful comments on earlier versions of this chapter. The studies described here were supported by grants from the NIH (MH63649 and DA015188).

References

Adolphs, R., Tranel, D., Koenigs, M. and Damasio, A. R. (2005) Preferring one taste over another without recognizing either. *Nat. Neurosci.* 8, 860–861.

Ahn, S. and Phillips, A. G. (1999) Dopaminergic correlates of sensory-specific satiety in the medial prefrontal cortex and nucleus accumbens of the rat. *J. Neurosci.* 19, B1–B6.

Aldridge, J. W. and Berridge, K. C. (this book) Neural coding of pleasure: "Rose-tinted glasses" of the ventral pallidum. In: *Pleasures of the Brain.* Eds. M. L. Kringelbach and K. C. Berridge. Oxford University Press: Oxford.

Anand, B. K. and Brobeck, J. R. (1951) Hypothalamic control of food intake in rats and cats. *Yale J. Biol. Med.* 24, 123–140.

Aragona, B. J., Liu, Y., Yu, Y. J., Curtis, J. T., Detwiler, J. M., Insel, T. R. and Wang, Z. (2006) Nucleus accumbens dopamine differentially mediates the formation and maintenance of monogamous pair bonds. *Nat. Neurosci.* 9, 133–139.

Bakshi, V. P. and Kelley, A. E. (1993) Striatal regulation of morphine-induced hyperphagia: an anatomical mapping study. *Psychopharmacology (Berl)* 111, 207–214.

Balleine, B. W. (2005) Neural bases of food-seeking: Affect, arousal and reward in corticostriatolimbic circuits. *Physiol. Behav.* 86, 717–730.

Bard, P. (1934) On emotional expression after decortication with some remarks on certain theoretical views. Part II. *Psychol. Rev.* 41, 424–449.

Barrett, L. F. and Wager, T. D. (2006) The structure of emotion: Evidence from neuroimaging studies. *Curr. Dir. Psychol. Sci.* 15, 79–83.

Bechara, A., Damasio, H. and Damasio, A. R. (2000) Emotion, decision making and the orbitofrontal cortex. *Cereb. Cortex* 10, 295–307.

Becker, J. B., Rudick, C. N. and Jenkins, W. J. (2001) The role of dopamine in the nucleus accumbens and striatum during sexual behavior in the female rat. *J. Neurosci.* 21, 3236–3241.

Beckstead, R. M., Morse, J. R. and Norgren, R. (1980) The nucleus of the solitary tract in the monkey: Projections to the thalamus and brain stem nuclei. *J. Comp. Neurol.* 190, 259–282.

Berridge, K. C. (1988) Brainstem systems mediate the enhancement of palatability by chlordiazepoxide. *Brain Res.* 447, 262–268.

Berridge, K. C. (1996) Food reward: Brain substrates of wanting and liking. *Neurosci. Biobehav. Rev.* 20, 1–25.

Berridge, K. C. (2000) Measuring hedonic impact in animals and infants: Microstructure of affective taste reactivity patterns. *Neurosci. Biobehav. Rev.* 24, 173–198.

Berridge, K. C. (2003a) Comparing the emotional brain of humans and other animals. In: *Handbook of Affective Sciences*, pp. 25–51. Eds. R. J. Davidson, H. H. Goldsmith and K. Scherer. Oxford University Press: Oxford.

Berridge, K. C. (2003b) Pleasures of the brain. *Brain Cogn* 52, 106–128.

Berridge, K. C. (2007a) Brain reward systems for food incentives and hedonics in normal appetite and eating disorders. In: *Appetite and Body Weight,* pp. 191–216. Eds. T. C. Kirkham and S. J. Cooper. Academic Press: London.

Berridge, K. C. (2007b) The debate over dopamine's role in reward: the case for incentive salience. *Psychopharmacology (Berl)* 191, 391–431.

Berridge, K. C., Flynn, F. W., Schulkin, J. and Grill, H. J. (1984) Sodium depletion enhances salt palatability in rats. *Behav. Neurosci.* 98, 652–660.

Berridge, K. C. and Peciña, S. (1995) Benzodiazepines, appetite, and taste palatability. *Neurosci. Biobehav. Rev.* 19, 121–131.

Berridge, K. C. and Robinson, T. E. (1998) What is the role of dopamine in reward: Hedonic impact, reward learning, or incentive salience? *Brain Res. Brain Res. Rev.* 28, 309–369.

Berridge, K. C. and Valenstein, E. S. (1991) What psychological process mediates feeding evoked by electrical stimulation of the lateral hypothalamus? *Behav. Neurosci.* 105, 3–14.

Berridge, K. C., Venier, I. L. and Robinson, T. E. (1989) Taste reactivity analysis of 6-hydroxydopamine-induced aphagia: Implications for arousal and anhedonia hypotheses of dopamine function. *Behav. Neurosci.* 103, 36–45.

Blood, A. J. and Zatorre, R. J. (2001) Intensely pleasurable responses to music correlate with activity in brain regions implicated in reward and emotion. *Proc. Natl Acad. Sci. USA* 98, 11818–11823.

Brauer, L. H., Cramblett, M. J., Paxton, D. A. and Rose, J. E. (2001) Haloperidol reduces smoking of both nicotine-containing and denicotinized cigarettes. *Psychopharmacology (Berl)* 159, 31–37.

Brauer, L. H. and de Wit, H. (1997) High dose pimozide does not block amphetamine-induced euphoria in normal volunteers. *Pharmacol Biochem Behav* 56, 265–272.

Breiter, H. C., Gollub, R. L., Weisskoff, R. M., Kennedy, D. N., Makris, N., Berke, J. D., Goodman, J. M., Kantor, H. L., Gastfriend, D. R., Riorden, J. P., Mathew, R. T., Rosen, B. R. and Hyman, S. E. (1997) Acute effects of cocaine on human brain activity and emotion. *Neuron* 19, 591–611.

Brown, S., Martinez, M. J. and Parsons, L. M. (2004) Passive music listening spontaneously engages limbic and paralimbic systems. *NeuroReport* 15, 2033–2037.

Burke, K. A., Franz, T., Miller, D. and Schoenbaum, G. (this book). Conditioned reinforcement and the specialized role of corticolimbic circuits in the pursuit of happiness and other more specific rewards. In: *Pleasures of the Brain.* Eds. M. L. Kringelbach and K. C. Berridge. Oxford University Press: Oxford.

Cabanac, M. (1971) Physiological role of pleasure. *Science* 173, 1103–1107.

Cabanac, M. (this book). The dialectics of pleasure. In: *Pleasures of the Brain.* Eds. M. L. Kringelbach and K. C. Berridge. Oxford University Press: Oxford.

Cagniard, B., Beeler, J. A., Britt, J. P., McGehee, D. S., Marinelli, M. and Zhuang, X. (2006) Dopamine scales performance in the absence of new learning. *Neuron* 51, 541–547.

Calder, A. J., Beaver, J. D., Davis, M. H., van Ditzhuijzen, J., Keane, J. and Lawrence, A. D. (2007) Disgust sensitivity predicts the insula and pallidal response to pictures of disgusting foods. *Eur J Neurosci* 25, 3422–3428.

Cardinal, R. N., Parkinson, J. A., Hall, J. and Everitt, B. J. (2002) Emotion and motivation: the role of the amygdala, ventral striatum, and prefrontal cortex. *Neurosci. Biobehav. Rev.* 26, 321–352.

Carelli, R. M. and Wightman, R. M. (2004) Functional microcircuitry in the accumbens underlying drug addiction: Insights from real-time signaling during behavior. *Curr. Opin. Neurobiol.* 14, 763–768.

Cooper, S. J. (1980) Benzodiazepines as appetite-enhancing compounds. *Appetite* 1, 7–19.

Cooper, S. J. and Estall, L. B. (1985) Behavioural pharmacology of food, water and salt intake in relation to drug actions at benzodiazepine receptors. *Neurosci. Biobehav. Rev.* 9, 5–19.

Cromwell, H. C. and Berridge, K. C. (1993) Where does damage lead to enhanced food aversion: the ventral pallidum/substantia innominata or lateral hypothalamus? *Brain Res.* 624, 1–10.

Damasio, A. R. (1999) *The Feeling of What Happens: Body and Emotion in the Making of Consciousness.* Harcourt Brace: New York.

Darwin, C. (1872) *The Expression of the Emotions in Man and Animals* (1998 edition: revised and with commentary by P. Ekman). Harper Collins–Oxford University Press: Oxford.

Davidson, R. J. and Irwin, W. (1999) The functional neuroanatomy of emotion and affective style. *Trends Cogn Sci* 3, 11–21.

Day, J. J. and Carelli, R. M. (2007) The nucleus accumbens and Pavlovian reward learning. *Neuroscientist* 13, 148–159.

de Araujo, I. E., Gutierrez, R., Oliveira-Maia, A. J., Pereira, A., Jr., Nicolelis, M. A. and Simon, S. A. (2006) Neural ensemble coding of satiety states. *Neuron* 51, 483–494.

de Araujo, I. E., Rolls, E. T., Kringelbach, M. L., McGlone, F. and Phillips, N. (2003) Taste-olfactory convergence, and the representation of the pleasantness of flavour, in the human brain. *Eur. J. Neurosci.* 18, 2059–2068.

Delgado, J. M. R. (1969) *Physical Control of the Mind: Toward a Psychocivilized Society.* Harper and Row: New York.

Dickinson, A. and Balleine, B. W. (this book). Hedonics: the cognitive–motivational interface. In: *Pleasures of the Brain.* Eds. M. L. Kringelbach and K. C. Berridge. Oxford University Press: Oxford.

Ekman, P. (1999) Facial expressions. In: *Handbook of Cognition and Emotion*, pp. 301–320. John Wiley & Sons Ltd: Chichester.

Evans, A. H., Pavese, N., Lawrence, A. D., Tai, Y. F., Appel, S., Doder, M., Brooks, D. J., Lees, A. J. and Piccini, P. (2006) Compulsive drug use linked to sensitized ventral striatal dopamine transmission. *Ann. Neurol.* 59, 852–858.

Faurion, A., Cerf, B., Le Bihan, D. and Pillias, A. M. (1998) fMRI study of taste cortical areas in humans. *Ann. N Y Acad. Sci.* 855, 535–545.

Fiorino, D. F., Coury, A. and Phillips, A. G. (1997) Dynamic changes in nucleus accumbens dopamine efflux during the Coolidge effect in male rats. *J. Neurosci.* 17, 4849–4855.

Firestone, L. L., Gyulai, F., Mintun, M., Adler, L. J., Urso, K. and Winter, P. M. (1996) Human brain activity response to fentanyl imaged by positron emission tomography. *Anesth Analg* 82, 1247–1251.

Fischman, M. W. and Foltin, R. W. (1992) Self-administration of cocaine by humans: a laboratory perspective. *Ciba Found Symp* 166, 165–173; discussion 173–180.

Francis, S., Rolls, E. T., Bowtell, R., McGlone, F., O'Doherty, J., Browning, A., Clare, S. and Smith, E. (1999) The representation of pleasant touch in the brain and its relationship with taste and olfactory areas. *NeuroReport* 10, 453–459.

Frijda, N. (this book). On the nature and function of pleasure In: *Pleasures of the Brain*. Eds. M. L. Kringelbach and K. C. Berridge. Oxford University Press: Oxford.

Frijda, N. H. (2006) *The Laws of Emotion*. Lawrence Erlbaum Associates: Mahwah, NJ.

Gallistel, C. R. (1980) *The Organization of Action: a New Synthesis*. Lawrence Erlbaum Associates: Hillsdale, NJ.

Goltz, F. (1892) Der hund ohne grosshirn. *Pflügers Arch. ges Physiol.* 51, 570.

Gottfried, J. A. (this book). Olfaction and its pleasures: Human neuroimaging perspectives. In: *Pleasures of the Brain*. Eds. M. L. Kringelbach and K. C. Berridge. Oxford University Press: Oxford.

Green, A. L., Pereira, E. and Aziz, T. Z. (this book). Deep brain stimulation and pleasure. In: *Pleasures of the Brain*. Eds. M. L. Kringelbach and K. C. Berridge. Oxford University Press: Oxford.

Grill, H. J. and Berridge, K. C. (1985) Taste reactivity as a measure of the neural control of palatability. In: *Progress in Psychobiology and Physiological Psychology*, Vol. 2, pp. 1–6. Eds. A. N. Epstein and J. Sprague. Academic Press: New York.

Grill, H. J. and Norgren, R. (1978a) The taste reactivity test. I. Mimetic responses to gustatory stimuli in neurologically normal rats. *Brain Res* 143, 263–279.

Grill, H. J. and Norgren, R. (1978b) The taste reactivity test. II. Mimetic responses to gustatory stimuli in chronic thalamic and chronic decerebrate rats. *Brain Res* 143, 281–297.

Grove, E. A. (1988a) Efferent connections of the substantia innominata in the rat. *J. Comp. Neurol.* 277, 347–364.

Grove, E. A. (1988b) Neural associations of the substantia innominata in the rat: Afferent connections. *J. Comp. Neurol.* 277, 315–346.

Gutierrez, H., Hernandez-Echeagaray, E., Ramirez-Amaya, V. and Bermudez-Rattoni, F. (1999) Blockade of *N*-methyl-D-aspartate receptors in the insular cortex disrupts taste aversion and spatial memory formation. *Neuroscience* 89, 751–758.

Haber, S. N., Groenewegen, H. J., Grove, E. A. and Nauta, W. J. (1985) Efferent connections of the ventral pallidum: Evidence of a dual striato pallidofugal pathway. *J Comp Neurol* 235, 322–335.

Hart, C. L., Ward, A. S., Haney, M., Comer, S. D., Foltin, R. W. and Fischman, M. W. (2002) Comparison of smoked marijuana and oral Delta(9)-tetrahydrocannabinol in humans. *Psychopharmacology (Berl)* 164, 407–415.

Hart, C. L., Ward, A. S., Haney, M., Foltin, R. W. and Fischman, M. W. (2001) Methamphetamine self-administration by humans. *Psychopharmacology (Berl)* 157, 75–81.

Heath, R. G. (1964) *The Role of Pleasure in Behavior; a Symposium by 22 Authors*. Hoeber Medical Division Harper & Row: New York.

Heath, R. G. (1972) Pleasure and brain activity in man. Deep and surface electroencephalograms during orgasm. *J Nerv and Ment Dis* 154, 3–18.

Heath, R. G. (1996) *Exploring the Mind–Brain Relationship*. Moran Printing Inc.: Baton Rouge, LA.

Heimer, L. and Wilson, R. D. (1975) The subcortical projections of allocortex: Similarities in the neural associations of the hippocampus, the periform cortex and the neocortex. In: *Proceedings of the Golgi Centennial Symposium*. Ed. M. Santini, pp. 173–193, New York: Raven Press.

Hernandez, G., Hamdani, S., Rajabi, H., Conover, K., Stewart, J., Arvanitogiannis, A. and Shizgal, P. (2006) Prolonged rewarding stimulation of the rat medial forebrain bundle: Neurochemical and behavioral consequences. *Behav. Neurosci.* 120, 888–904.

Herzog, J., Reiff, J., Krack, P., Witt, K., Schrader, B., Muller, D. and Deuschl, G. (2003) Manic episode with psychotic symptoms induced by subthalamic nucleus stimulation in a patient with Parkinson's disease. *Mov. Disord.* 18, 1382–1384.

Higgs, S. and Cooper, S. J. (1996) Hyperphagia induced by direct administration of midazolam into the parabrachial nucleus of the rat. *Eur. J. Pharmacol.* 313, 1–9.

Hoebel, B. G., Rada, P. V., Mark, G. P. and Pothos, E. N. (1999) Neural systems for reinforcement and inhibition of behavior: Relevance to eating, addiction, and depression. In: *Well-being: the Foundations of Hedonic Psychology*, pp. 558–572. Russell Sage Foundation: New York

Hollerman, J. R., Tremblay, L. and Schultz, W. (2000) Involvement of basal ganglia and orbitofrontal cortex in goal-directed behavior. *Prog Brain Res* 126, 193–215.

Houeto, J. L., Mesnage, V., Mallet, L., Pillon, B., Gargiulo, M., du Moncel, S. T., Bonnet, A. M., Pidoux, B., Dormont, D., Cornu, P. and Agid, Y. (2002) Behavioural disorders, Parkinson's disease and subthalamic stimulation. *J. Neurol. Neurosurg. Psychiatry* 72, 701–707.

Hughlings Jackson, J. (1958) *Selected writings of John Hughlings Jackson*, Vols. 1 and 2. Ed. J. Taylor. Staples Press: London.

Insel, T. R. and Fernald, R. D. (2004) How the brain processes social information: Searching for the social brain. *Annu Rev Neurosci* 27, 697–722.

Kalivas, P. W. and Nakamura, M. (1999) Neural systems for behavioral activation and reward. *Curr. Opin. Neurobiol.* 9, 223–227.

Kalivas, P. W. and Volkow, N. D. (2005) The neural basis of addiction: a pathology of motivation and choice. *Am. J. Psychiatry* 162, 1403–1413.

Kelley, A. E., Bakshi, V. P., Haber, S. N., Steininger, T. L., Will, M. J. and Zhang, M. (2002) Opioid modulation of taste hedonics within the ventral striatum. *Physiol. Behav.* 76, 365–377.

Kelley, A. E., Baldo, B. A., Pratt, W. E. and Will, M. J. (2005) Corticostriatal–hypothalamic circuitry and food motivation: Integration of energy, action and reward. *Physiol. Behav.* 86, 773–795.

Kirkham, T. C. (2005) Endocannabinoids in the regulation of appetite and body weight. *Behav. Pharmacol.* 16, 297–313.

Knutson, B. and Cooper, J. C. (2005) Functional magnetic resonance imaging of reward prediction. *Curr. Opin. Neurol.* 18, 411–417.

Knutson, B., Fong, G. W., Adams, C. M., Varner, J. L. and Hommer, D. (2001) Dissociation of reward anticipation and outcome with event-related fMRI. *NeuroReport* 12, 3683–3687.

Komisaruk, B. R., Whipple B. and Beyer, C. (this book). Sexual pleasure. In: *Pleasures of the Brain*. Eds. M. L. Kringelbach and K. C. Berridge. Oxford University Press: Oxford.

Komisaruk, B. R. and Whipple, B. (2005) Functional MRI of the brain during orgasm in women. *Annu. Rev. Sex Res.* 16, 62–86.

Kringelbach, M. L. (2005) The human orbitofrontal cortex: Linking reward to hedonic experience. *Nat. Rev. Neurosci.* 6, 691–702.

Kringelbach, M. L. (this book). The hedonic brain: a functional neuroanatomy of human pleasure. In: *Pleasures of the Brain*. Eds. M. L. Kringelbach and K. C. Berridge. Oxford University Press: Oxford.

Kringelbach, M. L., Jenkinson, N., Owen, S. L. and Aziz, T. Z. (2007) Translational principles of deep brain stimulation. *Nat. Rev. Neurosci.* 8, 623–635.

Kringelbach, M. L., O'Doherty, J., Rolls, E. T. and Andrews, C. (2003) Activation of the human orbitofrontal cortex to a liquid food stimulus is correlated with its subjective pleasantness. *Cereb. Cortex* 13, 1064–1071.

Landis, B. N., Leuchter, I., San Millan Ruiz, D., Lacroix, J. S. and Landis, T. (2006) Transient hemiageusia in cerebrovascular lateral pontine lesions. *J. Neurol. Neurosurg. Psychiatry* 77, 680–683.

LeDoux, J. (1996) *The Emotional Brain: the Mysterious Underpinnings of Emotional Life*. Simon & Schuster: New York.

Leknes, S. and Tracey, I. (this book). Pleasure and pain: Masters of mankind. In: *Pleasures of the Brain*. Eds. M. L. Kringelbach and K. C. Berridge. Oxford University Press: Oxford.

Leyton, M. (this book). The neurobiology of desire. Dopamine and the regulation of mood and motivational states in humans. In: *Pleasures of the Brain*. Eds. M. L. Kringelbach and K. C. Berridge. Oxford University Press: Oxford.

Leyton, M., Boileau, I., Benkelfat, C., Diksic, M., Baker, G. and Dagher, A. (2002) Amphetamine-induced increases in extracellular dopamine, drug wanting, and novelty seeking: a PET/[11C]raclopride study in healthy men. *Neuropsychopharmacology* 27, 1027–1035.

Leyton, M., Casey, K. F., Delaney, J. S., Kolivakis, T. and Benkelfat, C. (2005) Cocaine craving, euphoria, and self-administration: a preliminary study of the effect of catecholamine precursor depletion. *Behav. Neurosci.* 119, 1619–1627.

Lundy, R. F., Jr. and Norgren, R. (2004) Activity in the hypothalamus, amygdala, and cortex generates bilateral and convergent modulation of pontine gustatory neurons. *J. Neurophysiol.* 91, 1143–1157.

Mahler, S. V., Smith, K. S. and Berridge, K. C. (2007) Endocannabinoid hedonic hotspot for sensory pleasure: Anandamide in nucleus accumbens shell enhances "liking" of sweet reward. *Neuropsychopharmacology* 32(11), 2267–2278.

Mandat, T. S., Hurwitz, T. and Honey, C. R. (2006) Hypomania as an adverse effect of subthalamic nucleus stimulation: Report of two cases. *Acta Neurochir. (Wien)* 148, 895–897; discussion 898.

McCoy, A. N., Crowley, J. C., Haghighian, G., Dean, H. L. and Platt, M. L. (2003) Saccade reward signals in posterior cingulate cortex. *Neuron* 40, 1031–1040.

Menon, V. and Levitin, D. J. (2005) The rewards of music listening: Response and physiological connectivity of the mesolimbic system. *Neuroimage* 28, 175–184.

Miller, F. R. and Sherrington, C. S. (1915) Some observations on the bucco-pharyngeal stage of reflex deglutition in the Cat. *Q. J. Exp. Physiol.* 9, 147–186.

Miller, J. M., Vorel, S. R., Tranguch, A. J., Kenny, E. T., Mazzoni, P., van Gorp, W. G. and Kleber, H. D. (2006) Anhedonia after a selective bilateral lesion of the globus pallidus. *Am. J. Psychiatry* 163, 786–788.

Mobbs, D., Greicius, M. D., Abdel-Azim, E., Menon, V. and Reiss, A. L. (2003) Humor modulates the mesolimbic reward centers. *Neuron* 40, 1041–1048.

Mogenson, G. J. and Yang, C. R. (1991) The contribution of basal forebrain to limbic-motor integration and the mediation of motivation to action. *Adv. Exp. Med. Biol.* 295, 267–290.

Monahan, J. L., Murphy, S. T. and Zajonc, R. B. (2000) Subliminal mere exposure: Specific, general and diffuse effects. *Psychol. Sci.* 11, 462–466.

Morgan, J. C., diDonato, C. J., Iyer, S. S., Jenkins, P. D., Smith, J. R. and Sethi, K. D. (2006) Self-stimulatory behavior associated with deep brain stimulation in Parkinson's disease. *Mov. Disord.* 21, 283–285.

Naqvi, N. H., Rudrauf, D., Damasio, H. and Bechara, A. (2007) Damage to the insula disrupts addiction to cigarette smoking. *Science* 315, 531–534.

Norgren, R. (1995) Gustatory system. In: *The Rat Nervous System*, pp. 751–771. Ed. G. Paxinos. Academic Press: San Diego, CA.

O'Doherty, J., Rolls, E. T., Francis, S., Bowtell, R., McGlone, F., Kobal, G., Renner, B. and Ahne, G. (2000) Sensory-specific satiety-related olfactory activation of the human orbitofrontal cortex. *Neuroreport* 11, 399–403.

O'Doherty, J. P. (2004) Reward representations and reward-related learning in the human brain: Insights from neuroimaging. *Curr. Opin. Neurobiol.* 14, 769–776.

Olds, J. (1961) *Differential Effects of Drives and Drugs on Self-Stimulation at Different Brain Sites*. University of Texas Press: Austin, TX.

Olds, J. and Milner, P. (1954) Positive reinforcement produced by electrical stimulation of septal area and other regions of rat brain. *J. Comp. Physiol. Psychol.* 47, 419–427.

Panksepp, J. (1991) Affective neuroscience: a conceptual framework for the study of emotions. In: *International Reviews of Studies in Emotions*, Vol. 1, pp. 59–99. Ed. K. Strongman. Wiley: Chichester.

Peciña, S. and Berridge, K. C. (1996) Brainstem mediates diazepam enhancement of palatability and feeding: Microinjections into fourth ventricle versus lateral ventricle. *Brain Res* 727, 22–30.

Peciña, S. and Berridge, K. C. (2000) Opioid site in nucleus accumbens shell mediates eating and hedonic "liking" for food: Map based on microinjection Fos plumes. *Brain Res.* 863, 71–86.

Peciña, S. and Berridge, K. C. (2005) Hedonic hot spot in nucleus accumbens shell: Where do mu-opioids cause increased hedonic impact of sweetness? *J. Neurosci.* 25, 11777–11786.

Peciña, S., Berridge, K. C. and Parker, L. A. (1997) Pimozide does not shift palatability: Separation of anhedonia from sensorimotor suppression by taste reactivity. *Pharmacol. Biochem. Behav.* 58, 801–811.

Peciña, S., Cagniard, B., Berridge, K. C., Aldridge, J. W. and Zhuang, X. (2003) Hyperdopaminergic mutant mice have higher "wanting" but not "liking" for sweet rewards. *J. Neurosci.* 23, 9395–9402.

Peciña, S., Smith, K. S. and Berridge, K. C. (2006) Hedonic hot spots in the brain. *Neuroscientist* 12, 500–511.

Pessiglione, M., Schmidt, L., Draganski, B., Kalisch, R., Lau, H., Dolan, R. J. and Frith, C. D. (2007) How the brain translates money into force: a neuroimaging study of subliminal motivation. *Science* 316, 904–906.

Petrovic, P. (this book). Placebo analgesia and the brain. In: *Pleasures of the Brain*. Eds. M. L. Kringelbach and K. C. Berridge. Oxford University Press: Oxford.

Pickel, V. M., Chan, J., Kash, T. L., Rodriguez, J. J. and MacKie, K. (2004) Compartment-specific localization of cannabinoid 1 (CB1) and mu-opioid receptors in rat nucleus accumbens. *Neuroscience* 127, 101–112.

Pritchard, T. C., Hamilton, R. B. and Norgren, R. (2000) Projections of the parabrachial nucleus in the old world monkey. *Exp. Neurol.* 165, 101–117.

Rauch, S. L., Shin, L. M., Dougherty, D. D., Alpert, N. M., Orr, S. P., Lasko, M., Macklin, M. L., Fischman, A. J. and Pitman, R. K. (1999) Neural activation during sexual and competitive arousal in healthy men. *Psychiatry Res.* 91, 1–10.

Reynolds, S. M. and Berridge, K. C. (2002) Positive and negative motivation in nucleus accumbens shell: Bivalent rostrocaudal gradients for GABA-elicited eating, taste "liking"/ "disliking" reactions, place preference/avoidance, and fear. *J. Neurosci.* 22, 7308–7320.

Richardson, D. K., Reynolds, S. M., Cooper, S. J. and Berridge, K. C. (2005) Endogenous opioids are necessary for benzodiazepine palatability enhancement: Naltrexone blocks diazepam-induced increase of sucrose-'liking'. *Pharmacol. Biochem. Behav.* 81, 657–663.

Rios, C., Gomes, I. and Devi, L. A. (2006) mu opioid and CB1 cannabinoid receptor interactions: Reciprocal inhibition of receptor signaling and neuritogenesis. *Br. J. Pharmacol.* 148, 387–395.

Robbins, T. W. and Everitt, B. J. (1996) Neurobehavioural mechanisms of reward and motivation. *Curr. Opin. Neurobiol.* 6, 228–236.

Robinson, S., Sandstrom, S. M., Denenberg, V. H. and Palmiter, R. D. (2005) Distinguishing whether dopamine regulates liking, wanting, and/or learning about rewards. *Behav. Neurosci.* 119, 5–15.

Robinson, T. E. and Berridge, K. C. (1993) The neural basis of drug craving: an incentive-sensitization theory of addiction. *Brain Res. Brain Res. Rev.* 18, 247–291.

Robinson, T. E. and Berridge, K. C. (2003) Addiction. *Annu. Rev. Psychol.* 54, 25–53.

Roitman, M. F., Wheeler, R. A. and Carelli, R. M. (2005) Nucleus accumbens neurons are innately tuned for rewarding and aversive taste stimuli, encode their predictors, and are linked to motor output. *Neuron* 45, 587–597.

Rolls, E. T. (2000) The orbitofrontal cortex and reward. *Cerebral Cortex* 10, 384–394.

Rolls, E. T., Kringelbach, M. L. and de Araujo, I. E. (2003a) Different representations of pleasant and unpleasant odours in the human brain. *Eur J Neurosci* 18, 695–703.

Rolls, E. T., O'Doherty, J., Kringelbach, M. L., Francis, S., Bowtell, R. and McGlone, F. (2003b) Representations of pleasant and painful touch in the human orbitofrontal and cingulate cortices. *Cereb. Cortex* 13, 308–317.

Rolls, E. T., Sienkiewicz, Z. J. and Yaxley, S. (1989) Hunger modulates the responses to gustatory stimuli of single neurons in the caudolateral orbitofrontal cortex of the macaque monkey. *Eur. J. Neurosci.* 1, 53–60.

Schallert, T. and Whishaw, I. Q. (1978) Two types of aphagia and two types of sensorimotor impairment after lateral hypothalamic lesions: Observations in normal weight, dieted, and fattened rats. *J. Comp. Physiol. Psychol.* 92, 720–741.

Schlaepfer, T. E., Cohen, M. X., Frick, C., Kosel, M., Brodesser, D., Axmacher, N., Joe, A. Y., Kreft, M., Lenartz, D. and Sturm, V. (2008) Deep brain stimulation to reward circuitry alleviates anhedonia in refractory major depression. *Neuropsychopharmacology.* 33(2), 368–377.

Schoenbaum, G. and Roesch, M. (2005) Orbitofrontal cortex, associative learning, and expectancies. *Neuron* 47, 633–636.

Schoffelmeer, A. N., Hogenboom, F., Wardeh, G. and De Vries, T. J. (2006) Interactions between CB1 cannabinoid and mu opioid receptors mediating inhibition of neurotransmitter release in rat nuclues accumbens core. *Neuropharmacology* 51, 773–781.

Schooler, J. W. and Mauss, I. (this book). To be happy and to know it: the experience and meta-awareness of pleasure. In: *Pleasures of the Brain.* Eds. M. L. Kringelbach and K. C. Berridge. Oxford University Press: Oxford.

Schulkin, J. (1991) *Sodium Hunger: the Search for a Salty Taste.* Cambridge University Press: New York.

Schultz, W. (1998) Predictive reward signal of dopamine neurons. *J. Neurophysiol.* 80, 1–27.

Sem-Jacobsen, C. W. (1976) Electrical stimulation and self-stimulation with chronic implanted electrodes: Interpretation and pitfalls of results. In: *Brain–Stimulation Reward*, pp. 505–520. Eds. A. Wauquier and E. T. Rolls. Elsevier-North Holland: Amsterdam.

Shimura, T., Imaoka, H. and Yamamoto, T. (2006) Neurochemical modulation of ingestive behavior in the ventral pallidum. *Eur. J. Neurosci.* 23, 1596–1604.

Shizgal, P. (1999) On the neural computation of utility: Implications from studies of brain stimulation reward. In: *Well-being: the Foundations of Hedonic Psychology*, pp. 500–524. Eds. D. Kahneman, E. Diener and N. Schwarz. Russell Sage Foundation: New York.

Sienkiewicz-Jarosz, H., Scinska, A., Kuran, W., Ryglewicz, D., Rogowski, A., Wrobel, E., Korkosz, A., Kukwa, A., Kostowski, W. and Bienkowski, P. (2005) Taste responses in patients with Parkinson's disease. *J Neurol Neurosurg Psychiatry* 76, 40–46.

Simon, S. A., de Araujo, I. E., Gutierrez, R. and Nicolelis, M. A. (2006) The neural mechanisms of gustation: a distributed processing code. *Nat. Rev. Neurosci.* 7, 890–901.

Skov, M. (this book). Neuroaesthetics. In: *Pleasures of the Brain.* Eds. M. L. Kringelbach and K. C. Berridge. Oxford University Press: Oxford.

Small, D. M. (2006) Central gustatory processing in humans. *Adv. Otorhinolaryngol.* 63, 191–220.

Small, D. M., Gregory, M. D., Mak, Y. E., Gitelman, D., Mesulam, M. M. and Parrish, T. (2003) Dissociation of neural representation of intensity and affective valuation in human gustation. *Neuron* 39, 701–711.

Small, D. M., Zatorre, R. J., Dagher, A., Evans, A. C. and Jones-Gotman, M. (2001) Changes in brain activity related to eating chocolate: From pleasure to aversion. *Brain* 124, 1720–1733.

Smith, K. S. and Berridge, K. C. (2005) The ventral pallidum and hedonic reward: Neurochemical maps of sucrose "liking" and food intake. *J. Neurosci.* 25, 8637–8649.

Smith, K. S. and Berridge, K. C. (2007) Opioid limbic circuit for reward: Interaction between hedonic hotspots of nucleus accumbens and ventral pallidum. *J. Neurosci.* 27, 1594–1605.

Soderpalm, A. H. and Berridge, K. C. (2000a) Food intake after diazepam, morphine or muscimol: Microinjections In the nucleus accumbens shell. *Pharmacol. Biochem. Behav.* 66, 429–434.

Soderpalm, A. H. and Berridge, K. C. (2000b) The hedonic impact and intake of food are increased by midazolam microinjection in the parabrachial nucleus. *Brain Res.* 877, 288–297.

Spector, A. C. (2000) Linking gustatory neurobiology to behavior in vertebrates. *Neurosci. Biobehav. Rev.* 24, 391–416.

Steiner, J. E. (1973) The gustofacial response: Observation on normal and anencephalic newborn infants. *Symp. Oral Sensation Perception* 4, 254–278.

Steiner, J. E., Glaser, D., Hawilo, M. E. and Berridge, K. C. (2001) Comparative expression of hedonic impact: Affective reactions to taste by human infants and other primates. *Neurosci. Biobehav. Rev.* 25, 53–74.

Stellar, J. R., Brooks, F. H. and Mills, L. E. (1979) Approach and withdrawal analysis of the effects of hypothalamic stimulation and lesions in rats. *J. Comp. Physiol. Psychol.* 93, 446–466.

Stratford, T. R., Kelley, A. E. and Simansky, K. J. (1999) Blockade of GABAA receptors in the medial ventral pallidum elicits feeding in satiated rats. *Brain Res.* 825, 199–203.

Taha, S. A. and Fields, H. L. (2005) Encoding of palatability and appetitive behaviors by distinct neuronal populations in the nucleus accumbens. *J. Neurosci.* 25, 1193–1202.

Tanda, G., Pontieri, F. E. and Di Chiara, G. (1997) Cannabinoid and heroin activation of mesolimbic dopamine transmission by a common mu1 opioid receptor mechanism. *Science* 276, 2048–2050.

Teitelbaum, P. and Epstein, A. N. (1962) The lateral hypothalamic syndrome: Recovery of feeding and drinking after lateral hypothalamic lesions. *Psychol. Rev.* 69, 74–90.

Teitelbaum, P. and Stellar, E. (1954) Recovery from the failure to eat produced by hypothalamic lesions. *Science* 120, 894–895.

Tindell, A. J., Berridge, K. C. and Aldridge, J. W. (2004) Ventral pallidal representation of pavlovian cues and reward: Population and rate codes. *J. Neurosci.* 24, 1058–1069.

Tindell, A. J., Berridge, K. C., Zhang, J., Peciña, S. and Aldridge, J. W. (2005) Ventral pallidal neurons code incentive motivation: Amplification by mesolimbic sensitization and amphetamine. *Eur. J. Neurosci.* 22, 2617–2634.

Tindell, A. J., Smith, K. S., Peciña, S., Berridge, K. C. and Aldridge, J. W. (2006) Ventral pallidum firing codes hedonic reward: When a bad taste turns good. *J. Neurophysiol.* 96, 2399–2409.

Valenstein, E. S. (1974) *Brain Control.* Wiley: New York.

Veldhuizen, M. G., Rudenga, K. J. and Small, D. M. and (this book). The pleasure of taste, flavor and food. In: *Pleasures of the Brain.* Eds. M. L. Kringelbach and K. C. Berridge. Oxford University Press: Oxford.

Wang, Z. and Aragona, B. J. (2004) Neurochemical regulation of pair bonding in male prairie voles. *Physiol. Behav.* 83, 319–328.

Watson, K. K., Shepherd, S.V. and Platt, M. L. (this book). Neuroethology of pleasure. In: *Pleasures of the Brain.* Eds. M. L. Kringelbach and K. C. Berridge. Oxford University Press: Oxford.

Whalen, P. J., Rauch, S. L., Etcoff, N. L., McInerney, S. C., Lee, M. B. and Jenike, M. A. (1998) Masked presentations of emotional facial expressions modulate amygdala activity without explicit knowledge. *J. Neurosci.* 18, 411–418.

Will, M. J., Franzblau, E. B. and Kelley, A. E. (2003) Nucleus accumbens mu-opioids regulate intake of a high-fat diet via activation of a distributed brain network. *J. Neurosci.* 23, 2882–2888.

Williams, C. M. and Kirkham, T. C. (2002) Reversal of delta 9-THC hyperphagia by SR141716 and naloxone but not dexfenfluramine. *Pharmacol. Biochem. Behav.* 71, 333–340.

Winkielman, P., Berridge, K. C. and Wilbarger, J. L. (2005) Unconscious affective reactions to masked happy versus angry faces influence consumption behavior and judgments of value. *Pers Soc Psychol Bull* 31, 121–135.

Winkielman, P., Zajonc, R. B. and Schwarz, N. (1997) Subliminal affective priming resists attributional interventions. *Cognition Emotion* 11, 433–465.

Wirsig, C. R. and Grill, H. J. (1982) Contribution of the rat's neocortex to ingestive control: I. Latent learning for the taste of sodium chloride. *J. Comp. Physiol. Psychol.* 96, 615–627.

Wise, R. A. (1998) Drug-activation of brain reward pathways. *Drug Alcohol Depend.* 51, 13–22.

Wise, R. A. and Bozarth, M. A. (1985) Brain mechanisms of drug reward and euphoria. *Psychiatr. Med.* 3, 445–460.

Wyvell, C. L. and Berridge, K. C. (2000) Intra-accumbens amphetamine increases the conditioned incentive salience of sucrose reward: Enhancement of reward "wanting" without enhanced "liking" or response reinforcement. *J. Neurosci.* 20, 8122–8130.

Yamamoto, T. (2006) Neural substrates for the processing of cognitive and affective aspects of taste in the brain. *Arch. Histol. Cytol.* 69, 243–255.

Yaxley, S., Rolls, E. T. and Sienkiewicz, Z. J. (1988) The responsiveness of neurons in the insular gustatory cortex of the macaque monkey is independent of hunger. *Physiol. Behav.* 42, 223–229.

Yeomans, M. R. and Gray, R. W. (2002) Opioid peptides and the control of human ingestive behaviour. *Neurosci. Biobehav. Rev.* 26, 713–728.

Zahm, D. S. (2000) An integrative neuroanatomical perspective on some subcortical substrates of adaptive responding with emphasis on the nucleus accumbens. *Neurosci. Biobehav. Rev.* 24, 85–105.

Zhang, M. and Kelley, A. E. (2000) Enhanced intake of high-fat food following striatal mu-opioid stimulation: Microinjection mapping and fos expression. *Neuroscience* 99, 267–277.

2

Conditioned Reinforcement and the Specialized Role of Corticolimbic Circuits in the Pursuit of Happiness and Other More Specific Rewards

KATHRYN A. BURKE, THERESA FRANZ, DANIELLE MILLER,
AND GEOFFREY SCHOENBAUM

Normal behavior is guided by both primary reinforcers (such as food, drugs, sex, etc.) and secondary or conditioned reinforcers. Conditioned reinforcers are cues that have been repeatedly paired with primary rewards such that they acquire the ability to support behavior even in the absence of those primary rewards (Mackintosh, 1974). Our everyday lives are filled with examples of conditioned reinforcers that drive our behavior. These include examples such as money and corporate icons such as the McDonalds Golden Arches, which acquire an emotional/affective or hedonic value of their own, as well as examples with more specific associations, such as the song that was playing when we met that special someone. Indeed, conditioned reinforcement is pervasive in modern society.

Conditioned reinforcement depends upon a circuit that includes the basolateral amygdala, the orbitofrontal cortex, and the nucleus accumbens (Burns et al., 1993; Cador et al., 1989; Everitt and Robbins, 1992; Parkinson et al., 1999, 2001; Pears et al., 2001, 2003; Whitelaw et al., 1996). However, our understanding of how these areas interact to mediate conditioned reinforcement is hampered by our limited understanding of underlying associative representations that allow these cues to serve as conditioned reinforcers. Specifically, it is not certain whether conditioned reinforcers are effective because of the affective value they acquire or because of the specific outcomes that they represent and the value that those outcomes have to us. In other words, are the Golden Arches effective because they predict hamburgers and fries, which we desire, or are they also effective in part because they evoke a more general feeling of happiness directly? Moreover, if both are true, might different brain circuits be specialized to mediate our pursuit of happiness versus our pursuit of more specific rewards? In this chapter, we will outline a new approach we have taken to address these questions.

Basolateral Amygdala, Orbitofrontal Cortex, and Nucleus Accumbens in Reward Processing and Conditioned Reinforcement

Basolateral Amygdala

The basolateral amygdala receives sensory inputs from the environment and is commonly believed to associate neutral sensory cues with information regarding the primary rewarding or aversive outcomes that these cues come to predict (Quirk et al., 1995; Schoenbaum et al., 1998, 1999, 2003). This type of learning is important in adaptive behavior, in which an animal directs its behavior according to the current value of the predicted outcome.

The role of amygdala in this type of learning has been clearly demonstrated in reinforcer devaluation tasks (Hatfield et al., 1996; Malkova et al., 1997). In devaluation (Holland and Straub, 1979), the animal first learns the association between a neutral cue and a food outcome through a Pavlovian conditioning procedure. After the rat learns this association, it will readily respond to the food cup once the cue comes on in anticipation of the impending outcome. Following learning, the food is devalued by pairing it with illness (or by selective satiation). Subsequently, food cup responding to the cue is assessed in an unrewarded probe session. Normal animals spontaneously decrease conditioned responding to the cue after devaluation, indicating that the cue–outcome association has been formed and can be updated and used to guide food cup responding. Rats with damage to basolateral amygdala failed to alter conditioned responding after devaluation (Hatfield et al., 1996; Malkova et al., 1997). Instead, these rats continued to respond to the cue to the same extent as the rats in which the food is not devalued. This deficit was observed despite normal learning and normal devaluation in earlier training, as well as normal extinction during the probe test. Interestingly, damage to basolateral amygdala made after initial conditioning no longer affects changes in responding after devaluation (Pickens et al., 2003), suggesting that the original deficit reflects a critical role for amygdala in learning the original associations rather than in updating or using that information later in the actual probe test.

Importantly, similar deficits have been found in monkeys with neurotoxic lesions of amygdala (Malkova et al., 1997). In this experiment, monkeys were trained in an object discrimination task where visual objects were paired with different food rewards. Once the monkeys learned the object discriminations, they were allowed to choose between two objects that predicted two different foods. Prior to some sessions, the monkeys were fed to satiety on one of the two foods in order to devalue that particular outcome. Satiation caused normal monkeys to bias subsequent choices away from the object paired with the satiated food, whereas monkeys with amygdala lesions failed to change their choice behavior after selective satiation.

The role of basolateral amygdala in signaling information about predicted outcomes is also evident in a variety of other settings. For example, Dwyer and Killcross used a mediated conditioning procedure to show that rats with basolateral amygdala lesions were impaired in using outcome-specific information evoked by cues to guide their behavior (Dwyer and Killcross, 2006). In this experiment, rats were placed in a Y-maze where each distinctive arm was associated with water, sucrose, or maltodextrin solution. After learning the associations between the contextual cues contained in the maze arms and the different solutions, the rats were then trained to associate one of the maze arms with illness induced by lithium chloride (LiCl) injections. Subsequently, normal rats reduced their consumption of the solution associated with the target arm. This reduction occurred even though that solution was never directly paired with illness, indicating that rats had activated specific representations of the different solutions when they were exposed to the arms while ill. Rats with damage to basolateral amygdala showed normal behavior during training and devaluation but failed to change their consumption of the solutions in the probe test.

These studies demonstrate a critical role for basolateral amygdala in the process by which neutral cues are able to evoke representations of the outcomes they predict, particularly the value of those outcomes. However, basolateral amygdala has also been implicated in other cue-evoked behaviors, which do not appear to depend specifically upon the value of the predicted outcome. For example, basolateral amygdala supports second-order conditioning (Hatfield et al., 1996; Setlow et al., 2002a). In this Pavlovian procedure, a neutral cue is paired with a reward. Following this, another neutral cue, termed the second-order cue, is then paired with the conditioned stimulus. Although the second-order cue has never been paired directly with reward, a conditioned response develops to this cue. Basolateral amygdala lesions prevent the development of this second-order conditioned response. Interestingly second-order conditioning is not critically dependent on the value of the outcome predicted by the conditioned stimulus (Holland and Rescorla, 1975).

Basolateral amygdala has also been implicated in Pavlovian-to-instrumental transfer (PIT) (Corbit and Balleine, 2005). In this procedure, a cue that has been paired with an appetitive outcome, through Pavlovian conditioning, is able to increase a previously trained instrumental response. This increased responding is termed "transfer." Transfer occurs despite the lack of any prior pairing between the cue and the instrumental response. As a result, it has been suggested that the effect of the Pavlovian cue is due to its ability to evoke affective or motivational representations. Rats with damage to basolateral amygdala show normal Pavlovian and instrumental conditioning but fail to increase instrumental responding in the presence of the Pavlovian cue. Like second-order conditioning, mentioned above, transfer is not affected by devaluation of the outcome predicted by the conditioned stimulus (Holland, 2004).

These studies suggest that basolateral amygdala also plays a role in allowing cues to evoke representations of general affect that are independent of the specific properties of the predicted outcome, particularly its current value. Again this role seems to be particularly evident prior to learning, since lesions of basolateral amygdala after initial or first-order conditioning no longer affect performance in these settings (Pickens et al., 2003; Setlow et al., 2002a).

Basolateral Amygdala and Orbitofrontal Cortex

The involvement of basolateral amygdala in linking neutral cues to representations of the specific outcomes they predict, particularly the value of those outcomes, depends critically on interactions between basolateral amygdala and orbitofrontal cortex (Baxter et al., 2000; Schoenbaum et al., 2003). Orbitofrontal cortex–lesioned rats, like rats with basolateral amygdala lesions, fail to reduce conditioned responding after changes in the value of the predicted outcome (Gallagher et al., 1999). Similar deficits have been observed in monkeys with orbitofrontal lesions and in monkeys with crossed lesions of orbitofrontal cortex and amygdala (Baxter et al., 2000; Izquierdo et al., 2004). Impairments after crossed lesions demonstrate that proper responding requires serial processing in these two areas.

The role of amygdala and orbitofrontal cortex in signaling outcome representations has also been shown in a human imaging study using a devaluation procedure (Gottfried et al., 2003). In this experiment, subjects were trained to associate visual cues with odors of different foods—food odor "rewards." After training, the subjects were scanned during presentation of the visual cues before and after being fed to satiation on one of the foods. The cue that signaled the satiated food odor elicited a decreased blood oxygen level–dependent (BOLD) signal in orbitofrontal cortex as well as in the basal nuclei in the amygdala, while signals in these regions to a cue that signaled a nonsatiated food odor were not affected.

The roles of basolateral amygdala and orbitofrontal cortex in responding after devaluation can also be dissociated by manipulating the timing of the lesions. Recall that when lesions of basolateral amygdala are made after conditioning, there is no longer any effect on performance. By contrast, when orbitofrontal cortex lesions are made after initial conditioning, or even after devaluation, there continues to be an impact on devaluation-induced changes in conditioned responding (Pickens et al., 2003). These results suggest that orbitofrontal cortex plays an ongoing role in the use of cue-evoked representations of the outcome's current value to guide responding.

Whether outflow from basolateral amygdala to orbitofrontal cortex is also important for mediating behaviors based on affective representations is somewhat unclear. Post-training lesions of orbitofrontal cortex affect discriminative responding for second-order cues (Cousens and Otto, 2003). Whether this sort of learning is devaluation-insensitive like Pavlovian second-order conditioning is not certain. Post-training lesions of orbitofrontal cortex have also been reported to impair transfer-like damage to basolateral amygdala (Ostlund and Balleine, 2007). However, lesions made before training had no effect, suggesting that while orbitofrontal cortex may play a role in information acquired when it is intact, it does not contribute critically to the systems that mediate the fundamental processing required for transfer. Thus while the amygdala–orbitofrontal circuit is clearly required for behaviors guided by the cue-evoked value of the predicted outcome, it may not be critical for behaviors guided by cue-evoked devaluation-insensitive, general affective information.

Basolateral Amygdala, Central Nucleus, and Accumbens

The basolateral amygdala also sends strong projections to the central nucleus and to the nucleus accumbens (Kelley et al., 1982; Krettek and Price, 1978; Wright et al., 1996). These outflow pathways appear to be preferentially involved in allowing cues to drive behavior through representations of general affect that are resistant to devaluation of the predicted outcome. As we have described, the basolateral amygdala is implicated in behaviors such as second-order conditioning and PIT (Corbit and Balleine, 2005; Hatfield et al., 1996). As discussed earlier, these behaviors are typically resistant to devaluation of the predicted outcome. Normal performance in these tasks also often depends on central nucleus of the amygdala or the nucleus accumbens (Balleine and Corbit, 2005; Corbit and Balleine, 2005; Holland and Gallagher, 2003; Setlow et al., 2002b). Interestingly, it has been shown that crossed lesions of basolateral amygdala and nucleus accumbens, designed to disconnect the two regions, impair second-order conditioning even when made after first-order conditioning (Setlow et al., 2002b). This contrasts with basolateral amygdala (ABL) lesions, which are most effective when made before initial conditioning (Setlow et al., 2002a). These results suggests that basolateral amygdala plays a role in acquiring these representations, which nucleus accumbens then employs to guide responding. This is similar to the interactions between amygdala and orbitofrontal cortex in devaluation-sensitive behaviors.

Though deficits in second-order conditioning do not occur after lesions to the central nucleus of the amygdala (Hatfield et al., 1996), damage to this area has been found to affect PIT (Corbit and Balleine, 2005; Hall et al., 2001), as has damage to nucleus accumbens (Balleine and Corbit, 2005; de Borchgrave et al., 2002; Hall et al., 2001). The role of these downstream areas in transfer is complicated by the observation that transfer can be outcome-specific, where both the Pavlovian cue and the lever are associated with the same outcome or transfer can be general when the cue and lever are associated with different outcomes. Unfortunately, these two forms of transfer are typically not elicited by the same training procedures, thus most studies of this phenomenon report on only one or the other. However it appears that basolateral amygdala is required for outcome-specific but not general transfer, whereas central nucleus, and to a lesser extent accumbens, are more strongly implicated in general transfer (Balleine and Corbit, 2005; Corbit and Balleine, 2005; Holland and Gallagher, 2003). One interpretation of these results is that as affective information moves from basolateral amygdala to these downstream areas, it becomes progressively more independent of the specific features of the outcome with which it was originally associated. This idea is consistent with the observation that damage to nucleus accumbens does not impair learned responses after reinforcer devaluation (de Borchgrave et al., 2002) but does affect transfer (Corbit et al., 2001; de Borchgrave et al., 2002; Hall et al., 2001).

Contributions of Reward Circuits to Conditioned Reinforcement

Conditioned reinforcement is the process by which a neutral stimulus, which has been paired previously with a primary reinforcer, is able to support the acquisition and maintenance of a new instrumental response. A large and growing number of studies show that the brain regions involved in reward learning, reviewed earlier, are also critical for normal conditioned reinforcement. For example, Roberts and colleagues have tested the contributions of amygdala and orbitofrontal cortex in marmosets on a two-schedule, progressive ratio task (Parkinson et al., 2001; Pears et al., 2003). In this task, the marmosets have to execute X number of responses to obtain the conditioned reinforcer and primary reward was delivered after Y presentations of the conditioned reinforcer. The sizes of X and Y were progressively increased until the marmosets stopped responding. Marmosets with bilateral amygdala lesions responded less vigorously and stopped responding on lower schedules than controls (Parkinson et al., 2001). These results show clearly that amygdala—including basolateral amygdala—is important for maintaining responding for conditioned reinforcers. Consistent with this, performance in amygdala-lesioned marmosets, unlike controls, was unaffected by omission of the conditioned reinforcer, suggesting that presentations of the conditioned reinforcers did not maintain the responding on these higher order schedules. Marmosets with orbitofrontal cortex lesions also behaved abnormally in this task; however, unlike amygdala-lesioned animals, they actually responded more and at higher schedules than controls (Pears et al., 2003). This effect points to differences in the role of basolateral amygdala versus orbitofrontal cortex in mediating responding in this complex setting. However, as was the case for amygdala-lesioned animals, the performance of the orbitofrontal-lesioned marmosets was insensitive to omission of the conditioned reinforcer. Thus orbitofrontal cortex is also important for the process by which conditioned stimuli come to support instrumental responding.

Connections between basolateral amygdala and nucleus accumbens have also been implicated in conditioned reinforcement. For example, rats with bilateral basolateral amygdala lesions were impaired in a conditioned reinforcement task (Cador et al., 1989). In this experiment, thirsty rats were conditioned to associate a light/noise with water. After learning, these rats were presented with two levers, one leading to the light-noise cue and the other lever leading to nothing. Rats with basolateral amygdala lesions showed normal Pavlovian conditioning to the light but failed to respond on the lever for presentation of the light cue. This impairment was ameliorated dose-dependently by infusion of D-amphetamine into nucleus accumbens, suggesting that the deficit reflected the loss of normal excitatory drive from basolateral amygdala to accumbens. Additionally, these authors found that these infusions increased responding on the lever that produced the light cue in controls.

Dissociating the Associative and Circuit Basis of Conditioned Reinforcement

The results reviewed earlier suggest that different brain circuits might be involved in conditioned reinforcement due to their respective roles in reward learning. Thus basolateral amygdala and orbitofrontal cortex may support conditioned reinforcement because they allow Pavlovian cues to evoke representations of the outcomes they predict. Similarly projections from basolateral amygdala to central nucleus and nucleus accumbens may support conditioned reinforcement because they allow Pavlovian cues to evoke representations of general affect.

In this model, basolateral amygdala would play a central role in endowing cues with the ability to serve as conditioned reinforcers, since it is required for cues to acquire both affect and outcome representations. This is consistent with observations that although amygdala-lesioned marmosets show normal Pavlovian conditioning, which could be mediated by direct associations between cues and responses, they will not subsequently acquire novel conditioned instrumental responding for those cues. This hypothesis might also account for the somewhat different pattern of results in orbitofrontal-lesioned marmosets, which overresponded on the complex scheduled conditioned reinforcement task described earlier in this chapter. If orbital lesions cause a selective inability to use the conditioned reinforcer to signal the outcome while leaving affective representations largely intact, they might show general overresponding over time because they would be unable to recognize that the outcome is not actually delivered on most trials. Though admittedly one might have expected some impact on responding when the conditioned reinforcer was omitted, it is possible that the complex and prolonged training regime allowed the affective properties of the conditioned reinforcer to transfer to the response in orbitofrontal-lesioned marmosets.

However, whether or not our hypothesis fully explains existing data is to some extent premature, since we do not currently know whether conditioned reinforcers support behavior because of the outcomes they predict or due to some inherent value or "affect" that the cues have acquired. Intuitively one might expect it is both; however, to the best of our knowledge, with the exception of one particular study (Parkinson et al., 2005), this has not been empirically tested. To test this idea, it is necessary to utilize more specific Pavlovian training techniques to create cues that are biased to trigger or evoke either outcome or affect representations. These cues can then be used to assess conditioned reinforcement. Here we will describe our initial studies aimed at identifying the underlying representations of conditioned reinforcers.

Conditioned Reinforcement Mediated by Devaluation-insensitive Representations of General Affect

To show that conditioned reinforcement can be mediated by devaluation-insensitive representations of general affect, we tested the ability of rats to acquire a novel instrumental response for a Pavlovian cue *after* devaluation of the outcome predicted by the cue (Burke et al., 2007, 2008). The training procedure is shown in Table 2.1. Rats received presentations of a neutral cue paired with delivery of a food outcome in a Pavlovian conditioning procedure. Additionally, a control cue was also presented. For most of the rats, this cue was presented without reward; for some rats, it was "blocked" from forming any association with the outcome (see the section "Conditioned Reinforcement Mediated by Devaluation-sensitive Representations of Specific Outcomes" for explanation). After training, rats were assigned to one of the two groups. Rats in one group underwent reinforcer devaluation, in which the food was paired with illness induced by LiCl injection. Rats in the control group received similar exposure to the food and illness on alternate days. As in prior work (Gallagher et al., 1999; Pickens et al., 2003; Schoenbaum and Setlow, 2005), rats in the paired group significantly reduced their food consumption, while rats in the unpaired group showed no change in their food consumption.

We next tested the ability of these cues to support conditioned reinforcement. For rats in the paired group, responding on one lever/chain resulted in presentations of the cue associated with a devalued food while responding on the other lever/chain led to the control cue. For rats in the unpaired group, one response produced the cue associated with a nondevalued food and the other the control cue. As illustrated in Figure 2.1, rats responded significantly more for either the nondevalued or a devalued cue versus the control cue, with no significant effect of devaluation. This result indicates that a devalued cue was able to support apparently normal conditioned reinforcement. These data show, consistent with prior results (Parkinson et al., 2005), that devaluation-insensitive, general affective properties evoked by Pavlovian cues can support conditioned reinforcement.

Table 2.1 Outline of Events for the Devaluation Procedure Preceding Conditioned Reinforcement

Outline of Events		
Conditioning	Devaluation	Conditioned Reinforcement
Cue 1: Food pellet, Cue 2: Control cue	Paired group: Food pellet–LiCL	R1: Devalued cue, R2: Control cue
Cue 1: Food pellet, Cue 2: Control cue	Unpaired group: No pellet–LiCl	R1: Nondevalued cue, R2: Control cue

Cue 1 was either a cue light or house light, counterbalanced. Cue 2 was either a light cue (cue light or house light) or a noise cue (tone or white noise). Cue 1, in the paired group, becomes the "devalued cue" while Cue 1, in the unpaired group, becomes the "nondevalued cue." Cue 2 is a control cue for both groups. R1 and R2 were two identical levers/chains. The food pellet was a 45 mg sucrose pellet (Research Diets, New Brunswick, NJ). LiCl was given in two to three nonconsecutive injections of 0.3 M LiCl (5 mg/kg).

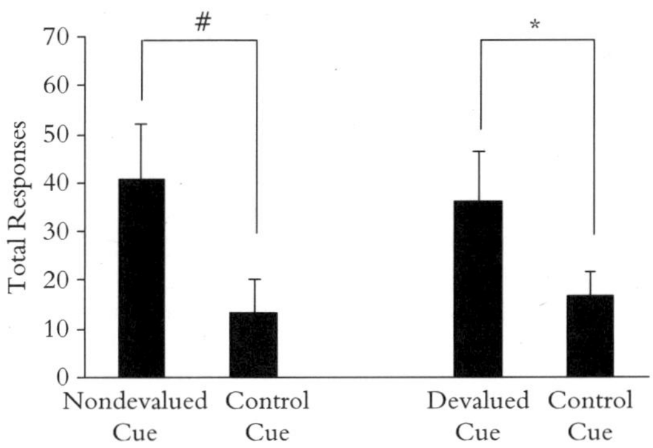

Figure 2.1. Acquisition of a new instrumental response for a cue associated with a devalued outcome. This graph shows the average total number of responses for nondevalued and devalued cues compared to the control cues over three days (i.e., three 30-min sessions) on a VR2 schedule. Responding on the left side of the graph shows responses from rats in the unpaired group, which did not experience food–LiCl pairings. Data on the right side of the graph represents responses from rats in the paired group that did experience food–LiCl pairings. In both groups, rats responded more for the cue associated with the food outcome (whether it was devalued or not) over the control cue, demonstrating no effect of devaluation. (★, $p < 0.05$; #, $p = 0.1$) (Graph modified from Burke et al., 2007.)

Conditioned Reinforcement Mediated by Devaluation-sensitive Representations of Specific Outcomes

To show that conditioned reinforcement can be mediated by outcome representations that are devaluation-sensitive, we used a Pavlovian training procedure termed transreinforcer blocking (Burke et al., 2007, 2008). Blocking refers to the observation that a cue that predicts reward will prevent the formation of associations between that reward and any other cues that are present. Thus, if a rat is trained that a light predicts food and is later presented with that same light and a tone, followed by the same food, the rat will not learn to associate the tone with any of the information evoked by presentation of that food (e.g., its sensory qualities and particular value and the more general affective properties it evokes, which are typically shared with many different outcomes). The light prevents or blocks the tone from forming associations with any of these representations.

Transreinforcer blocking varies this procedure by using two different but equally preferred outcomes. The light is initially presented alone followed by one outcome. Subsequently the light is presented in compound with the tone, followed by the second outcome. Because both outcomes are equally preferred, they trigger comparable emotional responses. These general affective properties are already predicted by the light, thus the tone is blocked from forming associations with them. However, the light does not predict information that is specific to the second outcome, such as its particular sensory properties and the value the animal attaches to these, so the tone is able to form associations with these outcome-specific properties. As a result, the tone becomes able to preferentially evoke representations of the outcome and its specific value and not of general affective representations triggered by the outcome. This assertion is supported by the demonstration that responding to a cue trained in this manner will not support general PIT (Rescorla, 1999) but is sensitive to devaluation of the specific outcome it predicts (see below). We will use this cue to test whether conditioned reinforcement is mediated by outcome-specific information.

The training procedure is shown in Table 2.2. For this task, four unique visual and auditory cues (A, B, X, and Y) and two differently flavored food pellet outcomes (O1 and O2) were used. These pellets, termed O1 and O2 (i.e., banana- and grape/chocolate-flavored sucrose pellets; Research Diets, New Brunswick, NJ) are equally preferred but still possess distinct, devaluation-sensitive properties. This is illustrated by preference and devaluation testing results, shown in the bar graph inset in Table 2.2.

In initial training, neutral cues, A and B, were each paired with one of the two food outcomes (A-O1 and B-O2). After this training, compound stimuli—AX and BY—were presented. AX was paired with O1 in a standard blocking procedure, while BY was paired with O1 in a transreinforcer blocking procedure. Because A predicts all features of O1, no learning occurs for X. By contrast, B does not predict the specific properties of O1 (properties that allowed the selective devaluation of one outcome but not the other in Table 2.2). As a result, Y acquires an association with these unique devaluation-sensitive features of the O1 outcome, while it is blocked from acquiring associations with the general affect shared between the two outcomes (properties that led to a similar preference between the two outcomes in Table 2.2).

At the end of testing, the effectiveness of this procedure was assessed in a single probe session, in which all four cues were presented alone, under extinction conditions. Consistent with predictions, rats responded most to the fully trained cue A/B, somewhat for the partially conditioned cue Y, and at levels comparable to the pre-CS period to the fully blocked cue X (data not shown; $p < 0.05$).

Subsequently, we tested the ability of the cues to support conditioned reinforcement. For some rats, pressing one lever resulted in presentation of the partially conditioned cue, Y, and another lever resulted in presentation of the blocked cue X. And for some rats, pressing one lever resulted in presentation of the fully conditioned cue A, while pressing on the other resulted in presentation of the blocked cue X. As illustrated in Figure 2.2, rats responded significantly more for either the fully or the partially conditioned cue when compared to the blocked cue, and there was no difference in responding for the fully or

Table 2.2 Outline of Events for the Transreinforcer Blocking Procedure Preceding Conditioned Reinforcement

Blocking and Conditioned Reinforcement		
Conditioning	Compound	Conditioned Reinforcement
A-O1, B-O2	AX-O1, BY-O1	Group 1: R1-A, R2-X Group 2: R1-Y, R2-X

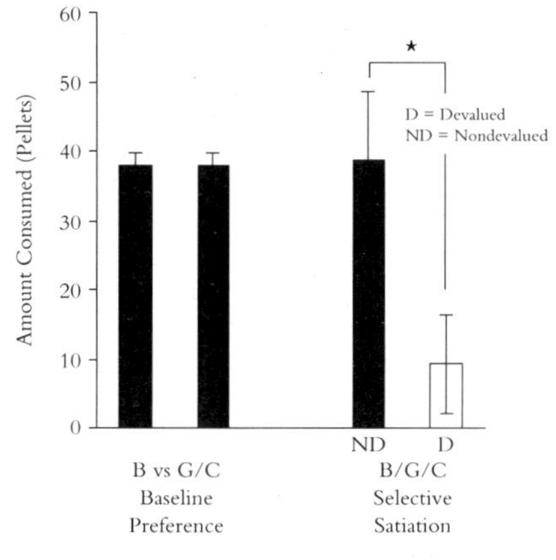

Amount consumed (pellets)
Nondevalued
Baseline preference
Selective satiation

Cues A and B were two different light cues (house light and cue light) and cues X and Y were two distinct noise cues (76 dB tone and 76 dB white noise, 4 kHz). Food pellets O1 and O2 were banana and grape or chocolate flavored sucrose pellets. R1 and R2 were levers/chains inserted into the walls of the behavioral chamber. All cues and food pellets were counterbalanced. (A) Taste preference testing for banana (B) versus grape (G) or chocolate (C) flavored sucrose pellets. Food-deprived rats were tested in a series of three preference tests. Rats were given 50 pellets of each flavor in their home cage for five minutes. Following these tests, rats were tested in a satiation procedure (three tests) where they were given the pellet to be satiated for 20 minutes in an unlimited quantity and immediately following, rats were presented with 100 pellets of the satiated and nonsatiated pellet. Consumption over all three days is shown as an average. (★,$p < 0.05$) (Graph modified from Burke et al., 2007.)

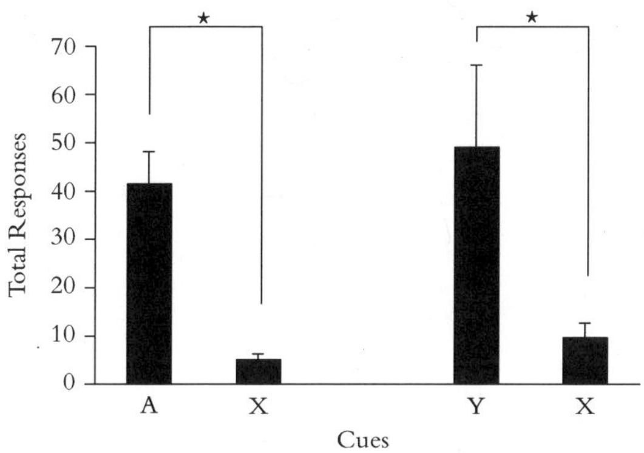

Figure 2.2 Acquisition of a new response mediated by a cue evoking an outcome representation. Figure shows the average total number of responses over two 30-min sessions on a VR2 schedule. Rats represented on the left side of the graph, had access to two levers/chains: one leading to the fully conditioned cue, A, and the other leading to the fully blocked cue, X. Rats, on the right side of the graph, had access to two levers/chains as well: one leading to the partially conditioned cue, Y, and the other to the fully blocked cue, X. Rats responded significantly more for the fully conditioned cue and the partially conditioned cue over the blocked cue (★,$p < 0.05$). (Graph modified from Burke et al., 2007.)

partially conditioned cues. These data show that representations of outcome-specific information evoked by Pavlovian cues are also sufficient to support conditioned reinforcement.

Roles of Basolateral Amygdala and Orbitofrontal Cortex in Conditioned Reinforcement Mediated by Affect and Outcome Representations

Data described earlier suggest that Pavlovian cues function as conditioned reinforcers due to their ability to evoke representations of the outcomes they predict and also due to their ability to signal the general affect normally evoked by those outcomes. The orbitofrontal cortex is known to play a role in signaling expected outcomes, thus the involvement of this brain region in conditioned reinforcement may reflect its role in outcome signaling. To test this hypothesis, we trained rats with lesions of orbitofrontal cortex in the transreinforcer and conditioned reinforcement task described earlier (Burke et al., 2008).

Neurotoxic lesions of orbitofrontal cortex were made prior to any training. After recovery from surgery, lesioned rats and controls were trained as shown in Table 2.2. Orbitofrontal lesions had no effect on Pavlovian conditioning, either in the initial training or during the compound training. However, when these cues were used as conditioned reinforcers to support the acquisition of a novel instrumental response, orbitofrontal-lesioned rats exhibited selective impairments in their ability to use outcome representations. These results are shown in the upper panels of Figure 2.3. Controls exhibited greater responding for either A or Y versus X, replicating the effect described in Figure 2.2, showing that outcome representations will support conditioned reinforcement. By contrast, orbitofrontal-lesioned rats showed normal responding for A, but failed to respond for Y. This pattern of selective impairment is consistent with the hypothesis that orbitofrontal cortex is required for conditioned reinforcement mediated by outcome representations.

To confirm that responding for Y in controls was mediated by outcome-specific information, we next devalued the outcome predicted by Y, by pairing O1 with illness. After several O1–illness pairings, we tested whether devaluation had any effect on the previously established conditioned responding. The results are shown in the lower panels of Figure 2.3. As predicted, the controls no longer responded to Y, indicating that responding for Y had been completely driven by devaluation-sensitive information about the O1 outcome.

Interestingly, there was no effect of devaluation on responding for A when the responding was considered as a ratio versus X. This was true both in controls and also in lesioned rats. The lack of any effect of devaluation on conditioned reinforcement mediated by a fully conditioned cue is consistent with data in Figure 2.1, showing that rats will respond for conditioned reinforcers even after devaluation of the predicted outcome. Thus these data confirm that devaluation-insensitive representations of general affect will support conditioned reinforcement. Indeed, such information appears to be sufficient to support apparently normal levels of responding. In addition, the lack of any effect of orbitofrontal cortex lesions on responding to A is consistent with the proposal that orbitofrontal cortex is not involved in conditioned reinforcement mediated by general affect. As a whole, these results are consistent with the proposal that orbitofrontal cortex supports conditioned reinforcement due to its already appreciated role in signaling expected outcomes.

The selective effects of orbitofrontal lesions on a specific form of conditioned reinforcement contrasts sharply with more general effects of basolateral

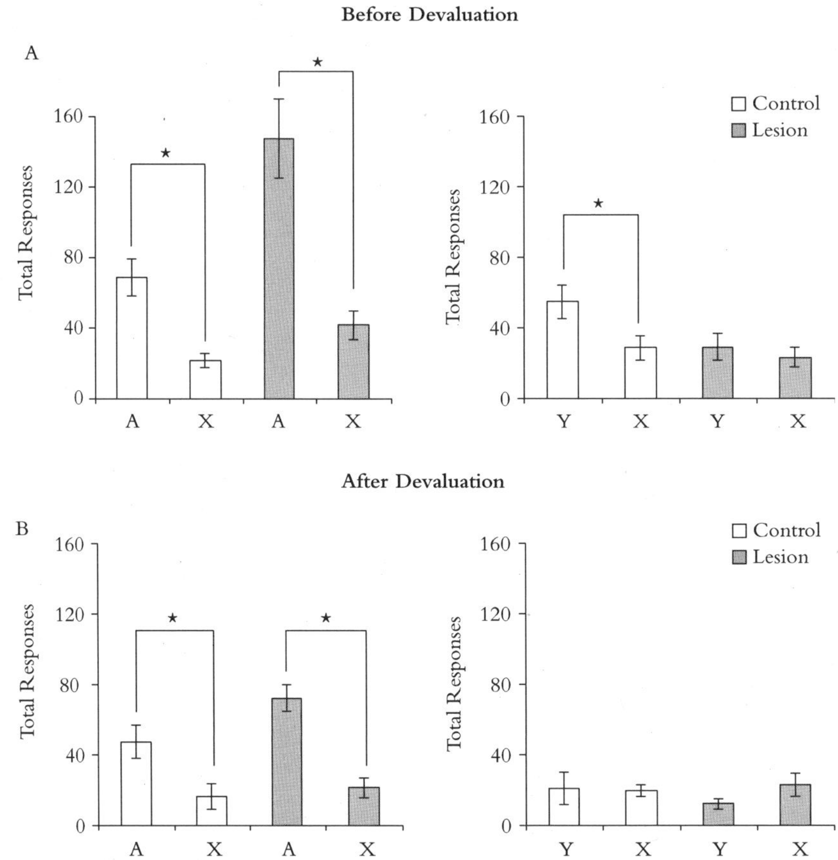

Figure 2.3 Effects of pretraining orbitofrontal cortex (OFC) lesions on conditioned reinforcement. Instrumental lever responding is shown before (3A) and after devaluation (3B) for cues A and X or Y and X. (A) Control rats responded significantly more for the fully conditioned cue, A, and partially conditioned cue, Y, over the fully blocked cue, X. On the other hand, as predicted, OFC lesioned rats showed conditioned reinforcement for the A cue but not for the Y cue. (B) Devaluation abolished responding for the Y cue in controls while having no effect on the A cue for either the controls or OFC lesioned rats (⋆, $p < 0.05$). (Modified from Burke et al., 2008.)

amygdala lesions on conditioned reinforcement (Burke et al., 2007, 2008). Rats with basolateral amygdala lesions were tested using the reinforcer devaluation and conditioned reinforcement task described in Table 2.1. Although basolateral amygdala lesions had no effect on Pavlovian conditioning, they completely abolished conditioned reinforcement. The results are shown in Figure 2.4. While responding for the cue that had been paired with food increased in controls across days of training, responding for these cues did not increase for the basolateral amygdala-lesioned rats. This effect was evident whether or not the food predicted by this cue had been devalued. These results are consistent with previously published reports that amygdala damage causes general deficits in conditioned reinforcement.

Conclusions

Despite their apparent involvement in different forms of associative learning, basolateral amygdala and its various outflow pathways through orbitofrontal cortex, central nucleus, and nucleus accumbens are each critical for normal conditioned reinforcement. Thus, instrumental responding for cues previously paired with food reward is sensitive to damage to amygdala, particularly basolateral amygdala (Burns et al., 1993; Cador et al.,

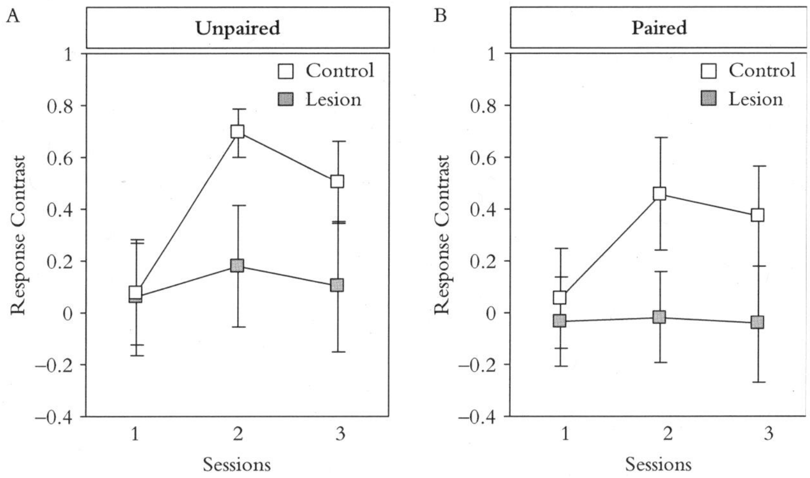

Figure 2.4 Effects of ABL lesions on conditioned reinforcement. For half of the rats, (A), the food outcome was not paired with LiCl. For the other half of the rats (B), the food outcome was paired with LiCl. The control unpaired and control paired rats (white boxes) showed an increase in their contrast after day 1, demonstrating conditioned reinforcement for the food-associated cue over the control cue with no effect of devaluation. ABL-lesioned rats (gray boxes), in both groups, did not show this increase of instrumental responding for these cues, compared to the controls, over the three sessions. (Graph modified from Burke et al., 2007.)

1989; Cousens and Otto, 2003; Hatfield et al., 1996; Parkinson et al., 2001; Setlow et al., 2002a), and also to the outflow pathways described earlier, including orbitofrontal cortex (but not other prefrontal areas) (Cousens and Otto, 2003; Pears et al., 2003), central nucleus of the amygdala, and regions of nucleus accumbens (Parkinson et al., 1999; Robledo et al., 1996; Setlow et al., 2002b; Taylor and Robbins, 1984). However, while damage to basolateral amygdala abolishes responding for these cues (Parkinson et al., 2001), manipulations elsewhere in these circuits have different effects. For example, in one report orbitofrontal-lesioned animals actually responded more for conditioned reinforcer cues compared to controls, as if their responding had become insensitive to some but not other aspects of the conditioned reinforcer (Pears et al., 2003). Similarly, central nucleus of the amygdala and nucleus accumbens seem to be important primarily for potentiating the control over behavior by conditioned reinforcers (Parkinson et al., 1999; Robledo et al., 1996; Taylor and Robbins, 1984). In addition, damage to basolateral amygdala is most effective when made before learning whereas damage to the outflow pathways—nucleus accumbens and orbitofrontal cortex—continues to be effective even when made after learning (Cousens and Otto, 2003; Pears et al., 2003; Setlow et al., 2002a,b).

One interpretation of these data is that conditioned reinforcement is not a unitary process but in fact reflects parallel activation of different types of associative information, mediated by these different circuits. Here, we have presented evidence in support of this proposal. Furthermore, at least for orbitofrontal cortex and basolateral amygdala, there appears to be some correspondence between the role these areas play in processing associative information evoked by Pavlovian cues and their roles in conditioned reinforcement. Thus basolateral amygdala is important in both settings for allowing cues to evoke representations of outcomes and also the affective information with which those outcomes are associated, potentially explaining why basolateral amygdala lesions cause a general deficit in conditioned reinforcement. By contrast, orbitofrontal cortex plays a more specific role in the process whereby cues evoke outcome representations in Pavlovian settings and orbitofrontal lesions cause a selective deficit in conditioned reinforcement when it is mediated by these representations. It will be of interest in the future to determine whether conditioned reinforcement mediated by other areas linked to basolateral amygdala, such as central nucleus or nucleus accumbens, is mediated primarily by affective information.

References

Balleine, B. W. and Corbit, L. H. (2005) Double dissociation of nucleus accumbens core and shell on the general and outcome-specific forms of pavlovian-instrumental transfer. *Soc. Neurosci. Abstr.* 71, 16.

Baxter, M. G., Parker, A., Lindner, C. C., Izquierdo, A. D. and Murray, E. A. (2000) Control of response selection by reinforcer value requires interaction of amygdala and orbitofrontal cortex. *J. Neurosci.* 20, 4311–4319.

Burke, K. A., Franz, T. M., Miller, D. N. and Schoenbaum, G. (2008)) The role of orbitofrontal in the pursuit of happiness and more specific rewards. *Nature.* 454, 340–344.

Burke, K. A., Franz, T. M., Miller, D. N. and Schoenbaum, G. (2007) Conditioned reinforcement can be mediated by either outcome-specific or general affective representations. *Front. Neurosci.* 1, 2. Epub 2007 Nov 2.

Burns, L. H., Robbins, T. W. and Everitt, B. J. (1993) Differential effects of excitotoxic lesions of the basolateral amygdala, ventral subiculum, and medial perfrontal cortex on responding with conditioned reinforcement and locomotor activity potentiated by intra-accumbens infusions of D-amphetamine. *Behav. Brain Res.* 55, 167–183.

Cador, M., Robbins, T. W. and Everitt, B. J. (1989) Involvement of the amygdala in stimulus-reward associations: Interactions with the ventral striatum. *Neuroscience* 30, 77–86.

Corbit, L. H. and Balleine, B. W. (2005) Double dissociation of basolateral and central amygdala lesions on the general and outcome-specific forms of pavlovian-instrumental transfer. *J. Neurosci.* 25, 962–970.

Corbit, L. H., Muir, J. L. and Balleine, B. W. (2001) The role of the nucleus accumbens in instrumental conditioning: Evidence of a functional dissociation between accumbens core and shell. *J. Neurosci.* 21, 3251–3260.

Cousens, G. A. and Otto, T. (2003) Neural substrates of olfactory discrimination learning with auditory secondary reinforcement. I. Contributions of the basolateral amygdaloid complex and orbitofrontal cortex. *Integr. Physiol. Behav. Sci.*38, 272–294.

de Borchgrave, R., Rawlins, J. N., Dickinson, A. and Balleine, B. W. (2002) Effects of cytotoxic nucleus accumbens lesions on instrumental conditioning in rats. *Exp. Brain Res.* 144, 50–68.

Dwyer, D. M. and Killcross, S. (2006) Lesions of the basolateral amygdala disrupt conditioning based on the retrieved representations of motivationally significant events. *J. Neurosci.* 26, 8305–8309.

Everitt, B. J. and Robbins, T. W. (1992) Amygdala-ventral striatal interactions and reward-related processes. In: *The Amygdala: Neurological Aspects of Emotion, Memory, and Mental Dysfunction*, pp. 401–429. Ed. J. P. Aggleton. John Wiley and Sons: Oxford, UK.

Gallagher, M., McMahan, R. W. and Schoenbaum, G. (1999) Orbitofrontal cortex and representation of incentive value in associative learning. *J. Neurosci.* 19, 6610–6614.

Gottfried, J. A., O'Doherty, J. and Dolan, R. J. (2003) Encoding predictive reward value in human amygdala and orbitofrontal cortex. *Science* 301, 1104–1107.

Hall, J., Parkinson, J. A., Connor, T. M., Dickinson, A. and Everitt, B. J. (2001) Involvement of the central nucleus of the amygdala and nucleus accumbens core in mediating Pavlovian influences on instrumental behavior. *Eur. J. Neurosci.* 13, 1984–1992.

Hatfield, T., Han, J. S., Conley, M., Gallagher, M. and Holland, P. (1996) Neurotoxic lesions of basolateral, but not central, amygdala interfere with Pavlovian second-order conditioning and reinforcer devaluation effects. *J. Neurosci.* 16, 5256–5265.

Holland, P. C. (2004) Relations between Pavlovian-Instrumental transfer and reinforcer devaluation. *J. Exp. Psychol. Anim. Behav. Process* 30, 104–117.

Holland, P. C. and Gallagher, M. (2003) Double dissociation of the effects of lesions of basolateral and central amygdala on conditioned stimulus-potentiated feeding and Pavlovian-instrumental transfer. *Eur. J. Neurosci.* 17, 1680–1694.

Holland, P. C. and Rescorla, R. A. (1975) The effects of two ways of devaluing the unconditioned stimulus after first and second-order appetitive conditioning. *J. Exp. Psychol. Anim. Behav. Process* 1, 355–363.

Holland, P. C. and Straub, J. J. (1979) Differential effects of two ways of devaluing the unconditioned stimulus after Pavlovian appetitive conditioning. *J. Exp. Psychol. Anim. Behav. Process* 5, 65–78.

Izquierdo, A., Suda, R. K. and Murray, E. A. (2004) Bilateral orbital prefrontal cortex lesions in rhesus monkeys disrupt choices guided by both reward value and reward contingency. *J. Neurosci.* 24, 7540–7548.

Kelley, A. E., Domesick, V. B. and Nauta, W. J. (1982) The amygdalostriatal projection in the rat—an anatomical study by anterograde and retrograde tracing methods. *Neuroscience* 7, 615–630.

Krettek, J. E. and Price, J. L. (1978) Amygdaloid projections to subcortical structures within the basal forebrain and brainstem in the rat and cat. *J. Comp. Neurol.* 178, 225–254.

Mackintosh, N. J. (1974) *The Psychology of Animal Learning.* Academic Press: London.

Malkova, L., Gaffan, D. and Murray, E. A. (1997) Excitotoxic lesions of the amygdala fail to produce impairment in visual learning for auditory secondary reinforcement but interfere with reinforcer devaluation effects in rhesus monkeys. *J. Neurosci.*17, 6011–6020.

Ostlund, S. B. and Balleine, B. W. (2007) Orbitofrontal cortex mediates outcome encoding in Pavlovian but not instrumental learning. *J. Neurosci.* 27, 4819–4825.

Parkinson, J.A., Roberts, A.C., Everitt, B.J. and Di Ciano, P. (2005) Acquisition of instrumental conditioned reinforcement is resistant to the devaluation of the unconditioned stimulus. *Q. J. Exp. Psychol.*, 1-12.

Parkinson, J. A., Crofts, H. S., McGuigan, M., Tomic, D. L., Everitt, B. J. and Roberts, A. C. (2001) The role of the primate amygdala in conditioned reinforcement. *J. Neurosci.* 21, 7770–7780.

Parkinson, J. A., Olmstead, M. C., Burns, L. H., Robbins, T. W. and Everitt, B. J. (1999) Dissociation of effects of lesions of the nucleus accumbens core and shell on appetitive Pavlovian approach behavior and the potentiation of conditioned reinforcement and locomotor activity by d-amphetamine. *J. Neurosci.* 19, 2401–2411.

Pears, A., Parkinson, J. A., Everitt, B. J. and Roberts, A. C. (2001) Effects of orbitofrontal cortex lesions on responding with conditioned reinforcement. *Brain Cogn.* 47, 44–46.

Pears, A., Parkinson, J. A., Hopewell, L., Everitt, B. J. and Roberts, A. C. (2003) Lesions of the orbitofrontal but not medial prefrontal cortex disrupt conditioned reinforcement in primates. *J. Neurosci.*23, 11189–11201.

Pickens, C. L., Saddoris, M. P., Setlow, B., Gallagher, M., Holland, P. C. and Schoenbaum, G. (2003) Different roles for orbitofrontal cortex and basolateral amygdala in a reinforcer devaluation task. *J. Neurosci.* 23, 11078–11084.

Quirk, G. J., Repa, C. and LeDoux, J. E. (1995) Fear conditioning enhances short-latency auditory responses of lateral amygdala neurons: Parallel recordings in the freely behaving rat. *Neuron* 15, 1029–1039.

Rescorla, R. A. (1999) Learning about qualitatively different outcomes during a blocking procedure. *Anim. Learn Behav.* 27, 140–151.

Robledo, P., Robbins, T. W. and Everitt, B. J. (1996) Effects of excitotoxic lesions of the central amygdaloid nucleus on the potentiation of reward-related stimuli by intra-accumbens amphetamine. *Behav. Neurosci.* 110, 981–990.

Schoenbaum, G. and Setlow, B. (2005) Cocaine makes actions insensitive to outcomes but not extinction: Implications for altered orbitofrontal–amygdalar function. *Cereb. Cortex* 15, 1162–1169.

Schoenbaum, G., Chiba, A. A. and Gallagher, M. (1998) Orbitofrontal cortex and basolateral amygdala encode expected outcomes during learning. *Nat. Neurosci.* 1, 155–159.

Schoenbaum, G., Chiba, A. A. and Gallagher, M. (1999) Neural encoding in orbitofrontal cortex and basolateral amygdala during olfactory discrimination learning. *J. Neurosci.* 19, 1876–1884.

Schoenbaum, G., Setlow, B., Saddoris, M. P. and Gallagher, M. (2003) Encoding predicted outcome and acquired value in orbitofrontal cortex during cue sampling depends upon input from basolateral amygdala. *Neuron* 39, 855–867.

Setlow, B., Gallagher, M. and Holland, P. (2002a) The basolateral complex of the amygdala is necessary for acquisition but not expression of CS motivational value in appetitive Pavlovian second-order conditioning. *Eur. J. Neurosci.* 15, 1841–1853.

Setlow, B., Holland, P. C. and Gallagher, M. (2002b) Disconnection of the basolateral amygdala complex and nucleus accumbens impairs appetitive Pavlovian second-order conditioned responses. *Behav. Neurosci.* 116, 267–275.

Taylor, J. R. and Robbins, T. W. (1984) Enhanced behavioral control by conditioned reinforcers following microinjections of d-amphetamine into the nucleus accumbens. *Psychopharmacology (Berl.)* 84, 405–412.

Whitelaw, R. B., Markou, A., Robbins, T. W. and Everitt, B. J. (1996) Excitotoxic lesions of the basolateral amygdala impair the acquisition of cocaine-seeking behaviour under a second-order schedule of reinforcement. *Psychopharmacology (Berl.)* 127, 213–224.

Wright, C. I., Beijer, A. V. J. and Groenewegen, H. J. (1996) Basal amygdaloid complex afferents to the rat nucleus accumbens are compartmentally organized. *J. Neurosci.* 16, 1877–1893.

3

Neural Coding of Pleasure: "Rose-tinted Glasses" of the Ventral Pallidum

J. WAYNE ALDRIDGE AND KENT C. BERRIDGE

Pleasure is not a sensation. What is it then? Nico Frijda's answer in the "pleasure questions" section of this book (which he suggested a number of years ago) epitomizes an emerging consensus among many psychologists and neuroscientists (Frijda, Chapter 6, this book). He notes that pleasure "is a 'pleasantness gloss' added to whatever is pleasant."

Other chapter authors in this book describe pleasure similarly in their answers to "pleasure questions" as "the subsequent valuation of sensory stimuli" (Kringelbach, Chapter 12, this book); "integrated with sensation" (Dickinson and Balleine, Chapter 4, this book); or "arises from a weighted combination of the sensory signals" (Kringelbach, Chapter 12, this book; Leknes and Tracey, Chapter 19 this book). Pleasure as a hedonic "gloss" (Frijda, Chapter 6, this book; Smith et al., Chapter 1, this book) on sensations is a succinct way to describe how brain signals representing mere sensations (or applied to behavior-generating signals, actions) become glazed by coincident hedonic neural activity that imbues them with pleasure, transforming the signals into hedonic stimuli (or hedonic actions). Thus, viewed through the brain's metaphoric "rose-tinted glasses of pleasure," ordinary sensations become pleasurable sensations.

Here we ask: how is a "pleasure gloss" encoded in brain activity? Where in the brain is this glossing operation performed and how does it work? Is it possible for neuroscientists to recognize the signature patterns of neural activity that represent a pleasure gloss? These are difficult questions that are only beginning to be addressed. The "pleasure gloss" metaphor, applied to the transformation of neural signals for a stimulus, is like a varnish that is applied on top of a dull object to transform it into a shiny one. Adding hedonic tone to the signal passed on to downstream structures, the neural gloss effectively gives the entire brain a "rose-tinted" hedonic perception of the stimulus as pleasant.

In the context of neural firing signals, our idea is that a particular pattern of neuronal spikes or action potentials in crucial neurons may apply a glaze of pleasure on what might otherwise be an ordinary sensation or action signal. At the moment of such a signal, neural activity related to a potentially hedonic sensation will be comingled with signals that specifically implement the pleasure gloss. It is the pleasure-generating neural pattern we wish to identify. A pleasure transformation might excite, inhibit, or vary the pattern of firing activity within the target structure it modulates and as a result dynamically recruits changes in activity profiles throughout an entire circuit. The particular pattern we will describe is an excitation in a large population of neurons within a hedonic hotspot of the ventral pallidum.

This chapter will focus on neural activity profiles in the ventral pallidum because we believe this brain structure is particularly important to applying a hedonic gloss (Figure 3.1). This structure in the subcortical forebrain is a nexus of circuits that process emotion information. We will describe reasons as to why we think the ventral pallidum is especially

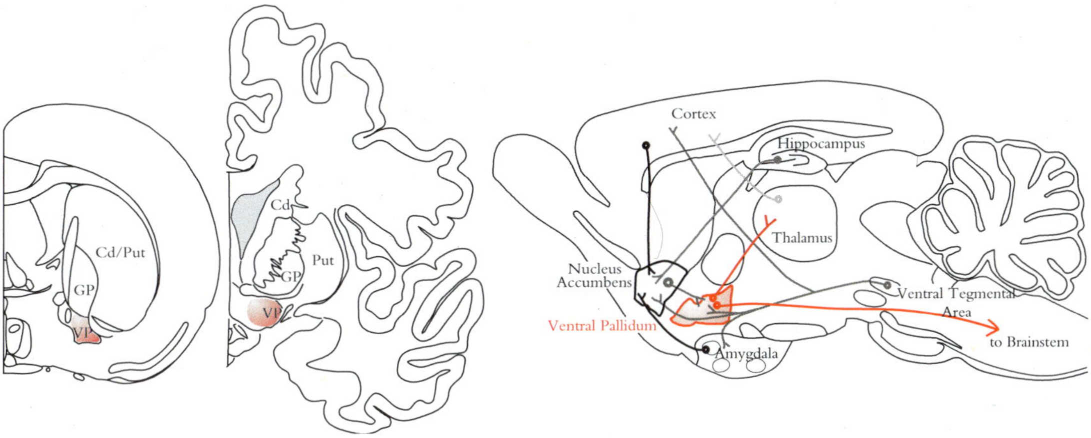

Figure 3.1 Ventral pallidum—sketches of rat and human coronal sections (left and middle panels). Schematic in right panel shows major input and output pathways of ventral pallidum.

likely to perform a hedonic transformation. It is worth noting that the ventral pallidum has only recently become appreciated as a distinct neuroanatomical entity, let alone one with special hedonic functions. Older literature did not usually discuss the ventral pallidum because it used to be considered as just the ventral part of the globus pallidus (a component of the basal ganglia, which are brain structures that include the neostriatum, globus pallidus, entopeduncular nucleus, subthalamic nucleus, and the substantia nigra in the midbrain ventral tegmentum). The basal ganglia were traditionally viewed as important to controlling movements, but now are also recognized as crucial for neural processing related to emotion, motivation, and reward too. Especially important to affective-motivational functions are particular components of basal ganglia such as ventral tegmentum and its mesolimbic dopamine projections, the nucleus accumbens (formerly known as the ventral portion of neostriatum), and the ventral pallidum.

We do not mean to exclude the importance of other brain structures. We focus on the ventral pallidum simply because it is a good place to start to understand pleasure. Besides the basal ganglia, of course, many other brain structures are also involved in assigning pleasure (Kringelbach, 2005) (many of the chapters, this book). Many of them might do so in cognitive and predictive ways that go beyond painting the basic pleasure gloss onto a sensation. Much of human pleasure has cognitive qualities that infuse uniquely human properties, and it is likely that abstract or higher pleasures depend on cortical brain areas for those qualities. For example, full pleasure derived by humans from tasty food, a humorous joke, or from listening to music requires learning and complex cognitive representations, such as in orbitofrontal cortex, insula cortex, and anterior cingulate cortex (Blood and Zatorre, 2001; Kringelbach, 2005; Watson et al., 2007). Thus, the particular pattern of coactivated cortical circuits would resolve the high level cognitive features of a pleasantness gloss on sensations or actions.

Ventral Pallidum: Applying a Pleasure Gloss to Sweetness?

Still, for seeing the basic glazing operation by which the pleasure gloss is actually generated and applied to sensations, the ventral pallidum has particular advantages. Here a special insight may be gained into why sugar tastes nice and how some other sensations can become as nice as sugar, at least when they get the same neuronal hedonic gloss.

Several reasons have led us to focus on the ventral pallidum, in particular, for adding a pleasure gloss to ongoing sensations via its neuronal firing patterns. First, the ventral pallidum contains a "hedonic hotspot" in its posterior half, a roughly cubic-millimeter brain site in which neuronal events can lead to amplifications of a sensory pleasure (Peciña and Berridge, 2005; Peciña et al., 2006; Smith and Berridge, 2005, 2007; Smith et al., Chapter 1, this book). In hedonic hotspots, microinjections of opioids and other neurochemicals are able to glaze an extra gloss of pleasure onto sweet sensations, enhancing 'liking' responses. Several hotspots have recently been mapped in the medial shell of the nucleus accumbens (sometimes called ventral striatum) and the ventral pallidum. In the ventral pallidum particularly, Kyle Smith, in his doctoral dissertation work in our laboratories, identified a 0.8 mm^3 hedonic hotspot in the posterior end of ventral pallidum in rats (Smith et al., Chapter 1, this book). If hotspots are scaled to overall brain size, the volume of a corresponding hotspot in humans might be closer to a cubic centimeter. In the ventral pallidum hotspot, for example, opioid stimulation caused over a doubling in the level of 'liking' reactions elicited by sucrose taste (Smith and Berridge, 2005; Tindell et al., 2006).

Another reason to focus on ventral pallidum for sensory pleasures is that electrophysiological recordings of neurons in its same hotspot, by Amy Tindell in her own dissertation study in the Aldridge laboratory, revealed vigorous firing in response to the taste of sugar (Tindell et al., 2004). With learning, ventral pallidal neurons shifted their activation pattern gradually to fire in response to predictive cues that were associated with sweet reward. Still, neurons continued to fire vigorously when the rats received their sugar pellet reward, suggesting that the neurons may encode persisting hedonic pleasure too.

A third reason is that neurons in the ventral pallidum appear to be especially crucial to normal hedonic 'liking' perhaps more so than in almost any other brain structure. That conclusion was suggested by an earlier lesion study in our laboratories by Casey Cromwell, who showed that destruction of neurons within the ventral pallidum hotspot eliminated normal 'liking' reactions to sweet tastes and replaced them with 'disliking' reactions (Cromwell and Berridge, 1993). Cromwell found that damage to ventral pallidum was more important to the loss of core 'liking' reactions than other nearby brain structures that had traditionally been associated with lesion-induced aversion (e.g., lateral hypothalamus). Indeed ventral pallidum might be the only brain structure known where discrete lesions can convert pleasure 'liking' to 'disliking'.

A final reason to think that this brain structure participates in applying a core 'liking' gloss is that the ventral pallidum in people has now been shown also to increase activity in response to diverse human rewards from food to money, even when the reward stimulus is so subliminally brief that it is not consciously perceived (Beaver et al., 2006; Childress et al., 2008; Pessiglione et al., 2007; Small et al., 2008). These considerations suggest that the ventral pallidum is a promising candidate to find out how the pleasure gloss looks from the point of view of the neurons that might apply it.

In terms of its neuroanatomical connections, the ventral pallidum receives convergent signals from the nucleus accumbens, a brain structure that is implicated generally in rewards (see Figure 3.1) (Baldo and Kelley, 2007; Burke et al., Chapter 2, this book; Carelli and Wightman, 2004; Day and Carelli, 2007; Garris and Rebec, 2002; Knutson et al., 2001a; Kringelbach, Chapter 12, this book; Leknes and Tracey, Chapter 19, this book; Salamone et al., 2007; Schultz, 2006; Smith et al., Chapter 1, this book; Wan and Peoples, 2006). Ventral pallidum also receives signals from other key limbic structures in the forebrain such as amygdala, orbitofrontal cortex, and insular cortex, as well as inputs from reward-related structures in the brainstem such as ventral tegmentum and parabrachial nucleus (Groenewegen and Trimble, 2007; Heimer and Van Hoesen, 2006; Kalivas and Volkow, 2005; Zahm, 2006). In return, the ventral pallidum sends output signals back to limbic areas of prefrontal cortex via the dorsomedial thalamus (Figure 3.1), including orbitofrontal, anterior cingulate, and insular regions of cortex. Ventral pallidum projects as well to many other brain structures involved in reward and thus may serve as a common focal point for integrating and distributing pleasure-related signals. Our studies exploit the ventral pallidum's natural pleasure responses to tasty foods to probe neural coding mechanisms of its hedonic gloss. The pleasure of foods is primary and is evolutionarily programmed into brains so much so that food pleasure may be one driving force behind the obesity epidemic today (Finlayson et al., 2007; Mela, 2006; Pelchat et al., 2004; Zheng and Berthoud, 2007). By recording firing patterns of neurons in the ventral pallidum hotspot at moments when rats are given pleasant and unpleasant tastes, we are able to probe hedonic coding mechanisms directly.

Amy Tindell, Kyle Smith, and others in our laboratory have begun the task of dissecting the hedonic qualities of food tastes in the pattern of firing activation of ventral pallidal neurons (Tindell et al., 2006). In the experiment described below, 'liking' and 'disliking' reactions were elicited from rats by the taste of salt, sugar, or plain water solutions infused directly into their mouths (painlessly and without disturbing them via tubes attached to oral cannula that had been previously implanted when they were anesthetized) (Tindell et al., 2006).

Sensory pleasure or aversion evoked by the tastes was determined by measuring facial affective 'liking' or 'disliking' reactions on video recorded in temporal synchrony with neural recordings on a computer. The natural positive 'liking' reactions elicited by

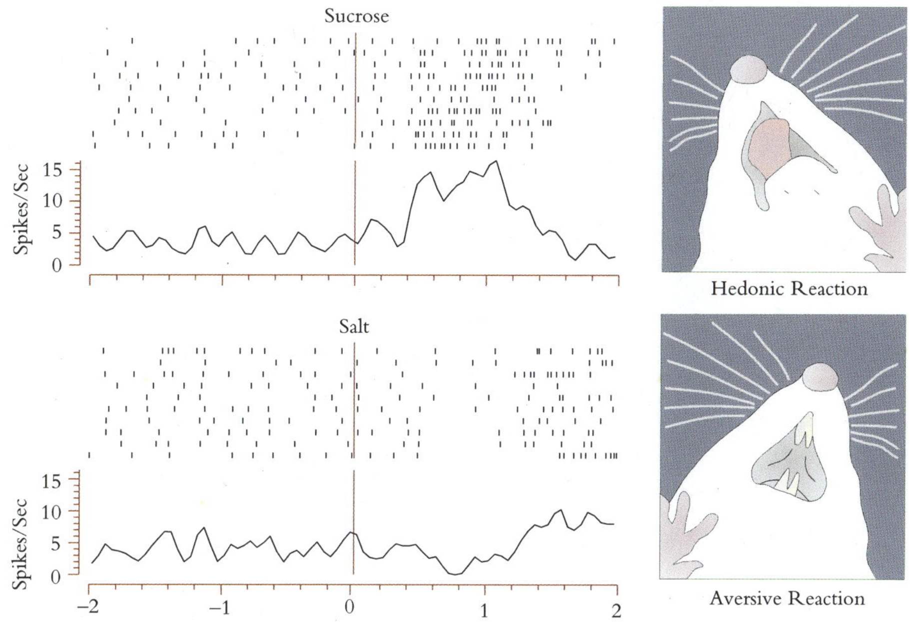

Figure 3.2 Neural responses in a single neuron to infused tastes of sucrose and salt. The perievent rasters on left consist of 10 trials aligned at time = 0 to the moment the infusion pump started. Each row represents a single trial and each dash represents a single spike. The histogram (line below each raster) shows the averaged response across all trials. Activity increased with the sucrose and decreased with concentrated salt tastes. The taste reactions are illustrated to the right. Sucrose evokes hedonic tongue protrusions while salt triggers gape movements.

pleasurable tastes, and 'disliking' reactions elicited by aversive tastes, reflect core pleasure/displeasure evaluations of taste in many different species, including infant humans, and nonhuman primates and rodents (Berridge, 2000). Positive hedonic taste reactions include anterior and lateral tongue protrusions (see Figure 3.2), and (in rats) paw licking movements. Aversive reactions, by contrast, include mouth gapes (see Figure 3.2) that spit out the taste solutions, along with headshaking, forelimb flailing, and (in rats) face washing and chin rubbing movements.

Isolating the Pleasure Code

We compared firing patterns in ventral pallidum during small infusions of pleasurable sugar (sucrose) solutions or aversively intense salt (NaCl) solutions. The crucial experimental strategy employed to dissociate sensory taste qualities (salty vs. sweet) from hedonic valence ('liked' vs. 'disliked') was to convert the unpleasant taste of a concentrated salt solution temporarily to one that was pleasant. The salt solution was three times more concentrated than seawater, or 10 times more concentrated than tears. Such an intense taste is normally too salty to most rats as well as to most people. The salty sensory stimulus itself did not change in our experiment; only the pleasurable affective quality of the taste changed dramatically and thus could be dissociated from sensation. That is, the *hedonic valence* was converted from *bad to good* by a physiological appetite, but the salt solution itself was the same on the tongue. (We note as slight caveat that salt appetite can produce minor sensory changes in brainstem gustatory nuclei, but those changes mostly make taste systems more sensitive to saltiness—which normally would make intense salt even worse. So those minor sensory changes could never explain why concentrated salt taste becomes pleasant.)

It is also important to note that this experimental strategy does not rely on learning in order to alter the hedonic value. Learning-based revaluations are themselves interesting and able to change hedonic valence, but they are complicated and introduce the possibility of alternative interpretations besides a change in core 'liking' reaction (for example, a change in reward expectations instead). Here we wished to focus on the basic hedonic reaction to taste, free of any other complication that we could avoid, and so a purely physiological method was used to change hedonic impact.

Converting the taste of concentrated salt solutions from negative to positive valence was effected by inducing a temporary physiological state of sodium deficiency (via furosemide/DOCA injections, which cause the kidneys to excrete sodium and stimulate brain systems of salt appetite) (Cabanac, 1971). Sodium deficiency induces a psychological appetite to ingest salt, which involves a selective increase in the hedonic perception of salty tastes (Schulkin, 2003). This phenomenon of a change in the hedonic quality of a sensation is called taste "alliesthesia" (Cabanac, 1971, Chapter 7, this book). People display salt appetite and alliesthesia too (Beauchamp et al., 1990). In fact, one of the first scientific reports of salt appetite was a case study of a young boy with an adrenal gland dysfunction, who insisted on eating handfuls of salt each day, and seemed to think of it as candy, to the alarm of his parents (Wilkins and Richter, 1940). In a physiological salt appetite state, 'disliking' gapes and other aversive reactions of rats to intensely salty tastes are diminished and replaced by hedonic 'liking' reactions (Berridge et al., 1984; Schulkin, 1991). Salt appetite exerts powerful effects on brain limbic processing. For example, inducing a salt appetite alters the morphology of nucleus accumbens neurons and changes dopamine and enkephalin release into the nucleus accumbens (Lucas et al., 2003; Roitman et al., 2002). Even previous exposures to salt appetite can have longlasting consequences, such as increasing the behavioral response to amphetamine through alterations in D2 dopamine receptor function (Clark and Bernstein, 2006). The powerful reach of salt appetite into brain limbic operations makes it a useful tool to probe the pleasure code in ventral pallidum recording experiments.

When Tindell and her colleagues put rats into a state of increased salt appetite, the aversive taste reactions to salt fell nearly to zero (Tindell et al., 2006), approaching the extremely low levels seen with sucrose. Most importantly, salt tastes were now 'liked' by rats as evidenced by increased numbers of positive hedonic facial reactions, such as tongue protrusions, that rose from near-zero to counts equaling those seen with sucrose tastes.

The neural representation of tastes was assessed by recording firing patterns of neurons in the hedonic hotspot of the ventral pallidum during behavioral testing. With this method, Tindell and her coworkers were able to show that firing rates of ventral pallidal neurons were directly correlated to 'liking' (hedonically positive behavioral reactions). Firing rates in response to sugar tastes were high while firing rates to tastes of salt solutions were low in the normal homeostatic physiological state (see Figure 3.3). After inducing a

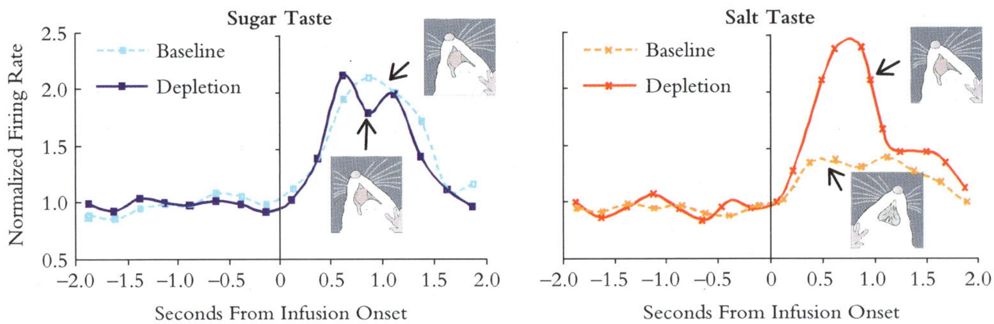

Figure 3.3 Average firing rates (normalized, relative to prestimulus baseline) of all neurons (n = 167 neurons) recorded in the ventral pallidum hotspot to sugar (left) or salt (right) taste infusions, which occurred at time = 0 on x-axis. Most neurons were excited by taste infusions and thus, on average the net change was an increase in firing. Neural responses to sugar did not change with salt depletion, but the response to salt increased dramatically during the salt appetite state, reaching peak amplitudes similar to sugar (solid line, right panel). In general, there were few inhibitory responses to salt tastes (6% before and 6% after depletion) and to sucrose tastes (1% before and 19% after depletion). Depletion (salt appetite) did not change the bias toward excitation significantly (Tindell et al., 2006). The insets indicate hedonic reactions to sucrose and aversive reactions to salt in baseline (dashed lines). After salt depletion (solid lines) the taste reactions to salt switched to hedonic.

physiological state of sodium deficiency, however, firing rates to infusions of salt tastes rose to equal or even exceed firing to sugar (Figure 3.3). In other words, ventral pallidal neurons fired faster when a taste was 'liked' and slower when a taste was 'disliked'. Thus the affective neural and facial responses to concentrated salt tastes were a function of the physiological state interacting with its sensation and not the sensory quality per se.

These findings suggested that firing rates of ventral pallidal neurons may be contributing to a neural code that implements the hotspot of hedonic value or sensory pleasure of taste. Neural firing changes in the ventral pallidum tracked the hedonic reversal of salt palatability specifically. Firing to sucrose tastes, which continued to evoke hedonic facial reactions in both physiological states, was relatively unchanged. Ventral pallidal neurons always fired at high rates to sucrose, and the rats always showed positive 'liking' reactions behaviorally, indicating the hedonic impact of sweetness remained constant. Since the sensory qualities of the stimuli did not change for either salt or sucrose, the changing activity of the ventral pallidum could be viewed as transforming the taste of salt with *a glaze of hedonic pleasure*. In this case, the pleasure gloss of neuronal firing was sufficient to convert a normally aversive sensation to one that is hedonically positive.

Separating the Code for Pleasure from Learning, Arousal, Movement...

Several features of our experiment were designed to help rule out alternative interpretations of the findings. For example, sucrose and salt stimuli were unchanged, and the physiologically induced salt deficient state had never been experienced by the rats before, which combine to rule out simple sensory or learning interpretations.

Other alternative explanations can be discounted in various ways. First, the experiment was designed to bypass all voluntary behavior or decisions by the rat that might have modulated its consumption of a taste and its resulting pleasure. All tastes were infused directly into the mouth in the same manner in every animal. Thus, rats could not "turn off" their experience of unpleasant salt, for example, by moving away from it. They could not change the amount of salt in their mouth to dilute it and so increase its pleasure. Further, acknowledging that behavioral electrophysiologists must always be alert to the possibility that movements might account for neural firing (as either neural commands to move or feedback from the movement), we took special steps to see whether movement coding could be ruled out.

We performed detailed analyses of movement reactions, and the results suggested that a simple motor explanation for the findings was unlikely. First, we found that the timing of neural firing to a taste differed substantially from the timing of motor reactions (assessed by a frame-by-frame video analysis of taste reactions). Ventral pallidal neurons fired nearly a second (800 ms) before taste reactions, which is much longer than one might expect in a neural circuit directly coupled to movement execution. Timing of this sort is much more likely to reflect a sensory-evoked psychological function, such as signaling hedonic impact. In addition, neuronal firing died away at the end of the taste stimulus, while the movements of 'liking' or 'disliking' reactions persisted for up to 10 s longer. These temporal dissociations suggest that neural firing was bound to sensory taste pleasure rather than to movements.

Finally, many movements of the mouth occurred when ventral pallidal neurons remained silent, especially when movements occurred spontaneously outside of a taste presentation. Still, we note in passing, for readers interested in the relation between affective and motor functions, that although hedonic activation patterns described above cannot be explained by simple movements, it does not completely exclude some role for the ventral pallidum in motor control, for example, perhaps a role in translating affect into movement.

The ventral pallidum is anatomically well-positioned to play an important role in higher levels of motor expression, consistent with the original Mogenson's proposal nearly 30 years ago that it may transform motivation to action (Mogenson et al., 1980). The basal ganglia have long been thought of as key structures in selecting appropriate movements from competing motor programs (Mink, 1996). Such ideas are entirely compatible with our view that ventral pallidum implements a pleasure gloss on stimuli and behavior. One idea we think worth considering is that the ventral pallidum could impart a motivational context function to movement selection, which is especially visible in the process of facilitating species-specific facilitatory consummatory actions that are often hedonically laden (eating, copulation, etc.) (Glickman and Schiff, 1967). It is axiomatic that pleasure mechanisms of the brain evolved to serve survival value (Frijda, Chapter 6, this book), and to the degree that engaging in particular actions promote survival, it seems plausible that some pleasurable actions might tap rather directly into ventral pallidal hotspot mechanisms, resulting in neural activity that occurs in tandem with movement.

Pleasure Versus Displeasure

In addition to pleasure, the brain must represent aversion or negative hedonic value. Does the ventral pallidum have this role too? A definitive answer is not possible from our experiments, but the data so far do not strongly support such a role at least in the caudal part of the ventral pallidum studied. In normal sodium balance, highly concentrated salt solutions evoked aversive reactions and although the corresponding firing rates were low, they still differed from no taste at all in the baseline period (see Figure 3.3). Could these lower (but still above baseline) firing rates represent a negative hedonic value of nasty salt? A bivalent coding scheme structured in this fashion would mean that baseline rates would represent no taste input while aversive tastes were represented by low elevations and hedonic tastes represented by high elevations of neural firing rates (and presumably neutral tastes might be represented by intermediate elevations that occur during an oral infusion). Although such a scheme is theoretically possible, it has logical difficulties and seems a potentially fragile mechanism. One problem with this idea, at least as stated so simply earlier, is that it means that with every "nice" taste encountered the trajectory of neural firing would be such that for a moment in time as the firing rate first passed through an intermediate range the signal would represent a "bad" taste and then a "good" taste. That seems delicate, and additional mechanisms would be needed to make this coding scheme work.

A related question, then, is just how the firing pattern of ventral pallidal neurons represents pleasure? The question remains unanswered, but we suspect the eventual answer will be more complicated than the mere firing *rate* code we have described so far. Of course, caution is always needed in assigning psychological processes in a one-to-one fashion to particular neuronal firing patterns. We think that neural firing rates alone, without taking other factors into consideration, are unlikely to fully account for a psychological affective reaction, whether it be pleasure or aversion.

There are other sites connected to ventral pallidum that are also involved in the hedonic code. There are also other possible coding mechanisms besides mere rates, for example, more complex patterns of spike timing and interspike intervals, and temporal and spatial patterns of selective activation that extend across multiple neurons to form coordinated networks. Dynamic recruitment of neuronal assemblies that form coordinated circuits within the ventral pallidum, and larger assemblies that stretch to other brain structures, including nucleus accumbens, brainstem, and orbitofrontal cortex, is likely to be crucial in generating pleasure or displeasure.

Even our own work has already shown that there is a likely population coding mechanism involved as well as rate coding. For example, we observe that more neurons are recruited during the presentation of 'liked' tastes than 'disliked' tastes, and even during conditioned stimuli that predict those 'liked' tastes. Neurons in related circuits that extend through globus pallidus and subthalamic nucleus are thought to have complex and irregular patterns of activity that defy notions of simple rate coding (Terman et al., 2002) and similar rules may hold for ventral pallidum. Future investigations will need to consider these possibilities, along with the organizational features between structures, to uncover interconnected firing patterns that might participate in codes for pleasure. Still, the identification of firing patterns in neurons of the ventral pallidum hotspot that appear to code pleasure, as described earlier, is an important step forward in the search for neural assemblies that use coded signals to add a gloss of pleasure to ongoing sensations and behaviors.

Beyond Pleasure: Other Reward Features Coded by Ventral Pallidum

Although ventral pallidal activity correlates to hedonic impact or pleasure gloss as described above, ventral pallidal circuits also code other additional features of reward alone besides pleasure, and neurons in the ventral pallidum hotspot may possibly be able to track each feature separately (Smith et al., 2007; Tindell et al., 2005). Besides pleasure or 'liking', reward also involves 'wanting' (motivation or incentive salience) and learning features (associations between predictive cues and hedonic rewards) (Berridge, 2007; Berridge and Robinson, 2003). Ventral pallidum firing carries encoded signals for these other aspects of reward in overlapping neuronal populations. For example, learning: after rats have learned that a Pavlovian conditioned stimulus (CS; an auditory tone) predicts sugar pellet rewards, some neurons in the ventral pallidum fire to the predictive CS as well as to the sweet reward itself, while other neurons respond to only predictive CS or to the sweet reward itself (i.e., unconditioned stimulus [UCS]) (Tindell et al., 2004, 2005, 2006). Neural activity in the ventral pallidum is associated with incentive 'wanting' properties of reward too (Lim et al., 2004; McFarland et al., 2004; Tindell et al., 2005). For instance, when incentive 'wanting' is amplified selectively and pulled apart from learning by sensitizing or pharmacologically activating mesolimbic dopamine systems that magnify incentive salience, ventral pallidal neurons fire higher bursts of action potentials to cues that carry the highest incentive salience (Tindell et al., 2005).

There is reason to think that the dopamine magnified 'wanting' signal instantiates a relatively pure code for incentive salience, because the Tindell experiment was arranged in way that separated the CS with the highest incentive salience from a different CS learned previously with the highest predictive value, and from the sweet reward (UCS) that gave the highest pleasure. The results of that experiment showed that dopamine magnification did not magnify either a predictive learning signal or a hedonic 'liking' signal, although both of those signals passed through ventral pallidal neurons too, but instead selectively magnified the incentive 'wanting' signal alone (Berridge and Aldridge, 2008; Tindell et al., 2005). Manipulations that alter the constellation of dopaminergic activity across limbic structures modulate the profiles of neural activity in ventral pallidum to separate 'wanting', 'liking'. and "predictive learning components" of reward (Tindell et al., 2005).

Thus, parsing the pleasure properties of food reward from 'wanting' and learning properties, as well as from sensation and movement properties, is an achievable experimental goal even if a challenging one. When it is done, results so far suggest that ventral pallidal neurons may keep track of each reward feature in a manner that allows tracking the features separately. It is an exciting task to try to understand the interactions among these various properties of rewards in coded signals as ventral pallidum processes pleasure information together with other sensory, motor, motivational, predictive, or cognitive aspects of reward.

Commingled 'Wanting' and 'Liking'

Neural codes for 'liking' pleasure are likely to commingle (i.e., be "multiplexed") with codes for 'wanting' and "learning" about those pleasures even on individual neurons. Multiplexed signals commingle in a manner akin to how wire and optical communications systems carry telephone or computer data signals from multiple telephone conversations, email communications, and internet web traffic over a single wire. Just as the different signals can be resolved at their destination by receivers that decode appropriately, we believe that multiple reward signals can be packed into the activity of single ventral pallidal neurons in much the same way, for potential unpacking downstream. The different signals represent different information about the reward at different moments, or perhaps multiple signals encoded in complex patterns of firing at the same moment. Decoding neural communication systems will involve parsing these multiple signals.

One reason we believe that multiple functional signals are carried by single neurons of the ventral pallidum is that we have observed a single neuron to encode all three signals, pleasure 'liking' motivation 'wanting', and "predictive learning" at various moments or in different ways (Smith et al., 2007; Tindell et al., 2005). The firing patterns produced in the same neuron by the three signals, however, are not necessarily identical even when the neuron is. For example, the same neuron may respond to the different signals with different temporal patterns of multispike activity. Individual neurons and the population as a whole often exhibit patterns of activity with highly complex *profiles* representing combinations of predictive information, reward value as well as incentive, in which different information signals can be discerned to be embedded in different aspects of the firing profile (Tindell et al., 2005). Further, these individual "information channels" can be manipulated experimentally such that the overall *profile* can change to selectively alter one signal more than others. For example, as mentioned above, mesolimbic activation by acute amphetamine injections or by sensitization results in a relative boost of the incentive component over reward value and predictive information in the overall activity profile of the same neurons (Tindell et al., 2005). Thus, it is conceivable that a pleasure gloss on behavior or sensations (i.e., 'liking') may also be represented by subtle and complex aspects of the firing profile that have yet to be uncovered.

Commingling together of 'liking' and 'wanting' signals in the same ventral pallidal neurons may be part of the reason why most rewards are both 'wanted' and 'liked' together. In normal circumstances, we like what we want and want what we like (see Question 4 in Pleasure Questions section of this book). The integrative nature of neural profiles in the ventral pallidum and the fact that this structure lies at an output nexus in the reward circuit suggest that the ventral pallidum may subserve a role for glossing sensations with pleasure as well as imbuing incentive value. It is the combination of a pleasure gloss with sensations or behavior and the ability of neurons in the ventral pallidum to have comingling signals that may allow the ventral pallidum to subserve a pleasure-glossing mechanism, and perhaps also help translate hedonic 'liking' into motivational 'wanting' for the same reward.

Reward in Other Brain Structures

We have focused here on the ventral pallidum as a hedonic hotspot that has special roles in coding and generating core 'liking' reactions to pleasure, but, of course, as mentioned in the beginning, many other brain regions participate with ventral pallidum to play roles in reward (see many other chapters in this book). For example, Morten Kringelbach has summarized cogent evidence that the orbitofrontal region of prefrontal cortex is especially important for cortical representations of reward (Kringelbach, 2005; Kringelbach, Chapter 12, this book). Additional cortical areas such as rostral insula, frontal operculum, and anterior cingulate cortex also respond to pleasant tastes, odors, sights, and textures of foodstuffs, and other sensory rewards, and participate in positive emotional reactions to diverse learned cultural rewards, ranging from attractive photographs and winning money to humor, art, and music (Blood and Zatorre, 2001; Burke et al., Chapter 2, this book; Feldman Barrett and Wager, 2006; Knutson et al., 2001b; Kringelbach, Chapter 12, this book; Rolls, 2005; Skov, Chapter 16, this book; Small and Veldhuizen, Chapter 9, this book; Watson et al., 2007).

A number of subcortical brain structures also code reward and interact with ventral pallidum. They include the mesolimbic dopamine system and its chief targets in the nucleus accumbens as well as the amygdala and other limbic structures. Wolfram Schultz and colleagues showed that mesolimbic dopamine neurons themselves are activated by food rewards, especially when they are surprising, as well as by cues that predict those rewards (Schultz, 2006). Many studies have shown the nucleus accumbens to be implicated in almost every type of reward (Everitt and Robbins, 2005; Green et al., Chapter 18, this book; Leyton, Chapter 13, this book; Smith et al., Chapter 1, this book), and the amygdala is also activated by many rewards including food, drugs, pleasant photographs, and sex (Burke et al., Chapter 2, this book; Komisaruk et al., Chapter 10, this book; Phan et al., 2002).

Several studies have demonstrated that cortical brain regions are especially activated by food rewards when individuals are hungry, but activation declines to the same reward after the individuals eat to satiety (Burton et al., 1975; Kringelbach, 2004, in press; Kringelbach et al., 2003; O'Doherty et al., 2000). These studies used an experimental logic similar to the salt appetite experiments we described above to show specifically that reward was coded by the brain activations they measured. Other studies have used learning to devalue a reward, for instance, by pairing the reward with an aversive event such as visceral illness in a learned taste aversion paradigm to similarly implicate reward or value predictions as the function embodied in the neural activation (Burke et al., Chapter 2, this book; Dickinson and Balleine, Chapter 4, this book). All such studies provide important insights into how brain systems code and process reward-related information.

Conclusion

The ventral pallidum is just one brain structure among many that code reward, but it may be special in applying the pleasure gloss to stimuli that makes them rewarding. It contains a hedonic hotspot that is both needed to generate normal core 'liking' reactions to pleasant stimuli and able to amplify 'liking' reactions to double normal levels when neurochemically stimulated. Neural firing in the ventral pallidum hotspot encodes 'liking' for the positive hedonic impact of sweet and salty taste rewards in a specific fashion. These coded 'liking' signals may reflect pleasure gloss generation from the neurons' point of view. The same neurons code other reward components too using different signal patterns (e.g., motivation 'wanting' and predictive learning), and link to other brain sites that participate in generating those components as well as hedonic 'liking'.

It seems plausible that ventral pallidum firing patterns mediate 'liking' for other natural sensory pleasures besides food and drug and other pleasures too. And, although our studies have used rats instead of people, we think our finding of pleasure coding in neuronal firing patterns of the ventral pallidum has direct implications for understanding the generation of the pleasure gloss in humans and for the way people perceive pleasure. As a result, conceivably, dysfunction in ventral pallidum firing and in its connections could be one reason why some people develop pathological problems that distort pleasure such as depression or that distort incentive salience and lead to pursuit or consumption of rewards, such as drug addiction, eating disorders, or compulsively excessive gambling. That is, if ventral pallidum signals code the normal hedonic impact of rewards, then pathological distortions of those hedonic signals might knock their pleasure response off balance, and contribute to hedonic dysfunction. Thus the ventral pallidum is an ideal focus for affective neuroscience studies to probe the causal brain mechanisms that gloss sensations with pleasure. It may be equally useful for understanding disorders of pleasure mechanisms and even perhaps eventually to develop more effective treatments.

Acknowledgments

We thank Morten Kringelbach, Kyle Smith, and Stephen Mahler for helpful comments on an earlier version of this manuscript. The research described was supported by grants from the NIH to J.W.A. (DA017752) and K.C.B. (MH063649, DA015188).

References

Baldo, B. A. and Kelley, A. E. (2007) Discrete neurochemical coding of distinguishable motivational processes: Insights from nucleus accumbens control of feeding. *Psychopharmacology (Berl.)* 191, 439–459.

Beauchamp, G. K., Bertino, M., Burke, D. and Engelman, K. (1990) Experimental sodium depletion and salt taste in normal human volunteers. *Am. J. Clin. Nutr.* 51, 881–889.

Beaver, J. D., Lawrence, A. D., van Ditzhuijzen, J., Davis, M. H., Woods, A. and Calder, A. J. (2006) Individual differences in reward drive predict neural responses to images of food. *J. Neurosci.* 26, 5160–5166.

Berridge, K. C. (2000) Measuring hedonic impact in animals and infants: Microstructure of affective taste reactivity patterns. *Neurosci. Biobehav. Rev.* 24, 173–198.

Berridge, K. C. (2007) The debate over dopamine's role in reward: the case for incentive salience. *Psychopharmacology (Berl.)* 191, 391–431.

Berridge, K. C. and Aldridge, J. W. (2008) Decision utility, incentive salience, and cue-triggered 'wanting'. In: *The Psychology of Action*, pp. 509–532. Eds. J. Bargh, E. Morsella. Oxford University Press: New York.

Berridge, K. C. and Robinson, T. E. (2003) Parsing reward. *Trends Neurosci.* 26, 507–513.

Berridge, K. C., Flynn, F. W., Schulkin, J. and Grill, H. J. (1984) Sodium depletion enhances salt palatability in rats. *Behav. Neurosci.* 98, 652–660.

Blood, A. J. and Zatorre, R. J. (2001) Intensely pleasurable responses to music correlate with activity in brain regions implicated in reward and emotion. *Proc. Natl. Acad. Sci. USA* 98, 11818–11823.

Burke, K. A., Miller, D. and Schoenbaum, G. (this book) Conditioned reinforcement and the specialized role of corticolimbic circuits in the pursuit of happiness and other more specific rewards. In: *Pleasures of the Brain*. Eds. M. L. Kringelbach, K. C. Berridge. Oxford University Press: Oxford.

Burton, M. J., Mora, F. and Rolls, E. T. (1975) Proceedings: Visual and taste neurones in the lateral hypothalamus and substantia innominata: Modulation of responsiveness by hunger. *J. Physiol.* 252, 50P–51P.

Cabanac, M. (1971) Physiological role of pleasure. *Science* 173, 1103–1107.

Cabanac, M. (this book) The dialectics of pleasure. In: *Pleasures of the brain*. Eds. M. L. Kringelbach, K. C. Berridge. Oxford University Press: Oxford.

Carelli, R. M. and Wightman, R. M. (2004) Functional microcircuitry in the accumbens underlying drug addiction: Insights from real-time signaling during behavior. *Curr. Opin. Neurobiol.* 14, 763–768.

Childress, A. R., Ehrman, R. N., Wang, Z., Li, Y., Sciortino, N., Hakun, J., Jens, W., Suh, J., Listerud, J., Marquez, K., Franklin, T., Langleben, D., Detre, J. and O'Brien, C. P. (2008) Prelude to passion: Limbic activation by "unseen" drug and sexual cues. *PLoS ONE* 3, e1506.

Clark, J. J. and Bernstein, I. L. (2006) Sensitization of salt appetite is associated with increased "wanting" but not "liking" of a salt reward in the sodium-deplete rat. *Behav. Neurosci.* 120, 206–210.

Cromwell, H. C. and Berridge, K. C. (1993) Where does damage lead to enhanced food aversion: the ventral pallidum/substantia innominata or lateral hypothalamus? *Brain Res.* 624, 1–10.

Day, J. J. and Carelli, R. M. (2007) The nucleus accumbens and Pavlovian reward learning. *Neuroscientist* 13, 148–159.

Dickinson, A. and Balleine, B. (this book) Hedonics: the cognitive–motivational interface. In: *Pleasures of the brain*. Eds. M. L. Kringelbach, K. C. Berridge. Oxford University Press: Oxford.

Everitt, B. J. and Robbins, T. W. (2005) Neural systems of reinforcement for drug addiction: From actions to habits to compulsion. *Nat. Neurosci.* 8, 1481–1489.

Feldman Barrett, L. and Wager, T. D. (2006) The structure of emotion: Evidence from neuroimaging studies. *Curr. Dir. Psychol. Sci.*15, 79–83.

Finlayson, G., King, N. and Blundell, J. E. (2007) Liking vs. wanting food: Importance for human appetite control and weight regulation. *Neurosci. Biobehav. Rev.* 31, 987–1002.

Frijda, N. (this book) On the nature and function of pleasure. In: *Pleasures of the Brain*. Eds. M. L. Kringelbach, K. C. Berridge. Oxford University Press: Oxford.

Garris, P. A. and Rebec, G. V. (2002) Modeling fast dopamine neurotransmission in the nucleus accumbens during behavior. *Behav.Brain Res.*137, 47–63.

Glickman, S. E. and Schiff, B. B. (1967) A biological theory of reinforcement. *Psychol. Rev.* 74, 81–109.

Green, A. L., Pereira, E. A. and Aziz, T. Z. (this book) Deep brain stimulation and pleasure. In: *Pleasures of the brain*. Eds. M. L. Kringelbach, K. C. Berridge. Oxford University Press: Oxford.

Groenewegen, H. J. and Trimble, M. (2007) The ventral striatum as an interface between the limbic and motor systems. *CNS Spectr.* 12, 887–892.

Heimer, L. and Van Hoesen, G. W. (2006) The limbic lobe and its output channels: Implications for emotional functions and adaptive behavior. *Neurosci. Biobehav. Rev.* 30, 126–147.

Kalivas, P. W. and Volkow, N. D. (2005) The neural basis of addiction: a pathology of motivation and choice. *Am. J. Psychiatry* 162, 1403–1413.

Knutson, B., Adams, C. M., Fong, G. W. and Hommer, D. (2001a) Anticipation of increasing monetary reward selectively recruits nucleus accumbens. *J. Neurosci.* 21, 159RC.

Knutson, B., Fong, G. W., Adams, C. M., Varner, J. L. and Hommer, D. (2001b) Dissociation of reward anticipation and outcome with event-related fMRI. *Neuroreport* 12, 3683–3687.

Komisaruk, B. R., Whipple, B. and Beyer, C. (this book) Sexual pleasure. In: *Pleasures of the Brain*. Eds. M. L. Kringelbach, K. C. Berridge. Oxford University Press: Oxford.

Kringelbach, M. L. (2004) Food for thought: Hedonic experience beyond homeostasis in the human brain. *Neuroscience* 126, 807–819.

Kringelbach, M. L. (2005) The human orbitofrontal cortex: Linking reward to hedonic experience. *Nat. Rev. Neurosci.* 6, 691–702.

Kringelbach, M. L. (this book) The hedonic brain: a functional neuroanatomy of human pleasure. In: *Pleasures of the brain*. Eds. M. L. Kringelbach, K. C. Berridge. Oxford University Press: Oxford.

Kringelbach, M. L., O'Doherty, J., Rolls, E. T. and Andrews, C. (2003) Activation of the human orbitofrontal cortex to a liquid food stimulus is correlated with its subjective pleasantness. *Cereb. Cortex* 13, 1064–1071.

Leknes, S. and Tracey, I. (this book) Pleasure and pain: Masters of mankind. In: *Pleasures of the Brain*. Eds. M. L. Kringelbach, K. C. Berridge. Oxford University Press: Oxford.

Leyton, M. (this book) The neurobiology of desire: Dopamine and the regulation of mood and motivational states in humans. In: *Pleasures of the brain*. Eds. M. L. Kringelbach, K. C. Berridge. Oxford University Press: Oxford.

Lim, M. M., Murphy, A. Z. and Young, L. J. (2004) Ventral striatopallidal oxytocin and vasopressin V1a receptors in the monogamous prairie vole (*Microtus ochrogaster*). *J. Comp. Neurol.* 468, 555–570.

Lucas, L. R., Grillo, C. A. and McEwen, B. S. (2003) Involvement of mesolimbic structures in short-term sodium depletion: In situ hybridization and ligand-binding analyses. *Neuroendocrinology* 77, 406–415.

McFarland, K., Davidge, S. B., Lapish, C. C. and Kalivas, P. W. (2004) Limbic and motor circuitry underlying footshock-induced reinstatement of cocaine-seeking behavior. *J. Neurosci.* 24, 1551–1560.

Mela, D. J. (2006) Eating for pleasure or just wanting to eat? Reconsidering sensory hedonic responses as a driver of obesity. *Appetite* 47, 10–17.

Mink, J. W. (1996) The basal ganglia: Focused selection and inhibition of competing motor programs. *Prog. Neurobiol.* 50, 381–425.

Mogenson, G. J., Jones, D. L. and Yim, C. Y. (1980) From motivation to action: Functional interface between the limbic system and the motor system. *Prog. Neurobiol.* 14, 69–97.

O'Doherty, J., Rolls, E. T., Francis, S., Bowtell, R., McGlone, F., Kobal, G., Renner, B. and Ahne, G. (2000) Sensory-specific satiety-related olfactory activation of the human orbitofrontal cortex. *Neuroreport* 11, 399–403.

Peciña, S. and Berridge, K. C. (2005) Hedonic hot spot in nucleus accumbens shell: Where do mu-opioids cause increased hedonic impact of sweetness? *J. Neurosci.* 25, 11777–11786.

Peciña, S., Smith, K. S. and Berridge, K. C. (2006) Hedonic hotspots in the brain. *Neuroscientist* 12, 500–511.

Pelchat, M. L., Johnson, A., Chan, R., Valdez, J. and Ragland, J. D. (2004) Images of desire: Food-craving activation during fMRI. *NeuroImage* 23, 1486–1493.

Pessiglione, M., Schmidt, L., Draganski, B., Kalisch, R., Lau, H., Dolan, R. and Frith, C. (2007) How the brain translates money into force: a neuroimaging study of subliminal motivation. *Science* 11, 904–906.

Phan, K. L., Wager, T., Taylor, S. F. and Liberzon, I. (2002) Functional neuroanatomy of emotion: a meta-analysis of emotion activation studies in PET and fMRI. *NeuroImage* 16, 331–348.

Roitman, M. F., Na, E., Anderson, G., Jones, T. A. and Bernstein, I. L. (2002) Induction of a salt appetite alters dendritic morphology in nucleus accumbens and sensitizes rats to amphetamine. *J. Neurosci.* 22, 1–5.

Rolls, E. (2005) *Emotions Explained*. Oxford University Press: Oxford.

Salamone, J. D., Correa, M., Farrar, A. and Mingote, S. M. (2007) Effort-related functions of nucleus accumbens dopamine and associated forebrain circuits. *Psychopharmacology (Berl.)* 191, 461–482.

Schulkin, J. (1991) *Sodium Hunger: the Search for a Salty Taste*. Cambridge University Press: New York.

Schulkin, J. (2003) *Rethinking Homeostasis: Allostatic Regulation in Physiology and Pathophysiology*. MIT Press: Cambridge, MA.

Schultz, W. (2006) Behavioral theories and the neurophysiology of reward. *Annu. Rev. Psychol.* 57, 87–115.

Skov, M. (this book) The pleasures of art. In: *Pleasures of the Brain*. Eds. M. L. Kringelbach, K. C. Berridge. Oxford University Press: Oxford.

Small, D. and Veldhuizen, M. (this book) Human crossmodal studies of taste and smell. In: *Pleasures of the Brain* . Eds. M. L. Kringelbach, K. C. Berridge. Oxford University Press: Oxford.

Small, D. M., Veldhuizen, M. G., Felsted, J., Mak, Y. E. and McGlone, F. (2008) Separable substrates for anticipatory and consummatory food chemosensation. *Neuron* 57, 786–797.

Smith, K. S. and Berridge, K. C. (2005) The ventral pallidum and hedonic reward: Neurochemical maps of sucrose "liking" and food intake. *J. Neurosci.* 25, 8637–8649.

Smith, K. S. and Berridge, K. C. (2007) Opioid limbic circuit for reward: Interaction between hedonic hotspots of nucleus accumbens and ventral pallidum. *J. Neurosci.* 27, 1594–1605.

Smith, K. S., Berridge, K. C. and Aldridge, J. W. (2007) Ventral pallidal neurons distinguish "liking" and "wanting" elevations caused by opioids versus dopamine in nucleus accumbens. Program No. 310.5, 2007 Neuroscience Meeting Planner. San Diego, CA: Society for Neuroscience, 2007. Online.

Smith, K. S., Mahler, S. V., Pecina, S. and Berridge, K. C. (this book) Hedonic hotspots: Generating sensory pleasure in the brain. In: *Pleasures of the Brain*. Eds. M. Kringelbach, K. C. Berridge. Oxford University Press: Oxford.

Terman, D., Rubin, J. E., Yew, A. C. and Wilson, C. J. (2002) Activity patterns in a model for the subthalamopallidal network of the basal ganglia. *J. Neurosci.* 22, 2963–2976.

Tindell, A. J., Berridge, K. C. and Aldridge, J. W. (2004) Ventral pallidal representation of Pavlovian cues and reward: Population and rate codes. *J Neurosci.* 24, 1058–1069.

Tindell, A. J., Berridge, K. C., Zhang, J., Peciña, S. and Aldridge, J. W. (2005) Ventral pallidal neurons code incentive motivation: Amplification by mesolimbic sensitization and amphetamine. *Eur. J. Neurosci.* 22, 2617–2634.

Tindell, A. J., Smith, K. S., Pecina, S., Berridge, K. C. and Aldridge, J. W. (2006) Ventral–pallidum firing codes hedonic reward: When a bad taste turns good. *J. Neurophysiol.* 96, 2399–2409.

Wan, X. and Peoples, L. L. (2006) Firing patterns of accumbal neurons during a pavlovian-conditioned approach task. *J. Neurophysiol.* 96, 652–660.

Watson, K. K., Matthews, B. J. and Allman, J. M. (2007) Brain activation during sight gags and language-dependent humor. *Cereb. Cortex* 17, 314–324.

Wilkins, L. and Richter, C. P. (1940) A great craving for salt by a child with cortico-adrenal insufficiency. *J. Am. Med. Assoc.* 114, 866–868.

Zahm, D. S. (2006) The evolving theory of basal forebrain functional–anatomical "macrosystems". *Neurosci. Biobehav. Rev.* 30, 148–172.

Zheng, H. and Berthoud, H. R. (2007) Eating for pleasure or calories. *Curr. Opin. Pharmacol.* 7, 607–612.

4

Hedonics: The Cognitive–Motivational Interface

ANTHONY DICKINSON AND BERNARD BALLEINE

From a biological perspective, understanding the function of mentality is compromised by the fact that it operates at a number of degrees removed from the ultimate evolutionary criterion of reproductive fitness. In other words, the biology of the mental is always filtered through the medium of behavior, a dislocation that immediately raises the issue of why evolution has bothered with the mental. If behavior is all that matters, why are we all not just Cartesian beast machines, simply performing the behavior required to get our genes into the next generation without an accompanying pageant being played out in the phenomenal theater of the conscious mind? This issue is raised most acutely in the case of hedonic and affective experience, which not only has phenomenal qualities but also carries strong motivational imperatives. What is the possible function of the relish or disgust elicited by a new, exotic flavor or of the erotic delight engendered by the caress of a lover?

At first sight, the answer appears obvious—the function of hedonic and noxious feelings is to motivate behavior. But this response just begs the question: If the processes underlying hedonic phenomenology carry all the motivational information about the utilities of the objects of our experience, why bother with pleasures and pains? Surely, all that is required is a direct transduction to behavioral mechanisms of the activity of taste and olfactory receptors, pain fibres, etc., when modulated by the appropriate hormonal, nutritional, osmotic, thermal, and other states, without the intervention of the panoply of phenomenal consciousness. It is, of course, this dilemma that has motivated claims that conscious experience is epiphenomenal and without any causal function with respect to behavior. However, for those of us committed to the view that psychology is biologically functional, this conclusion is most unsatisfactory.

Some years ago, we (Balleine and Dickinson, 1998b; Dickinson and Balleine, 2000a) argued that the function of hedonic and affective experience is to act as a motivational interface between cognitive and motivational systems, an interface that is required because these systems use incommensurate psychologies. To illustrate this hypothesis, we shall retell an episode that motivated our initial empirical examination of the theory (Dickinson & Balleine, 1993).

The Palermo Protocol

Many years ago on holiday in Sicily, one of the authors, A.D., came across a market square in Palermo with numerous fruit stalls filled with watermelons, a fact that he recalled later when, feeling thirsty in the Sicilian afternoon sun, he set off in search of watermelons. After some trial-and-error learning, he found the square again and had his first watermelon—delicious it was too, slaking his thirst with its distinctive and slightly perfumed flavor. So, he learned two things from that day's experience: first, the instrumental knowledge about how to find the market square and, second, a desire for watermelon when thirsty. So, a couple of days later, when thirsty again, he retraced the route to the square in search of watermelons. This time, however, the sight of the segments arrayed across the stalls produced a fleeting nausea, which he promptly dismissed with the memory of how delicious their taste and fluid texture had been. But that was the last time that A.D. knowingly tasted watermelon—it was now disgusting.

A.D. had no idea why his desire for watermelon changed so capriciously until a few years later, when, as a student of psychology, he learned that a single pairing of a novel flavor with gastric malaise can condition a profound aversion to the flavor even though an hour or more intervenes between the tasting of the flavor and the induction of nausea (Smith and Roll, 1967). This knowledge led him to reappraise the events of that first day in Palermo with the recollection that he had been horribly sick on the evening of the very day on which he had had first tasted watermelon, a malaise induced by a youthful overindulgence of the local red wine. The fact that aversion conditioning occurs preferentially to novel rather than familiar flavors, even when the temporal contiguity favors the familiar (Ahlers and Best, 1971), explains why its was the novel taste of watermelon, rather the familiar one of red wine, that had attracted the aversion.

Although this analysis identifies the origin of A.D.'s aversion, what it does not explain is the apparent disconnection between his desires and his basic affective reactions. At the time he retraced his steps to the Palermo square in search of watermelon on the second occasion, he, being thirsty, had a strong desire for watermelon while at the same time having a latent aversion to their taste, of which he was totally ignorant, at least at a conscious level. One way of understanding this disconnection is in terms of a dual psychology. As a cognitive creature, A.D.'s search for watermelon was a rational and intentional action controlled by a belief about the route to the square and his desire for watermelons. In contrast, the disgust that he felt (or, more strictly speaking, his orofacial reactions: gagging and spitting out) on re-tasting the melon was a manifestation of his nonrepresentational, reflex psychology. Consequently, during the intervening days between his two trips to the market square, these two psychologies appeared to remain radically disconnected and at variance with each other. As a cognitive creature, he represented watermelon as highly desirable and valued (when thirsty), whereas latent within his reflex machine psychology was an aversion of which of his cognitive creature was totally ignorant.

Importantly, what fused these two psychologies, thereby allowing them to interact in the control of his subsequent behavior, was his phenomenal experience on the second exposure to the melon. It was the experience of nausea and disgust, in conjunction with a perceptual-cognitive representation of the melon as the object of this powerful negative affect, that led to the loss of his desire. If he had not experienced nor cared about the feeling of disgust phenomenally—and, indeed, there was something it was like to experience that nausea—A.D. would probably still seek out watermelon on hot summers days.

Reflection on this episode led us to the central idea of hedonic interface theory, namely that the function for conscious hedonic and affective experience is to act as a motivational interface between the psychologies of the cognitive creature and the reflex machine. The function of this interface is to ground intentional desires, or in other words, cognitive representations of goal values in the biological responses of the reflex machine to motivationally relevant variables, such as nutritional and fluid depletion, poisoning, hormonal states, body temperature, and so on. This grounding occurs through the contiguous experience in phenomenal consciousness of the perception (or thought) of the target object or event (the melon) and the affect that it engenders (disgust) with the perception (or thought) being mediated by the cognitive psychology and the affect by the reflex machine psychology. The motivational imperative of this experience then changes the representation of goal value in the cognitive psychology so that subsequent goal-directed actions are controlled by this new value.

This claim about the function of hedonics and affect in motivation required a firmer empirical foundation than the recollection of one of A.D.'s life events, and so we decided to replicate his Palermo experience for our rats, which we already knew were capable of goal-directed behavior (Adams and Dickinson, 1981), to see whether they also acted in apparent ignorance of a latent aversion (Balleine and Dickinson, 1991). The rats were thirsty just as A.D. had been, but instead of learning the route from the beach to the square, we taught them to press a lever in a single session, not for watermelon, but for a novel sugar water solution. Immediately following this session, we made them mildly ill, not by intoxication with red wine, but by injecting a mild toxin that induces gastric malaise. In spite of these procedural variations, the basic structure of the experience was the same as that of A.D. on the day of his first experience of watermelon. Consequently, if the rats were anything like A.D., what they should have learned as cognitive creatures was first how to get sugar water by lever-pressing and, second, that sugar water is delicious when thirsty. But, as reflex machines, they should also have acquired a latent aversion to the sugar water, an aversion that should have been unknown to the cognitive psychology controlling their purposive actions. Consequently, when once again, given the opportunity to seek sugar water by lever-pressing, our rats should readily have done so. And, just like A.D., this is what they did—their propensity to lever-press was unaffected by the aversion treatment.

It is important to note that this test was conducted in the absence of the sugar water because it was designed to assess the rat's motivation to seek out the solution on the basis of their representation of its value, just as A.D. had sought out watermelons on that second occasion before he had re-tasted them. However, as soon as the rats earned the sugar water and tasted it, thereby discovering that it was now noxious, they immediately stopped pressing just as A.D. has avoided watermelon ever since his re-tasting. In all significant ways, the behavior of our rats was analogous to that of A.D. in Palermo—they too appeared to act in ignorance of their latent aversion, in this case, from sugar water.

By itself, however, the Palermo-like behavior of our rats is not very compelling evidence for a homology with A.D.'s dual psychology. Although A.D. is pretty convinced that his second trip to the market square was motivated by his desire for watermelon, memory is fallible, and perhaps he was in fact seeking some other goal entirely. Therefore, we needed to determine whether our rats were really lever-pressing in the extinction test in response to desire for the sugar water even following the aversion conditioning. Consequently, we considered what would have happened if on the day following A.D.'s initial experience with watermelon, he had been presented with a slice in a different context, for example, at dinner on the following evening. This episode should have allowed him to experience how disgusting the melon had become, thereby transforming his cognitive desire into an aversion and short-circuiting any future propensity to seek it out. So, if our rats were acting out of a desire for sugar water during the test, allowing them to re-taste it, following their initial experience with the solution during training should have enabled them to discover that the sugar water was now distasteful. Following this discovery, they should have been reluctant to seek it again in the test if they were acting as cognitive creatures.

To test this prediction, we gave a second group of rats the full Palermo protocol received by the first group. The only difference was that between the two opportunities to seek the sugar water by lever-pressing, we gave these rats a taste of the solution in a separate drinking cage. Our assumption was that at the outset of the re-tasting, the rats had a desire for sugar water and at the same time a latent aversion to it that was unknown to the cognitive psychology controlling their goal-directed instrumental behavior. Our claim was that re-tasting the sugar water would have two effects. First, it would engage a cognitive representation of the sugar water that would be consciously experienced as a perception of this solution. Second, the re-tasting would also activate a conditioned sickness response, which in turn would be experienced consciously as nausea. The perception of the sugar water conjointly with the strong aversive motivational imperative of the experienced nausea would then remove the cognitive desire for this solution. Therefore, our prediction was that these animals, unlike the standard Palermo rats, should have had no desire to seek the sugar water on the second occasion, and indeed they did not do so (Balleine and Dickinson, 1991).

Hedonic Interface Theory

In summary, what the Palermo experiment suggested is that the control of the goal-directed instrumental behavior of rats, like that of A.D., is not directly connected to the basic motivational processes that determine the values of goals. Rather, animals have to learn about goal values by direct experience with the goal object or event, a form of learning that we refer to as (instrumental) *incentive learning*. We have reviewed elsewhere the extensive evidence that incentive learning mediates not only the impact of aversion conditioning on goal value, but also that of primary motivational states, such as hunger and thirst, (Dickinson and Balleine, 1994, 2002). For example, if rats are trained to press a lever for a novel food reward while either hungry through food deprivation or nondeprived, a shift to the other motivational state has no effect on the propensity to lever-press unless, prior to the shift, they have had the opportunity to eat the food in the alternative state, and therefore, to learn about its incentive value in that state (Balleine, 1992). Therefore, the rats had to learn about the relative values of the food in the different motivational states, presumably by experiencing the fact that the food was more pleasant when hungry than when undeprived.

Moreover, we have also sketched an evolutionary just-so story for the origin of incentive learning (Dickinson and Balleine, 2000b). Our argument is as follows. When neural plasticity first evolved to support the benefits of learning, it took the form of a noncognitive, nonrepresentational, and nonconscious learning process, such as that envisaged by Thorndike (1911) in his classic stimulus–response (S–R)/reinforcement mechanism. The core idea of the S–R mechanism is that pairing a stimulus or response with an attractive biologically potent event strengthens or reinforces an associative connection between a current stimulus and an adaptive response so that the representation of that stimulus is more likely to elicit the response in future. This simple learning mechanism enables the animal to respond adaptively to predictive signals of biologically important reinforcers and to control the occurrence of such events through the acquisition of S–R habits, especially once it is recognized that the associated response can vary all the way from a change in the gain of the spinal stretch reflex in motor learning (Wolpaw, 1997) to the activation of complex motivational systems (Balleine and Killcross, 2006). Indeed, this form of learning has been deployed to great effect in the sophisticated "reinforcement" learning algorithms of artificial intelligence (Sutton and Barto, 1998).

A model S–R system is that mediating the orofacial ingestive and rejection reactions elicited by biologically significant flavors in many different mammalian species (Berridge, 2000). For example, a sweet sugar solution elicits a set of ingestive reactions, whereas aversive rejection reactions follow an intraoral infusion of a bitter quinine solution. There are two points to note about this response system. First, the responses can be modified by learning. Pairing an initially attractive sugar solution with gastric illness, as in the Palermo protocol, changes the reactions elicited by the solution from ingestive to aversive (Breslin et al., 1992). Second, these response systems are *directly* modulated by primary biological needs. A compelling example of this direct modulation is the response of rats to a strong salt solution. Normally, this solution elicits pronounced aversive reactions, but the very first time the solution is tasted after salt depletion, these reactions are transformed into the ingestive pattern (Berridge et al., 1984). The fact that this shift occurs on very first exposure to the salt after depletion shows the modulation by the need state is direct and does not depend upon learning.

The major constraint on an S–R machine is that, due to the paucity of its representations that control the responding, it is unable to adapt immediately to changes in goal value and therefore to plan a course of action in light of its current goals. This limitation can again be illustrated by the Palermo experiments. Recall that having taught our rats how to get sugar water by lever-pressing, we allowed them to learn about their aversion to this reward by tasting it outside the context of the lever-pressing. When given the opportunity to press the lever once again, they were now very reluctant to do so. In contrast to our rats, however, the experienced change in value of the sugar water would have had no direct impact on lever- pressing by an S–R machine. All that the S–R machine has acquired is a compulsion to press whenever it sees the lever; it has no knowledge that this response gives access to sugar water. Therefore, it cannot adapt its behavior directly to changes in the values of outcomes or reinforcers and, indeed, can only do so by direct punishment if the (now noxious) outcome is made contingent upon the response, thereby reengaging the S–R/reinforcement mechanism so that the change in the reinforcing property of the sugar water can impact on responding.

Our argument is that the capacity for goal-directed action could not be supported by an elaboration of the basic S–R psychology of the reflex machine[1] but rather required the evolution of capacity for some form of cognitive control. For example, Dickinson and colleagues (Dickinson, 1980; Heyes and Dickinson, 1990) have argued that a goal-directed action is mediated by a form of practical inference that takes as its arguments, beliefs about the causal outcomes of its actions, and the desirability or value of the outcomes. Although some may well be skeptical about attributing such causal reasoning to the rat, this attribution has received recent support from a demonstration that the instrumental behavior of the rat manifests a form of causal inference that appears to lie outside the scope of simple associative processes (Blaisdell et al., 2006; Clayton and Dickinson, 2006). However, this adaptation leaves the animal with two incommensurate psychologies. The S–R process is a psychological *mechanism* in that its processes—excitation, inhibition, and connection strengthening—are explanatory by analogy with physical processes. In contrast, the cognitive inference process and its arguments, the animal's beliefs and desires, are intentional or representational with the practical inference process itself constrained by cannons of rationality rather than by those of *psychological* mechanistic interactions.

This distinction at the psychological level is manifest in the implementation of these psychologies in brain processes. Being mechanistic, S–R learning is, at least in principle, transparent with respect to neural processes and can be implemented directly in the experience-dependent synaptic plasticity governing the connections between sensory and motor brain systems. By contrast, the cognitive inference process, being intentional and rational, is likely to be more opaque with respect to the underlying brain processes just as the processes operating on the symbolic variables of a high-level computer language are opaque with respect to the machine's hardware processes.

The evolution of a dual psychology in the service of goal-directed action inevitably brought with it the problem of connecting up or interfacing these two systems, a problem that is particularly acute in the case of motivation. As we have already noted for ingestive/rejection behavior, it is the basic response systems of the reflex machine that are directly sensitive to the biological motivational states and variables. By contrast, the values of or desires for goals are represented abstractly in the cognitive system where their function is both to cause and rationalize behavioral decisions. Therefore, in order to be effective, the cognitive control of goal-directed action must have coevolved with an interface between the two psychologies so that the values and desires of the cognitive creature could be grounded in the biologically relevant variables of the S–R psychology of the reflex machine.

In principle, the interface or transducer between the two psychologies could have taken many forms. For example, in a computer-controlled industrial plant, the interface between the pressure within a vessel and the symbolically encoded procedures for pressure control may take the form of an analog-to-digital pressure-sensitive transducer. By analogy, it is possible that evolution could have "solved" the interface problem by a nonconscious transducer. However, our claim is that Nature did, as a matter of fact, solve the interface problem by the coevolution of the capacity for phenomenal, first-order conscious experience along with the cognitive control of goal-directed action. Within the interface, the motor commands elicited with the S–R mechanisms, for example, the orofacial reactions to a flavor, are transformed by some (at present, mysterious) neural process into the phenomenal affective experience, for example, of disgust or delight, which in turn leads to assignment of an appropriate value or desirability to any flavor that is concurrently perceived. The crucial point about this transduction is that the affective or hedonic imperative of the experience is not epiphenomenal to the function of the interface—it is the fact that the flavor is consciously experienced as so disgusting, the toothache as so painful, the caress as so exciting that grounds the very assignment of value. In other words, the imperative for goal-directed action arises from the motivationally compelling nature of conscious hedonic experience.

In the remainder of this chapter, we shall unpack the implication of this hedonic interface theory (HIT) and consider the empirical evidence that bears on its major features.

Dual Psychology of Behavior

The central assumption of HIT is that behavior can be controlled by two radically different psychologies: the intentional psychology of the cognitive creature and the S–R psychology of the reflex machine. A similar assumption is common, if disputed, in the field of human cognitive psychology with the distinction between controlled or explicit and automatic or implicit processing and learning, although the distinction is rarely brought to bear in the context of the motivational control of goal-directed behavior (Gawronski and Bodenhausen, 2006). Over the last decade, however, it has become increasingly apparent that the two psychologies can be dissociated in the control of motivated instrumental behavior by brain lesions.

These dissociations have employed a variant of the outcome devaluation procedure for distinguishing between goal-directed and habitual S–R behavior that is based upon specific satiety. In an example of this procedure, we trained hungry rats to press a lever and pull a chain with one action earning a salty starch solution and the other a sour one (Balleine and Dickinson, 1998a; Colwill and Rescorla, 1985). Following this training, we fed the rats to satiety on one of the solutions in order to reduce its goal value before giving them the opportunity to press the lever and pull the chain. In this test, the rats performed the response trained with the devalued outcome less than that trained with the valued outcome. As in the case of the Palermo protocol, this test was conducted in the absence of the outcomes, or in other words in extinction, so that we could be sure that the selective choice of the valued action was based upon the rats' knowledge of the differential consequences of each action and therefore goal-directed. However, instrumental performance is not always goal-directed and sensitive to outcome devaluation. Many years ago, Adams (1982) demonstrated that overtraining can render instrumental responding impervious to outcome devaluation, a finding that suggests that one and the same action can be controlled by two different psychological processes depending upon the conditions of training (Dickinson, 1985). With extended training, a transition in control occurs from the practical inference processes of the cognitive creature to the habitual S–R mechanisms of the reflex machine.

This behavioral dissociation is now supported by extensive neurobiological evidence. We have known for sometime that the integrity of the prelimbic area of the prefrontal cortex is necessary for the acquisition of goal-directed behavior in that, although acquisition and terminal rates of performance are unaffected, instrumental responding by rats with pretraining lesions of this area is insensitive to outcome devaluation (Balleine and Dickinson, 1998c; Corbit and Balleine, 2003) and, indeed, the fact that lesions of the medial dorsal thalamus produced a similar deficit (Corbit et al., 2003) suggest that the acquisition cognitive control over instrumental behavior is mediated by the cortico-striatal loops via thalamus.[2] Further evidence for the role of cortico-striatal interactions comes from the fact that dysfunction of the striatal target areas of the prelimbic cortex, the nucleus accumbens (Corbit et al., 2001) and the dorsomedial striatum (Yin et al., 2005a; Yin et al., 2005b) also abolish sensitivity to goal devaluation.

In contrast to goal-directed behavior, another set of prefrontal and striatal structures appear to mediate habitual S–R instrumental responding. Killcross and Coutureau (2003) discovered that lesions of the infralimbic cortex prevented the loss of sensitivity to outcome devaluation with overtraining, thereby suggesting the lesion disrupted the habitual S–R mechanisms, and we now know that this effect is replicated by dysfunction of the dorsolateral striatum (Yin et al., 2004, 2006). Within the present context, our purpose in reviewing this neurobiological evidence is not to delineate the neural circuits underlying goal-directed and habitual behavior, but rather to argue that the cognitive and S–R systems of behavioral control are dissociable.

These dissociations are particularly compelling given that the target response, lever-pressing by rats for a food reward, their motivational state of hunger, and the stimulus context of an operant chamber are common features of the dissociations. However, perhaps the most compelling evidence comes from the finding that control can switched between these two systems by the temporary inactivation of the infralimbic cortex. Having overtrained their rats so that responding was impervious to outcome devaluation, Coutureau and Killcross (2003) found that the temporary inactivation of the infralimbic cortex restored goal-directed control, suggesting that there is some form of competitive interaction between the two systems for behavioral control (Balleine and Ostlund, 2007).

"(Dis)Liking"—the Hedonic Basis of Incentive Learning

The second main claim of HIT is that the desires or goal values of the cognitive psychology are based on hedonic experiences, which are the phenomenological manifestations of the responses of the reflex machine to motivational significant stimuli. Therefore, a straightforward prediction of the theory is that changing the hedonic response to the stimulus should produce a corresponding change in its goal value or desirability, a prediction that we have tested within the context of the Palermo protocol.

The logic of the experiment can be illustrated by considering what would have happened if A.D. had taken an antinausea drug immediately prior to his second tasting of the watermelon. If the drug had been sufficiently effective to have completely blocked any nauseous reaction, he would have again experienced them as delicious, an experience that would have confirmed his desire for them when thirsty, thereby prolonging the dissociation between his cognitive and S–R psychologies. We tested this prediction of HIT by bringing about this scenario for our Palermo rats (Balleine et al., 1995). Recall that in one condition, our rats learned to lever-press for sugar water before we gave our rats the opportunity to taste the sugar after they had acquired a latent aversion to it. This re-tasting had a profound effect when we subsequently tested their desire to search for the sugar water by lever-pressing. In the absence of the re-tasting, they behaved as though they were ignorant of their aversion by readily lever-pressing, whereas those that had re-tasted the sugar water showed little propensity to seek it out again. This finding we took as evidence that our rats, like A.D. and watermelons, based the goal value that they assigned to the sugar water on their affective reactions to it during the re-tasting.

If goal values are grounded on affect in the way that this analysis suggests, we should have been able to attenuate the loss of goal value produced by the re-tasting experience by ameliorating the experienced nausea with an antinausea drug. To do so, we used an antinausea drug, ondansetron, that reduces the aversive taste reactivity patterns elicited by a sweet solution after aversion conditioning without any detectable effect on consumption (Limebeer and Parker, 2000). As HIT assumes that affective experience originates from these reactions, the prediction is that when subsequently given the opportunity to seek out sugar water by lever-pressing, the rats would more readily do so if they had re-tasted the solution while they were under the influence of ondansetron, a prediction confirmed by Balleine et al. (1995).

Once again, the Palermo protocol is a fragile edifice on which to mount a general theory of the role of affective experience in the motivation of goal-directed behavior. There is, however, an array of evidence that treatments that enhance the hedonic reactions to a stimulus, as indexed by taste-reactivity patterns in the case of oral stimuli, also augment the goal value assigned to that stimulus as a result of experiencing these reactions. One such treatment is the administration of benzodiazepine drugs, for example, midazolam (Söderpalm and Berridge, 2000), which augment appetitive taste-reactivity (Berridge and Peciña, 1995) and, by implication, the hedonic reactions to foods. To determine whether this hedonic enhancement also leads to an increment in goal value, Balleine et al. (1994) trained rats to press a lever for a palatable food reward when they were not deprived of their maintenance diet. One group was then allowed to eat the food in separate feeding cages with or without an injection of midazolam. If the drug enhances the pleasure elicited by the food, then the experience of eating the food under the influence of the drug should have led to the assignment of a higher goal value, at least while in the drug state. To test whether this was so, Balleine et al. (1994) once again allowed the rats to seek the food by lever-pressing, although this test was conducted in extinction to assess the goal-directed component of the behavior. In accord with HIT, rats that had previously experienced the food under midazolam pressed more on test than the rats that did not experience food under midazolam.

In summary, there is a variety of evidence to suggest that the values of biologically significant goals are based on the hedonic reactions elicited by these goals and that the impact of primary motivational states on goal-directed behavior is mediated by the modulation of these hedonic reactions. We shall conclude our presentation of HIT by briefly contrasting it with three somewhat similar accounts of motivation: Damasio's somatic marker hypothesis, Berridge and Robinson's incentive salience theory, and Cabanac's common currency theory.

Autonomy of the Cognitive

The distinction between the somatic marker hypothesis and HIT is subtle, but important. According to the somatic marker hypothesis (Damasio, 1996), goal-directed behavior is regulated online by re-experiencing the affective and somatic consequences of contact with the goal. The operation of this process can be illustrated by a somatic marker account of why A.D. to this day avoids watermelons. The argument would be that the very act of contemplating the melon as a goal activates nauseous reactions, which serve to inhibit any propensity to seek them out. By contrast, HIT assumes that the representation of goal value, although initially grounded on affective and hedonic experience, is in fact purely cognitively and affect free. It is, of course, this assumption that is at the core of the radical disconnection between the cognitive and reflex machine psychologies. According to HIT, therefore, A.D. should be perfectly capable of avoiding watermelons without any feedback from conditioned nauseous reactions—this commodity simply has a low goal value for him and therefore will not rationalize the intention to seek melons. And, indeed, to this very day, A.D. does not in fact know whether or not he would still have an aversion to watermelon if he were to taste it again.

The empirical distinction between these two accounts once again can be illustrated by contemplating the effect of giving A.D. an antinausea drug at the time when he is deciding whether or not to seek watermelons. According to the somatic marker hypothesis, the presence of the drug, by attenuating the online nauseous reactions, should reduce his propensity to avoid the melon. By contrast, HIT predicts that the drug would have no impact because the cognitive representation of the goal value guiding his actions is affect-free. What evidence we have from the incentive learning studies with rats favors HIT (Balleine, 2005). For example, administering the antinausea drug, ondansetron, at the time of testing in the Palermo protocol did not affect the rats' propensity to seek sugar water (Balleine et al., 1995). So, although manipulating their affective experience at the time when they were assigning value or desirability during re-tasting modulated their subsequently instrumental choices, the same manipulation was ineffective when they performed the instrumental action itself. Therefore, it would appear that performing an intentional action can be independent of the affective experiences that grounded the goal value in the first.

By way of a caveat, however, we should note that although the Palermo design controls the interaction between the cognitive and S–R psychologies, the two psychologies are of course continually running in parallel. So, for example, if the context of the instrumental decision contains cues that are associated with an outcome from which the animal has an aversion, these cues could well elicit a conditioned affective response that could then modulate instrumental behavior. Incentive learning could occur at the time of an instrumental decision through the retrieval of a representation of the goal conjointly with a conditioned activation of the associated affective experience. Alternatively, the conditioned affective response could reinforce or punish the habitual responding over a series of instrumental decisions through the mechanisms of the S–R psychology. Therefore, in accord with the somatic marker hypothesis, HIT allows for such online interactions while at the same time maintaining that the goal representations that control instrumental actions can be affect free.

What About "Wanting"?

In many ways, HIT is complementary to incentive salience theory. It has long been recognized that signals for biologically important resources acquire motivational properties (Bindra, 1974; Dickinson and Dearing, 1979; Estes, 1943; Konorski, 1967; Rescorla and Solomon, 1967). A paradigmatic example of the motivational influence of such signals is the so-called general Pavlovian-to-instrumental transfer effect. If a stimulus that has been independently established as a signal for a reward is presented while an animal is engaged in the performance of an instrumental response, such as pressing a lever in search of food, the performance of the response is enhanced.[3] The presence of the signal appears to augment the motivation of the instrumental behavior. What Berridge and Robinson (Berridge, 2001; Robinson and Berridge, 1993) added to this basic motivational theory was twofold. First, the psychological contribution of incentive salience theory was to identify the state induced by the signal as one of 'wanting' that is distinct from any 'liking' or hedonic reactions elicited by the signal or the reward that it predicts. Second, on a neurobiological level, incentive salience theory identifies the mesolimbic dopamine system as the neural substrate for 'wanting'.

This is not the place to recapitulate the extensive evidence offered by Berridge and Robinson in favor of these two claims. For present purposes, it is sufficient to note the dissociation between indexes of 'wanting' and 'liking', such as that produced by dopamine antagonists. Most importantly, these antagonists attenuate the motivational impact of reward-predicting signals in the transfer paradigm (Dickinson et al., 2000), while having no effect on the ingestive response patterns indicative of a hedonic evaluation (Peciña et al., 1997). Given that HIT claims that hedonics supplies the motivational input into the control of cognitively mediated goal-direct action, 'wanting', at least as characterized by incentive salience theory, would appear to be the motivational process controlling the cue-directed habitual behavior of the S–R psychology.

This characterization is supported by the concordance between motivational indices of 'wanting' and conditions that establish habitual S–R control of behavior. Holland (2004) observed that the motivational influence of a reward-predicting signal increased in the transfer paradigm as instrumental responding became more habitual with training in the sense of being resistant to outcome devaluation. A similar concordance is also observed in response to a sensitization-inducing regime of injections of a dopamine agonist. Such a regime not only enhances motivational transfer (Wyvell and Berridge, 2001), but also facilitates the acquisition of habitual instrumental behavior (Nelson and Killcross, 2006).

In summary, we argue that incentive salience theory characterizes the motivational processes of the S–R psychology by focussing on the impact of reward-associated cues on habitual behavior. However, there is a lacuna at the core of the theory—although the distinction between 'wanting' and 'liking' is a central and important one, the theory leaves the hedonic component without any motivational function, and it remains a mystery as to why we should not only 'want' but also 'like'. It is in this sense that HIT complements incentive salience by articulating a psychological function for 'liking', namely that of grounding the incentive values underlying goal-directed actions in basic motivational states.

Hedonics as a "Common Currency"

HIT assumes that there is a commonality between the affective reactions to the many objects of (dis)pleasure, which takes the form of a common, phenomenologically experienced, bivalent motivational imperative. Cabanac (1992) pioneered the claim that pleasure provides a common motivational currency, at least in its contemporary form. In supporting this thesis, he noted that hedonic evaluations are directly modulated by primary motivation states, such as hunger or a thermic state, a process that he refers to as "alliesthesia"

(Cabanac, 1971) and, importantly, he has documented that behavioral choices can be predicted from independent hedonic evaluations (Cabanac, 1992). However, our discussion of somatic marker and incentive salience theories makes clear that HIT diverges from the common currency account in a number of respects. First, in accord with incentive salience theory, HIT assumes that hedonic processes play no role in the motivational control of the S–R habitual behaviors, such as the general behavioral activation produced by reward signals and the direct modulation of this activation by primary motivational states (Balleine, 1994; Dickinson and Dawson, 1987). To the extent that there is a commonality in the processes of S–R psychology, it is probably at the physiological rather than psychological level, possibly in terms of the role of dopamine as envisaged by incentive salience theory (Berridge, 2007).

Even at the level of cognitively mediated actions, HIT does not really view (dis)pleasure as a common currency. In contrasting HIT with somatic marker theory, we pointed out that once goal value has been assigned through hedonic experience of the goal object or event, cognitively mediated action and choice become independent of this experience in the sense that one does not need to experience (dis)pleasure at the time of action. Therefore, HIT assumes not only the affective but also a cognitive common currency in the form an abstract representation of value that can enter into and rationalize the practical inferences underlying action. It is through this cognitive common currency that nonhedonic determinants of value can enter into the control of action and choice (Higgins, 2006). Rather than viewing (dis)pleasure as a sole common currency, HIT argues that the biological determination of goal value occurs through an exchange between the affective and cognitive common currencies.

Cognitive Motivation and Consciousness

In summary, we view HIT as elaborating and/or complementing many of the ideas offered by the somatic marker and common currency hypotheses and, particularly, incentive salience theory. In HIT, however, we emphasize the motivational imperative endowed by the conscious nature of affective experience and, indeed, our original presentation of the core ideas of HIT was offered as an account of the function of primary consciousness (Balleine and Dickinson, 1998). HIT views primary consciousness as a mental medium for interfacing affect and cognition with the function of assigning cognitive values or desires to potential goal objects on the basis of the experienced affect. Once values or desires are assigned to goal–objects, however, the cognitive processes deploying these values in pursuit of the goals can function not only in the absence of affect but also unconsciously. In this respect, HIT recapitulates the Freudian view that "there are no unconscious affects as there are unconscious ideas" (Freud, 1915, p. 180) and, indeed, we (Dickinson and Balleine, 1994), following Tolman (1949a,b), have already noted that the process of goal-value assignment (instrumental incentive learning) in HIT has many of the characteristics of the Freudian process of *cathexis*.

Notes

1. Although many authors posit a role for response–outcome (R–O) associations in the performance of motivated instrumental responses (Colwill and Rescorla, 1986), there are no satisfactory accounts of how the R–O association interacts with motivational process to control behavior. There are a number of problems with the only viable mechanism that we know of, the associative-cybernetic model (Dickinson, 1994; Dickinson and Balleine, 1993), that are too technical to present here.
2. More recent evidence, however, shows that integrated prefrontal function is required only for the acquisition of goal-directed behavior in that post-training lesions of the prelimbic area leave responding sensitive to outcome devaluation (Ostlund and Balleine, 2005).
3. We refer here to general, rather than outcome specific transfer. General transfer is the capacity of a signal for a reward to augment instrumental performance even if the response was trained with a reinforcer that differs from the reward associated with the signal. By contrast, outcome-specific transfer refers to the ability of the signal to enhance selectively responses trained with the same reward. General and outcome-specific transfer can be dissociated within the amygdala (Corbit and Balleine, 2005). Moreover, general transfer is modulated by the relevance of the signalled reward to the animal's motivational state (Balleine, 1994; Dickinson and Dawson, 1987), whereas outcome-specific transfer is unaffected by devaluation of the outcome (Rescorla, 1994) suggesting that it is a response-priming rather than a motivational effect.

References

Adams, C. D. (1982). Variations in the sensitivity of instrumental responding to reinforcer devaluation. *Q. J. Exp. Psychol.* 34B, 77–98.

Adams, C. D. and Dickinson, A. (1981) Instrumental responding following reinforcer devaluation. *Q. J. Exp. Psychol.* 33B, 109–122.

Ahlers, R. H. and Best, P. J. (1971) Novelty vs temporal contiguity in learned taste aversions. *Psychon. Sci.* 25, 34–36.

Balleine, B. (1992) Instrumental performance following a shift in primary motivation depends upon incentive learning. *J. Exp. Psychol.: Anim. Behav. Process* 18, 236–250.

Balleine, B. (1994) Asymmetrical interactions between thirst and hunger in Pavlovian-instrumental transfer. *Quarterly J. Exp. Psychol.* 47B, 211–231.

Balleine, B., Ball, J. and Dickinson, A. (1994) Benzodiazepine-induced outcome revaluation and the motivational control of instrumental action in rats. *Behav. Neurosci.* 108, 573–589.

Balleine, B. and Dickinson, A. (1991) Instrumental performance following reinforcer devaluation depends upon incentive learning. *Q. J. Exp. Psychol.* 43B, 279–296.

Balleine, B. and Dickinson, A. (1998a) The role of incentive learning in instrumental outcome revaluation by sensory-specific satiety. *Anim. Learn. Behav.* 26, 46–59.

Balleine, B. W. (2005) Neural bases of food-seeking: Affect, arousal and reward in corticostriatolimbic circuits. *Physiol. Behav.* 86, 717–730.

Balleine, B. and Dickinson, A. (1998b) Consciousness—the interface between affect and cogntion. In: *Consciousness and Human Identity,* pp. 57–85. Ed. J. Cornwall. Oxford Univeristy Press: Oxford.

Balleine, B. W. and Dickinson, A. (1998c) Goal-directed instrumental action: Contingency and incentive learning and their cortical substrates. *Neuropharmacology* 37, 407–419.

Balleine, B. W., Garner, C. and Dickinson, A. (1995) Instrumental outcome devaluation is attenuated by the anti-emetic ondansetron. *Q. J. Exp. Psychol.* 48B, 235–251.

Balleine, B. W. and Killcross, S. (2006) Parallel incentive processing: An intergrated view of amygdala function. *Trends Neurosci.* 29, 272–279.

Balleine, B. W. and Ostlund, S. B. (2007) Action selection and initiation in instrumental conditioning. *Ann. N Y Acad. Sci.* 1104, 147–171.

Berridge, K. C. (2000) Measuring hedonic impact in animals and infants: Microstructure of affective taste reactivity patterns. *Neurosci. Biobehav Rev.* 24, 173–198.

Berridge, K. C. (2001) Reward leaning: Reinforcement, incentives and expectations. In: *The Psychology of Leaning and Motivation,* Vol. 40, pp. 223–278. Ed. D. L. Medin. Academic Press: New York.

Berridge, K. C. (2007) The debate over dopamine's role in reward: the case for incentive salience. *Psychopharmacology* 191, 391–431.

Berridge, K. C., Flynn, F. W., Schulkin, J. and Grill, H. J. (1984) Sodium depletion enhances salt palatablity in rats. *Behav. Neurosci.* 98, 652–660.

Berridge, K. C. and Peciña, S. (1995) Benzodiazepines, appetite, and taste palatability. *Neurosci. Biobehav. Rev.* 19, 121–131.

Bindra, D. (1974) A motivational view of learning, performance, and behavior. *Psychol. Rev.* 81, 199–213.

Blaisdell, A. P., Sawa, K., Leising, K. J. and Waldmann, M. R. (2006) Causal reasoning in rats. *Science* 311, 1020–1022.

Breslin, P. A., Spector, A. C. and Grill, H. J. (1992) A quantative comparison of taste reactivity behaviors to sucrose before and after lithium chloride pairings: a unidimensional account of palatability. *Behav. Neurosci.* 106, 820–836.

Cabanac, M. (1971) Physiological role of pleasure. *Science* 173, 1103–1107.

Cabanac, M. (1992) Pleasure: the common currency. *J. Theor. Biol.* 155, 173–200.

Clayton, N. and Dickinson, A. (2006) Rational rats. *Nat. Neurosci.* 9, 472–474.

Colwill, R. M. and Rescorla, R. A. (1985) Postconditioning devaluation of a reinforcer affects instrumental responding. *J. Exp. Psychol.: Anim. Behav. Process.* 11, 520–536

Colwill, R. M. and Rescorla, R. A. (1986) Associative structures in instrumental learning. In: *The Psychology of Learning and Motivation Vol. 20,* pp. 55–104. Ed. G. H. Bower. Academic Press: New York.

Corbit, L. H. and Balleine, B. W. (2003) The role of prelimbic cortex in instrumental conditioning. *Behav. Brain. Res.* 146, 145–157.

Corbit, L. H. and Balleine, B. W. (2005) Double dissociation of basolateral and central amygdala lesions on the general and outcome-specific forms of Pavlovian-instrumental transfer. *J. Neurosci.* 25, 962–970.

Corbit, L. H., Muir, J. L. and Balleine, B. W. (2001) The role of the nucleus accumbens in instrumental conditioning: Evidence of a functional dissociation between accumbens core and shell. *J. Neurosci.* 21, 3251–3260.

Corbit, L. H., Muir, J. L. and Balleine, B. W. (2003) Lesions of mediodorsal thalamus and anterior thalamic nuclei produce dissociable effects on instrumental conditioning in rats. *Eur. J. Neurosci.* 18, 1286–1294.

Coutureau, E. and Killcross, A. S. (2003) Inactivation of the infralimbic prefrontal cortex reinstates goal-directed responding in overtrained rats. *Behav. Brain Res.* 146, 167–174.

Damasio, A. R. (1996) The somatic marker hypothesis and the possble functions of the prefrontal cortex. *Philos. Trans. R Soc. Lond. B Biol. Sci.* 351, 1413–1420.

Dickinson, A. (1980). *Contemporary Animal Learning Theory.* Cambridge: Cambridge University Press.

Dickinson, A. (1985) Actions and habits: the development of behavioural autonomy. *Philos. Trans. R Soc. Lond. B Biol. Sci.* 308, 67–78.

Dickinson, A. (1994) Instrumental conditioning. In: *Animal Learning and Cognition*, pp. 45–79. Ed. N. J. Mackintosh. Academic Press: San Diego, CA.

Dickinson, A. and Balleine, B. (1993) Actions and responses: the dual psychology of behaviour. In: *Problems in the Philosophy and Psychology of Spatial Representation*, pp. 277–293. Eds. N. Eilan, R. A. McCarthy, and M. W. Brewer. Blackwell: Oxford.

Dickinson, A. and Balleine, B. (1994) Motivational control of goal-directed action. *Anim. Learn. Behav.* 22, 1–18.

Dickinson, A. and Balleine, B. W. (2000a) Causal cognition and goal-directed action. In: *The Evolution of Cognition*, pp. 185–204. Eds. C. Heyes and L. Huber. The MIT Press: Cambridge MA.

Dickinson, A. and Balleine, B. (2002) The role of learning in the operation of motivational systems. In: *Stevens' Handbook of Experimental Psychology*, pp. 497–533. Eds. H. Pashler and R. Gallistel. John Wiley & Sons: New York,

Dickinson, A. and Dawson, G. R. (1987) Pavlovian processes in the motivational control of instrumental performance. *Q. J. Exp. Psychol.* 39B, 201–213.

Dickinson, A. and Dearing, M. F. (1979) Appetitive-aversive interactions and inhibitory processes. In: *Mechanism of Learning and Motivation*, pp. 203–231. Eds. A. Dickinson and R. A. Boakes. Lawrence Erlbaum Associates: Hillsdale, NJ.

Dickinson, A., Smith, J. and Mirenowicz, J. (2000) Dissociation of Pavlovian and instrumental incentive learning under dopamine antagonists. *Behav. Neurosci.* 114, 468–483.

Estes, W. K. (1943) Discriminative conditioning. I. A discriminative property of conditioned anticipation. *J. Exp. Psychol.* 29, 390–400.

Gawronski, B. and Bodenhausen, G. V. (2006) Associative and propositional processes in evaluation: an integrative review of implicit and explicit attitude change. *Psychol. Bull.* 132, 692–731.

Heyes, C. and Dickinson, A. (1990). The intentionality of animal action. *Mind Lang., 5*, 87-104

Higgins, E. T. (2006) Value from hedonic experience and engagement. *Psychol. Rev.* 113, 439–460.

Holland, P. C. (2004) Relations between Pavlovian-instrumental transfer and reinforcer devaluation. *J. Exp. Psychol.: Anim. Behav. Process.* 30, 104–117.

Killcross, A. S. and Coutureau, E. (2003) Coordination of actions and habits in the medial prefrontal cortex of rats. *Cereb. Cortex* 13, 400–408.

Konorski, J. (1967) *Integrative Activity of the Brain: an Interdisciplinary Approach.* University of Chicago Press: Chicago.

Limebeer, C. L. and Parker, L. A. (2000) The antiemetic drug ondansetron interferes with lithium-induced conditioned rejection reactions, but not lithium-induced taste avoidance in rats. *J. Exp. Psychol.: Anim. Behav. Process.* 26, 371–384.

Nelson, A. and Killcross, K. (2006) Amphetamine exposure enhances habit formation. *J. Neurosci.* 26 , 3805–3812.

Ostlund, S. B. and Balleine, B. W. (2005) Lesions of medial prefrontal cortex disrupt the acquisition but not the expression of goal-directed learning. *J. Neurosci.* 25, 7763–7770.

Peciña, S., Berridge, K. C. and Parker, L. A. (1997) Pimozide does not shift palatability: Separation of anhedonia from sensorimotor suppression by taste reactivity. *Pharmacol. Biochem. Behav.* 58, 801–811.

Rescorla, R. A. (1994) Transfer of instrumental control mediated by a devalued outcome. *Anim. Learn. Behav.* 22, 27–33.

Rescorla, R. A. and Solomon, R. L. (1967) Two-process learning theory: Relationship between Pavlovian conditioning and instrumental learning. *Psychol. Rev.* 74, 151–182.

Robinson, T. E. and Berridge, K. C. (1993) The neural basis of drug craving: an incentive sensitization theory. *Brain Res. Rev.* 18, 247–291.

Smith, J. C. and Roll, D. L. (1967) Trace conditioning with X-rays as an aversive stmulus. *Psychon. Sci.* 11–12.

Söderpalm, A. H. V. and Berridge, K. C. (2000) The hedonic impact and intake of food are increased by midazolam microinjections in the parabrachial nucleus. *Brain Res.* 877, 288–297.

Sutton, R. S. and Barto, A. G. (1998) *Reinforcement Learning.* The MIT Press: Cambridge, MA.

Thorndike, E. L. (1911). *Animal Intelligence: Experimental Studies.* Macmillan: New York.

Tolman, E. C. (1949a) There is more than one kind of learning. *Psychol. Rev.* 56, 144–155.

Tolman, E. C. (1949b) The nature and functioning of wants. *Psychol. Rev.* 56, 357–369.

Wolpaw, J. R. (1997) The complex structure of memory. *Trends Neurosci.* 20, 588–594.

Wyvell, C. L. and Berridge, K. C. (2001) Incentive sensitization by previous amphetamine exposure: Increased cue-triggered "wanting" for sucrose reward. *J. Neurosci.* 21, 7831–7840.

Yin, H. H., Knowlton, B. J. and Balleine, B. B. (2004) Lesions of dorsolateral striatum preserve outcome expectancy but disrupt habit formation in instrumental learning. *Eur. J. Neurosci.* 19, 181–189.

Yin, H. H., Knowlton, B. J. and Balleine, B. W. (2005a) Blockade of NMDA receptors in the dorsomedial striatum prevents action outcome learning in instrumental conditioning. *Eur. J. Neurosci.* 22, 505–512.

Yin, H. H., Knowlton, B. J. and Balleine, B. W. (2006) Inactivation of dorsolateral striatum enhances sensitivity to changes in the action–outcome contingency in instrumental conditioning. *Behav. Brain Res.* 166, 189–196.

Yin, H. H., Ostlund, S. B., Knowlton, B. J. and Balleine, B. W. (2005b) The role of the dorsomedial striatum in instrumental conditioning. *Eur. J. Neurosci.* 22, 513–523.

5

Neuroethology of Pleasure

KARLI K. WATSON, STEPHEN V. SHEPHERD, AND MICHAEL L. PLATT

"Sur, Sura, Sundari."
"Víno, ženy a zpev."
"Vin, kvinder og sang."

Readers unacquainted with Hindi, Czech, or Danish may find these words unfamiliar, but all three sayings capture the same idea. In English, they translate as "wine, women, and song"—or, perhaps more familiarly, in the modern era "sex, drugs, and rock'n'roll." These clichés are relatively pancultural because they speak to all humans about our goals and joys. We pursue water to slake our thirst and wine to please the palate. We pursue the companionship of friends, the dalliances of lovers, and the bonds of matrimony. And we pursue beauty in music, landscape, ideas, and art.

Our pleasures are ingrained deeply in our ancestry. The motivation to climb four flights of stairs to retrieve a candy bar is mirrored by 7000-year-old petroglyph depictions of honey hunters enduring the wrath of a hive (Crane, 1999). In a hunter-gatherer world, where every calorie is hard won, sweet and fatty foods offer the most value for effort. Such temptations persist today when high-calorie foods are available on every street corner. The rewards we evolved to enjoy continue to motivate our behavior, however inappropriately, simply because obtaining them once determined whether our ancestors flourished or perished. Just as bipedal walking and enhanced cranial capacity allowed ancestral humans to outreproduce their competition, so too did the tendencies to savor a sweet, enjoy the presence of fellow individuals, or revel in cool water on a hot day. Viewing pleasure from an evolutionary perspective thus provides a framework within which we can understand our motivational drives and their relationship to the modern—as well as the primordial—world.

Many pleasures, far from being uniquely human, also motivate other animals. We cannot know what animals feel (Nagel, 1974; Quine, 1968) or even ask them what they enjoy, but we can observe the things they pursue and the way that specific patterns of brain activity mediate pursuit behavior (Skinner, 1938; Thorndike, 1898). Although the distinction between pleasure per se and motivation is explored elsewhere in this book (see Berridge, Chapter 1), we feel that in the colloquial sense of the word "pleasure," hedonic experiences and the drive to obtain them are often linked—at least in the natural world. Much of the literature we review here, therefore, concerns the circuitry and neural coding of preference, by which animals pursue the more favorable of outcomes. Such a proxy is indispensable because it provides a quantitative behavioral metric for what, if anything, a particular animal might find pleasurable. This is an important distinction, as the individual selective pressures bearing on various animal lineages may drive nonhuman species to take pleasure in sensations quite alien to our own. Cats, for example, lack sweet-taste receptors (Li et al., 2005) and are thus largely oblivious to the sort of orally applied sucrose solution that so many other animals, including ourselves, find pleasurable. Conversely, female mice may take great pleasure in the ultrasonic serenade of courting males, which we humans cannot even hear (Holy and Guo 2005).

Without a behavioral assay with which to link the diverse neuronal circuitry found in the animal kingdom to the distinct evolutionary pressures that shaped this circuitry, it would be much more difficult to put the "creature comforts" of humans in their proper context. Preference assays allow experimenters to infer what various animals find motivating, even when the utility of, or behavioral response to, these stimuli does not readily translate across species. For example, Mason and colleagues (2001) devised an experiment that allowed captive mink to express, via choice behavior, their preferences for various environmental stimuli. The animals were able to select different cage compartments, each equipped with something hypothesized to be pleasurable based on species-typical behavior: a water pool, a nesting site, and special toys. A "cost" was imposed on moving between compartments by attaching weights to the doors between them. The choice behavior of the mink demonstrated that water to splash in was the most valuable of the offered resources (Figure 5.1). Furthermore, when deprived of water, the mink showed elevated cortisol levels similar to those observed following food deprivation.

These results suggest that minks have an internal drive to splash just as they have a drive to eat. It can be difficult to distinguish whether such approach behavior is motivated by potential enjoyment, craving, avoidance of unpleasantness, or mere dispassionate reflex. Nonetheless, it seems no less reasonable to imagine these animals enjoy splashing than it does to imagine they enjoy a meal. Comparison of such preferences across species can suggest the adaptive significance and phylogenic origin of a behavior—answering two of Tinbergen's four questions—and provide us with animal models in which we can investigate the other two: biological mechanism and ontogenetic development. All four questions must be addressed to fully describe the causes and consequences of "pleasure," as well as to identify what is uniquely human about its experience. By comparing the neuroethology of pleasure in humans and other animals, we map out not only the human dimensions of happiness, but also the evolutionary forces that define them.

The network of brain areas and neurotransmitters involved in processing rewards and punishments is quite complex compared with our own seemingly effortless awareness of pleasure and pain (Halgren et al., 1978; Olds and Milner, 1954; Schoenbaum et al., 1998; Schultz, 2000; Schultz et al., 1997; Tremblay and Schultz, 2000). But, in fact, our "simple pleasures" are neither so easy nor so simple and are rooted in an evolutionary history rife with conflict. Our pleasures not only shepherd us, but lead us into

Measurement metric	Definition	Major field of usage
Elasticity of demand	Decline in visit rate as costs increase	Economics
Consumer surplus	Area under demand curve, e.g. visit price vs. visit number	Economics
Reservation price	Maximum price paid to reach resource	Experimental psychology
Total expenditure	Total price paid over the duration of each experiment	Behavioral ecology

Figure 5.1 The value an animal ascribes to a particular resource can be estimated experimentally. Left, a caged American mink (Mustela vison) with access to a water pool, a highly valued resource. Costs to pay were implemented by weighting doors to access various resources. Right, metrics used in this experiment to measure resource value. (Figure adapted, with permission, from Mason et al., 2001.)

temptation. By comparing the neural circuits activated in humans reporting pleasant experiences and animals performing identical tasks, we can shed some light onto the brain systems governing pleasure. Because homologous–neuronal circuits appear to govern similar types of decisions in both humans and animals, understanding reward processing in animals can serve as an effective and expedient model for how our own brains process pleasure. In this chapter, we examine in detail two different domains of reward and decision making in people and animals. Specifically, we focus on the allure of risk and the pleasures of social interaction.

Risky Pleasures—Gambling in Animals

Gambling is an ancient pastime and is common to all cultures in the modern world (McMillen, 1996). The prevalence of pathological gambling is estimated to be about 1% in the U.S. population (Shaffer et al., 1999), and over 80% of U.S. adults have participated in some form of commercial or state-sponsored gambling in their lifetimes (National Research Council, 1999). Various nonhuman animals also demonstrate distinct preferences when confronted with a gamble (Kacelnik and Bateson, 1996). Like humans, however, their preferences are influenced by the specific environmental context in which their decisions are made (see Platt and Huettel 2008 for a review).

Early ethological models of behavior assumed, for simplicity, that animals maintained complete knowledge of the environment and that reward contingencies were deterministic (see Stephens and Krebs 1986 for a review). In practice, however, uncertainty about environmental contingencies places strong constraints on behavior. The impact of uncertainty on choice has long been acknowledged in economics, which defines the spread of an outcome's known probability as risk. In the 18th century, Daniel Bernoulli proposed that the subjective (i.e., as determined by the economic agent) utility of a sum of money is a concave function of its face value (Figure 5.2). It follows directly from this idea that individuals will be generally risk-averse. The generality of risk aversion has been widely observed for both humans and animals of several species in several contexts. Importantly, however, this preference is not stable across all contexts (Kacelnik and Bateson, 1996; Weber et al., 2002) and often varies greatly amongst individuals (Weber et al., 2002).

For example, Caraco and colleagues (1980) observed that preferences of yellow-eyed juncos for risky and safe food options depend on their energetic state. In that study, birds were given the option of choosing a tray with a fixed number of millet seeds ("safe") or a tray with a probabilistically varying number of seeds with the same mean as the fixed option ("risky"). After 1 hour of food restriction, birds preferred the fixed option, but after 4 hours, they switched preference to the variable option.

One explanation for the switch from risk aversion to risk seeking is that, after the shorter interval of food deprivation, the gain from the fixed option was sufficient for the bird to maintain a positive energy budget. After the longer interval of food deprivation, however, energy stores were depleted, so the fixed option was no longer adequate to meet the animal's energy needs. When hungry, the bird's best chance for survival was to gamble on the risky option, since it offered at least the chance of a viable energetic yield. In an intriguing parallel, preference for a variable drug source over a fixed drug source was induced in drug addicts when they read scripts that simulated drug withdrawal, whereas this preference was reversed when the subjects read scripts that simulated drug satiation (Bickel et al., 2004). Similarly, the demographics of state lottery players in the United States—most prevalent amongst low-income adults lacking high-school diplomas (Clotfelter et al., 1999)—may reflect a facultative but adaptive response to economic hardship.

These findings suggest that risk preferences have evolved to motivate adaptive decision making. Neural circuits may generate contentment in order to promote

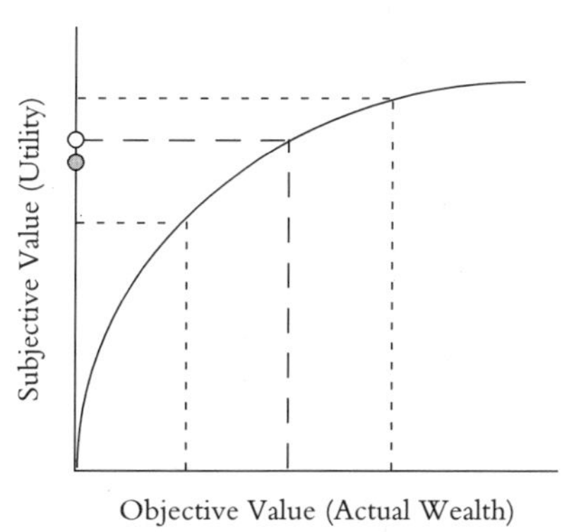

Figure 5.2 Concave utility curve. The subjective utility of a single fixed value (white dot, dashed line) is greater than the average utility from two variable values with the same mean objective value (gray dot, dotted lines). This indicates an aversion to risk.

stability within benign environmental conditions, or may generate restlessness, impulsivity, and risk-seeking when prospects are grim. According to this line of reasoning, animals under stress are more inclined to make risky decisions because the status quo presents a bleak outlook for evolutionary success. A key realization, in this context, is that economic models developed to explain the behavior of humans acting to maximize monetary returns can also describe the behavior of animals acting to maximize genetic fitness. These "neuroeconomic" models make explicit the common origins of and shared mechanisms underlying human and animal decision-making.

Neural Basis of Decision Making in Risky Contexts

A key idea in economics is that, for equal expected values, a strong preference for a risky or safe option implies a distinction between the objectively measured outcome (expected reward or punishment) and the subjectively experienced sensation (pleasure or discomfort). Similarly, the idiosyncratic preferences of individual organisms reflect a combination of inherited adaptations, learned experiences, and random variation. Thus, the precise way in which brains conflate or distinguish externally observable rewards and their subjective valuation is important for understanding the biological basis of behavioral decisions.

Recent studies have sketched out a basic portrait of the brain systems that contribute to decision making in risky contexts. For example, electrophysiological studies in animals indicate that tonic dopamine responses to reward cues vary with uncertainty; responses are maximal for a cue that is 50% valid and decrease with increasing reward predictability. Likewise, rapid phenylalanine/tyrosine depletion (RPTD), which reduces dopamine levels in ventral striatum (Mehta et al., 2005), significantly deflates the motivation of human subjects to bet money in a simulated gambling task (Roiser et al., 2005). These studies demonstrate that reward uncertainty, like reward prediction error, may influence dopaminergic signals that guide decision making.

Neuroimaging studies in humans have revealed that preference for a risky option is associated with increases in neuronal activity in the ventral striatum—a principal target of dopamine projections—and posterior parietal cortex (Kuhnen and Knutson, 2005; Preuschoff et al., 2006; Tom et al., 2007), and that the act of choosing a risky option activates the dorsal striatum and precuneus (Rustichini et al., 2005). Economists often distinguish between "risky" options, with known probabilities, and "ambiguous" options, in which the outcome probabilities are not known. The neural mechanisms mediating decision under risk or ambiguity were explored by Hsu and colleagues, and Huettel and colleagues (Hsu et al., 2005; Huettel et al., 2006). In the latter study, Huettel and colleagues probed choices between monetary gambles while human subjects were scanned using functional magnetic resonance imaging (fMRI). They found that subjects' preferences for ambiguity were correlated with activation in prefrontal cortex, while preferences for risk were correlated with activity in parietal cortex (Figure 5.3). This study was the first direct parametric link of economic preferences to activation of specific brain regions.

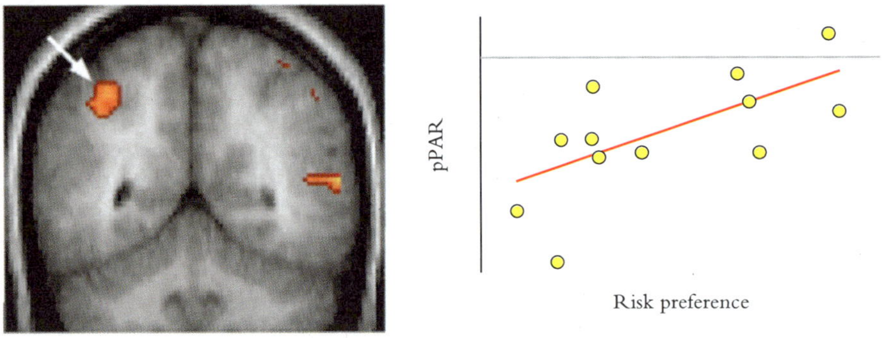

Figure 5.3 Subjective economic preferences predict changes in brain activation. Correlation between cortical activity in the posterior parietal cortex (see left, white arrow) and behavioral preference for risk. Right, each subject's mean preference value and fMRI parameter value (a measure of relative brain activity) is indicated by a single circle. Correlation = –0.62, $p < 0.00001$. (Figure adapted from Huettel et al., 2006.)

These studies, and others like them (see Platt and Huettel 2008 for a review), have identified brain areas that extract reward and probability information from the environment and translate this information into subjective value functions scaled by state variables such as satiety and perceived wealth. How such preferences are then mapped onto overt behavior, however, has received less attention from neurobiologists. Based on anatomical considerations, the posterior cingulate cortex (CGp), among other limbic areas, may contribute to the subjective evaluation of rewarding and punishing events by binding this information to action (Dean et al., 2004; McCoy et al., 2003). CGp is interconnected with visual, motor, and reward-related brain areas (Vogt et al., 1992). Moreover, CGp neurons respond not only after task events and actions (McCoy et al., 2003; Olson et al., 1996), but also after rewards are delivered or omitted, suggesting an evaluative role in behavior (McCoy et al., 2003).

To probe the role of CGp in behavioral evaluation and decision making, we studied the responses of CGp neurons when monkeys chose between a gamble and a sure bet with the same mean payoff (McCoy and Platt, 2005). Like people and other animals, rhesus monkeys show reliable risk preferences. Unlike people and many other animals, these monkeys favor a risky option even when it pays less, on average, than a safe option (Figure 5.4). CGp neurons closely reflect this behavioral bias by responding more strongly to risky options and following risky choices (McCoy and Platt, 2005). These data suggest that CGp plays a critical role in decisions made under uncertainty by binding subjective value to potential targets of action. Whether this binding process requires, or evokes, feelings of pleasure, as speculated above, remains unknown.

One question these data raise is why monkeys prefer the risky option, even when its objective value is less than the sure bet. One possibility is that monkeys focus on the large reward and ignore bad outcomes—a possibility suggested by prior behavioral studies in people and rats (Rachlin et al., 2000; Tversky and Kahneman, 1981), but heretofore unexplored in neurobiological studies of reward and decision making. We tested this hypothesis directly by examining the relationship between risk preference and delay between trials in monkeys (Hayden et al., 2007), and found that preference for a risky option declines with increasing delays between trials. These findings can be explained by "string theory" (Rachlin, 2000), which proposes that the salience of a large reward and the expected delay until that reward can be obtained together influence the valuation of a risky option. Similar processes have

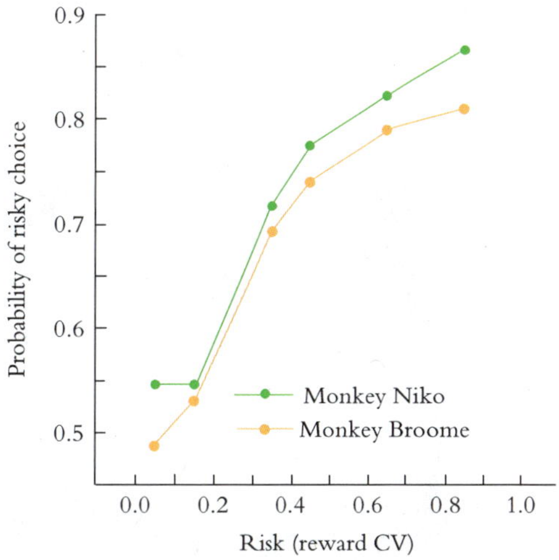

Figure 5.4 Macaque monkeys are sensitive to risk and prefer to gamble for a large juice reward rather than opt for a sure bet on a mid-sized reward. These two monkeys chose the risky target *more than* half the time, and the probability of choosing the risky target increased with increasing risk (logistic regression coefficients: Broome, 2.442, $p < 0.0000001$; Niko, 2.426, $p < 0.0000001$). CV = coefficient of variation of rewards associated with the risky target, a dimensionless measure of relative risk. (Figure adapted from McCoy and Platt, 2005.)

been proposed for humans who pursue the immediate intense "high" of drugs of abuse, while simultaneously discounting the delayed "low" of withdrawal (Bernheim and Rangel, 2004; Bickel et al., 1999).

Thus, assessment of affective responses to risk in humans may provide a basis for linking these neural processes to the pleasures and sorrows of gambling. Consistent with this interpretation, gambling addicts report greater levels of emotional arousal during baseline conditions than do casual gamblers and they also show greater increases in arousal while gambling (Brown et al., 2004). This general increase in arousal could potentially increase the incentive salience of the jackpot, thus inducing problem gamblers to persevere in the face of accumulating losses.

The Pleasure of Your Company: The Neurobiology of Friendship, Sex, and Companionship

As humans, we owe our evolutionary success, in no small part, to our drive to share resources with our companions. However, just as we have gained by cooperation with our conspecifics, we have also competed brutally with them. The human experience is thus quintessentially social, and just as the adaptive consequences of our social interactions have grown, so too must have the abilities of our brains to process the intrinsic rewards and risks of social interaction.

Despite the complexity of human social interaction, we share many fundamental elements of social behavior with our nonhuman kin. As discussed previously, the most thoroughly characterized mechanism for assigning outcome value to events and action is the mesolimbic dopamine system (Hyman and Malenka, 2001; Wise, 1987, 1998). Neurons in the ventral midbrain release dopamine into the ventral striatum and critically to the nucleus accumbens (NAcc). A vast literature shows that the activity of these dopaminergic cells, and the consequent release of dopamine in the NAcc, plays an essential role in mediating the hedonic value of primary reinforcers such as food or juice, as well as the motivation to respond to cues that these reinforcers will soon be available (Berridge and Robinson, 1998; Cromwell et al., 2005; McClure et al., 2003; Phillips et al., 2003; Roitman et al., 2004; Schultz et al., 1992; Wise, 2004).

Intriguingly, dopamine has also been shown to contribute to social reward and to motivate interaction with others. The best evidence for the role of dopamine in social reward processing comes from studies of long-term pair bonding—in small rodents called voles. In monogamous voles, males and females form strong pair bonds after mating, whereas in promiscuous voles, such pair bonds are not formed.

Pair bonding depends crucially on the action of dopamine within reward circuitry, including NAcc, medial prefrontal cortex (mPFC), and orbitofrontal cortex (OFC). Dopamine is associated with both affiliative and aggressive components of pair bonding by monogamous male voles (Keverne and Curley, 2004). In NAcc, D_2-type dopamine receptors are involved in affiliative behavior, whereas D_1-type receptors are involved in aggressive behavior (Aragona et al., 2006). Notably, monogamous voles, but not promiscuous voles, have high levels of oxytocin in NAcc and arginine vasopression in the ventral pallidum, the ventral forebrain, and the mesolimbic dopaminergic reward pathway. Dopamine receptors are also localized within the mPFC and OFC and these receptors appear to interact with oxytocin and vasopressin receptors during pair bonding (Smeltzer et al., 2006). Thus, work on rodents indicates that dopamine signaling in the NAcc, and possibly OFC, is critical for social motivation and social bonding. Similar findings hold for sheep (e.g. Kendrick, 2004), and several studies suggest that dopamine also makes specific contributions to reward circuitry mediating the complex social relationships of primates. Seeing an image of a loved one (Aron et al., 2005) or an attractive member of the opposite sex (Aharon et al., 2001; Kampe et al., 2001) activates dopaminergic brain areas in humans. Moreover, dysfunction of these circuits has been associated with both autism (Kendrick, 2004) and social anxiety (Schneier et al., 2002) disorders.

Although concrete goals such as food acquisition and mating clearly motivate behavior, abstract goals such as information gathering and affiliative interaction can be just as potent. Among group living species, including many primates, the dynamic social environment constitutes the behavioral context in which individuals pursue rewards, avoid punishments, evaluate risks, and make decisions. Many decisions are motivated by competitive and cooperative interactions with others in a social group; status not only constrains behavioral options, but constitutes a major target of goal-directed activity. In humans, recent research has revealed that cooperative transactions (Rilling et al., 2002) and opportunities to punish traitors (de Quervain et al., 2004) can be as motivating as primary rewards such as food and water. Indeed, the observation that humans will exert physical effort (Aharon et al., 2001) and forgo nonsocial rewards (Fehr and Gachter, 2002) to view and/or interact with others demonstrates that social stimuli have intrinsic reinforcement value.

The adaptive significance of navigating a complex social environment explains why social stimulation—an attractive smiling face, a cooperative transaction, the opportunity to punish a traitor—evokes neural activity in circuits that process other primary rewards. Physical features of the face, for example, provide information about genetic quality and affective attitude, and thus also about the quality and likelihood of potential alliances and mating opportunities. Similarly, grooming and poise can indicate both the intent and expectations of an observed individual. In a broad variety of situations, people use visually encoded information to assess the probability and tenor of potential interactions: for example, whether to expect avoidance, aid, or aggression from encountered individuals (e.g. Winston et al., 2002). Similarly, monkeys visually inspect members of their social group, preferentially investing in relationships with dominant individuals (Cheney and Seyfarth, 1990). Likewise, male primates often use visual cues to predict female mating receptivity (Hrdy and Whitten, 1987). These observations implicate a neural system linking social stimuli, such as faces, to the valuation functions guiding behavioral decision-making.

Consistent with such a system, a wealth of data indicates that primates find social stimuli to be intrinsically rewarding and that some types of social stimuli are more reinforcing than others (Anderson, 1998; Emery, 2000). Rewarding properties of social stimuli for both human and nonhuman primates can be inferred from studies of preferential looking. Adult monkeys spend more time looking at pictures of faces looking towards them than pictures with averted gaze (Keating and Keating, 1982), most likely because directed gaze signals elevated likelihood of interaction. Monkeys also appear to maximize information acquisition by directing their gaze more often toward higher-ranking than lower-ranking animals in group contexts (McNelis and Boatright-Horowitz, 1998) and, when viewing faces, by preferentially examining mobile, expressive features such as the eyes and mouth (Keating and Keating, 1982; Kyes et al., 1992). Similar social biases in orienting have been demonstrated in the visual scanning behavior of humans (Yarbus, 1967), with some biases arising quite early in development (e.g. Serrano et al., 1992).

These observations support the hypothesis that the primate brain has evolved mechanisms for attributing positive valence to social stimuli and that attention is guided by the reinforcing value of inspecting these stimuli. Deaner et al. (2005) explored this hypothesis quantitatively in the laboratory by developing an economic task in which male rhesus macaques were given a choice between two visual targets. Orienting to one target yielded fruit juice alone, while orienting to the other yielded both fruit juice and a picture of a familiar monkey. By systematically changing the juice payoffs for each target and the specific pools of images displayed, Deaner and colleagues were able to estimate the subjective value of different types of social and reproductive stimuli in a liquid currency. Their work revealed that male monkeys will forego larger juice rewards in order to view female sexual signals or the faces of high-ranking males, but require larger rewards to choose the faces of low-ranking males (Figure 5.5a).

The duration of looking at the images in each class hint at the complexity of social influences on decision making. Specifically, after drinking their juice, monkeys continued staring at female sexual signals for longer than they gazed at either high-ranking or low-ranking male faces (Figure 5.5b). Male monkeys may thus value the opportunity to view the faces of high-ranking males not because they are hedonically pleasing, but because they are potentially threatening and thus highly relevant for guiding behavior. That is to say, while looking may not be pleasurable, *not* looking may be even more uncomfortable. This implies that orienting decisions may activate reward machinery even when hedonic value is absent—another instance, perhaps, of distinction between "craving" and "enjoyment." It is important to note that negative affect has been associated with approach behavior in previous studies. For example, anger generates behavioral approach paired with strongly negative affect (Ohman et al., 2001).

A subsequent study by Klein et al. (2008) investigated how these social rewards influence orienting decisions. In this experiment, the authors recorded the responses of neurons in the macaque lateral intraparietal area (LIP) to images while monkeys performed the aforementioned "pay per view" task. Klein and colleagues found that LIP neurons responded most strongly when monkeys chose to view images of female sexual signals, less strongly when they chose to view images of the faces of dominant males, and least of all on the rare occasions when they chose to view the faces of subordinate males (Figure 5.5c). Thus, LIP neurons, which help determine where a monkey will next shift his gaze, represent the intrinsic value of the social information that can be gained by orienting toward a particular part of space. Interestingly, LIP represents this information in much the same way that it represents primary reinforcers when eye movements are explicitly rewarded with juice (e.g. see Platt and Glimcher, 1999). Together, these results endorse the idea that the primate reward system is organized, in part, to adaptively acquire valuable social information.

Recently, we have shown that humans, like monkeys, process social rewards in an economic fashion. Specifically, humans will pay more to view pictures of attractive members of the opposite sex than to view pictures of unattractive ones, even when the reward cues are implicit (Hayden et al., 2007). This paradigm also allowed us to estimate the subjective economic value of attending to an attractive member of the opposite sex. Specifically, men placed a value of around half a cent (U.S. dollar) for the opportunity to view an attractive woman, whereas the value women placed on the opportunity to view attractive men was not significantly different from zero (Figure 5.6). Importantly, we found that the opportunity to view members of the opposite sex was discounted temporally and was

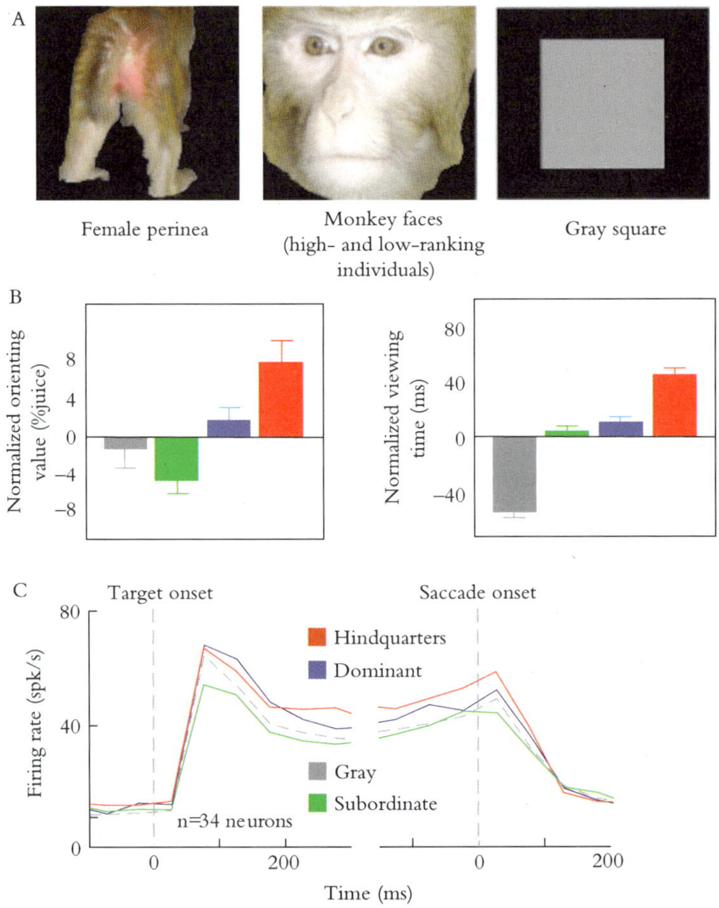

Figure 5.5 Monkeys value visual signals of status and sex, and parietal cortex signals the value of these images in the visual scene. (A) Example images shown to monkeys during a "pay per view" task used to assess valuation of socially-relevant visual images. (B) Mean normalized orienting values (left) and looking times (right) for various image classes. Orienting values are significantly higher for both the perinea (red bar) and high-status faces (blue bar) in contrast to either the low-status faces (green) or gray square (gray). Although the monkeys choose to orient more frequently to the high-status faces than the low-status faces, the length of time they gaze at either of these image classes are both shorter than the time they spend viewing the perinea. (Adapted from Deaner et al., 2005.) (C) Peristimulus time histrogram of 34 LIP neurons recorded during the "pay per view" task. Note that activity associated with high-value images, such as female perinea and dominant faces, is consistently greater than that associated with low-value subordinate face images. (Adapted from Klein and Platt, 2008.)

also exchangeable for effort. Together, these findings implicate a generic neural system that mediates decisions about abstract rewards, including social ones. Consistent with this idea, we have found that reward-related brain areas, including ventral striatum and OFC, signal the subjective value of social and monetary rewards in a similar fashion (Smith et al., submitted).

While the exact mechanisms relating social and nonsocial rewards remain obscure, their actions in any particular species surely reflect prior evolutionary pressures. Most nonhuman primates live in large, structured social groups, characterized by dominance hierarchies and extensive aggressive and affiliative interactions—a social organization that has become greatly elaborated in our own species. It seems plausible, given the structure of society, that human pleasure in some social interactions has been similarly expanded and enriched. While the neural mechanisms of such experiences as companionship, pride, and love

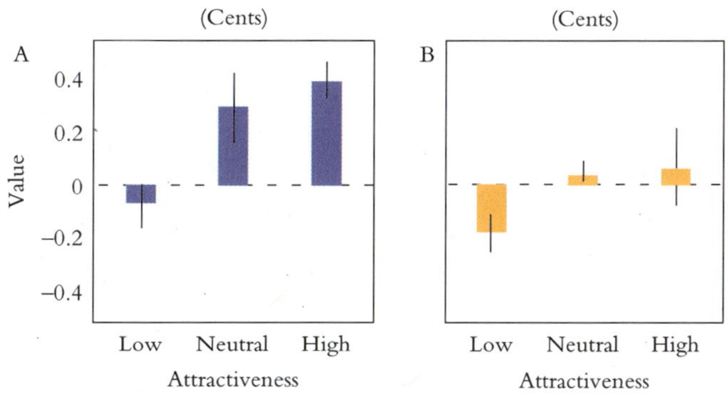

Figure 5.6 Men, but not women, implicitly value images of the opposite sex. (A) Average point of subjective equivalence (PSE) for all males, expressed in terms of equivalent monetary price (U.S. cents). Males valued the opportunity to view attractive and neutral females, but required a small (nonsignificant) payment to view unattractive females. Bars indicate 1 S.E.M. estimated by jackknife. (B) Average PSEs for all females in population. Females did not significantly value the opportunity to view attractive and neutral males, and required a payment to view unattractive males. (Figure adapted from Hayden et al., 2007.)

remain to be discovered, it seems assured that the sensitivity of human decisions to social context reflects an evolved specialization of primate reward systems for appropriately motivating and regulating behavior in complex social groups (Insel, 2003).

Conclusions

Human motivations reflect our evolutionary past and our personal experiences, and are deeply writ in the architecture of our brains. All these forces shape our experience of pleasure, and to a large extent, they are shared not only with other individuals of our own species, but with individuals of other closely related species. By understanding the neural mechanisms mediating preferences, and by extension pleasure, in humans and other animals, we can infer the way that evolution has shaped behavior and the biological mechanisms that govern it.

If one were to index the "simple pleasures" that characterize the human experience, we suspect the list would be surprisingly complex. But despite these complexities—or perhaps because of them—we believe it is productive to adopt a neuroethological perspective. Animal models are one of our most powerful means for understanding not only the mechanisms of behavior, decision, and learning, they are an integral part of understanding the broader biological landscape in which these phenomena arise.

It is for other chapters to elaborate, in more detail, the varieties of human pleasure, ranging from physical rewards like food and wine to abstract rewards like information and novelty. Our evolutionary history, however, has shaped all of these things—from the pleasures of sex to the pleasures of song—and it is outside ourselves, as much as inward, that we must look to fully understand them.

References

Aharon, I., Etcoff, N., Ariely, D., Chabris, C. F., O'Connor, E. and Breiter, H. C. (2001) Beautiful faces have variable reward value: fMRI and behavioral evidence. *Neuron* 32, 537–551.

Anderson, J. R. (1998) Social stimuli and social rewards in primate learning and cognition. *Behav.Processes* 42, 159–175.

Aragona, B. J., Liu, Y., Yu, Y. J., Curtis, J. T., Detwiler, J. M., Insel, T. R. and Wang, Z. X. (2006) Nucleus accumbens dopamine differentially mediates the formation and maintenance of monogamous pair bonds. *Nat. Neurosci.* 9, 133–139.

Aron, A., Fisher, H., Mashek, D. J., Strong, G., Li, H. and Brown, L. L. (2005) Reward, motivation, and emotion systems associated with early-stage intense romantic love. *J. Neurophysiol.* 94, 327–337.

Bernheim, B. D. and Rangel, A. (2004) Addiction and cue-triggered decision processes. *Am. Econ. Rev.* 94, 1558–1590.

Berridge, K. C. and Robinson, T. E. (1998) What is the role of dopamine in reward: Hedonic impact, reward learning, or incentive salience? *Brain Res. Brain Res. Rev.* 28, 309–369.

Bickel, W. K., Giordano, L. A. and Badger, G. J. (2004) Risk-sensitive foraging theory elucidates risky choices made by heroin addicts. *Addiction* 99, 855–861.

Bickel, W. K., Odum, A. L. and Madden, G. J. (1999) Impulsivity and cigarette smoking: Delay discounting in current, never, and ex-smokers. *Psychopharmacology* 146, 447–454.

Brown, S. L., Rodda, S. and Phillips, J. G. (2004) Differences between problem and nonproblem gamblers in subjective arousal and affective valence amongst electronic gaming machine players. *Addict. Behav.* 29, 1863–1867.

Caraco, T., Martindale, S. and Whittam, T. S. (1980) An empirical demonstration of risk-sensitive foraging preferences. *Anim. Behav.* 28, 820–830.

Cheney, D. L. and Seyfarth, R. M. (1990) *How Monkeys See the World: Inside the Mind of Another Species.* University of Chicago Press: Chicago, IL, 377pp.

Clotfelter, C. T., Cook, P. J., Edell, J. A. and Moore, M. (1999) State lotteries at the turn of the century. National Gambling Impact Study Commission.

Crane, E. (1999). *The World History of Beekeeping and Honey Hunting.* Routledge: New York.

Cromwell, H. C., Hassani, O. K. and Schultz, W. (2005) Relative reward processing in primate striatum. *Exp. Brain Res.* 162, 520–525.

de Quervain, D. J., Fischbacher, U., Treyer, V., Schellhammer, M., Schnyder, U., Buck, A. and Fehr, E. (2004) The neural basis of altruistic punishment. *Science* 305, 1254–1258.

Dean, H. L., Crowley, J. C. and Platt, M. L. (2004) Visual and saccade-related activity in macaque posterior cingulate cortex. *J. Neurophysiol.* 92, 3056–3068.

Deaner, R. O., Khera, A. V. and Platt, M. L. (2005) Monkeys pay per view: Adaptive valuation of social images by rhesus macaques. *Curr. Biol.* 15, 543–548.

Emery, N. J. (2000) The eyes have it: the neuroethology, function and evolution of social gaze. *Neurosci. Biobehav. Rev.* 24, 581–604.

Fehr, E. and Gachter, S. (2002) Altruistic punishment in humans. *Nature* 415, 137–140.

Halgren, E., Walter, R. D., Cherlow, D. G. and Crandall, P. H. (1978) Mental phenomena evoked by electrical-stimulation of human hippocampal formation and amygdala. *Brain* 101, 83–117.

Hayden, B. Y., Parikh, P. C., Deaner, R. O. and Platt, M. L. (2007) Economic principles motivating social attention in humans. *Proc. Biol. Sci.* 274, 1751–1756.

Holy, T.E., and Guo, Z.S. (2005). Ultrasonic songs of male mice. *PLoS. Biol.* 3, 2177–2186.

Hrdy, S. B. and Whitten, P. L. (1987) Patterning of sexual activity. In: *Primate Societies.* Ed. D. L. C. B.B. Smuts, R.M. Seyfarth, R.W. Wrangham and T.T. Struhsaker. University of Chicago Press: Chicago, IL.

Hsu, M., Bhatt, M., Adolphs, R., Tranel, D. and Camerer, C. F. (2005) Neural systems responding to degrees of uncertainty in human decision-making. *Science* 310, 1680–1683.

Huettel, S. A., Stowe, C. J., Gordon, E. M., Warner, B. T. and Platt, M. L. (2006) Neural signatures of economic preferences for risk and ambiguity. *Neuron* 49, 765–775.

Hyman, S. E. and Malenka, R. C. (2001) Addiction and the brain: the neurobiology of compulsion and its persistence. *Nat. Rev. Neurosci.* 2, 695–703.

Insel, T. R. (2003) Is social attachment an addictive disorder? *Physiol. Behav.* 79, 351–357.

Kacelnik, A. and Bateson, M. (1996) Risky theories—The effects of variance on foraging decisions. *Am. Zool.* 36, 402–434.

Kampe, K. K., Frith, C. D., Dolan, R. J. and Frith, U. (2001) Reward value of attractiveness and gaze. *Nature* 413, 589.

Keating, C. F. and Keating, E. G. (1982) Visual scan patterns of rhesus monkeys viewing faces. *Perception* 11, 211–219.

Kendrick, K. M. (2004) The neurobiology of social bonds. *J. Neuroendocrinol.* 16, 1007–1008.

Keverne, E. B. and Curley, J. P. (2004) Vasopressin, oxytocin, and social behaviour. *Curr. Opin. Neurobiol.* 14, 777–783.

Klein, J. T., Deaner, R.O., and Platt, M.L. (2008). Neural correlates of social target value in macaque parietal cortex. *Curr. Biol.* 18, 419–424.

Kuhnen, C. M. and Knutson, B. (2005) The neural basis of financial risk taking. *Neuron* 47, 763–770.

Kyes, R. C., Mayer, K. E. and Bunnell, B. N. (1992) Perception of stimuli presented as photographic slides in cynomolgus macaques (*Macaca fascicularis*). *Primates* 33, 407–412.

Li, X., Li, W., Wang, H., Cao, J., Maehashi, K., Huang, L., Bachmanov, A. A., Reed, D. R., Legrand-Defretin, V., Beauchamp, G. K. and Brand, J. G. (2005) Pseudogenization of a sweet receptor gene accounts for cats' indifference toward sugar. *PLoS Genet.* 1, e3.

Mason, G. J., Cooper, J. and Clarebrough, C. (2001) Frustrations of fur-farmed mink. *Nature* 410, 35–36.

McClure, S. M., Berns, G. S. and Montague, P. R. (2003) Temporal prediction errors in a passive learning task activate human striatum. *Neuron* 38, 339–346.

McCoy, A. N. and Platt, M. L. (2005) Risk-sensitive neurons in macaque posterior cingulate cortex. *Nat. Neurosci.* 8, 1220–1227.

McCoy, A. N., Crowley, J. C., Haghighian, G., Dean, H. L. and Platt, M. L. (2003) Saccade reward signals in posterior cingulate cortex. *Neuron* 40, 1031–1040.

McMillen, J. (1996) *Gambling Cultures: Studies in History and Interpretation.* Routledge: London.

McNelis, N. L. and Boatright-Horowitz, S. L. (1998) Social monitoring in a primate group: the relationship between visual attention and hierarchical ranks. *Anim. Cogn.* 1, 65–69.

Mehta, M. A., Gumaste, D., Montgomery, A. J., McTavish, S. F. and Grasby, P. M. (2005) The effects of acute tyrosine and phenylalanine depletion on spatial working memory and planning in healthy volunteers are predicted by changes in striatal dopamine levels. *Psychopharmacology (Berl.)* 180, 654–663.

Nagel, T. (1974) What is it like to be a bat? *Philos. Rev.* 83(4), 435–450.

National Research Council (1999) *Pathological Gambling: a Critical Review.* National Academy Press: Washington, D.C.

Ohman, A., Lundqvist, D. and Esteves, F. (2001) The face in the crowd revisited: a threat advantage with schematic stimuli. *J. Pers. Soc. Psychol.* 80, 381–396.

Olds, J. and Milner, P. (1954) Positive reinforcement produced by electrical stimulation of septal area and other regions of rat brain. *J. Comp. Physiol. Psychol.* 47, 419–427.

Olson, C. R., Musil, S. Y. and Goldberg, M. E. (1996) Single neurons in posterior cingulate cortex of behaving macaque: Eye movement signals. *J. Neurophysiol.* 76, 3285–3300.

Ophir, A. G., Phelps, S. M., Sorin, A.B. and Wolff, J.O. (2008). Social but not genetic monogamy is associated with greater breeding success in prairie voles. *Anim. Behav.* 75(3), 1143–1154.

Phillips, P. E., Stuber, G. D., Heien, M. L., Wightman, R. M. and Carelli, R. M. (2003) Subsecond dopamine release promotes cocaine seeking. *Nature* 422, 614–618.

Platt, M. L. and Glimcher, P. W. (1999) Neural correlates of decision variables in parietal cortex. *Nature* 400, 233–238.

Platt, M.L., and Huettel, S.A. (2008). Risky business: the neuroeconomics of decision making under uncertainty. *Nat. Neurosci.* 11, 398–403.

Preuschoff, K., Bossaerts, P. and Quartz, S. R. (2006) Neural differentiation of expected reward and risk in human subcortical structures. *Neuron* 51, 381–390.

Quine, W. V. (1968) *Ontological Relativity and Other Essays.* Columbia University Press: New York.

Rachlin, H. (2000) *The Science of Self-Control.* Harvard University Press: Cambridge, MA.

Rachlin, H., Brown, J. and Cross, D. (2000) Discounting in judgments of delay and probability. *J. Behav. Decis. Making* 13, 145–159.

Rilling, J., Gutman, D., Zeh, T., Pagnoni, G., Berns, G. and Kilts, C. (2002) A neural basis for social cooperation. *Neuron* 35, 395–405.

Rogers, R. D., Tunbridge, E. M., Bhagwagar, Z., Drevets, W. C., Sahakian, B. J. and Carter, C. S. (2003) Tryptophan depletion alters the decision-making of healthy volunteers through altered processing of reward cues. *Neuropsychopharmacology* 28, 153–162.

Roiser, J. P., McLean, A., Ogilvie, A. D., Blackwell, A. D., Bamber, D. J., Goodyer, I., Jones, P. B. and Sahakian, B. J. (2005) The subjective and cognitive effects of acute phenylalanine and tyrosine depletion in patients recovered from depression. *Neuropsychopharmacology* 30, 775–785.

Roitman, M. F., Stuber, G. D., Phillips, P. E., Wightman, R. M. and Carelli, R. M. (2004) Dopamine operates as a subsecond modulator of food seeking. *J. Neurosci.* 24, 1265–1271.

Rustichini, A., Dickhaut, J., Ghirardato, P., Smith, K. and Pardo, J. V. (2005) A brain imaging study of the choice procedure. *Games Econ. Behav.* 52, 257–282.

Schneier, F. R., Blanco, C., Antia, S. X. and Liebowitz, M. R. (2002) The social anxiety spectrum. *Psychiatr. Clin. North Am.* 25, 757–774.

Schoenbaum, G., Chiba, A. A. and Gallagher, M. (1998) Orbitofrontal cortex and basolateral amygdala encode expected outcomes during learning. *Nat. Neurosci.* 1, 155–159.

Schultz, W. (2000) Multiple reward signals in the brain. *Nat. Rev. Neurosci.* 1, 199–207.

Schultz, W., Apicella, P., Scarnati, E. and Ljungberg, T. (1992) Neuronal activity in monkey ventral striatum related to the expectation of reward. *J. Neurosci.* 12, 4595–4610.

Schultz, W., Dayan, P. and Montague, P. R. (1997) A neural substrate of prediction and reward. *Science* 275, 1593–1599.

Serrano, J. M., Iglesias, J. and Loeches, A. (1992) Visual discrimination and recognition of facial expressions of anger, fear, and surprise in 4- to 6-month-old infants. *Dev. Psychobiol.* 25, 411–425.

Shaffer, H. J., Hall, M. N. and Vander Bilt, J. (1999) Estimating the prevalence of disordered gambling behavior in the United States and Canada: a research synthesis. *Am. J. Public Health* 89, 1369–1376.

Skinner, B. F. (1938) *The Behavior of Organisms.* Appleton-Century: New York.

Smeltzer, M. D., Curtis, J. T., Aragona, B. J. and Wang, Z. (2006) Dopamine, oxytocin, and vasopressin receptor binding in the medial prefrontal cortex of monogamous and promiscuous voles. *Neurosci. Lett.* 394, 146–151.

Stephens, D. W. and J. R. Krebs (1986) *Foraging Theory.* Princeton University Press: New Jersey, NJ.

Thorndike, E. L. (1898) Animal intelligence: an experimental study of the associative processes in animals. In: *Psychological Review Monograph Supplement.* Macmillan: New York.

Tom, S. M., Fox, C. R., Trepel, C. and Poldrack, R. A. (2007) The neural basis of loss aversion in decision-making under risk. *Science* 315, 515–518.

Tremblay, L. and Schultz, W. (2000) Modifications of reward expectation-related neuronal activity during learning in primate orbitofrontal cortex. *J. Neurophysiol.* 83, 1877–1885.

Tversky, A. and Kahneman, D. (1981) The framing of decisions and the psychology of choice. *Science* 211, 453–458.

Vogt, B. A., Finch, D. M. and Olson, C. R. (1992) Functional heterogeneity in cingulate cortex: the anterior executive and posterior evaluative regions. *Cereb. Cortex* 2, 435–443.

Weber, E. U., Blais, A.-R. and Betz, N. E. (2002) A domain-specific risk-attitude scale: Measuring risk perceptions and risk behaviors. *J. Behav. Decis. Making* 15, 263–290.

Winston, J. S., Strange, B. A., O'Doherty, J. and Dolan, R. J. (2002) Automatic and intentional brain responses during evaluation of trustworthiness of faces. *Nat. Neurosci.* 5, 277–283.

Wise, R. A. (1987) The role of reward pathways in the development of drug dependence. *Pharmacol. Ther.* 35, 227–263.

Wise, R. A. (1998) Drug-activation of brain reward pathways. *Drug Alcohol Depend.* 51, 13–22.

Wise, R. A. (2004) Dopamine, learning and motivation. *Nat. Rev. Neurosci.* 5, 483–494.

Yarbus, A. L. (1967) *Eye Movements and Vision.* Plenum: New York.

PART II

HUMAN PLEASURES

6

On the Nature and Function of Pleasure

NICO H. FRIJDA

Pleasure is a fundamental notion in the description and explanation of emotion. Some investigators consider it, together with pain, as their core: All emotions can be regarded as variants of pleasure or pain, and with variations in activation (Russell, 2003). Also, human striving is often said to follow the "pleasure principle": all striving would aim at obtaining pleasure or at decreasing pain.

But what is pleasure? It is sometimes considered *ineffable*—impossible to be described in words. Yet, it evidently contains sufficient cues to enable children to learn using the word systematically.

In psychology, pleasure has become a technical term. It denotes a positive evaluation of sensations, objects, one's movements, people, and events that underlies use of a number of epithets like "pleasant," "agreeable," "liked," "likeable," "attractive," "nice," and more recently, "cool." It also allows considering a number of emotions as "positive emotions" (Fredrickson, 2001).

Further specification appears difficult and this is not surprising. Pleasure occurs in a number of different guises. One of these is that it occurs as a subjective state: that of having a "feeling of pleasure." Examples are the felt response to hearing that one's football team has won its match, and the feeling that arises when, after a day of physical exertion, one sinks down in a warm bath and finally relaxes: "it gave me a feeling of great pleasure." Such feelings clearly imply self-reflexivity.

Note that feelings of pleasure do not stand alone. They are about something. Pleasure and displeasure standing alone "are nothing more than convenient abstractions" (Ruckmick, 1936). This is even more empathically so in the second guise of pleasure. Most pleasure occurs in the form of an experienced property of sensations, objects, actions, people, or events (Arnold, 1960; Ruckmick, 1936; Wundt, 1896). Introducing his summary of the introspective feeling research, Beebe-Center (1932, p.1) wrote: "When confronted with actual objects, people rarely use the terms pleasantness and unpleasantness. Instead, they use the corresponding adjectives: "this perfume is very pleasant"; "this weather is unpleasant." Pleasure generally occurs as a "niceness gloss" attached to its object. Such a gloss can also be attached to body feelings: that agreeable sense of relaxation, for instance, or that nice sense of bodily vigor. Displeasure likewise mostly occurs as "badness gloss." Beebe-Center (1932) caught both together under the heading of the "hedonic tone" of perceived stimuli.

In a third guise, pleasure is embedded in emotions; the same holds for displeasure. It is useful to distinguish between affects (hedonic tone, pleasure, and pain) and emotions. The distinction is common (Arnold, 1960; Dumas, 1948; Gemelli, 1949; Scherer, 2005). Affects are states or experiences of pleasure or pain (or of mixed feeling). Emotions are multicomponential responses (Scherer, 2005) to stimulus events as appraised on the basis of spatiotemporal context, and include action tendency or some other modification of action readiness (Frijda, 2007). Niceness or badness gloss of stimuli, actions, or events co-occurs with awareness of one's action impulse or loss of motivation. An experience of joy can be described as awareness of a pleasant event merged with a felt urge for exuberant, enthusiastic, and often gratuitous interaction with the environment (Davitz, 1969; Frijda, 1986); jumping, shouting, singing, and other playful behavior illustrate the latter. Inclination to interact may expand into readiness to "broaden and build" relationships with others and the environment (Fredrickson, 2001). Feedback from exuberant action and interaction colors the feeling. Hence the prominence of body feelings in introspective accounts of many pleasures. However, the experience is not of the body only. It is experience of bodily action that modulates one's relation with an external object or an object of thought.

The modulation of experienced relationship applies to different "positive emotions" in different fashions. Pride includes pleasant awareness of being exposed to the appreciating glance and judgment of others and proudly exposing oneself to it. Admiration and awe include turning receptive attention to a target that gives pleasure by its value and for the moment humbling oneself before it (Keltner and Haidt, 2003). And so forth.

Whatever the guise, the question recurs: what does "pleasure," "pleasantness," or "niceness" mean? What is the nature of the gloss given those names? Is it a kind of experience that differs from sensory awareness? And if so, what properties distinguish it from neutral awareness and pleasantness from unpleasantness?

These questions have been thoroughly examined and hotly debated at the time that conscious experience was regarded as composed of mental elements—the time of Wundt (1896) and Titchener (1908). Should a further kind of element—feelings—be added to sensations, images, and (somewhat later) thoughts?

Experiments consisted of presenting tones, colors, odors, pieces of music, or more complex stimuli. The subjects, trained in introspection, were asked to abstract from the sensory contents as such and to concentrate on the accompanying experiences. They described the stimuli as pleasant or unpleasant. These judgments came easily and naturally (Alechsieff, 1907). Yet, it proved difficult to specify them. As one observer said: "I cannot find anything I left out, and I know just as little as at the beginning why I like some tones better than others" (Nakashima, 1909, p.166). Still, the observed experiences did have properties that set them apart from sensory impressions. I list some of them, as included in a listing by Koch (1913) (cited from Beebe-Center, 1932, p. 94): (1) Feelings could not be localized in or outside the body. (2) They were independent of sense organs. (3) They could not be referred to specifiable properties of the stimuli. (4) They did not stand alone but were experienced as properties of the stimuli. (5) They differed from body sensations that often accompanied them, but not always.(6) They were "evanescent" (I think the term was introduced by Titchener), that is, they tended to disappear when attention was directed at them rather than at the stimulus to which they adhered.

Discussion arose between Wundt and Titchener whether there existed three kinds of feeling (pleasant/unpleasant ones, calm/excited ones, and tense/relaxed ones), or only the first kind; the discussion was never resolved.

A blow to the notion of "affect" or "feeling" as a special kind of experience was delivered by Nafe (1924), a pupil of Titchener. In his experiments, "from first to last, the reports are made in terms of touch" (Nafe, 1924, p. 517). Pleasure, he concluded, is "bright pressure," and displeasure is "dull pressure." Others (e.g., Young, 1927) could not generally confirm these observations. Subjects from Cornell University did confirm, but from other laboratories they did not.

Nafe's work strengthened the opinion of those who thought that hedonic tone could be identified with body sensations. Note, however, that James (1894) was not of the same opinion. Emotional feeling may consist of peripheral feedback, but, in his view, pleasure and displeasure as such do not (James, 1894, p.523). Anyway, conclusions in introspective research kept on differing. They appeared to depend strongly on the observers' prior opinions (Arnold, 1960; Beebe-Center, 1932) and on their attentional set: "In order to experience these pressures, one 'must maintain a psychological nonperceptive attitude"' (Nafe, 1924, p. 540).

This bit of history serves to clarify what the experience of pleasure consists of. Most of the conclusions to follow are given by Arnold (1960) in her very careful survey of the summarized research:

First, pleasure is not some kind of sensation. Second, it never stands alone but always is a "gloss" to some other experience, including one's general condition. Third, pleasure is certainly not a *quale*, a specific irreducible experience. It does not even have a characteristic phenomenology, as has the color red or the experience of bright pressure. Fourth, even an experience of "bright pressure" has something of this gloss. "Bright" and "dull" distinctly produce different semantic differential ratings. Fifth, examination of feelings of pleasure is severely hampered by the feeling's evanescence. Attention directed to it rather than to the pleasant object or sensation makes it vanish.

Is this a surprising conclusion for a notion so generally held to be the core of emotions? I don't know. Contrary to what one might believe, "pleasure" is not a lexical universal. It does not occur in all languages (Wierzbicka, 1999). "Good" and "bad" do. So do "feeling good," "feeling bad," and "X feels good (or bad)." "Pleasure" in the generic sense that the word has acquired in English does not have an equivalent in a number of languages. Wierzbicka reports (personal communication) that it does exist in Polish, but not in Russian. Neither, I think, does it exist in French, where *plaisir* has a more or less specific overtone of pleasurable gustatory or sexual experience, more akin to the English "lust."

In this chapter, I take "pleasure" to denote "feeling good" or the experiential property that some object feels good. Many different words are used for such valued experience besides "pleasure," such as "nice" or "cool." They all are used as epithets for objects, persons, or events that psychology treats as giving rise to pleasure.

The Pleasure Process

But still, what do these various words denote? In fact, some subjects in the mentioned experiments did give a clue. "When I say 'pleasant' it doesn't stand for anything more than 'I would smell it more if I could.' I am not able to discriminate anything beside that in most cases" reported one subject. A different subject, when shown a 52-color spectrum band reported: "I perhaps found some pleasure in inspecting the colors although I don't know what that means when I make use of that expression except that it is the verbal equivalent for continuing to look" (Young, 1927, pp. 180–181).

These remarks are consonant with the main reasons for ascribing pleasure or feeling good about to someone else and even to some animal. These reasons are several. First, the individual welcomes the stimulus. He or she allows the stimulus to affect him or her and may prolong the interaction. Stimuli rated as interesting are watched for a longer time than uninteresting ones, for instance (Berlyne, 1971).

Second, he or she shows "acceptance wriggles": movements (usually classed among "consummatory responses") that intensify or expand perception of the stimulus, its appearance or taste, or whatever of the stimulus. In Berridge's (2000) experiments with mice, their eyes twinkle, tongues protrude, and mouths open when faced with preferred foods; the movements are interpreted as signs of 'liking'. Each sort of pleasure generates its own acceptance wriggles that can form extensive repertoires. Take relishing wine: it includes sipping, letting the wine circulate in one's mouth, curving one's tongue so as to maximize the sensation. Acceptance wriggles explore the variety of stimulus aspects contained in the stimulus and its context: the flavors of the wine, its "robe," its bouquet, and the structure of its perfume. Exploration renders each kind of pleasure different from any other in further aspects than the sensory ones only, because acceptance wriggles generate different subject–object relationships. Touching implies intimate proximity which smelling does not, for instance.

The notion of acceptance wriggles may be extended to turning an event over and over in one's mind. The wriggles blend over in active efforts to continue interaction and to continuing to engage in interaction—as in "optimal experiences" when working on engaging intellectual, artistic, manual tasks, or manual skills, which shows in eagerness to return to work after interruption, the impatience to have one's next thought completed, the engagement with which one pens down one's next word when writing. These are similar to the eagerness with which one may gulp down one's glass of wine, or, by contrast, withhold gulping it down for savoring it.

Third, the individual may indicate an acceptance stance by signals that perhaps are derived from signals of social acceptance. In humans, this is the smile. The smile is not only, or perhaps not even mainly, an expression of pleasure. The smile or some variant of it (Garotti et al., 1993) is a sign of social acceptance or a request for being accepted (Van Hooff, 1972). In dogs, tail-wagging appears to have the same functions (Coren, 2001) as do the trembling whiskers of a mouse (Berridge, 2000).

Fourth, pleasure facilitates the actions that engender it, turning encountered difficulties to welcomed challenges such as the pleasures of dancing, solving cross word puzzles, and encountering novel objects. Extensive research has shown that it facilitates cognitive and social actions generally, such as helping others and cognitive fluency (Isen, 2004).

Many or all of the earlier mentioned behaviors are prominent during savoring: actions for which maximizing pleasure forms the explicit goal and extends to reflexive awareness of the obtained pleasure (Frijda and Sundararajan, 2007).

Fifth, feeling intensity, attention, speed and amplitude of reactions, and duration of commerce with the pleasant object vary in magnitude; pleasure is a graded property. The magnitude of sensory pleasure depends on stimulus intensity in nonmonotonous fashion. It follows the "Wundt curve" (Berlyne, 1971) and increases to a maximum and then decreases, flipping over into unpleasantness. Nonsensory pleasures may show a "sawtooth" curve (Brehm, 1999; Miron et al., 2007). Their magnitude varies with the number of concerns or meanings involved (Frijda, 2007), with unexpectedness (Ortony et al., 1988), reference standards (Parducci, 1995), degree of habituation (Scitovsky, 1976), and motive state (e.g., hunger, thirst, specific nutritional deficits, duration since last sexual satisfaction).

These various phenomena lead to a functional definition of pleasure, or of feeling good or feeling good about something. Pleasure is the sense of accepting some stimulus, event, action, interaction, or personal state. Pleasantness is the perceived property of that stimulus or other object of inviting you to accept it, or one's sensed willingness to have that stimulus or other object continue.

Accepting and acceptance and willingness to continue imply that pleasure is a stable state. It forms a state of harmony, in which there is no felt or operative need for change. During pleasure—that is, during unmitigated pleasure—things are good as they are. This, of course, represents its major contrast to pain. The latter is defined by willingness to change. As such, change would be welcomed. I add "as such," because pain may be sought, for the subsequent satisfactions for which pain is a means, such as honoring one's God.

The feeling of pleasure, for all its ineffability, reflects the absence of need to change. It is a pointer to stability and actual acceptance, with its preparedness for acceptance wriggles.

Conceiving of feelings in this way may be unusual. It conflicts with a general sensory bias in conceiving them. I think there is some implicit presupposition that all experience should be of some sort of sensory nature. That, I think, is sheer dogma. No theory of conscious awareness posits it as necessarily true, for the simple reason that no such theory exists. How any neural process, including those from sense impressions, turns into conscious awareness is as yet simply a mystery (Chalmers, 1996).

Such sensory bias is manifestly incorrect. It was hotly debated in the discussions about "imageless thought" in the early 20th century (see Humphrey, 1951). There is manifold evidence. One can be aware of "affordances" for action (Gibson, 1979); before any motor involvement actually takes place (Nöe, 2004); depth perception and distance perception are cases in point. One also can be aware of inclinations and desires, long before they generate motor nerve innervations. Some central processes appear as such to give rise to conscious experience. Nonverbal awareness of what someone else feels, through activation of one's mirror neurons, also appears to be contingent on that latter central activation only. The notion of feelings as pointers, moreover, occurs elsewhere than with respect to affective feelings or perceived hedonic qualities (and is the core of Dennett's conception of awareness). There exist "feelings of knowing." "Oxymoron? Just wait a minute, I do know what that word means, yes, here it is. . . ." (see Bühler, 1907 for evidence with regard to the awareness of what, for instance, a proverb means; Frijda, 2007 for a very brief summary, Humphrey, 1951, p.287, for an extensive one).

The word "pointer" indicates that the information that feelings reflect can become manifest under appropriate conditions. One such manifestation is the action that results from activated dispositions for acceptance wriggles and other actions. Another is image-like or propositional explicitation as in "Oh, yes, *idiot savant* indeed is one" or, further, when you say: "Oxymoron? Well, eh, that, eh, is...a figure of speech that combines two normally contradictory terms." Similarly, affective feelings point to the facts and acts of acceptance. Recall the quote: "When I say 'pleasant' it doesn't stand for anything more than 'I would smell it more if I could.' "

The processes that feelings are pointers too can occur, whether or not feeling is aroused. Evidence for this assertion consists of the priming effects exerted upon subsequent stimuli by hedonic stimuli presented under the threshold of awareness, as in backward masking experiments. Other evidence comes from the occurrence of acceptance wriggles in individuals with no, or merely very elementary conscious experience, such as anencephalic infants (Steiner, 1973, p. 18). Another kind again comes from the many actions that are undertaken and to which one returns again and again—mountain climbing, solving intellectual problems, exploring unknown information—that give experience of pleasure only after looking back on them. When asked why these actions are undertaken, the answer is: "because I like it" (Csikszentmihalyi, 1990).

Feeling of pleasure thus results from a pleasure process that as such is nonconscious. Berridge (2007) calls it "core pleasure" or "core liking." The pleasure process consists of acceptance tuning: being set to accept oncoming stimuli and events, including one's own state; readiness to continue commerce with the object or event; readiness to deploy acceptance wriggles such as fit the currently perceived situation; and, in the event, to experience the object or event as pleasant. Acceptance tuning thus appears to include lowering the threshold for pleasant feelings, or feelings of liking. It includes detection of acceptable stimuli at lower levels of incoming information than that of indifferent stimuli, and it includes facilitation of cognitive and social processes that involve information acceptance.

Pleasure processes may emerge and exist, as acceptance tuning, independently of an object, for instance by biochemical means such as morphine. They may attach to events that had not elicited them, such as pictures shown after backward masking or other nonconscious priming (Berridge and Winkielman, 2003; Zajonc, 1980). They thus may be "attributed" to arbitrary cognitive contents (Russell, 2003), although "attribution" would not appear to be the proper designation of the process involved.

Different Pleasures

The above analysis of pleasure, or feeling good, was modeled mainly after sensory pleasures. Sensory pleasures often lead to the assumption that neural sensitivities exist that, when activated by stimuli, directly lead on to pleasure. Evolution proposes a rationale for this assumption: recognizing objects (foods, mates) as pleasant promotes reproductive fitness, just as recognition of harmful objects as aversive does (e.g., Bradley et al., 2001). However, there exist other kinds of pleasure for which the link to particular sensory stimuli is not so clear. Yet, the characterization of "acceptance and acceptance tuning" also applies to them, as some of the examples given indicated.

Different sorts of pleasure can be distinguished (e.g., Kubovy, 1999; Rozin, 1999). I think it meaningful to distinguish the following (Frijda, 2007): (1) Nonsensory likings: one likes particular people, objects, or possessions; one likes money; one likes familiar stimuli more than unfamiliar ones (Zajonc, 2004). Familiar stimuli obviously have no sensory signature. (2) Pleasures of gain and relief. They are caused by changes over previous states or expectations, and by disappearance of what was unpleasant rather than by pleasant sensory stimuli. (3) Achievement pleasures, notably those of achieving desired goals and obtaining concern satisfactions, such as having gained social prestige or having restored social harmony after discord. Pleasures of achievement include pleasures contingent upon achievement in the more narrow sense of mastering some skill, or arriving at some goal after effort. (4) Social pleasures: chatting, watching one's children, participating in a group, and interacting intimately with a loved one. They all form objects of pleasure that are difficult to characterize in stimulus terms nor as predicting particular sensory sensations. (5) Activity pleasures, where the pleasure is in doing rather than in obtaining an outcome, as in dancing, singing, going for a walk, and solving crossword puzzles. (6) Esthetic pleasures: pleasures derived from just watching or from hearing things that one cannot or does not consume otherwise, and that again are not primarily characterized by particular sensory aspects, formal theories of beauty notwithstanding, but by suggested meanings and empathy.

These pleasures all involve objects, events, or actions that one is inclined to remain with, seek, or return to. They all can induce acceptance wriggles that are specific for the kind of objects etc. at hand: celebrating escape from danger or oppression, turning one's achievement over in thought, smiling and slapping shoulders when meeting old friends, contended looking around one's group, and so on.

Even many sensory pleasures are not simple responses to sensory events alone. They figure within contexts of action and interpersonal interaction that add to the pleasure or underlie it in the first place. Pleasure passes beyond that of just receiving the stimuli. Take the pleasures of skin upon skin: those of stroking and being stroked. They not only calm a crying infant. They also delight two adults because they are transformed from what simply was touch into elements of interpersonal intimacy. Of course, lip upon lip is pleasant for a similar reason.

Many sensory pleasures fully derive from a context of action. They represent affordances, in the Gibsonian sense (Gibson, 1979). Smooth skin of a potential mate is pleasant to watch because smooth skin is good to touch and stroke. It shows caressability. Such interpenetration of perception and action is general in visual and tactile perception (Nöe, 2004).

Many pleasant stimuli correspond to sensitivities that are the entry points of motivational systems. For sex, this again is obvious: faces and bodies of potential mates are appraised as attractive. Perceiving them is a pleasure. People can spend hours looking for and at them. Ratings of attractive female faces and figures are generally quite positive, in particular by male subjects (Bradley et al., 2001). They also cause activation in the nucleus accumbens (NAcc) (Aharon et al., 2001). Just perceiving a female rat functions as reinforcement for a male rat, even when there is no occasion for further interaction. Physical appearance and the smell of sex pheromones, and hearing mating calls all are sensory stimuli with incentive value. They form unconditioned stimuli, to a large part derived from innate sensitivities (Buss, 1994), which when activated confer incentive value upon the stimulus objects in mature organisms. They largely do so even in naive animals that have never yet experienced orgasm, and elicit search for where the smell comes from without representation of what will appear at the end and will elicit acceptance wriggles of their own. A face or body is attractive when it attracts, that is, alerts the sexual motivational system. This also applies to affiliation—a pleasant manner evokes the affiliative system, puts at ease, and may evoke a pleasant manner in return. Nice looking or smelling food makes the mouth water, which is part of the ingestive system. The close links of such pleasures with the respective systems are suggested by the covariation between hunger and the apparent palatability of foods (Berridge, 1999; Dickinson and Balleine, 2002), as it is by the susceptibility for friendliness in prisoners. The latter forms a familiar technique in interrogations.

Contextual and other neutral sensory stimuli can obtain incentive value by evaluative conditioning (De Houwer et al., 2001) and by conditioning to the pleasures of final satisfaction (Dickinson and Balleine, 2002). Such signal stimuli do not appear to regularly elicit pleasure. They merely elicit wanting: directed or undirected desire (Berridge, 1999; Dickinson and Balleine, 2002). They do not usually elicit acceptance wriggles and inclinations to prolong interaction with them. However, on occasion, they may do so when desire is suspended and the stimulus is consciously taken as a signal for pleasure to come.

Sensory pleasures can expand beyond the strictly sensory because all or most such stimuli appear within some spatiotemporal context. As in an earlier example, being stroked can be perceived as a caress, and the caress as being given affection, and may project further onto the nonsensory pleasure of intimate interaction. Such expansion of pleasure occurs by expansion of meaning. It can likewise occur with what was nonsensory pleasure to begin with. It is not only that a caress may lead to intimacy; intimacy can lead to a caress and onto sex, and come to mean a gateway to the latter, as it often does in dependency and trust relationships, and under shared hardships. That considerably enhances pleasure. "Why is sex so little fun on your own?" did Fonagy (2008) wonder.

All this has considerable theoretical implications. It shows that even many sensory pleasures are determined by more complex configurations than the mere appearance of some stimulus. It applies to the amorous pleasure from caresses, the social pleasures of chatting with others at the village pump or on the Internet.

Elicitation by complex configurations generally applies to the pleasures that are embedded in positive emotions. Emotions (as defined by action or changes in action readiness, in addition to pleasure or pain and autonomic arousal) are produced by appraisals that involve presence or absence of some pleasant or unpleasant core event, within a spatiotemporal context, and relevance of that event and context to the individual's concerns or major goals. Most of the mentioned nonsensory pleasures are embedded in emotions: in hope, relief, pride, joy, joyous anticipation, enjoyment, and so forth. The moderate role of sensory stimuli has been emphasized in the preceding discussion when describing the various kinds of pleasure. By contrast, entertaining interpersonal relationships and interactions, enablement, performance, and smooth progress of actions appear to carry more of their explanation.

The role of the nonsensory pleasures is exemplified by the just-mentioned expansions of pleasure by expansions of meaning. Such expansions can begin even in affectively neutral events. Viewing a bird in flight can generate a nostalgic sense of freedom: the pleasure of being delivered from the constraints of gravity, even if only in imagination. A cute infant, or even a clumsy child in rags, can become moving, an object of delight, when viewed as exemplifying how vulnerability or loneliness forms the core of self-sufficiency in hard surroundings (Tan and Frijda, 1999). Pleasures of this kind—filling events with meanings that relate to concerns—form the content of much of the appreciation of art, literature, and cinema (Tan, 1996). They also form the target of a mode of attending to objects, landscapes, words, lines in poetry, pictures, that has been called "savoring," notably when meant as the translation of the Chinese word *pin wei* (Frijda and Sundararajan, 2007). *Pin wei* refers to attending, grasping context, letting one's imagination go, and being aware of one's feelings as well as of the object and its meanings.

Where do the pleasures of savoring come from? What is the pleasure, mixed as it is in nostalgia and the sadness of a moving film, in recognizing vulnerability and loneliness as modes of existence and fate? Exploring this issue is a worthwhile enterprise, the work in which I am not very familiar with. I would guess, though, that they mainly come from the enriched understanding of the objects, the enlargement of the scope of what one can participate in through empathy, and the concomitant widened scope of one's being at home and orientated in the world. If true, this exemplifies nonsensory pleasures that hinge on doing things, and being able to do things, rather than with more "extrinsic" rewards.

With these differences in kinds of pleasures, is the word "pleasure" more than merely a common label for different things? For instance, a distinction in kind has frequently been made between "lower" and "higher" pleasures (e.g., Dumas, 1948; Hutcheson, 1728, 1972; Paulhan, 1887). Are all pleasures the same or are they not?

The question is like that of whether a glass is half full or half empty. Certainly, the pleasure experiences differ importantly. Their phenomenology is inseparable from the nature of the pleasant object or event, and what one can or cannot do with it. Different experiences are shaped by their very different acceptance wriggles, their interpersonal meanings, and their implications for further conduct.

But at the same time, they all share the central features of acceptance, acceptance tuning, and stability. Their basic mechanisms of motivating acceptance and prolonged commerce may well be the same, for the same reasons as those of sensory pain and mental anguish appear to do (Panksepp, 1998). It appears a safe bet that all pleasures share a neurohumoral final common pathway, as several findings (e.g., Knutson and Peterson, 2005; Knutson and Wimmer, in press), and contributions in the present volume suggest.

What Elicits Pleasure?

And are all pleasures caused by some general common constellation, or are they not? Perhaps indeed they are, their very different sensory and nonsensory eliciting conditions notwithstanding. The elicitors of nonsensory pleasures and at least a large number of sensory ones do indeed appear to share something.

I will argue that many diverse pleasures are elicited by opportunity for exerting some function of the organism or person, and by some function's actual successful exertion, when such opportunity or exertion is not a matter of routine. Perhaps this even applies to all pleasures. This holds for functions for which a criterion of adequate, smooth, or optimal performance exists.

Most or all pleasure, I propose, results from opportunities for exerting such functions, from progress in their successful exertion, and from their successful completion, when such pleasure is not inhibited by simultaneous processes that are aversive and generate pain. The latter proviso is in line with the Konorski model (Dickinson and Balleine, 2002). Pain, unpleasantness, conversely, comes from frustration, obstruction, or disturbance of opportunity, progress, or completion.

We have seen that many sensory pleasures represent the entry points of activation of some motivational system. The sensitivity that recognizes the stimulus as pleasant—a caress, a genital sensation, a friendly smile—is part of a motivational system. The stimulus sets it going.

Motivational systems consist of sets of functions or competences that include appetitive and consummatory action programs, and sensitivity for particular outcomes, in addition to sensitivities for activating it. The sexual system, for instance, contains competences for responding to certain smells or movements and for recognizing sexual objects (the "sensitivities"), for evoking interest in a potential mate, for accepting bodily contacts, becoming sexually aroused, sexually arousing the mate, sexual intercourse proper, and obtaining terminal sexual satisfaction.

The latter produces pleasure: the rewarding outcome. But, as discussed before, so does recognition of sexual object, the first phase of the system's cycle. Making progress along the chains of the system's competences also gives pleasure, at least in the human species. Each phase is liked. Flirting is a pleasure and so is foreplay. There is no good reason to doubt that the same occurs during animal courtship, and perceiving a target's appropriate courtship dance. The pleasures may not extinguish when no outcome pleasure occurs, as suggested by being attracted and persistence of interest in sexually dysfunctional men (Geer and Janssen, 2000).

The affiliation system likewise represents a set of competences: those for trusting another individual, for moving closer in a way that does not evoke distrust or discomfort, for exchanging affection and information, for understanding his or her affiliation signals, and for emitting them oneself: smiling, uttering noises or words in friendly tones. Every phase again brings pleasure: recognizing a potential partner, actual friend, or one's child, and interacting in the various ways proposed by one's endowments (e.g., speaking, empathy) and one's culture. Here too, relevant incentive stimulus values appear unlearned, with occasion to cling, being held, cooing, and, in humans, being smiled at as examples. Probably, the rewards that may maintain responsiveness to the latter are not themselves of a truly sensory nature, but stem from synchronized interaction (Fogel, 1993). Here too, it is doubtful that the pleasantness of eliciting cues derives from some simple or single final sensory pleasantness, such as being held. What reinforces may rather be finding and having occasion to cling, as in Harlow's rhesus monkey infants.

The three above constellations—encountering stimuli that activate a motivational system, progress in the exertion of the system's skills, and achieving the system's built-in termination—have great generality as elicitors of pleasure. The third constellation is achieving the system's termination: goal achievement as the major cause of pleasure represents a core assumption of many current emotion theories (Arnold, 1960; Frijda, 1986; Scherer, 2005; Stein and Trabasso, 1992). In questionnaire studies, it turns up as one of the main antecedents of joy, together with positive changes in personal relationships (Scherer et al., 1986). The link between goal achievement and pleasure can be generalized over everything that can be considered a "goal": optimal body temperature in temperature regulation, gaining or regaining self-esteem and social esteem; achieving one's standards or expectations of performance in work, or maintaining social harmony.

The second constellation is progress: pleasure is greatly enhanced by previous unexpectedness of a positive outcome, or uncertainty about obtaining one's motivational goal. The point was mentioned before. Relief and pleasant surprise probably provide the clearest examples (Gregory, 1924). By contrast, when pleasurable stimuli arrive as a matter of course (e.g., opening the fridge and finding food, for us, in the West), no noticeable pleasure response occurs. Carver and Scheier (1990) accordingly specified the condition for pleasure as a rate of progress toward goal achievement that is faster than standard, or faster than expected.

With regard to opportunity for exerting motivational systems or competences: "Some objects will permit a given function to work according to its natural design" (Arnold, 1960, p.83). They not only permit it, but may actually activate it. Sweet substances make the mouth water and prompt swallowing; pictures of sexual intercourse indeed alert components of the sexual system such as genital arousal in both men and women (Laan and Janssen, in press). Perceived "cuteness" actually cues the caring-for system. Many people may in passing stroke the hair of a perfectly strange cute little child or animal, or at least smile at it. And when the system is less easily aroused, pleasures go down, as attested by alliesthesia and reward devaluation after being sated (Dickinson and Balleine, 2002; Kringelbach et al., 2003).

Mere opportunity for exercising motivational systems only evokes pleasure, though, in the absence of simultaneous aversive influences that inhibit such exercise and pleasure. For instance, in a study in which female participants watched male-produced sex movies, most participants showed genital arousal but felt the movies to be disgusting and did not experience pleasure or even desire (Laan and Janssen, 2007). Moreover, in women, correlations between genital arousal and experienced desire or excitement are generally low (Laan and Janssen, 2007).

Pleasure by opportunity for exertion not only shows in welcoming relevant stimuli. Opportunities for exercise are often avidly sought, in what have been called "competence motivation" (White, 1959) and "intrinsic motivation" (Deci and Ryan, 1985). Achievement and mastery pleasures do not only or even primarily result from action outcomes. Reaching the summit in mountain climbing is not their pleasure's major source, but only seals it. Grasping the skills of holding on to ridges and getting in the pins, improving on one's techniques, and the skills of supporting suspense and controlling nervousness all contribute importantly. The applause and pay for stunt riding round off the thrill of risk-taking and performance that are described by their actors as "feeling out of this world" (Piët, 1987). The doing rather than the outcomes constitute the enthrallment of skilled action that Csikszentmihalyi (1990) describes under the headings of "flow" and "optimal experience."

Activity pleasures too come from doing rather than achieving. Achieving, in so far as relevant, just proves that the doing was good and smooth. The paradigm of activity pleasure is the fun of rough and tumble play. It is motivated by no other direct goal than the game of exerting social and physical skills, and stretching the limits of these competences.

Pleasures of skill exertion are manifest even in simple skills, when such exertion is not a matter of routine. Examples are the great pleasure of just plain walking after having been for some time confined to bed or of manipulating objects after having had one's hand bandaged (personal communications). Pleasure is generated by exercising still more elementary functions. Lying in the sunshine can be a delight. So, again, can lying in a warm bath after effort be. The pleasures again not only or mainly come from warmth. They may well result from successfully exercising a recuperation function (e.g., Ulrich et al., 1990). They involve the "skills" to relax and let things go, with their presuppositions of having appraised safety and security. At a similar elementary level is the satisfaction from overcoming difficulties in completing simple motor tasks: writing on an uneven surface, finishing one's spoken sentence regardless of being interrupted—both when satisfaction is not blotted out by irritation.

Illuminating instantiations of pleasure from skill exertion come from experiments by Premack (1965). Occasion for exerting a skill that has not been exerted for some time (for instance, playing a pinball machine, looking at a movie cartoon, eating small chocolate candies, and hammering on a wooden pegboard for a kindergarten child) can serve as a reinforcer to perform another of these actions that had been performed recently. Evidently, performing skills is liked when adaptation level has not flattened that liking out. Adaptation levels generally shift liking of skill exertion to a higher level: for a mountain climber, mastering a given ascent drives him or her to try a more challenging one. But all kinds of everyday actions also appear enjoyable or, at least, appear to be exerted with enjoyment. Lykken (2000) gives convincing accounts, both of his own activities and of those of others he observed.

Exertion of very elementary skills appears pleasurable. Cognitive assimilation provides an instance; it drives the emotion of curiosity. Curiosity engages the skills of detecting gaps between novel information and existing knowledge (Loewenstein, 1994), appreciating the gap as manageable, cognitively assimilating the novel information to the existing knowledge, and accommodating that existing knowledge to what came in new (Piaget, 1936). "Interest" and "curiosity" both indicate generally pleasant emotions that center on challenges to cognitive assimilation with a perspective of success. The relative ease of cognitive assimilation appears to render stimuli more pleasurable. Winkielman and Cacioppo (2001) showed that facilitation of visual processing, by priming the contours of briefly presented pictures, generates positive affect, as indicated by ratings and by recorded zygomatic muscle activity, smiling. Winkielman et al. (2006) demonstrated that stimulus prototypicality (of dot patterns and common geometric patterns) enhances both categorization speed and liking for the stimuli. This might explain, for instance, the observed preference for symmetrical faces (Langlois and Roggeman, 1990), as well as the preference for familiar as opposed to novel stimuli (Zajonc, 2004). The "hedonic fluency model" of cognitive processing, based on these findings, appears an instance of a more general model of the hedonic effects of successful skill exertion (Winkielman and Cacioppo, 2001).

"Function pleasure" has been a designation for pleasures of being actively engaged in children (Bühler, 1931); it characterizes much of play and is applicable to a much wider array of pleasures. It extends to events and actions that in their consequences are not pleasant at all. One kind is that of viewing destruction. For instance, after the September 11, 2001 destruction of Twin Towers, many people watched the TV pictures again and again, with a fascination that is not explained by horror (Rimé et al., 2005). Related are the pleasures of war (Broughton, 1993), of which self-report accounts exist at least from the Middle Ages onward (Elias, 1939 [1969]). A mild form of the same are the pleasures derived from violent entertainment (Goldstein, 1998), some of which are experienced as "the sublime" (Burke, 1757, 1990). So far, there is no satisfactory explanation of those pleasures. Simpler than assuming an aggressive instinct (Lorenz, 1963) is interpreting those pleasures as paradigms of the pleasures of skill exertion: actual or imaginary exertion of destructive power and of the skills of tolerating fear and suspense.

Sensory pleasures have formed a problem for approaches to pleasure like the present one. There are, in fact, more connections between pleasant sensory stimuli and competences than appear at first sight. Erotic attractiveness may derive from manifest affordances. Smooth skin is good to the touch, for members of species that have physical provisions for caressing, and caressing acts in their repertoire. Grains may look attractive to a newborn chick because they appear peckable; they fit the chick's pecker. As indicated, such intertwinings of perception and action are general in perception (Nöe, 2004; Varela et al., 1991).

And take the case of aversive tastes or smells. Do they evoke disgust because they are hedonically aversive, or are they aversive because they—bitter taste and the smell of sulfuric acid, for instance— happen to be rejected by the system anyway, because they make the tongue cringe or the nose contract, just as (if I am correct) sweet tastes elicit salivation anyway (and perhaps even at aversive concentrations)? We may be made by nature to accept subtle sweet flavors and smells, and thus like them, just as we are made to accept acrid and pungent tastes by early training (if we were an Indonesian or Mexican baby), and come to delight in them.

The problem thus may not be large for all sensory pleasures; still, some are not readily accommodated. Plato, in his *Philebus* essay, wondered what striving might possibly underlie his pleasure from the smell of his roses when entering his garden in the early morning.

Monitoring Well-Functioning

This latter problem apart, the various observations lead to a generalization. Pleasure reflects well-functioning. It reflects well-functioning of one's functions, skills, and competences: bodily, motor, perceptual, cognitive, emotional, and social. "Well-functioning" includes meeting opportunities for exercise, smooth progress, and natural or successful termination of motivational systems and goal-directed action structures. Well-functioning can be assessed at various levels of functioning: that of one's respiration or moving one's finger, and that of one's understanding and assimilation of events, and of one's social competence in general, and of one's obtaining and maintaining life satisfaction.

Pleasure reflects well-functioning when such well-functioning is not routine. It reflects positive changes in well-functioning: reestablishment of functioning well after failure, threat or difficulty, or functioning better than before or than expected (Carver and Scheier, 1990). Routinely functioning well is not affectively noticed, or noticed at all. We do not notice breathing smoothly, walking normally, or having a stable unproblematic marriage that has lasted for years, until one looks around at others.

This, in turn, implies that pleasure results from a process and a provision for that process: that of function monitoring. It produces pain or unpleasantness as its output on detecting malfunctioning or on detecting threat to functioning well.

Pleasure and pain processes result from meta-monitoring, as Reisenzein (1988) and Carver and Scheier (1990) called it: "They [affects or emotions] represent an organismic monitoring of 'how things are going' with respect to those values" (goal values, Carver and Scheier, 1990, p.33). Along with others (e.g., Scherer, 2004), I suggest that such monitoring is more general. It extends beyond belief-desire or current state-goal value concordance or discordance, to functioning that reaches monitoring generally, as in the example of relaxing in the sunshine. It can give rise to nonconscious pleasure process responses and can given rise to articulate monitoring in conscious feelings and felt qualities (Scherer, 2004) that instigate and guide actions (Dickinson and Balleine, 2002).

This interpretation of pleasure (and of unpleasantness) is not novel. On the contrary, it is very old. Aristotle gives the interpretation in the *Nicomachean Ethics*: pleasure, *hedonè*, is the sense of unimpeded functioning (Aristotle, 1976). His formulation only suffers from defining pleasure by way of what it is not. Spinoza (1677; 1989) improved on it: "Pleasure (*laetitia*) is man's transition from a less state of perfection to a greater" (*Definitions of the Emotions, I* (1989, p. 128), in which "perfection" simply means: in possession of the full power of action that one's nature is endowed with. Many investigators advanced similar hypotheses (e.g., Arnold, 1960; MacDougall, 1905; Ruckmick, 1936). Simon (1967) also proposed a similar principle, formulated by Winkielman and Cacioppo (2001, p. 991) as "affective feedback is one of the ways of monitoring changes in cognitive processing." The present interpretation extends it beyond cognitive processing back to the old philosophers.

There remains the somewhat problematic position of sensory pleasures. One might argue that attractiveness of mates and smiles are tricks shaped by evolution. That no doubt is true, but by itself provides no explanation of how the processes operate here and now, in each individual and individual encounter. My hunch is that evolution has not made each sexually relevant event pleasant. Rather, it has shaped a sexual motivational system that by necessity includes sensitivities for stimuli that are proper elicitors of each phase or section of the system: for recognition of mates, for initiating courtship and response to the partner's courtship displays, for engaging in and completion of intercourse, and so forth, in the case of the sexual system. Likewise, evolution has shaped the competences for affiliation, for aggression and dominance, for cognitive assimilation and the assessment of cognitive gaps. Exercise of all of them then became pleasant as evidences of smooth progress and functioning.

It still is plausible to assume that sensory pleasure—the sensory pleasure process—represents the evolutionary original format of pleasure. Even insects have it, since they detect and approach food and mates. But more general monitoring appears to come not much later. I once caught a wild turtle and put it in a terrarium. It kept me awake for the night by its persistent effort to get out and the sound of its nails against the glass. When brought back to its lake the next day, it at once swam away in a straight unhesitating line. Perhaps sensory pleasure forms a preadaptation (see also Panksepp, 1998; Rozin, 1999) for more encompassing process monitoring.

Anyway, under this perspective, pleasantness thus is not a property of the relevant stimulus events as such. Smells and sights from members of the other sex, and victory over attackers, are not liked because they directly access the liking system. Appearance of a lovely person is a pleasure because it activates erotic approach propensity and indirectly the entire sexual behavioral system.

This inversion of liking and relevance to a motivational system even corresponds to some mentioned empirical observations. One considers lovely whom one loves. Smooth skin fits caressing. Sweet substances are readily chewed and swallowed. Relaxing enables restoring functions. The magnitude of pleasantness can be considered an index of the degree and extent of fit between the events and the function or motivational system or systems at stake, as pleasantness is an index of processing fluency.

The inversion of liking and functioning relevance is also not implausible when considering the disconnections between behavior system activation and felt pleasure that also occur. As has been mentioned, backward masking may increase amounts of drink consumed in the absence of felt pleasantness of the prime, but also in the absence of increase in felt liking of the drink (Berridge and Winkielman, 2003). More telling is the mentioned evidence for partial sexual system activation in women by images that also evoked felt disgust.

The hypothesis may have implications for, say, neurophysiological research. Are pleasure circuits best understood as linked to particular stimuli or to widely distributed circuits for function monitoring? Their location in the frontal brain may suggest the latter, perhaps (Kringelbach et al., 2003).

Functions of Pleasure

The nature of pleasure as a state of acceptance and a process of acceptance tuning suggests an overall function of pleasure. Pleasure leads to inclination to continue the present interaction with its object or situation or to persist in the present kind of action.

This applies not only to the deployment of acceptance wriggles, but equally to more complex dealings, as well as to relaxation and rest and to mere liking through mere exposure (Zajonc, 1968), that is, to remain in the neighborhood of familiar objects since these serve as safety signals (Zajonc, 2004).

This emphasis on interaction and action forms, I think, the proper perspective. The nonsleeping animal is continuously active and any action or interaction just represents a modulation of that continuous activity. Affect does not emerge mainly upon external stimuli, but upon the progress of ongoing processing.

To the extent that the encompassing hypothesis is correct—pleasure as a well-functioning index—this function of the index extends to nonsensory pleasures. It applies to well-functioning at all levels of functional integration to which central or "meta-monitoring" applies. This includes the failure of automatic regulation of body warmth and respiration, and of food supply by one's internal clock and meal habits, so that the unpleasant feeling of one's stomach grumbling has to intervene.

Pleasantness, like unpleasantness, does not arise only by events that directly impinge on sensitivities and ongoing processing. Recalled pleasantness—remembered utility, as Kahneman (1999) called it—extends this function in a further way. Remembered utility, in a way, can arise automatically by evaluative conditioning (De Houwer et al., 2001). For instrumental learning through pleasant response–outcome contingencies, conscious pleasure and contingency awareness appear required (Dickinson and Balleine, 2002). In other words, while not being the sole or even primary function of pleasure, one of its function is the detection of concern-relevant objects and that of pleasant feeling includes serving instrumental learning.

Pleasure and Desire

This chapter began with stressing the role of pleasure in emotions and striving, as laid down in the pleasure principle.

Yet, I concluded that pleasure, in contrast to pain, is best described as a stable state. It is the state that brings the individual to accept the state, situation, or action he or she is in, and to persist in that state to prolong it or to return to it. This description was based both on self-reports from studies examining feelings of pleasure or pleasantness, from behavior in pleasant states, and from the occurrence of calm and serene pleasures. These latter pleasures conflict with the link that has often been made between pleasure and approach tendency (e.g., Cacioppo et al., 2004). This indeed cannot be the primary attribute of pleasure.

There thus is a contradiction between pleasure as a stable state and pleasure as that which instigates pursuing it. The contradiction is not complete though. Pleasure as such instigates acceptance wriggles, which include actions to extend one's interaction with the pleasant object or situation in space and time, and often the desire to immediately repeat the experience. There is regret that the ice cream is finished and that the movie ends.

By itself, the contradiction agrees with the distinction between liking and wanting. This distinction is primarily based on neurohumoral findings. It is strengthened by the many ordinary empirical observations of liking that does not induce lifting a finger or making a detour. One sees attractive people and continues on one's way; one enjoys the natural beauty around the speedway one is speeding on.

There also is an apparent contradiction in the close link between the incentive value of signals for possible pleasant outcomes and those outcomes themselves. The signals predict those outcomes as a result of contingency learning, often combined with anticipation of that pleasure. But why do these elicit wanting, since pleasure by itself does not motivate other than its own continuation? How to bridge the paradox? What motivates the motivation of wanting?

Theoretically, wanting implies some gap to be filled. Appraisal theory predicts that the constellation for wanting contains cues that indicate or contain a gap, and of course it does so. Anticipation of pleasure is not presence of pleasure, or occasion to immediately obtain it. Anticipation of pleasure implies inability to deploy relevant acceptance wriggles and further consummatory actions. Anticipation of pleasure and desire posses an inherent ambivalence, because pleasure is near, but there also is uncertainty whether and when it will be obtained. Salience of either nearness or being out of reach cause anticipated pleasure and desire to hover themselves between being pleasant or painful. At any moment balance may shift from going to obtain something pleasant towards getting rid of a painful urge. Anticipated pleasure is therefore a different emotion from joy or enjoyment, as their neural representations confirm (Knutson and Cooper, 2005; Knutson and Peterson, 2005). Being in love represents the paradigmatic illustration of this ambivalence. It usually is not a pleasant emotion. One is filled with anxiety, shot through with moments of delight. In other words, pleasure itself serves as an attractor, as befits it being a stable state. Pleasure being an attractor may underlie one's pursuit when moved away from the point of attraction by viewing further possibility or by the attractor valley having been leveled by habituation.

This analysis, of course, represents a hypothesis about the transition from liking to wanting. Data are probably needed to truly obtain insight. I do not know how much data already exist about the mechanics of wanting outside addictions. It does occur outside it: in the salted peanut phenomenon; in the mysteries of greed—of wanting more under abundance; in the cravings when dieting and faced with cookies, or the thought of them.

References

Aharon, I., Etcoff, N., Ariely, D., Chabris, C. F., O'Connor, E. and Breiter, H. C. (2001) Beautiful faces have variable reward value: fMRI and behavioral evidence. *Neuron* 32, 537–51.

Alechsieff, N. (1907) Die Grundformen der Gefühle. *Psychologische Studien* 3, 156–271.

Aristotle (1976) *The Nicomachean Ethics, Book 10* (transl. J.A.K. Thomson). Penguin Books: London.

Arnold, M. B. (1960) *Emotion and Personality*, Vol. I. Columbia University Press: New York.

Beebe-Center, J. G. (1932, 1965) *The Psychology of Pleasantness and Unpleasantness* (reissued 1965). Russell & Russell: New York.

Berlyne, D. E. (1971) *Aesthetics and Psychobiology*. Appleton-Century-Crofts: New York.

Berridge, K. C. (1999) Pleasure, pain, desire, and dread: Hidden core processes of emotion. In: *Foundations of Hedonic Psychology: Scientific Perspectives on Enjoyment and Suffering*, pp. 525–557. Eds. D. Kahneman, E. Diener and N. Schwarz. Sage: New York.

Berridge, K. C. (2000) Measuring hedonic impact in animals and infants: Microstructure of affective taste reactivity patterns. *Neurosci. Biobehav. Rev.* 24, 173–198.

Berridge, K. C. (2007) The debate over dopamine's role in reward: the case for incentive salience. *Psychopharmacology (Berl.)* 191, 391–431.

Berridge, K. C. and Winkielman, P. (2003) What is an unconscious emotion? (The case for unconscious "liking"). *Cogn. Emot.* 17, 181–211.

Bradley, M. M., Codispoti, M., Cuthbert, B. N. and Lang, P. J. (2001) Emotion and motivation I: Defensive and appetitive reactions in picture processing. *Emotion* 1, 276–298.

Brehm, J. (1999) The intensity of emotion. *Pers. Soc. Psychol. Rev.* 3, 2–22.

Broughton, J. M. (1993) The pleasures of the Gulf war. *Recent trends in theoretical psychology* III, 231–246.

Bühler, C. (1931) *Kindheit und Jugend*. Hirzel: Leipzig, Germany.

Bühler, K. (1907) Tatsachen und Problemen zu einer Psychologie der Denkvorgänge. I. Über Gedanken. *Arch. Psychol.* 9, 297–365.

Burke, E. (1757;1990) *A Philosophical Enquiry into the Origin of Our Ideas of the Sublime and Beautiful*. Oxford University Press: Oxford.

Buss, D. M. (1994) *The Evolution of Desire*. Basic Books: New York.

Cacioppo, J. T., Larsen, J. T., Smith, N. K., & Berntson, G. G. (2004). The affect system: What lurks below the surface of feelings? In: *Feelings and Emotions: the Amsterdam Symposium*, pp. 223–242. Eds. A. R. S. Manstead, N. Frijda and A. Fischer. Cambridge University Press: Cambridge.

Carver, C. S. and Scheier, M. F. (1990) Origins and functions of positive and negative affect: a control-process view. *Psychol. Rev.* 97, 19–35.

Chalmers, D. (1996) *The Conscious Mind: In Search of a Fundamental Theory*. Oxford University Press: New York.

Coren, S. (2001) *How to Speak Dog: Mastering the Art of Dog–Human Communication*. Simon & Schuster: New York.

Csikszentmihalyi, M. (1990) *Flow: the Psychology of Optimal Experience*. Harper Collins: New York.

Davitz, J. R. (1969) *The Language of Emotion*. Academic Press: New York.

De Houwer, J., Thomas, S. and Baeyens, F. (2001) Associative learning of likes and dislikes: a review of 25 years of research on human evaluative conditioning. *Psychol. Bull.* 128, 853–825.

Deci, E. L. and Ryan, R. M. (1985) *Intrinsic Motivation and Self-Determination in Human Behavior.* Plenum Press: New York.

Dickinson, A. and Balleine, B. (2002) The role of learning in the operation of motivational systems. In: *Stevens' Handbook of Experimental Psychology,* Vol. 3, pp. 497–562. Eds. R. Gallistel and H. Pashler. Wiley: New York.

Dumas, G. (1948) *La vie affective*. Presses Universite France: Paris.

Elias, N. (1939 [1969]) *Der Prozess der Zivilization [The Civilizing Process]*. Urizen Books: New York.

Fogel, A. (1993) *Developing through Relationships* Harvester: New York.

Fonagy, P. (2008) A genuinely developmental theory of sexual enjoyment and its implications for psychoanalytic technique. *J. Am. Psychoanal. Assoc.* 56, 11–36.

Fredrickson, B. L. (2001) The role of positive emotions in positive psychology: the broaden-and-build theory of positive emotions. *Am. Psychol.* 56, 218–226.

Frijda, N. H. (1986) *The Emotions*. Cambridge University Press: Cambridge.

Frijda, N. H. (2007) *The Laws of Emotion*. Lawrence Erlbaum Associates: Mahwah, NJ.

Frijda, N. H. & Sundararajan, L. (2007). Emotion refinement. *Persp. Psychol. Sci.,2,* 227–241.

Garotti, P. L., Caterina, R., Brighetti, G., Giberti, E. and Ricci-Bitti, P. E. (1993) Il ruolo del sorriso come indicatore di stati emotivi differenziati [The role of the smile as indicating differentiated emotional states]. *Rassegna di Psicologia* 10, 29–53.

Geer, J. H. and Janssen, E. (2000) The sexual response system. In: *Handbook of Psychophysiology*, pp. 315–341. Eds. J. T. Cacioppo, L. Tassinari and G. Berntson. Cambridge University Press: New York.

Gemelli, A. (1949) Orienting concepts in the study of affective states. *J. Nerv. Mental Dis.* 110(I), 198–214; (II), 299–314.

Gibson, J. J. (1979) *The Ecological Approach to Visual Perception*. Houghton Mifflin: Boston, MA.

Goldstein, J. (1998) *Why We Watch: The Attractions of Violent Entertainment*. Oxford University Press: New York.

Gregory, J. C. (1924) *The Nature of Laughter.* Kegan Paul: London.

Humphrey, G. (1951) *Thinking: an Introduction to its Experimental Psychology.* Methuen: London.

Hutcheson, F. (1728; 1972) *An Essay on the Nature and Conduct of the Passions and Affections*. Scolar Press: Menston, UK.

Isen, A. M. (2004) Some perspectives on positive feelings and emotions: Positive affect facilitates thinking and problem solving. In: *Feelings and Emotions: the Amsterdam Symposium*, pp. 263–281. Eds. A. R. S. Manstead, N. H. Frijda and A. Fischer. Cambridge University Press: Cambridge.

James, W. (1894) The physical basis of emotions. *Psychol. Rev.* 1, 516–529.

Kahneman, D. (1999) Objective happiness. In: *Foundations of Hedonic Psychology: Scientific Perspectives on Enjoyment and Suffering*, pp. 3–25. Eds. D. Kahneman, E. Diener and N. Schwartz. Sage: New York.

Keltner, D. and Haidt, J. (2003) Approaching awe, a moral, spiritual, and aesthetic emotion. *Cogn. Emot.* 17, 297–314.

Knutson, B. and Cooper, J. C. (2005) Functional magnetic resonance imaging of reward prediction. *Curr. Opin. Neurol.* 18, 411–7.

Knutson, B. and Peterson, R. (2005) Neurally reconstructing expected utility. *Games Econ. Behav.* 52, 305–315.

Knutson, B. and Wimmer, G. E. (in press) Reward: Neural circuitry for social valuation. In: *Fundamentals of Social Neuroscience*. Eds. E. Harmon-Jones and P. Winkielman. Guilford Press: New York.

Koch, B. (1913) *Experimentelle Untersuchungen über die elementare "Gefühlsqualitäten." [Experimental Investigations into the Elementary "Feeling Qualities"]*. Quelle & Mayer: Heidelberg, Germany.

Kringelbach, M. L., O'Doherty, J., Rolls, E. T. and Andrews, C. (2003) Activation of the human orbitofrontal cortex to a liquid food stimulus is correlated with its subjective pleasantness. *Cereb. Cortex* 13, 1064–71.

Kubovy, M. (1999) On the pleasures of the mind. In: *Foundations of Hedonic Psychology: Scientific Perspectives on Enjoyment and Suffering*, pp. 134–154. Eds. D. Kahneman, E. Diener and N. Schwartz. Russell Sage: New York.

Laan, E. and Janssen, E. (2007) How do men and women feel? Determinants of subjective experience of sexual arousal. In: *The Psychophysiology of Sex*, pp. 278–290. Ed. E. Janssen. Indiana University Press: Bloomington, IN.

Langlois, J. H. and Roggeman, L. A. (1990) Attractive faces are only average. *Psychological Science* 1, 115–121.

Loewenstein, G. (1994) The psychology of curiosity: a review and reinterpretation. *Psychol. Bull.* 116, 75–98.

Lorenz, K. (1963) *Das sogenannte Böse [On Aggression]*. G.Borotha-Schoeler: Wien, Austria.

Lykken, D. T. (2000) *The Nature of Happiness: a Psychologist Explores the Truth Behind the World's Most Elusive Emotion*. St. Martin's Press: New York.

MacDougall, W. (1905) *Physiological Psychology.* Aldine: London.

Miron, A. M., Parkinson, S. K. and Brehm, J. W. (2007) Does happiness function like a motivational state? *Cogn. Emot.* 21, 248–267.

Nafe, J. P. (1924) An experimental study of the affective qualities. *Am. J. Psychol.* 35, 507–544.

Nakashima, T. (1909) Contributions to the study of the affective processes. *Am. J. Psychol.* 20, 157–181.

Nöe, A. (2004) *Action in Perception.* MIT Press: Cambridge, MA.

Ortony, A., Clore, G. and Collins, A. (1988) *The Cognitive Structure of Emotions*. Cambridge University Press: Cambridge.

Panksepp, J. (1998) *Affective Neuroscience*. Oxford University Press: Oxford.

Parducci, A. (1995) *Happiness, Pleasure, and Judgment.* Erlbaum: Mahwah, N. J.

Paulhan, F. (1887) *Les phénomènes affectifs et les lois de leur apparition. Essai de psychologie générale.* Alcan: Paris.

Piaget, J. (1936) *La naissance de l'intelligence chez l'enfant.* Delachaux & Niestlé: Neuchatel, Switzerland.

Piët, S. (1987) What motivates stuntmen? *Motiv. Emot.* 11, 195–213.

Premack, D. (1965) Reinforcement theory. In: *Nebraska Symposium on Motivation*, pp. 123–180. Ed. M. R. Jones. University of Nebraska Press: Lincoln, NE.

Reisenzein, R. (1988) A theory of emotional feelings as metarepresentational states of mind. In: ISRE '98:

Proceedings of the 10th Conference of the ISRE, pp. 186–190. Ed. A. Fischer. ISRE Publications: Amsterdam.

Rimé, B., Delfosse, C. and Corsini, S. (2005) Emotional fascination: Responses to viewing pictures of September 11 attacks. *Cogn. Emot.* 19, 923–932.

Rozin, P. (1999) Preadaptation and the puzzles and properties of pleasure. In: *Well-being: The Foundations of Hedonic Psychology: Scientific Perspectives on Enjoyment and Suffering*, pp. 109–133. Eds. D. Kahneman, E. Diener and N. Schwartz. Russell Sage: New York.

Ruckmick, C. A. (1936) *The Psychology of Feeling and Emotion.* McGraw-Hill: New York.

Russell, J. A. (2003) Core affect and the psychological construction of emotion. *Psychol.l Rev.* 110, 145–172.

Scherer, K. R. (2004) Feelings integrate the central representation of appraisal-driven response organization in emotion. In: *Feelings and Emotions: the Amsterdam Symposium.* Eds. A. R. S. Manstead, N. Frijda and A. Fischer. Cambridge University Press: New York.

Scherer, K. R. (2005) What are emotions? And how can they be measured? . *Social Science Information* 44, 695–729.

Scherer, K. R., Walbott, H. G. and Summerfield, A. B. (1986) *Experiencing Emotions: a Cross-Cultural Study.* Cambridge University Press: Cambridge.

Scitovsky, T. (1976) *The Joyless Economy: The Psychology of Human Satisfaction.* Oxford University Press: New York.

Simon, H.A. (1967). Motivational and emotional controls of cognition. *Psychol. Rev. 74,* 29–39.

Spinoza, B. (1677;1989) *Ethica* [English translation, G.H.R. Parkinson]. Everyman's Library: London.

Stein, N. and Trabasso, T. (1992) The organization of emotional experience: Creating links between emotion, thinking, and intentional action. *Cogn. Emot.* 6, 225–244.

Steiner, J.E. (1973). The gustofacial response: Observation on normal and anencephalic newborn infants. *Symp. Oral Sensation Percept., 4,* 254–278.

Tan, E. S. H. (1996) *Emotion and the Narrative Film: Film as an Emotion Machine.* Erlbaum: Hillsdale, NJ.

Tan, E. S. H. and Frijda, N. H. (1999) Sentiment in film viewing. In: *Passionate Views: Film, Cognition, and Emotion*, pp. 48–64. Eds. C. Plantinga and G. Smith. Johns Hopkins University Press: Baltimore, MD.

Titchener, E. B. (1908) *Lectures on the Elementary Psychology of Feeling and Attention.* McMillan: New York.

Ulrich, R. S., Dimberg, U. and Driver, B. L. (1990) Psychophysiological indicators of leisure consequences. *J. Leisure Res.* 22, 154–166.

Van Hooff, J. A. R. A. M. (1972) A comparative approach to the phylogeny of laughter and smiling. In: *Non-Verbal Communication*, pp. 209–237. Ed. R. A. Hinde. Cambridge University Press: Cambridge.

Varela, F. J., Thompson, E. and Rosch, E. (1991) *The Embodied Mind.* MIT Press: Cambridge, MA.

White, R. W. (1959) Motivation reconsidered: the concept of competence. *Psychol. Rev.* 66, 297–333.

Wierzbicka, A. (1999) *Emotions across Languages and Cultures.* Cambridge University Press: Cambridge.

Winkielman, P. and Cacioppo, J. T. (2001) Mind at ease puts a smile on the face: Psychophysiological evidence that processing facilitation elicits positive affect. *J. Pers. Soc. Psychol.* 81, 989–1000.

Winkielman, P., Halberstadt, J., Fazendeiro, T. and Catty, S. (2006) Prototypes are attractive because they are easy on the mind. *Psychol. Sci.* 17, 799–806.

Wundt, W. (1896) *Grundriss der Psychologie.* Engelmann: Stuttgart, Germany.

Young, P. T. (1927) Studies in affective psychology. *Am. J. Psychol.* 38, 157–193.

Zajonc, R. B. (1968) Attitudinal effects of mere exposure. *J. Pers. Soc. Psychol. Monograph Suppl.* 9(2, Pt. 2), 1–27.

Zajonc, R. B. (1980) Feeling and thinking. Preferences need no inferences. *Am. Psychol.* 35, 151–175.

Zajonc, R. B. (2004) Exposure effects: an unmediated phenomenon. In: *Feelings and Emotions: the Amsterdam Symposium*, pp. 194–203. Eds. A. R. S. Manstead, N. H. Frijda and A. Fischer. Cambridge University Press: Cambridge.

7

The Dialectics of Pleasure

MICHEL CABANAC

In the 17th century, the mathematician–philosopher Blaise Pascal, while considering the world of his time, recognized anxiously that each human was balanced between two abysses: the infinitely small abyss of atomic particles, *raccourci d'atome*, and the infinitely large abyss of the cosmos, *l'univers* (Pascal, 1623–1662).

However, in the middle of the 20th century, the palaeontologist–philosopher Pierre Teilhard de Chardin, in his attempt to reconcile biology with physics and astronomy[1] removed the Pascalian anguish. His reconciliation was the result of his reckoning that the human phenomenon represents also an infinite. He considered that matter is organized in living beings and, as a rough estimate of this organization, he counted the number of atoms coordinated in autonomous entities such as the human body.

In this organism, the number reached 10^{25}, that is, a number similar to the magnitude of the positive and negative exponents for the size of the cosmos and the atom measured in centimetres (Teilhard de Chardin, 1965) (Figure 7.1).

The Brain and the Infinite Complexity

Gell-Mann has underlined the difficulty of defining and, furthermore, quantifying complexity (Gell-Mann, 1994). The quantitative comparing of biological complexity to the dimensions of universe and quarks makes little sense in terms of numbers because complexity has no dimension, and therefore, cannot be compared to units of length. However, it is of interest to obtain numbers to illustrate the astronomic complexity reached in living beings. The number of atoms in the human body, used by Teilhard de Chardin, as a gross estimate of complexity can no longer be accepted; the number of atoms organized in a Sequoia or a whale is much larger than in a human body and nothing indicates that these entities are more complex than humans. As suggested by Teilhard de Chardin himself, the number of atoms is not the best index of complexity. Among other candidate indexes of complexity, such as the genetic code, the central nervous system is more likely to be the locus of the highest and most organized complexity in nature.

The human central nervous system is generally accepted as containing about 10^{11} to 10^{12} neurones organized in a single whole entity. Each neurone is capable of communicating synaptically with other neurones. Rather than the raw number of cells contained in the brain, synapses are a better indicator of the brain complexity. It is possible to make a rough estimate of the number of synapses accumulated in the human central nervous system (Kandel and Schwartz, 1985). The number of synapses per neurone is extremely variable from 2 in the highly specialized bipolar retinal cells, to 150,000 in Purkinje cells. A conservative figure may be estimated at ca. 10,000 as the average number of synapses per neurone. The resulting number of synapses would thus reach ca. 10^{15}. This is a conservative

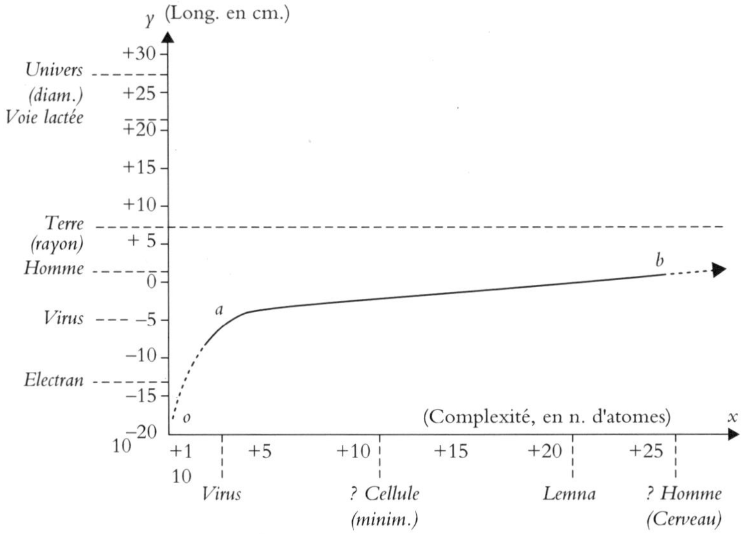

Figure 7.1 Teilhard de Chardin's natural curve of complexity. In ordinates the length of identifiable entity-objects measured in centimetres (*long. en cm.*). Human (*Homme*); earth radius (*Terre rayon*); galaxy (*Voie lactée*); universe (*Univers*); in abscissa complexity (*Complexité, en n. d'atomes*) as estimated in number of atoms. Cell (*Cellule*); water lentil, duckweed (*Lemna*); human brain (*Homme cerveau*). "a", appearance of life; "b", appearance of Homo. (With permission from Teilhard de Chardin, 1965.)

estimate since 10^{14} to 10^{15} was proposed as the number of synapses in the human cortex only (Changeux, 1983). Such a number defies imagination, but can be compared to other phenomena to illustrate its magnitude: if the universe is 10 to 20 billion years old, the number of synapses in a human brain is the same as the number of minutes since the Big Bang.

Yet, this is not the ultimate of brain complexity. The mechanism of synaptic transmission involves neurotransmitter vesicles that operate as quanta of synaptic transmission. The number of vesicles in a given synaptic button is variable but may be estimated conservatively as 10^2. In turn, the index we have selected to estimate complexity now reaches 10^{17}. This figure is given to illustrate the brain complexity. Other criteria would provide numbers of similar or higher magnitude. For example, Penrose (1994) estimated that the brain is able to perform 10^{24} operations per second. Thus, Teilhard de Chardin's solution to Pascal's anguish may be transposed from the human body to the single human brain, the complexity of which may be considered as a third infinite, the infinite complex. A somewhat similar line was developed in what is commonly designated as the Anthropic principle (Barrow and Tipler, 1986; Bertola and Curi, 1993; Breuer, 1991). However, the argument defended in the present chapter is independent of the anthropic principle. Here, while studying "pleasure," I address the question of "how" agents are brought into motion, not the question of "why" they are motioned. We may now consider what can move this infinite complex and what force can induce the brain, taken as an individual whole entity, to behave.

Emergence

Before going further, we must acknowledge that the brain is the locus of the second emergence of evolution. After emergence of life from matter, thinking and consciousness emerged from complex nervous systems. We should not be arrogant enough to shun the possibility that other forms of thinking complexity may exist in the cosmos but, in the present state of our knowledge, the human brain is the most advanced locus of thinking. In the following, let us restrict the infinite complex to the thinking brain and concentrate on its emerging properties, sensation and consciousness.

A behaviorist approach of psychology ignores the mind and works according to Figure 7.2A. In Figure 7.2A, the brain is a 'black box' receiving stimuli from the subject's environment and the *milieu intérieur* via afferent nervous pathways and produces behavioral

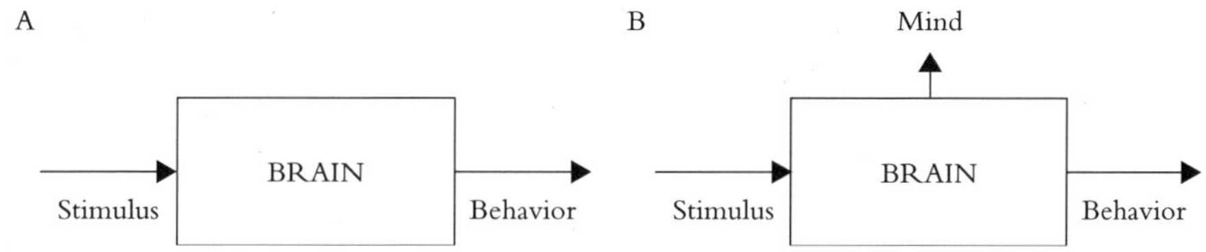

Figure 7.2 The brain as a black-box (see text).

responses adapted to these stimuli. According to this attitude, a true scientist may and should deal only with public data that can be shared with anybody. Mental objects are considered as 'private' and, therefore, unreachable scientifically because they cannot be shared by several observers. It is true that the study of many behaviors may be limited to the measurement of tropisms, tactisms, or reflexes and do not need any mental explanation. When Limneas are placed in a gradient of temperature or light, they all move and gather to stay at the same temperature and the same light intensity. One may accept that such a behavior is due to a simple tropism, similar to the growing of a plant toward light. When a toad adapts its posture and turns to keep facing a worm-shaped target that moves around it, but remains indifferent to the same target moving identically if the 'worm' is vertical (Ewert, 1968), we are entitled to think that the toad's behavior is purely reflexive and similar to the multisynaptic spinal reflexes well studied by Sherrington.

Yet, there exist other, more complex, behaviors that are contemporaneous with mental events. The conscious motivations for these behaviors take place in the brain, in the black box of Figure 7.1B, between the two arrows. To take these events into account is sometimes designed somewhat derogatory as a "mentalist" approach. Yet, the mental sphere exists and it is possible to study mental events scientifically. The founders of experimental psychology, Weber and Fechner, studied scientifically the sensations—mental events—aroused by sounds and lights, and modern psychophysicists keep on this track. The principle consists in presenting a well-measured stimulus and asking the subject to describe behaviorally or verbally the sensation aroused. It is, therefore, possible to study sensation scientifically and to extend the method to all mental events.

Figure 7.3 is a model of any sensation aroused by a stimulus in a psychophysics experiment. The mental event may be described with four dimensions. The quality of sensation is analogous to the nature of the stimulus: vibration in the inner ear, wave length on the retina, gas on the olfactory epithelium, liquid solution on the tongue, and so on. Physiology and psychology textbooks limit sensations to the five senses: vision, audition, olfaction, taste, and skin sensation (touch and pain). However, it would be more appropriate to accept that any afferent fiber to the brain can arouse a sensation (Cabanac, 1994). The quality dimension of sensation, therefore, describes the nature of the stimulus. The intensity of sensation describes the magnitude of the stimulus: amplitude of the wave length of the sound, concentration of odorant or solutes, area of stimulated skin, and so on.

The third dimension is hedonicity, which is the focus of this chapter and this book: the displeasure of a bitter taste, the pleasure of a caress, and so on. This dimension motivates the stimulated subject to seek or to avoid the stimulus. There are sound reasons to accept that the pleasure of a sensation indicates the usefulness of the stimulus and that the displeasure indicates uselessness or noxiousness (Cabanac, 1971; 1994). It follows that the seeking of sensory pleasure is beneficial to the subject and optimizes behavior, as seen from a physiological point of view.

The fourth dimension is time and indicates the duration of the exposure to the stimulus. Note that quality, intensity, and duration cannot be nil or negative, but hedonicity can be.

Pleasure and the Philosophers

Moralists have always recognized that the seeking of pleasure is the motor of action. Here is what but a sample of a few thinkers wrote: "mankind have nothing better under the sun than to eat and drink and rejoice" (Ecclesiastes 900 B.C.); "one may also think that, if all humans seek pleasure, that is because they desire to live" (Aristotle, 384–322 B.C.); "nature gives us organs to inform us through pleasure about what we should seek and through pain what we should avoid" (Condillac, 1754); "pleasure is the spring of action" (Bentham [1748–1832], 1823 in Bowring,

1962); "pleasure is always useful" (Dostoievski: *Alexis Ivanovitch, The Player*); "the Creator[...] gave us the inducement of appetite, the encouragement of taste, and the reward of pleasure" (Brillat-Savarin, 1828); "pleasures and pains represent the sole genuine basis for understanding human motives" (Jevons, 1871); "from every point of view the affective process must be regarded as motivational in nature" (Young, 1959). Finally, the Fathers of the American Constitution also acknowledged that the pursuit of happiness is the main aim of the human life. In the following discussion, it will be argued that the seeking of pleasure is the motor of human behavior. I do not claim that the seeking of pleasure and the avoidance of displeasure explain brain function, but, as foreseen by philosophers, that they explain behavior. Yet, this knowledge now will be based, not on intuition and introspection, like the moralists of the past, but on scientific evidence. We must acknowledge that proximal explanation of behavior may seem unnecessary when examined from the point of view of natural selection. However, proximal explanations of behavior must also be provided. In addition, the emergence of thinking, consciousness, and awareness must and can be viewed from an evolutionary perspective (Cosmides et al., 1992). The problem is not that we can understand evolution phenomenologically without consciousness; the problem is that consciousness is there, we must deal with it, and understand its role in nature.

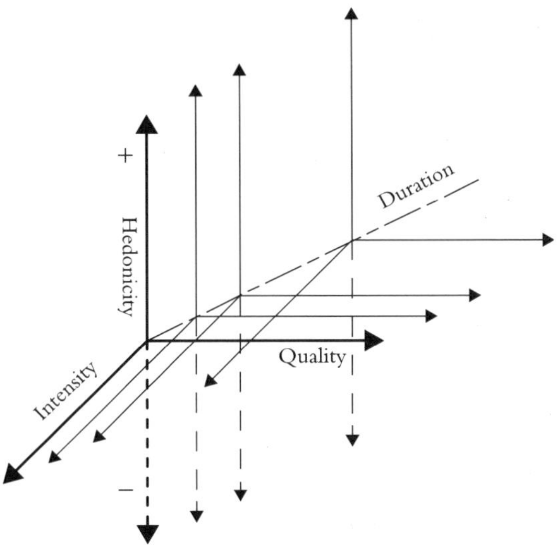

Figure 7.3 Sensation seen as a multidimensional event in response to a stimulus. Quality is non parametric and describes the nature of the stimulus; quantity/intensity, describes the intensity of the stimulus; affectivity/hedonicity (pleasure or displeasure), describes the physiological usefulness (survival value) of the stimulus. Duration is added to the three other fundamental axes of sensation. It is argued, here, that this model describes equally sensation and consciousness. (With permission from Cabanac, 1996.)

Physiological Role of Pleasure

The starting point of the chain of reasoning is the observation that sensory pleasure tags stimuli that are useful for optimal physiological functioning and survival and, obversely, that sensory displeasure indicates useless or noxious stimuli (Cabanac, 1971, 1979).

Sensation is the mental representation of a stimulus that reaches a sensory organ. Whenever a stimulus excites a sensory ending, the sensation aroused is a four-dimensional phenomenon (Figure 7.3). The first dimension, the quality of the sensation indicates the nature of the stimulus. The second dimension, the intensity of the sensation, is non-parametric and indicates the magnitude of the stimulus. The third dimension, the hedonicity of the sensation (pleasure/displeasure), indicates the potential usefulness of the stimulus. The fourth dimension, the duration of the sensation, indicates the duration of the stimulus.

Let us examine the third dimension. The pleasure we experience when stimulated with an agreeable stimulus is not a constant but rather a variable and depends on our internal state. Let us take an example: we are fasted and hungry; we reach for a delicious candy. We eat another candy, then another one, and after 5, and then 10, the 11th candy bar is still chemically the same as the first one but is no longer nearly as delicious. Yet, when our hunger returns, the candy will again arouse intense pleasure. The same phenomenon of pleasure alteration can be found with olfactory and thermal sensations. Then, the same stimulus can make an individual feel either pleasant or unpleasant according to the internal state of the individual and the resulting sensation can move up and down the affectivity axis. Indeed, this is especially obvious in the determinism of thermoregulatory behavior (Attia, 1984) and of the control of food intake (Fantino 1984, 1995; McBride 1997).

When a subject is hypothermic, a warm skin feels pleasant and a cool one unpleasant. When the subject is already hyperthermic, a warm skin feels unpleasant and a cool one pleasant. The efficacy of this mechanism is almost limitless: the simplest behavior, such as the seeking of pleasure by a hyperthermic subject who immerses only his hand in stirred 20°C water, can extract as much as 70 W from the hand, *i.e.*, as much as the subject's basal heat production (Cabanac *et al.*, 1972). The adaptation of sensory pleasure and displeasure to physiological need is such that very similar equations describe thermal preference (Bleichert *et al.*, 1973; Cabanac *et al.*, 1972), and the autonomic regulatory responses, evaporative heat loss (Stolwijk *et al.*, 1968), or chemical thermogenesis (Nadel *et al.*, 1970). In addition, thermal pleasure and displeasure is adapted to the defense not only of core temperature but to the difference T_{set}–T_{core}, the error signal of temperature regulation. The case of alimentary pleasure is similar, with pleasure occurring from food stimuli when applied to hungry subjects, and displeasure when applied to satiated subjects (Fantino, 1984, 1995).

This, therefore, seems to be a general case: when the *milieu intérieur* varies, pleasure changes according to the stimulus usefulness for the body. The word *alliesthesia* was coined to describe the fact that a same given stimulus can make an individual feel pleasant or unpleasant according to the internal state of the individual and that the resulting sensation can move up and down the affectivity axis. Usefulness being estimated by the body regarding physiological regulatory set points. The similarity with temperature regulation was recognized when alliesthesia to sweet stimuli disappeared in subjects whose body weight was reduced with dieting, and recovered when body weight returned to control value. Originally, I thought that only those stimuli susceptible of modifying the *milieu intérieur* would be a source of alliesthesia. However, it was more recently shown that the hedonicity can change with all sensations. For example, the functional role of pleasure may be found in even in postural adjustments: the most precise gestures are reached with the help of the affective capability of musculo-articular sensation (Rossetti *et al.*, 1994). Finally, all sensations are likely to display alliesthesia, including those like vision or audition that are not 'plugged' directly on the *milieu intérieur* (Brondel and Cabanac, 2007) contrary to what I believed initially.

Sensory pleasure is thus the axis of sensation that allows optimization of behavior, as seen from the physiologist point of view, that of immediate improvement of physiological function, and survival of the subject.

The Common Currency

Let us accept as a postulate that at each instant, the animals satisfy behaviorally their most urgent motivation. When they theorized on the "behavioural final common path," McFarland and Sibly stated that the central nervous system responsible for behavior must rank on a priority scale all motivations simultaneously present according to their urgency and that, to achieve such a ranking, the brain must choose among competing motives based on a common currency (McFarland and Sibly, 1975). Thus, a trade-off will be able to take place among simultaneous motivations, one being suppressed to allow the manifestation of another one that will rank first. Animals occasionally renounce to feed because of the presence of a predator or of a reproductive partner. In that case, immediate survival or reproductive necessity are ranked higher than the necessity to feed. The common currency makes possible such a trade-off between motivations and their ranking according to the urgency of the need to access to behavior.

The reading of McFarland and Sibly's "behavioural final common path" and later of other published works of McFarland's gave me a strong intellectual excitement because these texts succeed in unifying biology to economical science through their borderline disciplines, ethology, and microeconomics; microeconomics is the branch of economical science dealing with the behavior of individual persons or agents and the principles of their decision making. Yet, the situation was still more stimulating because it led me to have an inkling of the implication of a third order of science, psychology. Sensory pleasure might be the common currency postulated by McFarland and Sibly. Let us examine this hypothesis further.

Ethology is the science of behavior stemming from zoology. Two great postulates form the basis of ethology. According to the first postulate, animal behavior is optimal by definition, since it was produced by natural selection over millions of years. For the ethologists, optimality is estimated from the reproductive efficiency of an individual and a species. Such a definition of optimality is therefore defined on the long-term privileged viewpoint for the zoologists, but does not exclude the short-term province of the physiologists.

Behavioral motivations are numerous as survival from predators, reproduction, alimentary needs, most of temperature regulation, and need to sleep are all satisfied behaviorally. The British ethologists McFarland and Sibly used the expression "behavioural final common path" to describe and underline that many different motivations converge toward a single means of expression (McFarland and Sibly, 1975). Such a term is especially eloquent for a physiologist. Indeed, it points toward the beautiful homology between behavior and the basic organization of the nervous system responsible for the motor response. The great English physiologist, Sir Charles Sherrington, was first to name "final common path," the alpha motoneurone whose cellular body is located in the ventral horn of the spinal chord and on which multiple excitatory and inhibitory paths converge (Sherrington, 1906). In the same way as the alpha-motoneurone integrates these influences and generates only one command, either inhibition or excitation, to the peripheral muscle, behavior integrates simultaneously contradictory motivations that cannot be satisfied simultaneously in a similar way (e.g., the need to sleep with the need to eat) and must manage a large mass of additional signals before acting. This illuminating image leads to the second postulate.

Another basic postulate of ethology is that behavior is the product of natural selection and thus is optimal (Baerends, 1956; Tinbergen, 1950). However, ethologists seldom question the proximate cause of behavior and optimization of behavior. McFarland and Sibly (1975) have theorized that since behavior is a *final common path,* there must exist a common currency within the brain–mind to allow trade-offs between motivations and thus rank them by order of priority (McFarland, 1985; McFarland and Sibly, 1975; McNamara and Houston, 1986). Over the last 30 years, a series of experiments led to the conclusion that pleasure is this common currency (Cabanac, 1995). More generally, I suggest the common currency is the hedonic axis of consciousness. The rationale is explained in the following text.

	Resulting hedonic experience	Action
Behavior 1	$a \rightarrow A$	yes
Behavior 2	$B \rightarrow b$	no
Behavior 1 + Behavior 2	$a + B \rightarrow A + b$	yes
with, $a < A$, and $B > b$ and with $a + B < A + b$		

Figure 7.4 Mechanism by which a behavior (Behavior 2) that produces displeasure (B to b) can be chosen by a subject if another behavior (Behavior 1) that produces pleasure ("a" to "A") is simultaneously chosen. The necessary and sufficient condition for the Behavior 2 to occur (action) is that the algebraic sum of affective experience (pleasure) of the yoked behaviors is positive (a+B<A+b). Capital letters "B" and "A" indicate larger pleasure than respective small letters "a" and "b." Theoretical presentation of two behaviors "1" and "2" and on how a behavior that arouses displeasure may be chosen by a subject. A subject can choose a behavior (Behavior 2) that produces displeasure if another behavior (Behavior 1) that produces pleasure is simultaneously chosen. The necessary and sufficient condition for the behavior "2" to occur (action) is that the algebraic sum of affective experience (pleasure) of the yoked behaviors is positive (P+D>p+d). Capital letters "P" and "D" indicate larger affective experience than respective small letters "p" and "d." (With permission from Cabanac, 1992.)

Since there is need for a common currency, if sensory pleasure/displeasure is the mental signal that allows us to produce useful behaviors adapted to their physiological goals, then pleasure/displeasure could be this common currency. This was confirmed in several experimental conflicts of motivations involving physiological functions such as temperature regulation, fatigue aroused by muscular exercise, and taste. Useful behaviors were selected by the subjects who simply maximized the algebraic sum of pleasures/displeasures aroused simultaneously in both sensory modalities involved. The same mechanism works with purely mental motivations. When subjects were placed in situations where a motivation involving a sensory modality (e.g., thermal discomfort, or hunger) was confronted with a purely mental motivation (e.g., the pleasure of playing a video game or the displeasure of losing money), they behaved in the same way. They maximized the algebraic sum of pleasures/displeasures aroused simultaneously in both sensory and mental modalities involved. Let us take an example: if we ask subjects to rate independently the hedonic magnitude of two kinds of stimuli such as the pleasure of earning money and the displeasure of experiencing pain, then if we present simultaneously money/pleasure AND pain/displeasure, the participants will end the session when the magnitude of the pain rating becomes higher than the pleasure of money earning. The principle of their decisions can be described by Figure 7.4. Finally, it was demonstrated in rats that motivations for acquiring rewards as different from one another as intracranial electrical stimulation and the sweet taste of sugar activated the same area of the brain (Shizgal, 1997; Shizgal and Conover, 1996). Thus, pleasure was also a common currency in animals.

Consciousness

The paramount role of pleasure and displeasure can be recognized not only in sensation and physiology but in the whole realm of consciousness (McKenna, 1996; Warburton, 1996). An evolutionary psychology perspective (Cosmides and Tooby, 1995; Cosmides et al., 1992) is useful here to extrapolate from sensation to consciousness and leads to a postulate. It may be postulated that consciousness evolved from sensation, including experiencing pleasure (Cabanac, 1996). A corollary of this postulate is that when consciousness evolved from sensation, consciousness inherited the properties of sensation, that is, Figure 7.3 describes not only sensation but also the four-dimensional structure of any conscious event.

We may accept the above conclusions as a result of mere introspection: I can analyze any thought I have and recognize a quality, an intensity, an hedonicity, and a duration. Furthermore, the postulate entails its corollary according to the following steps:

1. The hedonic axis of sensation, pleasure and displeasure, is the sign of the physiological usefulness of a stimulus. Pleasure is both the tag of a useful stimulus and the force that orients behavior to approach and eventually consume this stimulus.
2. Since behavior is a final common path, the brain needs a common currency to rank the motivations for access to behavior in a time-sharing pattern (McFarland and Sibly, 1975; McNamara and Houston, 1986).
3. Since maximization of sensory pleasure is the motivation for behaviors adapted to physiological goals and since motivations with physiological goals compete with other motivations such as playing, esthetic, and social ones for access to the behavioral final common path, pleasure is this common currency (Cabanac, 1992). In the same way as sensory pleasure tags the usefulness of a stimulus, joy tags the usefulness of any other conscious experience.[2]
4. The hedonic axis is thus the motivational capability of consciousness.
5. If consciousness possesses the hedonic axis, then the other axes of sensation also were inherited by consciousness.

The reader will have recognized in these five steps the same rationale as used earlier. The behavioral final common path, and the common currency, lead to the challenging conclusion that all conscious events bear the structure sketched in Figure 7.3. It follows that affectivity/hedonicity might be the general motivation for any behavior and that all the properties of sensory pleasure also would belong to the affective axis of consciousness. Joy is, in consciousness, congruent with pleasure in sensation. Sensory pleasure possesses several characteristics: pleasure is contingent, pleasure is the sign of a useful stimulus, pleasure is transient, and pleasure motivates behavior (Cabanac, 1971; 1979). Therefore, it follows from the same premises that joy keeps with global consciousness the same properties as pleasure with sensation. Joy possesses the same characteristics as sensory pleasure: joy is contingent, joy is the sign of a useful thought, joy is transient, and joy motivates behavior (Cazeneuve 1962; Cabanac 1986, 1996; Warburton 1996).

It follows that pleasure/joy is the ineluctable motor of action of the human mind and bears the properties of an impelling force, the application of which will actuate the human being. This conclusion does not exclude that animals are susceptible to obey also the law of maximization of pleasure; interested readers will find elements of discussion in Griffin (1992) and Dawkins (1993). The evolutionary advantage of sensory pleasure, for the animals that first had it, is that it saved the nervous system from storing astronomical numbers of potentially useful or noxious stimulus–response reflexes. Sensory pleasure added flexibility to the behavioral pattern of those which possessed it and passed it to their offspring. Similarly, joy saves the brain from storing an infinite number of rules of thumb and provides a formidable advantage both in amount of information stored and in flexibility of the behavioral responses. The major advantages of hedonic decision process as compared to rational (or reflexive) prehedonic decision process (in living or inanimate agents) is to facilitate the ranking of priorities and to add flexibility to this process. The hedonic dimension of consciousness may be what makes a conscious agent different from an algorithmic Türing machine, as pointed out by Penrose (1994).

McFarland and Sibly's theoretical analysis is satisfying for the ethologists. Such an analysis shows its excellency when, and by, permitting a mathematical description of behavior. However, if the theory describes from outside, phenomenologically how decisions are made, it does not inform on the mental cognition of the decision-making process. Yet, the process of decision making is a mental one. McFarland and Sibly explain the *how* of behavioral decisions, but there remains a problem to understand the *why* of behavior. Now, we just saw that sensory pleasure appeared to be linked to the physiological needs of the body and that the pursuit of sensory pleasure leads to behaviors adapted to satisfying the body's physiological needs. Sensory pleasure might answer the question left open by the ethologists in the theorization on the behavioral final common path and the common currency. Might it be that sensory pleasure would be the common currency that allows the trade-off among motivations and their ranking by order of priority? From a theoretical point of view, it is permitted to accept the hypothesis.

The reasoning goes as follows: sensory pleasure permits the optimization of behavior by triggering useful behaviors. Yet, if the first postulate of ethology, according to which behavior is always optimal is true, sensory pleasure should optimize any function that ranks first on the ladder of the motivations ranked by order of priority. Pleasure thus should optimize various motivations. If this it so, then the next fundamental question is: might pleasure be the common currency postulated as necessary on theoretical grounds?

According to such a hypothesis, it may be accepted that the trade-off between motivations takes place simply by comparing the magnitude of the various pleasures aroused if satisfying the various competing motivations. The motivation whose satisfaction will fulfill the strongest pleasure will be satisfied first. The second rank will be occupied by the motivation whose satisfaction will provide the highest remaining pleasure, and so on.

The theory permits to render count also of the selection of behaviors that procure displeasure. The affective dimension of any perception being the common currency would allow associating a behavior-producing pleasure to another behavior-producing displeasure. The decision would take place according to the result of the simple algebraic sum of pleasures and displeasures aroused by various behaviors, as in the theoretical example of Figure 7.4.

If a behavior "1" arouses a pleasure that increases from "p" to "P," the subject will tend to achieve such a behavior that procures pleasure.

If a behavior "2" arouses a displeasure that increases from "d" to "D," the subject will tend to avoid such a behavior that procures displeasure.

However, if behavior "1" and "2" are associated, yoked, the procurement is both "P" and "D," the subject will tend to achieve these yoked behaviors if in the affective result, the pleasure of "P"+ "D" is larger than the pleasure of "p"+ "d" (or if the displeasure is smaller). In the opposite case, of course the subject will abstain.

This association of a positive element providing pleasure with a negative element lacking pleasure evokes irresistibly other but similar processes taking place in living beings. It is common knowledge that the living beings are capable of synthesizing energy-rich molecules in order to store this energy for future use. Such a synthesis is spontaneously improbable because by costing energy to the environment, it apparently violates the second principle of thermodynamics. Living cells nevertheless synthesize routinely energy-rich molecules.[3] The transfer of energy and its storage in energy-rich molecules is possible only because a larger amount of energy is liberated simultaneously from the destruction of other energy-rich bonds. It is the yoking of an exothermic reaction with an endothermic reaction that allows the storage of energy to take place. (However, it should be recalled that the overall balance of the yoked reactions nevertheless augments total entropy of the system in its environment, as nature does not violate the second principle of thermodynamics.)

One can recognize some analogy between cellular thermodynamics and behavioral decision-making processes. In a manner similar to energy transfer, the yoking of two behaviors, one arousing pleasure and the other displeasure, permits behaviors that arouse displeasure and therefore that are spontaneously improbable to occur because the net balance is positive in favor of pleasure.

The law of the behavioral final common path is at the center of this reasoning because it forces motivations to be compared to one another by way of the common currency. In the same way as we went from physiology to psychology, then from physiology to microeconomics, the ineluctable implication of the behavioral final common path will now lead us to extrapolate once again and probe the very nature of happiness. For this, we must first learn the lesson from the teachings of the simplest case, that of sensory motivation. From the behavioral final common path, it follows that the teachings received from the studies in the field of sensory motivation should necessarily apply also to more complex motivations.

Let us return to the simplest case, thermal sensation. In Figure 7.5, we recognize the butterfly wing-type sensory response of a hypothermic subject, who feels cold stimuli as unpleasant and warm stimuli pleasant, and, while hyperthermic, feels the opposite in response to the same stimuli. Hyperthermia and hypothermia responses are completed with the responses of a subject in a state of normothermia. Let us start the analysis of the results here with the response during normothermia.

The normothermic subject does not feel pleasure from cold nor warm stimulations. Extreme temperatures, beyond the pain thresholds, arouse displeasure but all other temperatures arouse indifferent sensations. Thus, pleasure takes place only in situations where an internal trouble has to be corrected, either hyperthermia or hypothermia. Yet, the very consumption of pleasurable stimuli corrects the internal trouble: the warming of the skin will tend to eliminate hypothermia, its cooling down will suppress hyperthermia. It follows that pleasure will only last as long as the internal temperature default is not corrected by the stimulus source of pleasure. Sensory pleasure in response to a stimulus is hence a dynamic element arising as long as the stimulus is useful, but generating eventually its own extinction. In the state of normothermia, all thermal stimuli are indifferent. The absence of need eliminates pleasure. The definition of thermal comfort follows from this establishment.

Before 1987, thermal comfort was defined by International Union of Physiological Sciences' (IUPS). Commission on Thermal Physiology as "the subjective state of satisfaction towards the environment." Actually, this definition is ambiguous because it incorporates situations where the environment is pleasurable and when it is indifferent. As situations where the environment is pleasurable are eminently unstable, I proposed in 1987 to redefine thermal comfort as "the state of subjective indifference of a subject towards its thermal environment" (Commission for Thermal Physiology [IUPS], 1987). Thus, a comfortable environment can remain indefinitely so. Thermal comfort has nothing exciting. We can easily follow an identical reasoning in the case of gustatory sensation. Pleasure takes place with stimuli that correct an internal deficit in energy and metabolites. In a state of satiety, these stimuli turn indifferent.

We can more schematically represent the experimental results of Figure 7.5 in Figure 7.6. In this double entry table, the thermal states are reduced to three rows: hypothermia, normothermia, and hyperthermia. The skin temperature stimuli are also reduced to three columns: cold, tepid, and hot. The table thus defines nine states where the symbols U, I, and P indicate whether the stimulus is perceived as unpleasant, indifferent, or pleasant. For example, the block on the top left presents the case of a cold stimulus on a hypothermic subject. The sensation, which is unpleasant in this case, is indicated by a U. We can see that most situations are recognized as unpleasant or indifferent.

To be able to extend to other motivations, and follow in the general case the steps that have led us to distinguish sensory pleasure from comfort, we must now return our syllogism. The concept of behavioral common final path leads us naturally to look for a homologue scheme for the all other behaviors that satisfy other motivations, for example, ludic play, social, and so on. If, in the physiological domain, the motivation is the pursuit of sensory pleasure and the avoidance of displeasure, what is the motivation in the case

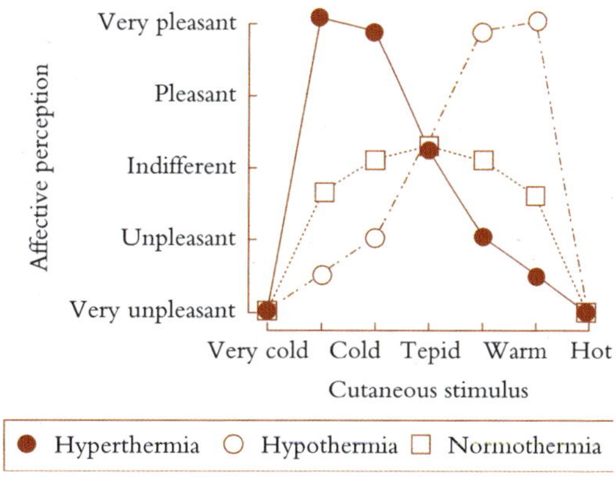

Figure 7.5 Pleasure (positive ratings) and displeasure (negative ratings) reported by a subject in response to thermal stimuli presented for 30 s on his left hand. During normothermia all stimuli between the threshold for cold and hot pains are experienced as indifferent. (With permission from Mower, 1976.)

Internal state \ Stimulus	Cold	Tepid	Hot
Hypothermia	U	I	P
Normothermia	U	I	U
Hyperthermia	P	I	U

Figure 7.6 The results of Figure 7.5 simplified, here, into a 3×3 matrix. The hedonic dimension of thermal sensation depends on the subject's internal state. A thermal stimulus feels unpleasant (U), indifferent (I) or pleasant (P) depending on body core temperature. Pleasure occurs only in dynamic situations when a stimulus tends to correct an internal trouble. (With permission from Cabanac, 1981.)

of other behaviors? If in physiology, behavior leads to the indifference of comfort, where will the other behaviors lead? If the brain works identically for physiological motivations and all other motivations—the behavioral final common path and the common currency make this a prerequisite—then, when physiological behavior tends to comfort, all other behaviors will tend to a state of satisfaction, a common denominator for all motivations. What is that state, could it be what common language calls happiness?

Ambiguous Happiness

What is happiness? Happiness has been defined as "the state of complete interior satisfaction"[4] or "the state of consciousness of complete satisfaction."[5] Actually, these definitions contain the same ambiguity as the former American Society of Heating, Refrigerating, and Air-Conditioning Engineers (ASHRAE) definition of thermal comfort, which was discussed in the preceding section, incorporating under the same word, a stable and indifferent element, comfort, and two dynamical, corrective elements, cold and warm pleasures. That ambiguity of the word happiness will be reflected in innumerable contradictory citations that we can list from literature attributed to several authors. Flaubert expressed it facetiously: "we must never think of happiness, this brings out the devil because it is him who invented this concept to enrage mankind." It is clear that the word *happiness* as it is usually used comprises two entities, the first one stable, the other transient. The use of a single vocable maintains the confusion found between Proust's writing: "this extension, this possible multiplication of ourselves that is happiness," on the one hand, and Jules Renard's, on the other hand: "we are not happy; our happiness is the silence of our unhappiness." These two citations are not necessarily wrong, even though they are opposed, but their opposition itself leads to the false conclusion that happiness is fallacious and does not exist, or not enough. Let us judge: "Man cannot and does not want to be happy" (Thomas Aquinas); "neither man nor any animal is happy" (David Hume); "God has not anticipated happiness for His creatures" (J. Giraudoux); "the great hindrance to happiness is to expect too great a happiness" (B. Fontenelle); "we are not happy, happiness does not exist" (A.P. Tchekov); "I learnt that happiness is the knowledge that happiness doesn't exist" (R. Fallet); finally, to limit an enumeration that could be much longer, "I'll never know happiness on earth" (R. Queneau).

Yet, happiness does exist. But, as we had to dissociate comfort in comfort-indifference toward the environment, and in pleasure, transient and dynamic component, strongly leading to pleasurable stimuli, we can dissociate happiness into two components, one indifferent and the other transient dynamic. The first one is indifference, nirvana-happiness (or, simply, happiness). The second one is joy-happiness (or, simply, joy).

According to the analogy happiness/comfort, happiness is a state of indifference toward all motivations; we find this concept of renunciation suggested in nirvana the Hinduist and especially Buddhist aim. The nirvana, "state of supreme serenity"[6] or "last state of contemplation characterized by the absence of pain,"[7] is actually an indifference toward all motivations; this ataraxia is recommended by the sceptics[8] toward all motivations Whereas comfort is only in the realm of physiological motivation. "Happiness is empty" for Victor Hugo and "is never grand" for Aldous Huxley. It is actually a stable state that can be durable, obtained from the absence of motivation or after satisfaction: "Happiness must be something solid and permanent" (Samuel Johnson). Happiness, like comfort, is a sign that everything is going well: "We cannot be happy for a wrong reason; we cannot be wrong of being happy" (H. Petit).

Joy is to happiness what pleasure is to comfort. Joy, the "pleasurable and profound emotion, extolling feeling perceived by the whole consciousness" (Dict. Robert), is indeed the dynamic, transient, and sought analog of sensory pleasure. "Joy is a burn that you cannot relish" (A. Camus). Joy is dynamic but short. "The pleasure of the other specie affects no portion of our body in particular, we name it pleasure of the spirit and I call it joy" (T. Hobbes). Like pleasure, joy is sought and, hence motivates innumerable behaviors. The more integrating the behavior—joy being both a cause and a result—the more lively the joy. "Joy is the passage of man from one smaller to a greater perfection" (Spinoza). Like pleasure, joy disappears with the completion of the given behavior, with the satisfaction of the given motivation and the reaching of happiness. A survey done on 467 subjects showed that the events considered the most wanted were all transient, such as a birth, a promotion, or a reconciliation (Henderson et al., 1984). From its characteristics, intensity and transience, joy generates disappointment if we expect it to be a perennial, a property that it lacks, from its very nature. We would like to grab joy, but it is already gone. One may paraphrase the old Latin adagio applied to sexual pleasure and generalize it to joy: *omnia anima post gaudium triste*.[9]

	Sensation	Consciousness
Pleasant	pleasure	joy
Indifferent	comfort	happiness

Figure 7.7 Joy is to happiness what pleasure is to comfort, and thus, the transience of pleasure can be found also with joy, whereas comfort and happiness are stable but indifferent.

To distinguish happiness and joy lifts the ambiguities (Cazeneuve, 1962) usually created by the use of the word happiness. It would seem reasonable, therefore, to accept that the match comfort/pleasure is only a particular case of the general match happiness/joy. These two couples are indeed homothetic. (Figure 7.7).

We can easily recognize in our day-to-day lives and that of our contemporaries the characteristics of joy and happiness. The success in an examination, a promotion, an honorific reward is all quickly forgotten. After a short transient while, of 24 to 48 hours, the new state appears as normal. We slip into the new situation: in the same way as sensory pleasure is dynamic but transient, joy of a promotion or success is brisk, hot, but transient.

Hunters and seducers declare often that the process of conquest brings them more satisfaction and motivates them more than the consumption of their capture. It is also commonplace that money does not bring happiness. On the contrary, one may say that wealth brings happiness, happiness being defined now as the absence of disagreement and discomfort. It is a sad happiness. Wealth does not bring dynamic and transient joy, which is independent of it.

This reasoning that we justly followed, a simple extrapolation of physiological data and knowledge, leads us to a utilitarian conclusion: joy, like pleasure, is a sign that the chosen behavior at a given time is useful for the individual at that time. It is then tempting to go one step further and to make it the key of the optimization of all individual decision making. The same conclusions, of course, can be reached from introspection alone (Cazeneuve, 1966).

Thus, the hedonic dimension of consciousness is what renders decisions simple. To minimize displeasure and to maximize pleasure is both the result and the aim of behavior. That pleasure dialectics is a dynamic phenomenon that serves to optimize behavior. It would also be of interest to explore the zoology of pleasure to discover when this property emerged in phylogeny and became so successful that it was maintained up to the human, but that is another story...

Notes

1. It is often considered that Teilhard's main thrust was an attempt to reconcile science and faith because he was a scientist and a priest. However, in this precise book, his aim was indeed what I state here.
2. This was confirmed in a series of experiments where pleasure/joy tagged efficacious mental activities in grammar Balaskó, M. and Cabanac, M. (1998), ethics and mathematics Cabanac M., et als. (2002) aggressiveness Ramírez J. M., Bonniot-Cabanac M.-C., Cabanac M. (2005), Politics and gambling Bonniot-Cabanac M.-Cl., Cabanac M. (2009) and actually even rational decisions Cabanac M., Bonniot-Cabanac M.-Cl. (2007).
3. See for example A. L. Lehninger (1965) *Bioenergetics. The Molecular Basis of Biological Energy Transformations.* New York: W. A. Benjamin Inc.
4. Petit dictionnaire Larousse, 1989.
5. Dictionnaire Robert. The psychologist M. Argyle gives a similar definitiuon: state resulting from satisfaction with life (*The Psychology of Happiness.* London: Methuen, 1987).
6. Dictionnaire Robert.
7. Dictionnaire Larousse.
8. Pyrrhon of Athens, 365 B.C.–275 B.C.
9. *omnia anima post coitum triste.*

References

Aristotle (384 B.C.–322 B.C.) *De l'âme.* Librairie philosophique J. Vrin: Paris.

Attia, M. (1984) Thermal pleasantness and temperature regulation in man. *Neurosci. Biobehav. Rev.* 8, 335–343.

Baerends, G. P. (1956) Aufbau des tierischen Verhaltens. In: *Handbuch der Zoologie Vol. VIII Mammalia*, pp. Teil 10 (Lfg 7). Eds. W. Kükenthal and T. Krumbach. De Gruyter & Co.: Berlin.

Balaskó, M. and Cabanac, M. (1998) Grammatical choice and affective experience in a second-language test. *Neuropsychobiology* 37, 205–210.

Barrow, J. D. and Tipler, F. T. (1986) *The Anthropic Cosmological Principle.* Oxford University Press: Oxford.

Bentham, J. (1823 [1748–1832]) *An Introduction to the Principles of Morals and Legislation (réédition, 1948).* Hafner Publishing Co.: New York.

Bertola, F. and Curi, U. (1993) *The Anthropic Principle: Proceedings of the Second Venice Conference on Cosmology and Philosophy.* Cambridge University Press: Cambridge.

Bleichert, A., Behling, K., Scarperi, M. and Scarperi, S. (1973) Thermoregulatory behavior of man during rest and exercise. *Pflüg. Arch.* 338, 303–312.

Bonniot-Cabanac M.-Cl., Cabanac M. (2009) Pleasure in decision making situations: Politics and gambling *J. Risk Res.* (in press).

Bowring, J. (1962) *The Works of Jeremy Bentham Published Under the Superintendance of his Executor John Bowring 1833–1844*. Russell & Russell Inc.: New York.
Breuer, R. (1991) *The Anthropic Principle. Man as the Focal Point of Nature*. Birkhäuser: Boston.
Brillat-Savarin, A. (1828) *Physiologie du goût, ou Méditations de gastronomie transcendante; ouvrage théorique, historique et à l'ordre du jour*. A. Sutelet et Cie, Libraires: Paris.
Brondel, L. and Cabanac, M. (2007) Alliesthesia in visual and auditory sensations from environmental signals. *Physiol. Behav.* 91, 196–201.
Cabanac, M. (1971) Physiological role of pleasure. *Science* 173, 1103–1107.
Cabanac, M. (1979) Sensory pleasure. *Q. Rev. Biol.* 54, 1–29.
Cabanac, M. (1986) Du confort au bonheur. *Psychiatrie Française* 17, 9–15 [*English version*: Cabanac, M. (2000) What is happiness? In *Fechner Day 2000*, pp. 29–34. Ed. C. Bonnet. Amalgame Impression: Strasbourg.]
Cabanac, M. (1992) Pleasure: the common currency. *J Theor Biol* 155, 173–200.
Cabanac, M. (1994) What is sensation? In: *Biological Perspectives on Motivated Activities*, pp. 409–428. Ed. R. Wong. Ablex: Norwood, NJ.
Cabanac, M. (1995) *La quête du plaisir*. Liber: Montréal.
Cabanac, M. (1996) On the origin of consciousness, a postulate and its corrollary. *Neurosci. Biobehav. Rev.* 20, 33–40.
Cabanac M., Bonniot-Cabanac M.-Cl. (2007) Decision making: rational or hedonic? *Behav. Brain Funct.* 3, 1–45.
Cabanac, M., Guillaume, J., Balasko, M. and Fleury, A. (2002) Pleasure in decision-making situations. *BiomedCentral*. http://www.biomedcentral.com/imedia/9977974521128571_ARTICLE.PDF
Cabanac, M., Massonnet, B. and Belaiche, R. (1972) Preferred hand temperature as a function of internal and mean skin temperatures. *J. Appl. Physiol.* 33, 699–703.
Cazeneuve, J. (1962) *Psychologie de la joie*. Michel Brient: Paris.
Cazeneuve, J. (1966) *Bonheur et civilisation*. Gallimard: Paris.
Changeux, J.-P. (1983) *L'homme neuronal*. Fayard: Paris.
Commission for Thermal Physiology (IUPS) (1987) Glossary of terms for thermal physiology. *Pflüg. Arch.* 410, 567–587.
Condillac, E. B. de (1754) *Traité des sensations*. Arthème Fayard: Paris.
Cosmides, L. and Tooby, J. (1995) From evolution to adaptations of behavior. In: *Biological Perspectives on Motivated Activities*, pp. 11–74. Ed. R. Wong. Ablex Publishing Co.: Northwood, NJ.
Cosmides, L., Tooby, J. and Barkow, J. H. (1992) Introduction: Evolutionary psychology and conceptual integration. In: *The Adapted Mind*, pp. 3–15. Eds. J. H. Barkow, L. Cosmides and J. Tooby. Oxford University Press: New York.
Dawkins, M. S. (1993) *Through Our Eyes Only?* W. H. Freeman & Co.: Oxford.
Ewert, J. P. (1968) Der Einfluss von Zwischenhirndetekten auf die Visuomotorik im Beute-und Fluchtverhalten der Erdkröte (*Bufo bufo* L.). *Z. Physiol.* 61, 41–70.
Fantino, M. (1984) Role of sensory input in the control of food intake. *J. Auton. Nerv. Syst.* 10, 326–347.
Fantino, M. (1995) Nutriments et alliesthésie alimentaire. *Cah. Nutrit. Diétét.* 30, 14–18.
Gell-Mann, M. (1994) *The Quark and the Jaguar*. W. H. Freeman & Co.: New York.
Griffin, D. R. (1992) *Animal Minds*. University of Chicago Press: Chicago.
Henderson, M., Argyle, M. and Furnham, A. (1984) The assesment of positive life events. *Unpublished*.
Jevons, W. S. (1871) *The Theory of Political Economy*. MacMillan: London.
Kandel, E. R. and Schwartz, J. H. (1985) *Principles of Neural Science*. Elsevier Science: Amsterdam.
McBride, R. L. (1997) *On the Relativity of Everyday Pleasures. The Value of Pleasures and the Question of Guilt*. Ed. D. M. Warburton Arise: Rome.
McFarland, D. J. (1985) *Animal Behaviour*. Pitman: London.
McFarland, D. J. and Sibly, R. M. (1975) The behavioural final common path. *Philos. Trans. R Soc. Lond.* 270, 265–293.
McKenna, F. (1996) Understanding risky choices. In: *Pleasure and Quality of Life*, pp. 157–170. Eds. D. M. Warburton and N. Sherwood. John Wiley & Sons Ltd.: Chichester.
McNamara, J. M. and Houston, A. I. (1986) The common currency for behavioural decisions. *Am. Nat.* 127, 358–378.
Nadel, E. R., Horvath, S. M., Dawson, C. A. and Tucker, A. (1970) Sensitivity to central and peripheral thermal stimulation in man. *J. Appl. Physiol.* 29, 603–609.
Pascal, B. (1623–1622) *Pensées*. Livre de poche 1969: Paris.
Penrose, R. (1994) *Shadows of the Mind*. Oxford University Press: Oxford.
Ramírez J. M., Bonniot-Cabanac M.-C., Cabanac M. (2005) Can aggression provide pleasure? *Eur. Psychol.*, 10, 136–145.
Rossetti, Y., Meckler, C. and Prablanc, C. (1994) Is there an optimal arm posture? Deterioration of finger localization precision and comfort sensation in extreme arm-joint postures. *Exp. Brain Res.* 99, 131–136.
Sherrington, C. S. (1906) *The Integrative Action of the Nervous System*. Yale University Press: New Haven, CT.
Shizgal, P. (1997) Neural basis of utility estimation. *Curr. Opin. Neurobiol.* 7, 198–208.
Shizgal, P. and Conover, K. (1996) On the neural computation of utility. *Curr. Dir. Psychol. Sci.* 5, 37–43.
Stolwijk, J. A. J., Saltin, B. and Gagge, A. P. (1968) Physiological factors associated with sweating during exercise. *Aerosp. Med.* 39, 1101–1105.
Teilhard de Chardin, P. (1965) *La place de l'homme dans la nature*. Éditions du Seuil: Paris.
Tinbergen, N. (1950) The hierarchical organization of mechanisms underlying instinctive behaviour. *Exp. Biol.* 4, 305–312.
Warburton, D. M. (1996) The functions of pleasure. In: *Pleasure and Quality of Life*, pp. 1–10. Eds. D. M. Warburton and N. Sherwood. John Wiley & Sons: Chichester.
Young, P. T. (1959) The role of affective processes in learning and motivation. *Psychol. Rev.* 66, 104–123.

8

Olfaction and Its Pleasures: Human Neuroimaging Perspectives

JAY A. GOTTFRIED

I like to think that the editors of this book made a highly calculated decision to insert this chapter on smell immediately following "On the Nature and Function of Pleasure" (Frijda, Chapter 6, this volume) and "The Dialectics of Pleasure" (Cabanac, Chapter 7, this volume). After all, the sense of smell has long played a critical role in hedonic experience, emotion, and behavior. Indeed, the emotional potency of smell can be traced back to unicellular organisms swimming the seas many hundred-millions of years ago. By virtue of chemical sensors on their cell membranes, these species were able to move sensibly through aqueous chemical gradients, using their "sense of smell" to pinpoint objects of pleasure, i.e., friends and food sources. The same chemical sense was equally critical for avoiding objects of displeasure, such as poisons, waste products, and predators that threatened survival. The emotional primacy of smells remains a key "function" of olfactory systems, evolutionarily conserved from molds to mammals in order to mediate biologically important behaviors. Although the human sense of smell is often considered inferior to the faculties of sight and sound, it is worth mentioning that the human nose can routinely (a) detect trace amounts of odorants in the range of parts-per-billion, (b) distinguish among odorants that only differ by a single carbon atom, (c) sense certain odorants even more acutely than rodents and dogs, and (d) discriminate tens of thousands of different smells, if not more.

The power of smell has been recognized since antiquity. In Greek mythology, it is said that the women of Lemnos, failing to make worship and offerings to the goddess Aphrodite, were cursed with such a foul stench that their husbands were forced to take new wives (Apollodorus, c. 1st century B.C.). The philosopher Theophrastus (370 B.C. to c. 285 B.C.) wrote about the medicinal properties of certain perfumes, chief among them "megaleion," a costly mixture of burnt resin, oil of balanos, cassia, cinnamon, and myrrh, which was thought to be an excellent wound anti-inflammatory. Also available in the ancient pharmacopeia were rose-perfume, which relieved strangury, a painful (perhaps common?) bladder syndrome of frequent urination, and iris-perfume, which was useful as a laxative (Theophrastus, c. 300 B.C., in Hort, 1926). In describing 5th century B.C. Scythian customs, the historian Herodotus (484 B.C. to c. 430 B.C.) noted that the battle-hardy Scythian nomads never washed with water, preferring instead to plaster their bodies with a paste of cypress, cedar, and frankincense woods that imparted a cleansing "sweet odor" and smooth texture to the skin (Herodotus, c. 1450 B.C., in Godolphin, 1974).

Well into the 17th century A.D., odors were believed to be the direct agent of diseases as well as their remedy. It was widely thought that odorous vapors were the carrier of bubonic plague, and sweetly scented pomanders were worn to ward off the illness. The ominous mask of the plague doctor, with its long pointed beak, was filled with a potpourri of spices to purify the air, providing an extra level of protection against the scourge (Jackson, 2003). Throughout the 17th and 18th centuries, numerous exotic odors with alleged healing powers abounded. For example, a musk-like odor ("civet") obtained from the perineal glands of civet cats was used to treat hysteria (when applied to the navel), vertigo and apoplexy (when applied to the nostrils and temple), pain, deafness, and epilepsy (Jackson, 2003). Another legendary substance was ambergris ("gray amber" in French), extracted from the bowels of sperm whales and prized as an essential component of love potions, among other sundry cosmetic and medicinal uses (Murphy, 1933).

The common-sense idea that human behavior is under the powerful sway of odor has even filtered into 20th century comic book culture. Working in the Golden Age of Comic Books (dating roughly from the first appearance of Superman in 1938 to the initiation of comic censorship regulations in 1954), the artist Jack Cole created a superhero known as Plastic Man, whose powers of elasticity enabled him to stretch into an infinite variety of shapes in his combat against crime. In one such episode (see *Police Comics* #76, March 1948; Cole, 1948), a disgruntled perfumer assembles a perfume out of ambergris, swamp fog, weeping willow sap, attar of roses, musk, onion juice, tear gas, and a half ounce of widow's tears (see Figure 8.1). This somber scent causes such pathological melancholy that saddened smellers willingly relinquish money, jewelry, and even oil deeds to the crafty perfumer, as he amasses a tidy fortune. (The sad tale ends on a happy note when Plastic Man, sprayed with the stench, falls into a river, thereby washing off the perfume, eliminating its grip on the crime fighter.)

Each of these anecdotes is meant to illustrate the sacred regard that humans have long maintained for smells. Recent behavioral data confirm these impressions by showing that odorous stimuli have significant effects on mood (Lundstrom et al., 2003; Lundstrom and Olsson, 2005; Villemure and Bushnell, 2007),

Figure 8.1 Illustrating the emotional potency of smell. In *Police Comics* #76 (March 1948) by Jack Cole, a disgruntled perfumer concocts a perfume so melancholy (top panels) that a mere whiff is sufficient to cause merchants, jewelers, and oil prospectors to relinquish their valuables (bottom panels). Eventually Plastic Man is forced to contend with this olfactory master of crime (panels not shown). (Reprinted with permission of DC Comics.)

physiology (Alaoui-Ismaili et al., 1997; Bensafi et al., 2002, 2004), and behavior (Herz and Schooler, 2002; Holland et al., 2005; Jacob et al., 2001a; Kovacs et al., 2004), even during implicit odor exposure (Degel and Koster, 1999; Koster et al., 2002) and even when odor concentrations are so low that they cannot be consciously perceived (Li et al., 2007). With these considerations in mind, the remainder of the chapter will focus on the anatomy and function of the human olfactory system, based largely on functional imaging studies over the last 15 years, with an emphasis on recent research involving odor affect and olfactory emotional learning. It should be noted that a variety of neuroimaging approaches, including functional magnetic resonance imaging (fMRI), 2-deoxy-glucose uptake, and optical imaging, have been successfully used to characterize the olfactory bulb in rodent models (Johnson et al., 1999; Malnic et al., 1999; Meister and Bonhoeffer, 2001; Mori et al., 1992; Rubin and Katz, 1999; Xu et al., 2003), but this topic is beyond the scope of the present chapter.

Neuroanatomy of the Human Olfactory System

A brief anatomical overview of human olfaction will help provide a framework for subsequent sections of this chapter on olfactory neuroimaging. Unfortunately, there have been very few careful anatomical studies of the human olfactory system, and none of these has extended much beyond a descriptive characterization at the gross level (Eslinger et al., 1982; Heimer, 1995; Mai et al., 1997). In particular, the anatomical boundaries of primary olfactory cortex, including piriform cortex, amygdala, and entorhinal cortex, are not well established in humans, and the relevance of animal models (mainly rodent and monkey) to the human system is not at all clear. What follows is a consensus view of how an aroma detected at the nose gains access to subcortical and cortical olfactory structures in the human brain.

When a volatile chemical (the "odorant") is inhaled through the nose, it binds to receptors of olfactory sensory neurons distributed throughout the olfactory epithelium in the absence of any systematic spatial organization. These receptors are in the family of G-protein-coupled receptors, the endings of which are submerged in the mucus layer of the nasal mucosa. The necessary diffusion of the odorant (much of which may be water-insoluble) through the aqueous mucus introduces one of many temporal delays in the chemical transduction of the odor signal. It is important to note that even though each sensory neuron expresses one type of olfactory receptor tuned to a unique chemical attribute of an odorant (e.g., molecular functional group), there is no simple one-to-one correspondence between odorant and receptor. This is because each neuron is capable of binding to multiple different odorants that share common chemical features, and because each odorant is capable of binding to multiple different sensory receptors with affinity for different chemical moieties of the native stimulus. In rodents there are approximately 1000 different receptors, though in humans, it is estimated that only about 350 receptors are functionally active, due to a high proportion of pseudogenes in the human olfactory genome (Gilad et al., 2003; Mombaerts, 1999; Niimura and Nei, 2005).

All of the sensory neurons expressing the same receptor project through the olfactory nerve (Cranial Nerve I) onto a small number of loci (glomeruli) in the olfactory bulb. Within glomeruli of the olfactory bulb, axon terminals of the sensory neurons contact the dendrites of mitral and tufted cells. Thus, there is a massive convergence of inputs at this first synapse, providing an opportunity for integration of odor information. Animal imaging studies indicate that specific chemical attributes of an odorant, such as carbon chain length or functional group, are coded in specific clusters or "modules" within olfactory bulb glomeruli (Johnson et al., 1999; Malnic et al., 1999; Meister and Bonhoeffer, 2001; Mori et al., 1992; Rubin and Katz, 1999; Xu et al., 2003), but whether these organizational rules apply to the human olfactory bulb remains to be determined. The human olfactory bulb is exceedingly slender, on the order of 50–60 mm^3 (per side) (Mueller et al., 2005; Turetsky et al., 2000), lying in the olfactory sulcus just lateral to the gyrus rectus on the medial orbital surface of the brain. Its small size and position beside the sinus cavity makes it highly prone to MRI signal loss and distortion, precluding its investigation using human functional imaging techniques.

Afferents from the olfactory bulb coalesce into the olfactory tract, which extends posterolaterally through the lateral olfactory tract and terminates on several regions collectively referred to as the "primary olfactory cortex" (POC) (see Figure 8.2). It is worth bearing in mind that there is probably very little that is "primary" about POC, with anatomical (Haberly, 1985; Illig and Haberly, 2003; Johnson et al., 2000) and functional features more resembling sensory association cortex, distinguishing it from other "primary" sensory areas such as visual striate cortex (V1) in the occipital lobe or somatosensory cortex (S1) in

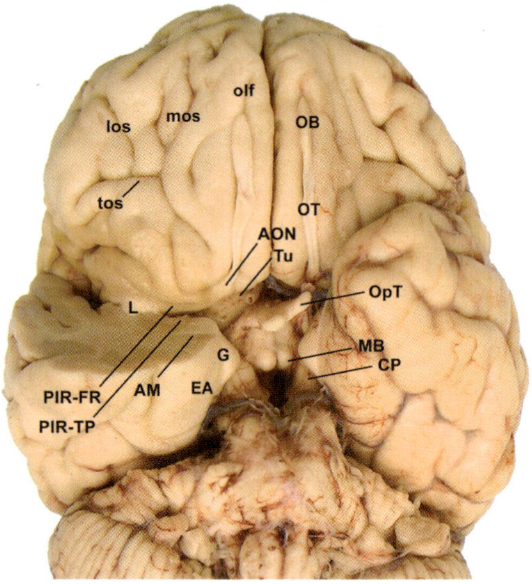

Figure 8.2 Neuroanatomy of the human olfactory system. This view of the human ventral forebrain and medial temporal lobes depicts the olfactory tract, its principal projections, and surrounding structures. The right medial temporal lobe (left side of figure) has been resected coronally through the mid-amygdala in order to reveal key olfactory cortical structures. AON, anterior olfactory nucleus; CP, cerebral peduncle; EA, entorhinal area; G, gyrus ambiens; L, limen insula; los, lateral olfactory sulcus; MB, mammillary body; mos, medial olfactory sulcus; olf, olfactory sulcus; PIR-FR, frontal piriform cortex; OB, olfactory bulb; OpT, optic tract; OT, olfactory tract; tos, transverse olfactory sulcus; Tu, olfactory tubercle; PIR-TP, temporal piriform cortex. (Image prepared with the help of Dr. Eileen H. Bigio, Department of Pathology, Northwestern University Feinberg School of Medicine, Chicago, IL. This figure was published as Fig. 1 in Gottfried and Zald (2005). Copyright © Elsevier 2005.)

the postcentral gyrus. It is also important to note that afferent olfactory projections remain ipsilateral, at least as far as cortex, which is unique among the senses.

The principal POC target of the olfactory bulb is the piriform cortex ("pear shaped"), a three-layer paleocortex situated at the junction of the frontal and temporal lobes. Animal evidence suggests that the piriform cortex is anatomically and physiologically heterogeneous, including anterior and posterior subdivisions, and human imaging studies appear to confirm functional piriform dissociations along an anterior–posterior axis. Other major projection sites of the olfactory bulb (from rostral to lateral) include the anterior olfactory nucleus (AON), the olfactory tubercle (OTu), the amygdala (specifically, the anterior and posterior cortical nuclei, medial nucleus, the nucleus of the lateral olfactory tract, and the periamygdaloid area), and a rostral portion of the entorhinal cortex. In humans, the AON and OTu are particularly small, and the lack of easily recognizable anatomical landmarks makes these structures very difficult to visualize without the aid of a microscope. On the basis of animal data, it is generally presumed that the AON provides a route for contralateral olfactory projections by way of the anterior commissure, and the OTu appears to be cytologically affiliated with ventral striatum and nucleus accumbens, regions rich in dopamine terminals that underlie reward-related processing. Most of POC, with the exception of the OTu, contains dense feedback projections to the olfactory bulb. All POC regions have substantial interconnections via associational intracortical fibers.

Downstream projections from the various sites in POC synapse upon numerous structures involved in emotion, learning, and memory, including basolateral amygdala, hypothalamus, hippocampus, insula, and the basal ganglia. One critical output from the piriform cortex projects directly to the orbitofrontal cortex (OFC) (Carmichael et al., 1994). Thus an odor input is no more than three synapses removed from prefrontal cortex (receptor neuron-bulb, bulb-piriform, piriform-OFC), providing an intimate link to neocortical brain regions that mediate multisensory integration, feeding, and adaptive behavior. Interestingly, the location of the putative human olfactory OFC (based on a meta-analysis of 13 imaging studies) is 15–20 cm rostral to the proposed monkey olfactory OFC (based on anatomy and electrophysiology), suggesting that attempts to infer human orbitofrontal function from animal models may not always be appropriate (Gottfried and Zald, 2005).

Regardless of interspecies homologies, it is clear from this anatomical arrangement that odor information has access to prefrontal cortex without a requisite thalamic intermediary, another distinguishing hallmark of olfaction. This observation implies that either (a) upstream structures (such as the olfactory bulb or piriform cortex) are able to carry out many of the processing functions otherwise reserved for sensory thalamic nuclei or (b) the olfactory system has no need to filter odor information through a thalamic module. Interestingly, research in rodents (Price and Slotnick, 1983) and nonhuman primates (Yarita et al., 1980) has demonstrated a second "indirect" pathway involving the mediodorsal thalamus (MDT) by which an odor may gain prefrontal access (from piriform to MDT to OFC). The precise functional role of this thalamic pathway is poorly understood, though some lesions of MDT impair odor discrimination learning in rats (Staubli et al., 1987).

Neuroimaging of the Human Olfactory System

Until the arrival of modern neuroimaging techniques, most of our knowledge of olfactory function in the human brain was based on lesion studies of patients with damage to the medial temporal and basal frontal lobes (e.g., Eichenbaum et al., 1983; Jones-Gotman and Zatorre, 1988). While such investigations provided important information about the central basis of human olfaction, the proximity of so many critical olfactory structures within a coarsely circumscribed lesional area limited efforts to relate anatomy to function in detail. Beginning with positron emission tomography (PET), and shortly thereafter with functional magnetic resonance imaging (fMRI), these imaging approaches have revolutionized analysis of the human olfactory system by permitting simultaneous whole-brain data acquisition at relatively high (millimeter) spatial resolution (reviewed in Sobel et al., 2003). Both PET and fMRI detect local activity-induced changes in hemodynamic state (specifically, regional cerebral blood flow [rCBF] in the case of PET; blood oxygenation level-dependent [BOLD] contrast in the case of fMRI), providing surrogate measures of neural activity that are constrained by the intrinsic time lag of neurovascular coupling. For this reason, the spatial virtues of PET and fMRI come at the expense of temporal resolution, generally on the order of seconds.

The first neuroimaging investigation of human olfaction was conducted in 1992. In this PET study, Zatorre et al. (1992) asked young healthy volunteers to inhale through their noses during presentation of eight different odorants absorbed onto cotton wands (experimental condition) or during presentation of an odorless wand (control condition) (see Figure 8.3). A comparison of odor to no-odor scans demonstrated significant neural responses in piriform cortex bilaterally as well as in the right OFC. Activity in the left OFC was also detected at slightly reduced statistical threshold. This groundbreaking study helped validate the use of functional imaging to characterize human

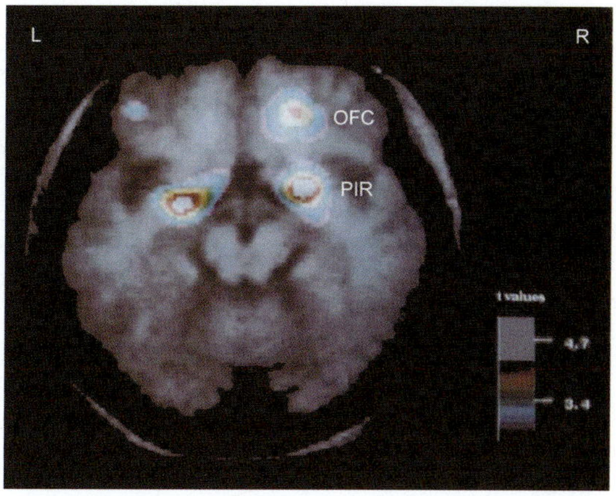

Figure 8.3 The first functional imaging study of human olfaction. In this PET study by Zatorre et al. in 1992, odor stimulation was associated with significant neural activation in piriform cortex (PIR) bilaterally and in orbitofrontal cortex (OFC) on the right side. Responses are overlaid on an axial section of the subject-averaged anatomical MRI scan. (Reprinted and modified from Zatorre et al. (1992) by permission from Macmillan Publishers Ltd.)

olfactory processing, and since this time, numerous investigators using both PET and fMRI have tested an impressive variety of olfactory-related paradigms, including manipulations of odor stimulus duration, odor intensity, odor hedonics, odor quality, odor attention, odor memory, odor localization, multisensory processing, appetite and motivation, higher-order cognitive judgments, and gender and age effects. What follows is by no means an exhaustive review of human olfactory neuroimaging, but is meant to highlight several important themes that have emerged over recent years, with particular emphasis on hedonic aspects of odor processing in the human brain, in keeping with the general themes of this book.

The Elusiveness of Primary Olfactory (Piriform) Cortex

A notable feature apparent across many of the initial olfactory studies was the unpredictable activation of POC (reviewed in Zald and Pardo, 2000), a perplexing and substantial problem. (Imagine for example: if the original fMRI studies of human vision showed inconsistent activation of primary visual cortex, the whole technique would have quickly fallen out of favor.) Work by Sobel et al. (2000) demonstrated that much of the response inconstancy in piriform cortex was due to sensory habituation. Because most imaging experiments at this time were "block designs," with prolonged odor delivery over a period of 30–60 s, piriform cortex underwent marked response decline (habituation), compromising efforts to measure neural activity. By modeling their odor block with an exponential decay function, Sobel and colleagues were able to record fMRI signal reliably from human piriform cortex (see Figure 8.4). This important finding demonstrated the value of limiting the exposure time of individual odorants and as such has helped motivate the use of event-related olfactory fMRI designs. It has also highlighted important physiological links with rodent piriform cortex, which exhibits analogous response properties of sensory habituation (Wilson, 1998). This result has since been confirmed by other investigators (Poellinger et al., 2001), and interestingly, recent work from our laboratory has revealed the odor specificity of sensory habituation, such that neural activity in human piriform cortex is reduced only to the exposed odor, but not to other odors, even for those that share qualitative or structural (molecular) features (Li et al., 2006).

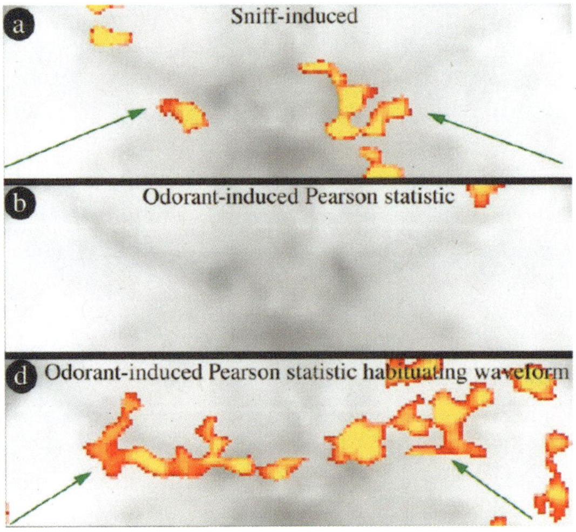

Figure 8.4 Odor habituation in human piriform cortex. In this fMRI study, odor and no-odor conditions alternated every 40 s, and subjects sniffed every 8 s. Sniffing induced neural activity in piriform cortex irrespective of odor presence (a), but odor stimulation (averaged across odor blocks) did not elicit significant piriform activity (b). However, when the time course of odor stimulation was modeled with an exponential reference waveform to account for sensory habituation, robust responses in piriform cortex were observed (d). (Reprinted and modified from Sobel et al., 2000.)

Another important factor compromising signal detection in piriform cortex is specific to fMRI. Conventional fMRI sequences are prone to considerable signal loss and signal distortion (susceptibility artifact) at air–tissue interfaces of the brain (Ojemann et al., 1997). As a result, these artifacts can reduce image quality in critical olfactory areas located in MRI-susceptible regions of the medial temporal and basal frontal lobes. Over the past few years, numerous techniques have been developed to minimize signal dropout in these regions (e.g., Deichmann et al., 2003; Gu et al., 2002; Wilson et al., 2002; Wilson and Jezzard, 2003) with moderate success. In our laboratory (using a Siemens 3-T whole-body MRI scanner), we have found that the combination of parallel imaging, an eight-channel head coil, an expanded spatial resolution (field-of-view, 220 × 220 mm; in-plane resolution, 1.72 × 1.72 mm; slice thickness, 2 mm, with 1-mm gap), and an oblique angle of data acquisition (30° to the intercommissural line, rostral > caudal) yields excellent fMRI signal recovery in piriform and orbitofrontal cortices (Li et al., 2006). Nevertheless, because this susceptibility artifact is not observed with PET, some investigators favor the PET method when piriform cortex is the focus of interest. In the opinion of this author, however, fMRI has important advantages over PET, provided that the imaging sequence has been optimized. The benefits of fMRI include the following: (a) it is safe and noninvasive, not requiring the intravenous delivery of radioactive tracers; (b) it is compatible with "event-related" studies, permitting the design of more complex experiments; and (c) the data can be collected continuously over an extended period of time, which is useful for studies involving learning and other time-dependent effects.

Functional Complexity of Human Piriform Cortex

As mentioned above, the conception of human piriform cortex as a "primary" olfactory region, functionally akin to other primary sensory structures, is really a misnomer. Initial fMRI studies of piriform cortex showed that this structure is not only odor-responsive, but also sniff-responsive (Sobel et al., 1998a). Sniffing of odorless air, as well as artificial sniffing induced by air puffs into the nostrils, activated piriform cortex, whereas partial physical occlusion of the nostrils, or topical anesthesia to the nasal passages, reduced piriform activation. These elegant results indicated that sniff-induced piriform activity is not simply due to the motor act of sniffing, but rather to the physical sensation of airflow across the nasal mucosa, and are compatible with animal data suggesting that the sniff may prime piriform cortex for optimal reception of a smell (Adrian, 1942; Ueki and Domino, 1961).

Based on recent fMRI studies over the last few years, however, it has become clear that piriform cortex supports a large variety of complex functions, and the idea of piriform cortex as a basic sensory throughput or relay is no longer tenable. The following subsections highlight some of these major functions.

Odorant Structure and Odor Quality. Understanding how the brain transforms an olfactory sensation at the nose into an integrated percept of odor quality is a research question of great importance that even preoccupied the minds of the ancient Greek philosophers. To Aristotle, olfactory perceptual experience could be reduced to Fire, one of the four Elements (with a capital "E"), reflecting humankind's intimate link to the universe (Aristotle, 350 B.C.). The general consensus, based largely on animal models, still champions reductionism, of a molecular genetic sort: it is presumed that the perceptual quality of an odorant necessarily follows from its chemical composition, such that the neural signature of odor quality is embodied in ensemble spatial activity across discrete clusters or "modules" in the olfactory bulb (Buck, 2004; Firestein, 2001; Leon and Johnson, 2003). There is even behavioral evidence in rodents to suggest that the degree of overlap between spatial activity patterns in the bulb may account for how well an animal can discriminate between different odorants (Cleland et al., 2002; Linster et al., 2001, 2002). However, this research is somewhat at odds with the finding that structurally related odorants can smell different, and structurally unrelated odorants can smell similar (Cain and Polak, 1992; Polak, 1973). It also conflicts with electrophysiological data showing synthetic or integrative odor coding in rodent piriform cortex, in marked distinction to more elemental or configural coding in the olfactory bulb. Such findings indicate that knowledge about an odorant's physical and chemical properties is not always sufficient to predict its odor quality.

In an attempt to explore the relationship between structure and quality coding in human piriform cortex, our laboratory used an olfactory version of fMRI cross-adaptation (Buckner et al., 1998; Grill-Spector and Malach, 2001; Kourtzi and Kanwisher, 2001; Winston et al., 2004) with the selection of odorant pairs that systematically differed in chemical structure (alcohol or aldehyde functional group) or perceptual quality ("lemon" or "vegetable") (Gottfried et al., 2006). As used here, the term "odor quality" refers to the perceptual character or identity emanating from an odorous object (in contrast to other perceptual features such as intensity or valence). Critically, this experimental design was fully balanced, controlling for the possibility that the findings could be confounded by variations in intensity, hedonics, or other perceptual dimensions.

Across 16 healthy subjects, we found that structure and quality were encoded in separable subregions of piriform cortex. Anterior piriform cortex was sensitive to odorant chemical structure, independently of odor quality, whereas posterior piriform cortex responded preferentially to odor quality irrespective of the underlying functional group. Odor quality coding also extended into other regions including OFC and hippocampus, consistent with the idea that odor quality representations may be distributed across a network of olfactory-related regions, in keeping with prior monkey (Tanabe et al., 1975) and human (Eichenbaum et al., 1983; Gottfried and Dolan, 2003; Royet et al., 1999; Savic et al., 2000) studies of olfactory discrimination and semantic processing. The identification of structure-based codes in anterior piriform would ensure stimulus constancy of the original input, which may be particularly important for a sensory system with high stimulus variability due to changes in wind speed or direction, respiratory cycling, and the presence of background odors. In turn, odor quality coding in posterior piriform cortex, irrespective of functional group (as one critical determinant of structure), implies that more synthetic mechanisms may govern the neural coding of smell at levels downstream of the olfactory bulb.

Learning-induced Plasticity. Neural representations of odor quality are not only maintained in human piriform cortex, but they can also be updated and modified through learning and experience. In our laboratory, Li et al. (2006) used an fMRI paradigm of perceptual learning (Gibson, 1991) to investigate how sensory experience alters behavioral and neural correlates of odor quality perception. In this study, subjects smelled four different stimuli: a target odorant (destined for habituation), an odorant related in perceptual quality, an odorant related in chemical structure (functional group), and an unrelated odorant, both before and after being exposed to the target odorant continuously for 3.5 min. Behaviorally, as a result of continuous odor exposure, subjects were better able to discriminate among quality- and functional group-related odorants. For example, a subject exposed to L-carvone (a minty ketone) became a mint "expert" and simultaneously developed expertise in distinguishing among ketone odorants. A comparison of pre- to post-habituation fMRI sessions revealed neural plasticity in piriform cortex in response to qualitatively similar odorants. Thus, the same (invariant) odor input was capable of evoking significantly different piriform responses depending on prior odor experience even in the absence of explicit training or feedback. Interestingly, neural plasticity was also observed in OFC and the magnitude of experience-dependent change in this region correlated directly with the degree of learning-induced change in behavioral discrimination, on a subject-by-subject basis (see Figure 8.5), suggesting that OFC critically mediates the effects of perceptual learning.

In a more recent experiment (Li et al., 2008), we implemented olfactory multivariate (pattern-based) fMRI analysis techniques to test the impact of aversive olfactory conditioning on perceptual and neural discrimination of a set of odor enantiomers (mirror-image molecules) that are perceptually indistinguishable. As a result of pairing one of these chiral odorants (the conditioned stimulus or CS+) with an electric shock (the unconditioned stimulus), perceptual discrimination between CS+ and its chiral (unshocked) counterpart was significantly enhanced. In parallel, fMRI pattern correlations between these two odorants (compared to a control nonconditioned enantiomer pair) significantly decreased in posterior piriform cortex, signifying greater pattern discriminability. These findings indicate that aversive learning induces piriform plasticity with corresponding gains in perceptual odor discrimination. That completely indiscriminable smells can be transformed into discriminable percepts further accentuates the potency of learning and experience to enhance olfactory perception.

Memory. Anatomical and computational models of olfactory processing have long suggested that piriform cortex may be a repository for odor memories (Haberly, 1985, 2001; Hasselmo and Bower, 1993). Several imaging studies have begun to confirm this in humans. One PET study comparing odor recognition scans to baseline scans revealed enhanced piriform cortex activity (Savic et al., 2000), and a second study also demonstrated participation of piriform cortex in memory processing at both short (minutes) and long (days) delays (Dade et al., 2002). Similar patterns of activation have been observed during an event-related fMRI study of olfactory contextual memory (Gottfried et al., 2004). Here, in an initial study phase, subjects were given combinations of smells and pictures and asked to covertly imagine a link or association between the two stimuli. The

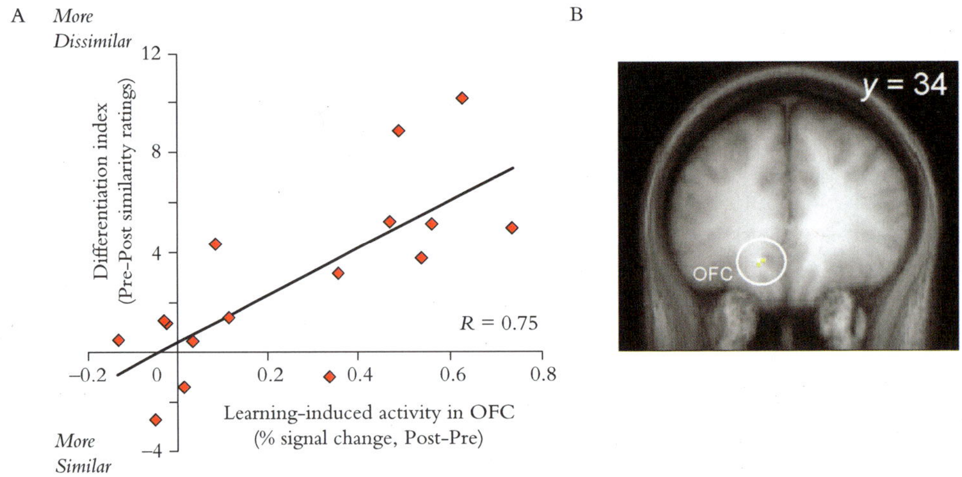

Figure 8.5 Neural activity in OFC relates to experience-dependent improvements in perceptual expertise. (A) Regression analysis demonstrates that the magnitude of learning-induced change in OFC (abscissa) directly correlates with the degree of perceptual enhancement (ordinate), on a subject-by-subject basis. Each diamond represents a different subject. (B) The mean effect of the behavioral correlation in OFC ($N = 16$ subjects) is overlaid on a coronal section of the group-averaged T1-weighted structural scan (threshold, $p < 0.005$ uncorrected). [Reprinted and modified from Li et al (2006). Copyright (2006), with permission from Elsevier.]

aim of this session was to encourage episodic memory encoding of odor-picture pairs. In a subsequent recognition memory test, subjects had to decide whether they were viewing a study (old) or novel (new) picture in the absence of any odor cues. Comparison of correctly remembered items to correct rejections ("old/new" effect) highlighted significant memory-related activity in piriform cortex even without sniffing or smelling. It is therefore tempting to suggest that the memory-related responses in piriform cortex provide evidence for the retrieval of olfactory context (experienced at encoding), in keeping with relational models of memory function (Cohen et al., 1997; Marr, 1971; Mishkin et al., 1997). Such an arrangement would help maintain sensory fidelity of the original memory trace (Mesulam, 1998) and facilitate reactivation of the multisensory memory by numerous sensory channels.

Attention. An fMRI investigation by Zelano et al. (2005) provided some of the first evidence to show that attention modulates neural processing in human piriform cortex. During odor detection blocks, subjects were asked to sniff and decide whether an odor was present or absent, throughout which attention to odor was maintained. During control sniff blocks, subjects were asked to sniff even though they were informed that no odor would be delivered on any of the trials. The critical comparison was between (a) odor detection trials in which subjects sniffed but *no odor* was delivered and (b) control trials in which subjects sniffed knowing that odor would not be delivered. This contrast ensured that any attention-dependent differences in neural activation could not confounded by the presence of odor. In this manner, it was found that attention to odor selectively increased responses in anterior piriform cortex, but not in posterior piriform cortex. More recent data from our laboratory (Plailly et al., 2008) indicate that when subjects are presented with a combined olfactory-auditory stimulus, attention to the odor (compared to attention to the tone) is associated with an increase in network coherence between posterior piriform cortex and mediodorsal thalamus, and between mediodorsal thalamus and OFC, suggesting that the so-called "indirect pathway" to olfactory OFC is functionally active in humans and may be critically involved in mediating conscious analysis of smell.

Odor Imagery. Piriform cortex is also activated during mental imagery of smells. In a clever PET study by Djordjevic and colleagues (2005), subjects were first given the name of a smell (e.g., rose) and asked to imagine its smell for 5 s. Subjects were then presented with a weak odor (that might or might not match the imagined odor) or with odorless air for 2 s, indicating via pushbutton whether an odor was present or absent. Behaviorally odor detection was more accurate when the imagined odor matched the presented odor in comparison to a mismatched condition. When odor imagery trials were directly compared to control trials (in which subjects were not asked to imagine odor), significant neural activity was identified in left piriform cortex. These results nicely coincide with our odor contextual memory study (Gottfried et al., 2004) and validate the idea that sensory-specific neural representations of odor are maintained in piriform cortex.

Imaging Olfactory "Pleasure" in the Human Brain

Neuroimaging studies have capitalized on the affective primacy of olfaction as a means of examining two different aspects of odor valence. One set of studies (described in "Odor As a Hedonic State") has focused on the hedonic properties of the odors themselves, in an effort to delineate the brain regions activated in response to pleasant or unpleasant smells. A second set of studies (described in "Odor As a Behavioral Reinforcer") has used pleasant and unpleasant odors as behavioral reinforcers in a variety of associative learning (Pavlovian conditioning) paradigms. Insofar as "pleasant" smells are rewards capable of provoking biologically important behaviors, putative human pheromones fulfill this operational definition, but such imaging studies (some of which implicate gender-dissociable activity in the hypothalamus) are beyond the scope of the chapter. Interested readers can consult the following articles for more information: Jacob et al. (2001b), Savic et al. (2001, 2005), and Sobel et al. (1999).

Odor As a Hedonic State. The first human imaging study to investigate neural responses to emotionally salient odors was performed by Zald and Pardo (1997). Using PET imaging, these investigators found that highly aversive sulfides elicited neural activity in left and right amygdala and left posterolateral OFC, in comparison to an odorless control (see Figure 8.6). In contrast, presentation of slightly less aversive (and less intense) odorants failed to evoke amygdala activity, though responses were still observed in OFC, in a position posterior to that identified with sulfide exposure. Moreover, the magnitude of activity in amygdala

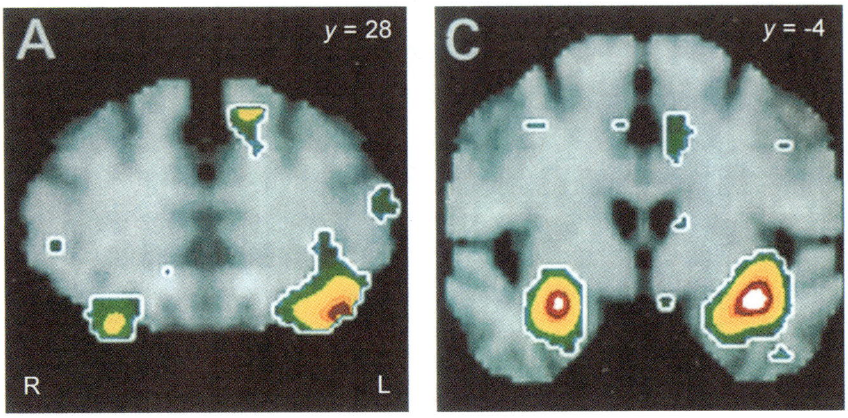

Figure 8.6 Aversive olfactory processing. Neural activations in the orbitofrontal cortex (a) and amygdala (b) were evoked during stimulation with unpleasant odor. Activations superimposed on coronal sections of frontal and temporal lobes. [Reprinted and modified from Zald and Pardo (1997). Copyright © 1997 National Academy of Sciences USA.]

and OFC correlated with subject-specific ratings of odor aversiveness. Together these findings strongly implicated human amygdala and left OFC in the processing of negative odor valence.

Subsequent fMRI studies, testing the effects of both pleasant and unpleasant odors within a single experiment, have provided mixed support for the idea that amygdala and left lateral OFC are sensitive to aversive odorants. One event-related study (Gottfried et al., 2002a) showed an absence of valence-specific effects in amygdala, such that it responded similarly to a pleasant odor (vanillin, "vanilla"), an unpleasant odor (4-methyl-pentanoic acid, "sweaty socks"), and a hedonically neutral odor (phenethyl alcohol, "rose"), raising questions about the valence specificity of this region. On the other hand, OFC (as well as anterior piriform cortex) did show variable activation patterns in response to valence (pleasant odor in right medial OFC, unpleasant odor in left lateral OFC). Another fMRI study tested a wider set of odorants, including three pleasant and three unpleasant compounds (Rolls et al., 2003). Odor pleasantness correlated with medial OFC activity; odor unpleasantness, with more lateral OFC activity (but not amygdala). In turn, odor intensity was associated with activity in piriform, entorhinal, and anterior cingulate cortices.

It is evident from the above studies that OFC is sensitive to odor valence, but whether amygdala is equally involved is less clear. One possible reason for this discrepancy may be due to the fact that odor valence frequently covaries with odor intensity. In other words, the perceived pleasantness of an odor often increases or decreases with changes in its intensity. To dissociate these perceptual factors, Anderson et al. (2003) conducted a study in which subjects were presented with high- and low-intensity versions of one pleasant odor (citral, "lemon") and one unpleasant odor (valeric acid, "sour cheese") during fMRI scanning. Critically, the pleasant and unpleasant odors were matched for their respective levels of intensity (high or low), and high-intensity and low-intensity odors were matched for their respective valence (pleasant or unpleasant), enabling these factors to be distinguished. Odor intensity was associated with selective activity in amygdala, whereas odor valence was associated with activity in OFC (right medial, pleasant; left lateral, unpleasant). Similar functional dissociations of intensity and valence have also been described with gustatory stimuli (Small et al., 2004).

Although these findings seemed to establish a clear functional dissociation of intensity and valence in amygdala and OFC, respectively, one important limitation of the study was that only the *extremes* of odor valence were examined. Follow-up work employing a broader valence range of odors suggests a more complex story. Winston and colleagues extended the Anderson et al. study by including hedonically neutral odors (anisole, "phenolic, gasoline, ethereal"; 2-heptanol, "earthy, oily") at low and high intensity, in addition to the pleasant citral odorant and the unpleasant valeric acid odorant (Winston et al., 2005). Here the hypothesis was that if amygdala simply coded intensity, then high-intensity versions of pleasant, unpleasant, and neutral odors should all evoke similar levels

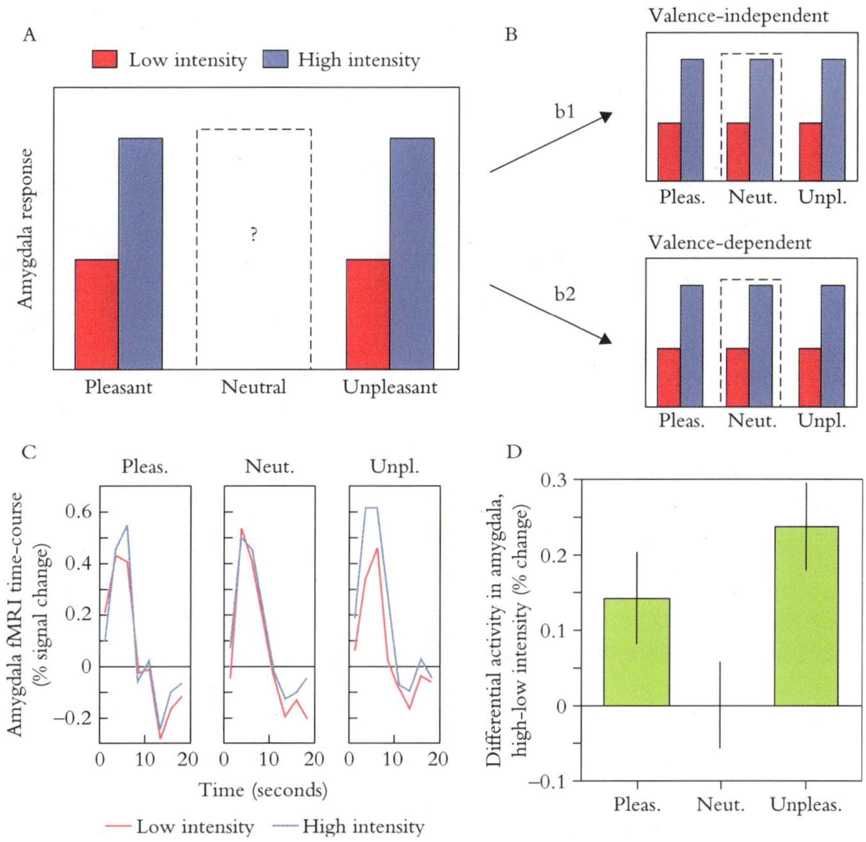

Figure 8.7 Intensity coding in the amygdala is valence-dependent. (a) Prior studies suggested that amygdala is preferentially responsive to odor intensity, though only the extremes of valence (pleasant and unpleasant) were tested, leaving open two competing hypotheses: in b1, amygdala codes intensity across the full valence spectrum, including neutral odor; alternatively, in b2, amygdala responds to intensity only for emotionally salient odors, reflecting a valence by intensity interaction. (c and d) fMRI results are consistent with the latter hypothesis (b2). (c) Response time courses show that high-intensity odors evoke greater activity in amygdala for pleasant and unpleasant odors, but not for neutral odor. (d) Differential response magnitudes in amygdala (high vs. low intensity) for each level of valence again reveal effects of intensity only at valence extremes. [Reprinted and modified from Winston et al. (2005). Copyright © Society for Neuroscience.]

of activity in amygdala (see Figure 8.7). Alternatively, if amygdala coded intensity only in the presence of behaviorally meaningful smells, then high-intensity versions of neutral odors should not elicit amygdala activation. Our data support the latter hypothesis: the amygdala selectively responded only to high (vs. low) pleasant and unpleasant odors, whereas for neutral odors there was no differential effect of intensity. The implication is that the amygdala is sensitive to an integrated combination of odor intensity and affect. From an ecological perspective, this stands to reason. To a carnivore searching for lunch, the scent of prey (a "pleasant" odor) or predator (an "unpleasant" odor) is likely to provoke a more robust behavioral response as that scent builds in intensity on the wind. In contrast, the scent of a termite mound (a "neutral" odor), also rising in the breeze, is unlikely (and ought not) to stir the animal. Ultimately, it is the combined emotional salience of intensity and affect (the overall emotional value of the smell) that matters most to the amygdala.

Interestingly, a role for amygdala in olfactory hedonic processing has emerged in an elegant fMRI study involving autobiographical odor memories (Herz et al., 2004). In a prescanning interview phase, subjects were identified for whom a particular perfume evoked a specific pleasant personal memory. Then, during scanning, subjects were presented with this personally meaningful perfume ("experimental odor" EO), a non-salient control perfume ("control odor" CO), and also visual counterparts of the two perfumes (pictures of the bottles: EV, CV). Comparison of the salient perfume odor (EO) to the control odor (CO) was associated with significant activity in amygdala and parahippocampal gyrus; moreover, this effect was specific to the olfactory context, as medial temporal activation was observed in the comparison of EO to EV. These intriguing findings are among the first to demonstrate the sensory potency of olfactory (vs. visual) cues on human amygdala activity, a mechanism that may underlie the purported emotional and mnemonic power of smells.

Characterization of olfactory hedonics in the human brain has also been studied using a wholly different approach. In humans, when a food is eaten to satiety, the perceived pleasantness of the sated food declines more than that of non-sated foods. This phenomenon has been termed sensory-specific satiation and is thought to help regulate feeding and ensure a nutritionally diverse food intake (Rolls et al., 1981a,b). In monkeys, a similar effect can be demonstrated, whereby sated foods are rejected, but non-sated foods continue to be accepted, and it has been shown that the responses of taste neurons within caudolateral OFC are modulated in parallel with these behavioral satiety effects (Rolls et al., 1989). Appetite states have also been shown to modulate OFC single-unit activity in the olfactory domain (Critchley and Rolls, 1996). These findings indicate that neuronal responses in primate OFC reflect sensitivity to the current reward value of foods and their associated smells.

Using selective satiety as a behavioral manipulation, O'Doherty et al. (2000) were able to demonstrate that information about olfactory hedonics (reward value) is also represented in portions of human OFC. In this fMRI study, subjects were scanned as they smelled either banana or vanilla odor, both before and after eating bananas to satiety. Postsatiety odor-evoked activations to the banana smell decreased in OFC in 5/6 subjects without any corresponding OFC signal decline in response to the vanilla smell, consistent with the behavioral effect of selective satiety. Similar findings in the gustatory domain have been demonstrated by other investigators. Responses in human OFC decrease as chocolate (a complex food stimulus with prominent olfactory components) is consumed to satiety (Small et al., 2001). Sensory-specific satiety effects have also been documented in OFC when subjects drink a flavorful food liquid (either tomato juice or chocolate milk) until they are sated (Kringelbach et al., 2003). Together, these findings not only extend data suggesting that OFC represents the reward value of an olfactory stimulus, but also corroborate an interspecies preservation of OFC function between monkeys and humans.

An unresolved issue concerns whether a functional dissociation of pleasant and unpleasant odor exists in OFC. As discussed above, several fMRI studies converge on the idea that right medial OFC encodes pleasantness and left lateral OFC encodes unpleasantness (Anderson et al., 2003; Gottfried et al., 2002a; Rolls et al., 2003). These findings also concur with imaging data showing valence-specific medial-lateral dissociations to a variety of non-olfactory stimuli, including tastes (Small et al., 2004), money (O'Doherty et al., 2001), and faces (O'Doherty et al., 2003). To evaluate these functional differences more carefully, Gottfried and Zald (2005) plotted the OFC activation peaks for pleasant and unpleasant odors from six studies in which odor valence was a main focus (Anderson et al., 2003; Gottfried et al., 2002a; Gottfried et al., 2004; Rolls et al., 2003; Zald and Pardo, 1997, 1998). Although a limited sample, the data reveal some interesting basic patterns (see Figure 8.8). First, aversive odorants tend to engage the left OFC, while pleasant odorants have a more bilateral representation. Second, neural responses evoked by pleasant or unpleasant odor do not localize to discrete OFC areas. Third, there is a tendency for unpleasant odors to activate more lateral areas of OFC compared to pleasant odors, but it is important to note that the activated voxels are equally situated in medial OFC (Walker areas 11 and 13) and lateral OFC (Walker area 47/12). This analysis is in partial agreement with a recent review by Kringelbach and Rolls (2004), who emphasize a medial-positive, lateral-negative distinction in the OFC. However, the identification of several medially positioned voxels in response to unpleasant odor suggests that a medial-positive, lateral-negative model of OFC function cannot completely account for the processing of odor hedonics.

Neuroimaging studies of explicit hedonic judgments have also been used to explore the emotional underpinnings of olfactory processing. Royet and colleagues (1999, 2000, 2001) have emphasized the importance of the left OFC in the conscious hedonic evaluation of odorants. This premise arose out of their initial finding that hedonic judgments produced left OFC activation that exceeded those produced by judgments of familiarity, edibility, intensity, or detection.

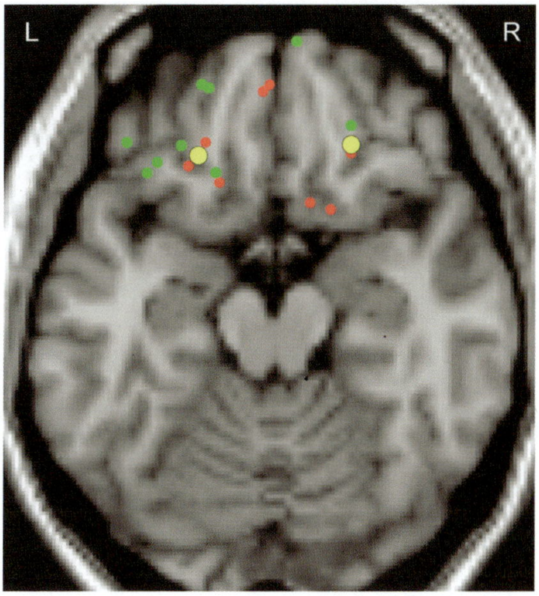

Figure 8.8 Localization of odor hedonics in the human orbitofrontal cortex. An imaging meta-analysis of six olfactory imaging studies demonstrates that coding of odor valence in OFC does not dissociate neatly across hemispheres or along medial-lateral orbitofrontal subdivisions. Pleasant odor, red dots. Unpleasant odor, green dots. The yellow dots indicate the location of putative human olfactory OFC. [Reprinted and modified with permission from Gottfried et al. (2006). Copyright © Oxford University Press 2006.]

However, two studies suggest that hedonic judgments are not exclusively lateralized to the left OFC. First, in a study by Zatorre et al. (2000), both intensity and pleasantness judgments were observed to activate the right OFC, and there was no preferential activation of the left OFC during pleasantness relative to intensity judgments. An additional concern may be raised regarding the earlier Royet studies in that the hedonic judgment tasks are commonly made in the exclusive presence of emotionally salient odors, whereas other judgments were not. Therefore, any task-related activations could be confounded by mere stimulus properties of the odorants themselves. In an effort to overcome this issue, Royet et al. (2003) conducted an fMRI study in which subjects smelled pleasant and unpleasant odors during alternating blocks of either hedonic judgments or a control task (random button press). A comparison of hedonic and control tasks (thereby controlling for mere stimulus effects) revealed significant activity in OFC bilaterally, partially dispelling the idea that left OFC is preferentially involved in olfactory emotional processing and supporting the possibility of right (as well as left) OFC involvement. On the basis of the current imaging data, evidence for hemispheric specialization of olfactory hedonics awaits further confirmation.

Odor as a Behavioral Reinforcer. Another set of imaging studies has focused on the use of emotionally salient odors as behavioral reinforcers in paradigms of associative learning, thus complementing the studies on odor affect and hedonics described above. Importantly, the use of pleasant and unpleasant odors as appetitive (reward-based) or aversive reinforcers, respectively, makes it possible to examine both valence types of learning within a single sensory modality and even within a single experiment. The design of these experiments followed closely from previous classical (Pavlovian) conditioning fMRI paradigms in which electric shock (Buchel et al., 1998; LaBar et al., 1998) or salty or sweet tastes (O'Doherty et al., 2002) were used as unconditioned stimuli (UCS). (Note that the designation of odor as "UCS" is an operational definition only; it does mean to imply that odor affect is necessarily an unconditioned (innate, hard-wired) property of the chemical stimulus. It is equally possible that the hedonic value of a smell is learned (conditioned), which could nevertheless serve as an effective secondary reinforcer in conditioning paradigms.)

In one of the first fMRI conditioning studies using olfactory reinforcers (Gottfried et al., 2002a,b), a series of neutral-expression faces (the to-be-conditioned stimuli, CS+) was repetitively paired with pleasant (vanillin; "vanilla"), neutral (phenethyl alcohol, "floral"), or unpleasant (4-methyl-pentanoic acid, "sour, sweaty") odor (the UCS stimuli). The face-odor pairings followed a 50% partial reinforcement schedule, such that one-half of all face presentations was paired with their corresponding odors, permitting the dissociation of CS+-specific effects from confounds related to the UCS. A fourth face was also presented that was never paired to odor (the non-conditioned stimulus, or CS–). To minimize task-related confounds, subjects performed the same gender judgment task (male or female) on each trial, which also ensured that they attended consistently throughout the experiment.

Behaviorally, as visual-olfactory learning proceeded, subjects responded faster to the appetitive CS+ (appCS+, i.e., the CS+ face paired to the pleasant odor UCS) and to the aversive CS+ (avCS+, i.e., the CS+ face paired to the unpleasant odor UCS) in comparison to the CS– stimulus. These findings provided a behavioral index of associative learning, suggesting that the repeated pairing of pictures and odors increased the emotional salience of the appCS+ and avCS+, drawing subjects' attentional resources more effectively toward these stimuli and resulting in more rapid processing of the faces. Correspondingly, the neural correlates of appetitive and aversive olfactory learning were identified primarily in the OFC. Interestingly, learning-related activity in the amygdala was observed preferentially in response to the appCS+, in contrast to prior fMRI studies of aversive conditioning (Buchel et al., 1998; LaBar et al., 1998). Functional dissociations between the appCS+ and avCS+ were seen in discrete subregions of OFC (appCS+: medial OFC; avCS+: central-lateral OFC) and in the medial temporal lobe. Together these findings helped confirm the idea that the hedonic properties of smells were sufficiently robust to be used as Pavlovian reinforcers. Moreover, the participation of OFC in human olfactory conditioning extended prior studies involving gustatory (O'Doherty et al., 2002) or abstract (monetary) (O'Doherty et al., 2001) reinforcers, implying that learning-induced OFC responses are at least partially independent of sensory modality.

Follow-up olfactory fMRI studies indicate that amygdala and OFC are specifically involved in encoding the predictive reward value of olfactory reinforcers. We used a paradigm known as reinforcer devaluation as a way to tease apart the various central representations that a CS+ may engage (Gottfried et al., 2003). Such approaches have been used in the context of animal studies of appetitive learning, which show that damage to amygdala and OFC impair the effects of reinforcer devaluation (Baxter et al., 2000; Gallagher et al., 1999; Hatfield et al., 1996; Malkova et al., 1997). In our study, 13 hungry subjects were scanned during learning and anticipation of two food-based olfactory rewards (vanilla and peanut butter odors) both before and after selective satiation. One odor was destined for reinforcer devaluation (target UCS), whereas the other underwent no motivational shift (nontarget UCS). Abstract visual images comprised target and non-target CS+ stimuli, which were paired with the corresponding UCS. In between pre- and post-satiety fMRI sessions, subjects were taken out of the scanner and fed a lunch containing the flavor matching the target odor UCS (i.e., vanilla ice cream or peanut butter sandwiches). The main hypothesis was that if amygdala and OFC maintain representations of predictive reward value, then neural responses evoked by a CS+ should be sensitive to selective satiety manipulations that devalue predicted reward.

Behavioral ratings of odor pleasantness revealed a robust effect of selective satiation: pleasantness of the target UCS declined significantly from pre- to post-satiety, whereas pleasantness of the non-target UCS changed very little. The neural correlates of this satiety effect were detected principally in amygdala and OFC. In these brain regions, neural activity evoked by the target CS+ selectively decreased in response to satiation in the absence of changes evoked by the non-target CS+. Other areas including anterior cingulate cortex, ventral striatum, and insula also showed a selective decline in response to the target CS+, as well as response increase to the non-target CS+, perhaps reflecting shifts in relative reward preference between target and non-target choices. These results underscore the selective impact of hunger and satiety on the neural correlates of reward prediction. For example, when a food changes from delectable to distasteful, the brain responses evoked by a predictive cue are attenuated in areas that maintain responses to predictors of other palatable stimuli.

A related fMRI study from our laboratory also considered the opposite effect of UCS value modulation. Rather than *decreasing* the current value of an olfactory UCS (via selective satiety), Gottfried and Dolan (2004) used a paradigm of UCS inflation (Rescorla, 1974), whereby the motivational value of an aversive odor was *increased*. In this design, two neutral faces (the CS+) were repetitively paired with two different unpleasant smells (4-methyl-pentanoic acid, "sweaty socks"; and ammonium sulfide, "rotten eggs"). Following an initial conditioning phase in which the face-odor contingencies were learned, subjects were exposed to a higher intensity of one of the two UCS stimuli (in the absence of the CS+) in order to increase or "inflate" its reinforcement value, while the other odor UCS was presented at the same intensity. In a final scanning phase, the CS+ faces were presented at extinction without further delivery of the UCS. Behavioral indices of UCS inflation indicate that the target CS+ face (associated with the inflated UCS) was rated as more aversive than the non-target CS+ face, and on a subjectwise basis, the magnitude of this behavioral effect linearly correlated with neural activity in lateral OFC. Together, the studies on reinforcer devaluation and reinforcer inflation demonstrate that an odor-predictive cue has direct access to internal representations of value in OFC, and that these representations are flexibly updated according to an individual's motivational state and recent sensory experience.

Human Olfactory Orbitofrontal Cortex: Localization, Function, and Other Considerations

The earliest imaging study of human olfaction (Zatorre et al., 1992) revealed significant neural activity in bilateral piriform cortex and right OFC. The right OFC activation was located centrally within the orbital cortex, in between the medial and lateral orbital sulci, but more rostral than would be predicted from the monkey data. Odor-evoked OFC activity was also observed on the left side, albeit at reduced statistical threshold. Since then, numerous investigators using both PET and fMRI techniques have further examined the location of an olfactory representation in OFC. A recent meta-analysis of 13 olfactory imaging studies (Gottfried and Zald, 2005) demonstrates that olfactory stimulation consistently activates a bilateral area along the medial orbital sulcus (i.e., the medial posterior limb of the "H"-shaped sulcus) close to the transverse orbital sulcus (i.e., the horizontal limb of the "H"-shaped sulcus). Comparison of these activations to monkey (Carmichael et al., 1994) and human (Ongur et al., 2003) architectonic maps suggests that this putative human olfactory OFC roughly corresponds to the posterior part of area 11, a location that is strikingly more anterior than one would predict on the basis of anatomical and electrophysiological data from animals (e.g., Carmichael et al., 1994; Morecraft et al., 1992; Yarita et al., 1980).

Regardless of specific cytological subdivisions, it is clear that the putative human region is situated at least 2 cm rostral to the monkey counterpart (in the agranular insula, along the posterior edge of the OFC). Indeed, the posterior-most agranular segment of the orbital surface is rarely activated in human olfactory neuroimaging. A complete survey of the olfactory imaging literature shows that, if anything, agranular insula is preferentially activated during higher-order tasks involving cross-modal integration (De Araujo et al., 2003; Small et al., 2004). Possibly this anatomical discrepancy is the result of methodological differences (as discussed in Gottfried and Zald, 2005). Alternatively, cognitive or attentional factors might alter fMRI response patterns such that odor stimulation evokes more anterior regions of OFC than would be predicted from the primate studies. However, it is interesting to speculate that biological differences in olfactory processing between monkeys and humans might account for some of the cross-species differences in functional anatomy, perhaps reflecting behavioral differences in the role that the sense of smell plays in these two species. Of particular note in this regard, the amygdala provides strong input to agranular OFC, but very scant projections to anterior central OFC, implying that limbic (amygdala) influences on odor processing in OFC may be highly limited in the human brain.

The human and monkey OFC data are in closer agreement when it comes to multisensory integration. The role of OFC as a site of sensory convergence has been well-documented in both species. In nonhuman primates, the OFC is a recipient of projections not only from primary olfactory structures, but also from gustatory, visual, and visceral centers, and it has been proposed that these inputs are together assembled into unitary flavor percepts that can guide feeding-related behavior (Carmichael et al., 1994). In humans, multisensory integration of odors and tastes (Cerf-Ducastel and Murphy, 2001; De Araujo et al., 2003; Small et al., 2004) or odors and pictures (Gottfried and Dolan, 2003) has been demonstrated in OFC. Moreover, through manipulations of semantic correspondence between odors and tastes (Small et al., 2004), or between odors and pictures (Gottfried and Dolan, 2003), OFC activity increased with increasing subjective congruency ratings. These observations underscore the idea that prior experience and sensory context can profoundly modulate the central processing of olfactory information. Such mechanisms may also help to resolve the inherent ambiguity in olfactory perception.

Regional dissociations within OFC have been observed to depend on both the route of odorant administration and the type of odorant (food vs. nonfood) (Small et al., 2005). External ("orthonasal") delivery of a chocolate odor elicited activation in caudolateral OFC, insula, and opercula, whereas delivery of the same smell via the mouth ("retronasal") elicited activity in medial OFC and perigenual cingulate cortex. These findings were interpreted to suggest that orthonasal food odor signifies reward availability (of a not-yet-consumed food), whereas retronasal food odor signifies reward receipt (of a food in the mouth), each associated with its own network of activations, consistent with related studies on gustatory processing in the human brain (O'Doherty et al., 2002) and akin to models of reward 'wanting' versus 'liking' (Berridge, 1996). Interestingly, route of odorant administration had little effect on neural responses evoked by nonfood odors, presumably since these stimuli signal neither anticipation nor consumption of reward.

It should be mentioned that many different cognitive tasks influence odor-evoked neural responses in the OFC. Judgments of intensity (Zatorre et al., 2000), familiarity (Royet et al., 1999, 2001), edibility (Royet et al., 1999, 2001), and pleasantness (described above in "Odor as a Hedonic State") have each been associated with activation of the OFC in PET studies. Some of these activations are situated near the purported olfactory projection area in OFC, though others are found more rostrally. It is worth noting that many judgment-related foci lie outside of the OFC altogether: olfactory judgment tasks typically evoke responses in large portions of frontal, temporal, parietal, and occipital cortex (Royet et al., 1999, 2000, 2003; Savic et al., 2000), and even in cerebellum (Ferdon and Murphy, 2003; Sobel et al., 1998b). For example, in an fMRI study that dissociated the "where" versus "what" systems of olfaction, the superior temporal gyrus was preferentially activated when subjects attempted to localize an odor, whereas occipital gyrus and paracentral lobule were more active when subjects attempted to identify an odor (Porter et al., 2005). These findings indicate that numerous areas tertiary to the olfactory system help mediate higher-order olfactory decision-making and are consistent with the idea of both serial hierarchical and parallel distributed modes of olfactory information processing that vary according to task demands (Royet and Plailly, 2004; Savic et al., 2000). However, it remains unclear how these non-olfactory brain areas specifically interact with the OFC and "primary" olfactory areas (e.g., piriform, amygdala) in mediating these tasks.

Out of these disparate findings, it is apparent that a single common function cannot be easily ascribed to human olfactory OFC. However, a brief glance through evolution sheds some light on this issue (for further discussion, see Gottfried, 2007). Prefrontal cortex (including OFC) is a distinctly mammalian region of the brain, emerging 175 million years ago in an ancestor of modern-day mammals (Jerison, 1997). Thus, it follows that many thousands of *non-mammalian* vertebrates (including bony fish, reptiles, amphibians, and birds) can smell quite well without the use of an OFC, effectively using olfactory signals to optimize feeding, mating, and maternal bonding, among other behaviors critical for survival. So where does that leave the function of olfactory OFC? Based on the OFC data described in the preceding sections of this chapter, it seems reasonable to speculate that the OFC is critical in situations where experience and learning have an opportunity to modify behavior adaptively. The ability to suppress natural response tendencies, to form new predictions about old stimuli, and to update information about sensory inputs, particularly for "emotionally" (biologically) important events, are some of the unique features that OFC contributes to olfactory processing.

Conclusions

Neuroimaging of olfaction has provided a unique glimpse into the "pleasures of the human brain" and has substantiated what was long suggested by animal studies: the human olfactory system intimately overlaps many limbic regions of the brain underlying emotion, reward, learning, and motivation. This is surely no coincidence. After all, chemical sensation has been a primary reinforcer of behavior throughout much of evolution. Unicellular protozoa, ciliated or flagellated or cilio-flagellated, relied on a sense of "smell" to detect chemical gradients in aqueous media, enabling them to find food, evade predators, and keep clear of toxic waste (see Dusenbery, 1992). Several hundred-million years later, olfactory sensory systems remained finely tuned to environmental signals carrying information essential for survival, perhaps reaching its apogee in rodents and canines. Nature being expedient, the human brain has since appropriated these olfactory structures for use in a wider variety of (non-olfactory) sensory contexts and behavioral contingencies, ensuring greater biological adaptability.

Thus the olfactory pleasures of the human brain coincide with many of the neurobiological systems described elsewhere in this book. The amygdala is preferentially sensitive to emotionally salient, highly orienting, behaviorally informative odor events: for example, intense pleasant and unpleasant, but not valence-neutral, smells; novel or unexpected sensory contingencies between olfactory and non-olfactory cues; odor-triggered personal, autobiographical memories, in the *Proustian* manner. It is tempting to speculate that the intimate connection of the olfactory system to the amygdala, a mere two synapses removed from the odor periphery, accounts for the emotional potency of smells, but empirical evidence to support this idea remains scant.

The OFC also plays an important role in the processing of odor valence and odor reward value (both intrinsic and predicted), multisensory interactions, and flavor integration. However, its true strengths emerge in times of biological strain, when natural or learned response predispositions need to be modified. As discussed above, there are thousands of vertebrates contending with odor valence, odor reward, and multisensory phenomena quite successfully, without the advantages of an orbitofrontal or prefrontal cortex. What such creatures lack, and what the OFC confers upon mammals so endowed, is the ability to unleash themselves from stimulus-bound behavior (Gottfried, 2007). Examples of OFC-impoverished *human* vertebrates, usually the result of head trauma, tumor, or vascular disease, abound in the neurological literature and highlight the difficulties such patients have in suppressing their instincts. In a classic series of articles, Lhermitte characterized the stimulus dependency typical of patients with frontal lobe lesions (Lhermitte, 1983, 1986; Lhermitte et al., 1986). These syndromes included "utilization behavior" in which patients felt compelled to grasp and utilize objects even when not instructed to do so, and "environmental dependency" in which patients acted upon social and physical cues despite the inappropriateness of the situation (like undressing and climbing into Prof. Lhermitte's bed or picking up the Prof's electric razor and proceeding to shave), as if they "were powerless in the face of influences from the outside world" (Lhermitte, 1986, p. 342). For the purposes of smell, the olfactory OFC is a critical locus where learning-induced neural plasticity helps to refine sensory perception and odor experience.

At the center of all things olfactory is the piriform cortex. Available data indicate that this brain region is less a "primary" sensory structure and more an "associative" cortical site, akin to inferotemporal cortex for higher-order visual processing. With massive converging and distributed afferent projections from the olfactory bulb, dense reciprocal feedback from amygdala and OFC, and extensive intracortical (intra-piriform) connectivity (Haberly, 1985, 2001), the piriform cortex is architecturally suited to house odor "objects" and to update these representations through expectation, experience, learning, and context. Olfactory traces in piriform cortex are highly mutable. Mere sensory exposure to one odorant continuously for 3 min is sufficient to modify neural codes of odor quality and enhance perceptual discrimination of related odorants (Li et al., 2006), and aversive olfactory learning can enhance perceptual and piriform discrimination of formerly indistinguishable odors (Li et al., 2008).

In summary, human neuroimaging studies of olfaction have confirmed and extended a deep-rooted belief, namely, that smells are intimately linked to hedonics, pleasure, and emotion. This notion has endured since antiquity: to Aristotle and his disciples in the 4th century B.C. (Aristotle, 350 B.C.), it was abundantly clear that the sense of smell was allied to Fire, the most passionate and temperamental of the four classical Elements. Research of a more contemporary mode has echoed this ancient hypothesis, for example, in the form of multidimensional scaling studies showing that human subjects primarily categorize odors along a hedonic (pleasant–unpleasant) axis (Schiffman, 1974). However, until the recent advent of robust olfactory imaging protocols, there was little neuroscientific evidence directly connecting human olfaction to brain systems mediating hedonics, emotional learning, and motivation. Many of the imaging studies described in this chapter lend biological credence to these links, demonstrating that olfactory pleasures (and displeasures) have potent access to limbic emotional structures. With the ever-accelerating scientific interest in human emotion and social cognition, it is worth emphasizing that odorous stimuli can provide highly effective functional probes of limbic processing in the human brain.

Acknowledgments

I would like to thank past and present members of the Gottfried laboratory, including J. Howard, W. Li, and J. Plailly, for helpful suggestions, and I would also like to give credit to H.M. Zankel for unwavering support. Portions of the research described here were funded by the National Institute on Deafness and Communication Disorders and by the Illinois Department of Public Health.

References

Adrian, E. D. (1942) Olfactory reactions in the brain of the hedgehog. *J. Physiol.* 100, 459–473.

Alaoui-Ismaili, O., Vernet-Maury, E., Dittmar, A., Delhomme, G. and Chanel, J. (1997) Odor hedonics: Connection with emotional response estimated by autonomic parameters. *Chem. Senses* 22, 237–248.

Anderson, A. K., Christoff, K., Stappen, I., Panitz, D., Ghahremani, D. G., Glover, G., Gabrieli, J. D. and Sobel, N. (2003) Dissociated neural representations of intensity and valence in human olfaction. *Nat. Neurosci.* 6, 196–202.

Apollodorus (c. 1st century B.C.) (1960) *The Library, Book 1 [1.9.17]. Sir J.G. Frazer, Transl. Loeb Classical Library Volumes 121 & 122.* Harvard University Press: Cambridge, MA.

Aristotle (350 B.C.) *On Sense and the Sensible.* Transl. J. I. Beare. [Electronic version] Retrieved on December 28, 2006 from http://classics.mit.edu/Aristotle/sense.html.

Baxter, M. G., Parker, A., Lindner, C. C., Izquierdo, A. D. and Murray, E. A. (2000) Control of response selection by reinforcer value requires interaction of amygdala and orbital prefrontal cortex. *J. Neurosci.* 20, 4311–4319.

Bensafi, M., Rouby, C., Farget, V., Bertrand, B., Vigouroux, M. and Holley, A. (2002) Autonomic nervous system responses to odours: the role of pleasantness and arousal. *Chem. Senses* 27, 703–709.

Bensafi, M., Tsutsui, T., Khan, R., Levenson, R. W. and Sobel, N. (2004) Sniffing a human sex-steroid derived compound affects mood and autonomic arousal in a dose-dependent manner. *Psychoneuroendocrinology* 29, 1290–1299.

Berridge, K. C. (1996) Food reward: Brain substrates of wanting and liking. *Neurosci. Biobehav. Rev.* 20, 1–25.

Buchel, C., Morris, J., Dolan, R. J. and Friston, K. J. (1998) Brain systems mediating aversive conditioning: an event-related fMRI study. *Neuron* 20, 947–957.

Buck, L. B. (2004) Olfactory receptors and odor coding in mammals. *Nutr. Rev.* 62, S184–S188; discussion S224–S241.

Buckner, R. L., Goodman, J., Burock, M., Rotte, M., Koutstaal, W., Schacter, D., Rosen, B. and Dale, A. M. (1998) Functional-anatomic correlates of object priming in humans revealed by rapid presentation event-related fMRI. *Neuron* 20, 285–296.

Cabanac, M. (this volume) The dialetics of pleasure. In: *Pleasures of the Brain*. Eds. M. L. Kringelbach and K. C. Berridge. Oxford University Press: Oxford, UK.

Cain, W. S. and Polak, E. H. (1992) Olfactory adaptation as an aspect of odor similarity. *Chem. Senses* 17, 481–491.

Carmichael, S. T., Clugnet, M. C. and Price, J. L. (1994) Central olfactory connections in the macaque monkey. *J. Comp. Neurol.* 346, 403–434.

Cerf-Ducastel, B. and Murphy, C. (2001) fMRI activation in response to odorants orally delivered in aqueous solutions. *Chem. Senses* 26, 625–637.

Cleland, T. A., Morse, A., Yue, E. L. and Linster, C. (2002) Behavioral models of odor similarity. *Behav. Neurosci.* 116, 222–231.

Cohen, N. J., Poldrack, R. A. and Eichenbaum, H. (1997) Memory for items and memory for relations in the procedural/declarative memory framework. *Memory* 5, 131–178.

Cole, J. R. (1948) The Melancholy Mr. Morbid. *Police Comics*, 76 (March issue) Quality Comics Group/DC Comics: New York.

Critchley, H. D. and Rolls, E. T. (1996) Hunger and satiety modify the responses of olfactory and visual neurons in the primate orbitofrontal cortex. *J. Neurophysiol.* 75, 1673–1686.

Dade, L. A., Zatorre, R. J. and Jones-Gotman, M. (2002) Olfactory learning: Convergent findings from lesion and brain imaging studies in humans. *Brain* 125, 86–101.

De Araujo, I. E., Rolls, E. T., Kringelbach, M. L., McGlone, F. and Phillips, N. (2003) Taste–olfactory convergence, and the representation of the pleasantness of flavour, in the human brain. *Eur. J. Neurosci.* 18, 2059–2068.

Degel, J. and Koster, E. P. (1999) Odors: Implicit memory and performance effects. *Chem. Senses* 24, 317–325.

Deichmann, R., Gottfried, J. A., Hutton, C. and Turner, R. (2003) Optimized EPI for fMRI studies of the orbitofrontal cortex. *Neuroimage* 19, 430–441.

Djordjevic, J., Zatorre, R. J., Petrides, M., Boyle, J. A. and Jones-Gotman, M. (2005) Functional neuroimaging of odor imagery. *Neuroimage* 24, 791–801.

Dusenbery, D. B. (1992) *Sensory Ecology: How Organisms Acquire and Respond to Information*. W. H. Freeman: New York, NY.

Eichenbaum, H., Morton, T. H., Potter, H. and Corkin, S. (1983) Selective olfactory deficits in case H.M. *Brain* 106, 459–472.

Eslinger, P. J., Damasio, A. R. and Van Hoesen, G. W. (1982) Olfactory dysfunction in man: Anatomical and behavioral aspects. *Brain Cogn.* 1, 259–285.

Ferdon, S. and Murphy, C. (2003) The cerebellum and olfaction in the aging brain: a functional magnetic resonance imaging study. *Neuroimage* 20, 12–21.

Firestein, S. (2001) How the olfactory system makes sense of scents. *Nature* 413, 211–218.

Frijda, N. (this volume) On the nature and function of pleasure. In: *Pleasures of the Brain*. Eds. M. L. Kringelbach and K. C. Berridge. Oxford University Press: Oxford, UK.

Gallagher, M., McMahan, R. W. and Schoenbaum, G. (1999) Orbitofrontal cortex and representation of incentive value in associative learning. *J. Neurosci.* 19, 6610–6614.

Gibson, E. J. (1991) *An Odyssey in Learning and Perception*. MIT Press: Cambridge, MA.

Gilad, Y., Man, O., Paabo, S. and Lancet, D. (2003) Human specific loss of olfactory receptor genes. *Proc. Natl. Acad. Sci. USA* 100, 3324–3327.

Gottfried, J. A. (2007) What can an orbitofrontal cortex-endowed animal do with smells? *Ann. NY Acad. Sci.* 1121, 102–120.

Gottfried, J. A. and Dolan, R. J. (2003) The nose smells what the eye sees: Crossmodal visual facilitation of human olfactory perception. *Neuron* 39, 375–386.

Gottfried, J. A. and Dolan, R. J. (2004) Human orbitofrontal cortex mediates extinction learning while accessing conditioned representations of value. *Nat. Neurosci.* 7, 1144–1152.

Gottfried, J. A. and Zald, D. H. (2005) On the scent of human olfactory orbitofrontal cortex: Meta-analysis and comparison to non-human primates. *Brain Res. Brain Res. Rev.* 50, 287–304.

Gottfried, J. A., Deichmann, R., Winston, J. S. and Dolan, R. J. (2002a) Functional heterogeneity in human olfactory cortex: an event-related functional magnetic resonance imaging study. *J. Neurosci.* 22, 10819–10828.

Gottfried, J. A., O'Doherty, J. and Dolan, R. J. (2002b) Appetitive and aversive olfactory learning in humans studied using event-related functional magnetic resonance imaging. *J. Neurosci.* 22, 10829–10837.

Gottfried, J. A., O'Doherty, J. and Dolan, R. J. (2003) Encoding predictive reward value in human amygdala and orbitofrontal cortex. *Science* 301, 1104–1107.

Gottfried, J. A., Smith, A. P., Rugg, M. D. and Dolan, R. J. (2004) Remembrance of odors past: Human olfactory cortex in cross-modal recognition memory. *Neuron* 42, 687–695.

Gottfried, J. A., Winston, J. S. and Dolan, R. J. (2006) Dissociable codes of odor quality and odorant structure in human piriform cortex. *Neuron* 49, 467–479.

Grill-Spector, K. and Malach, R. (2001) fMR-adaptation: a tool for studying the functional properties of human cortical neurons. *Acta Psychol.(Amst.)* 107, 293–321.

Gu, H., Feng, H., Zhan, W., Xu, S., Silbersweig, D. A., Stern, E. and Yang, Y. (2002) Single-shot interleaved z-shim EPI with optimized compensation for signal losses due to susceptibility-induced field inhomogeneity at 3 T. *Neuroimage* 17, 1358–1364.

Haberly, L. B. (1985) Neuronal circuitry in olfactory cortex: Anatomy and functional implications. *Chem. Senses* 10, 219–238.

Haberly, L. B. (2001) Parallel-distributed processing in olfactory cortex: New insights from morphological and physiological analysis of neuronal circuitry. *Chem. Senses* 26, 551–576.

Hasselmo, M. E. and Bower, J. M. (1993) Acetylcholine and memory. *Trends Neurosci.* 16, 218–222.

Hatfield, T., Han, J. S., Conley, M., Gallagher, M. and Holland, P. (1996) Neurotoxic lesions of basolateral, but not central, amygdala interfere with Pavlovian second-order conditioning and reinforcer devaluation effects. *J. Neurosci.* 16, 5256–5265.

Heimer, L. (1995) *The Human Brain and Spinal Cord: Functional Neuroanatomy and Dissection Guide.* Springer-Verlag: New York.

Herodotus (c. 1450 B.C.) On the Scythians. In: *The Metropolitan Museum of Art Bulletin,* New Series, Vol. 32, No. 5, Ed. F. R. B. Godolphin, *From the Lands of the Scythians: Ancient Treasures from the Museums of the U.S.S.R. 3000 B.C.–100 B.C. (1973–1974),* pp. 129–149.

Herz, R. S. and Schooler, J. W. (2002) A naturalistic study of autobiographical memories evoked by olfactory and visual cues: Testing the Proustian hypothesis. *Am. J. Psychol.* 115, 21–32.

Herz, R. S., Eliassen, J., Beland, S. and Souza, T. (2004) Neuroimaging evidence for the emotional potency of odor-evoked memory. *Neuropsychologia* 42, 371–378.

Holland, R. W., Hendriks, M. and Aarts, H. (2005) Smells like clean spirit. Nonconscious effects of scent on cognition and behavior. *Psychol. Sci.* 16, 689–693.

Illig, K. R. and Haberly, L. B. (2003) Odor-evoked activity is spatially distributed in piriform cortex. *J. Comp. Neurol.* 457, 361–373.

Jackson, W. (2003) The story of civet. *Pharmaceut. J.* 271, 859–861.

Jacob, S., Hayreh, D. J. and McClintock, M. K. (2001a) Context-dependent effects of steroid chemosignals on human physiology and mood. *Physiol. Behav.* 74, 15–27.

Jacob, S., Kinnunen, L. H., Metz, J., Cooper, M. and McClintock, M. K. (2001b) Sustained human chemosignal unconsciously alters brain function. *Neuroreport* 12, 2391–2394.

Jerison, H. J. (1997) Evolution of prefrontal cortex. In: *Development of the Prefrontal Cortex: Evolution, Neurobiology, and Behavior.* pp. 27–47. Eds. N. A. Krasnegor, G. R. Lyon, P. S. Goldman-Rakic. Paul H. Brookes Publishing: Baltimore, MD.

Johnson, B. A., Woo, C. C., Hingco, E. E., Pham, K. L. and Leon, M. (1999) Multidimensional chemotopic responses to *n*-aliphatic acid odorants in the rat olfactory bulb. *J. Comp. Neurol.* 409, 529–548.

Johnson, D. M., Illig, K. R., Behan, M. and Haberly, L. B. (2000) New features of connectivity in piriform cortex visualized by intracellular injection of pyramidal cells suggest that "primary" olfactory cortex functions like "association" cortex in other sensory systems. *J. Neurosci.* 20, 6974–6982.

Jones-Gotman, M. and Zatorre, R. J. (1988) Olfactory identification deficits in patients with focal cerebral excision. *Neuropsychologia* 26, 387–400.

Koster, E. P., Degel, J. and Piper, D. (2002) Proactive and retroactive interference in implicit odor memory. *Chem. Senses* 27, 191–206.

Kourtzi, Z. and Kanwisher, N. (2001) Representation of perceived object shape by the human lateral occipital complex. *Science* 293, 1506–1509.

Kovacs, G., Gulyas, B., Savic, I., Perrett, D. I., Cornwell, R. E., Little, A. C., Jones, B. C., Burt, D. M., Gal, V. and Vidnyanszky, Z. (2004) Smelling human sex hormone-like compounds affects face gender judgment of men. *Neuroreport* 15, 1275–1277.

Kringelbach, M. L. and Rolls, E. T. (2004) The functional neuroanatomy of the human orbitofrontal cortex: Evidence from neuroimaging and neuropsychology. *Prog. Neurobiol.* 72, 341–372.

Kringelbach, M. L., O'Doherty, J., Rolls, E. T. and Andrews, C. (2003) Activation of the human orbitofrontal cortex to a liquid food stimulus is correlated with subjective pleasantness. *Cereb. Cortex* 13, 1064–1071.

LaBar, K. S., Gatenby, J. C., Gore, J. C., LeDoux, J. E. and Phelps, E. A. (1998) Human amygdala activation during conditioned fear acquisition and extinction: a mixed-trial fMRI study. *Neuron* 20, 937–945.

Leon, M. and Johnson, B. A. (2003) Olfactory coding in the mammalian olfactory bulb. *Brain Res. Brain Res. Rev.* 42, 23–32.

Lhermitte, F. (1983) 'Utilization behaviour' and its relation to lesions of the frontal lobes. *Brain* 106 (Pt 2), 237–255.

Lhermitte, F. (1986) Human autonomy and the frontal lobes. Part II: Patient behavior in complex and social situations: the "environmental dependency syndrome". *Ann. Neurol.* 19, 335–343.

Lhermitte, F., Pillon, B. and Serdaru, M. (1986) Human autonomy and the frontal lobes. Part I: Imitation and utilization behavior: a neuropsychological study of 75 patients. *Ann. Neurol.* 19, 326–334.

Li, W., Howard, J. D., Parrish, T. B. and Gottfried, J. A. (2008) Aversive learning enhances perceptual and cortical discrimination of indiscriminable odor cues. *Science* 319, 1842–1845.

Li, W., Luxenberg, E., Parrish, T. and Gottfried, J. A. (2006) Learning to smell the roses: Experience-dependent neural plasticity in human piriform and orbitofrontal cortices. *Neuron* 52, 1097–1108.

Li, W., Moallem, I., Paller, K. A. and Gottfried, J. A. (2007) Subliminal smells can guide social preferences. *Psychol. Sci.* 18, 1044–1049.

Linster, C., Johnson, B. A., Yue, E., Morse, A., Xu, Z., Hingco, E. E., Choi, Y., Choi, M., Messiha, A. and Leon, M. (2001) Perceptual correlates of neural representations evoked by odorant enantiomers. *J. Neurosci.* 21, 9837–9843.

Linster, C., Johnson, B. A., Morse, A., Yue, E. and Leon, M. (2002) Spontaneous versus reinforced olfactory discriminations. *J. Neurosci.* 22, 6842–6845.

Lundstrom, J. N., Goncalves, M., Esteves, F. and Olsson, M. J. (2003) Psychological effects of subthreshold exposure to the putative human pheromone 4,16-androstadien-3-one. *Horm. Behav.* 44, 395–401.

Lundstrom, J. N. and Olsson, M. J. (2005) Subthreshold amounts of social odorant affect mood, but not behavior, in heterosexual women when tested by a male, but not a female, experimenter. *Biol. Psychol.* 70, 197–204.

Mai, J. K., Assheuer, J. and Paxinos, G. (1997) *Atlas of the Human Brain.* Academic Press: San Diego, CA.

Malkova, L., Gaffan, D. and Murray, E. A. (1997) Excitotoxic lesions of the amygdala fail to produce impairment in visual learning for auditory secondary reinforcement but interfere with reinforcer devaluation effects in rhesus monkeys. *J. Neurosci.* 17, 6011–6020.

Malnic, B., Hirono, J., Sato, T. and Buck, L. B. (1999) Combinatorial receptor codes for odors. *Cell* 96, 713–723.

Marr, D. (1971) Simple memory: a theory for archicortex. *Phil. Trans. R. Soc. Lond.* 262, 23–81.

Meister, M. and Bonhoeffer, T. (2001) Tuning and topography in an odor map on the rat olfactory bulb. *J. Neurosci.* 21, 1351–1360.

Mesulam, M. M. (1998) From sensation to cognition. *Brain* 121, 1013–1052.

Mishkin, M., Suzuki, W. A., Gadian, D. G. and Vargha-Khadem, F. (1997) Hierarchical organization of cognitive memory. *Philos. Trans. R. Soc. Lond. B Biol. Sci.* 352, 1461–1467.

Mombaerts, P. (1999) Odorant receptor genes in humans. *Curr. Opin. Genet. Dev.* 9, 315–320.

Morecraft, R. J., Geula, C. and Mesulam, M. M. (1992) Cytoarchitecture and neural afferents of orbitofrontal cortex in the brain of the monkey. *J. Comp. Neurol.* 323, 341–358.

Mori, K., Mataga, N. and Imamura, K. (1992) Differential specificities of single mitral cells in rabbit olfactory bulb for a homologous series of fatty acid odor molecules. *J. Neurophysiol.* 67, 786–789.

Mueller, A., Rodewald, A., Reden, J., Gerber, J., von Kummer, R. and Hummel, T. (2005) Reduced olfactory bulb volume in post-traumatic and post-infectious olfactory dysfunction. *Neuroreport* 16, 475–478.

Murphy, R. C. (1933) Floating gold: the romance of ambergris, Part One. *Nat. History* March–April. From http://www.naturalhistorymag.com/index_archivepicks.html

Niimura, Y. and Nei, M. (2005) Comparative evolutionary analysis of olfactory receptor gene clusters between humans and mice. *Gene* 346, 13–21.

O'Doherty, J., Rolls, E. T., Francis, S., Bowtell, R., McGlone, F., Kobal, G., Renner, B. and Ahne, G. (2000) Sensory-specific satiety-related olfactory activation of the human orbitofrontal cortex. *Neuroreport* 11, 893–897.

O'Doherty, J., Kringelbach, M. L., Rolls, E. T., Hornak, J. and Andrews, C. (2001) Abstract reward and punishment representations in the human orbitofrontal cortex. *Nat. Neurosci.* 4, 95–102.

O'Doherty, J., Deichmann, R., Critchley, H. D. and Dolan, R. J. (2002) Neural responses during anticipation of a primary taste reward. *Neuron* 33, 815–826.

O'Doherty, J., Winston, J. S., Critchley, H., Perrett, D., Burt, D. M. and Dolan, R. J. (2003) Beauty in a smile: the role of medial orbitofrontal cortex in facial attractiveness. *Neuropsychologia* 41, 147–155.

Ojemann, J. G., Akbudak, E., Snyder, A. Z., McKinstry, R. C., Raichle, M. E. and Conturo, T. E. (1997) Anatomic localization and quantitative analysis of gradient refocused echo-planar fMRI susceptibility artifacts. *Neuroimage* 6, 156–167.

Ongur, D., Ferry, A. T. and Price, J. L. (2003) Architectonic subdivision of the human orbital and medial prefrontal cortex. *J. Comp. Neurol.* 460, 425–449.

Plailly, J., Howard, J. D., Gitelman, D. R. and Gottfried, J. A. (2008) Attention to odor modulates thalamocortical connectivity in the human brain. *J. Neurosci.* 28, 5257–5267.

Poellinger, A., Thomas, R., Lio, P., Lee, A., Makris, N., Rosen, B. R. and Kwong, K. K. (2001) Activation and habituation in olfaction—an fMRI study. *Neuroimage* 13, 547–560.

Polak, E. H. (1973) Multiple profile-multiple receptor site model for vertebrate olfaction. *J. Theor. Biol.* 245, 175–184.

Porter, J., Anand, T., Johnson, B., Khan, R. M. and Sobel, N. (2005) Brain mechanisms for extracting spatial information from smell. *Neuron* 47, 581–592.

Price, J. L. and Slotnick, B. M. (1983) Dual olfactory representation in the rat thalamus: an anatomical and electrophysiological study. *J. Comp. Neurol.* 215, 63–77.

Rescorla, R. A. (1974) Effect of inflation of the unconditioned stimulus value following conditioning. *J. Comp. Physiol. Psychol.* 86, 101–106.

Rolls, B. J., Rolls, E. T., Rowe, E. A. and Sweeney, K. (1981a) Sensory specific satiety in man. *Physiol. Behav.* 27, 137–142.

Rolls, B. J., Rowe, E. A., Rolls, E. T., Kingston, B., Megson, A. and Gunary, R. (1981b) Variety in a meal enhances food intake in man. *Physiol. Behav.* 26, 215–221.

Rolls, E. T., Sienkiewicz, Z. J. and Yaxley, S. (1989) Hunger modulates the responses to gustatory stimuli of single neurons in the caudolateral orbitofrontal cortex of the macaque monkey. *Eur. J. Neurosci.* 1, 53–60.

Rolls, E. T., Kringelbach, M. L. and De Araujo, I. E. (2003) Different representations of pleasant and unpleasant odours in the human brain. *Eur. J. Neurosci.* 18, 695–703.

Royet, J. P. and Plailly, J. (2004) Lateralization of olfactory processes. *Chem. Senses* 29, 731–745.

Royet, J. P., Koenig, O., Gregoire, M. C., Cinotti, L., Lavenne, F., Le Bars, D., Costes, N., Vigouroux, M., Farget, V., Sicard, G., Holley, A., Mauguiere, F., Comar, D. and Froment, J. C. (1999) Functional anatomy of perceptual and semantic processing for odors. *J. Cogn. Neurosci.* 11, 94–109.

Royet, J. P., Zald, D., Versace, R., Costes, N., Lavenne, F., Koenig, O. and Gervais, R. (2000) Emotional responses to pleasant and unpleasant olfactory, visual, and auditory stimuli: a positron emission tomography study. *J. Neurosci.* 20, 7752–7759.

Royet, J. P., Hudry, J., Zald, D. H., Godinot, D., Gregoire, M. C., Lavenne, F., Costes, N. and Holley, A. (2001) Functional neuroanatomy of different olfactory judgments. *Neuroimage* 13, 506–519.

Royet, J. P., Plailly, J., Delon-Martin, C., Kareken, D. A. and Segebarth, C. (2003) fMRI of emotional responses to odors: Influence of hedonic valence and judgment, handedness, and gender. *Neuroimage* 20, 713–728.

Rubin, B. D. and Katz, L. C. (1999) Optical imaging of odorant representations in the mammalian olfactory bulb. *Neuron* 23, 499–511.

Savic, I., Gulyas, B., Larsson, M. and Roland, P. (2000) Olfactory functions are mediated by parallel and hierarchical processing. *Neuron* 26, 735–745.

Savic, I., Berglund, H., Gulyas, B. and Roland, P. (2001) Smelling of odorous sex hormone-like compounds causes sex- differentiated hypothalamic activations in humans. *Neuron* 31, 661–668.

Savic, I., Berglund, H. and Lindstrom, P. (2005) Brain response to putative pheromones in homosexual men. *Proc. Natl. Acad. Sci. USA* 102, 7356–7361.

Schiffman, S. S. (1974) Physicochemical correlates of olfactory quality. *Science* 185, 112–117.

Small, D. M., Zatorre, R. J., Dagher, A., Evans, A. C. and Jones-Gotman, M. (2001) Changes in brain activity related to eating chocolate: From pleasure to aversion. *Brain* 124, 1720–1733.

Small, D. M., Voss, J., Mak, Y. E., Simmons, K. B., Parrish, T. and Gitelman, D. (2004) Experience-dependent neural integration of taste and smell in the human brain. *J. Neurophysiol.* 92, 1892–1903.

Small, D. M., Gerber, J. C., Mak, Y. E. and Hummel, T. (2005) Differential neural responses evoked by orthonasal versus retronasal odorant perception in humans. *Neuron* 47, 593–605.

Sobel, N., Prabhakaran, V., Desmond, J. E., Glover, G. H., Goode, R. L., Sullivan, E. V. and Gabrieli, J. D. (1998a)

Sniffing and smelling: Separate subsystems in the human olfactory cortex. *Nature* 392, 282–286.

Sobel, N., Prabhakaran, V., Hartley, C. A., Desmond, J. E., Zhao, Z., Glover, G. H., Gabrieli, J. D. and Sullivan, E. V. (1998b) Odorant-induced and sniff-induced activation in the cerebellum of the human. *J. Neurosci.* 18, 8990–9001.

Sobel, N., Prabhakaran, V., Hartley, C. A., Desmond, J. E., Glover, G. H., Sullivan, E. V. and Gabrieli, J. D. (1999) Blind smell: Brain activation induced by an undetected air-borne chemical. *Brain* 122, 209–217.

Sobel, N., Prabhakaran, V., Zhao, Z., Desmond, J. E., Glover, G. H., Sullivan, E. V. and Gabrieli, J. D. (2000) Time course of odorant-induced activation in the human primary olfactory cortex. *J. Neurophysiol.* 83, 537–551.

Sobel, N., Johnson, B. N., Mainland, J. and Yousem, D. M. (2003) Functional neuroimaging of human olfaction. In: *Handbook of Olfaction and Gustation*. pp. 251–273. Ed. R. L. Doty. Marcel Dekker: New York.

Staubli, U., Schottler, F. and Nejat-Bina, D. (1987) Role of dorsomedial thalamic nucleus and piriform cortex in processing olfactory information. *Behav. Brain Res.* 25, 117–129.

Tanabe, T., Iino, M. and Takagi, S. F. (1975) Discrimination of odors in olfactory bulb, pyriform-amygdaloid areas, and orbitofrontal cortex of the monkey. *J. Neurophysiol.* 38, 1284–1296.

Theophrastus (c. 300 B.C.) *Treatise on Odours*. In *Enquiry into Plants*, Vol. II, Books 6–9 (Transl. Sir A. F. Hort). Loeb Classical Library, 1916 Edition. Harvard University Press: Cambridge, MA.

Turetsky, B. I., Moberg, P. J., Yousem, D. M., Doty, R. L., Arnold, S. E. and Gur, R. E. (2000) Reduced olfactory bulb volume in patients with schizophrenia. *Am. J. Psychiatry* 157, 828–830.

Ueki, S. and Domino, E. F. (1961) Some evidence for a mechanical receptor in olfactory function. *J. Neurophysiol.* 24, 12–25.

Villemure, C. and Bushnell, M. C. (2007) The effects of the steroid androstadienone and pleasant odorants on the mood and pain perception of men and women. *Eur. J. Pain* 11, 181–191.

Wilson, D. A. (1998) Habituation of odor responses in the rat anterior piriform cortex. *J. Neurophysiol.* 79, 1425–1440.

Wilson, J. L. and Jezzard, P. (2003) Utilization of an intra-oral diamagnetic passive shim in functional MRI of the inferior frontal cortex. *Magn Reson.Med.* 50, 1089–1094.

Wilson, J. L., Jenkinson, M., de, A. I., Kringelbach, M. L., Rolls, E. T. and Jezzard, P. (2002) Fast, fully automated global and local magnetic field optimization for fMRI of the human brain. *Neuroimage* 17, 967–976.

Winston, J. S., Henson, R. N., Fine-Goulden, M. R. and Dolan, R. J. (2004) fMRI-adaptation reveals dissociable neural representations of identity and expression in face perception. *J. Neurophysiol.* 92, 1830–1839.

Winston, J. S., Gottfried, J. A., Kilner, J. M. and Dolan, R. J. (2005) Integrated neural representations of odor intensity and affective valence in human amygdala. *J. Neurosci.* 25, 8903–8907.

Xu, F., Liu, N., Kida, I., Rothman, D. L., Hyder, F. and Shepherd, G. M. (2003) Odor maps of aldehydes and esters revealed by functional MRI in the glomerular layer of the mouse olfactory bulb. *Proc. Natl. Acad. Sci. USA.* 100, 11029–11034.

Yarita, H., Iino, M., Tanabe, T., Kogure, S. and Takagi, S. F. (1980) A transthalamic olfactory pathway to orbitofrontal cortex in the monkey. *J. Neurophysiol.* 43, 69–85.

Zald, D. H. and Pardo, J. V. (1997) Emotion, olfaction, and the human amygdala: Amygdala activation during aversive olfactory stimulation. *Proc. Natl. Acad. Sci. USA* 94, 4119–4124.

Zald, D. H. and Pardo, J. V. (1998) Unpublished observations.

Zald, D. H. and Pardo, J. V. (2000) Functional neuroimaging of the olfactory system in humans. *Int. J. Psychophysiol.* 36, 165–181.

Zatorre, R. J., Jones-Gotman, M., Evans, A. C. and Meyer, E. (1992) Functional localization and lateralization of human olfactory cortex. *Nature* 360, 339–340.

Zatorre, R. J., Jones-Gotman, M. and Rouby, C. (2000) Neural mechanisms involved in odor pleasantness and intensity judgments. *Neuroreport* 11, 2711–2716.

Zelano, C., Bensafi, M., Porter, J., Mainland, J., Johnson, B., Bremner, E., Telles, C., Khan, R. and Sobel, N. (2005) Attentional modulation in human primary olfactory cortex. *Nat. Neurosci.* 8, 114–120.

9

The Pleasure of Taste, Flavor, and Food

MARIA G. VELDHUIZEN, KRISTIN J. RUDENGA, AND DANA M. SMALL

Food is an important sensory pleasure. The aim of this chapter is to outline what is currently known about the neural basis for the hedonic representation of taste and flavor in lean healthy humans. The word "taste" is often used to describe the sensation of food in the mouth. However, what we colloquially refer to as "taste" is more accurately labeled "flavor," which is the unitary perception that results from the integration of multiple distinct sensory inputs, including taste, smell, and oral somatosensation (Small and Prescott, 2005). Here we use the word "taste" to refer to the sensations of sweet, sour, salty, bitter, and savory, and the word "flavor" to refer to a stimulus that has at least a gustatory and an olfactory component. Taste and flavor are both proximal sensations, in that the stimulus must come into contact with the body in order for sensation to occur. Information about food, in particular its availability, is also obtained through the distal senses of olfaction and vision. Inasmuch as sensory experiences are rewarding, we argue that it is important to consider the two categories of sensory representation of food (proximal and distal) within the context of anticipatory and consummatory phases of food reward (Berridge, 1996; Berridge and Robinson, 2003). This is because it is difficult to draw a line between where sensory coding ends and affective coding begins (Carmichael and Price, 1996; Small et al., 2007), and a comprehensive understanding of both will most certainly require sensitivity to both sensory and affective components.

We begin the chapter by describing the basic brain anatomy of taste and flavor. We then explore what is known about determinants of the perceived pleasantness of taste and flavor from the psychophysical and neural perspectives. Specifically, we highlight the role of (1) sensory factors, (2) experience, (3) internal state, (4) expectation and beliefs, and (5) unconscious affect in determining perceived pleasantness of taste, flavor, and food. Finally, we consider the sensory and affective coding of food within the framework of anticipatory and consummatory food reward and argue for the existence of separable but overlapping substrates representing these phases of food reward in humans.

Brain Anatomy of Taste Processing

Given the multifaceted nature of taste processing in the brain, we first describe the anatomical pathways for pure taste and then for flavor. Our current understanding has come both from animal research and subsequently from neuroimaging studies in humans. It is important to realize that there are critical differences between various species, in particular between primates and rodents.

Pure Taste

The sensation of taste occurs when molecules interact with taste receptors in the oral cavity. Common examples of taste sensations include sweet, salty, sour,

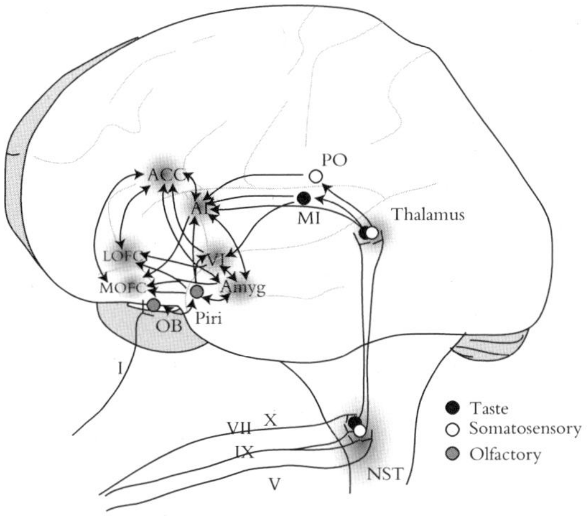

Figure 9.1 A glass brain schematic depiction of taste (black circles), somatosensory (white circles), and olfactory (gray circles) pathways. Anatomical locations are only approximate and connectivity is not exhaustive. Information from taste receptors on the tongue is conveyed via the chorda tympani (VII), the glossopharyngeal nerve (IX), and the vagus nerve (X) to the rostral nucleus tractus solitarius (NTS), which then projects to the thalamus. From here taste information projects to the midinsula (MI) and anterior insula and overlying frontal operculum (AI). AI also projects to the ventral insula (VI), medial orbitofrontal cortex (MOFC), and lateral orbitofrontal cortex (LOFC). Somatosensory input reaches the NTS via the glossopharyngeal nerve (IX) and the trigeminal nerve (V), which then projects to the thalamus. Oral somatosensory information is then relayed to the opercular region of the postcentral gyrus (PO). Olfactory information is conveyed via cranial nerve I to the olfactory bulb, which projects to primary olfactory cortex, including piriform cortex (piri). Piriform projects to VI and orbital regions. The anterior cingulate cortex (ACC) and amygdala (Amyg) are also heavily interconnected with the insula and orbital regions representing taste, smell, and oral somatosensation.

bitter, and savory. In humans, the taste signal is conveyed from the taste receptor cells and/or presynaptic cells (Chandrashekar et al., 2006; Tomchik et al., 2007) via the cranial nerves VII (the chorda tympani branch of the facial nerve), IX (the glossopharyngeal nerve), and X (the vagus nerve) to the nucleus tractus solitarius (NTS) in the hindbrain, the first gustatory relay in the brain (Beckstead et al., 1980) (Figure 9.1). Second-order gustatory fibers ascend ipsilaterally from the NTS to join the central tegmental tract and project to the parvicellular part of the ventroposteromedial nucleus of the thalamus (VPMpc) (Beckstead et al., 1980). The primary efferent projection from VPMpc is located in ipsilateral insular/opercular cortex adjacent to the superior limiting sulcus (anterior insula and overlying frontal operculum) and extending rostrally to the caudolateral orbitofrontal cortex (clOFC) (Mufson and Mesulam, 1984; Ogawa et al., 1985; Pritchard et al., 1986). A second, less extensive, projection terminates in areas 3a, 3b, 2, and 1 along the lateral margin of the precentral gyrus (Pritchard et al., 1986). Thus, anatomical studies have identified two "primary" gustatory regions within the insula/operculum.

Neuroimaging studies of taste consistently report activation of the human homologues of the two primary areas identified in monkeys (Small et al., 1999; Verhagen and Engelen, 2006). However, activation often extends beyond these regions to include the ventral (Kinomura et al., 1994; Small et al., 2003a) and posterior insula (Ogawa et al., 2005), as well as overlying rolandic, frontal, and parietal operculum areas of cortex (Cerf-Ducastel et al., 2001; Faurion et al., 1998). This suggests that taste information from the primary regions likely synapses with adjacent insular and opercular neurons. The result is that there is a relatively extensive region of insula/operculum with taste-responsive neurons. However, an important feature of so-called gustatory cortex is that only a small portion of cells in these areas actually respond to taste (Scott and Plata-Salaman, 1999). Thus taste representation is extensive but sparse in cortical areas.

Information from primary taste cortex in the anterior insula and overlying operculum (AI/FO) also projects to caudomedial orbitofrontal cortex (cmOFC) and clOFC (Baylis et al., 1995; Carmichael and Price, 1995a; Pritchard et al., 2005). The caudolateral taste region receives inputs from the frontal opercular taste area, anterior dorsal insula extending into more ventral regions of the insula, and the amygdala, mediodorsal thalamus, rhinal sulcus, and substantia nigra (Baylis et al., 1995). The cmOFC receives input from AI/FO and projects to clOFC (Carmichael and Price, 1996; Pritchard et al., 2005). Because of its functional and anatomical properties, this area is thought to constitute gustatory cortex which functions as an intermediate area between the AI/FO and clOFC (Pritchard et al., 2005). Notably, the caudomedial taste region has the highest proportion of taste-responsive neurons in primates yet identified, with 20% of the neurons responding to gustatory stimulation. Figure 9.1 provides a cartoon description of the taste pathway in humans. Human imaging studies frequently report brain activation responses to taste stimuli in caudomedial, caudolateral, and more anterior regions of OFC (O'Doherty et al., 2001, 2002a; Small et al., 1999, 2003a).

Although not formally considered part of the gustatory system, the amygdala has reciprocal connections with virtually every level of the gustatory pathway (Amaral and Price, 1984; Carmichael and Price, 1995b; Mufson et al., 1981; Norgren, 1976; Price and Amaral, 1981; Turner et al., 1980) and taste-responsive neurons are present within several amygdaloid nuclei (Scott et al., 1993). In the monkey, gustatory information reaches the amygdala via direct pathways from cortical taste neurons within the insula and operculum (Aggleton et al., 1980; Mufson et al., 1981) and OFC (Carmichael et al., 1994; Carmichael and Price, 1995a,b). The amygdala also sends projections back to the NTS (Price and Amaral, 1981), where it may exert an influence on taste processing at this early level of the primate gustatory neuroaxis. In humans, resection of the amygdala for the treatment of pharmaçological intractable epilepsy leads to changes in taste quality and intensity perception (Henkin et al., 1977; Small et al., 1997b, 2001b, c, 2005a). The anterior cingulate cortex (ACC) is also not typically considered gustatory cortex, but has connections with the thalamus, insula, OFC, and amygdala (Carmichael and Price, 1996; Vogt et al., 1987) and shows a very consistent response to taste in human neuroimaging studies (Small et al., 2003a; Verhagen and Engelen, 2006) (Figure 9.2). Additionally, the area shows supra-additive responses to congruent taste–odor mixtures (Small et al., 2004) and appears to play a role in encoding other sensory components of food (de Araujo and Rolls, 2004; de Araujo et al., 2003).

We note that it is very likely that interactions between affective value and sensory coding of taste differ between rodents and primates (Scott and Plata-Salaman, 1999; Scott et al., 1995). This is most probably because of interspecies differences in the anatomy of the gustatory system. In rodents, the pons contains an important secondary gustatory relay, the parabrachial nucleus (Di Lorenzo and Monroe, 1995, 1997; Spector, 1995a,b;Yamamoto et al., 1994) within which much centrifugal modulation occurs to guide complex feeding behaviors (Lundy and Norgren, 2001, 2004). From the pons, two separate projections

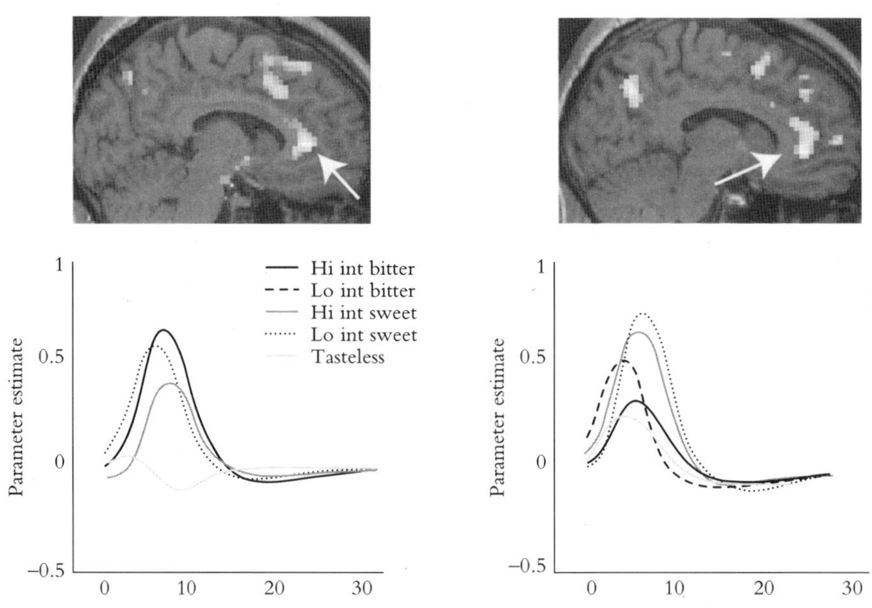

Figure 9.2 Sagittal sections from two representative subjects who participated in an fMRI study of taste perception (Small et al., 2003). The images illustrate the robust response to two concentrations of sweet and two concentrations of bitter taste that are observed in the anterior cingulate cortex (ACC) (indicated by the arrows). The graphs depict the response to these four taste stimuli plus a tasteless solution at the location in the ACC indicated by the arrow.

arise, one which synapses in the parvicellular medial tip of the VPMpc, which in turn sends axons to taste cortex. This pathway is thought to be largely sensory in nature. A second pathway projects widely to the ventral forebrain areas including hypothalamus and amygdala and is thought to be largely affective in nature. Recent support for this dissociation comes from Hajnal and Norgren (2005) who found that lesions of the pons, but not the thalamus, disrupted dopamine overflow in the accumbens during sucrose licking. Remarkably, the pontine taste relay does not appear to exist in human (Topolovec et al., 2004) and nonhuman (Beckstead et al., 1980; Norgren, 1984, 1990; Pritchard et al., 2000) primates, or if it does exist, its role is much reduced. Based upon the potential importance of these interspecies differences, we have elected to focus upon the human and nonhuman primate literature.

Flavor and Food

Taste sensations rarely occur in isolation. Rather, when we taste we also touch the food or drink and smell its odor via retronasal olfaction. The term flavor describes this multimodal experience of taste, touch, and smell. Taste and oral somatosensation are particularly intimately related (Green, 2003; Simon et al., 2008). Taste receptors lie side-by-side in the oral cavity with thermo-, mechano-, and noci-receptors. Every time we taste, we simultaneously stimulate taste receptors and touch receptors. Consequently, even "pure taste" sensations have an oral somatosensory component. Indeed, stimulation of mechanoreceptors can, under some circumstances, lead to taste perceptions. For example, Cruz and Green (2000) showed that heating and cooling the tongue can lead to sweet, sour, and salty taste perceptions in most people.

Gustatory neurons from cranial nerves VII, IX, and X are joined in the NTS by oral somatosensory projections from the spinal trigeminal nucleus. The precise locations of the trigeminal projections vary across species, but there is evidence (including in humans) of overlap with gustatory areas (Whitehead, 1990; Whitehead and Frank, 1983) and of tracts that run within the NTS that may facilitate cross-modal integration (Travers, 1988) (Figure 9.1). In addition, somatosensory input also reaches the NTS via the glossopharyngeal nerve, which contains taste-sensitive, as well as mechano- and thermo-sensitive neurons (Bradley et al., 1992). Overlapping representation of gustatory and somatosensory information also occurs in the thalamus (Pritchard et al., 1989) and at the cortical level (Cerf-Ducastel et al., 2001;Pritchard et al., 1986). For example, the so-called primary gustatory cortex contains nearly as many somatosensory-specific as taste-specific neurons, in addition to bimodal neurons responding to both somatosensory and taste stimulation (Kadohisa et al., 2004; Plata-Salaman et al., 1992,1996; Smith-Swintosky et al., 1991; Yamamoto et al., 1985).

By chewing, swallowing, and the interaction of saliva with food, volatile molecules are released during eating into the oral cavity. The volatile molecules traverse the nasopharynx to interact with receptors on the olfactory epithelium. This is referred to as retronasal olfaction (see Figure 9.3). Via cranial nerve I, the olfactory signal travels to the olfactory bulb, which projects to the anterior olfactory nucleus, the olfactory tubercle, the piriform cortex, several amygdaloid subnuclei, and rostral entorhinal cortex, which in turn project to the insula, additional amygdala subnuclei, the entorhinal cortex, and OFC (de Olmos et al., 1978; Price, 1973; Turner et al., 1978) (Figure 9.1). Odors sensed retronasally produce distinct sensory and neural responses compared to odors sensed orthonasally (i.e.,

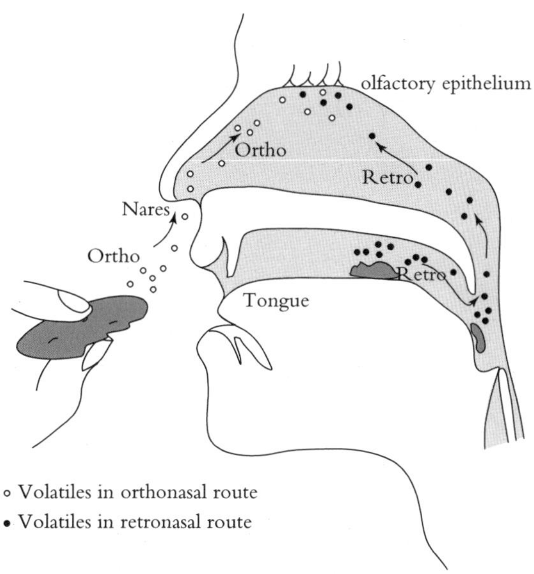

Figure 9.3 Schematic depiction of the two routes of olfactory perception: orthonasal and retronasal. Odors sensed orthonasally enter the body through the nose (nares) and travel directly to the olfactory epithelium in the nasal cavity. Odors sensed retronasally enter the mouth during eating and drinking. Volatiles are released from the food or drink, which pass through the nasophyarynx, at the back of the oral cavity, to enter the nasal cavity and reach the olfactory epithelium.

via the nose rather than the mouth) (Heilmann and Hummel, 2004; Rozin, 1982; Small et al., 2005b).

Although flavor is multimodal, the independent sensory signals fuse into a single percept, which we tend to ascribe to the gustatory sense, as when we say something "tastes" delicious. This fusion occurs for two reasons. First, odors and tastes share common characteristics. For example, both odors and tastes can be perceived as sweet. Therefore, the subjective boundary between what is taste and what is smell is blurred (Auvray and Spence, 2008). Second, two separate illusions cause taste, touch, and retronasal olfaction to appear to originate from a common location (Small, 2008). Taste sensations are localized to the site where the food or drink contacts the mouth because the concurrently sensed touch "captures" the taste (Green, 2003; Todrank and Bartoshuk, 1991). For example, Todrank and Bartoshuk showed that although taste cells are located in discrete patches on the tongue, when a cotton swab soaked in a sweet solution was "painted" across the tongue, sweetness appeared to arise along the entire path of the swab. In other words, taste sensations were perceived as coming from areas where there were no taste cells at all; touch had "captured" the taste.

The second illusion is an olfactory referral illusion, whereby olfactory sensations like strawberry odor, which are dependent on stimulation of olfactory receptors, come to be referred to the mouth. To demonstrate this illusion, hold your nose, place a strawberry jelly bean in your mouth, and chew. You should perceive taste—sweetness and a little sourness—along with the touch of the jelly bean. With your nose held, you do not perceive the strawberry odor because holding the nose prevents volatiles from reaching the olfactory epithelium. However, when you let go of your nose, the odor molecules are allowed to travel through the nasal cavity to the olfactory receptors and suddenly the jelly bean "tastes" like strawberry (Figure 9.3). This demonstrates that the perception of strawberry "taste" is dependent upon the volatiles reaching the olfactory receptors, even though it is perceived as coming from the oral cavity. It is thought that these illusions facilitate the binding of the different sensations together into a fused unitary flavor percept (Green, 2003; Small, 2008).

Small has recently proposed that processing in the somatomotor mouth area of the brain cortex is responsible for perceptual binding of taste, touch, and smell into a unitary flavor percept (Small, 2008). This brain region displays greater response to retronasal compared to orthonasal odors and is consistently activated during the perception of flavorful compared to tasteless solutions (Cerf-Ducastel and Murphy, 2001; de Araujo and Rolls, 2004; Marciani et al., 2006). Further, given the proposed importance of oral referral, it is logical that the cross-modal binding mechanism is localized in the cortical representation of the mouth region.

In addition to orchestrating binding, Small proposed that the neural computations in the somatomotor mouth area play a "permissive" role in enabling the sculpting of multimodal neurons—specifically, that unimodal taste and unimodal smell neurons located in the piriform and anterior dorsal insula sculpt the profiles of bimodal taste/smell neurons located in the ventral anterior insula and OFC only when there is concurrent activation of the binding substrate (and associated oral referral). Thus, the two functions of the binding mechanism in the cortical somatomotor mouth area are to mediate oral referral and to bind together responses across a distributed network of cortical regions. This network was proposed to comprise unimodal and multimodal representations of taste, smell, and oral somatosensation that arise when a stimulus is in the mouth. Certainly odor representations in the piriform cortex (Gottfried et al., 2006b; Wilson and Stevenson, 2004) and OFC (Gottfried et al., 2006b; Schoenbaum and Eichenbaum, 1995a,b) are included in this network. There is also evidence for overlapping responses to taste, smell, and oral somatosensation in the insula and overlying operculum (Cerf-Ducastel and Murphy, 2001; de Araujo and Rolls, 2004; Marciani et al., 2006; Poellinger et al., 2001; Savic et al., 2000; Small et al., 1999, 2003b; Verhagen et al., 2004; Zald et al., 1998), and OFC (Francis et al., 1999; Frank et al., 2003; Gottfried et al., 2002, 2006a; Marciani et al., 2006; O'Doherty et al., 2000; Poellinger et al., 2001; Savic et al., 2000; Small et al., 1997a, 1999, 2003b; Sobel et al., 1998; Zald and Pardo, 1997; Zald et al., 1998; Zatorre et al., 1992).

In accordance with these findings in humans, single-cell recording studies in monkeys have identified both taste- and smell-responsive cells in the insula/operculum (Scott and Plata-Salaman, 1999) and OFC (Rolls and Baylis, 1994; Rolls et al., 1996). The ACC also receives direct projections from the insula and the OFC (Carmichael and Price, 1996b; Vogt and Pandya, 1987), responds to taste and smell (de Araujo and Rolls, 2004; de Araujo et al., 2003; Marciani et al., 2006; O'Doherty et al., 2000; Royet et al., 2003; Savic et al., 2000; Small et al., 2001a, 2003b; Zald and Pardo, 1997; Zald et al., 1998), and shows supra-additive responses to congruent taste-odor pairs

(Small et al., 2004), and is therefore included in the proposed flavor network.

In addition, prior work has shown that the perception of congruent, but not incongruent, taste–odor solutions gives rise to responses in the AI/FO, anteroventral insula/caudal OFC, frontal operculum, and ACC which are greater than the summed responses to independent stimulation with the taste or odor (McCabe and Rolls, 2007;Small et al., 2004). Such supra-additive responses are thought to be a hallmark of multisensory integration (Calvert, 2001; Stein, 1998). The fact that the supra-additive responses in these regions are experience-dependent strongly supports the possibility that these areas are key nodes of the distributed representation of the flavor object.

Unlike flavor, which is a proximal sensation, distal information about food is obtained through orthonasal olfaction and vision. Orthonasal and retronasal olfactory information combine with oral touch and taste in the ventral insula and OFC (de Araujo and Rolls, 2004; McCabe and Rolls, 2007; Small et al., 2004). Visual information projects from primary visual cortex to inferotemporal areas, which in turn project to amygdala, OFC, and perirhinal cortex (Barbas, 1988; Jones and Powell, 1970; Rolls and Deco, 2002; Van Essen and De Yoe, 1995). Simmons and colleagues (2005) showed that the OFC and insula respond preferentially to pictures of food compared to nonfood objects, and suggested that in addition to flavor, the object category "food" may be represented in these regions.

Taste and Flavor Pleasantness

The perceived pleasantness of a taste or a flavor is dependent upon a number of variables. These include: (1) sensory factors, such as stimulus intensity and quality; (2) internal state, such as whether an animal is hungry or nutrient deficient; (3) experience, such as whether a flavor is novel or has been previously paired with a taste or with calories; (4) beliefs and expectations, such as whether the stimulus is expected; and (5) unconscious affect, such as those nonconscious or implicit representations that may not be readily accessible to conscious introspection, but which nevertheless exert powerful influences over behavior.

Sensory Factors: Perception

An indirect and complex relationship exists between perceived pleasantness and the perceptual characteristics of a gustatory stimulus that lead to its perceived pleasantness (Moskowitz et al., 1975b). One of these characteristics is perceived intensity, which is a function of stimulus concentration. The pleasantness of a stimulus can be described as a single-peaked function of perceived intensity. Wilhelm Wundt was the first to describe this relationship, which is now generally known as the "Wundt-curve" (Wundt, 1897). This function posits that, as subjective intensity increases, pleasantness increases up to a certain maximum, where it reaches a plateau, after which pleasantness will decrease. This "single-peaked" hedonic function has been demonstrated for sweet, sour, salty, and bitter taste (Ekman and Åkesson, 1965; Moskowitz et al., 1974, 1975b; Pfaffmann, 1980) and can also be understood as the expression of an optimal liking point (Moskowitz, 1971). Recently, there has been considerable interest in individual differences in optimal liking points. For example, it has been shown that those who prefer more concentrated sweet solutions are more likely to have a family history of alcoholism (Kampov-Polevoy et al., 2001; Pepino and Mennella, 2007). This suggests that taste hedonics may provide a marker for underlying genetically determined differences in the neurophysiology of reward.

Although the psychohedonic function of gustatory stimuli tends to conform to the inverted-U shape of the Wundt-curve, there are differences between tastants (Moskowitz et al., 1974, 1975b; Pfaffmann, 1980). In other words, stimulus quality interacts with intensity to shape pleasantness. For example, the pleasantness of sucrose tends to increase relatively slowly initially, plateau at a high pleasantness for moderate to strong intensities, after which it decreases, but does not necessarily become negative. In contrast, bitter substances increase minimally in pleasantness or start at the optimal point at a weak subjective intensity and decrease thereafter. Psychohedonic functions may also be conceived of as consisting of two underlying functions, one of pleasantness and one of unpleasantness (Coombs and Avrunin, 1977). This proposal is in line with the dual affective architecture of a negative and positive neural network as proposed by Cacioppo and Berntson (1999). The central tenant of this theory is that initially, appetitive and aversive information are processed in separable neural substrates, to be later integrated into a unidimensional subjective feeling that may be accompanied by approach or avoidance behavior. The dual affective architecture theory also argues that each system has its own activation function. The positive system has a weak positive output at zero input (positivity offset), whereas the activation function for negativity is argued to have a lower offset and rise more quickly, referred to as the negativity bias. This formulation is consistent with observations of taste perception. For example, in 1897 Wundt observed that "for equal (perceived) intensities, sour, and more especially, bitter (taste) produce 'much stronger feelings' than sweet" (Wundt, 1897). Later, Kocher and Fisher (1969) showed that tastes rated as unpleasant tended to receive higher intensity ratings.

A unique feature of the gustatory system is that there is a very stable relationship between the quality of a taste (sweet, sour, salty, bitter, and savory), its physiological significance, and its perceived pleasantness. Most researchers agree that the five major categories of taste quality sweet, sour, salty, bitter, and savory- are tuned to identify a specific nutrient or physiological threat, and that each is associated with a particular physiological function, namely, ensuring energy reserves (sweet), maintaining electrolyte balance (salty), guarding pH (sour, bitter), motivating protein intake (savory), and avoiding toxins (bitter) (Bartoshuk, 1991; Scott and Plata-Salaman, 1991). Thus, the primary function of the gustatory sense is to identify substances that may promote or disrupt homeostasis. Accordingly, toxicity is a better predictor of central neural response than physical structure (Scott and Mark, 1987), and taste elicits perceptual and autonomic responses (Norgren, 1985). In sum, the perceived pleasantness of a taste varies with intensity and the nature of this relationship is related to quality, likely because of the intimate coupling between quality and physiological significance.

Sensory Factors: Neural Representation

The tight coupling between quality and physiological significance is reflected in the physiological organization of the gustatory system. There is evidence for separate substrates for the so-called attractive tastes of umami and sweet, and the aversive taste of bitter, at virtually every level of the neuroaxis. Beginning with the receptors, both sweet and umami receptors are heterodimers and share a common subunit (T1R2). The T1R2 subunit combines with the T1R1 subunit to form the heterodimeric G-protein coupled receptor for sweet, or with the T1R3 subunit to form the heterodimeric G-protein-coupled receptor for umami (Chandrashekar et al., 2006). Bitter substances, on the other hand, are transduced by a family of homodimeric G-protein coupled receptors known as the T2Rs (Chandrashekar et al., 2000). It is thought that taste receptor cells express only one type of receptor and thus that taste receptor cells are tuned to specific qualities (Chandrashekar et al., 2006) (though see Tomchik et al., 2007). However, if a sweet receptor is expressed in a bitter taste cell, the animal will learn to avoid the chemical that binds to that receptor. In contrast, if a T2R is expressed in a sweet taste cell, then the animal will learn to ingest the chemical that binds to it (Mueller et al., 2005). This suggests that labeled lines exist at the level of the taste cell, not the receptor, with independent lines for attractive versus aversive tastes. Whether taste quality is coded in a labeled line beyond the taste receptor cell continues to be a subject of considerable debate in the field (Di Lorenzo et al., 2003a, b; Erickson et al., 1994; Katz et al., 2002a,b; Scott and Plata-Salaman, 1991; Smith and Scott, 2003; Smith and St. John, 1999; Smith et al., 2000; Tomchik et al., 2007).

Outlining this debate is beyond the scope of the current chapter. However, there is fairly strong evidence to suggest that there is at least some segregation between the pathways and neural representation of attractive versus aversive taste in rats (Accolla et al., 2007; Sugita and Shiba, 2005) and in humans (O'Doherty et al., 2001; Small et al., 2003a; Zald et al., 2002). For example, Acolla and colleagues used calcium dye imaging to show spatially segregated responses for different tastes in the rodent insula. They observed that two tastants of similar hedonic value activated areas with more common regions than two tastants with the opposite hedonic value (Accolla et al., 2007). In humans, Small and colleagues (2003a) used fMRI to examine neural response to two concentrations of a sweet taste and two concentrations of a bitter taste. Different regions of the insula and OFC responded preferentially to the sweet or the bitter stimuli.

According to the dual architecture model, the separate signals for pleasant (i.e., attractive tastes) and unpleasant (i.e., aversive tastes) stimuli must integrate at some point so that a single feeling and behavior is expressed (Cacioppo and Berntson, 1999). In addition, in the gustatory system, the affective code must be integrated with information about intensity and quality.

Although there has been some debate over whether the insula encodes the affective value of taste, we believe that there are enough independent observations of value-related responses to conclude that affective value is represented in the insula. For example, insular response to the ingestion of chocolate changes as a function of satiety and the magnitude of the neural change correlates with changes in pleasantness ratings (Small et al., 2001a; Smeets et al., 2006); preferential response in the insula is noted to unpredicted preferred compared to nonpreferred solutions (Berns et al., 2001); cognitive manipulations that change the perceived pleasantness of pure taste stimuli also modulate insular response to the taste (Nitschke et al., 2006); and insular lesions result in changes to taste intensity perception, but the magnitude of this effect is related to the hedonic value of the taste (Mak et al., 2005). In addition to playing a role in affective coding, the insula has also been shown to play a role in encoding taste intensity (Grabenhorst and Rolls, 2008; Mak et al., 2005; Scott et al., 1986; Small et al., 2003a; Yaxley et al., 1990) and quality (Adolphs et al., 2005; Pritchard et al., 1999; Schoenfeld et al., 2004; Smith and Scott, 2003; Smith-Swintosky et al., 1991).

Similarly, although amygdala response does not correlate with ratings of the perceived pleasantness of taste or flavor (de Araujo et al., 2003; Kringelbach et al., 2003; McCabe and Rolls, 2007; Small et al., 2001a), the structure nevertheless likely plays an important role in affective coding. Consistent with the known role of the amygdala in encoding fear (LeDoux, 2000), early chemosensory studies suggested that the amygdala encoded aversive tastes, flavors, and odors (Small et al., 1997a; Zald and Pardo, 1997; Zald et al., 1998). For example, Zald and colleagues reported preferential response of the amygdala to an aversive odor compared to a pleasant odor (Zald and Pardo, 1997), and an aversive saline solution compared to a pleasant chocolate flavor (Zald et al., 1998). Small and colleagues then reported that patients with removal of the amygdala for the treatment of pharmacologically intractable epilepsy rated unpleasant sour, salty, and bitter taste stimuli as more intense than did control subjects (Small et al., 2001c). In contrast, there were no group differences in pleasant sweet intensity ratings. Based on these findings, it was proposed that processing in the amygdala reflects an interaction between valence and intensity, and as such, that encoding here may underlie the negativity bias for unpleasant tastes (Small et al., 2001c).

In a subsequent study, we reasoned that if this was true, then if the concentration of a sweet solution is increased so that it matched the intensity perception of a bitter taste, then the amygdala should respond equally to pleasant sweet taste and unpleasant bitter taste. As predicted, this is exactly what was found; the amygdala response was similarly strong to sweet and bitter solutions that were matched for intensity (Small et al., 2003a). However, in keeping with the fact that it is the interaction between valence and intensity that drives amygdala response, Winston and colleagues (2005) showed that when amygdala response is measured during sensation of two concentrations of pleasant, unpleasant, and neutral odors, concentration-dependent responses are only observed for the pleasant and unpleasant stimuli.

Like the insula, the amygdala also plays a role in taste quality coding (Henkin et al., 1977; Small et al., 1997b, 2005a). Unilateral resection of the amygdala for the treatment of pharmacologically intractable epilepsy results in increases in taste recognition thresholds (Henkin et al., 1977; Small et al., 1997b), and bilateral lesions of the amygdala have been associated with gustatory agnosia, which is the complete inability to identify taste quality despite intact ability to discriminate stimuli based on intensity or affective differences (Small et al., 2005a).

In contrast to the amygdala and insula, the OFC appears to play a more specialized role in encoding value. Valence-specific taste responses in the OFC are independent of intensity (Small et al., 2003a) and likely of quality (Gottfried et al., 2006a). In addition, when subjects are asked to evaluate the pleasantness of a taste, the OFC is selectively engaged (Bender et al., 2005; Grabenhorst and Rolls, 2008; Grabenhorst et al., 2007; Small et al., 2007). For example, Grabenhorst and Rolls (2008) showed that response to a savory taste and flavor in the pregenual cingulate cortex and medial OFC correlates with pleasantness ratings and that these regions respond preferentially when subjects evaluate taste pleasantness compared to the intensity of these stimuli. Small and colleagues showed that a region of lateral OFC was preferentially engaged when subjects judged the pleasantness of sweet, sour, salty, and tasteless solutions compared to when they performed a detection task, judged quality, or sampled passively (Bender et al., 2005; Small et al., 2007). Although the response occurred to the tasteless solution, as well as to the solutions containing a tastant, preferential connectivity was observed between the lateral OFC and the medial OFC, ventral striatum, ventral and anterior insula when subjects received a taste versus a tasteless solution. This suggests that the lateral OFC plays a general role in judging pleasantness and selectively modulates sensory-specific cortex depending on the modality being judged. Consistent with this possibility, Royet and colleagues (2001, 2003) found that a similar area of lateral OFC responded when subjects judged odor pleasantness.

In sum, the pleasantness of taste is influenced by quality and intensity. The insula and amygdala appear to encode quality and intensity but the magnitude or precise location of the response appears dependent on value. Both of these regions project to regions of the OFC, in which response to pure taste has been shown to vary as a function of pleasantness (Baylis et al., 1995; Carmichael and Price, 1995a,b, 1996a; Pritchard et al., 2005). Thus we propose that processing within and between these regions integrates separable affective signals with information about quality and intensity to compute a single affective representation, which is realized within the OFC.

Experience: Perception

As discussed above, there is a very stable relationship between taste quality and affective value. There is even evidence that these associations are innate (Steiner et al., 2001b). Babies show prototypical responses to different taste qualities. For example, they will gag when given something bitter and lick their lips when presented with sweet. Here we refer to this aspect of gustatory hedonics as the "intrinsic affective value" of the quality. The presence of intrinsic affective values has led some investigators to propose that taste perceptions are primary reinforcers (Rolls, 1999).

However, if this were true, then affective values should be immutable. They are not. In rodents, diet has been shown to have a profound effect on gustatory organization (Hendricks et al., 2004; Hill and Przekop, 1988; Krimm and Hill, 1999; Pittman and Contreras, 2002; Shuler et al., 2004), which may in turn be related to preferences (Hill et al., 1986). There is also an extensive literature examining taste-nutrient conditioning in rodents (Ackroff and Sclafani, 1991, 1994; Ackroff et al., 1993; Azzara and Sclafani, 1998; Drucker and Sclafani, 1997; Drucker et al., 1993; Perez and Sclafani,1990; Perez et al., 1998; Sclafani, 1988, 2004, 2007a; Sclafani and Ackroff, 1994; Sclafani and Glendinning, 2005; Sclafani and Nissenbaum, 1987; Sclafani et al., 1993a,b, 1996, 1998, 2007).

One particularly important finding from this literature is that postingestive effects play an important role in preference learning for pure tastes (Sclafani and Ackroff, 2003). This postingestive signal may be mediated in part by sweet taste signaling in the gut (Sclafani, 2007b) but a recent study found that trpm5 knock-out mice, who do not taste sweet, savory, or bitter, learn to prefer caloric versus noncaloric sweet drinks, indicating that postingestive effects can influence preference learning in the absence of taste sensation (de Araujo et al., 2008).

In humans, diet has been shown to affect the pleasantness perception of taste qualities (Moskowitz et al., 1975a) and the importance of being able to adapt preferences to the food available in the environment has been emphasized (Glendinning, 1994). The influence of diet upon taste pleasantness perception and preference suggests that although the relationship between the intrinsic affective value and quality is generally very stable, learning can occur.

In contrast to pure tastes, the perceived pleasantness of flavors is largely determined by experience (Cardello, 1996; Laing et al., 1994; Mela, 2001; Prescott et al., 1996). Specifically, taste–odor "congruency," which refers to the extent to which a taste and an odor appear to match or be harmonious, is directly related to prior experience with the taste–odor pair and plays an important role in determining the perceived pleasantness of a taste–odor mixture (Schifferstein and Verlegh, 1996).

Behavioral studies of taste–odor integration show that odors can enhance perceived taste intensity, but only if they have previously been experienced with that taste (Frank and Byram, 1988; Frank et al., 1993; Sakai et al., 2001; Schifferstein and Verlegh, 1996; van de Klauuw and Frank, 1996). For example, strawberry odor will enhance the perceived intensity of a sweet but not a salty taste solution.

Similarly, Dalton and colleagues (2000) demonstrated that detection thresholds for an odorant were significantly reduced while subjects held a taste in the mouth, but only if the taste was perceptually congruent (Dalton et al., 2000). It is thought that perceptual congruency, which clearly influences the nature of integration, develops in part because odor quality changes as a result of being experienced in solution with a taste. Specifically, odors experienced in solution with a sweet or a salty taste, come to be perceived as sweeter or saltier after the exposure (Stevenson and Boakes, 2003; Stevenson and Tomiczek, 2007; Stevenson et al., 1995). This process is also proposed to lead to the formation of a flavor engram, which is encoded via configural learning (Stevenson and Boakes, 2004) as a distributed pattern of response across multiple regions of insula and overlying operculum, OFC, and ACC (Small, 2008).

In sum, simple exposure to taste–odor pairs is thought to lead to the formation of a flavor engram via configural learning, which is associated with the development of perceptual congruency. Perceptual congruency is, in turn, postulated to be an important determinant of flavor pleasantness.

Although there is a clear positive association between taste–odor congruency and flavor pleasantness, taste–odor sensory learning (i.e., the acquisition of taste-like properties by odors) is not always accompanied by changes in odor liking (Yeomans et al., 2006). Rather, Yeomans and colleagues showed that sweet taste exposure increased the sweetness ratings of an odor but only increased the perceived pleasantness of the odor if the participant was a sweet liker. They suggested that the existence of this dissociation indicates that hedonic and sensory learning occur via independent mechanisms.

One important distinction may relate to flavor-nutrient learning, which is known to play an important role in the development of flavor preference learning (Birch et al., 1990; Johnson et al., 1991; Zellner et al., 1983). For example, children liked a previously unfamiliar flavor better, and consumed more of it, after it was paired with a high-caloric solution for a week during their morning snack (Birch et al., 1990). Similarly, the liking of a novel drink increased in adults if caffeine was consumed afterwards (Richardson et al., 1996). Accordingly, Yeomans and Mobini (Yeomans et al. 2006) have now shown that hunger influences whether taste–odor exposure results in hedonic shifts (i.e., learning) in odors, while sensory properties like acquired sweetness or bitterness are insensitive to hunger. That is, odors were rated as sweeter as a function of exposure, but were only rated as more pleasant if the exposure occurred while subjects were hungry. Notably, hunger had no effect on sensory or hedonic learning for the bitter taste. This dissociation again highlights the fact that there appear to be separate pathways representing bitter and sweet taste.

Experience: Neural Representation

We are unaware of studies examining the neural correlates of taste hedonics as a function of diet in human or nonhuman primates. One neuroimaging study by Faurion and colleagues (2002) found increased response in the insula and operculum to pure taste as a function of exposure. This was linked to increased sensitivity but it is unknown if exposure also changed perceived pleasantness.

There is a relatively larger literature examining the neural correlates of flavor perception in human and nonhuman primates. In accordance with the psychophysical literature, these studies highlight the importance of congruency in integration and affective representation (de Araujo et al., 2003; McCabe and Rolls, 2007; Small et al., 1997a, 2004). In nonhuman primates, bimodal neurons in the ventral insula and caudal OFC respond selectively to odors and tastes that have previously been experienced together (Rolls and Baylis, 1994). For example, a cell may respond to the presentation of glucose and banana odor, but not glucose and an onion odor.

In humans, evoked potential latency to discrete odor exposure is significantly shorter, and of greater amplitudes, when presented with congruent but not incongruent tastes (Welge-Lussen et al., 2004). Activations to retronasally presented odors (usually in aqueous solution) have been observed in insula, OFC and ACC (Cerf-Ducastel and Murphy, 2001; de Araujo et al., 2003; Marciani et al., 2006; McCabe and Rolls, 2007; Okamoto et al., 2006; Small et al., 2004, 2005b). Bimodal responses in humans to retronasally presented odors and tastes have been observed in frontal operculum and anterior insula, amygdala, ACC, and caudal OFC (de Araujo et al., 2003; McCabe and Rolls, 2007).

In a study by Small et al. (2004), supra-additive responses in ventral anterior insula extending into OFC and ACC were observed for pleasant congruent taste/odor pairs compared to the sum of the response to independent stimulation of their components. In addition, both regions showed significantly greater activation to congruent compared to incongruent mixtures (Small et al., 2004). Notably, a supra-additive response to an unpleasant and incongruent taste–odor pair was also observed in an adjacent but not overlapping region of ventral insula. This region receives projections from the anterior insula (Baylis et al., 1995) and the piriform cortex (Carmichael et al., 1994) and likely plays an important role in taste–odor learning. A similar conclusion was reached by de Araujo et al. (2003) who found that pure tastes, retronasally presented odors, and taste/odor mixtures all produce activation in this region. However, they only observed supra-additive responses (to the combination of sucrose and strawberry odor) in the left anterior OFC. Notably, response in this region of OFC also correlated with flavor pleasantness ratings (de Araujo et al., 2003).

However, in a subsequent study, supra-additive response, (which correlated with perceived pleasantness) to combined vegetable odor and MSG taste was observed in the medial OFC and ACC (McCabe and Rolls, 2007). Taken together, the data suggest that the ventral insula, as well as multiple regions of OFC and ACC, are likely important for flavor learning and encoding flavors. However, the contribution of congruency to perceived pleasantness of flavors is likely a function of specific processing in the ACC and OFC. It is also noteworthy that in addition to the ventral insula, unfamiliar and unpleasant flavors preferentially recruit the basal forebrain and amygdala but not the

OFC. This suggests that hedonic and aversive representations of flavors have separable substrates (Small et al., 1997a).

We are currently not aware of any published studies examining the neural correlates of flavor-nutrient conditioning in human or nonhuman primates. However, this topic has been extensively investigated in rodents (Sclafani, 2004; Sclafani et al., 2001) and deserves to be the focus of future investigations in humans.

Internal State: Perception

The process by which homeostatic signals influence perceived pleasantness of sensory stimuli is known as "alliesthesia" (Cabanac, 1971). As reviewed above, homeostatic signals interact with both sensory factors and experience to influence preference development. The pleasantness of taste, flavor, and indeed any sensory cue that is associated with food, also changes as a function of fluctuations in internal state related to the feeding cycle (Cabanac, 1971; Gottfried et al., 2003; LaBar et al., 2001; Rolls et al., 1981). For example, after consumption of sugar, ratings of the pleasantness of a sweet stimulus decrease. In contrast, injection of insulin, which leads to subjective hunger, causes pleasantness ratings of a sweet stimulus to increase (Cabanac et al., 1968; Jacobs, 1958). Similar effects have been shown for salty taste (Beauchamp and Cowart, 1985; Jacobs et al., 1988). There is also a considerable literature showing that long-term deprivation, or acute depletion, of particular nutrients leads to the development of specific appetites in rodents (McCaughey and Tordoff, 2002; McCaughey et al., 2005; Johnson, 2007) For example, NaCl deprivation has been shown to lead to increased palatability of saltiness (Berridge et al., 1984).

In humans, a sodium-specific appetite has only been demonstrated for severe, prolonged sodium depletion (Cotterill and Cunliffe, 1973; Knowles, 1958; Wilkins and Richter, 1940). However, short-term experimental manipulations of sodium intake in healthy subjects can change sensitivity and preference. Beauchamp and colleagues (1990) fed subjects a low-sodium meal for 10 days. Following this manipulation, detection thresholds for salt decreased and preference judgments of salty foods increased. In addition, an increased preference for salty foods has been demonstrated after subjects exercise, presumably reflecting the need to replenish depleted sodium stores and regulate hydration (Leshem et al., 1999). However, these effects are considerably more modest than the specific appetites observed in rodents (Mattes, 1997), possibly reflecting the fact that in humans the coupling between palatability and need state is less straightforward than in rodents (Yeomans et al., 2004).

Alliesthesia also occurs for foods and flavors (Cabanac, 1971). At least part of the modulatory effects of internal state are thought to be sensory-specific (Hetherington et al., 1989; Rolls, 1986; Rolls et al., 1981, 1982, 1983). In a seminal experiment, Barbara Rolls and colleagues demonstrated that pleasantness reduction is maximal to the color, shape, and flavor of the food on which subjects are fed. They showed that after eating chocolates of one color, the pleasantness of the taste of the eaten color declined more than of the noneaten colors, although these chocolates differed only in appearance. These data unequivocally demonstrated that sensory factors interact with homeostatic signals in determining perceived pleasantness.

Internal State: Neural Representation

At the neurophysiological level, unimodal gustatory and olfactory responses to specific foods are greatly attenuated by satiety in the OFC (Critchley and Rolls, 1996; Nakano et al., 1984; Rolls et al., 1989), while leaving responses to other foods in tact. Sensory-specific satiety related declines have also been seen for visual responses to food items, confirming the multimodal nature of the phenomenon (Critchley and Rolls, 1996; Nakano et al., 1984; Rolls et al., 1989). In contrast to the OFC, single-cell recordings of the insula-operculum primary taste region of cortex do not show evidence of a sensory-specific satiety effect, which led Rolls and colleagues to conclude that this process first emerges at the level of the OFC.

Data from humans suggest that insular responses to foods do actually decrease with feeding (Gautier 2001; Small et al. 2001a), but this may reflect the more general process of alliesthesia, rather than sensory-specific satiety. Single-unit responses to taste also decrease as a function of satiety in the amygdala (Yan and Scott, 1996), caudomedial OFC (Tom Pritchard, personal communication), lateral hypothalamus, and substantia innominata (Rolls, 1981a). Notably, the effect of satiety is much more pronounced in the OFC and lateral hypothalamus than it is in the amygdala (Smith and Scott, 2003).

In humans, fMRI has been used to track changes in brain response following, or over the course of, a glucose challenge (Liu et al., 2000; Matsuda et al., 1999; Smeets et al., 2005a,b). All of these studies have focused upon response in the hypothalamus and all show that response here decreases as a function of satiety. This effect appears to require the presence of a sweet taste and calories (Smeets et al., 2005a), is associated with changes in plasma insulin levels (Liu et al., 2000; Smeets et al., 2005a, b), peaks at approximately 10 min postingestion (Liu et al., 2000), and is blunted in overweight subjects (Matsuda et al., 1999).

To our knowledge, there is no published work examining how satiety effects changes in neural response to pure tastes in other brain regions in humans. We also know of no study that has examined the neural correlates of specific nutrient depletion in human or nonhuman primates. However, there are a number of comprehensive reviews on this topic in rodents (Johnson, 2007; McCaughey and Scott, 1998; McCaughey and Tordoff, 2002; McCaughey et al., 2005).

In contrast, there is now a fairly well developed literature examining whole brain response to changes in response to flavors and foods as a function of changes in internal state. This literature includes studies of participants tasting flavors or foods (Kringelbach et al., 2003; Small et al., 2001b; Smeets et al., 2006), or being exposed to food cues (Gottfried et al., 2003; Hinton et al., 2004; LaBar et al., 2001; Mohanty et al., 2008; Morris and Dolan, 2001; O'Doherty et al., 2000).

The human neuroimaging literature accords with single-unit work in primates to highlight a role for the OFC in sensory specific satiety for flavors and food cues. Two studies have reported on brain response to flavored food or drink as subjects ate or drank that substance to satiety or beyond. Responses in the medial OFC to palatable food or drink decreased as that substance was eaten to satiety (and beyond), and these decreases correlated with declining pleasantness ratings (Small et al., 2001a; Kringelbach et al., 2003). In contrast, activity in the lateral OFC, extending into ventrolateral prefrontal cortex, increased with satiety and correlated inversely with pleasantness ratings (Small et al., 2001a). These data suggest that there are both decreases and increases in activity with satiety, and that a medial/lateral segregation distinguishes these responses.

These findings accord with data from Tataranni and colleagues who have used positron emission tomography (PET) to measure resting brain response after a 36-h fast (condition 1: hungry), followed by a 2-mL infusion of a liquid meal (condition 2: taste) and then a satiating amount of a liquid meal (condition 3: satiety) (Del Parigi et al., 2002a, b, c, d, 2003, 2004; Gautier et al., 1999, 2001; Tataranni and Del Parigi, 2003; Tataranni et al., 1999). In lean subjects, activity in the caudal OFC and insula is greater during hunger compared to satiety, while activity in the dorsal and ventral prefrontal cortex, extending into the lateral most part of the OFC (areas 8, 9, 10, 46, 47), is greater in satiety compared to hunger (Tataranni and Del Parigi, 2003; Tataranni et al., 1999). Both effects are significantly larger in obese subjects (Del Parigi et al., 2002c; Gautier et al., 2001; Tataranni and Del Parigi, 2003). The authors have proposed that the prefrontal cortex signals satiation by sending inhibitory inputs to limbic/paralimibic areas, thus suppressing hunger. They further suggest that since both responses are greater in obese subjects, the prefrontal inhibitory signal may be working harder to suppress chronically hyperactive areas of limbic brain structures that promote eating (Tataranni and Del Parigi, 2003).

A dissociation in the functional role between medial and lateral prefrontal cortex in food reward also accords with data from a study in which brain response was measured as subjects perused menus of high and low incentive value (Arana et al., 2003). Medial OFC activity was observed when subjects viewed preferred versus nonpreferred menus (Arana et al., 2003; Hinton et al., 2004). In contrast, lateral OFC activity was observed when participants had to suppress responses to alternative desirable items to select their most preferred item (Arana et al., 2003).

A follow-up study was then designed to look at the interaction between what the authors termed intrinsic and extrinsic determinants of food reward (Hinton et al., 2004). Intrinsic factors were defined as contributions by homeostatic mechanisms (e.g., hunger, satiety) and extrinsic factors as contributions by variables related to the food, such as food preferences and the associated experience-based incentive properties of food. The follow-up study was identical to the earlier experiment except that subjects performed the task while hungry and while full. Consequently, they were able to evaluate the effect of intrinsic factors (internal state) upon extrinsic factors evaluated by their task. In addition to replicating their initial findings, and in accordance with previous studies comparing hunger and satiety, they found that the lateral OFC was more responsive when viewing and choosing items from a menu when full as compared to when hungry. More importantly, they found that intrinsic factors (i.e., hunger vs. satiety) had a far greater effect upon neural response than extrinsic factors (imagining high vs. low incentive foods). The only region of the brain in which an interaction was observed between intrinsic and extrinsic factors was in a region of the mid-anterior OFC (Chiavaras and Petrides, 2000), which was preferentially activated by viewing and choosing food from high- versus low-incentive menus when hungry compared to when full. The implication is that there is a high degree of overlap of neural representation of the two facets of food reward measured, but that some specializations do exist, and these can be found in the OFC.

The effect of satiety on brain response to olfactory and visual food cues has also been investigated. O'Doherty and colleagues (2000) presented subjects orthonasally with a banana and a vanilla odor before and after eating bananas to satiety. In all six subjects, activity evoked in the caudal and medial OFC by banana but not by vanilla odor decreased after eating. Several subjects also showed decreases in the amygdala and/or insula. In a subsequent study, Gottfried et al. (2003) paired abstract visual designs with food odors and then fed subjects to satiety with a food associated with one of the experimental odors. They found that activity in the OFC, amygdala, insula, anterior cingulate, and ventral striatum decreased in response to visual cues that predicted odors associated with the food eaten to satiety. No change was observed in response to visual stimuli predicting the odor not associated with the food eaten to satiety.

More recently, Mohanty and colleagues asked participants, in a fasted or sated state, to perform a target detection task in which central cues signaled visual locations of motivationally relevant (food) and irrelevant (tool) attentional targets that were presented peripherally (Mohanty et al., 2008). The subject's task was to press a button as soon as the target (food or tool) appeared. Response in the amygdala, posterior cingulate cortex, locus coeruleus, and substantia nigra showed selective sensitivity to food-related cues when hungry but not when sated, an effect that did not generalize to tools. In addition, responses in the visual spatial attention network and OFC correlated with the speed of attentional shifts, which were sensitive not just to motivational state, but also to the value of the target. This indicates that OFC neurons are sensitive to internal state and the reward value of the target, and that this information is used to successfully bias attention in the service of goal directed behavior (i.e., faster target detection). Interestingly, the authors observed dissociable responses in the OFC, with the increased medial OFC activity associated with faster reaction times for food-related targets while hungry and increased lateral OFC activity associated with slower reaction times while hungry and faster reaction times to food related-stimuli while full.

Sensory-specific satiety has been used extensively to study reward devaluation in nonhuman animals (Holland, 2004; Holland and Gallagher, 2004; Rescorla, 1987). The findings from this literature accord with human neuroimaging work in highlighting a role for the amygdala (specifically the basolateral nucleus) and OFC in sensory-specific satiety. Additionally, specialized roles for individual regions within this network are beginning to be defined. For example, the amygdala is critical for the process by which initially neutral events gain associative access to the incentive value of reinforcers, whereas the OFC is critical for the use of the expectancy induced by these incentive cues to guide behavior (Holland and Gallagher, 2004; Pickens et al., 2003; Schoenbaum et al., 2003).

Finally, we note that there is evidence for dissociable circuits representing sensory-specific satiety and alliesthesia and devaluation of food receipt versus food cues. Deactivation, which is observed in the ventromedial OFC/subcallosal area and hypothalamus, may result from a combination of homeostatic and sensory factors, whereas activity on the ventral surface of the OFC is associated specifically with sensory devaluation. Additionally, the amygdala and ventral striatum appear to be selectively sensitive to devaluation of visual and olfactory food cues. These regions are not sensitive to devaluation of food or drink (Kringelbach et al., 2003; Small et al., 2001a). We have proposed that the explanation for this discrepancy can be found by differentiating neural responses to satiety of sensory cues predicting food reward versus sensory cues associated with the experience of food reward (Gottfried et al., 2006a; Small et al., 2005b, 2008).

Expectation and Beliefs

Neuroimaging investigations of the influence of expectations and beliefs on affective perceptual and neural responses to food-related stimuli can be divided into those that examine the influence of temporal predictability and those that manipulate beliefs about the stimulus identity or quality (Berns et al., 2001; de Araujo et al., 2005; McClure et al., 2003, 2004; Nitschke et al., 2006; O'Doherty et al., 2002a, 2003a, 2004, 2006; Pagnoni et al., 2002; Plassmann et al., 2008; Sarinopoulos et al., 2006; Veldhuizen et al., 2007).

Studies examining the importance of temporal predictability in encoding of taste/flavor arose in response to work conducted by Shultz and colleagues in the mid and late 1990s (Schultz, 1998; Schultz et al., 1997). In brief, Schultz and colleagues showed that dopamine responses change as a function of learning and that the nature of these changes is consistent with the possibility that they signal a prediction error required for reward learning to occur (Schultz et al., 1997). A cornerstone of this theory is the finding that there is greater dopamine release, and responses in dopamine target regions, to sensation of unpredicted food rewards. There are now a handful of studies that have used fMRI to investigate Schultz's hypotheses in humans (Berns et al., 2001; McClure et al., 2003; O'Doherty, 2004; O'Doherty et al., 2002b, 2003b, 2004; Pagnoni et al., 2002). Since this work is aimed primarily toward understanding reward learning, we highlight only those findings relevant to understanding taste and flavor pleasure and preference. The reader is referred to a review by John O'Doherty (2004) for a more thorough discussion of these results and their theoretical underpinnings.

The first neuroimaging study of temporal predictability was conducted by Berns and colleagues (2001). Based upon Schultz's work, they reasoned that predictability may influence brain response to reward in humans and that this would be detectable with fMRI. In support of the prediction, they found greater response to fruit juice and water when these stimuli were delivered in an unpredictable as compared to a predictable pattern in the nucleus accumbens and medial OFC. Neither response was related to the subject's stated preference (i.e., for water or juice). Rather, preference was reflected by response in the somatosensory cortex. However, they identified an interaction between preference and predictability in a region of right anterior insula; that is, this region responded maximally to the preferred unpredictable solution. Taken together, the findings suggest that the two signals are separable but that the dopaminergic reward signal is integrated with sensory information in the anterior insula (Berns et al., 2001).

Subsequent studies confirmed the importance of the striatum and OFC in encoding both positive and negative prediction errors (McClure et al., 2003; O'Doherty, 2004; O'Doherty et al., 2003a; Pagnoni et al., 2002). More recent data from O'Doherty and colleagues shows that responses to cues that predict pleasant drinks in the ventral midbrain and ventral putamen are also related to behavioral food preferences (O'Doherty et al., 2006). Notably, in contrast to Berns and colleagues, they reported that flavor-related responding in the anterior insula did not reflect flavor preference (O'Doherty et al., 2006). However, in contrast to the original study by Berns and colleagues, the flavors were predicted by abstract visual cues. This is consistent with the suggestion that the insular response is a function of an interaction between predictability and preference.

Studies examining the influence of expectations and beliefs about stimulus quality and identity are based on the idea that top-down influences can shape the encoding of incoming sensory signals. Many of the studies of this topic reflect a growing interest in the new field of neuroeconomics. In the first study of its kind, McClure and colleagues (2004) asked whether information about brand could influence stated preferences and choices, and if so, how such influences might be represented in the brain. When beverages were delivered without providing knowledge about the brand, response in the medial OFC/ventromedial prefrontal cortex tracked anonymous behavioral preference. Providing information about brand shifted behavioral preference toward the stated favored brand and altered brain response so that response in the hippocampus, dorsolateral prefrontal cortex, and midbrain reflected preference. These data are compelling because they show that brand information can override sensory determinants of preference to guide behavior.

Later work showed that the medial OFC/ventromedial prefrontal cortex response could also be enhanced by labels that cause an increase in ratings of perceived pleasantness (de Araujo et al., 2005; Plassmann et al., 2008). De Araujo and colleagues showed that when an odor was labeled as "cheddar cheese" it was rated as significantly more pleasant and preferentially activated the medial OFC/ventromedial prefrontal cortex compared to when it was labeled "body odor" (de Araujo et al., 2005). In contrast, although response in the amygdala differed as a function of label, the magnitude was not related to pleasantness rating. Plassman and colleagues assigned differing prices to the same wine and showed that increasing the price increased pleasantness ratings and response in the medial OFC/ventromedial prefrontal cortex (de Araujo et al., 2005; Plassmann et al., 2008).

The medial OFC/ventromedial prefrontal cortex is also implicated in label-induced reductions in aversiveness (Nitschke et al., 2006; Sarinopoulos et al., 2006). In these studies, subjects were given two concentrations of sweet and bitter stimuli. Solution delivery was preceded by a cue indicating whether the stimulus was aversive (strong-bitter) or mildly aversive (weak-bitter) or pleasant (strong-sweet) or mildly pleasant (weak-sweet). In addition, in some trials, the strong-bitter solution was labeled as mildly aversive and the strong-sweet as mildly pleasant. These "mislabelings" resulted in the strong-bitter being rated as less aversive, and the strong-sweet as being rated as less pleasant. Although no neural modulation of sweet taste encoding was observed as a function of label, the authors found less response in the amygdala and insula when the strong-bitter was preceded by the "mildly aversive" label than when it was preceded the "aversive" label. Further response in the dorsolateral prefrontal cortex and medial OFC/ventromedial prefrontal cortex predicted modulation in these regions, implicating the prefrontal cortex as a source of top-down modulation of incoming sensory signals.

Taken together, these findings suggest that sensory encoding of taste, flavor, and odor in the insula and OFC can be modulated by dopaminergic reward circuits as a function of predictability and by higher-order prefrontal regions as a function of cognitive manipulations of stimulus identity. The data also highlight a role for the striatum in mediating the effects of temporal expectation and the medial OFC/ventromedial prefrontal cortex in mediating shifts in pleasantness associated with labeling.

Unconscious Affect

In this chapter, we have focused primarily on understanding factors that determine the perceived pleasantness of taste and flavor. However, we acknowledge that there are also nonconscious or implicit representations that may not be readily accessible to conscious introspection, but which nevertheless exert powerful influences over behavior (Berridge and Robinson, 2003). Affective responses to tastes can occur without conscious perception. Anencephalic newborns, who are born without a cortex, but with an intact brainstem, display the same reactions as normal newborns to taste stimulation; they smile or smack after sweet and grimace and gape in response to bitter tastes (Steiner, 1973; Steiner et al., 2001a). This indicates that neither conscious sensation nor sensory cortex is a necessary condition for affective responses to taste to occur.

The exceptional case of patient B who lost almost all of his paralimbic brain structures (including the insula, OFC, anterior cingulate gyrus, and amygdala) provides another indication that taste affective responses can be orchestrated without sensory cortex (Adolphs et al., 2005). Patient B was given a single glass with either sugar water or saline; tastes perceived as pleasant and unpleasant respectively by normal controls. Patient B, however, responded indiscriminately positive to either of the stimuli: he displayed a pleased facial expression, said the drink tasted delicious and would drink it readily. Yet in a slightly different situation, when given a choice between the two, patient B would consistently prefer sugar water over saline. Moreover, when encouraged to try, he would refuse to drink the saline, even though he was unable to express why beyond stating that he just liked the other solution better. This example shows that complex choice behavior can exist without consciousness of taste pleasantness.

Anticipatory Versus Consummatory Chemosensation

It is widely acknowledged that reward processing is multifaceted (e.g., Bindra, 1978; Dickinson and Balleine, 1994; Robbins and Everitt, 1996; Schultz, 2000; Wise, 1985, 2002; Wise, 2002; White, 1989). Theories of the neural mechanisms of food reward generally distinguish between homeostatic mechanisms versus sensory mechanisms (Sclafani, 2004; Sclafani and Ackroff, 2004) and between regions encoding anticipation versus receipt (Berridge, 1996; Berridge and Robinson, 2003; Small, 2002). We have discussed contributions of sensory and homeostatic mechanisms in determining taste and flavor pleasantness. We conclude this chapter by drawing a connection between what Berridge has termed 'wanting' versus 'liking' (Berridge, 1996) and what we have termed anticipatory and consummatory chemosensation (Small et al., 2008).

As highlighted in the introduction, food is sensed proximally and distally; that is, at a distance and in contact with the body. For example, perception of the sight and smell of a food often precedes and predicts its ingestion and perception of its flavors. There is now good evidence to indicate that dissociable networks encode proximal and distal sensations of food. In a direct comparison between blood oxygen level–dependent (BOLD) response to taste anticipation and receipt O'Doherty and colleagues (2002b) reported that the amygdala, midbrain, and ventral striatum respond preferentially to abstract visual stimuli that predicted the arrival of sugar water compared to neutral cues and compared to the receipt of the sugar water. They also found distinct responses in the OFC to the cue and the receipt of the taste. In a subsequent study, we showed that the amygdala, thalamus, midbrain, ventral striatum, and ventral pallidum also respond more to smell of a food aroma that predicted its taste compared to the actual receipt of that taste (Small et al., 2008). In contrast, the insula responded to both cues and receipt. A unique feature of our study was that, in addition to being ecologically relevant, the aroma and drink were perceptually similar (i.e., the aromas and the flavors represented the same food object) and hedonically equivalent. Consequently, the design allowed us to examine differential responses to cues and receipt while minimizing differences in sensory, semantic, and hedonic features between stimuli. We were therefore able to rule out the possibility that differences in sensory, semantic, or hedonic features lead to the differential responses. This is consistent with data presented above that indicates that the amygdala does not represent the perceived pleasantness of tastes or odors.

Although we believe that the amygdala does not play a role in representing the perceived pleasantness of taste, flavor, or food cues, it does appear to play a critical role in encoding the incentive value of stimuli associated with, or predictive of, food (Arana et al., 2003; Gottfried et al., 2003; LaBar et al., 2001; Morris and Dolan, 2001). This possibility is consistent with studies showing that the amygdala is differentially sensitive to the devaluation of proximal and distal food sensations. Specifically, amygdala response to pleasant foods and flavors is not modulated by eating to satiety, and it does not correlate with ratings of perceived pleasantness, which change as a function of satiety (Kringelbach et al., 2003; Small et al., 2001a). In contrast, amygdala response to food aromas (O'Doherty et al., 2000), pictures of foods (LaBar et al., 2001; Morris and Dolan, 2001), and visual stimuli predicting food aromas (Gottfried et al., 2003) does change with satiety. Collectively, these findings highlight an important and perhaps preferential role for the amygdala in encoding anticipatory chemosensation and food reward.

These issues are important not only for understanding neural encoding of reward and chemosensation but also because subtle differences in the representation of food reward may prove important for understanding how individual differences may contribute to overeating and the current obesity epidemic. For example, it has been shown that obese compared to lean individuals will work harder to obtain food even though they do not find the food more palatable (Saelens and Epstein, 1996).

Conclusions

Taste and flavor pleasantness are determined by a variety of factors including the sensory properties of the stimulus, internal state, prior experience and beliefs, and expectations about the stimulus. There is very good evidence to suggest that multiple separable systems convey signals about these factors to the OFC, which then computes the value of the stimulus in the service of goal directed behaviors. Perceived pleasantness is a conscious correlate of this process. Also, in keeping with the proposal that there are separable substrates representing food 'wanting' and 'liking', we suggest that separable circuits represent anticipatory and consummatory chemosensation of foods.

References

Accolla, R., Bathellier, B., Petersen, C. C. and Carleton, A. (2007) Differential spatial representation of taste modalities in the rat gustatory cortex. *J. Neurosci.* 27, 1396–1404.

Ackroff, K. and Sclafani, A. (1991) Sucrose to polycose preference shifts in rats: the role of taste, osmolality, and the fructose moiety. *Physiol. Behav.* 49, 1047–1060.

Ackroff, K. and Sclafani, A. (1994) Flavor preferences conditioned by intragastric infusions of dilute polycose solutions. *Physiol. Behav.* 55, 957–962.

Ackroff, K., Manza, L. and Sclafani, A. (1993) The rat's preference for sucrose, polycose, and their mixtures. *Appetite* 21, 69–80.

Adolphs, R., Tranel, D., Koenigs, M. and Damasio, A. R. (2005) Preferring one taste over another without recognizing either. *Nat. Neurosci.* 8, 860–861.

Aggleton, J. P., Burton, M. J. and Passingham, R. E. (1980) Cortical and subcortical afferents to the amygdala of the rhesus monkey (*Macaca mulatta*). *Brain Res.* 190, 347–368.

Amaral, D. G. and Price, J. L. (1984) Amygdalo-cortical projections in the monkey (*Macaca fascicularis*). *J. Comp. Neurol.* 230, 465–496.

Arana, F. S., Parkinson, J. A., Hinton, E., Holland, A. J., Owen, A. M. and Roberts, A. C. (2003) Dissociable contributions of the human amygdala and orbitofrontal cortex to incentive motivation and goal selection. *J. Neurosci.* 23, 9632–9638.

Auvray, M. and Spence, C.(2008) The multisensory perception of flavor. *Conscious. Cogn.* 17, 1016–1031.

.Azzara, A. V. and Sclafani, A. (1998) Flavor preferences conditioned by intragastric sugar infusions in rats: Maltose is more reinforcing than sucrose. *Physiol. Behav.* 64, 535–541.

Barbas, H. (1988) Anatomic organization of basoventral and mediodorsal visual recipient prefrontal regions in the rhesus monkey. *J. Comp. Neurol.* 276, 313–342.

Bartoshuk, L. M. (1991) Taste, smell, and pleasure. In: *The Hedonics of Taste and Smell*, pp. 15–28. Ed. R. C. Bolles. Lawrence Erlbaum Associates: Mahwah, NJ.

Baylis, L. L., Rolls, E. T. and Baylis, G. C. (1995) Afferent connections of the caudolateral orbitofrontal cortex taste area of the primate. *Neuroscience* 64, 801–812.

Beauchamp, G. K. and Cowart, B. J. (1985) Congenital and experiential factors in the development of human flavor preferences. *Appetite* 6, 357–372.

Beauchamp, G. K., Bertino, M., Burke, D. and Engelma, K. (1990) Experimental sodium depletion and salt taste in normal human volunteers. *Am. J. Clin. Nutr.* 51, 881–889.

Beckstead, R. M., Morse, J. R. and Norgren, R. (1980) The nucleus of the solitary tract in the monkey: Projections to the thalamus and brain stem nuclei. *J. Comp. Neurol.* 190, 259–282.

Bender, G., Mak, Y. E. and Small, D. M. (2005) The interaction between evaluative and passive response to taste in the human brain. *Chem. Senses* 30, A188.

Berns, G. S., McClure, S. M., Pagnoni, G. and Montague, P. R. (2001) Predictability modulates human brain response to reward. *J. Neurosci.* 21, 2793–2798.

Berridge, K. C. (1996) Food reward: Brain substrates of wanting and liking. *Neurosci. Biobehav. Rev.* 20, 1–25.

Berridge, K. C. and Robinson, T. E. (2003) Parsing reward. *Trends Neurosci.* 26, 507–513.

Berridge, K. C., Flynn, F. W., Schulkin, J. and Grill, H. J. (1984) Sodium depletion enhances salt palatability in rats. *Behav. Neurosci.* 98, 652–660.

Bindra, D. (1978) How adaptive behavior is produced: a perceptual-motivation alternative to response reinforcement. *Behavioral Brain Science* 28, 41–91.

Birch, L. L., McPhee, L., Steinberg, L. and Sullivan, S. (1990) Conditioned flavor preferences in young children. *Physiol. Behav.* 47, 501–505.

Bradley, R. M., Smoke, R. H., Akin, T. and Najafi, K. (1992) Functional regeneration of glossopharyngeal nerve through micromachined sieve electrode arrays. *Brain Res.* 594, 84–90.

Cabanac, M. (1971) Physiological role of pleasure. *Science* 173, 1103–1107.

Cabanac, M., Minaire, Y. and Adair, E. R. (1968) Influence of internal factors on the pleasantness of a gustative sweet sensation. *Commun. Behav. Biol.* 1, 77–82.

Cacioppo, J. T. and Berntson, G. G. (1999) The affect system: Architecture and operating characteristics. *Curr. Dir. Psychol. Sci.* 8, 133–137.

Calvert, G. A. (2001) Crossmodal processing in the human brain: Insights from functional neuroimaging studies. *Cereb. Cortex.* 11, 1110–1123.

Cardello, A. V. (1996) The role of the human senses in food acceptance. In: *Food Choice Acceptance and Consumption.* Eds. H. L. Meiselman, H. J. H. Macfie. Blackie Academic & Professional: London.

Carmichael, S. T. and Price, J. L. (1995a) Limbic connections of the orbital and medial prefrontal cortex in macaque monkeys. *J. Comp. Neurol* 363, 615–641.

Carmichael, S. T. and Price, J. L. (1995b) Sensory and premotor connections of the orbital and medial prefrontal cortex of Macaque monkeys. *J. Comp. Neurol.* 363, 642–664.

Carmichael, S. T. and Price, J. L. (1996) Connectional networks within the orbital and medial prefrontal cortex of macaque monkeys. *J. Comp. Neurol.* 371, 179–207.

Carmichael, S. T., Clugnet, M. C. and Price, J. L. (1994) Central olfactory connections in the macaque monkey. *J. Comp. Neurol.* 346, 403–434.

Cerf-Ducastel, B. and Murphy, C. (2001) fMRI activation in response to odorants orally delivered in aqueous solutions. *Chem. Senses* 26, 625–637.

Cerf-Ducastel, B., Van de Moortele, P. F., MacLeod, P., Le Bihan, D. and Faurion, A. (2001) Interaction of gustatory and lingual somatosensory perceptions at the cortical level in the human: a functional magnetic resonance imaging study. *Chem. Senses.* 26, 371–383.

Chandrashekar, J., Hoon, M. A., Ryba, N. J. and Zuker, C. S. (2006) The receptors and cells for mammalian taste. *Nature* 444, 288–294.

Chandrashekar, J., Mueller, K. L., Hoon, M. A., Adler, E., Feng, L., Guo, W., Zuker, C. S. and Ryba, N. J. (2000) T2Rs function as bitter taste receptors. *Cell* 100, 703–711.

Chiavaras, M. M. and Petrides, M. (2000) Orbitofrontal sulci of the human and Macaque monkey brain. *J. Comp. Neurol.* 422, 35–54.

Coombs, C. H. and Avrunin, G. S. (1977) Single-peaked functions and the theory of preference. *Psychol. Rev.* 84, 216–230.

Cotterill, J. A. and Cunliffe, W. J. (1973) Self-medication with liquorice in a patient with Addison's disease. *Lancet* ii, 294–295.

Critchley, H. D. and Rolls, E. T. (1996) Hunger and satiety modify the responses of olfactory and visual neurons in the primate orbitofrontal cortex. *J. Neurophysiol.* 75, 1673–1686.

Cruz, A. and Green, B. G. (2000) Thermal stimulation of taste. *Nature* 403, 889–892.

Dalton, P., Doolittle, N., Nagata, H. and Breslin, P. A. (2000) The merging of the senses: Integration of subthreshold taste and smell. *Nat. Neurosci.* 3, 431–432.

de Araujo, E. and Rolls, E. T. (2004) Representation in the human brain of food texture and oral fat. *J. Neurosci.* 24, 3086–3093.

de Araujo, E., Rolls, E. T., Kringelbach, M. L., McGlone, F. and Phillips, N. (2003) Taste-olfactory convergence, and the representation of the pleasantness of flavour in the human brain. *Eur. J. Neurosci.* 18, 2059–2068.

de Araujo, I. E., Rolls, E. T., Velazco, M. I., Margot, C. and Cayeux, I. (2005) Cognitive modulation of olfactory processing. *Neuron* 46, 671–679.

de Araujo, I. E., Oliveira-Maia, A. J., Sotnikova, T. D., Gainetdinov, R. R., Caron, M. G., Nicolelis, M. A. and Simon, S. A. (2008) Food reward in the absence of taste receptor signaling. *Neuron* 57, 930–941.

de Olmos, J., Hardy, H. and Heimer, L. (1978) The afferent connections of the main and the accessory olfactory bulb formations in the rat: an experimental HRP-study. *J. Comp. Neurol.* 181, 213–244.

Del Parigi, A., Chen, K., Salbe, A. D., Reiman, E. M. and Tataranni, P. A. (2003) Are we addicted to food? *Obes. Res.* 11, 493–495.

Del Parigi, A., Chen, K., Hill, D. O., Wing, R. R., Reiman, E. and Tataranni, P. A. (2004) Persistence of abnormal neural responses to a meal in postobese individuals. *Int. J. Obes.* 28, 370–377.

Del Parigi, A., Chen, K., Salbe, A. D., Gautier, J. F., Ravussin, E., Reiman, E. M. and Tataranni, P. A. (2002b) Tasting a liquid meal after a prolonged fast is associated with preferential activation of the left hemisphere. *Neuroreport.* 13, 1141–1145.

Del Parigi, A., Gautier, J. F., Chen, K., Salbe, A. D., Ravussin, E., Reiman, E. and Tataranni, P. A. (2002c) Neuroimaging and obesity: Mapping the brain responses to hunger and satiation in humans using positron emission tomography. *Ann. NY Acad. Sci.* 967, 389–397.

Del Parigi, A., Chen, K., Gautier, J. F., Salbe, A. D., Pratley, R. E., Ravussin, E., Reiman, E. M. and Tataranni, P. A. (2002a) Sex differences in the human brain's response to hunger and satiation.[erratum appears in *Am. J. Clin. Nutr.* 2002, 76(2), 492]. *Am. J. Clin. Nutr.* 75, 1017–1022.

Di Lorenzo, P. M. and Monroe, S. (1995) Corticofugal influence on taste responses in the nucleus of the solitary tract in the rat. *J. Neurophysiol.* 74, 258–272.

Di Lorenzo, P. M. and Monroe, S. (1997) Transfer of information about taste from the nucleus of the solitary tract to the parabrachial nucleus of the pons. *Brain Res.* 763, 167–181.

Di Lorenzo, P. M., Hallock, R. M. and Kennedy, D. P. (2003a) Temporal coding of sensation: Mimicking taste quality with electrical stimulation of the brain. *Behav. Neurosci.* 117, 1423–1433.

Di Lorenzo, P. M., Lemon, C. H. and Reich, C. G. (2003b) Dynamic coding of taste stimuli in the brainstem: Effects of brief pulses of taste stimuli on subsequent taste responses. *J. Neurosci.* 23, 8893–8902.

Dickinson, A. and Balleine, B. (1994) Motivational control of goal-directed action. *Anim. Learn. Behav.* 22, 1–18.

Drucker, D. B. and Sclafani, A. (1997) The role of gastric and postgastric sites in glucose-conditioned flavor preferences in rats. *Physiol. Behav.* 61, 351–358.

Drucker, D. B., Ackroff, K. and Sclafani, A. (1993) Flavor preference produced by intragastric polycose infusions in rats using a concurrent conditioning procedure. *Physiol. Behav.* 54, 351–355.

Ekman, G. and Åkesson, C. (1965) Saltness, sweetness, and preference. A study of quantitative relations in individual subjects. *Scand. J. Psychol.* 6, 241–253.

Erickson, R. P., Di Lorenzo, P. M. and Woodbury, M. A. (1994) Classification of taste responses in brain stem: Membership in fuzzy sets. *J. Neurophysiol.* 71, 2139–2150.

Faurion, A., Cerf, B., Le Bihan, D. and Pilliasa, A.-M. (1998) fMRI study of taste cortical areas in humans. *Ann. NY Acad. Sci.* 855, 535–545.

Faurion, A., Cerf, B., Pillias, A. M. and Boireau, N. (2002) Increased taste sensitivity by familiarization to novel stimuli: Psychophysics, fMRI, and electrophysiological techniques suggest modulations at peripheral and central levels. In: *Olfaction, Taste and Cognition*. pp. 350–366. Eds. C. Rouby, B. Schaal, D. Dubois, R. Gervais, A. Holley. Cambridge University Press: Cambridge.

Francis, S., Rolls, E. T., Bowtell, R., McGlone, F., O'Doherty, J., Browning, A., Clare, S. and Smith, E. (1999) The representation of pleasant touch in the brain and its relationship with taste and olfactory areas. *Neuroreport* 10, 435–459.

Frank, G. K., Kaye, W. H., Carter, C. S., Brooks, S., May, C., Fissell, K. and Stenger, V. A. (2003) The evaluation of brain activity in response to taste stimuli – a pilot study and method for central taste activation as assessed by event-related fMRI. *J. Neurosci. Methods* 131, 99–105.

Frank, R. A. and Byram, J. (1988) Taste-smell interactions are tastant and odorant dependent. *Chem. Senses* 13, 445–455.

Frank, R. A., van der Klaauw, N. J. and Schifferstein, H. N. (1993) Both perceptual and conceptual factors influence taste-odor and taste-taste interactions. *Percept. Psychophys.* 54, 343–354.

Gautier, J. F., Del Parigi, A., Chen, K., Salbe, A. D., Bandy, D., Pratley, R. E., Ravussin, E., Reiman, E. M. and Tataranni, P. A. (2001) Effect of satiation on brain activity in obese and lean women [comment]. *Obes. Res.* 9, 676–684.

Gautier, J. F., Chen, K., Uecker, A., Bandy, D., Frost, J., Salbe, A. D., Pratley, R. E., Lawson, M., Ravussin, E., Reiman, E. M. and Tataranni, P. A. (1999) Regions of the human brain affected during a liquid-meal taste perception in the fasting state: a positron emission tomography study. *Am. J. Clin. Nutr.* 70, 806–810.

Glendinning, J. I. (1994) Is the bitter rejection response always adaptive? *Physiol. Behav.* 56, 1217–1227.

Gottfried, J. A., O'Doherty, J. and Dolan, R. J. (2003) Encoding predictive reward value in human amygdala and orbitofrontal cortex. *Science* 301, 1104–1107.

Gottfried, J. A., Small, D. M. and Zald, D. H. (2006a) The chemical senses. In: *The Orbitofrontal Cortex*. pp. 125–171. Eds. D. H. Zald, S. L. Rauch. Oxford University Press: New York.

Gottfried, J. A., Winston, J. S. and Dolan, R. J. (2006b) Dissociable codes of odor quality and odorant structure in human piriform cortex. *Neuron* 49, 467–479.

Gottfried, J. A., Deichmann, R., Winston, J. S. and Dolan, R. J. (2002) Functional heterogeneity in human olfactory cortex: an event-related functional magnetic resonance imaging study. *J. Neurosci.* 22, 10819–10828.

Grabenhorst, F. and Rolls, E. T. (2008) Selective attention to affective value alters how the brain processes taste stimuli. *Eur. J. Neurosci.* 27, 723–729.

Grabenhorst, F., E. T. Rolls and A. Bilderbeck (2008). How cognition modulates affective responses to taste and flavor: Top-down influences on the orbitofrontal and pregenual cingulate cortices. *Cerebral Cortex* 18(7), 1549–1559.

Green, B. G. (2003) Studying taste as a cutaneous sense. *Food Qual. Prefer.* 14, 99–109.

Hajnal, A. and Norgren, R. (2005) Taste pathways that mediate accumbens dopamine release by sapid sucrose. *Physiol. Behav.* 84, 363–369.

Heilmann, S. and Hummel, T. (2004) A new method for comparing orthonasal and retronasal olfaction. *Behav. Neurosci.* 118, 412–419.

Hendricks, S. J., Brunjes, P. C. and Hill, D. L. (2004) Taste bud cell dynamics during normal and sodium-restricted development. *J. Comp. Neurol.* 472, 173–182.

Henkin, R. I., Comiter, H., Fedio, P. and O'Doherty, D. (1977) Defects in taste and smell recognition following temporal lobectomy. *Trans. Am. Neurol. Assoc.* 102, 146–150.

Hetherington, M., Rolls, B. J. and Burley, V. J. (1989) The time course of sensory-specific satiety. *Appetite* 12, 57–68.

Hill, D. L. and Przekop, P. R., Jr. (1988) Influences of dietary sodium on functional taste receptor development: a sensitive period. *Science* 241, 1826–1828.

Hill, D. L., Mistretta, C. M. and Bradley, R. M. (1986) Effects of dietary NaCl deprivation during early development on behavioral and neurophysiological taste responses. *Behav. Neurosci.* 100, 390–398.

Hinton, E. C., Parkinson, J. A., Holland, A. J., Arana, F. S., Roberts, A. C. and Owen, A. M. (2004) Neural contributions to the motivational control of appetite in humans. *Eur. J. Neurosci.* 20, 1411–1418.

Holland, P. C. (2004) Relations between Pavlovian-instrumental transfer and reinforcer devaluation. *J. Exp. Psychol. Anim. Behav. Process.* 30, 104–117.

Holland, P. C. and Gallagher, M. (2004) Amygdala–frontal interactions and reward expectancy. *Curr. Opin. Neurobiol.* 14, 148–155.

Jacobs, H. L. (1958) The preference for glucose solutions and its modification by injections of insulin. *J. Comp. Physiol. Psychol.* 51, 304–310.

Jacobs, K. M., Mark, G. P. and Scott, T. R. (1988) Taste responses in the nucleus tractus solitarius of sodium-deprived rats. *J. Physiol.* 406, 393–410.

Johnson, A. K. (2007) The sensory psychobiology of thirst and salt appetite. *Med. Sci. Sports & Exerc.* 39, 1388–1400.

Johnson, S. L., McPhee, L. and Birch, L. L. (1991) Conditioned preferences: Young children prefer flavors associated with high dietary fat. *Physiol. Behav.* 50, 1245–1251.

Jones, E. G. and Powell, T. P. (1970) An anatomical study of converging sensory pathways within the cerebral cortex of the monkey. *Brain* 93, 793–820.

Kadohisa, M., Rolls, E. T. and Verhagen, J. V. (2004) Orbitofrontal cortex: Neuronal representation of oral temperature and capsaicin in addition to taste and texture. *Neuroscience* 127, 207–221.

Kampov-Polevoy, A. B., Tsoi, M. V., Zvartau, E. E., Nezmanov, N. G. and Kalitov, E. (2001) Sweet liking and a familiy history of alcoholism in hopitalized alcoholic and non-alcoholic patients. *Alcohol* 36, 165–170.

Katz, D. B., Nicolelis, M. A. and Simon, S. A. (2002a) Gustatory processing is dynamic and distributed. *Curr. Opin. Neurobiol.* 12, 448–454.

Katz, D. B., Simon, S. A. and Nicolelis, M. A. (2002b) Taste-specific neuronal ensembles in the gustatory cortex of awake rats. *J. Neurosci.* 22, 1850–1857.

Kinomura, S., Kawashima, R., Yamada, K., Ono, S., Itoh, M., Yoshioka, S., Yamaguchi, T., Matsui, H., Miyazawa, H., Itoh, H. and et al. (1994) Functional anatomy of taste perception in the human brain studied with positron emission tomography. *Brain Res.* 659, 263–266.

Knowles, J. P. (1958) Addison's disease with glycyrrhizophilia. *Proc. R. Soc. Med.* 51, 178.

Kocher, E. C. and G. L. Fisher (1969). Subjective intensity and taste preference. *Percept. Motor Skills* 28(3), 735–740.

Krimm, R. F. and Hill, D. L. (1999) Early dietary sodium restriction disrupts the peripheral anatomical development of the gustatory system. *J. Neurobiol.* 39, 218–226.

Kringelbach, M. L., O'Doherty, J., Rolls, E. T. and Andrews, C. (2003) Activation of the human orbitofrontal cortex to a liquid food stimulus is correlated with its subjective pleasantness. *Cereb. Cortex* 13, 1064–1071.

LaBar, K. S., Gitelman, D. R., Parrish, T. B., Kim, Y. H., Nobre, A. C. and Mesulam, M. M. (2001) Hunger selectively modulates corticolimbic activation to food stimuli in humans. *Behav. Neurosci.* 115, 493–500.

Laing, D. G., Prescott, J., Bell., G. A., Gillmore, R., Allen, S., Best, D. J., Yoshida, M., Yamazaki, K. and Ishii-Mathews, R. (1994) Responses of Japanese and Australians to sweetness in the context of different foods. *J. Sens. Stud.* 9, 131–155.

LeDoux, J. E. (2000) Emotion circuits in the brain. *Ann. Rev. Neurosci.* 23, 155–184.

Leshem, M., Abutbul, A. and Eilon, R. (1999) Exercise increases the preference for salt in humans. *Appetite* 32, 251–260.

Liu, Y., Gao, J. H., Liu, H. L. and Fox, P. T. (2000) The temporal response of the brain after eating revealed by functional MRI. *Nature* 405, 1058–1062.

Lundy, R. F., Jr. and Norgren, R. (2001) Pontine gustatory activity is altered by electrical stimulation in the central nucleus of the amygdala. *J. Neurophysiol.* 85, 770–783.

Lundy, R. F., Jr. and Norgren, R. (2004) Activity in the hypothalamus, amygdala, and cortex generates bilateral and convergent modulation of pontine gustatory neurons. *J. Neurophysiol.* 91, 1143–1157.

Mak, Y. E., Simmons, K. B., Gitelman, D. R. and Small, D. M. (2005) Taste and olfactory intensity perception changes following left insular stroke. *Behav. Neurosci.* 119, 1693–1700.

Marciani, L., Pfeiffer, J. C., Hort, J., Head, K., Bush, D., Taylor, A. J., Spiller, R. C., Francis, S. and Gowland, P. A. (2006) Improved methods for fMRI studies of combined taste and aroma stimuli. *J. Neurosci. Methods* 158, 186–194.

Matsuda, M., Liu, Y., Mahankali, S., Pu, Y., Mahankali, A., Wang, J., DeFronzo, R. A., Fox, P. T. and Gao, J. H. (1999) Altered hypothalamic function in response to glucose ingestion in obese humans. *Diabetes* 48, 1801–1806.

Mattes, R. D. (1997) The taste for salt in humans. *Am. J. Clin. Nutr.* 65, 692S-697S.

McCabe, C. and Rolls, E. T. (2007) Umami: a delicious flavor formed by convergence of taste and olfactory pathways in the human brain. *Eur. J. Neurosci.* 25, 1855–1864.

McCaughey, S. A. and Scott, T. R. (1998) The taste of sodium. *Neurosci. Biobehav. Rev.* 22, 663–676.

McCaughey, S. A. and Tordoff, M. G. (2002) Magnesium appetite in the rat. *Appetite* 38, 29–38.

McCaughey, S. A., Forestell, C. A. and Tordoff, M. G. (2005) Calcium deprivation increases the palatability of calcium solutions in rats. *Physiol. Behav.* 84, 335–342.

McClure, S. M., Berns, G. S. and Montague, P. R. (2003) Temporal prediction errors in a passive learning task activate human striatum. *Neuron* 38, 339–346.

McClure, S. M., Li, J., Tomlin, D., Cypert, K. S., Montague, L. M. and Montague, P. R. (2004) Neural correlates of behavioral preference for culturally familiar drinks. *Neuron* 44, 379–387.

Mela, D. J. (2001) Why do we like what we like? *J. Sci. Food Agri.* 81(1), 10–16

Mohanty, D. R., Gitelman, D. R., Small, D. M. and Mesulam, M. (2008) The spatial attention network interacts with limbic and monaminergic systems to modulate motivation-induced attentional shifts. *Cereb. Cortex,* 18(11), 2604–2613.

Morris, J. S. and Dolan, R. J. (2001) Involvement of human amygdala and orbitofrontal cortex in hunger-enhanced memory for food stimuli. *J. Neurosci.* 21, 5304–5310.

Moskowitz, H. R. (1971) The sweetness and pleasantness of sugars. *Am. J. Psychol.* 84, 387–405.

Moskowitz, H. W., Kumaraiah, V. and Sharma, K. N. (1975a) Cross-cultural differences in simple taste preferences. *Science* 190, 1217–1218.

Moskowitz, H. R., Kluter, R. A., Westerling, J. and Jacobs, H. L. (1974) Sugar sweetness and pleasantness: Evidence for different psychological laws. *Science* 184, 583–585.

Moskowitz, H. W., Kumaraiah, V., Sharma, K. N., Jacobs, H. L. and Sharma, S. D. (1975b) Cross-cultural differences in simple taste preferences. *Science* 190, 1217–1218.

Mueller, K. L., Hoon, M. A., Erlenbach, I., Chandrashekar, J., Zuker, C. S. and Ryba, N. J. (2005) The receptors and coding logic for bitter taste. *Nature* 434, 225–229.

Mufson, E. J. and Mesulam, M. M. (1984) Thalamic connections of the insula in the rhesus monkey and comments on the paralimbic connectivity of the medial pulvinar nucleus. *J. Comp. Neurol.* 227, 109–120.

Mufson, E. J., Mesulam, M. M. and Pandya, D. N. (1981) Insular interconnections with the amygdala in the rhesus monkey. *Neurosci.* 6, 1231–1248.

Nakano, Y., Oomura, Y., Nishino, H., Aou, S., Yamamoto, T. and Nemoto, S. (1984) Neuronal activity in the medial orbitofrontal cortex of the behaving monkey: Modulation by glucose and satiety. *Brain Res. Bull.* 12, 381–385.

Nitschke, J. B., Dixon, G. E., Sarinopoulos, I., Short, S. J., Cohen, J. D., Smith, E. E., Kosslyn, S. M., Rose, R. M. and Davidson, R. J. (2006) Altering expectancy dampens neural response to aversive taste in primary taste cortex. *Nat. Neurosci.* 9, 435–442.

Norgren, R. (1976) Taste pathways to hypothalamus and amygdala. *J. Comp. Neurol.* 166, 17–30.

Norgren, R. (1984) Central neural mechanisms of taste. In: *Handbook of Physiology*. Ed. I. Darien-Smith. American Physiology Society: Washington, D. C.

Norgren, R. (1985) Taste and the autonomic nervous system. *Chem. Senses* 10, 143–161.

Norgren, R. (1990) Gustatory system. In: *The Human Nervous System*. pp. 845–861. Ed. G. Paxinos. Academic Press: San Diego, CA.

O'Doherty, J. P. (2004) Reward representations and reward-related learning in the human brain: Insights from neuroimaging. *Curr. Opin. Neurobiol.* 14, 769–776.

O'Doherty, J. P., Buchanan, T. W., Seymour, B. and Dolan, R. J. (2006) Predictive neural coding of reward preference involves dissociable responses in human ventral midbrain and ventral striatum. *Neuron* 49, 157–166.

O'Doherty, J. P., Deichmann, R., Critchley, H. D. and Dolan, R. J. (2002a) Neural responses during anticipation of a primary taste reward. *Neuron* 33, 815–826.

O'Doherty, J. P., Dayan, P., Friston, K., Critchley, H. and Dolan, R. J. (2003a) Temporal difference models and reward-related learning in the human brain. *Neuron* 38, 329–337.

O'Doherty, J., Rolls, E. T., Francis, S., Bowtell, R. and McGlone, F. (2001) Representation of pleasant and aversive taste in the human brain. *J. Neurophysiol.* 85, 1315–1321.

O'Doherty, J., Dayan, P., Schultz, J., Deichmann, R., Friston, K. and Dolan, R. J. (2004) Dissociable roles of ventral and dorsal striatum in instrumental conditioning. *Science* 304, 452–454.

O'Doherty, J., Rolls, E. T., Francis, S., Bowtell, R., McGlone, F., Kobal, G., Renner, B. and Ahne, G. (2000) Sensory-specific satiety-related olfactory activation of the human orbitofrontal cortex. [republished in *Neuroreport* 2000, 11(4), 893–897]. *Neuroreport* 11, 399–403.

Ogawa, H., Ito, S. and Nomura, T. (1985) Two distinct projection areas from tongue nerves in the frontal operculum of macaque monkeys as revealed with evoked potential mapping. *Neurosci. Res.* 2, 447–459.

Ogawa, H., Wakita, M., Hasegawa, K., Kobayakawa, T., Sakai, N., Hirai, T., Yamashita, Y. and Saito, S. (2005) Functional MRI detection of activation in the primary gustatory cortices in humans. *Chem. Senses* 30, 583–592.

Okamoto, M., Dan, H., Singh, A. K., Hayakawa, F., Jurcak, V., Suzuki, T., Kohyama, K. and Dan, I. (2006) Prefrontal activity during flavor difference test: Application of functional near-infrared spectroscopy to sensory evaluation studies. *Appetite* 47(2), 220–232.

Pagnoni, G., Zink, C. F., Montague, P. R. and Berns, G. S. (2002) Activity in human ventral striatum locked to errors of reward prediction. *Nat. Neurosci.* 5, 97–98.

Pepino, M. Y. and Mennella, J. A. (2007) Effects of cigarette smoking and family history of alcoholism on sweet taste perception and food cravings in women. *Alcohol. Clin. Exp. Res.* 31, 1891–1899.

Perez, C. and Sclafani, A. (1990) Developmental changes in sugar and starch taste preferences in young rats. *Physiol. Behav.* 48, 7–12.

Perez, C., Lucas, F. and Sclafani, A. (1998) Increased flavor acceptance and preference conditioned by the postingestive actions of glucose. *Physiol. Behav.*64, 483–492.

Pfaffmann, C. (1980) Wundt's schema of sensory affect in the light of research on gustatory preferences. *Psychol. Res.* 42, 165–174.

Pickens, C. P., Saddoris, M. P., Setlow, B., Gallagher, M., Holland, P. C. and Schoenbaum, G. (2003) Different roles for orbitofrontal cortex and basolateral amygdala in a reinforcer devaluation task. *J. Neurosci.* 23, 11078–11084.

Pittman, D. W. and Contreras, R. J. (2002) Dietary NaCl influences the organization of chorda tympani neurons projecting to the nucleus of the solitary tract in rats. *Chem. Senses* 27, 333–341.

Plassmann, H., O'Doherty, J., Shiv, B. and Rangel, A. (2008) Marketing actions can modulate neural representations of experienced pleasantness. *Proc. Natl. Acad. Sci. USA* 105, 1050–1054.

Plata-Salaman, C. R., Scott, T. R. and Smith-Swintosky, V. L. (1992) Gustatory neural coding in the monkey cortex: L-amino acids. *J. Neurophysiol.* 67, 1552–1561.

Plata-Salaman, C. R., Smith-Swintosky, V. L. and Scott, T. R. (1996) Gustatory neural coding in the monkey cortex: Mixtures. *J. Neurophysiol.* 75, 2369–2379.

Poellinger, A., Thomas, R., Lio, P., Lee, A., Makris, N., Rosen, B. R. and Kwong, K. K. (2001) Activation and habituation in olfaction—an fMRI study. *Neuroimage* 13, 547–560.

Prescott, J., Bell, G. A., Gillmore, R., Yoshida, M., O'Sullivan, M., Korac, S. and Yamazaki, K. (1996) Cross-cultural comparisons of Japanese and Australian responses to manipulations of sweetness in foods. *Food Qual. Pref.* 8, 45–55.

Price, J. L. (1973) An autoradiographic study of complementary laminar patterns of termination of afferent fibers to the olfactory cortex. *J. Comp. Neurol.* 150, 87–108.

Price, J. L. and Amaral, D. G. (1981) An autoradiographic study of the projections of the central nucleus of the monkey amygdala. *J. Neurosci.* 1, 1242–1259.

Pritchard, T. C., Hamilton, R. B. and Norgren, R. (1989) Neural coding of gustatory information in the thalamus of Macaca mulatta. *J. Neurophysiol.* 61, 1–14.

Pritchard, T. C., Hamilton, R. B. and Norgren, R. (2000) Projections of the parabrachial nucleus in the Old World monkey. *Exp. Neurol.* 165, 101–117.

Pritchard, T. C., Macaluso, E. and Eslinger, P. J. (1999) Taste perception in patients with insular cortex lesions. *Behav. Neurosci.*. 113, 663–671.

Pritchard, T. C., Hamilton, R. B., Morse, J. R. and Norgren, R. (1986) Projections of thalamic gustatory and lingual areas in the monkey, Macaca fascicularis. *J. Comp. Neurol.* 244, 213–228.

Pritchard, T. C., Edwards, E. M., Smith, C. A., Hilgert, K. G., Gavlick, A. M., Maryniak, T. D., Schwartz, G. J. and Scott, T. R. (2005) Gustatory neural responses in the medial orbitofrontal cortex of the old world monkey. *J. Neurosci.* 25, 6047–6056.

Rescorla, R. A. (1987) A Pavlovian analysis of goal-directed behavior. *Am. Psychol.* 42, 119–129.

Richardson, N. J., Rogers, P. J. and Elliman, N. A. (1996) Conditioned flavour preferences reinforced by caffeine consumed after lunch. *Physiol. Behav.* 60, 257–263.

Robbins, T. W. and Everitt, B. J. (1996) Neurobehavioural mechanisms of reward and motivation. *Curr. Opin. Neurobiol.* 6, 228–236.

Rolls, B. J. (1986) Sensory-specific satiety. *Nutr. Rev.* 44, 93–101.

Rolls, B. J., Rowe, E. A. and Rolls, E. T. (1982) How sensory properties of foods affect human feeding behavior. *Physiol. Behav.* 29, 409–417.

Rolls, B. J., Rolls, E. T., Rowe, E. A. and Sweeney, K. (1981) Sensory specific satiety in man. *Physiol. Behav.* 27, 137–142.

Rolls, E. T. (1981a) Central nervous mechanisms related to feeding and appetite. *Br. Med. Bull* 37, 131–134.

Rolls, E. T. (1981b) Processing beyond the inferior temporal visual cortex related to feeding, memory, and striatal function. In: *Brain Mechanisms of Sensation*. pp. 241–269. Eds. Y. Katsuki, R. Norgren, M. Sato. Wiley: New York.

Rolls, E. T. (1999) *The Brain and Emotion*. Oxford University Press: Oxford.

Rolls, E. T. and Baylis, L. L. (1994) Gustatory, olfactory, and visual convergence within the primate orbitofrontal cortex. *J. Neurosci.* 14, 5437–5452.

Rolls, E. T. and Deco, G. (2002) *Computational Neuroscience of Vision*. Oxford University Press: Oxford.

Rolls, E. T., Critchley, H. D. and Treves, A. (1996) Representation of olfactory information in the primate orbitofrontal cortex. *J. Neurophysiol.* 75, 1982–1996.

Rolls, E. T., Rolls, B. J. and Rowe, E. A. (1983) Sensory-specific and motivation-specific satiety for the sight and taste of food and water in man. *Physiol. Behav.* 30, 185–192.

Rolls, E. T., Sienkiewicz, Z. J. and Yaxley, S. (1989) Hunger modulates the responses to gustatory stimuli of single neurons in the caudolateral orbitofrontal cortex of the macaque monkey. *Eur. J. Neurosci.* 1, 53–70.

Royet, J. P., Plailly, J., Delon-Martin, C., Kareken, D. A. and Segebarth, C. (2003) fMRI of emotional responses to odors: Influence of hedonic valence and judgment, handedness, and gender. *Neuroimage* 20, 713–728.

Royet, J. P., Hudry, J., Zald, D. H., Godinot, D., Gregoire, M. C., Lavenne, F., Costes, N. and Holley, A. (2001) Functional neuroanatomy of different olfactory judgments. *Neuroimage*. 13, 506–519.

Rozin, P. (1982) "Taste-smell confusions" and the duality of the olfactory sense. *Percept. Psychophys.* 31, 397–401.

Saelens, B. E. and Epstein, L. H. (1996) Reinforcing value of food in obese and non-obese women. *Appetite* 27, 41–50.

Sakai, N., Kobayakawa, T., Gotow, N., Saito, S. and Imada, S. (2001) Enhancement of sweetness ratings of aspartame by a vanilla odor presented either by orthonasal or retronasal routes. *Percept. Mot. Skills* 92, 1002–1008.

Sarinopoulos, I., Dixon, G. E., Short, S. J., Davidson, R. J. and Nitschke, J. B. (2006) Brain mechanisms of expectation associated with insula and amygdala response to aversive taste: Implications for placebo. *Brain Behav. Immun.* 20, 120–132.

Savic, I., Gulyas, B., Larsson, M. and Roland, P. (2000) Olfactory functions are mediated by parallel and hierarchical processing. *Neuron* 26, 735–745.

Schifferstein, H. N. J. and Verlegh, P. W. J. (1996) The role of congruency and pleasantness in odor-induced taste enhancement. *Acta Psychol.* 94, 87–105.

Schoenbaum, G. and Eichenbaum, H. (1995a) Information coding in the rodent prefrontal cortex. I. Single-neuron activity in orbitofrontal cortex compared with that in pyriform cortex. *J. Neurophysiol.* 74, 733–750.

Schoenbaum, G. and Eichenbaum, H. (1995b) Information coding in the rodent prefrontal cortex. II. Ensemble activity in orbitofrontal cortex. *J. Neurophysiol.* 74, 751–762.

Schoenbaum, G., Setlow, B., Saddoris, M. P. and Gallagher, M. (2003) Encoding predicted outcome and acquired value in orbitofrontal cortex during cue sampling depends upon input from basolateral amygdala. [comment]. *Neuron*. 39, 855–867.

Schoenfeld, M. A., Neuer, G., Tempelmann, C., Schussler, K., Noesselt, T., Hopf, J. M. and Heinze, H. J. (2004) Functional magnetic resonance tomography correlates of taste perception in the human primary taste cortex. *Neurosci.* 127, 347–353.

Schultz, W. (1998) Predictive reward signal of dopamine neurons. *J. Neurophysiol.* 80, 1–27.

Schultz, W. (2000) Multiple reward signals in the brain. *Nat. Rev. Neurosci.* 1, 199–207.

Schultz, W., Dayan, P. and Montague, P. R. (1997) A neural substrate of prediction and reward. *Science* 275, 1593–1599.

Sclafani, A. (1988) Carbohydrate appetite in rats: Taste and postingestive factors. *Appetite* 11, 20–25.

Sclafani, A. (2004) Oral and postoral determinants of food reward. *Physiol. Behav.* 81, 773–779.

Sclafani, A. (2007a) Fat and sugar flavor preference and acceptance in C57BL/6J and 129 mice: Experience attenuates strain differences. *Physiol. Behav.* 90, 602–611.

Sclafani, A. (2007b) Sweet taste signaling in the gut [comment]. *Proc. Natl Acad. Sci. USA* 104, 14887–14888.

Sclafani, A. and Ackroff, K. (1994) Glucose- and fructose-conditioned flavor preferences in rats: Taste versus postingestive conditioning. *Physiol. Behav.* 56, 399–405.

Sclafani, A. and Ackroff, K. (2003) Reinforcement value of sucrose measured by progressive ratio operant licking in the rat. *Physiol. Behav.* 79, 663–670.

Sclafani, A. and Ackroff, K. (2004) The relationship between food reward and satiation revisited. *Physiol. Behav.* 82, 89–95.

Sclafani, A. and Glendinning, J. I. (2005) Sugar and fat conditioned flavor preferences in C57BL/6J and 129 mice: Oral and postoral interactions. *Am. J. Physiol. Regul. Integr. Comp. Physiol.* 289, R712–720.

Sclafani, A. and Nissenbaum, J. W. (1987) Oral versus postingestive origin of polysaccharide appetite in the rat. *Neurosci. Biobehav. Rev.* 11, 169–172.

Sclafani, A., Lucas, F. and Ackroff, K. (1996) The importance of taste and palatability in carbohydrate-induced overeating in rats. *Am. J. Physiol.* 270, R1197–1202.

Sclafani, A., Thompson, B. and Smith, J. C. (1998) The rat's acceptance and preference for sucrose, maltodextrin, and saccharin solutions and mixtures. *Physiol. Behav.* 63, 499–503.

Sclafani, A., Weiss, K., Cardieri, C. and Ackroff, K. (1993b) Feeding response of rats to no-fat and high-fat cakes. *Obes. Res.* 1, 173–178.

Sclafani, A., Zukerman, S., Glendinning, J. I. and Margolskee, R. F. (2007) Fat and carbohydrate preferences in mice: the contribution of alpha-gustducin and Trpm5 taste-signaling proteins. *Am. J. Physiol. Regul. Integr.Comp. Physiol.* 293, R1504–1513.

Sclafani, A., Azzara, A. V., Touzani, K., Grigson, P. S. and Norgren, R. (2001) Parabrachial nucleus lesions block taste and attenuate flavor preference and aversion conditioning in rats. *Behav. Neurosci.* 115, 920–933.

Sclafani, A., Cardieri, C., Tucker, K., Blusk, D. and Ackroff, K. (1993a) Intragastric glucose but not fructose conditions robust flavor preferences in rats. *Am. J. Physiol.* 265, R320–325.

Scott, T. R. and Mark, G. P. (1987) The taste system encodes stimulus toxicity. *Brain Res.* 414, 197.

Scott, T. R. and Plata-Salaman, C. R. (1991) Coding of taste quality. In: *Smell and Taste in Health and Disease*. Eds. T.V. Getchell, R.L. Doty, L.M. Bartoshuk and J.B. Snow, Jr. Raven Press: New York.

Scott, T. R. and Plata-Salaman, C. R. (1999) Taste in the monkey cortex. *Physiol. Behav.* 67, 489–511.

Scott, T. R., Yan, J. and Rolls, E. T. (1995) Brain mechanisms of satiety and taste in macaques. *Neurobiology (Bp.)* 3, 281–292.

Scott, T. R., Yaxley, S., Sienkiewicz, Z. J. and Rolls, E. T. (1986) Gustatory responses in the frontal opercular cortex of the alert cynomolgus monkey. *J. Neurophysiol.* 56, 876–890.

Scott, T. R., Karadi, Z., Oomura, Y., Nishino, H., Plata-Salaman, C. R., Lenard, L., Giza, B. K. and Aou, S. (1993) Gustatory neural coding in the amygdala of the alert macaque monkey. *J. Neurophysiol.* 69, 1810–1820.

Shuler, M. G., Krimm, R. F. and Hill, D. L. (2004) Neuron/target plasticity in the peripheral gustatory system. *J. Comp. Neurol.* 472, 183–192.

Simmons, W. K., Martin, A. and Barsalou, L. W. (2005) Pictures of appetizing foods activate gustatory cortices for taste and reward. *Cereb. Cortex* .15, 1602–1608.

Simon, S. A., de Araujo, I., Stapleton, J. R. and Nicolelis, M. A. (2008) Multisensory processing of gustatory stimuli. *Chemosens. Percept.* 1, 95–102.

Small, D. M. (2002) Toward an understanding of the brain substrates of reward in humans. *Neuron.* 33, 668–671.

Small, D. M. (2008) Flavor and the formation of category-specific processing in olfaction. *Chemosens. Percept.*, 1, 136–146.

Small, D. M. and Prescott, J. (2005) Odor/taste integration and the perception of flavor. *Exp. Brain Res.* 166, 345–357.

Small, D. M., Zatorre, R. J. and Jones-Gotman, M. (2001b) Changes in taste intensity perception following anterior temporal lobe removal in humans. *Chem. Senses* 26, 425–432.

Small, D. M., Zatorre, R. J. and Jones-Gotman, M. (2001c) Increased intensity perception of aversive taste following right anteromedial temporal lobe removal in humans. *Brain* 124, 1566–1575.

Small, D. M., Gerber, J. C., Mak, Y. E. and Hummel, T. (2005b) Differential neural responses evoked by orthonasal versus retronasal odorant perception in humans. *Neuron* 47, 593–605.

Small, D. M., Bernasconi, N., Bernasconi, A., Sziklas, V. and Jones-Gotman, M. (2005a) Gustatory agnosia. *Neurology* 64, 311–317.

Small, D. M., Jones-Gotman, M., Zatorre, R. J., Petrides, M. and Evans, A. C. (1997a) Flavor processing: More than the sum of its parts. *Neuroreport* 8, 3913–3917.

Small, D. M., Jones-Gotman, M., Zatorre, R. J., Petrides, M. and Evans, A. C. (1997b) A role for the right anterior temporal lobe in taste quality recognition. *J. Neurosci.* 17, 5136–5142.

Small, D. M., Veldhuizen, M. G., Felsted, J., Mak, Y. E. and McGlone, F. (2008) Separable substrates for anticipatory and consummatory food chemosensation. *Neuron* 57, 786–797.

Small, D. M., Zatorre, R. J., Dagher, A., Evans, A. C. and Jones-Gotman, M. (2001a) Changes in brain activity related to eating chocolate: From pleasure to aversion. *Brain* 124, 1720–1733.

Small, D. M., Bender, G., Veldhuizen, M. G., Rudenga, K., Nachtigal, D. and Felsted, J. (2007) The role of the human orbitofrontal cortex in taste and flavor processing. *Ann. N Y Acad. Sci.,* 1121, 136–151.

Small, D. M., Gregory, M. D., Mak, Y. E., Gitelman, D., Mesulam, M. M. and Parrish, T. (2003a) Dissociation of neural representation of intensity and affective valuation in human gustation. *Neuron* 39, 701–711.

Small, D. M., Voss, J., Mak, Y. E., Simmons, K. B., Parrish, T. and Gitelman, D. (2004) Experience-dependent neural integration of taste and smell in the human brain. *J. Neurophys.* 92, 1892–1903.

Small, D. M., Zald, D. H., Jones-Gotman, M., Zatorre, R. J., Pardo, J. V., Frey, S. and Petrides, M. (1999) Human cortical gustatory areas: a review of functional neuroimaging data. *Neuroreport* 10, 7–14.

Smeets, P. A., de Graaf, C., Stafleu, A., van Osch, M. J. and van der Grond, J. (2005a) Functional magnetic resonance imaging of human hypothalamic responses to sweet taste and calories. *Am. J. Clin. Nutr* .82, 1011–1016.

Smeets, P. A., de Graaf, C., Stafleu, A., van Osch, M. J. and van der Grond, J. (2005b) Functional MRI of human hypothalamic responses following glucose ingestion. *Neuroimage* 24, 363–368.

Smeets, P. A., de Graaf, C., Stafleu, A., van Osch, M. J., Nievelstein, R. A. and van der Grond, J. (2006) Effect of satiety on brain activation during chocolate tasting in men and women. *Am. J. Clin. Nutr.* 83, 1297–1305.

Smith, D. V. and Scott, T. R. (2003) Gustatory neural coding. In: *Handbook of Olfaction and Gustation.* pp. 731–758. Ed. R. L. Doty. Marcel Dekker: Philedelphia, PA.

Smith, D. V. and St John, S. J. (1999) Neural coding of gustatory information. *Curr. Opin. Neurobiol.* 9, 427–435.

Smith, D. V., John, S. J. and Boughter, J. D. (2000) Neuronal cell types and taste quality coding. *Physiol. Behav.* 69, 77–85.

Smith-Swintosky, V. L., Plata-Salaman, C. R. and Scott, T. R. (1991) Gustatory neural coding in the monkey cortex: Stimulus quality. *J. Neurophysiol.* 66, 1156–1165.

Sobel, N., Prabhakaran, V., Desmond, J. E., Glover, G. H., Goode, R. L., Sullivan, E. V. and Gabrieli, J. D. (1998) Sniffing and smelling: Separate subsystems in the human olfactory cortex. *Nature* 392, 282–286.

Spector, A. C. (1995a) Gustatory function in the parabrachial nuclei: Implications from lesion studies in rats. *Rev. Neurosci.* 6, 143–175.

Spector, A. C. (1995b) Gustatory parabrachial lesions disrupt taste-guided quinine responsiveness in rats. *Behav. Neurosci.* 109, 79–90.

Stein, B. E. (1998) Neural mecahnisms for synthesizing sensory information and producing adaptive behaviors. *Exp. Brain Res.* 123, 124–135.

Steiner, J. E. (1973) The gustofacial response: Observation on normal and anencephalic newborn infants. *Symp. Oral Sens. Percept.* 4, 254–278.

Steiner, J. E., Glaser, D., Hawilo, M. E. and Berridge, K. C. (2001a) Comparative expression of hedonic impact: Affective reactions to taste by human infants and other primates. *Neurosci. Biobehav. Rev.* 25, 53–74.

Stevenson, R. J. and Boakes, R. A. (2003) A mnemonic theory of odor perception. *Psychol. Rev.* 110, 340–364.

Stevenson, R. J. and Boakes, R. A. (2004) Sweet and sour smells: Learned synesthesia between the senses of taste and smell. In: *The Handbook of Multisensory Processes.* pp. 69–83. Eds. G. A. Calvert, C. Spence, B. E. Stein. MIT Press: Boston, MA.

Stevenson, R. J. and Tomiczek, C. (2007) Olfactory-induced synesthesias: a review and model. *Psychol. Bull.* 133, 294–309.

Stevenson, R. J., Prescott, J. and Boakes, R. A. (1995) The acquisition of taste properties by odors. *Learn. Motiv.* 26, 433–455.

Sugita, M. and Shiba, Y. (2005) Genetic tracing shows segregation of taste neuronal circuitries for bitter and sweet. *Science* 309, 781–785.

Tataranni, P. A. and Del Parigi, A. (2003) Functional neuroimaging: a new generation of human brain studies in obesity research. *Obes. Rev.* 4, 229–238.

Tataranni, P. A., Gautier, J. F., Chen, K., Uecker, A., Bandy, D., Salbe, A. D., Pratley, R. E., Lawson, M., Reiman, E. M. and Ravussin, E. (1999) Neuroanatomical correlates of hunger and satiation in humans using positron emission tomography. *Proc. Nat. Acad. Sci.. U S A* 96, 4569–4574.

Todrank, J. and Bartoshuk, L. M. (1991) A taste illusion: Taste sensation localized by touch. *Physiol. Behav.* 50, 1027–1031.

Tomchik, S. M., Berg, S., Woul Kim, J., Chaudhar, N. and Roper, S. D. (2007) Breadth of tuning and taste coding in mammalian taste buds. *J. Neurosci.*. 27, 10840–10848.

Topolovec, J. C., Gati, J. S., Menon, R. S., Shoemaker, J. K. and Cechetto, D. F. (2004) Human cardiovascular and gustatory brainstem sites observed by functional magnetic resonance imaging. *J. Comp. Neurol.* 471, 446–461.

Travers, J. B. (1988) Efferent projections from the anterior nucleus of the solitary tract of the hamster. *Brain Res.* 457, 1–11.

Turner, B. H., Gupta, K. C. and Mishkin, M. (1978) The locus and cytoarchitecture of the projection areas of the olfactory bulb in Macaca mulatta. *J. Comp. Neurol.* 177, 381–396.

Turner, B. H., Mishkin, M. and Knapp, M. (1980) Organization of the amygdalopetal projections from modality-specific cortical association areas in the monkey. *J. Comp. Neurol.* 191, 515–543.

van de Klauuw, N. J. and Frank, R. A. (1996) Scaling component intensities of complex stimuli: the influence of response alternatives. *Environ. Int.l* 22, 21–31.

Van Essen, D. C. and De Yoe, E. A. (1995) Concurrent processing in the primate visual cortex. In: *The Cognitive Neurosciences*. pp. 383–400. Ed. M. Gazzaniga. MIT Press: Cambridge, MA.

Veldhuizen, M. G., Bender, G., Constable, R. T. and Small, D. M. (2007) Trying to detect taste in a tasteless solution: Modulation of early gustatory cortex by attention to taste. *Chem. Senses* 32, 569–581.

Verhagen, J. V. and Engelen, L. (2006) The neurocognitive bases of human multimodal food perception: Sensory integration. *Neurosci. Biobehav. Rev.* 30, 613–650.

Verhagen, J. V., Kadohisa, M. and Rolls, E. T. (2004) Primate insular/opercular taste cortex: Neuronal representations of the viscosity, fat texture, grittiness, temperature, and taste of foods. *J. Neurophysiol.* 92, 1685–1699.

Vogt, B. A. and Pandya, D. (1987) Cingulate cortex of the rhesus monkey: II. Cortical afferents. *J. Comp. Neurol.* 262, 271–289.

Vogt, B. A., Pandya, D. N. and Rosene, D. L. (1987) Cingulate cortex of the rhesus monkey: I. Cytoarchitecture and thalamic afferents. *J. Comp. Neurol.* 262, 256–270.

Welge-Lussen, A., Drago, J., Wolfensberger, M. and Hummel, T. (2004) Interaktion von schmecken und riechen. *HNO-Infromationen* 29, 123.

White, N. M. (1989) Reward or reinforcement: What's the difference? *Neurosci. Biobehav. Rev.* 13, 181–186.

Whitehead, M. C. (1990) Subdivisions and neuron types of the nucleus of the solitary tract that project to the parabrachial nucleus in the hamster. *J. Comp. Neurol.* 301, 554–574.

Whitehead, M. C. and Frank, M. E. (1983) Anatomy of the gustatory system in the hamster: Central projections of the chorda tympani and the lingual nerve. *J. Comp. Neurol.* 220, 378–395.

Wilkins, L. and Richter, C. P. (1940) A great craving for salt by a child with cortico-adrenal insufficiency. *J. Am. Med. Assoc.* 114, 866–868.

Winston, J. S., Gottfried, J. A., Kilner, J. M. and Dolan, R. J. (2005). Integrated neural representations of odor intensity and affective valence in human amygdala. *J. Neurosci.* 25(39), 8903–8907.

Wilson, D. A. and Stevenson, R. J. (2004) The fundamental role of memory in olfactory perception. *Trends Neurosci.* 25, 243–247.

Wise, R. A. (1985) The anhedonia hypothesis: Mark III. *Behav. Brain Sci.* 8.

Wise, R. A. (2002) Brain reward circuitry: Insights from unsensed incentives. *Neuron.* 36, 229–240.

Wundt, W. (1897) *Outlines of Psychology*. Translated by C.H. Judd. Wilhelm Engelmann: Leipzig (Reprinted Thoemmes: Bristol, 1999); first published in German as Wundt, W. (1896). *Grundriss der Psychologie*. Wilhelm Engelmann: Leipzig.

Yamamoto, T., Shimura, T., Sakai, N. and Ozaki, N. (1994) Representation of hedonics and quality of taste stimuli in the parabrachial nucleus of the rat. *Physiol. Behav.* 56, 1197–1202.

Yamamoto, T., Yuyama, N., Kato, T. and Kawamura, Y. (1985) Gustatory responses of cortical neurons in rats. II. Information processing of taste quality. *J. Neurophysiol.* 53, 1370–1386.

Yan, J. and Scott, T. R. (1996) The effect of satiety on responses of gustatory neurons in the amygdala of alert cynomolgus macaques. *Brain Res.* 740, 193–200.

Yaxley, S., Rolls, E. T. and Sienkiewicz, Z. J. (1990) Gustatory responses of single neurons in the insula of the macaque monkey. *J. Neurophysiol.* 63, 689–700.

Yeomans, M. R. and Mobini, S. (2006). Hunger alters the expression of acquired hedonic but not sensory qualities of food-paired odors in humans. *J. Exp. Psychol. Anim. Behav. Process.* 32(4), 460–466.

Yeomans, M. R., Blundell, J. E. and Leshem, M. (2004) Palatability: Response to nutritional need or need-free stimulation of appetite? *Br. J. Nutr.* 92, S3–14.

Yeomans, M. R., Mobini, S., Elliman, T. D., Walker, H. C. and Stevenson, R. J. (2006) Hedonic and sensory characteristics of odors conditioned by pairing with tastants in humans. *J. Exp. Psychol. Anim. Behav. Process.* 32, 215–228.

Zald, D. H. and Pardo, J. V. (1997) Emotion, olfaction, and the human amygdala: Amygdala activation during aversive olfactory stimulation. *Proc. Nat. Acad. Sci.U S A* 94, 4119–4124.

Zald, D. H., Hagen, M. C. and Pardo, J. V. (2002) Neural correlates of tasting concentrated quinine and sugar solutions. *J. Neurophysiol.*. 87, 1068–1075.

Zald, D. H., Lee, J. T., Fluegel, K. W. and Pardo, J. V. (1998) Aversive gustatory stimulation activates limbic circuits in humans. *Brain* 121, 1143–1154.

Zatorre, R. J., Jones-Gotman, M., Evans, A. C. and Meyer, E. (1992) Functional localization and lateralization of human olfactory cortex. *Nature* 360, 339–340.

Zellner, D. A., Rozin, P., Aron, M. and Kulish, C. (1983) Conditioned enhancement of human's liking for flavor by pairing with sweetness. *Learn. Motiv.* 14, 338–350.

10

Sexual Pleasure

BARRY R. KOMISARUK, BEVERLY WHIPPLE, AND CARLOS BEYER

Like is like like. By definition, we like pleasure. That is, the origin of the word, "pleasure" has a common root with the word, "like." This is schematized in Figure 10.1. What pleases us (i.e., what we like—what suits us) is that which gives us pleasure. There is a commonality between pleasure and a match (a fit) with what we like. Thus, there is a commonality between likeness, in the sense of similarity on the one hand, and liking, as in pleasure, on the other. (In the diagram, the transition is eased by the concept of fitting ends together: [affinity–liking], which produces a smooth [uninterrupted, flat] surface.) *Placer* in Spanish means "pleasure." (Please as a request implies that the response will match [be like] what the requester would like to be done, and will hence "please" the requester, thereby providing pleasure.)

The unified sense of these two meanings of "like" may reflect a fundamental psychological feeling and understanding of commonality between the two concepts. This could be an example of how the two meanings came to have their origin in language—a sort of archae-psychology.

Neurologically, perhaps there is a specific role for memory in the experience of pleasure, such that for a stimulus or event to give us pleasure, an underlying brain mechanism provides a matching mechanism between the present event and a prior comparable event, that is, requiring a memory process—(very) short and/or long term—to provide a likeness between the present and the prior moments, which gives us the feeling of liking or pleasure. The involvement of a memory process would not account for pleasurable experiences that occur for the first time, although a test of the generality of a link between memory and pleasure could be provided in the hypothetical case of a person with total amnesia; could such a person experience pleasure?

We can also trace an intimate connection between "pleasure" and "love" in their etymological and archae-psychological roots. This is schematized in Figure 10.2. "Like" and "taste" have a common origin in "*gustare*," both of which have a common origin in "*delectare*." The word "*lubh*," which means "to desire" in Sanskrit generated "lubet" (it delights), which generated "love" and "delectare." From "*lubh*" also originated "libido," which means "desire"; this incorporates the sense of lacking or yearning, on the basis that it means "to be away from one's star."

This archae-psychological duality and tension between desire and pleasure remains unresolved today, as represented in the ongoing debate over whether addictive substances such as cocaine that act via the dopaminergic system, produce 'wanting' or "pleasure" (Berridge, 2006).

Sexual Pleasure is Good for Our Health

Numerous studies have suggested the health benefits of sexual activity. Many of the studies are correlational,

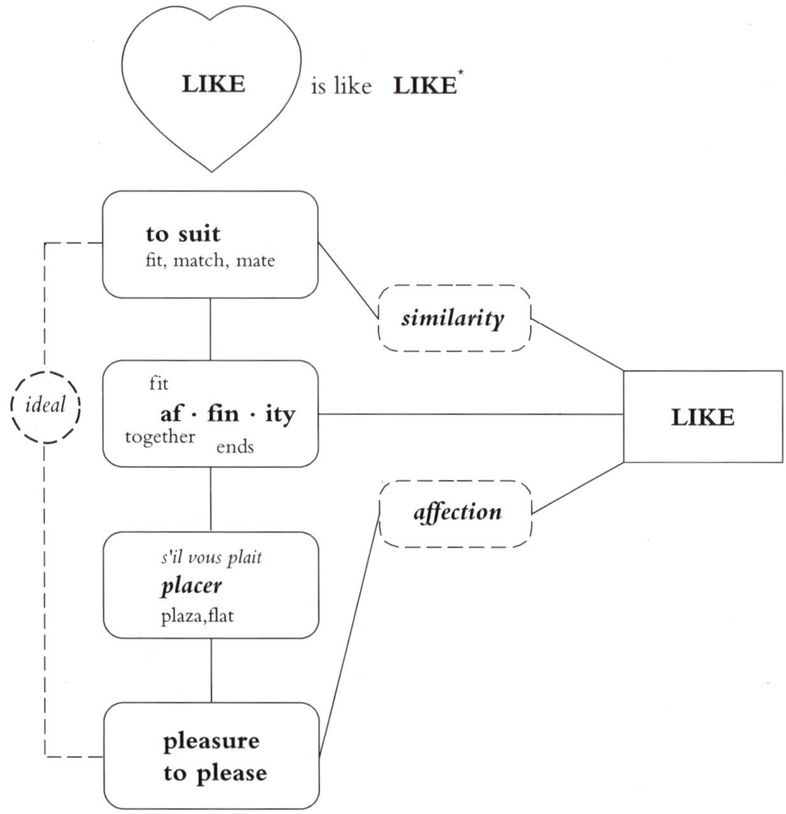

"What we like we think is just so."

Shipley, Dictionary of Word Origins

* For lexophiles:
"Time flies like an arrow; fruit flies like a banana."

Figure 10.1 Like is like like. The two meanings of "like"—that is, affection and similarity, are seen to have a common sense, concisely stated by (Shipley, 1945) as shown. In other words, what gives one pleasure may match one's preexisting template or expectation.

and hence do not inform the direction of causality. It is likely a reciprocal interaction between sexual activity and health, each promoting the other. Further research is necessary to ascertain which comes first.

Of particular interest is the finding that sexual *pleasure,* specifically, is correlated with longevity in women. Palmore (1982) found, in a 25-year longitudinal study of 252 men and women, that longevity was greater in women who reported past enjoyment of intercourse. By contrast with men, the frequency of sexual intercourse in women was not predictive of their longevity; for the men, frequency of intercourse was the only significant predictor of longevity. The incidence of heart attack (myocardial infarction) in women was significantly correlated with "sexual frigidity [defined by the author as 'a partial or complete inability of the female to be aroused sexually or to achieve orgasm'], and dissatisfaction," which was found among 65% of the coronary patients as compared with 24% of the comparison group, who were age-matched but admitted to the hospital for other illnesses, for example, respiratory, urinary, gastrointestinal, hypertension, diabetes, and so on (Abramov, 1976).

Historical Note

Using "medical massage," doctors from the time of Hippocrates to Freud stimulated orgasms in women to treat "hysteria," which was believed to be due to

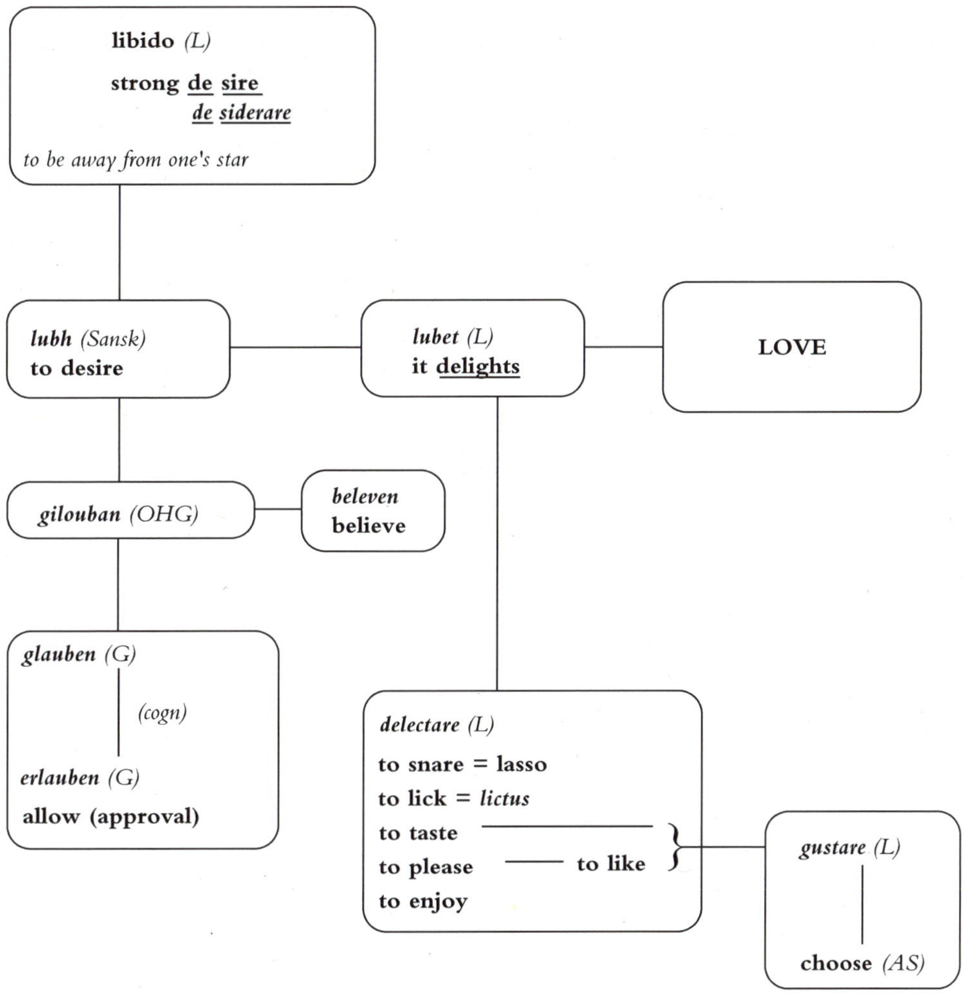

Figure 10.2 The word, "love" is seen to be etymologically related to the words, "like," "desire," and "libido." The languages of origin are abbreviated as: G: German, L: Latin, OHG: Old High German, Sansk: Sanskrit; and cogn: cognate. [Sources for both figures: Shipley (1945) and Onions (1966).]

a uterine reaction to sexual deprivation ("hyster" = womb in Greek). After introduction of electromechanical vibrators as a medical appliance in the 1880s, mechanical stimulation gained cultural popularity for its ability to produce sexual pleasure. This led the embarrassed medical establishment to abandon its use as a therapy in the 1930s (Komisaruk et al., 2006).

In a 10-year follow-up study of 918 men who were initially between the ages of 45 and 59 years, the mortality risk was found to be 50% lower among the men who had frequent orgasms (defined in the study as two or more per week) than among men who had orgasms less than monthly. Even when age, social class, and smoking status were controlled for, a strong and significant inverse relationship was found between orgasm frequency and risk of death. The authors concluded that "sexual activity seems to have a protective effect on men's health" (Davey Smith et al., 1997).

A relatively high frequency of ejaculations over the years was reported to correlate with a reduced incidence of prostate cancer (Giles et al., 2003). In a study of more than 2000 men under the age of 70 years, those who recalled having had an average of four or more ejaculations per week in their twenties, thirties, and forties had a significantly lower (by one-third) risk of developing prostate cancer than men who reported an average of fewer than three ejaculations per week during the same age period. There was no association of prostate cancer with the number of sexual partners, suggesting that infectious factors did not account for the difference. A questionnaire study of 50,000 men, aged 40 to 75 years, was confirmatory (Leitzmann et al., 2004). Men who reported at least 21 ejaculations per month (via intercourse or masturbation) had a significantly lower risk of prostate cancer than men who reported seven or fewer ejaculations per month. The authors speculated that ejaculations may clear the prostate of potential carcinogenic substances and that psychological stress reduction resulting from ejaculation could reduce the release of prostate cell-stimulating factors from the nerves that supply the prostate.

In another study of cancer, the incidence of breast cancer *in men* was reported to be inversely related to the frequency of orgasm, most dramatically in men aged 50 to 59 years. Of 21 cases of breast cancer in this age group, 13 occurred in men who reported up to six orgasms per month, but only two cases in men who reported at least 14 orgasms per month (Petridou et al., 2000).

In our own research, we measured the effect of specifically pleasurable vaginal self-stimulation on perception of pain. Pain thresholds were measured as the force at which a gradually increasing compression applied to the finger tips was reported as feeling painful. The women applied vaginal self-stimulation with a nonvibrating sterilized smooth plastic cylinder. When the 10 women in the study applied steady pressure to the anterior vaginal wall, their pain thresholds increased significantly by a mean of over 40%. However, when they were asked to apply the vaginal self-stimulation in a way that felt pleasurable to them, their pain thresholds increased significantly by over 80%. Four of the women experienced orgasm during this pleasurable self-stimulation, at which time their mean pain thresholds increased by over 100%. Their tactile thresholds, measured via von Frey fibers applied to the back of the hand, showed no significant change during any of these events (Whipple and Komisaruk,1985).

A separate study provided further evidence that sexual pleasure, per se, counteracts at least this type of experimentally induced pain. External genital (clitoral) self-stimulation produced a significantly greater elevation in pain thresholds when the women applied it in a way that felt pleasurable than when it was applied as constant pressure (Whipple and Komisaruk, 1988).

Abundant evidence points to dopamine as a neurotransmitter essential to orgasm in humans. Administration of the dopamine precursor, levo-DOPA, or of the dopaminergic agonist, apomorphine, are reported to facilitate orgasm in both men and women. Both cocaine, a dopamine reuptake inhibitor, and amphetamine, a dopamine releaser, administered before sexual intercourse are reported to facilitate orgasm and enhance the pleasurable quality of this experience (Kall, 1992; Miller and Gold, 1988). Kall (1992) studied 29 young men who used amphetamine more than once per week for at least 6 months. Of 27 of these men using amphetamine during sexual intercourse, 21 reported intensified orgasms and 23 reported that the drug prolonged intercourse. Conversely, antipsychotics and many antidepressants that impair orgasm possess the common property of blocking postsynaptic D2 or D4 receptors. Chronic use of cocaine or amphetamines induces sexual disorders and anorgasmia in a high proportion of addicts, often producing the side effect of symptoms similar to paranoid schizophrenia.

Role of Dopamine in Sexual Pleasure

The anatomical distribution of dopaminergic neurons in the brain and their projection sites are well known. Most dopaminergic neuron cell bodies are located in the mesencephalon, and three fiber systems ascend from this region to the forebrain. These fiber systems connect the ventral tegmental area in the rostral mesencephalon (A10, A8) and substantia nigra (A9) with three forebrain areas: (1) the nigrostriatal pathway: caudate nucleus and putamen; (2) the mesocortical pathway: cingulate cortex, enthorhinal cortex, and medial prefrontal cortex; and (3) the mesolimbic pathway: nucleus accumbens, amygdala, septum, and olfactory tubercle (Majovski et al., 1981).

The ventral tegmental area, the origin of the mesocortical and mesolimbic pathways, is reported to be activated during ejaculation in men, as measured by positron emission tomography (PET) (Holstege et al., 2003), and its projection target region in the nucleus accumbens is activated during orgasm in women, as measured by functional magnetic resonance imaging (fMRI) in our own studies (Komisaruk et al., 2002, 2004; see Figure 10.3). Studying rats, Pfaus et al. (1995) reported that dopamine is released in the nucleus accumbens during mating. Aron et al. (2005), using fMRI, found that men and women who were "intensely in love," when observing pictures of their beloved, showed activation in the ventral midbrain and the caudate nucleus. These regions comprise components of the mesolimbic

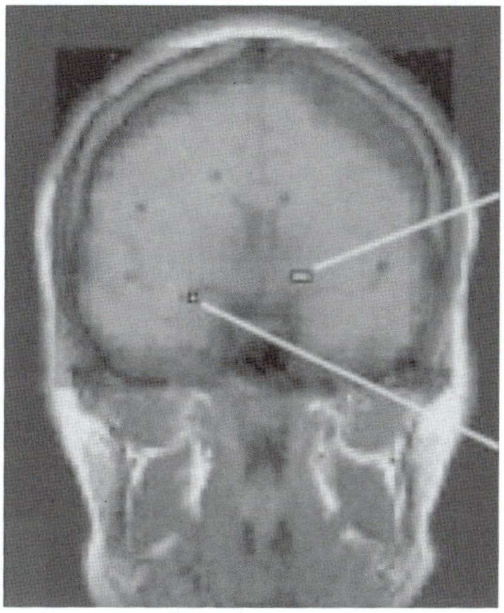

Figure 10.3 The nucleus accumbens and amygdala were two of the brain regions activated during orgasm in this woman in our study, as registered by fMRI. The activated voxels are highlighted for clarity.

dopaminergic system. The mesocortical dopaminergic system was implicated in a pharmacological study of brain areas activated during penile erection. Using PET, while men with erectile dysfunction who were receiving the dopaminergic agonist, apomorphine, also watched a sexually stimulating video sequence, a significant correlation was observed between the measured degree of penile rigidity (using the commercial RigiScan device) and increased activity in the anterior cingulate cortex and right prefrontal cortex (Bornhovd et al., 2002).

By contrast, the nigrostriatal dopaminergic pathway is evidently not related to orgasm, for antidopaminergic drugs that do not block this system, for example, clozapine, produce orgasmic dysfunction, most likely by blocking activity in the mesolimbic and/or mesocortical systems (Pilowsky et al., 1997). This is consistent with the report (Melis et al., 1998) that clozapine *increases* the firing rate of nigrostriatal neurons in unanesthetized rats. Thus, even if the same is true for humans, the increased dopamine release does not counteract the inhibitory effect of clozapine on producing orgasmic dysfunction, further supporting a role for the mesolimbic, rather than the nigrostriatal, dopaminergic system in the production of orgasm.

Neurophysiological studies indicate that dopamine acts more as a neuromodulator of sensory inputs rather than as an excitatory neurotransmitter. The failure of dopaminergic drugs to produce orgasm by themselves is probably due to the fact that dopamine does not "turn on an orgasm switch." It is more likely that dopamine augments the flow of sensory impulses generated by genital stimulation and other sexually relevant stimuli to activate the relevant limbic system circuits. Recent studies support this idea. Thus, Hagemann et al., (2003) found, in patients with erectile dysfunction, that the dopamine agonist, apomorphine, enhanced the response in the frontal cortex and the rostral anterior cingulate cortex (using PET) to the presentation of a sexually stimulating video. The authors concluded that apomorphine enhances the brain response to sexual stimuli. Conversely, somatosensory discrimination is significantly impaired in patients with the dopaminergic deficiency that characterizes Parkinson's disease, compared with normal volunteers (Weder et al., 2000). This is consistent with the syndrome termed "sensory neglect" that is characterized by deficits in orienting to somatosensory stimuli in rats with lesions in lateral hypothalamus; these interrupt the ascending dopaminergic pathways (Marshall et al., 1971). The modulatory role of dopamine on sensory activity is also manifest in the observation that acute administration of drugs that increase dopaminergic activity (agonists, releasers, or uptake inhibitors) rarely induce orgasm in the absence of other factors. A possible exception is cocaine, which, when rapidly incorporated into the blood circulation, can induce the cocaine "rush" (Nelson et al., 2006; Seecof and Tennant, 1986) that some individuals report as feeling similar to genital orgasm.

Berridge (2006) has suggested that the mesolimbic pathway determines the incentive salience or wanting of a prospective reward rather than the pleasurable experience of the reward itself—an expectation of pleasure, rather than 'liking'. He states, "... the best short answer to the question of what dopamine does in reward is that it causes 'wanting' for rewards but not learning or 'liking' for the same rewards" (p. 45).

This concept is consistent with findings of Phillips et al., (2008) who measured dopamine release into the nucleus accumbens by microdialysis in awake rats. They report that "First, a significant increase in DA (dopamine) efflux is observed in anticipation of reward when the rat is separated from the food reward or receptive female by a perforated wire screen. Second, a decline in DA efflux parallels the development of satiety for both the initial food and copulation with the first receptive female. Furthermore, exposure to a novel food or female and the reinstatement of motivated behavior is accompanied by a rebound increase in DA efflux." This may also be consistent with the report by Glick et al., (1994) using in vivo microdialysis of dopamine, that when levels were low, rats pressed a lever to self-administer cocaine intravenously more than when dopamine levels were high, although an alternative interpretation in this case is that dopamine at high levels could magnify aversive effects of cocaine. For additional discussion of this tricky issue, see the chapters by Berridge and by Leyton (Chapter 13) in this book.

To apply this concept to sexual response in humans, perhaps the release of dopamine during sexual response and orgasm (as evidenced by the brain imaging findings of activation of ventral tegmentum and nucleus accumbens in humans) is that it generates the 'wanting' of orgasm. But why then do sexual stimulation and orgasm feel pleasurable? Perhaps it is the intense multiple-modality sensory stimulation generated by the interpersonal interaction plus the muscular activity, that is, self- or proprioceptive stimulation—rather than dopamine levels per se—that is in itself pleasurable and rewarding. There is extensive evidence that sensory stimulation per se and its developmental elaborations (e.g., cognitive and symbolic social interaction) is rewarding and pleasurable, for when it is deprived, humans generate behavioral and physiological stimulation to provide it, and if such stimulation is thwarted or otherwise blocked, the result may be "psychosomatic" disease (for further discussion, see Komisaruk (1982) and Komisaruk and Whipple (1998)).

An fMRI study by Breiter et al., (1997) of the cocaine rush reported a pattern of activation of the mesolimbic dopaminergic system that may at first seem at odds with our report of activation of the nucleus accumbens during orgasm. They reported that the activation of the ventral tegmentum correlated with the perception of the cocaine rush. The post-rush period of craving was more closely correlated with activation of the nucleus accumbens, subcallosal cortex, and other regions. Perhaps the apparent discrepancy can be reconciled in that the rush is an immediate, "blast" effect of cocaine, whereas sexual stimulation leading to orgasm may contain a craving element leading up to the rush. A finer-grained temporal analysis of brain activity leading up to, during, and immediately after orgasm is necessary to adequately reconcile these findings.

In addition to cocaine and amphetamine, other drugs of abuse, for example, opiates, marijuana, caffeine, and nicotine, also increase the release of dopamine intrinsic to the mesolimbic system (Adinoff, 2004) although by different mechanisms. While cocaine blocks dopamine reuptake by blocking the dopamine transporter, and amphetamines increase the release of dopamine from the synaptic vesicles, tetrahydrocannabinol binds to G protein-coupled receptors. The opiates, for example, heroin and morphine, act indirectly by inhibiting the GABA neurons that in turn inhibit the dopamine neurons in the ventral tegmental area, thereby disinhibiting the dopamine neurons and increasing their dopamine-releasing activity in at least the nucleus accumbens (Johnson and North, 1992).

There are recent preliminary reports that "sexual desire" can be increased in women by the intranasal administration of a small synthetic peptide, bremelanotide, which acts presumably by stimulating melanocortin receptors in the brain (Perelman, 2007; Shadiack et al., 2007).

Pleasurable Vaginal Self-Stimulation Counteracts Pain

We have found that in women, pain thresholds are more than doubled during orgasm (Whipple and Komisaruk, 1985). In our earlier studies in rats, we had concluded, through the use of the opiate antagonist, naloxone, and receptor blocking agents, that endogenous opioids, norepinephrine, serotonin, GABA, and glycine each contribute to the ability of vaginal stimulation to potently suppress a wide variety of responses to noxious stimulation, indicative of the multiple roles served by this organ (Komisaruk and Whipple, 2000).

In our more recent studies in women using fMRI, we found that the insular cortex and anterior cingulate cortex in the forebrain, among other areas, are activated during orgasm (Komisaruk and Whipple, 2000; Komisaruk et al., 2004). Other investigators report that these cortical regions are activated during painful stimulation (Casey et al., 2001). These findings, considered together, suggest that a significant (active inhibitory) interaction occurs between orgasm and pain in the insular and anterior cingulate cortices, indicating that they are involved in both pain and pleasure. Could these brain regions have some property that is common to both pain and pleasure, perhaps intense emotional *expression*—controlling the contorted facial expression that occurs both during painful anguish and similarly during impending orgasm—separate from the actual different *feelings* of pain versus pleasure? It seems possible that (at least female) genital stimulation and orgasm, which we have shown attenuates the aversive component of pain, may not attenuate the *arousing* quality of pain. That is, these two properties of pain—aversion and arousal—may be independent and differentiable. This might help account for some individuals' practice of willingly receiving what would appear to be pain-inducing stimulation in a sexually stimulating context, a combination that apparently intensifies their pleasure. In that case, the sexual stimulation may attenuate the aversive, but not the arousing, component of the stimulation.

What is the adaptive significance of the pleasure produced by vaginal stimulation and its effect on suppressing pain? By reducing the potentially aversive intensities of sensory stimulation that may occur during sexual intercourse, perhaps the positive-reinforcing effects of the pleasurable component of the genital sensory stimulation are increased. In this view, the pleasurable component of vaginal and cervical sensory input would play a significant reinforcing role in intercourse, and the pain-attenuating effect of this sensory input could facilitate that reinforcing effect by attenuating any aversive components of the stimulation.

Speculation on Possible Adaptive Significance of Sexual Pleasure

In a different context, the vaginal and cervical stimulation-produced analgesia mechanism may also be activated during labor and delivery. We have found that indeed, the threshold to experimental, externally applied pain (finger compression) increases significantly during labor (Whipple et al., 1990), and in rats, pain threshold increased momentarily as each pup emerged from the birth canal (Toniolo et al., 1987). (In that study, pain threshold was measured as the latency for the parturient females to withdraw the tail, hanging down from the grid floor, from 50°C water). The endogenous contractions and the fetus-produced distention of the birth canal (uterus, cervix, and vagina) generate afferent activity via the sensory nerves of the reproductive tract as the fetus passes through the birth canal. It is possible that by activating this pain-attenuating mechanism during childbirth, the stress of parturition might be reduced, which might thereby promote the maternal–infant bond.

In the case of sexual intercourse, it seems plausible that pleasurable sensation produced by sexual intercourse would provide a mechanism that reinforces it, thereby promoting repeat performances. Viewed in this evolutionarily adaptive context, it would appear that the pleasurable feelings generated by sexual intercourse and orgasm would be selected-for, evolutionarily. Consequently, sexual pleasure can be viewed as a potently effective and important adaptive mechanism, the function of which is to ensure the procreation of the species.

It is important to recognize that not all women fit into the male linear model of sexual function, that is, desire, arousal, and orgasm, in that order. Women can experience sexual arousal, orgasm, and satisfaction without sexual desire, and they can experience desire, arousal, and satisfaction without orgasm (Whipple, 2002). Leiblum (2001) has noted that "many women can experience adequate sexual arousal and even orgasm without experiencing any genuine satisfaction, pleasure, or even the inclination to repeat the experience" (p. 165). It is important to consider reports of pleasure and satisfaction in women as characteristics of normal sexual function, regardless of their experiencing orgasm (Sugrue and Whipple, 2001).

Future research on dissociations between pleasure and orgasm would be illuminating—for example, of regional brain activation during reports of pleasure with, versus without, orgasm, as well as orgasm with, versus without pleasure, the latter as in the persistent genital arousal disorder (e.g., Leiblum et al., 2007).

Conclusion

We have presented evidence that activation of the dopaminergic system is a necessary condition for the experience of sexual pleasure. While we have focused on the role of dopamine, we realize that other neuronal systems, for example, serotoninergic, peptidergic, aminergic, among others, play significant roles in sexual response. However, this limitation in the present discussion pales in comparison with the much more difficult question of how these dopaminergic neurons, or *any* neurons for that matter, generate the cognitive experience of pleasure. Of course, we do not understand how any bit of cognitive experience is transduced from any neuronal activity—light, heat, color, taste, odor, touch, sound, pain. We cannot do much better than recognize that activity in some neuron groups (e.g., cortical) apparently can generate cognitive experience unlike others, such as the spinal cord if separated from the brain. We can also say that at any given moment, cognitive experience may be related to the activity of certain cortical neuron groups but not to others. However, we are at a total loss to account for the difference between how our neurons can generate our experience of pain versus our experience of pleasure, except to point to some (hopefully relevant) differences in neuronal systems and neurotransmitters.

By observing what we can, we have the sense that we are getting somewhere in understanding how our brain generates pleasure. The trouble is, we do not really know where "there" is. We know from our cognitive experience that pleasure exists, and that it differs from other cognitive experience such as pain. What we lack is a concept of how any neurons produce any bit of cognitive experience. To paraphrase Supreme Court Justice Potter Stewart in 1964, "I can't define pornography, but I know it when I see it" we know pleasure when we feel it; we just do not know which or how our neurons generate it.

References

Abramov, L. A. (1976) Sexual life and sexual frigidity among women developing acute myocardial infarction. *Psychosom. Med.* 38, 418–425.

Adinoff, B. (2004) Neurobiologic processes in drug reward and addiction. *Harvard Rev. Psychiatry* 12, 305–320.

Aron, A., Fisher, H., Mashek, D. J., Strong, G., Li, H. and Brown, L. L. (2005) Reward, motivation, and emotion systems associated with early-stage intense romantic love. *J. Neurophysiol* 94, 327–337.

Berridge, K. C. (2006) The debate over dopamine's role in reward: the case for incentive salience. *Psychopharmacology (Berl)* 10, 1007–1077.

Bornhovd, K., Quante, M., Glauche, V., Bromm, B., Weiller, C. and Buchel, C. (2002) Painful stimuli evoke different stimulus-response functions in the amygdala, prefrontal, insula and somatosensory cortex: a single-trial fMRI study. *Am. J. Gastroenterol.* 97, 654–661.

Breiter, H. C., Gollub, R. L., Weisskoff, R. M., Kennedy, D. N., Makris, N., Berke, J. D., Goodman, J. M., Kantor, H. L., Gastfriend, D. R., Riorden, J. P., Mathew, R. T., Rosen, B. R. and Hyman, S. E. (1997) Acute effects of cocaine on human brain activity and emotion. *Neuron* 19, 591–611.

Casey, K. L., Morrow, T. J., Lorenz, J. and Minoshima, S. (2001) Temporal and spatial dynamics of human forebrain activity during heat pain: Analysis by positron emission tomography. *J. Neurophysiol.* 85, 951–959.

Davey Smith, G., Frankel, S. and Yarnell, J. (1997) Sex and death: Are they related? Findings from the Caerphilly Cohort Study. *BMJ* 315, 1641–1644.

Giles, G. G., Severi, G., English, D. R., McCredie, M. R., Borland, R., Boyle, P. and Hopper, J. L. (2003) Sexual factors and prostate cancer. *Br. J. Urol.* 92, 211–216.

Glick, S. D., Raucci, J., Wang, S., Keller, R. W. Jr. and Carlson, J. N. (1994) Neurochemical predisposition to self-administer cocaine in rats: Individual differences in dopamine and its metabolites. *Brain Res.* 653, 148–154.

Hagemann, J. H., Berding, G., Bergh, S., Sleep, D. J., Knapp, W. H., Jonas, U. and Stief, C.G. (2003) Effects of visual sexual stimuli and apomorphine SL on cerebral activity in men with erectile dysfunction. *Eur. Urol.* 43, 412–420.

Holstege, G., Georgiadis, J. R., Paans, A. M. J., Meiners, L. C., van der Graaf, F. H. C. E. and Reinders, A. A. T. S. (2003) Brain activation during human male ejaculation. *J. Neurosci.* 23, 9185–9193.

Johnson, S. W. and North, R.A. (1992) Opioids excite dopamine neurons by hyperpolarization of local interneurons. *J. Neurosci.* 12, 483–486.

Kall, K. I. (1992) Effects of amphetamine on sexual behavior of male i.v. drug users in Stockholm—a pilot study. *AIDS Edu. Prev.* 4, 6–17.

Komisaruk, B., Whipple, B., Crawford, A., Grimes, S., Liu, W.-C., Kalnin, A. and Mosier, K. (2004) Brain activation during vaginocervical self-stimulation and orgasm in women with complete spinal cord injury: fMRI evidence of mediation by the Vagus nerves. *Brain Res.* 1024, 77–88.

Komisaruk, B. R. (1982) Visceral-somatic integration in behavior, cognition, and "psychosomatic" disease. In: *Advances in the Study of Behavior*, Vol. 12, pp. 107–140. Ed. J. S. Rosenblatt, R. A. Hinde, C. Beer and M.-C. Busnel. Academic Press, New York, NY.

Komisaruk, B. R., Beyer, C. and Whipple, B. (2006) *The Science of Orgasm.* The Johns Hopkins University Press: Baltimore, MD.

Komisaruk, B. R., Whipple, B., Crawford, A., Grimes, S., Kalnin, A. J., Mosier, K., Liu, W.-C. and Harkness, B. (2002) Brain activity (fMRI and PET) during orgasm in women, in response to vaginocervical self-stimulation. In: *Society for Neuroscience* Program No. 841.817, Washington, DC.

Komisaruk, B. R. and Whipple, B (1998) Love as sensory stimulation: Physiological effects of its deprivation and expression. *Psychoneuroendocrinology* 23, 927–944.

Komisaruk, B. R. and Whippple,, B. (2000) How does vaginal stimulation produce pleasure, pain, and analgesia? In: *Sex, Gender, and Pain*, pp. 109–134. Ed. R. B. Fillingim. IASP Press, Seattle, WA.

Leiblum, S. (2001) Critical overview of the new consensus-based definitions and classification of female sexual dysfunction. *J. Sex Marital Ther.* 27, 158–168.

Leiblum, S., Seehuus, M., Goldmeier, D. and Brown, C. (2007) Psychological, medical, and pharmacological correlates of persistent genital arousal disorder. *J. Sex. Med.* 4, 1358–1366.

Leitzmann, M. F., Platz, E. A., Stampfer, M. J., Willett, W. C. and Giovannucci, E. (2004) Ejaculation frequency and subsequent risk of prostate cancer. *J. Am. Med. Assoc.* 7, 1578–1586.

Majovski, L. V., Jacques, S., Hartz, G. and Fogwell, L. A. (1981) Dopaminergic (DA) systems: Their role in pathological neurobehavioral symptoms. *Neurosurgery* 9, 751–757.

Marshall, J. F., Turner, B. H. and Teitelbaum, P. (1971) Sensory neglect produced by lateral hypothalamic damage. *Science* 174, 523–525.

Melis, M., Gessa, G. L. and Diana, M. (1998) Clozapine does activate nigrostriatal dopamine neurons in unanesthetized rats. *Eur. J. Pharmacol.* 363, 135–138.

Miller, N. S. and Gold, M. S. (1988) The human sexual response and alcohol and drugs. *J. Subst. Abuse Treat.* 5, 171–177.

Nelson, R. A., Boyd, S. J., Ziegelstein, R. C., Herning, R., Cadet, J. L., Henningfield, J. E., Schuster, C. R., Contoreggi, C. and Gorelick, D. A. (2006) Effect of rate of administration on subjective and physiological effects of intravenous cocaine in humans. *Drug Alcohol Depend.* 82, 19–24.

Onions, C. T. (1966) *The Oxford Dictionary of English Etymology.* The Oxford University Press: New York.

Palmore, E. B. (1982) Predictors of the longevity difference: a 25-year follow-up. *Gerontologist* 22:, 513–518.

Perelman, M. A. (2007) Clinical application of CNS-acting agents in FSD. *J. Sex. Med.* 4, 280–290.

Petridou, E., Giokas, G., Kuper, H., Mucci, L. A. and Trichopoulos, D. (2000) Endocrine correlates of male breast cancer risk: a case-control study in Athens, Greece. *Br. J. Cancer* 83, 1234–1237.

Pfaus, J. G., Damsma, G., Wenkstern, D. and Fibiger, H. C. (1995) Sexual activity increases dopamine transmission in the nucleus accumbens and striatum of female rats. *Brain Res* 693, 21–30.

Phillips, A. G., Vacca, G. and Ahn, S. (2008) A top-down perspective on dopamine, motivation and memory. *Pharmacol. Biochem. Behav.*. 90, 236–249.

Pilowsky, L. S., Mulligan, R. S., Acton, P. D., Ell, P. J., Casta, D. C. and Kerwin, R. W. (1997) Limbic selectivity of clozapine. *Lancet* 350, 490–491.

Seecof, R. and Tennant, F. S. Jr. (1986) Subjective perceptions to the intravenous "rush" of heroin and cocaine in opioid addicts. *Am. J. Drug Alcohol Abuse* 12, 79–87.

Shadiack, A. M., Sharma, S. D., Earle, D. C., Spana, C. and Hallam, T. J. (2007) Melanocortins in the treatment of male and female sexual dysfunction. *Curr. Top. Med. Chem.* 7, 1137–1144.

Shipley, J. T. (1945) *Dictionary of Word Origins*. Philosophical Library: New York.

Sugrue, D. P. and Whipple, B. (2001) The consensus-based classification of female sexual dysfunction: Barriers to universal acceptance. *J Sex Marital Ther* 27, 221–226.

Toniolo, M. V., Whipple, B. and Komisaruk, B. R. (1987) Spontaneous maternal analgesia during birth in rats. In: *NIH Centennial MBRS–MARC Symposium*. p. 100.

Weder, B., Azari, N. P., Knorr, U., Seitz, R. J., Keel, A., Nienhusmeier, M., Maguire, R. P., Leenders, K. L. and Ludin, H.-P. (2000) Disturbed functional brain interactions underlying deficient tactile object discrimination in Parkinson's disease. *Hum. Brain Mapp.* 11, 131–145.

Whipple, B. (2002) Women's sexual pleasure and satisfaction. *Female Patient* 27, 44–47.

Whipple, B. and Komisaruk, B. R. (1985) Elevation of pain threshold by vaginal stimulation in women. *Pain* 21, 357–367.

Whipple, B. and Komisaruk, B. R. (1988) Analgesia produced in women by genital self-stimulation. *J. Sex Res.* 24, 130–140.

Whipple, B., Josimovich, J. B., and Komisaruk, B. R. (1990) Sensory thresholds during the antepartum, intrapartum, and postpartum periods. *Int. J. Nurs. Stud.* 27, 213–221.

11

The Sweetest Taboo: Functional Neurobiology of Human Sexuality in Relation to Pleasure

JANNIKO R. GEORGIADIS AND RUDIE KORTEKAAS

Every organism that is alive today is a direct descendant of one or two ancestors: one in the case of asexual reproduction and two in the case of sexual reproduction. This self-evident truth tells us two things about those ancestors: (i) they reached sexual maturity in a sufficiently healthy state and (ii) they actually replicated. These two achievements, individual survival and species survival, are essential for the continuity of life and, as a result, any trait that increases the ability to execute them is favored by natural selection. It is tempting to call these achievements the goals of life, but apparent goals is a safer term to use.

Dawkins (1976) remarked that the two apparent goals of life can easily be simplified by regarding the gene as the "atom" of natural selection and replication. Two seemingly rivaling apparent goals of life now become one: individual survival and species survival can both be seen as two different expression forms of the conservation of an organism's own genes (abbreviated as COG).

The fact that life forms across phyla are as diverse as they are proves that there are many different solutions to optimize COG. Our own species employs sexual reproduction, a useful COG-optimizing strategy in a variable environment. Another adaptation to a variable environment is a complicated nervous system. In our case, this system is extremely expensive in a metabolic sense: roughly, one-fourth of the total body glucose consumption is bestowed on the nervous system, which constitutes only 2% of the total body weight (see, e.g., Raichle and Mintun, 2006). A system that uses these amounts of energy would never have evolved unless it contributes enormously to COG.

The mere fact that both sexual reproduction and a complex nervous system can be regarded as adaptations to achieve COG in an unpredictable environment leads us to the hypothesis that the brain in general and higher brain functions in particular play a decisive role in sexual behavior. These functions include attributing valence, wanting, approach, liking, the development of satiety, and cognition (memory). The peculiar acuteness and intensity of the hedonia associated with successful sexual encounters lead us to choose sex as a model for COG-relevant hedonic/motivated behaviors.

A recent review summarized that the neuronal circuit for pleasure includes the ventral pallidum, the nucleus accumbens, the amygdala, and the orbitofrontal cortex (OFC) (Leknes and Tracey, 2008). A large number of neuromodulators and hormones are involved in these processes, most notably dopamine and opioids (Berridge, 1996).

In this chapter, we will propose a neurobiology of human sexuality by reviewing functional neuroimaging studies and neuropsychological and pharmacoendocrinological anomalies that affect human sexuality. Special reference will be given to the question of which brain parts make sexual activity feel good.

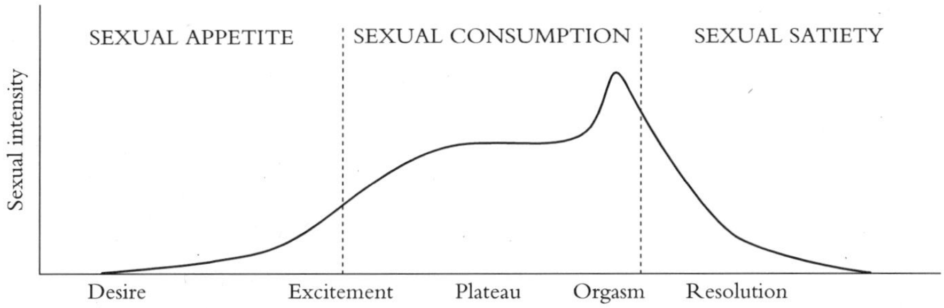

Figure 11.1 The human sexual response cycle.

The Human Sexual Response

In the 1960s and 1970s, the rapidly growing scientific interest in human sexuality led to the development of two models of the human sexual response: a physiology-based four-phase model by Masters and Johnson (1966) and a clinical-based three-phase model by Kaplan (1974). Both models may be fused into a five-phase model: (1) desire; (2) excitement; (3) plateau; (4) orgasm; (5) resolution (Figure 11.1) (Stoeckart et al., 2000). It is also customary to adopt terminology commonly used for feeding behavior: If the cycle is completed, the motivation to engage in sexual activity (sexual appetite) is followed by the physical "consumption" of the sexual need (sexual consumption), after which there is a period of sexual quiescence (sexual satiety).

Neuroimaging Studies on Sexual Appetite

Sexual appetite can be elicited from different sources—visual, auditory, olfactory, somatosensory, and even cognitive (imagery). By far the most used stimulus to elicit sexual desire and arousal in positron emission tomography (PET) and functional magnetic resonance imaging (fMRI) studies is visual sexual stimulation (VSS), using either still (photos) or dynamic (film excerpts) pornographic material. Studies investigating male brain responses to VSS have focused on healthy heterosexual (Arnow et al., 2002; Beauregard et al., 2001; Bocher et al., 2001; Ferretti et al., 2005; Hamann et al., 2004; Karama et al., 2002; Kim et al., 2006; Miyagawa et al., 2007; Moulier et al., 2006; Park et al., 2001b; Redouté et al., 2000; Stark et al., 2005; Stoleru et al., 1999; Tsujimura et al., 2006) and homosexual men (Paul et al., 2007; Safron et al., 2007), on male patients suffering from hyposexuality (Hagemann et al., 2003; Montorsi et al., 2003; Redouté et al., 2005; Stoleru et al., 1999; Yang, 2004) and on men with specific paraphilic preferences (Schiffer et al., 2008; Stark et al., 2005; Walter et al., 2007). In sharp contrast to the abundant interest for male sexuality, relatively few studies have investigated the female counterpart. Basically, two groups of "female" studies exist: (1) studies focussing on hormonal influences on female sexual appetite and associated brain responses (Archer et al., 2006; Gizewski et al., 2006; Jeong et al., 2005; Yang et al., 2008) and (2) studies that aimed to compare male and female brain responses during sexual appetite (Hamann et al., 2004; Karama et al., 2002; Park et al., 2001a).

Finding Common Ground

To set the stage for discussion of the reported brain responses, we quickly summarize the experimental set up used in this class of studies: (1) Most subjects are male, heterosexual, sexually healthy, and college students. (2) VSS type and length varies substantially between studies, ranging from still pictures of nude people presented for 3 s to XXX-rated film excerpts shown for 30 min. (3) Considerable variability is also seen with respect to the "control stimuli" that have been used to compare the VSS task with, that is, the stimuli that should match the VSS for elements like luminance, number of subjects depicted in the excerpt, and emotionality. (4) Subjective (feelings, questionnaires, visual analogue scales) and objective (e.g., penile erection) measures of sexual arousal are only occasionally collected. (5) Almost never are separate measures of pleasure performed. It is needless to say that these factors have to be taken into account when interpreting brain responses related to the processing of VSS in general and sexual pleasure in particular.

Because healthy men represent the vast majority of subjects who have been included in VSS-paradigms, this group of subjects is best suited as a starting point for a review. Next we will provide an overview of brain regions involved in male sexual appetite by identifying the most consistent effects across studies.

Male Brain Responses to VSS: Amygdala and Hypothalamus

Reigning theories about how the brain governs sexual behavior are derived from investigations in experimental animals. The picture painted by these investigations is that of a circuitry involving four main subcortical brain areas: the midbrain, the hypothalamus, the thalamus, and the (extended) amygdala (Baum and Everitt, 1992; Coolen et al., 1996; Kollack-Walker and Newman, 1997; Marson and Murphy, 2006; Marson et al., 1993). Although researchers disagree in some respects, there seems to be general consensus that sexual salience is recognized in the (extended) amygdala, which, via its projections to the hypothalamus and midbrain, "activates" a sexual response. Sensory consequences of sexual behavior are registered by the thalamus, especially at the moment of ejaculation.

Involvement of the amygdala and hypothalamus could be demonstrated by VSS studies, but in most instances lenient statistical approaches, especially region of interest analyses, were necessary to bring out these activations. Amygdala activation was not related to the degree of sexual arousal (Ferretti et al., 2005) and was observed only in the fMRI studies, which typically used VSS of shorter duration (Beauregard et al., 2001; Ferretti et al., 2005; Gizewski et al., 2006; Hamann et al., 2004; Karama et al., 2002). This is in agreement with the function of the amygdala in sexual behavior proposed by the animal literature, namely identifying distal olfactory and visual sexual signals (Newman, 1999; Parfitt and Newman, 1998).

Unlike the amygdala, the level of hypothalamus activation was associated with the degree of sexual arousal as indicated by penile tumescence measurements (Arnow et al., 2002; Ferretti et al., 2005; Karama et al., 2002; Redouté et al., 2000). Heightened subjective sexual arousal in response to apomorphine (a dopamine agonist) treatment (Montorsi et al., 2003), or in relation to elevated plasma testosterone levels (Redouté et al., 2005) also increased hypothalamic activity. Ferretti demonstrated that the hypothalamus was activated during the onset of penile erection, but not when sustaining it (Ferretti et al., 2005). In other words, once penile erection has been established, it can be sustained without further marked activation of the hypothalamus, provided that there is (visual) erotic stimulation.

The hypothalamus may be an important mediator of sexual pleasure: Plasma levels of the pituitary hormones oxytocin and prolactin are tightly linked to sexual pleasure and satiety. Nasal administration of oxytocin enhances sexual pleasure in men (Burri et al., 2008), whereas prolactin released during orgasm inhibits sexual desire during the postorgasmic phase (Krüger et al., 2003), probably by modulating promotivational dopaminergic systems (Krüger et al., 2005).

Male Brain Responses to VSS: Extrastriate Cortex, Anterior Cingulate Cortex, Insula, and Inferior Parietal Lobule

Activation of extrastriate visual regions in the occipitotemporal cortex (ES), anterior cingulate cortex (ACC), and insula is reported in many, if not most, neuroimaging studies of visually induced emotional states. ES activation is a well-known phenomenon in visual paradigms that reflects visual attention processes (Corbetta et al., 1990) and that occurs most readily when emotionally laden stimuli are used (Lane et al., 1997; Lang et al., 1998). Underscoring the susceptibility of VSS paradigms to side effects is the fact that ES activation is usually much stronger when film clips are presented rather than still images (Ferretti et al., 2005; Redouté et al., 2000).

In the studies that used penile tumescence as an objective marker of sexual arousal (Arnow et al., 2002; Ferretti et al., 2005; Karama et al., 2002; Moulier et al., 2006; Redouté et al., 2000, 2005; Stoleru et al., 1999), there was a strong association between the degree of penile erection and the activity in the insula and ACC. However, these responses do not necessarily constitute a unique network for sexual arousal, but are likely to reflect processing of sensory information. The insula monitors the internal state of the body and receives (predominantly) visceral afferent information from all parts of the body (Craig, 2002; Saper, 2002). The distal end of the penis (glans) has a viscera-like innervation with free nerve endings providing sensory information to the central nervous system via small-diameter Aδ fibers, at least in the rat (Halata and Munger, 1986; Johnson and Halata, 1991). It is therefore not surprising that responses in the insula were found when the internal state of the body changed due to penile erection and enhanced sexual arousal. Cell clusters within the

ACC are also known to respond to many different stimuli (for review, see Vogt, 2005). There is a large body of evidence showing that the rostral ACC is involved in processing both negative (Kulkarni et al., 2005; Vogt et al., 1996) and positive (Aron et al., 2005; Bartels and Zeki, 2000; Petrovic et al., 2002) emotions. Caudal parts of the ACC are mostly activated by cognitively demanding tasks, like reward-related decision-making and response selection (Bush et al., 2002; Devinsky et al., 1995).

Stoléru, the first researcher to implement a VSS paradigm for the purpose of neuroimaging, tried to explain the caudal ACC remarking that "subjects viewing erotica in a scanner environment are simultaneously confronted with the urge to act and with the impossibility to do so" (Stoleru et al., 1999). In a later study, he performed a VSS paradigm with healthy men and men suffering from hypoactive sexual desire disorder (HSDD). As expected, the HSDD group reported significantly lower sexual arousal than the healthy controls. Both groups activated the rostral ACC, but only the healthy men also activated the caudal ACC. The authors concluded that healthy men had an urge to physically respond to the VSS that was absent in the sexual arousal–deprived HSDD group (Stoleru et al., 2003). A region where metabolism was strongly linked to the degree of penile tumescence was the inferior parietal lobule (IPL) (; Ferretti et al., 2005; Moulier et al., 2006; Redouté et al., 2000, 2005). This region contains the secondary somatosensory cortex (SII), indicating higher-order somatosensory processing associated with the sensation of penile erection. The central processing of genital sensory information will be discussed in more detail in the section on sexual consumption.

Male Brain Responses to VSS: Basal Ganglia, Orbitofrontal Cortex, and Temporal Lobe

ES, rostral ACC, insula, and IPL are representatives of a neural network underlying heightened arousal and attention in response to VSS, as well as awareness and perception of penile erection, albeit within the boundaries of a neuroimaging experiment. Amygdala and hypothalamus are part of a basic system, which is also found in animals for identifying sexual salience and establishing sexual motivation and genital responses. In fact, the hypothalamus may stimulate sexual pleasure via oxytocin release from the pituitary gland (Burri et al., 2008). However, sexual pleasure can still not be fully explained at this point.

As indicated in the introduction, dopamine (DA) and opioids are associated with 'wanting' and 'liking', respectively (Berridge, 1996; Leknes and Tracey, 2008). The basal ganglia are crucial elements of DA-sensitive corticosubcortical loops. In VSS studies, the basal ganglia are consistently involved, particularly the head of caudate (Arnow et al., 2002; Redouté et al., 2000, 2005; Schiffer et al., 2008; Stoleru et al., 2003; Walter et al., 2008) and the ventral striatum and/or pallidum (Hamann et al., 2004; Karama et al., 2002; Miyagawa et al., 2007; Rauch et al., 1999; Redouté et al., 2000, 2005; Stark et al., 2005; Schiffer et al., 2008). The head of caudate is instrumental for the evaluation of one's own actions (Lau and Glimcher, 2007) and processing performance feedback (Tricomi and Fiez, 2008).

The ventral striatum (including the nucleus accumbens) and ventral pallidum are part of a distributed neuronal network underlying attribution of valence and such functions as wanting, approach, liking, and the development of satiety (Berridge, 1996; Leknes and Tracey, 2008; Schultz et al., 2000). DA release in the ventral striatum functions to signal the incentive salience—or 'wanting'—of a stimulus or object (Berridge and Robinson, 1998). For example, activity in the ventral striatum was correlated positively with subjective ratings of cocaine craving in drug addicts (Risinger et al., 2005; Volkow et al., 1996). The ventral striatum has also been found to be recruited by multiple forms of positive affect, ranging from romantic and maternal love (Bartels and Zeki, 2000, 2004) and shivers down the spine when listening to intensely pleasurable music (Blood and Zatorre, 2001), to competitive arousal (Rauch et al., 1999) and experiencing the effects of morphine (Becerra et al., 2006). Animal studies have shown that these hedonic experiences are stimulated by endogenous μ-opioids that modulate activity levels in the ventral striatum (Szechtman et al., 1981; Zubieta et al., 2001).

Using an interesting design with sexually arousing and disgusting visual material and people with and without sadomasochistic (SM) preferences, Stark and coworkers showed that, across groups and conditions, the ventral striatum was only activated when stimuli valenced as pleasant were compared with those valenced as unpleasant (Stark et al., 2005). This provides strong evidence that the ventral striatum mediates feelings of pleasure during VSS-induced sexual arousal. We therefore propose that both processes, ascribing sexual salience and experiencing hedonia during sexual arousal, are mediated by a network including the ventral striatum.

The OFC is a target of the mesocortical DA projection system and has been strongly implicated in reward processing (Kringelbach, 2005; Rolls, 2000; Schultz et al., 2000). However, the OFC was not consistently reported in these VSS studies. One reason for this could be, at least in the studies that employed blood level oxygenated–dependent (BOLD) fMRI, that the close proximity of air-filled sinuses caused signal dropout in the OFC (difficulties investigating the human OFC are discussed in Kringelbach and Rolls, 2004).

Alternatively, the paradigms were perhaps not pleasurable enough to elicit a response in the OFC. Shedding a different light on the function of the OFC during the sexual response is the finding that the lateral OFC was activated when subjects voluntarily inhibited VSS-induced sexual desire (Beauregard et al., 2001). In agreement with this are reports of decreased lateral OFC activity in relation to increased subjective arousal induced either by apomorphine administration (Montorsi et al., 2003) or elevated plasma testosterone levels (Redouté et al., 2005). Nonsexual paradigms have shown that prefrontal metabolism decreases with increased striatal DA (Volkow et al., 2007), suggesting that sexually enhanced DA levels decrease prefrontal metabolic activity. Similarly, it has been postulated that reduced neocortical input to subcortical structures constitutes part of the neurobiological substrate for euphoric mental states or pleasure (Volkow et al., 1996).

Viewed from a neuroimaging perspective, this means that such a mechanism would correspond to deactivations or negative correlations for certain cortical areas during euphoric mental states, which indeed seemed to be the case for prefrontal and temporal areas (Bartels and Zeki, 2004; Georgiadis et al., 2006, 2009; Small et al., 2001). Indeed, temporal deactivation was found in relation to VSS-induced sexual arousal in a number of studies (Arnow et al., 2002; Bocher et al., 2001; Redouté et al., 2000; Stoleru et al., 1999).

Female Brain Responses to VSS

In women, hormonal fluctuations during the menstrual cycle influence sexual interest (Clayton et al., 1999), which complicates the design of experimental paradigms to decipher the neuronal substrates of female sexual arousal. As a consequence, most VSS studies in women have focused on identifying brain regions that mediate the relationship between menstrual phase and sexual arousal. In general, women showed the same network of VSS-induced activation as men (Gizewski et al., 2006; Hamann et al., 2004; Karama et al., 2002; Park et al., 2001a; Yang et al., 2008).

VSS-induced brain activity varied as a function of menstrual phase, as shown by increased activity in the ACC, the insula, and the OFC during the mid-luteal phase compared to the menses phase (Gizewski et al., 2006). Likewise, in postmenopausal women, estrogen or testosterone treatment reinstated their sexual interest, which coincided with "increased limbic brain activity" (Archer et al., 2006). Together with similar central effects of testosterone observed in men (Park et al., 2001a; Redouté et al., 2005; Stoleru et al., 2003), this underscores the vital influence of gonadal hormones in mediating the sexual response.

In our opinion, reliable gender differences in brain responses to VSS were found only by Hamann, because only in this study gender groups were carefully matched for levels of sexual arousal perceived during the VSS (Hamann et al., 2004). Gender differences were found in the hypothalamus and amygdala, indicating a neuronal basis for differential processing of visual erotic stimuli in men and women. One might argue that in men visual stimuli have greater access to basic systems underlying the sexual response, and that this reflects the presumed greater role of visual stimuli in human male sexual behavior (Laumann et al., 1994).

Sexual Orientation and Brain Responses to VSS

The overall picture for VSS-induced brain activity did not differ greatly for homosexual and heterosexual men (Paul et al., 2007). However, a tendency toward greater involvement of parietal areas was found in homosexuals, whereas in heterosexuals there was stronger activation of the right caudate nucleus and the ACC. Weaker correlations between the perceived level of sexual arousal and fMRI signal in the hypothalamus were found in homosexual men. With the earlier mentioned gender differences in the hypothalamus demonstrated by Hamann (2004) in mind, this could signify female-like brain responses to VSS in homosexual men. In support of this, it has been demonstrated in human subjects exposed to androgen- and estrogen-like pheromones that hypothalamic metabolic activity depends on sexual orientation (Berglund et al., 2006; Savic et al., 2005).

VSS Paradigms With Hyposexual and Paraphilic Patient Groups

Hyposexual patients generally show decreased sexual interest and arousal in response to VSS, which can be restored by drug or hormone treatment. Patients with low testosterone levels can teach us the brain regions where activity is testosterone dependent (Park et al., 2001b; Redouté et al., 2005; Stoleru et al., 2003).

Although their results did not fully overlap, there was agreement between these research groups that the amount of activity that was found, especially in subcortical and limbic areas, varied as a function of testosterone levels. The study by Redouté made a strong case for specific testosterone-mediated activity in the left lat-OFC (Redouté et al., 2005). Likewise, apomorphine treatment in patients with psychogenic erectile dysfunction normalized VSS-induced activity in the left lat-OFC (Hagemann et al., 2003) and in subcortical areas like the midbrain and nucleus accumbens (Montorsi et al., 2003).

Subjects with exclusive pedophilic and SM preferences have also been investigated. Pedophilia-related neuronal responses pointed to altered subcortical activity, even if it was increased (in striatum, globus pallidus, thalamus) in one study (Schiffer et al., 2008) and decreased (in hypothalamus) in another study (Walter et al., 2007). The SM study demonstrated that similar brain responses were found in SM subjects watching SM videos as in control subjects watching "regular"' VSS (Stark et al., 2005). As mentioned before, the ventral striatum was activated across groups when stimuli eliciting high positive affect were compared with stimuli eliciting strong negative affect.

Conclusion from VSS Studies

In the sheer mosaic of brain regions that have been reported to be involved in the central processing of VSS, few regions can be convincingly linked to sexual pleasure. For any emotion to be fully established a network of brain regions must be recruited, including regions involved in attention, somatosensory processing, valence, and arousal processes. Such effects were indeed reported by most VSS studies. However, we feel that most of these studies have not been able to nail down what is the essence of the sexual human brain and how sexual arousal makes us feel good. This is partly due to the constraints of the neuroimaging techniques used, like restrictions in temporal and spatial resolution, and the fact that the experimental setting of a neuroimaging experiment is not necessarily compatible with sexual pleasure. Another important issue is that seemingly opposite emotional states, such as pleasure and pain (Leknes and Tracey, 2008), or sexual pleasure and disgust (Stark et al., 2005), show remarkable neurobiological overlap. For VSS-based neuroimaging experiments, which often make use of rather rigid paradigms, this may be particularly relevant: under those circumstances, sexual arousal does not necessarily imply sexual pleasure, but could just as easily cause feelings of guilt, shame, or frustration.

We conclude that increased subcortical activity, especially in the ventral striatum and hypothalamus, and (less convincingly) associated decreased OFC activity are hallmarks of sexual arousal–related euphoric or pleasurable feelings. In addition, DA, opioids, and sex steroids regulate sexual behavior through action on these regions.

Human Brain Studies on Sexual Consumption

Of the five stages of human sexual behavior as defined, *plateau* and *orgasm* are part of *sexual consumption*. Normally, the plateau phase consists of some form of sexual genital stimulation. Feelings of pleasure and euphoria are perhaps the main characteristics of the sexual consumption phase, especially during orgasm. As such, sexual consumption constitutes an ideal model to investigate brain mechanisms underlying (sexual) pleasure. When reading this section, it is important to keep in mind that popular fascination for sexual activity is not nearly equaled by the amount of neuroscientific research, and that our research group is one of only two groups in the world that is currently engaged in a systematic investigation this phase of the sexual response in humans.

Early Studies

In several studies over the past 60 years, electroencephalography (EEG) has been used to investigate changes in the electrical activity of the brain during orgasm. Mosovich and Tallaferro (1954) reported high-voltage paroxysmal activity during orgasm in men and women. Around this time, the psychiatrist Robert Heath performed deep brain stimulations to improve his patients' condition (Heath, 1963). He discovered that injections of acetylcholine into the septal area, located between the lateral ventricles, induced intense pleasure, including multiple sexual orgasms lasting as long as 30 min. When Heath recorded from electrodes in the septal area, he found slow wave and spike activity with interspersed fast activity during orgasm (Heath, 1972). These waveforms resembled paroxysmal discharges seen with epileptic seizures (Calleja et al., 1988). Results from more recent surface EEG studies have suggested that during orgasm, but not during a "faked orgasm" electrical activity was right-lateralized (Cohen et al., 1976), and that orgasm was associated with a depression of alpha power, a well-known but nonspecific marker of attentional shifts (Graber et al., 1985).

Two neuroimaging studies have reported on orgasm and the human brain. The first was conducted in men by Tiihonen (1994). Using single photon emission computed tomography (SPECT), they were able to show that there were mainly decreases in tracer uptake during ejaculation. In particular, they showed that all neocortical areas were deactivated except for the right prefrontal cortex. However, the SPECT technique used by Tiihonen did not allow for analysis of subcortical regions. The other study, an fMRI study in spinal cord–lesioned women, was conducted by Komisaruk (2004). Many regions were reported to be related to vaginocervical stimulation and orgasm, including the hypothalamus, the amygdala, the insula, and the caudal brainstem. However, their experimental setup was not comparable with ours (different imaging technique, stimulation type, and subject group). Also, the BOLD fMRI acquisition type that was used in combination with the relatively long periods of continuous scanning in their paradigm most likely caused signal drift and, associated therewith random artifactual brain activity (Smith et al., 1999).

Recent Advances: PET Experiments on Sexual Consumption

In a series of PET experiments conducted in Groningen from 2000 to 2004, healthy male and female subjects underwent genital stimulation and attempted to reach orgasm. The radiotracer $H_2^{15}O$ was used as a measure of regional cerebral blood flow (rCBF). Male subjects received manual penile stimulation by their female partners, and female subjects received clitoral stimulation by their male partners. During scanning, the subjects kept their eyes closed. One of the advantages of including the subjects' partners was that it ensured a sexually salient context with some degree of familiarity for the subject, while maximizing the chances for pleasurable experiences for the subjects. The clitoris was chosen as locus of stimulation for two reasons: (1) Women reach orgasm most easily through clitoral stimulation, and not via penetration (Hite, 1976; Lloyd, 2005). (2) The clitoris is embryonically equivalent to the penis. As such, these organs have the same peripheral innervation, which, at least in theory, reduces experimental variance when both groups are analysed together.

rCBF maps were calculated for sexual genital stimulation relative to a nonsexual resting state ("Stimulation"), and for orgasm relative to the nonsexual resting state and relative to sexual genital stimulation ("Orgasm").

Figure 11.2 depicts the major stimulation and orgasm-related rCBF changes. These effects were investigated separately within each gender group (Georgiadis and Holstege, 2005; Georgiadis et al., 2006, 2007; Holstege et al., 2003). Recently, we performed an analysis to determine gender differences and commonalities with respect to stimulation and orgasm (Georgiadis et al., 2009). The following paragraphs give an overview of the most important effects reported in our studies.

Sexual Genital Stimulation: Somatosensory Areas

One of the most influential discoveries in the history of neuroscience is that of cerebral somatotopic maps, or homunculi (Penfield and Rasmussen, 1950). On the Penfield homunculus of the primary somatosensory cortex (SI), the location of the genitals is on the interhemispheric surface, directly dorsal to the callosomarginal sulcus in the paracentral lobule. For the penis, this location has been corroborated by some investigators (Allison et al., 1996; Foerster et al., 1936; Guerit and Opsomer, 1991; Mäkelä et al., 2003; Nakagawa et al., 1998), but disputed by others (Bradley et al., 1998; Kell et al., 2005, Pfeifer, 1920) who have put forward a location on the dorsal convexity of the postcentral gyrus. We concur with the finding that the genital representation is more dorsal on SI than Penfield originally described (Georgiadis and Holstege, 2005; Georgiadis et al., 2006, 2009) and we support the call for revision of the somatosensory homunculus (Kell et al., 2005).

As we have shown in the previous section, SII is a higher-order somatosensory area the metabolism of which is correlated to the degree of penile tumescence (Ferretti et al., 2005; Moulier et al., 2006; Redouté et al., 2000, 2005), most likely as a result of altered sensory input provided by an erect penis. Our result of SII activation during sexual genital stimulation in men and women supports this explanation (Georgiadis and Holstege, 2005; Georgiadis et al., 2006, 2009). Although SII was also activated by nonsexual genital stimulation, like stimulation of the dorsal penile nerve (Mäkelä et al., 2003) or penile skin (Kell et al., 2005), and, in women, of the vaginal vestibulum (Pukall et al., 2005), SII was activated more strongly when

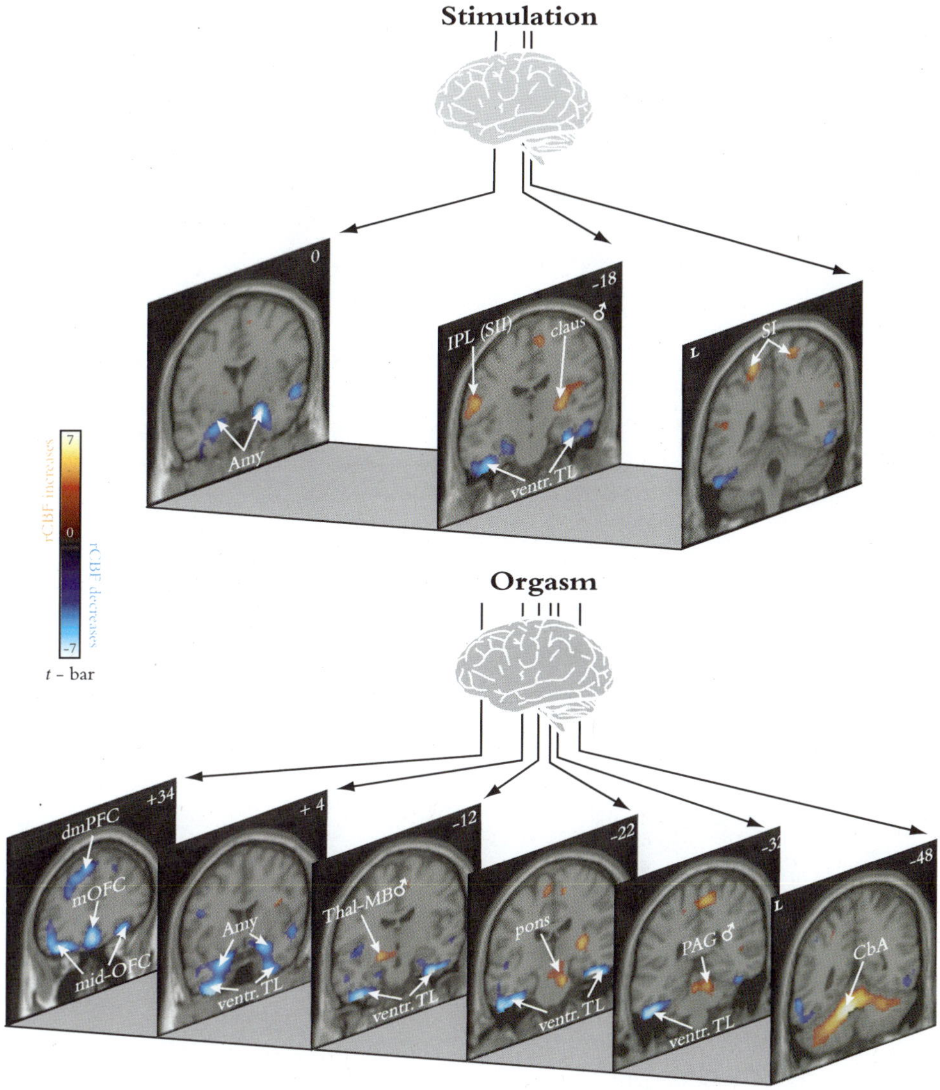

Figure 11.2 rCBF changes related to sexual genital stimulation and orgasm. Scans of sexual genital stimulation and orgasm were compared with scans of a nonsexual resting state, resulting in patterns of rCBF changes representing sexual genital stimulation ("stimulation"; top) and orgasm (bottom), respectively. Increased rCBF is depicted in red shading, decreased rCBF in blue shading. The threshold for all rCBF changes depicted in the figure is $p<0.001$ (uncorrected). Sections are lined up from anterior (left section) to posterior (right section), with the distance to the anterior commissure (in mm) indicated in the right top corner. Abbreviations: Amy, amygdala; CbA, cerebellum, anterior lobe; claus, claustrum; dmPFC, dorsomedial prefrontal cortex; IPL, inferior parietal lobule; L, left hemisphere; lat-OFC, lateral orbitofrontal cortex; mid-OFC, medial orbitofrontal cortex; PAG, periaqueductal gray matter; rCBF, regional cerebral blood flow; SI, primary somatosensory cortex; SII, secondary somatosensory cortex; Thal-Mb, thalamus-midbrain transition zone; T, T value; ventr. TL, ventral temporal lobe; ♂, activated more in men than in women; ♀, activated more in women than in men.

genital (vaginal) stimulation was perceived as painful (Pukall et al., 2005). Thus, it appears that SII activation is also a reflection of the context or salience of the sensory stimulus, which is in agreement with numerous pain paradigms reporting SII activation (see, e.g., Ploner et al., 2002). SII has strong reciprocal connections with temporal lobe limbic structures like the amygdala (Friedman et al., 1986), which most likely serve to promote the integration of somatosensory and salient aspects of tactile stimuli. Here, the somatosensory perception of genital stimulation could receive a sexual "color."

Sexual Genital Stimulation: Amygdala and Ventromedial Temporal Lobe

Animal studies (see, e.g., Coolen et al., 1996) and VSS neuroimaging studies in humans (Beauregard et al., 2001; Ferretti et al., 2005; Gizewski et al., 2006; Hamann et al., 2004;Karama et al., 2002) have convincingly advocated a prosexual function for the amygdala, arguing that the amygdala is invaluable for the identification of sexual salience. However, we found clear and significant deactivation in the amygdala in response to sexual penile and clitoral stimulation (Georgiadis and Holstege, 2005; Georgiadis et al., 2006, 2009). Supporting our finding are studies showing that euphoric mental states such as cocaine rush (Breiter et al., 1997) and romantic love (Bartels and Zeki, 2004) are characterized by attenuated amygdala and temporal lobe activity. These "rush" states share characteristics with the rush experienced during sexual genital stimulation.

We propose a double duty for the amygdala in sexual behavior. It is conceivable that activation of the amygdala is important for identifying sexual salience, and that this is a prerequisite for sexual desire. Likewise, the identification of objects or events that are detrimental for survival is also a function of the amygdala, and this is a prerequisite for vigilance and fear behavior (Amaral, 2002; Davis and Whalen, 2001; Rauch et al., 2000). Once sexual desire has been established, however, amygdala activity decreases. This change in amygdala activity may be crucial for prolonged sexual arousal during sexual consumption. Examplifying this are patients with chronically enhanced amygdala activity, for instance, as a result of war- and combat-related posttraumatic stress disorder (Rauch et al., 2000), who experience more sexual difficulties than healthy controls (Letourneau et al., 1997).

The amygdala deactivation was joined by bilateral deactivation of the ventromedial temporal lobe, most notably the fusiform gyrus. Unlike amygdala deactivation, there was already some indication of temporal lobe deactivation during VSS-induced sexual appetite (Arnow et al., 2002; Bocher et al., 2001; Redouté et al., 2000; Stoleru et al., 1999). Temporal lobe lesions may cause hypersexuality (Aloni and Katz, 1999; Baird et al., 2002), indicating that temporal lobes are generally inhibitory to sexual behavior. In that sense, release of neocortical inhibition is necessary for sexual desire.

Sexual Genital Stimulation: Claustrum

The right claustrum was significantly more activated in men than in women (Georgiadis and Holstege, 2005; Georgiadis et al., 2009). This result corresponded with other reports showing that claustrum activity levels correlated with the degree of penile turgidity (Arnow et al., 2002; Redouté et al., 2000, 2005). This strongly suggests that the claustrum is specifically related to penile stimulation and not to genital stimulation per se.

The claustrum has extensive and mostly reciprocal connections with the neocortex (for review, see Crick and Koch, 2005) and functionally has been linked to cross-modal matching (Hadjikhani and Roland, 1998; Horster et al., 1989) and multisensory integration (Naghavi et al., 2007). During VSS, claustrum activation may reflect cross-modal transfer of visual input to imagined tactile (penile) stimulation (Arnow et al., 2002). Conversely, during sexual penile stimulation with eyes closed, claustrum activation could reflect cross-modal transfer of penile information to a visually imagined situation.

This raises the possibility that the men employed visual imagery more than the women did. Supporting this idea is the fact that, at least in cat (Narkiewicz, 1964; Olson and Graybiel, 1980) and monkey (Pearson et al., 1982), the posterior portion of the claustrum is connected to visual cortical areas. We also found activation of the right ventral occipitotemporal region in men, a visual area that is activated during visual imagery (Ishai et al., 2000; Roland and Gulyas, 1995). Next to these neuroscientific arguments, there is the plain observation that men seem to have a greater interest in visual sexual stimuli than women (Laumann et al., 1994).

Sexual Genital Stimulation: Frontoparietal Areas

A robust finding was that women had larger activation in the left parietal lobe and adjacent posterior part of the left frontal lobe (Georgiadis et al., 2006, 2009). Included were the "hand-area" of the primary motor cortex (MI), somatosensory area 2 (SA2), the posterior parietal cortex (PPC), as well as the right premotor cortex. A possible explanation for gender differences in somatosensory areas is that the stimulation was not identical for both genders due to the different shape of penis and clitoris. For instance, in women, often the entire vaginal vestibulum was stimulated, and not only the clitoris. The strongest gender difference was found in the PPC, but this multisensory association area (for review, see Andersen et al., 1997) does not fit a strictly somatosensory explanation.

A different kind of explanation is provided by the mirror neuron concept (for review, see Rizzolatti and Craighero, 2004). Connections between sensory and motor areas provide the basis for understanding another person's actions by simulating that person's actions onto one's own sensory-motor representations (Gazzola et al., 2006; Rizzolatti and Craighero, 2004). Transferring this concept to our experiment, it could be that female subjects were creating a mental representation of their partner's hand performing the clitoral stimulation. The left-lateralized MI and parietal activation is consistent with this and with the fact that the stimulation was always right-handed. Women are stronger empathizers than men (Baron-Cohen et al., 2005), and it is also known that individuals who are better at taking the perspective of other people show stronger activation in their mirror neuron areas, including premotor and somatosensory areas (Gazzola et al., 2006). We want to stress that in the absence of any data on perspective taking this explanation should be regarded as tentative.

Orgasm: Cerebellum

The famous neuroscience pioneer Gall (1758–1828) claimed that the instinct of reproduction was located in the cerebellum (Gall, 1822). In the early 1970s, Heath reported increased electrical activity in the cerebellum during orgasm (Heath, 1972), while recently Komisaruk also reported activity increases in the cerebellum as their most robust finding (Komisaruk et al., 2004). We add new information by showing specific cerebellar sites, namely the left anterior lobe of the cerebellum and adjacent deep cerebellar nuclei, that can be functionally connected to orgasm in men and women (Georgiadis et al., 2006, 2007, 2009).

The role of the cerebellum in the coordination of movement is well-established, and the anterior lobe of the vermis contains a sensorimotor representation of the body with the body axis located in the midline of the vermis. During orgasm, the axial muscles are predominantly active, examplified by pelvic muscular contractions (van Netten et al., 2008). In accordance with the concept of Schmahmann, we found midline anterior vermis activation in both sexes, along with a strong positive association in this area in women between blood flow and rectal pressure fluctuations (Georgiadis et al., 2006), a reliable measure of pelvic muscular activity (van Netten et al., 2008). We and others have found similar cerebellar activation related to pelvic motor function in women performing voluntary pelvic motor contractions (Blok et al., 1997; Georgiadis et al., 2006; Seseke et al., 2006).

The cerebellum also plays a significant role in the modulation of nonmotor aspects of behavior, like cognition, autonomic regulation, affect, and emotionally important memory (Middleton and Strick, 1994; Schmahmann, 2004; Snider, 1950; Snider and Maiti, 1976). Neuroimaging studies have confirmed the role of the cerebellum in autonomic regulation by showing increased activity, particularly in the anterior lobe of the vermis, during heightened cardiovascular and respiratory arousal induced by hypercapnia (Parsons et al. 2001), and physical or mental exercise (Critchley et al., 2000). Indeed, orgasms are characterized by substantial cardiovascular and respiratory arousal (Carmichael et al., 1994). Emotionally, clinical studies have revealed that large lesions of the vermis in humans may cause blunting of affect and inappropriate behavior (Schmahmann and Sherman, 1998; Stone et al., 2001). Conversely, electrical stimulation of the cerebellum may alleviate violent behavior (Heath, 1977), and cerebellar activation was found in association with intensely pleasurable music (Blood and Zatorre, 2001) and the perceived funniness of a joke (Berns, 2004).

Orgasm: Temporal Lobe

Relative to the nonsexual resting state, there were profound, widespread, and bilateral rCBF decreases in the temporal lobe during orgasm in both gender groups (Georgiadis et al., 2006, 2009). In particular, there were additional deactivations in the left parahippocampal gyrus and anterior temporal pole compared to the preceding stage of sexual genital stimulation, which were most remarkable in women (Georgiadis et al., 2006). No such additional deactivations were found for the amygdala. Similar temporal lobe deactivations were found by Tiihonen (1994). The observation of a gradual decrease of temporal lobe activity as the sexual response goes along is confirmed by subjective measures of sexual pleasure obtained in the female subject group: There was a highly significant inverse relationship between sexual pleasure and temporal lobe activity levels (Georgiadis et al., 2006). Conversely, increased activity in the parahippocampal gyrus was associated with unpleasantness as a result of music dissonance (Blood et al., 1999). Thus, we have shown that temporal lobe deactivation was uncertain during VSS-induced sexual arousal, but became manifest during sexual consumption, with the lowest activity levels during orgasm.

Abnormal activity in the temporal lobe, for example, during an epileptic seizure, may in fact cause sexual or even orgasmic auras (Fadul et al., 2005; Janszky et al., 2004). There seems to be a preponderance for an onset in the anterior temporal pole (Janszky et al., 2004), which fits nicely with our finding of significantly decreased rCBF in this part of the left anterior temporal pole during orgasm.

Orgasm: Orbitofrontal Cortex

In men and women alike, there was profound deactivation throughout the prefrontal cortex (Georgiadis et al., 2006, 2007), but especially in the anterior part of the OFC. As mentioned above when we discussed VSS-induced sexual appetite, the OFC is part of the brain's reward system (Leknes and Tracey, 2008; Rolls, 2000; Schultz et al., 2000). The OFC receives input from several sensory modalities (Carmichael and Price, 1995a, b), suggesting that the OFC is involved in the analysis of sensory stimuli, based on which behavior can be adapted. There is evidence that the OFC is made up of functionally distinct portions (Kringelbach and Rolls, 2004; Öngür et al., 2003). More specifically, a tripartite model (medial, middle, lateral) has been proposed for the anterior OFC (Kringelbach, 2005).

The medial OFC (mOFC) is thought to constitute a crucial part of a neural network underlying self-monitoring (Kringelbach, 2005) and self-referential thought (Gusnard et al., 2001; Northoff et al., 2006). Other parts of this network include the amygdala and the cingulate cortex (Gusnard and Raichle, 2001). The mOFC has strong reciprocal connections with the amygdala, and joint deactivation of these regions is the main neural mechanism underlying cognitive reappraisal of fearful stimuli, which alleviates fearful feelings (Hariri et al., 2000). A joint element of this study and ours is that mOFC deactivation seems to promote a more carefree state of mind.

The weight of the orgasm-related OFC deactivation was located more laterally in the middle and lateral part of the left anterior OFC. Strikingly, the middle OFC (mid-OFC) appears to specifically encode hedonic experience, at least with respect to food intake. In this part of the OFC, activity increased with increasing satiation and subjective pleasantness, and decreased when cues were presented that were associated with feelings of satiety (Kringelbach et al., 2003). If orgasm is thought of as a sexual variant of satiation, one would expect activity levels in the mid-OFC to be very high. Immediately after orgasm, a profound feeling of sexual satiety and sometimes even sexual aversion is established, which would then be associated with very low levels of mid-OFC activity. Following on from this, the orgasm-related mid-OFC deactivation could signify ongoing sexual satiety processes. Unfortunately, our PET data did not provide the necessary means to dissociate between the neuronal events related to sexual satiety and the satiation that preceded it. This rapid behavioral change is the topic of a set of fMRI experiments we are currently running.

The lateral-most part of the OFC (lat-OFC) has been strongly linked to decision making and evaluations leading to behavioral change (Kringelbach, 2005). Again, this notion fits nicely with the changing lat-OFC activity during orgasm, because orgasm marks the transition from intense sexual motivation and pleasure to sexual satiety and associated feelings of aversion, which clearly represent two different behavioral states. Urge suppression seems to be another prominent function of this part of the OFC, as evidenced by both by functional neuroimaging studies (Beauregard et al., 2001; Dougherty et al., 1999; Small et al., 2001), and by studies evaluating the effects of OFC damage. The latter may result in antisocial, impulsive, and disinhibited behavior, including promiscuity and sexual disinhibition (Damasio et al., 1994; Lapierre et al., 1995; Yang et al., 2005).

Conversely, hyperactivity of the OFC is a feature of disorders like obsessive compulsive disorder (Saxena and Rauch, 2000) and these patients exhibit excessive self-control and hyposexuality (Aksaray et al., 2001). Strikingly, most of our participants reported (off the record) that during sexual genital stimulation, they often had to "hold back" in order to prevent orgasm. Given the logic described above, this conscious control over a sexual urge may likely be reflected by the activation level of the lat-OFC: increased rCBF during sexual genital stimulation may represent high behavioral control, whereas decreased rCBF during

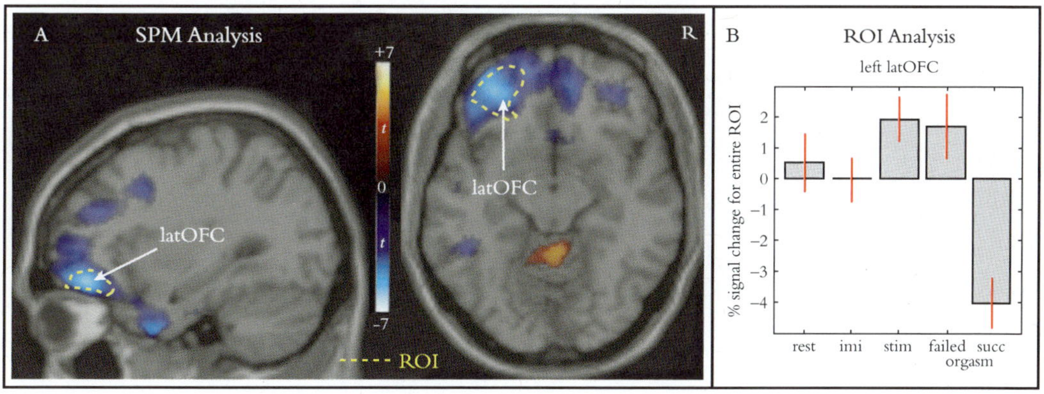

Figure 11.3 Female orbitofrontal function during sexual consumption. (A) statistical parametric mapping (SPM) analysis: Two-tailed t-map showing the group result for the comparison Successful Orgasm versus Clitoral Stimulation ($p<0.05$, corrected for multiple comparisons). Red scaling refers to rCBF increases, blue to decreases. The yellow dashed line indicates the form and extent of a region of interest (ROI) used to further analyze rCBF in the left lateral orbitofrontal cortex (LOFC). (B) ROI analysis for the left LOFC. We used a standard ROI from the ROI analysis software package MarsBAR (http://marsbar.sourceforge.net). The mean activity for the entire ROI was set to zero. For each parameter the percentage signal change is displayed. Abbreviations: rest, passive rest; imi, imitation of orgasm; stim, stimulation of the clitoris; failed, failed orgasm; succ, successful orgasm.

orgasm may represent sexual disinhibition. The finding that during failed orgasm attempts activity in the left lat-OFC also seemed to be enhanced illustrates the power of the OFC during the sexual response (see Figure 11.3). Deactivation of the lat-OFC possibly reflects the "loss of conscious control" or the "release of tension" that people report when they describe an orgasmic experience (Mah and Binik, 2001).

Orgasm: Insula

The insula is a key brain region in affective neuroscience and, as we have argued above, VSS studies showed clear insula involvement during male sexual appetite. During sexual consumption, however, involvement of the insula was less conclusive. A small effect was found in the right posterior insula during sexual genital stimulation, which should probably be ascribed to stimulation of the erect penis (Georgiadis and Holstege, 2005). One explanation for differential involvement of the insula during different stages of the sexual response could be that during the emergence of sexual arousal and associated onset of penile erection, there is a clear change of the body state, which the insula would readily respond to (Saper, 2002). Sexual genital stimulation, on the other hand, reflects the relatively stable plateau phase of the sexual response.

Strikingly, the same insular region showed higher activity in women than in men during orgasm (Georgiadis et al., 2009). This gender difference in orgasm-related insular activity is potentially interesting, especially because a recent fMRI study in women has specifically linked activity levels in the insula to self-reports of orgasm frequency, ease, and satisfaction (Ortigue et al., 2007).

Orgasm: Subcortical Regions

Subcortical regions seem to be crucial for positive affect (Berridge and Robinson, 1998; Burgdorf and Panksepp, 2006; Leknes and Tracey, 2008). During orgasm, perhaps the most pleasurable of all human experiences under physiological conditions, there appeared to be more subcortical effects than during the preceding phase of sexual genital stimulation (Figure 11.2). For some of these regions, there was a clear gender bias (Georgiadis et al., 2009).

First of all, there was activation in the pons: This brainstem region contains cerebellum-projecting nuclei, but it is also the site for cardiovascular and respiratory control centers, as well as for centers controlling sympathetic tone. This corresponds well with both the peak state of cardiovascular arousal (Exton et al., 1999; Krüger et al., 1998) and the high level of norepinephrine in the cerebrospinal fluid (Krüger et al., 2006) that is found during orgasm. Also in agreement with this is a study on the neuronal correlates of mental and physical exercise in humans, which showed that the pons and anterior lobe of the cerebellar vermis were positively correlated with cardiovascular arousal (Critchley et al., 2000).

Increased rCBF was also found in the lateral part of the left transition zone of midbrain and thalamus, and this effect was mainly present during orgasm in men (Georgiadis et al., 2007). The parvocellular subparafascicular nucleus of the thalamus has been extensively linked to the sensory processing of ejaculation, at least in rodents (Baum and Everitt, 1992; Coolen et al., 2003). This might also explain why this cluster of activation was not found during female orgasm.

We collected subjective measures of subjective arousal only in women and found a significant positive correlation between the subjects' reported perceived level of sexual arousal and rCBF in the ventral rostral midbrain (Georgiadis et al., 2006) and the right head of caudate. Because in rodents increased DA release has been demonstrated during sexual behavior (Paredes and Agmo, 2004), it is quite possible that the positive correlation found in our study represents increased activation of dopaminergic cell groups in the midbrain and their striatal target cells. The head of caudate was also reported by VSS studies (Arnow et al., 2002; Redouté et al., 2000, 2005; Schiffer et al., 2008; Stoleru et al., 2003; Walter et al., 2008).

Men had stronger activation than women in the periaqueductal gray matter (PAG). The PAG has a prominent role in the control of reproductive behavior in experimental animals (Sakuma and Pfaff, 1979a, b), and our data for the first time indicate that the PAG may also be involved in human sexual activity. Besides its role in reproductive reflexes, the PAG is also known to be involved in vocalization, cardiovascular and respiratory arousal, and pain suppression (for review, see Holstege et al., 1996). In human opiate addicts, the PAG was activated after intravenous administration of heroin (Sell et al., 1999). These people commonly report "orgasmic pleasure" with opiate use, but they also experience a severe lack of sexual interest (De Leon and Wexler, 1973; Mintz et al., 1974; Mirin et al., 1980; Seecof and Tennant, 1986).

This is in agreement with the recent idea that, neurochemically, pleasure, but not motivation, may be subserved primarily by opioid mechanisms (Berridge, 1996; Leknes and Tracey, 2008). Opiates presumably stimulate oxytocin release (Murphy et al., 1990), which, in turn, has been shown to enhance orgasmic pleasure (Murphy et al., 1987; Burri et al., 2008). Male-biased endogenous opioid release might explain the differential activation of the PAG during orgasm, because morphine activates a greater percentage of antinociceptive PAG-medulla projecting neurons in male compared with female rats (Loyd et al., 2007). There is no evidence, however, for greater orgasmic pleasure in men (Vance and Wagner, 1976).

In animal models of the neurobiology of sexual behavior, the hypothalamus plays a major part in every stage of the sexual response (McKenna, 1999). In humans, pituitary hormones, most notably oxytocin and prolactin, are crucially involved in mediating orgasmic pleasure and installing sexual satiety, respectively (Exton et al., 1999, 2001; Krüger et al., 2003, 2005). However, our studies did not show as much as a hint of hypothalamic involvement. It is possible that the neuronal events that might occur in the hypothalamus during orgasm (e.g. controlling hormone release) do not require a metabolic increase that sufficiently increases rCBF compared to a control condition. Alternatively, the temporal resolution of our PET experiment may have been too limited to detect short-lasting events occurring in the hypothalamus, or there may have been high intersubject variability in the hypothalamic response.

Conclusions about Orgasmic Pleasure

We have shown that deactivation of the OFC and certain temporal lobe regions is the main feature of orgasm in men and women. Thus, we were able to substantially refine and specify Tiihonen's work in men, who showed much more widespread neocortical deactivation (Tiihonen et al., 1994). What is interesting is that, at least in nonhuman primates, the regions that were deactivated during sexual consumption seem to be heavily interconnected. Strong connections exist between the lat-OFC and the temporal lobe, especially the parahippocampal gyrus (Kondo et al. 2005) and the anterior temporal pole (Kondo et al., 2003). Increased activity in these paralimbic areas is found with negative emotions such as guilt (Shin et al., 2000) and unpleasant feelings elicited by dissonant music (Blood et al., 1999). This suggests that these areas constitute a functional neocortical network promoting positive or negative affect by deactivation or activation of these areas, respectively. Supporting this idea is the suggestion of Volkow (1996) that decreased cortical inhibition of subcortical areas, especially the striatum, is one of the neural prerequisites for the experience of pleasure. Subcortical involvement was more prominent in men than in women during orgasm. Whether this reflects more orgasmic pleasure in men is unlikely: Written orgasmic experiences of male and female students are so similar that they could not be differentiated by gynaecologists, medical students, and psychologists (Vance and Wagner, 1976). Further investigation into the role of the hypothalamus in mediating orgasmic pleasure is also needed, as our methodological setup was perhaps not sensitive enough to measure activity in this region.

Recent Advances: Ongoing fMRI Experiments

Pleasure Systems during Sexual Genital Stimulation

The pattern of activations and deactivations found during sexual genital stimulation in men and women does not immediately point to involvement of pleasure or reward brain systems, like the nucleus accumbens, the ventral pallidum, the OFC, and the amygdala (Leknes and Tracey, 2008). Unlike most of the studies exploring VSS-induced sexual appetite, we found no effects in the basal ganglia during sexual genital stimulation. Komisaruk and colleagues, using fMRI, did report basal ganglia involvement associated with vaginocervical self-stimulation in a small group of spinal cord–injured women (Komisaruk et al., 2004), but, as we have mentioned earlier, their fMRI design was prone to artifacts.

To fully benefit from fMRI's advantages, and, at the same time, avoid its pitfalls, we recently developed an arterial spin labeling (ASL) fMRI acquisition type for the study of human sexual behavior. ASL-fMRI produces a stable signal that permits reliable investigation of brain responses on a long timescale (Wang et al., 2003). Fourteen healthy male subjects were scanned under this protocol. Better sensitivity of fMRI versus PET was suggested by strong bilateral activation in the posterior ventral pallidum during stimulation of the erect penis, which was not found with our PET experiments (Georgiadis et al., 2008). These activations may be linked to pleasurable feelings experienced during penile stimulation, because bilateral lesions of the globus pallidus cause anhedonia (Miller et al., 2006). Together with the previously discussed euphoria-related amygdala deactivation (Bartels and Zeki, 2000; Breiter et al., 1997), and deactivation of the ventromedial temporal lobe that is associated with increased sexual arousal and behavior (Aloni and Katz, 1999; Baird et al., 2002; Janszky et al., 2004), this may represent a neuronal framework for the pleasurable sexual experience during sexual genital stimulation.

The Brain and Anomalous Sexuality

Changes in sexual function are common in the clinic as results of trauma, cancer, surgery, epilepsy, vasculopathy, and the use of both prescription and non-prescription drugs. Many reviews have been written about altered sexual function due to neurological disorders (Rees et al., 2007) and drugs (Aldridge, 1982; Galbraith, 1991; Smith, 1982). Here, we focus on those changes that apparently have a primary neuronal cause.

Hyposexuality: Loss of Sexual Desire and Erectile Impotence

Many prescription drugs may cause reduced sexual desire, including antiepileptic drugs (Bergen et al., 1992) and centrally acting antihypertensive agents (Weiss, 1991). Selective serotonin reuptake inhibiting (SSRI) antidepressants are well known to disturb erectile function (Makhlouf et al., 2007), but bupropion, an antidepressant that increases dopaminergic tone, does not have this disadvantage (Rowland et al., 1997). Drugs that reduce the action of DA seem to be most detrimental to sexual desire. DA receptor antagonists that are used as antipsychotics, such as risperidone (Jayaram et al., 2007), have erectile impotence as a side effect. Neuroleptics are antipsychotics that block dopamine receptors and they commonly cause reversible impotence, loss of sexual desire, and anorgasmia (Martin-Du, 1978). The mediating hormone for this effect could well be prolactin because neuroleptics induce a reversible hyperprolactinaemia (Bouloux and Grossman, 1987). In agreement with this loss of libido is that the recreational drug 3,4-et hylenedioxymethamphetamine (MDMA) ("ecstasy") also increases prolactin levels and impairs sexual drive and function (Passie et al., 2005). These studies underscore the importance of dopamine in hypothalamic function for regulation of sexual desire and genital responses.

The peripheral vasoactive compounds sildenafil (also known as Viagra) and to a lesser extent phentolamine are well-known proerectile drugs. More interesting from the neurobiological point of view are the centrally active proerectile drugs apomorphine (dopamine agonist), yohimbine, trazodone (5-HT_2 antagonist), and testosterone (Sonda et al., 1990; Vitezic and Pelcic, 2002). Of these compounds, it appears that dopaminergic agonists have the most robust proerectile effects in human and nonhuman males. Apomorphine, a nonselective dopamine receptor agonist, is erectogenic in normal subjects (Heaton, 2000; Lal et al., 1984), in men with Parkinson's disease (O'Sullivan and Hughes, 1998), and even in men with psychogenic impotence (Segraves et al., 1991).

Orgasmic Dysfunction

Contrary to dopamine, which generally enhances sexual behavior, serotonin (5-HT) is inhibitory to sexual behavior. The inhibitory properties of selective serotonin reuptake inhibitors (SSRI) on orgasm are particularly notorious, causing delayed ejaculation and anorgasmia (Seidman, 2006). This side effect has, however, been put to therapeutic use as SSRI treatment delays ejaculation in men suffering from premature ejaculation (Waldinger et al., 2001, 2003). Conversely, a 5-HT_2 receptor antagonist (cyproheptadine) reversed anorgasmia induced by the SSRI citalopram (Lauerma, 1996). In the rat, serotonin released in the lateral hypothalamus (Veening et al., 1982) inhibits dopamine release in the nucleus accumbens and prohibits sexual motivation (Lorrain et al., 1999). Thus, serotonin interacts with dopamine to mediate sexual desire, most likely to establish post-orgasmic sexual inactivity. Impaired ejaculation may also be caused by centrally acting antihypertensive agents (Weiss, 1991), probably as a result of decreased sympathetic tone. Normal ejaculation and orgasm generally go together with high norepinephrine levels (Krüger et al., 2006), indicative of high sympathetic tone.

Orgasmic dysfunction is fairly common in multiple sclerosis (MS), and in these patients anorgasmia correlated significantly with brain stem and pyramidal abnormalities (Barak et al., 1996). Large cerebral defects like traumatic brain injury and stroke certainly influence orgasmic ability, but the results are contradictory. Postinjury orgasmic ability may be decreased or increased compared to the preinjury situation (Aloni and Katz, 1999; Hibbard et al., 2000; Sandel et al., 1996).

Hypersexuality

Hypersexuality in nursing care facilities includes "cuddling, touching of the genitals, sexual remarks, propositioning, grabbing and groping, use of obscene language, and masturbating without shame" (Nagaratnam and Gayagay, 2002). In many cases, these behavioral aberrations are probably due to various forms of functional or structural brain abnormalities like frontal lobe lesions (Sandel et al., 1996), right capsulolenticular hematoma (Donnet et al., 1997), MS plaques in the right hypothalamus and mesencephalon extending into the right side of the brainstem (Frohman et al., 2002), and lesions of the septal nuclei (Cavazos et al., 1997; Gorman and Cummings, 1992). Diencephalic injury may also result in a syndrome involving hypersexuality. In particular, thalamic infarction is associated with frontal hypoperfusion (Mutarelli et al., 2006) and the mediodorsal thalamic nucleus might be the mediating nucleus in this respect (Spinella, 2004). Nonpharmacological antiparkinsonian therapies, like high-frequency stimulation of the subthalamic nucleus (Romito et al., 2002) or globus pallidus (Roane et al., 2002), can also induce hypersexuality.

Temporal lobe lesions cause true hypersexuality (Aloni and Katz, 1999; Baird et al., 2002), which is reminiscent of the original description of Klüver-Bucy syndrome in monkeys. Strikingly, postencephalitic Klüver-Bucy syndrome in young children involves frequent holding of genitals, intermittent pelvic thrusting movements, and rubbing the genitals to the bed (Pradhan et al., 1998). Epileptic events in the temporal lobes can cause hypersexual paroxysms with masturbation (Gautier-Smith, 1980).

Sexual hallucinations are also not uncommon in certain types of epilepsy and may include tactile genital sensations, sexual feelings, and even orgasm (Fadul et al., 2005; Janszky et al., 2004). Epileptic orgasmic auras are more prominent in women than in men (Janszky et al., 2004; Remillard et al., 1983), but whether this represents a greater contribution of temporal lobe function to the orgasmic experience is uncertain.

Drug-induced hypersexuality is also quite common. Most commonly hypersexuality is caused by dopaminergic drugs. Sexual arousal, excessive masturbation, and hypersexual behavior may occur with methylphenidate treatment (Bilgic et al., 2007), and because dopaminergic drugs are the first treatment for Parkinson's disease (PD), hypersexuality is commonly induced in PD patients due to these compounds (Uitti et al., 1989).

Altered Sexual Preference

Altered sexual preference can follow damage to the diencephalon and medial-basal forebrain (Miller et al., 1986), and lesions of the temporal lobe, which contains the limbic structures amygdala and hippocampal formation, may lead to transvestite and transsexual behavior (Gautier-Smith, 1980). The importance of the temporal lobes for sexual preference and normal sexuality is illustrated by a study describing two cases of temporal lobe hypometabolism in association with late-life homosexual pedophilia and concomitant increases in sexual behavior (Mendez et al., 2000). Perhaps, a predisposition to pedophilia was "unmasked" by brain disease or the patients' ability to recognize fitting sex partners was disabled. Similarly, pedophilia may follow the onset of cognitive impairment (Regestein and Reich, 1978).

Conclusions from Anomalous Sexuality in Patients

It is apparent from the described neuropsychiatric and pharmacoendocrinological cases that behavioral control over sexual behavior consists of at least three relatively autonomous and spatially segregated neuronal processes: (1) Sexual desire is mediated by upper brainstem and limbic structures such as hypothalamus, septum, and hippocampus. Dopamine and prolactin are involved in erectile function and sexual desire and have opposite behavioral and emotional effects, suggestive of functional antagonism. (2) The interpretation of stimuli as sexual and the sexual preference that results from this process is a temporal lobe function. (3) Behavioral inhibition is a prefrontal cortical function.

Disorders of all of these processes can in their own way lead to hyper- and hyposexual behavior. Another conclusion is that sexual behavioral patterns such as genital manipulation and rubbing and pelvic thrusting are already present in very young, sexually naive children. There appear to be suppressing neuronal functions that can become damaged in disease, thus releasing these behaviors, which then present as sexual stereotypes.

Conclusions

Most people alive today are capable of procreative sexuality because this capability is largely determined by the genome inherited from the parents. The same logic applies to sexual interest, sexual motivation, and sexual pleasure. In general, behaviors that increase the ability to conserve one's own genes (COG) are recognizable by pleasure.

Our thesis that the brain plays a decisive role in reproduction is confirmed by brain studies that indicate very prominent changes in cerebral metabolism during sexual behavior. An awkward vacuity in these studies is the infrequent use of subjective hedonic endpoints, which makes it relatively venturesome to implicate brain areas in sexual pleasure. Also, some of the likely areas are metabolically too minute or too constant to be detected with in vivo neuroimaging methods.

Figure 11.4 summarizes the main findings: Sexual partner selection and sexual preference, as well as recognizing sexual cues, appear to be functions of the temporal lobe, hypothalamus, and amygdala. Sexual arousal is mediated primarily by the hypothalamus and by downregulation of temporal lobe activity. Hedonic properties of sexual arousal are regulated by a network that includes at least ventral striatum and hypothalamus. Dopamine and gonadal hormones are crucial neuromodulators for establishing sexual interest and arousal.

When physical sexual activity commences by stimulation of the external genitalia, metabolic activity in the ventromedial and anterior temporal lobe gradually decreases until orgasm has been reached. The amygdala, which was activated during sexual appetite, becomes deactivated during sexual consumption. Recent and ongoing fMRI advances strongly suggest that the posterior ventral pallidum, a pleasure-related brain area, is also substantially activated during stimulation of the erect penis. Downregulation of the OFC is specific for orgasm and not for the preceding phase of sexual genital stimulation with high sexual arousal. Especially deactivation of the left mid-, and lat-OFC is a prime neuronal candidate for orgasmic bliss, as it most likely reflects decreased behavioral control and release of tension. The subcortical contribution to orgasm was inconclusive, but prime candidates for regulating orgasmic pleasure are the PAG and the hypothalamus. Opioids and oxytocin may be particularly important neuromodulators for orgasmic pleasure, whereas serotonin and prolactin released during orgasm promote the termination of sexual behavior by inhibiting promotivational brain systems.

A recent review mentioned the ventral striatum, ventral pallidum, the amygdala, and the OFC as prime areas where hedonic value is coded (Leknes and Tracey, 2008). As we have demonstrated in this chapter, all these regions are heavily involved in sexual behavior. In general one could summarize: if it is good for COG, it is pleasant.

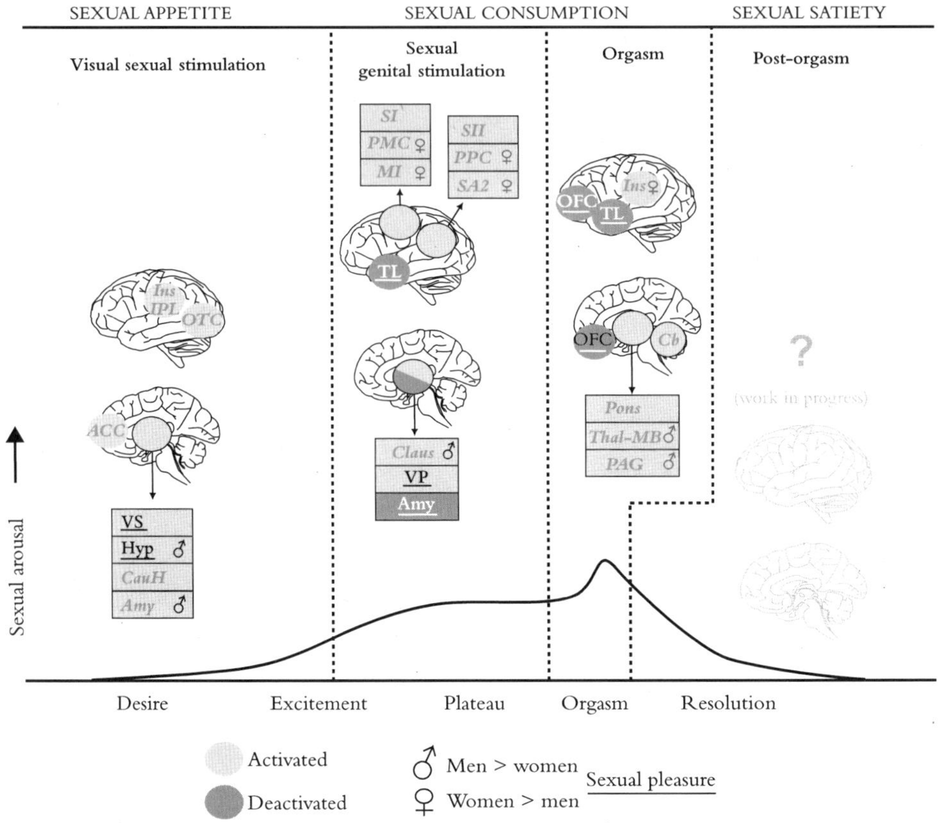

Figure 11.4 Conceptual overview of brain regions involved in different stages of the human sexual response based on neuroimaging evidence. The circles represent activated (light gray) and deactivated (dark gray) brain areas. The circle in the center of the bottom brain (mesial view) represents all subcortical areas. Brain regions that can be linked to sexual pleasure are depicted in bold and are underlined. Please also note the complete lack of knowledge regarding the human postorgasmic stage. This phase is currently being investigated by our research group. Abbreviations: ACC, anterior cingulate cortex; Amy, amygdala; Cb, cerebellum, anterior lobe; CauH, head of caudate; claus, claustrum; Hyp, hypothalamus; Ins, insula; IPL, inferior parietal lobule; L, left hemisphere; MI, primary motor cortex; OFC, orbitofrontal cortex; OTC, occipitotemporal cortex; PAG, periaqueductal gray matter; PMC, premotor cortex; PPC, posterior parietal cortex; rCBF, regional cerebral blood flow; SI, primary somatosensory cortex; SII, secondary somatosensory cortex; SA2, somatosensory area 2; Thal-Mb, thalamus-midbrain transition zone; TL, temporal lobe; VP, ventral pallidum; VS, ventral striatum; ♂, activated more in men than in women; ♀, activated more in women than in men.

References

Aksaray, G., Yelken, B., Kaptanoglu, C., Oflu, S. and Ozaltin, M. (2001) Sexuality in women with obsessive compulsive disorder. *J.Sex Marital Ther.* 27, 273–277.

Aldridge, S. A. (1982) Drug-induced sexual dysfunction. *Clin.Pharm.* 1, 141–147.

Allison, T., McCarthy, G., Luby, M., Puce, A. and Spencer, D. D. (1996) Localization of functional regions of human mesial cortex by somatosensory evoked potential recording and by cortical stimulation. *Electroencephalogr. Clin. Neurophysiol.* 100, 126–140.

Aloni, R. and Katz, S. (1999) A review of the effect of traumatic brain injury on the human sexual response. *Brain Inj.* 13, 269–280.

Amaral, D. G. (2002) The primate amygdala and the neurobiology of social behavior: Implications for understanding social anxiety. *Biol.Psychiatry* 51, 11–17.

Andersen, R. A., Snyder, L. H., Bradley, D. C. and Xing, J. (1997) Multimodal representation of space in the posterior parietal cortex and its use in planning movements. *Annu. Rev. Neurosci.* 20, 303–330.

Archer, J. S., Love-Geffen, T. E., Herbst-Damm, K. L., Swinney, D. A. and Chang, J. R. (2006) Effect of estradiol versus estradiol and testosterone on brain-activation patterns in postmenopausal women. *Menopause* 13, 528–537.

Arnow, B. A., Desmond, J. E., Banner, L. L., Glover, G. H., Solomon, A., Polan, M. L., Lue, T. F. and Atlas, S. W. (2002) Brain activation and sexual arousal in healthy, heterosexual males. *Brain* 125, 1014–1023.

Aron, A., Fisher, H., Mashek, D. J., Strong, G., Li, H. and Brown, L. L. (2005) Reward, motivation, and emotion systems associated with early-stage intense romantic love. *J. Neurophysiol.* 94, 327–337.

Baird, A. D., Wilson, S. J., Bladin, P. F., Saling, M. M. and Reutens, D. C. (2002) Hypersexuality after temporal lobe resection. *Epilepsy Behav.* 3, 173–181.

Barak, Y., Achiron, A., Elizur, A., Gabbay, U., Noy, S. and Sarova-Pinhas, I. (1996) Sexual dysfunction in relapsing-remitting multiple sclerosis: Magnetic resonance imaging, clinical, and psychological correlates. *J. Psychiatry Neurosci.* 21, 255–258.

Baron-Cohen, S., Knickmeyer, R. C. and Belmonte, M. K. (2005) Sex differences in the brain: Implications for explaining autism. *Science* 310, 819–823.

Bartels, A. and Zeki, S. (2000) The neural basis of romantic love. *Neuroreport* 11, 3829–3834.

Bartels, A. and Zeki, S. (2004) The neural correlates of maternal and romantic love. *Neuroimage* 21, 1155–1166.

Baum, M. J. and Everitt, B. J. (1992) Increased expression of c-fos in the medial preoptic area after mating in male rats: Role of afferent inputs from the medial amygdala and midbrain central tegmental field. *Neuroscience* 50, 627–646.

Beauregard, M., Levesque, J. and Bourgouin, P. (2001) Neural correlates of conscious self-regulation of emotion. *J. Neurosci.* 21, RC165.

Becerra, L., Harter, K., Gonzalez, R. G. and Borsook, D. (2006) Functional magnetic resonance imaging measures of the effects of morphine on central nervous system circuitry in opioid-naive healthy volunteers. *Anesth. Analg.* 103, 208–216.

Bergen, D., Daugherty, S. and Eckenfels, E. (1992) Reduction of sexual activities in females taking antiepileptic drugs. *Psychopathology* 25, 1–4.

Berglund, H., Lindstrom, P. and Savic, I. (2006) Brain response to putative pheromones in lesbian women. *Proc. NatlAcad. Sci .USA* 103, 8269–8274.

Berns, G. S. (2004) Something funny happened to reward. *Trends Cogn Sci.* 8, 193–194.

Berridge, K. C. (1996) Food reward: Brain substrates of wanting and liking. *Neurosci. Biobehav. Rev.* 20, 1–25.

Berridge, K. C. and Robinson, T. E. (1998) What is the role of dopamine in reward: Hedonic impact, reward learning, or incentive salience? *Brain Res. Brain Res. Rev.* 28, 309–369.

Bilgic, A., Gurkan, K. and Turkoglu, S. (2007) Excessive masturbation and hypersexual behavior associated with methylphenidate. *J. Am. Acad. Child Adolesc. Psychiatry* 46, 789–790.

Blok, B. F., Sturms, L. M. and Holstege, G. (1997) A PET study on cortical and subcortical control of pelvic floor musculature in women. *J. Comp. Neurol.* 389, 535–544.

Blood, A. J. and Zatorre, R. J. (2001) Intensely pleasurable responses to music correlate with activity in brain regions implicated in reward and emotion. *Proc. Natl Acad.Sci. USA* 98, 11818–11823.

Blood, A. J., Zatorre, R. J., Bermudez, P. and Evans, A. C. (1999) Emotional responses to pleasant and unpleasant music correlate with activity in paralimbic brain regions. *Nat. Neurosci.* 2, 382–387.

Bocher, M., Chisin, R., Parag, Y., Freedman, N., Meir Weil, Y., Lester, H., Mishani, E. and Bonne, O. (2001) Cerebral activation associated with sexual arousal in response to a pornographic clip: a 15O-H2O PET study in heterosexual men. *Neuroimage* 14, 105–117.

Bouloux, P. M. and Grossman, A. (1987) Hyperprolactinaemia and sexual function in the male. *Br. J. Hosp. Med.* 37, 503–510.

Bradley, W. E., Farrell, D. F. and Ojemann, G. A. (1998) Human cerebrocortical potentials evoked by stimulation of the dorsal nerve of the penis. *Somatosens. Mot. Res.* 15, 118–127.

Breiter, H. C., Gollub, R. L., Weisskoff, R. M., Kennedy, D. N., Makris, N., Berke, J. D., Goodman, J. M., Kantor, H. L., Gastfriend, D. R., Riorden, J. P., Mathew, R. T., Rosen, B. R. and Hyman, S. E. (1997) Acute effects of cocaine on human brain activity and emotion. *Neuron* 19, 591–611.

Burgdorf, J. and Panksepp, J. (2006) The neurobiology of positive emotions. *Neurosci. Biobehav. Rev.* 30, 173–187.

Burri, A., Heinrichs, M., Schedlowski, M. and Kruger, T. H. (2008) The acute effects of intranasal oxytocin administration on endocrine and sexual function in males. *Psychoneuroendocrinology* 33, 591–600.

Bush, G., Vogt, B. A., Holmes, J., Dale, A. M., Greve, D., Jenike, M. A. and Rosen, B. R. (2002) Dorsal anterior cingulate cortex: a role in reward-based decision making. *Proc. Natl Acad. Sci. USA* 99, 523–528.

Calleja, J., Carpizo, R. and Berciano, J. (1988) Orgasmic epilepsy. *Epilepsia* 29, 635–639.

Carmichael, M. S., Warburton, V. L., Dixen, J. and Davidson, J. M. (1994) Relationships among cardiovascular, muscular, and oxytocin responses during human sexual activity. *Arch. Sex Behav.* 23, 59–79.

Carmichael, S. T. and Price, J. L. (1995a) Limbic connections of the orbital and medial prefrontal cortex in macaque monkeys. *J. Comp. Neurol.* 363, 615–641.

Carmichael, S. T. and Price, J. L. (1995b) Sensory and premotor connections of the orbital and medial prefrontal cortex of macaque monkeys. *J. Comp. Neurol.* 363, 642–664.

Cavazos, J. E., Wang, C. J., Sitoh, Y. Y., Ng, S. E. and Tien, R. D. (1997) Anatomy and pathology of the septal region. *Neuroimaging Clin. N. Am.* 7, 67–78.

Clayton, A. H., Clavet, G. J., McGarvey, E. L., Warnock, J. K. and Weiss, K. (1999) Assessment of sexual functioning during the menstrual cycle. *J. Sex Marital Ther.* 25, 281–291.

Cohen, H. D., Rosen, R. C. and Goldstein, L. (1976) Electroencephalographic laterality changes during human sexual orgasm. *Arch. Sex Behav.* 5, 189–199.

Coolen, L. M., Peters, H. J. and Veening, J. G. (1996) Fos immunoreactivity in the rat brain following consummatory elements of sexual behavior: a sex comparison. *Brain Res.* 738, 67–82.

Coolen, L. M., Veening, J. G., Petersen, D. W. and Shipley, M. T. (2003) Parvocellular subparafascicular thalamic nucleus in the rat: Anatomical and functional compartmentalization. *J. Comp. Neurol.* 463, 117–131.

Corbetta, M., Miezin, F. M., Dobmeyer, S., Shulman, G. L.and Petersen, S. E. (1990) Attentional modulation of neural processing of shape, color, and velocity in humans. *Science* 248, 1556–1559.

Craig, A. D. (2002) How do you feel? Interoception: the sense of the physiological condition of the body. *Nat. Rev. Neurosci.* 3, 655–666.

Crick, F. C. and Koch, C. (2005) What is the function of the claustrum? *Philos. Trans. R. Soc. Lond. B. Biol. Sci.* 360, 1271–1279.

Critchley, H. D., Corfield, D. R., Chandler, M. P., Mathias, C. J. and Dolan, R. J. (2000) Cerebral correlates of autonomic cardiovascular arousal: a functional neuroimaging investigation in humans. *J. Physiol.* 523 Pt 1:259–70., 259–270.

Damasio, H., Grabowski, T., Frank, R., Galaburda, A. M. and Damasio, A. R. (1994) The return of Phineas Gage: Clues about the brain from the skull of a famous patient. *Science* 264, 1102–1105.

Davis, M. and Whalen, P. J. (2001) The amygdala: Vigilance and emotion. *Mol. Psychiatry* 6, 13–34.

Dawkins, R. (1976) *The Selfish Gene*. Oxford University Press: Oxford, UK.

De Leon, G. and Wexler, H. K. (1973) Heroin addiction: Its relation to sexual behavior and sexual experience. *J.Abnorm.Psychol.* 81, 36–38.

Devinsky, O., Morrell, M. J. and Vogt, B. A. (1995) Contributions of anterior cingulate cortex to behaviour. *Brain* 118 (Pt 1), 279–306.

Donnet, A., Schmitt, A., Poncet, M., Graziani, N. and Grisoli, F. (1997) [Hallucinations of supernumerary limbs, left hemineglect and hypersexuality in a case of right capsulo-lenticular hematoma]. *Rev.Neurol.(Paris)* 153, 587–590.

Dougherty, D. D., Shin, L. M., Alpert, N. M., Pitman, R. K., Orr, S. P., Lasko, M., Macklin, M. L., Fischman, A. J. and Rauch, S. L. (1999) Anger in healthy men: a PET study using script-driven imagery. *Biol.Psychiatry* 46, 466–472.

Exton, M. S., Bindert, A., Kruger, T., Scheller, F., Hartmann, U. and Schedlowski, M. (1999) Cardiovascular and endocrine alterations after masturbation-induced orgasm in women. *Psychosom. Med.* 61, 280–289.

Exton, M. S., Kruger, T. H., Koch, M., Paulson, E., Knapp, W., Hartmann, U. and Schedlowski, M. (2001) Coitus-induced orgasm stimulates prolactin secretion in healthy subjects. *Psychoneuroendocrinology* 26, 287–294.

Fadul, C. E., Stommel, E. W., Dragnev, K. H., Eskey, C. J. and Dalmau, J. O. (2005) Focal paraneoplastic limbic encephalitis presenting as orgasmic epilepsy. *J.Neurooncol.* 72, 195–198.

Ferretti, A., Caulo, M., Del Gratta, C., Di Matteo, R., Merla, A., Montorsi, F., Pizzella, V., Pompa, P., Rigatti, P., Rossini, P. M., Salonia, A., Tartaro, A. and Romani, G. L. (2005) Dynamics of male sexual arousal: Distinct components of brain activation revealed by fMRI. *Neuroimage.* 26, 1086–1096.

Foerster, O., Bumke, O. and Foerster, O. (1936) Sensible corticale felder. In: *Handbuch der Neurologie*. pp. 358–449. Eds. O. Bumke, O. Foerster. Springer: Berlin.

Friedman, D. P., Murray, E. A., O'Neill, J. B. and Mishkin, M. (1986) Cortical connections of the somatosensory fields of the lateral sulcus of macaques: Evidence for a corticolimbic pathway for touch. *J. Comp. Neurol.* 252, 323–347.

Frohman, E. M., Frohman, T. C. and Moreault, A. M. (2002) Acquired sexual paraphilia in patients with multiple sclerosis. *Arch. Neurol.* 59, 1006–1010.

Galbraith, R. A. (1991) Sexual side effects of drugs. *Drug Ther.(NY)* 21, 38–40, 45.

Gall, F. J. (1822) *Sur les fonctions du cerveau et sur celles de chacune de ses parties, avec des observations sur la possibilit, de reconnoitre les instincts, les penchans, les talens, ou les dispositions morales et intellectuelles des hommes et des animaux, par la configuration de leur cerveau et de leur t^te*. J.B. Balliere: Paris.

Gautier-Smith, P. C. (1980) [Cerebral dysfunction and disorders of sexual behaviour (author's translation)]. *Rev. Neurol.(Paris)* 136, 311–319.

Gazzola, V., Aziz-Zadeh, L. and Keysers, C. (2006) Empathy and the somatotopic auditory mirror system in humans. *Curr. Biol.* 16, 1824–1829.

Georgiadis, J. R. and Holstege, G. (2005) Human brain activation during sexual stimulation of the penis. *J. Comp. Neurol.* 493, 33–38.

Georgiadis, J. R., Reinders, A. A., van der Graaf, F. H., Paans, A. M. and Kortekaas, R. (2007) Brain activation during human male ejaculation revisited. *Neuroreport* 18, 553–557.

Georgiadis, J. R., Kortekaas, R., Kuipers, R., Nieuwenburg, A., Pruim, J., Reinders, A. A. T. S. and Holstege, G. (2006) Regional cerebral blood flow changes associated with clitorally induced orgasm in healthy women. *Eur J. Neurosci.* 24, 3305–3316.

Georgiadis, J. R., Farrell, M., Boessen, R., Denton, D. A., Gavrilescu, M., Kortekaas, R., Renken, R. J., Hoogduin, J. M. and Egan, G. F. (2008) The feasibility of PULSAR arterial spin labeling in the investigation of the male sexual response. OHBM Annual Meeting, Melbourne, Australia.

Georgiadis, J.R., Reinders, A.A., Paans, A.M., Renken, R., and Kortekaas, R. (2009) Men versus women on sexual brain function: Prominent differences during tactile genital stimulation, but not during orgasm. *Hum. Brain Mapp.* [Epub ahead of print] DOI 10.1002/hbm.20733

Gizewski, E. R., Krause, E., Karama, S., Baars, A., Senf, W. and Forsting, M. (2006) There are differences in cerebral activation between females in distinct menstrual phases during viewing of erotic stimuli: a fMRI study. *Exp. Brain Res.* 174, 101–108.

Gorman, D. G. and Cummings, J. L. (1992) Hypersexuality following septal injury. *Arch. Neurol.* 49, 308–310.

Graber, B., Rohrbaugh, J. W., Newlin, D. B., Varner, J. L. and Ellingson, R. J. (1985) EEG during masturbation and ejaculation. *Arch. Sex Behav.* 14, 491–503.

Guerit, J. M. and Opsomer, R. J. (1991) Bit-mapped imaging of somatosensory evoked potentials after stimulation of the posterior tibial nerves and dorsal nerve of the penis/clitoris. *Electroencephalogr. Clin. Neurophysiol.* 80, 228–237.

Gusnard, D. A. and Raichle, M. E. (2001) Searching for a baseline: Functional imaging and the resting human brain. *Nat. Rev. Neurosci.* 2, 685–694.

Gusnard, D. A., Akbudak, E., Shulman, G. L. and Raichle, M. E. (2001) Medial prefrontal cortex and self-referential mental activity: Relation to a default mode of brain function. *Proc. Natl Acad. Sci. USA* 98, 4259–4264.

Hadjikhani, N. and Roland, P. E. (1998) Cross-modal transfer of information between the tactile and the visual representations in the human brain: a positron emission tomographic study. *J. Neurosci.* 18, 1072–1084.

Hagemann, J. H., Berding, G., Bergh, S., Sleep, D. J., Knapp, W. H., Jonas, U. and Stief, C. G. (2003) Effects of visual sexual stimuli and apomorphine SL on cerebral activity in men with erectile dysfunction. *Eur.Urol.* 43, 412–420.

Halata, Z. and Munger, B. L. (1986) The neuroanatomical basis for the protopathic sensibility of the human glans penis. *Brain Res.* 371, 205–230.

Hamann, S., Herman, R. A., Nolan, C. L. and Wallen, K. (2004) Men and women differ in amygdala response to visual sexual stimuli. *Nat. Neurosci.* 7, 411–416.

Hariri, A. R., Bookheimer, S. Y. and Mazziotta, J. C. (2000) Modulating emotional responses: Effects of a neocortical network on the limbic system. *Neuroreport* 11, 43–48.

Heath, R. G. (1963) Electrical self-stimulation of the brain in man. *Am. J. Psychiatry* 120, 571–577.

Heath, R. G. (1972) Pleasure and brain activity in man. Deep and surface electroencephalograms during orgasm. *J. Nerv. Ment. Dis.* 154, 3–18.

Heath, R. G. (1977) Modulation of emotion with a brain pacemamer. Treatment for intractable psychiatric illness. *J. Nerv. Ment. Dis.* 165, 300–317.

Heaton, J. P. (2000) Central neuropharmacological agents and mechanisms in erectile dysfunction: the role of dopamine. *Neurosci. Biobehav. Rev.* 24, 561–569.

Hibbard, M. R., Gordon, W. A., Flanagan, S., Haddad, L. and Labinsky, E. (2000) Sexual dysfunction after traumatic brain injury. *NeuroRehabilitation* 15, 107–120.

Hite, S. (1976) *The Hite Report: A Nationwide Survey of Female Sexuality.* MacMillan: New York.

Holstege, G., Bandler, R. and Saper, C. (1996) The emotional motor system. *Prog. Brain Res.* 107, 3–6.

Holstege, G., Georgiadis, J. R., Paans, A. M., Meiners, L. C., van der Graaf, F. H. and Reinders, A. A. (2003) Brain activation during human male ejaculation. *J.Neurosci.* 23, 9185–9193.

Horster, W., Rivers, A., Schuster, B., Ettlinger, G., Skreczek, W. and Hesse, W. (1989) The neural structures involved in cross-modal recognition and tactile discrimination performance: an investigation using 2-DG. *Behav. Brain Res.* 33, 209–227.

Ishai, A., Ungerleider, L. G. and Haxby, J. V. (2000) Distributed neural systems for the generation of visual images. *Neuron* 28, 979–990.

Janszky, J., Ebner, A., Szupera, Z., Schulz, R., Hollo, A., Szucs, A. and Clemens, B. (2004) Orgasmic aura—a report of seven cases. *Seizure* 13, 441–444.

Jayaram, M. B., Hosalli, P. M. and Stroup, T. S. (2007) Risperidone versus olanzapine for treatment of schizophrenia. *Schizophr. Bull.* 33, 1274–1276.

Jeong, G. W., Park, K., Youn, G., Kang, H. K., Kim, H. J., Seo, J. J. and Ryu, S. B. (2005) Assessment of cerebrocortical regions associated with sexual arousal in premenopausal and menopausal women by using BOLD-based functional MRI. *J. Sex Med.* 2, 645–651.

Johnson, R. D. and Halata, Z. (1991) Topography and ultrastructure of sensory nerve endings in the glans penis of the rat. *J. Comp. Neurol.* 312, 299–310.

Kaplan, H. S. (1974) *The New Sex Therapy: Active Treatment of Sexual Dysfunctions.* Routledge: New York.

Karama, S., Lecours, A. R., Leroux, J. M., Bourgouin, P., Beaudoin, G., Joubert, S. and Beauregard, M. (2002) Areas of brain activation in males and females during viewing of erotic film excerpts. *Hum. Brain Mapp.* 16, 1–13.

Kell, C. A., von Kriegstein, K., Rosler, A., Kleinschmidt, A. and Laufs, H. (2005) The sensory cortical representation of the human penis: Revisiting somatotopy in the male homunculus. *J. Neurosci.* 25, 5984–5987.

Kim, S. W., Sohn, D. W., Cho, Y. H., Yang, W. S., Lee, K. U., Juh, R., Ahn, K. J., Chung, Y. A., Han, S. I., Lee, K. H., Lee, C. U. and Chae, J. H. (2006) Brain activation by visual erotic stimuli in healthy middle aged males. *Int. J. Impot. Res.* 18(5), 452–457.

Kollack-Walker, S. and Newman, S. W. (1997) Mating-induced expression of c-fos in the male Syrian hamster brain: Role of experience, pheromones, and ejaculations. *J. Neurobiol.* 32, 481–501.

Komisaruk, B. R., Whipple, B., Crawford, A., Liu, W. C., Kalnin, A. and Mosier, K. (2004) Brain activation during vaginocervical self-stimulation and orgasm in women with complete spinal cord injury: fMRI evidence of mediation by the vagus nerves. *Brain Res.* 1024, 77–88.

Kondo, H., Saleem, K. S. and Price, J. L. (2003) Differential connections of the temporal pole with the orbital and medial prefrontal networks in macaque monkeys. *J. Comp. Neurol.* 465, 499–523.

Kondo, H., Saleem, K. S., and Price, J. L. (2005) Differential connections of the perirhinal and parahippocampal cortex with the orbital and medial prefrontal networks in macaque monkeys. *J. Comp. Neurol.* 493, 479–509.

Kringelbach, M. L. (2005) The human orbitofrontal cortex: Linking reward to hedonic experience. *Nat. Rev. Neurosci.* 6, 691–702.

Kringelbach, M. L. and Rolls, E. T. (2004) The functional neuroanatomy of the human orbitofrontal cortex: Evidence from neuroimaging and neuropsychology. *Prog. Neurobiol.* 72, 341–372.

Kringelbach, M. L., O'Doherty, J., Rolls, E. T. and Andrews, C. (2003) Activation of the human orbitofrontal cortex to a liquid food stimulus is correlated with its subjective pleasantness. *Cereb. Cortex* 13, 1064–1071.

Krüger, T., Exton, M. S., Pawlak, C., zur Muhlen, A., Hartmann, U. and Schedlowski, M. (1998) Neuroendocrine and cardiovascular response to sexual arousal and orgasm in men. *Psychoneuroendocrinology* 23, 401–411.

Krüger, T. H., Hartmann, U. and Schedlowski, M. (2005) Prolactinergic and dopaminergic mechanisms underlying sexual arousal and orgasm in humans. *World J. Urol.* 23, 130–138.

Krüger, T. H., Schiffer, B., Eikermann, M., Haake, P., Gizewski, E. and Schedlowski, M. (2006) Serial neurochemical measurement of cerebrospinal fluid during the human sexual response cycle. *Eur. J. Neurosci.* 24, 3445–3452.

Krüger, T. H., Haake, P., Chereath, D., Knapp, W., Janssen, O. E., Exton, M. S., Schedlowski, M. and Hartmann, U. (2003) Specificity of the neuroendocrine response to orgasm during sexual arousal in men. *J. Endocrinol.* 177, 57–64.

Kulkarni, B., Bentley, D. E., Elliott, R., Youell, P., Watson, A., Derbyshire, S. W., Frackowiak, R. S., Friston, K. J. and Jones, A. K. (2005) Attention to pain localization and unpleasantness discriminates the functions of the medial and lateral pain systems. *Eur. J. Neurosci.* 21, 3133–3142.

Lal, S., Ackman, D., Thavundayil, J. X., Kiely, M. E. and Etienne, P. (1984) Effect of apomorphine, a dopamine receptor agonist, on penile tumescence in normal subjects. *Prog. Neuropsychopharmacol. Biol. Psychiatry* 8, 695–699.

Lane, R. D., Reiman, E. M., Bradley, M. M., Lang, P. J., Ahern, G. L., Davidson, R. J. and Schwartz, G. E. (1997) Neuroanatomical correlates of pleasant and unpleasant emotion. *Neuropsychologia* 35, 1437–1444.

Lang, P. J., Bradley, M. M., Fitzsimmons, J. R., Cuthbert, B. N., Scott, J. D., Moulder, B. and Nangia, V. (1998) Emotional arousal and activation of the visual cortex: an fMRI analysis. *Psychophysiology* 35, 199–210.

Lapierre, D., Braun, C. M. and Hodgins, S. (1995) Ventral frontal deficits in psychopathy: Neuropsychological test findings. *Neuropsychologia* 33, 139–151.

Lau, B. and Glimcher, P. W. (2007) Action and outcome encoding in the primate caudate nucleus. *J. Neurosci.* 27, 14502–14514.

Lauerma, H. (1996) Successful treatment of citalopram-induced anorgasmia by cyproheptadine. *Acta Psychiatr. Scand.* 93, 69–70.

Laumann, E. O., Gagnon, J. H., Michael, R. T. and Michaels, S. (1994) *The Social Organization of Sexuality: Sexual Practices in The United States*. The University of Chicago Press: Chicago, IL.

Leknes, S. and Tracey, I. (2008) A common neurobiology for pain and pleasure. *Nat. Rev. Neurosci.* 9, 314–320.

Letourneau, E. J., Schewe, P. A. and Frueh, B. C. (1997) Preliminary evaluation of sexual problems in combat veterans with PTSD. *J. Trauma Stress.* 10, 125–132.

Lloyd, E. A. (2005) *The Case of the Female Orgasm: Bias in the Science of Evolution*. Harvard University Press: Cambridge, MA.

Lorrain, D. S., Riolo, J. V., Matuszewich, L. and Hull, E. M. (1999) Lateral hypothalamic serotonin inhibits nucleus accumbens dopamine: Implications for sexual satiety. *J. Neurosci.* 19, 7648–7652.

Loyd, D. R., Morgan, M. M. and Murphy, A. Z. (2007) Morphine preferentially activates the periaqueductal gray-rostral ventromedial medullary pathway in the male rat: a potential mechanism for sex differences in antinociception. *Neuroscience* 147, 456–468.

Mah, K. and Binik, Y. M. (2001) The nature of human orgasm: a critical review of major trends. *Clin. Psychol. Rev.* 21, 823–856.

Mäkelä, J. P., Illman, M., Jousmaki, V., Numminen, J., Lehecka, M., Salenius, S., Forss, N. and Hari, R. (2003) Dorsal penile nerve stimulation elicits left-hemisphere dominant activation in the second somatosensory cortex. *Hum. Brain Mapp.* 18, 90–99.

Makhlouf, A., Kparker, A. and Niederberger, C. S. (2007) Depression and erectile dysfunction. *Urol. Clin.North Am.* 34, 565–574.

Marson, L. and Murphy, A. Z. (2006) Identification of neural circuits involved in female genital responses in the rat: a dual virus and anterograde tracing study. *Am. J. Physiol Regul. Integr. Comp. Physiol.* 291, R419-R428.

Marson, L., Platt, K. B. and McKenna, K. E. (1993) Central nervous system innervation of the penis as revealed by the transneuronal transport of pseudorabies virus. *Neuroscience* 55, 263–280.

Martin-Du, P. R. (1978) [Neuroleptics and sexual dysfunction in man. Neuroendocrine aspects]. *Schweiz. Arch. Neurol. Neurochir. Psychiatr.* 122, 285–313.

Masters, W. H. and Johnson, V. E. (1966) *Human Sexual Response*. Little, Brown and Company: Boston, MA.

McKenna, K. (1999) The brain is the master organ in sexual function: Central nervous system control of male and female sexual function. *Int. J. Impot.Res.* 11 (Suppl 1), S48–55., S48-S55.

Mendez, M. F., Chow, T., Ringman, J., Twitchell, G. and Hinkin, C. H. (2000) Pedophilia and temporal lobe disturbances. *J. Neuropsychiatry Clin. Neurosci.* 12, 71–76.

Middleton, F. A. and Strick, P. L. (1994) Anatomical evidence for cerebellar and basal ganglia involvement in higher cognitive function. *Science* 266, 458–461.

Miller, B. L., Cummings, J. L., McIntyre, H., Ebers, G. and Grode, M. (1986) Hypersexuality or altered sexual preference following brain injury. *J. Neurol. Neurosurg. Psychiatry* 49, 867–873.

Miller, J. M., Vorel, S. R., Tranguch, A. J., Kenny, E. T., Mazzoni, P., van Gorp, W. G. and Kleber, H. D. (2006) Anhedonia after a selective bilateral lesion of the globus pallidus. *Am. J. Psychiatry* 163, 786–788.

Mintz, J., O'Hare, K., O'Brien, C. P. and Goldschmidt, J. (1974) Sexual problems of heroin addicts. *Arch. Gen. Psychiatry* 31, 700–703.

Mirin, S. M., Meyer, R. E., Mendelson, J. H. and Ellingboe, J. (1980) Opiate use and sexual function. *Am. J. Psychiatry* 137, 909–915.

Miyagawa, Y., Tsujimura, A., Fujita, K., Matsuoka, Y., Takahashi, T., Takao, T., Takada, S., Matsumiya, K., Osaki, Y., Takasawa, M., Oku, N., Hatazawa, J., Kaneko, S. and Okuyama, A. (2007) Differential brain processing of audiovisual sexual stimuli in men: Comparative positron emission tomography study of the initiation and maintenance of penile erection during sexual arousal. *Neuroimage* 36, 830–842.

Montorsi, F., Perani, D., Anchisi, D., Salonia, A., Scifo, P., Rigiroli, P., Deho, F., De Vito, M. L., Heaton, J., Rigatti, P. and Fazio, F. (2003) Brain activation patterns during video sexual stimulation following the administration of apomorphine: Results of a placebo-controlled study. *Eur. Urol.* 43, 405–411.

Mosovich, A. and Tallaferro, A. (1954) Studies on EEG and sex function orgasm. *Dis. Nerv. Syst.* 15, 218–220.

Moulier, V., Mouras, H., Pelegrini-Issac, M., Glutron, D., Rouxel, R., Grandjean, B., Bittoun, J. and Stoleru, S. (2006) Neuroanatomical correlates of penile erection evoked by photographic stimuli in human males. *Neuroimage* 33, 689–699.

Murphy, M. R., Checkley, S. A., Seckl, J. R. and Lightman, S. L. (1990) Naloxone inhibits oxytocin release at orgasm in man. *J. Clin. Endocrinol. Metab.* 71, 1056–1058.

Murphy, M. R., Seckl, J. R., Burton, S., Checkley, S. A. and Lightman, S. L. (1987) Changes in oxytocin and vasopressin secretion during sexual activity in men. *J. Clin. Endocrinol. Metab.* 65, 738–741.

Mutarelli, E. G., Omuro, A. M. and Adoni, T. (2006) Hypersexuality following bilateral thalamic infarction: Case report. *Arq. Neuropsiquiatr.* 64, 146–148.

Nagaratnam, N. and Gayagay, G., Jr. (2002) Hypersexuality in nursing care facilities—a descriptive study. *Arch. Gerontol.Geriatr.* 35, 195–203.

Naghavi, H. R., Eriksson, J., Larsson, A. and Nyberg, L. (2007) The claustrum/insula region integrates conceptually related sounds and pictures. *Neurosci. Lett.* 422, 77–80.

Nakagawa, H., Namima, T., Aizawa, M., Uchi, K., Kaiho, Y., Yoshikawa, K., Orikasa, S. and Nakasato, N. (1998) Somatosensory evoked magnetic fields elicited by dorsal penile, posterior tibial and median nerve stimulation. *Electroencephalogr. Clin. Neurophysiol.* 108, 57–61.

Narkiewicz, O. (1964) Degenerations in the claustrum after regional neocortical ablations in the cat. *J. Comp. Neurol.* 123, 335–355.

Newman, S. W. (1999) The medial extended amygdala in male reproductive behavior. A node in the mammalian social behavior network. *Ann. NY Acad. Sci.* 877, 242–257.

Northoff, G., Heinzel, A., de, G. M., Bermpohl, F., Dobrowolny, H. and Panksepp, J. (2006) Self-referential processing in our brain—a meta-analysis of imaging studies on the self. *Neuroimage* 31, 440–457.

O'Sullivan, J. D. and Hughes, A. J. (1998) Apomorphine-induced penile erections in Parkinson's disease. *Mov Disord.* 13, 536–539.

Olson, C. R. and Graybiel, A. M. (1980) Sensory maps in the claustrum of the cat. *Nature* 288, 479–481.

Öngür, D., Ferry, A. T. and Price, J. L. (2003) Architectonic subdivision of the human orbital and medial prefrontal cortex. *J. Comp. Neurol.* 460, 425–449.

Ortigue, S., Grafton, S. T. and Bianchi-Demicheli, F. (2007) Correlation between insula activation and self-reported quality of orgasm in women. *Neuroimage.* 37, 551–560.

Paredes, R. G. and Agmo, A. (2004) Has dopamine a physiological role in the control of sexual behavior? A critical review of the evidence. *Prog. Neurobiol.* 73, 179–226.

Parfitt, D. B. and Newman, S. W. (1998) Fos-immunoreactivity within the extended amygdala is correlated with the onset of sexual satiety. *Horm. Behav.* 34, 17–29.

Park, K., Kang, H. K., Seo, J. J., Kim, H. J., Ryu, S. B. and Jeong, G. W. (2001a) Blood-oxygenation-level-dependent functional magnetic resonance imaging for evaluating cerebral regions of female sexual arousal response. *Urology* 57, 1189–1194.

Park, K., Seo, J. J., Kang, H. K., Ryu, S. B., Kim, H. J. and Jeong, G. W. (2001b) A new potential of blood oxygenation level dependent (BOLD) functional MRI for evaluating cerebral centers of penile erection. *Int. J. Impot. Res.* 13, 73–81.

Parsons, L.M., Egan, G., Liotti, M. Brannan, S., Denton, D., Shade, R., Robillard, R., Madden, L., Abplanalp, B., and Fox, P.T. (2001) Neuroimaging evidence implicating cerebellum in the experience of hypercapnia and hunger for air. *Proc. Natl Acad. Sci. USA* 98, 2041–2046.

Passie, T., Hartmann, U., Schneider, U., Emrich, H. M. and Kruger, T. H. (2005) Ecstasy (MDMA) mimics the post-orgasmic state: Impairment of sexual drive and function during acute MDMA-effects may be due to increased prolactin secretion. *Med. Hypotheses* 64, 899–903.

Paul, T., Schiffer, B., Zwarg, T., Kruger, T. H., Karama, S., Schedlowski, M., Forsting, M. and Gizewski, E. R. (2007) Brain response to visual sexual stimuli in heterosexual and homosexual males. *Hum. Brain Mapp.* 29, 726–735.

Pearson, R. C., Brodal, P., Gatter, K. C. and Powell, T. P. (1982) The organization of the connections between the cortex and the claustrum in the monkey. *Brain Res.* 234, 435–441.

Penfield, W. and Rasmussen, T. (1950) *The Cerebral Cortex of Man. A Clinical Study of Localization of Function.* Macmillan: New York.

Petrovic, P., Kalso, E., Petersson, K. M. and Ingvar, M. (2002) Placebo and opioid analgesia—Imaging a shared neuronal network. *Science* 295, 1737–1740.

Pfeifer, B. (1920) Die Storungen der Sensibilitat im Gebiete der Genito-Analhaut bei Hirnverletzten. *Zeitschrift der gesamten Neurologie und Psychiatrie* 53, 5–38.

Ploner, M., Gross, J., Timmermann, L. and Schnitzler, A. (2002) Cortical representation of first and second pain sensation in humans. *Proc. Natl. Acad. Sci.USA* 99, 12444–12448.

Pradhan, S., Singh, M. N. and Pandey, N. (1998) Kluver Bucy syndrome in young children. *Clin. Neurol. Neurosurg.* 100, 254–258.

Pukall, C. F., Strigo, I. A., Binik, Y. M., Amsel, R., Khalife, S. and Bushnell, M. C. (2005) Neural correlates of painful genital touch in women with vulvar vestibulitis syndrome. *Pain* 115, 118–127.

Raichle, M. E. and Mintun, M. A. (2006) Brain work and brain imaging. *Annu. Rev. Neurosci.* 29, 449–476.

Rauch, S. L., Whalen, P. J., Shin, L. M., McInerney, S. C., Macklin, M. L., Lasko, N. B., Orr, S. P. and Pitman, R. K. (2000) Exaggerated amygdala response to masked facial stimuli in posttraumatic stress disorder: a functional MRI study. *Biol. Psychiatry* 47, 769–776.

Rauch, S. L., Shin, L. M., Dougherty, D. D., Alpert, N. M., Orr, S. P., Lasko, M., Macklin, M. L., Fischman, A. J. and Pitman, R. K. (1999) Neural activation during sexual and competitive arousal in healthy men. *Psychiatry Res.* 91, 1–10.

Redouté, J., Stoleru, S., Gregoire, M. C., Costes, N., Cinotti, L., Lavenne, F., Le Bars, D., Forest, M. G. and Pujol, J. F. (2000) Brain processing of visual sexual stimuli in human males. *Hum.Brain Mapp.* 11, 162–177.

Redouté, J., Stoleru, S., Pugeat, M., Costes, N., Lavenne, F., Le Bars, D., Dechaud, H., Cinotti, L. and Pujol, J. F. (2005) Brain processing of visual sexual stimuli in treated and untreated hypogonadal patients. *Psychoneuroendocrinology* 30, 461–482.

Rees, P. M., Fowler, C. J. and Maas, C. P. (2007) Sexual function in men and women with neurological disorders. *Lancet* 369, 512–525.

Regestein, Q. R. and Reich, P. (1978) Pedophilia occurring after onset of cognitive impairment. *J. Nerv. Ment. Dis.* 166, 794–798.

Remillard, G. M., Andermann, F., Testa, G. F., Gloor, P., Aube, M., Martin, J. B., Feindel, W., Guberman, A. and Simpson, C. (1983) Sexual ictal manifestations predominate in women with temporal lobe epilepsy: a finding suggesting sexual dimorphism in the human brain. *Neurology* 33, 323–330.

Risinger, R. C., Salmeron, B. J., Ross, T. J., Amen, S. L., Sanfilipo, M., Hoffmann, R. G., Bloom, A. S., Garavan, H. and Stein, E. A. (2005) Neural correlates of high and craving during cocaine self-administration using BOLD fMRI. *Neuroimage* 26, 1097–1108.

Rizzolatti, G. and Craighero, L. (2004) The mirror-neuron system. *Annu. Rev. Neurosci.* 27, 169–192.

Roane, D. M., Yu, M., Feinberg, T. E. and Rogers, J. D. (2002) Hypersexuality after pallidal surgery in Parkinson disease. *Neuropsychiatry Neuropsychol. Behav. Neurol.* 15, 247–251.

Roland, P. E. and Gulyas, B. (1995) Visual memory, visual imagery, and visual recognition of large field patterns by the human brain: Functional anatomy by positron emission tomography. *Cereb. Cortex* 5, 79–93.

Rolls, E. T. (2000) The orbitofrontal cortex and reward. *Cereb. Cortex* 10, 284–294.

Romito, L. M., Raja, M., Daniele, A., Contarino, M. F., Bentivoglio, A. R., Barbier, A., Scerrati, M. and Albanese, A. (2002) Transient mania with hypersexuality after surgery for high frequency stimulation of the subthalamic nucleus in Parkinson's disease. *Mov. Disord.* 17, 1371–1374.

Rowland, D. L., Myers, L., Culver, A. and Davidson, J. M. (1997) Bupropion and sexual function: a placebo-controlled prospective study on diabetic men with erectile dysfunction. *J. Clin. Psychopharmacol* 17, 350–357.

Safron, A., Barch, B., Bailey, J. M., Gitelman, D. R., Parrish, T. B. and Reber, P. J. (2007) Neural correlates of sexual arousal in homosexual and heterosexual men. *Behav. Neurosci.* 121, 237–248.

Sakuma, Y. and Pfaff, D. W. (1979a) Facilitation of female reproductive behavior from mesensephalic central gray in the rat. *Am. J. Physiol* 237, R278–R284.

Sakuma, Y. and Pfaff, D. W. (1979b) Mesencephalic mechanisms for integration of female reproductive behavior in the rat. *Am. J. Physiol* 237, R285–R290.

Sandel, M. E., Williams, K. S., Dellapietra, L. and Derogatis, L. R. (1996) Sexual functioning following traumatic brain injury. *Brain Inj.* 10, 719–728.

Saper, C. B. (2002) The central autonomic nervous system: Conscious visceral perception and autonomic pattern generation. *Annu. Rev. Neurosci.* 25, 433–469.

Savic, I., Berglund, H. and Lindstrom, P. (2005) Brain response to putative pheromones in homosexual men. *Proc. Natl. Acad. Sci.USA* 102, 7356–7361.

Saxena, S. and Rauch, S. L. (2000) Functional neuroimaging and the neuroanatomy of obsessive-compulsive disorder. *Psychiatr. Clin. North Am.* 23, 563–586.

Schiffer, B., Krueger, T., Paul, T., de, G. A., Forsting, M., Leygraf, N., Schedlowski, M. and Gizewski, E. (2008) Brain response to visual sexual stimuli in homosexual pedophiles. *J.Psychiatry Neurosci.* 33, 23–33.

Schmahmann, J. D. (2004) Disorders of the cerebellum: Ataxia, dysmetria of thought, and the cerebellar cognitive affective syndrome. *J. Neuropsychiatry Clin. Neurosci.* 16, 367–378.

Schmahmann, J. D. and Sherman, J. C. (1998) The cerebellar cognitive affective syndrome. *Brain* 121, 561–579.

Schultz, W., Tremblay, L. and Hollerman, J. R. (2000) Reward processing in primate orbitofrontal cortex and basal ganglia. *Cereb. Cortex* 10, 272–284.

Seecof, R. and Tennant, F. S., Jr. (1986) Subjective perceptions to the intravenous "rush" of heroin and cocaine in opioid addicts. *Am. J. Drug Alcohol Abuse* 12, 79–87.

Segraves, R. T., Bari, M., Segraves, K. and Spirnak, P. (1991) Effect of apomorphine on penile tumescence in men with psychogenic impotence. *J. Urol.* 145, 1174–1175.

Seidman, S. (2006) Ejaculatory dysfunction and depression: Pharmacological and psychobiological interactions. *Int. J. Impot. Res.* 18 Suppl 1, S33–S38.

Sell, L. A., Morris, J., Bearn, J., Frackowiak, R. S., Friston, K. J. and Dolan, R. J. (1999) Activation of reward circuitry in human opiate addicts. *Eur. J. Neurosci.* 11, 1042–1048.

Seseke, S., Baudewig, J., Kallenberg, K., Ringert, R. H., Seseke, F. and Dechent, P. (2006) Voluntary pelvic floor muscle control-an fMRI study. *Neuroimage* 31, 1399–1407.

Shin, L. M., Dougherty, D. D., Orr, S. P., Pitman, R. K., Lasko, M., Macklin, M. L., Alpert, N. M., Fischman, A. J. and Rauch, S. L. (2000) Activation of anterior paralimbic structures during guilt-related script-driven imagery. *Biol. Psychiatry* 48, 43–50.

Small, D. M., Zatorre, R. J., Dagher, A., Evans, A. C. and Jones-Gotman, M. (2001) Changes in brain activity related to eating chocolate: From pleasure to aversion. *Brain* 124, 1720–1733.

Smith, A. M., Lewis, B. K., Ruttimann, U. E., Ye, F. Q., Sinnwell, T. M., Yang, Y., Duyn, J. H. and Frank, J. A. (1999) Investigation of low frequency drift in fMRI signal. *Neuroimage* 9, 526–533.

Smith, C. G. (1982) Drug effects on male sexual function. *Clin. Obstet. Gynecol.* 25, 525–531.

Snider, R. S. (1950) Recent contributions to the anatomy and physiology of the cerebellum. *Arch. Neurol. Psychiatry* 64, 196–219.

Snider, R. S. and Maiti, A. (1976) Cerebellar contributions to the Papez circuit. *J. Neurosci. Res.* 2, 133–146.

Sonda, L. P., Mazo, R. and Chancellor, M. B. (1990) The role of yohimbine for the treatment of erectile impotence. *J. Sex Marital Ther.* 16, 15–21.

Spinella, M. (2004) Hypersexuality and dysexecutive syndrome after a thalamic infarct. *Int. J. Neurosci.* 114, 1581–1590.

Stark, R., Schienle, A., Girod, C., Walter, B., Kirsch, P., Blecker, C., Ott, U., Schafer, A., Sammer, G., Zimmermann, M. and Vaitl, D. (2005) Erotic and disgust-inducing pictures—differences in the hemodynamic responses of the brain. *Biol. Psychol.* 70, 19–29.

Stoeckart, R., Slob, A. K., Moors-Mommers, M. A. C. T., Slob, A. K., Vink, C. W., Moors, J. P. C. and Everaerd, W. (2000) Biologische aspecten van de seksuele respons. In: *Leerboek Seksuologie*. Bohn Stafleu Van Loghum: Houten, Netherlands.

Stoleru, S., Redouté, J., Costes, N., Lavenne, F., Bars, D. L., Dechaud, H., Forest, M. G., Pugeat, M., Cinotti, L. and Pujol, J. F. (2003) Brain processing of visual sexual stimuli in men with hypoactive sexual desire disorder. *Psychiatry Res.* 124, 67–86.

Stoleru, S., Gregoire, M. C., Gerard, D., Decety, J., Lafarge, E., Cinotti, L., Lavenne, F., Le Bars, D., Vernet-Maury, E., Rada, H., Collet, C., Mazoyer, B., Forest, M. G., Magnin, F., Spira, A. and Comar, D. (1999) Neuroanatomical correlates of visually evoked sexual arousal in human males. *Arch. Sex Behav.* 28, 1–21.

Stone, J., Smith, L., Watt, K., Barron, L. and Zeman, A. (2001) Incoordinated thought and emotion in spinocerebellar ataxia type 8. *J. Neurol.* 248, 229–232.

Szechtman, H., Hershkowitz, M. and Simantov, R. (1981) Sexual behavior decreases pain sensitivity and stimulated endogenous opioids in male rats. *Eur. J. Pharmacol.* 70, 279–285.

Tiihonen, J., Kuikka, J., Kupila, J., Partanen, K., Vainio, P., Airaksinen, J., Eronen, M., Hallikainen, T., Paanila, J. and Kinnunen, I. (1994) Increase in cerebral blood flow of right prefrontal cortex in man during orgasm. *Neurosci. Lett.* 170, 241–243.

Tricomi, E. and Fiez, J. A. (2008) Feedback signals in the caudate reflect goal achievement on a declarative memory task. *Neuroimage* epub ahead of print.

Tsujimura, A., Miyagawa, Y., Fujita, K., Matsuoka, Y., Takahashi, T., Takao, T., Matsumiya, K., Osaki, Y., Takasawa, M., Oku, N., Hatazawa, J., Kaneko, S. and Okuyama, A. (2006) Brain processing of audiovisual sexual stimuli inducing penile erection: a positron emission tomography study. *J. Urol.* 176, 679–683.

Uitti, R. J., Tanner, C. M., Rajput, A. H., Goetz, C. G., Klawans, H. L. and Thiessen, B. (1989) Hypersexuality with antiparkinsonian therapy. *Clin. Neuropharmacol.* 12, 375–383.

van Netten, J. J., Georgiadis, J. R., Nieuwenburg, A. and Kortekaas, R. (2008) 8–13 Hz fluctuations in rectal pressure are an objective marker of clitorally-induced orgasm in women. *Arch. Sex Behav.* 37, 279–285.

Vance, E. B. and Wagner, N. N. (1976) Written descriptions of orgasm: a study of sex differences. *Arch. Sex Behav.* 5, 87–98.

Veening, J. G., Swanson, L. W., Cowan, W. M., Nieuwenhuys, R. and Geeraedts, L. M. (1982) The medial forebrain bundle of the rat. II. An autoradiographic study of the topography of the major descending and ascending components. *J. Comp. Neurol.* 206, 82–108.

Vitezic, D. and Pelcic, J. M. (2002) Erectile dysfunction: Oral pharmacotherapy options. *Int. J. Clin. Pharmacol. Ther.* 40, 393–403.

Vogt, B. A. (2005) Pain and emotion interactions in subregions of the cingulate gyrus. *Nat. Rev. Neurosci.* 6, 533–544.

Vogt, B. A., Derbyshire, S. and Jones, A. K. (1996) Pain processing in four regions of human cingulate cortex localized with co-registered PET and MR imaging. *Eur. J. Neurosci.* 8, 1461–1473.

Volkow, N. D., Wang, G. J., Fowler, J. S., Gatley, S. J., Ding, Y. S., Logan, J., Dewey, S. L., Hitzemann, R. and Lieberman, J. (1996) Relationship between psychostimulant-induced "high" and dopamine transporter occupancy. *Proc. Natl Acad. Sci.USA* 93, 10388–10392.

Volkow, N. D., Wang, G. J., Telang, F., Fowler, J. S., Logan, J., Jayne, M., Ma, Y., Pradhan, K. and Wong, C. (2007) Profound decreases in dopamine release in striatum in detoxified alcoholics: Possible orbitofrontal involvement. *J. Neurosci.* 27, 12700–12706.

Waldinger, M. D., Zwinderman, A. H. and Olivier, B. (2001) Antidepressants and ejaculation: a double-blind, randomized, placebo-controlled, fixed-dose study with paroxetine, sertraline, and nefazodone. *J. Clin. Psychopharmacol.* 21, 293–297.

Waldinger, M. D., Zwinderman, A. H. and Olivier, B. (2003) Antidepressants and ejaculation: a double-blind, randomized, fixed-dose study with mirtazapine and paroxetine. *J. Clin. Psychopharmacol.* 23, 467–470.

Walter, M., Stadler, J., Tempelmann, C., Speck, O. and Northoff, G. (2008) High resolution fMRI of subcortical regions during visual erotic stimulation at 7 T. *Mag. Resonance Mater. Phys. Biol. Med.* 21, 103–111.

Walter, M., Witzel, J., Wiebking, C., Gubka, U., Rotte, M., Schiltz, K., Bermpohl, F., Tempelmann, C., Bogerts, B., Heinze, H. J. and Northoff, G. (2007) Pedophilia is linked to reduced activation in hypothalamus and lateral prefrontal cortex during visual erotic stimulation. *Biol. Psychiatry* 62, 698–701.

Wang, J., Aguirre, G. K., Kimberg, D. Y., Roc, A. C., Li, L. and Detre, J. A. (2003) Arterial spin labeling perfusion fMRI with very low task frequency. *Magn Reson. Med.* 49, 796–802.

Weiss, R. J. (1991) Effects of antihypertensive agents on sexual function. *Am. Fam. Physician* 44, 2075–2082.

Yang, J. C. (2004) Functional neuroanatomy in depressed patients with sexual dysfunction: Blood oxygenation level dependent functional MR imaging. *Korean J. Radiol.* 5, 87–95.

Yang, J. C., Park, K., Eun, S. J., Lee, M. S., Yoon, J. S., Shin, I. S., Kim, Y. K., Chung, T. W., Kang, H. K. and Jeong, G. W. (2008) Assessment of Cerebrocortical Areas Associated with Sexual Arousal in Depressive Women Using Functional MR Imaging. *J. Sex Med.* 5, 602–609.

Yang, Y., Raine, A., Lencz, T., Bihrle, S., LaCasse, L. and Colletti, P. (2005) Volume reduction in prefrontal gray matter in unsuccessful criminal psychopaths. *Biol. Psychiatry* 57, 1103–1108.

Zubieta, J. K., Smith, Y. R., Bueller, J. A., Xu, Y., Kilbourn, M. R., Jewett, D. M., Meyer, C. R., Koeppe, R. A. and Stohler, C. S. (2001) Regional mu opioid receptor regulation of sensory and affective dimensions of pain. *Science* 293, 311–315.

12

The Hedonic Brain: A Functional Neuroanatomy of Human Pleasure

MORTEN L. KRINGELBACH

Pleasure is central to our lives and intimately linked to emotional and reward processing in the brain. In general, hedonic experience is arguably at the heart of what makes us human, but at the same time it is also one of the most important factors that is keeping us from staying healthy (Kringelbach, 2004b; Saper et al., 2002). Understanding the underlying brain mechanisms can therefore help us understand and potentially treat the serious problems of affective disorders, such as, for example unipolar depression, bipolar disorder, chronic pain, and the worldwide epidemic of obesity.

The central premise of this chapter is that in order to more effectively treat affective disorders, we need to develop a better understanding of hedonic processing—that is the affective component of sensory processing—in the human brain (Kringelbach, 2005, 2009). Pleasure is here defined as one of the positive dimensions of the more general category of hedonic processing, which also includes other negative and unpleasant dimensions such as pain (see Leknes and Tracey, Chapter 19, this book). Importantly, malignant affective disorders such as depression, chronic pain, and eating disorders are characterized by the lowered or missing ability to experience pleasure, anhedonia. Thus, in order to help with these disorders, we will have to further our understanding of the cortical and subcortical mechanisms involved in pleasure and hedonic processing in general (Berridge and Kringelbach, 2008).

This review explores the evidence for underlying brain mechanisms and principles of pleasure and hedonic processing in the human brain. This evidence comes from human neuroimaging, neuropsychology, and neurosurgery. In particular, this chapter concentrates on the evidence linking the human orbitofrontal cortex (OFC) to general hedonic processing (see Figure 12.1).

Of Pleasures Past and Present

Pleasure must serve a central role in fulfilling the Darwinian imperative of survival and procreation (Darwin, 1872). This means that for all animals, the sensory pleasures linked to food intake as well as sex are likely to be basic pleasures (Berridge, 1996; Kringelbach, 2004b). Common to both survival and procreation are the social interactions with conspecifics, which may potentially lead to the propagation of genes. This has probably been selected for in evolution, which means that social pleasures are also likely to be part of our repertoire of basic pleasures (Kringelbach and Rolls, 2003). In the development of the social pleasures, the early attachment bonds between parents and infants are likely to be extremely important (Lorenz, 1943; Stein et al., 1991). In fact, in social species such as humans, it might well be that the social pleasures are at least as pleasurable as the sensory and the sexual pleasures (Watson et al., Chapter 5, this book).

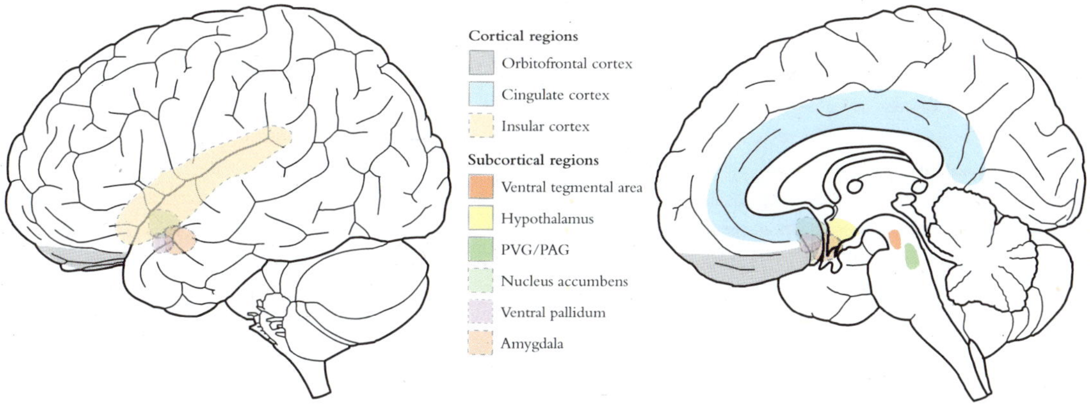

Figure 12.1 Brain regions involved in human hedonic processing. The figure shows the human brain seen from the side (top) and split in the middle (bottom) overlaid with the approximate location of the important brain structures of the pleasure brain. These include cortical areas such as the orbitofrontal (gray), the cingulate (light blue), and the insular cortices (buried between the prefrontal and temporal lobes, orange), as well as subcortical areas such as the ventral tegmental area in the brainstem (light red), hypothalamus (yellow), periventricular gray/periacqueductal gray (PVG/PAG, green), nucleus accumbens (light green), ventral pallidum (light purple), and the amygdala (light red).

In addition to these basic sensory and social pleasures, there are a large number of higher order pleasures, including monetary, artistic, musical, altruistic, and transcendent pleasures (Skov, Chapter 16, this book; Vuust and Kringelbach, Chapter 15, this book). Such higher order pleasures might be conceptualized as higher-dimensional combinations of the basic pleasures and as such may reuse some of the same brain mechanisms.

During the last century, a large corpus of animal experimentation has investigated reward processing in the brain. Many people have subsequently defined pleasure to be the conscious experience of reward but it is questionable whether such a narrow definition is meaningful or indeed useful. Such a definition would rather limit pleasure to conscious organisms, which is problematic for a number of reasons, given that we do not have a good definition of consciousness.

Pleasure is not a sensation (Frijda, Chapter 6, this book), since it does not fit most common definitions of sensations, as pointed out by Ryle (1954). Instead, pleasure would appear to be part of the subsequent valuation of sensory stimuli needed in decision making, including most importantly the hedonic valence, and as such may well be present in many species.

While the pleasure—or hedonic impact—of a reward such as sweetness can be measured by verbal reports in conscious humans, this hedonic processing is not dependent on the presence of language. In most nonlinguistic mammals, pleasure will also elicit "acceptance wriggles" that adds a hedonic gloss to the sensation that we experience as conscious pleasure (Frijda, Chapter 6, this book). Pleasure-elicited behaviors such as protruding tongue movements to sweet foods are present in other animals including rodents and have been proposed as an objective measure of the pleasure elicited (Steiner et al., 2001). While human infants initially exhibit similar kinds licking of their lips for sweet foods, these stereotyped behaviors disappear after a while.

Humans do, however, exhibit many pleasure behaviors, from the carefree smiles and laughter of pleasant social interactions to the deep groans of sensory and sexual pleasure (James, 1890). Most people instinctively feel that our pleasures would somehow not be quite the same without these pleasure-elicited behaviors.

At the same time, much of our brain activity is not available for conscious introspection and the neuroscientific evidence from humans and other animals has made it clear that nonconscious brain activity is essential for controlling our behavior. Some of this nonconscious brain activity is related to hedonic processing and may lead to hedonic reactions, where we are not conscious of their origin but are nevertheless happy to confabulate about the causes (Kringelbach, 2004c).

In a similar way to how it has proven useful to divide emotion into the nonconscious and conscious subcomponents of emotions and feelings, it might be more useful and meaningful to divide pleasure into both nonconscious and conscious subcomponents of evaluative hedonic processing (Kringelbach, 2004a). Such a definition would hold that while pleasure plays a central role for emotions and conscious feelings, it is not itself a conscious feeling.

Reward and hedonic processing are closely linked with motivation and emotion. Historically, early drive theories of motivation proposed that need potentiated previously learned habits and that need reduction strengthened new stimulus–response habit bonds (Hull, 1951). This was then taken to mean that hedonic behavior is controlled by need states. But these theories do not, for example, explain why people still continue to eat when sated. This led to theories of incentive motivation where hedonic behavior is mostly determined by the incentive value of a stimulus or its capacity to function as a reward (Bindra, 1978). Need states, such as hunger, are still important but only work indirectly on the stimulus' incentive value. Alliesthesia is the principle of modulation of the hedonic value of a consummatory sensory stimulus by homeostatic factors (Cabanac, 1971, Chapter 7, this book).

A useful distinction has been proposed between two aspects of reward: hedonic impact and incentive salience, where the former refers to the 'liking' or pleasure related to the reward and the latter to the 'wanting' or desire for the reward (Berridge, 1996; Berridge and Robinson, 1998). In order to provide hedonic evaluation of stimuli, the brain regions implicated in hedonic assessment must receive salient information about stimulus identity from the primary and secondary sensory cortices.

Neuroimaging offers a powerful way to investigate both the 'liking' and 'wanting' components in the human brain. One way to investigate 'liking' is to take subjective hedonic ratings throughout a human neuroimaging experiment and then correlate these ratings with changes in activity in the human brain (De Araujo et al., 2003b,c; Kringelbach et al., 2003). This allows for a unique window on the hedonic processes evaluating the pleasantness of salient stimuli. Such measures are only correlational in nature and will need to be combined with experiments offering causal inferences. This chapter will therefore discuss one promising way of getting the best of both worlds, which is to use deep brain stimulation (DBS) in patients in conjunction with neuroimaging methods such as magnetoencephalography (MEG) (Kringelbach et al., 2007a,b).

The Sensory Pleasures: Food Intake

The essential energy to sustain life is obtained from food intake and food consumption and is in turn associated with reward and a sensation of subjective pleasure—at least in humans. Although the necessary homeostatic regulation and consummatory behavior are hardwired in even brainless species, the challenges of regulating feeding are much greater for mammals who must maintain a stable body temperature in a wide variety of hostile climates, which in turn requires intricate neural circuits. The relative sophistication of foraging in higher primates compared to other mammals indicates that significant parts of our large brains are dedicated to the required motivational, emotional, and cognitive processing, and that mental processes related to food intake may indeed underlie other higher functions (Kringelbach, 2004b).

Food intake relies on our brain to obtain sensory information about a food, to evaluate for desirability, and to choose the appropriate behavior. Part of this process is closely linked to homeostatic regulation, which has been elucidated in great detail in animal models, with mammals including humans sharing many subcortical circuits and molecules (such as leptin and ghrelin) as, for example, outlined elsewhere (Saper et al., 2002).

However, food intake in humans is not only regulated by homeostatic processes as is illustrated by our easy overindulgence on sweet foods beyond homeostatic needs and the epidemic proportions of obesity, which has become a major health problem. Instead, food intake relies on the interaction between homeostatic regulation and hedonic processing. This complex subcortical and cortical processing involves higher order processes such as learning, memory, planning, and prediction, and gives rise to conscious experience of not only the sensory properties of the food (such as the identity, intensity, temperature, fat contents, and viscosity) but also the valence elicited by the food (including, most importantly, the pleasure experienced).

Food intake involves crucial decisions, where the brain must compare and evaluate the predicted reward value of various behaviors. This processing can be complex as the estimations will vary in quality depending on the sampling rate of the behavior and the variance of reward distributions. It is hard to provide a reliable estimate of the reward value of a food that appears to be highly desirable and is high in nutritional value, but is only rarely available and varies significantly in quality. This raises the classic problem in animal learning of how to optimize behavior such that the amount of exploration is balanced with the amount of exploitation, where exploration is the time spent sampling the outcome of different behaviors and exploitation is the time spent using existing behaviors with known reward values.

Food-related behaviors have to be precisely controlled because the decision to swallow toxins, microorganisms, or nonfood objects on the basis of erroneously determining the sensory properties of the food can be fatal. Humans and other animals have therefore developed elaborate food-related behaviors to balance conservative risk-minimizing and life-preserving strategies (exploitation) with occasional novelty seeking (exploration) in the hope of discovering new, valuable sources of nutrients (Rozin, 2001).

Pleasure and hedonic processing in general is central to this balancing act between exploitation and exploration. The evidence from neuroimaging studies has linked regions of the human brain—and in particular the OFC—to various aspects of food intake and especially to the representation of the subjective pleasantness of foods (Kringelbach, 2004b).

These findings appear for the first time to provide a solid basis for the further exploration of the brain systems involved in the conscious experience of pleasure and reward and provide a unique method for studying the hedonic quality of human experience. This hedonic experience is related to *qualia*, which has been described as "the hard problem of consciousness" (Chalmers, 1995) and which some philosophers believe will never become amendable to scientific analysis. Yet, recent neuroimaging of the neural mechanisms behind various aspects of food intake suggest that this line of scientific inquiry may still yield important insights into the core of subjective experience.

From Sensory Processing to Hedonic Experience of Food

All of the classic five senses (vision, hearing, smell, taste, and touch) are involved in the regulation of food intake. However, in addition to these sensory systems there are also other sensory receptors such as those in the digestive tract that are sensitive to gastric distension or those in the circulatory system that are sensitive to changes in blood pressure or level of carbon dioxide gas in the blood.

The two most important senses involved in food intake are smell and taste. In the following, it is shown how they interact to facilitate decision making and hedonic experience. Four computational principles have been proposed for the interaction between sensory and hedonic processing in humans: (1) motivation-independent processing of identity and intensity; (2) formation of learning-dependent multimodal sensory representations; (3) reward representations using state-dependent mechanisms including selective satiation, and (4) representations of hedonic experience, monitoring/learning, or direct behavioral change (Kringelbach, 2006).

Motivation-independent Processing of Identity and Intensity

The primary taste area in humans has been found to be located in the anterior insula/frontal operculum (Kinomura et al., 1994; O'Doherty et al., 2001b; Small et al., 1997, 1999). These neuroimaging studies of taste are agreement with the findings from human lesion studies and the neurophysiological findings in primates (Bornstein, 1940a,b; Scott et al., 1986; Veldhuizen et al., Chapter 9, this book).

The largest functional magnetic resonance imaging (fMRI) study of taste processing to date used 40 data sets from 38 right-handed subjects (13 women and 25 men, of which 2 subjects participated in two experiments) in four taste investigations that used: (1) identical delivery of the taste stimuli; (2) the same control procedure in which a tasteless solution was delivered after every taste stimulus; and (3) event-related interleaved designs (Kringelbach et al., 2004). A total of eight unimodal and six multimodal taste stimuli (oral stimuli that produce typically taste, olfactory, and somatosensory stimulation) ranging from pleasant to unpleasant were used in the four experiments. The results of the main analysis included both unimodal and multimodal taste stimuli, which was then confirmed in a separate analysis using only unimodal taste stimuli.

Stringent random effects analysis of taste activity across the forty datasets revealed three cortical activity foci to the main effects of taste in the human brain (which were corrected for multiple comparisons). Bilateral activity in the anterior insular/frontal opercular cortex was found with a slightly stronger response on the right side. This slight asymmetry in bilateral taste processing fits with an early meta-analysis of gustatory responses gathered from neuroimaging studies suggesting that the preponderance of activity peaks to taste, fall in the right hemisphere (Small et al., 1999).

Taste stimuli also produced activity in the medial caudal OFC, which is likely to coincide with the secondary taste cortex. This fits well with subsequent neurophysiological recordings in medial parts of the macaque OFC (Pritchard et al., 2005).

Similar to taste stimuli, pure olfactory stimuli have given dissociable brain areas for motivation-independent representations of reinforcer identity and hedonic representations. Neuroimaging studies have found representations of olfactory identity in primary olfactory cortices (Anderson et al., 2003; Gottfried et al., 2002, Chapter 8, this book; O'Doherty et al., 2000; Rolls et al., 2003a; Royet et al., 2001; Zald and Pardo, 1997), which are distinct from hedonic representations in other brain areas including the orbitofrontal cortex.

In general, the experiments in primates including humans have clearly demonstrated that the primary sensory areas for taste and smell are not modulated by motivational state, and that hedonic processing is generally thought to occur in higher-order, multimodal areas such as the OFC.

Formation of Learning-dependent Multimodal Sensory Representations

In addition to multimodal information from taste and smell, decisions about food intake also integrate, for example, somatosensory information, which is sensed by receptors in the oral and nasal cavity. This sensory information includes temperature, viscosity, fat contents, pungency, and irritation and is mediated by a large variety of neural systems. This integrated information is processed and made available for the crucial decision of ingestion or rejection of a potentially poisonous food—although simple brainstem mechanisms also exist (Grill and Norgren, 1978).

Neuroimaging studies have investigated the multimodal integration involved in food intake and found that one of the central brain regions is the human OFC. Auditory (Frey et al., 2000), gustatory (Small et al., 1999), olfactory (Zatorre et al., 1992), somatosensory (Rolls et al., 2003b), and visual (Aharon et al., 2001) inputs as well as information from the visceral sensory system (Critchley et al., 2002) have been shown to produce activity in the human OFC. This is in agreement with neurophysiological recordings finding that the nonhuman primate OFC receives input from all of the five senses (Rolls, 1999).

These sensory inputs enter the OFC mostly through its posterior parts. The interaction between taste and smell have been studied with neuroimaging and significant activity has been found in slightly more anterior parts of the OFC and nearby agranular insula for the combination of taste and smell (De Araujo et al., 2003c; Kringelbach et al., 2003; Small et al., 1997).

A good example of multimodal integration is how subjective olfactory experience appears different depending on whether a smell reaches the nasal cavity through the nose (orthonasal) or mouth via the posterior nares of the nasopharynx (retronasal) (Pierce and Halpern, 1996). These are likely to be related to differences in somatosensory influences (e.g., mastication). Several neuroimaging studies have found corresponding differences in cortical activity patterns between ortho- and retronasal olfaction in the OFC and related brain regions (Cerf-Ducastel and Murphy, 2001; De Araujo et al., 2003a; Small et al., 2005).

Reward Representations of Sensory Stimuli

In contrast to these motivation-independent representations of uni- and multimodal reinforcer identities, neuroimaging studies have found that the valence is encoded in a network of other brain regions.

In a neuroimaging taste study, a dissociation was found between the brain regions responding to taste intensity and taste affective valence (Small et al., 2003). Brain regions responding to intensity regardless of valence were found in the cerebellum, pons, middle insular cortex, and amygdala, while valence-specific responses were observed in the OFC with the right caudolateral OFC responding preferentially to pleasant compared to unpleasant taste, irrespective of intensity.

Another neuroimaging study found that the subjective ratings of taste pleasantness (but not intensity) correlate with activity in the medial OFC and in the anterior cingulate cortex (De Araujo et al., 2003b) (Figure 12.2). Moreover, in this study investigating the effects of thirst and subsequent replenishment, it was found that the medial OFC and a region of mid-insula were correlated with subjective pleasantness ratings of water across the whole experiment (De Araujo et al., 2003b).

Further evidence of neural correlates of subjective experience of pure taste was found in an experiment investigating true taste synergism, which is the phenomenon whereby the intensity of a taste is dramatically enhanced by adding minute doses of another taste. The results of this neuroimaging experiment showed that the strong subjective enhancement of umami taste occurring when 0.005 M inosine 5'-monophosphate is added to 0.5 M monosodium glutamate (compared to both delivered separately) was correlated with

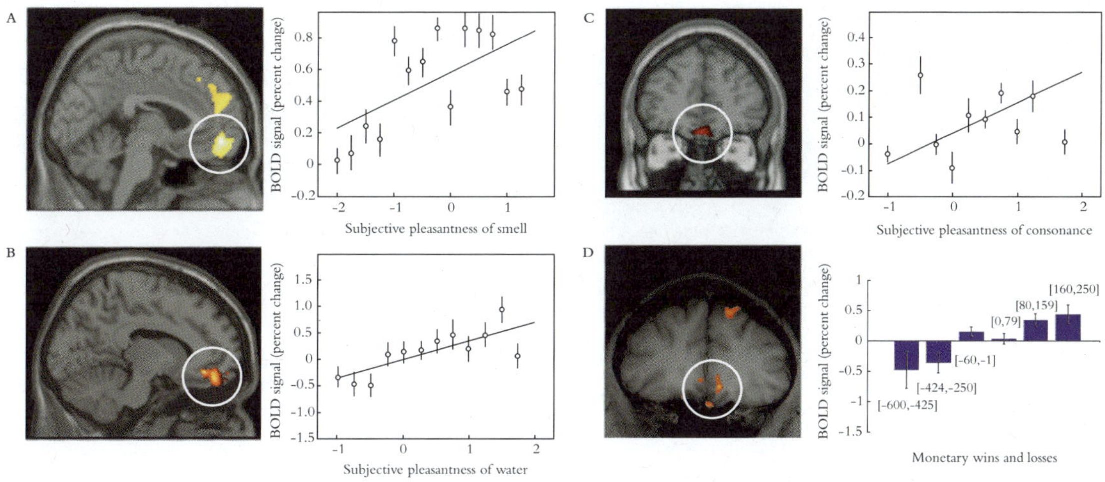

Figure 12.2 Valence coding in medial orbitofrontal cortex (mOFC). (A) The activity in mOFC correlates with the subjective ratings of pleasantness in an experiment with three pleasant and three unpleasant odors (Rolls et al., 2003a). (B) Similarly, the activity in mOFC was also correlated with the subjective pleasantness ratings of water in a thirst experiment (De Araujo et al., 2003b). A correlation in a very similar part of mOFC was found with the pleasantness of other pure tastants used in the experiment (not shown). (C) This corresponded to the findings in an experiment investigating taste and smell convergence and consonance, which found that activity in the mOFC was correlated to subjective consonance ratings (De Araujo et al., 2003c). (D) Even higher order rewards such as monetary reward was found to correlate with activity in the mOFC (O'Doherty et al., 2001a).

increased activity in a mid-anterior part of the OFC (Figure 12.3) (De Araujo et al., 2003a).

Several neuroimaging olfaction studies have found dissociable encoding of olfactory stimuli with the intensity encoded in the amygdala and nearby regions, and the pleasantness correlated with activity in the medial OFC and anterior cingulate cortex (Anderson et al., 2003; Gottfried et al., 2002; Rolls et al., 2003a). This is consistent with studies that have found that hedonic judgments are correlated with activity in the medial OFC (Royet et al., 2001) and that the unpleasantness of aversive odours correlates with activity in the lateral OFC (Zald and Pardo, 1997). Furthermore, it has been found that the OFC represents the sensory-specific decrease of smell (O'Doherty et al., 2000), showing that the reward value of olfactory stimuli is represented in the OFC.

Other recent strong evidence for the role of the OFC in the representation of the reward value of olfactory stimuli comes from an appetitive conditioning neuroimaging experiment, which measured the brain activity related to two arbitrary visual stimuli both before and after olfactory devaluation (Gottfried et al., 2003). In the amygdala and the OFC, responses evoked by a predictive target stimulus were decreased after devaluation, whereas responses to the nondevalued stimulus were maintained. It would thus appear that differential activity in the amygdala and the OFC encodes the current value of reward representations accessible to predictive cues. It should also be noted that the affect and intensity judgments of odour in a paired discrimination task correlates with activity in the OFC (Zatorre et al., 2000).

This evidence is compatible with studies in nonhuman primates where monkeys with lesions to the OFC responded normally to associations between food and conditioners but failed to modify their behavior to the cues when the incentive value of the food was reduced (Butter et al., 1963), and where lesions to the OFC altered food preferences in monkeys (Baylis and Gaffan, 1991). Similarly, unilateral crossed lesions between the OFC and the basolateral part of the amygdala in monkeys disrupted devaluation effects in a procedure in which the incentive value of a food was reduced by satiation on that specific food (Baxter et al., 2000).

Representations of Hedonic Experience

The evidence from neuroimaging studies of pure taste and smell cited above shows that the OFC is consistently correlated with the subjective pleasantness ratings of

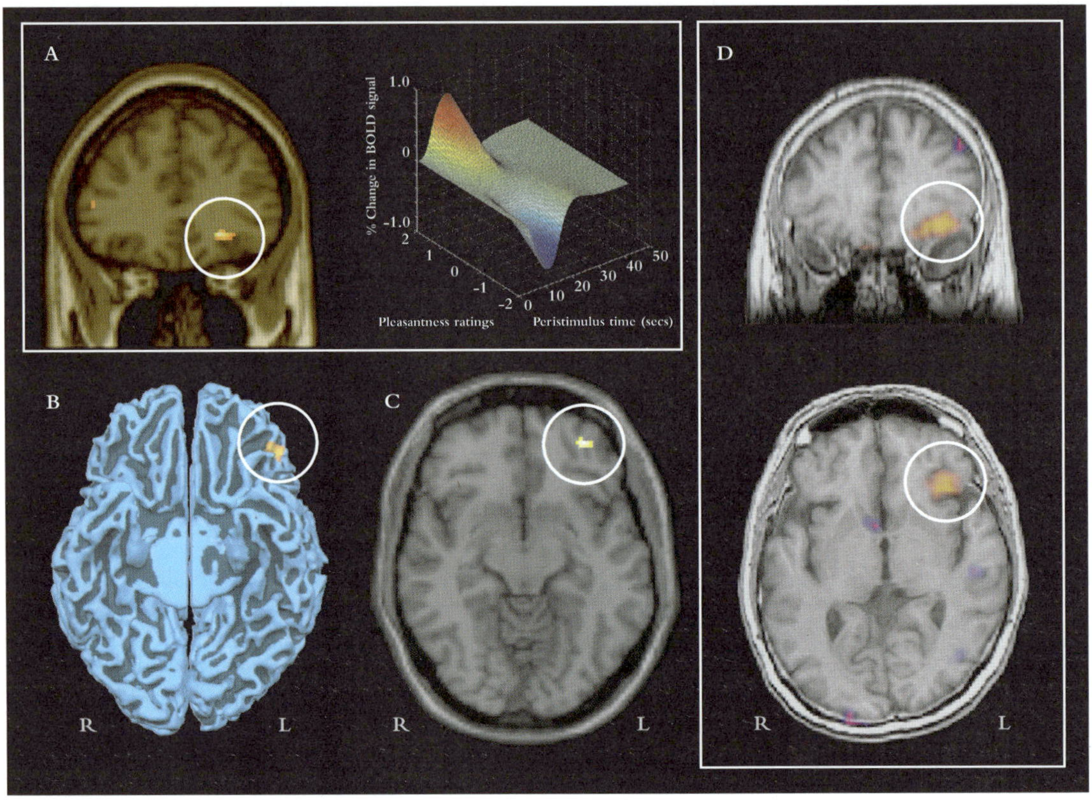

Figure 12.3 Hedonic experiences. (A) A neuroimaging study using selective satiation found that mid-anterior parts of the OFC are correlated with the subjects' subjective pleasantness ratings of the foods throughout the experiment (Kringelbach et al., 2003). On the right is shown a plot of the magnitude of the fitted haemodynamic response from a representative single subject against the subjective pleasantness ratings (on a scale from –2 to +2) and peristimulus time in seconds. (B) Additional evidence for the role of the OFC in subjective experience comes from another neuroimaging experiment investigating the supra-additive effects of combining the umami tastants monosodium glutamate and inosine monophosphate (De Araujo et al., 2003a). The figure shows the region of mid-anterior OFC showing synergistic effects (rendered on the ventral surface of human cortical areas with the cerebellum removed). The perceived synergy is unlikely to be expressed in the taste receptors themselves and the activity in the OFC may thus reflect the subjective enhancement of umami taste, which must be closely linked to subjective experience. (C) Adding strawberry odor to a sucrose taste solution makes the combination significantly more pleasant than the sum of each of the individual components. The supralinear effects, reflecting the subjective enhancement were found to significantly correlate with the activity in a lateral region of the left anterior OFC, which is remarkably similar to that found in the other experiments (De Araujo et al., 2003c). (D) These findings were strengthened by findings using deep brain stimulation (DBS) and magnetoencephalography (MEG) (Kringelbach et al., 2007a). Pleasurable subjective pain relief for chronic pain in a phantom limb in a patient was causally induced by effective deep brain stimulation in the PVG/PAG part of the brainstem. When using MEG to directly measure the concomitant changes in the rest of the brain, a significant change in power was found in the mid-anterior orbitofrontal. Note that while the results may seem to demonstrate a laterality effect in the left OFC, the effects were also present in the right OFC.

the stimuli. Therefore, it is to be expected that studies using multimodal combinations of taste and smell as well as state-dependent changes in pleasantness should find correlations between subjective pleasantness and activity in these brain regions. Compelling evidence that this is indeed the case comes from a sensory-specific satiety neuroimaging study, which has shown that a region of the left OFC showed not only a sensory-specific decrease in the reward value to the whole food eaten to satiety (and not to the whole food not eaten), but also a correlation with pleasantness ratings (see Figure 12.3a) (Kringelbach et al., 2003). This result strongly indicates that the reward value of the taste, olfactory, and somatosensory components of a whole food are represented in the OFC and that the subjective pleasantness of food thus might be represented here.

It is an open but very interesting question whether the OFC and perhaps even subregions thereof are both necessary and sufficient for the experience of sensory and social pleasure. The evidence from the psychosurgery experiments of last century is not clear since the lack of neuroimaging methods and post-mortem investigations meant that the surgical lesions were not adequately described (Heath, 1954). One interpretation, consistent with more recent surgical circumscribed lesions to the OFC (Hornak et al., 2003), would suggest that disconnections of the white matter of the OFC can lead to serious emotional changes. Direct tests of anhedonia linked to lesions to the OFC have, however, not been carried out.

Further evidence comes from a study investigating the nonspecific satiation effects of chocolate (with both olfactory and gustatory components), which found a correlation between the decrease in pleasantness and activity in the OFC (Small et al., 2001). Another multimodal study investigating the link between olfaction and vision found activity in the anterior OFC for semantically congruent trials (Gottfried and Dolan, 2003).

When investigating the synergistic enhancement of a matched taste and retronasal smell, it was again found that a region of the OFC was significantly active (Figure 12.3c) (De Araujo et al., 2003c). This region was located very near to the region of the OFC active by the synergistic combinations of umami (see Figure 12.3b) (De Araujo et al., 2003a).

Other Sensory Pleasures

In addition to these food-related results, a number of other neuroimaging studies have investigated the hedonic processing involved in other sensory pleasures such as sex, drugs, and rock'n roll. Sex has been remarkably little studied with neuroimaging, although some progress has recently been made (Georgiadis and Kortekaas, Chapter 11, this book; Komisaruk et al., Chapter 10, this book). Studies of sexual orgasms and excitement in both normal male and female volunteers found an important involvement of the mid-anterior OFC (Holstege et al., 2003; Georgiadis et al., 2006). Interestingly, the active brain regions were remarkably similar to those found in a study of fluid analogies and creativity (Geake and Hansen, 2005).

Drugs such as cocaine have been studied to a much greater extent (Breiter et al., 1997) although it is not entirely clear how these drugs come to influence the blood oxygen level–dependent (BOLD) signals measured with fMRI. One such study found a correlation between a reliable index of the rush of intravenous met-amphetamine in drug-naïve subjects and activity in the medial OFC (Völlm et al., 2004).

Rock'n roll is somewhat hard to study in the noisy environment of the fMRI scanner but PET studies have found activity in the OFC, which correlates with the negative dissonance (pleasantness) of musical chords (Blood et al., 1999) and intensely pleasurable responses, or "chills", that are elicited by music are correlated with activity in the OFC, ventral striatum, cingulate, and insula cortex (Blood and Zatorre, 2001). The brain correlates of the social aspects of music such as participation in making music and dancing to music have, however, hardly been studied at all (Vuust and Kringelbach, Chapter 15, this book), which is somewhat surprising given that these are some of the most pleasurable aspects of music (Neveu Kringelbach, 2005).

The Social Pleasures: Face Processing

Humans are intensely social, and experiments have shown time and again that our preferred route to health, pleasure, and perhaps even happiness is through social relationships with other people (Layard, 2005). Human social relationships are very rich and complex, and we have only begun to understand some of the underlying brain processes (Adolphs, 2003; Watson et al., Chapter 5, this book).

In humans and other primates, facial expressions act as important social cues to regulate behavior (Darwin, 1872; Ekman and Friesen, 1971). Much is known about the neural correlates of the decoding of face expressions from neurophysiological studies in nonhuman primates (Bruce et al., 1981; Desimone and Gross, 1979; Hasselmo et al., 1989; Perrett et al., 1982), and from human lesion (Adolphs et al., 1994; Bodamer, 1947; Sergent and Villemure, 1989) and imaging studies (Haxby et al., 1994), but very little is known about the neural correlates of how face expressions govern human social behavior. In addition, it is clear that infant faces play an important role in the early attachment between parents and children, which is the foundation of our hedonic brain.

Infant and Infantile Faces as a Tool for Understanding Social Attachment

The scientific interest in the cuteness of infant faces started with Charles Darwin who pointed out that in order for infants to survive and to perpetuate the human species, adults need to respond and care for their young (Darwin, 1872). The Nobel Prize winner Konrad Lorenz proposed that it is the specific structure of the infant face that serves to elicit these parental responses (Lorenz, 1971), but the biological basis for this has remained elusive. Lorenz argued that infantile features serve as "innate releasing mechanisms" for affection and nurturing in adult humans and that most of these features are evident in the face, including a relatively large head, predominance of the brain capsule, large and low-lying eyes, and bulging cheek region (Lorenz, 1971). Thus it is argued that these "babyish" features of infants increase the infant's chance of survival by evoking parental responses (Bowlby, 1957, 1969) and the parents' ability to respond is important for the survival of the species (Darwin, 1872).

While a considerable body of research has focused on how the human brain processes adult faces, much less research has investigated the processing of infant faces (Frith, 2006). A number of studies have used fMRI to examine parental responses to children's faces, which has advanced our understanding of some of the underlying neural circuitry (Swain et al., 2007). Most studies have compared parental responses to their own children with their responses to other children. It has been found that there is stronger activity to one's own children compared to other infants in striate and extrastriate visual areas and in reward-related areas such as the nucleus accumbens, anterior cingulate, and amygdala (Ranote et al., 2004; Swain et al., 2007).

While these studies have substantially increased our general knowledge of the parental neural responses to children faces, there are a number of reasons why a substantial test of the Lorenz's theory of the specificity of infant faces requires a direct comparison between matched adult and infant faces from the first year of life; preferably where the faces are unfamiliar and using neuroimaging techniques that permit the temporal progression of brain activity to be studied.

We used MEG to investigate the temporal and spatial distribution of the underlying neural systems for these facial responses in 12 adult human participants (Kringelbach et al., 2008). Consistent with previous findings, we found that face processing of both adult and infant faces elicits a wave of activity starting in the striate cortices and spreading along ventral and dorsal pathways (Blair, 2003).

In addition, however, we found that at around 130 ms after presentation of the infant faces, activity occurred in the medial OFC identifying for the first time a neural basis for this vital evolutionary process (see Figure 12.4). This was not evident in response to the adult faces. Since the infant and adult faces used in

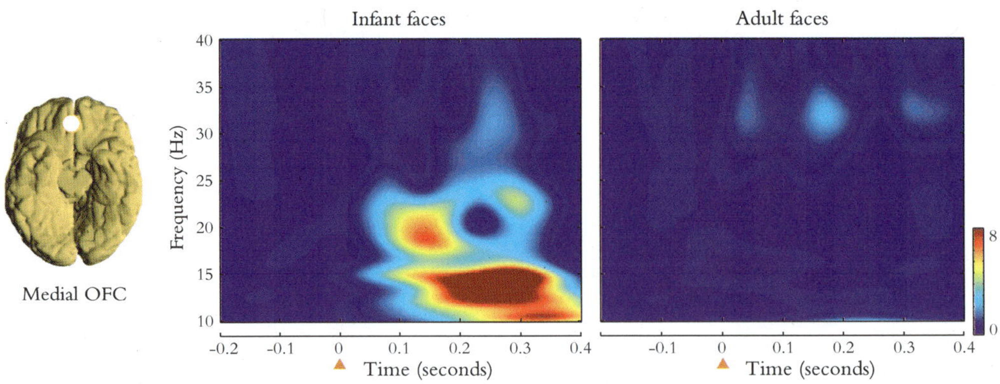

Figure 12.4 Infant faces. Significant activity was present from around 130 ms in the mOFC (leftmost panel) when viewing infant faces (middle panel) but not when viewing adult faces (rightmost panel). Time–frequency representations of the normalized evoked average group responses to baby and adult faces from the virtual electrodes show that the initial response to infant faces is present in the 12–20 Hz band from around 130 ms—and not present to adult faces (Kringelbach et al., 2008).

this study were carefully matched by independent panels of participants for emotional valence and arousal, and attractiveness, the findings provide evidence that it is the distinct features of the infant faces compared to adult faces which are important, rather than evaluative subjective processing such as attractiveness or emotional valence.

These specific responses to unfamiliar infant faces occur so fast that they are almost certainly quicker than anything under conscious control, suggesting that they are automatized. The findings are therefore potentially of interest in that they suggest a temporally earlier role than previously thought for the medial OFC in guiding affective reactions, which may be even nonconscious.

The medial OFC may thus provide the necessary attentional—and perhaps hedonic—tagging of infant faces that predisposes humans to treat infant faces as special and elicits caring, as suggested by Lorenz.

There is a potentially important clinical application of these findings in relation to postnatal depression. Postnatal depression is common, occurring in approximately 13% of mothers and 3% of fathers after birth (O'Hara and Swain, 1996) and often within 6 weeks (Cooper and Murray, 1998). Postnatal depression has been associated with a range of adverse child outcomes including attachment, behavioral and emotional disturbances, and there is also some evidence for poorer cognitive outcomes. There is increasing evidence that certain features of the behavior of depressed mothers are associated with adverse outcome, in particular their lack of responsiveness to the infant, the reduced ability to perceive their infant's signals and less mimetic behavior (Cohn and Tronick, 1983; Reck et al., 2004) with a resultant lack of contingency between the infant's actions and the mother's responses (Murray et al., 1996). Furthermore, it has been shown experimentally that infants respond adversely with distress, crying, increased arousal, and then avoidance to an unresponsive maternal face (the still face paradigm) (Reck et al., 2004; Tronick et al., 1977).

It is possible that changes to activity in the medial OFC may thus be secondary to depression and adversely affect parental responsivity. Further research could identify whether these early and specific medial orbitofrontal responses to infant faces (own and others) are affected and even suppressed by depression, thereby helping to explain this lack of maternal responsiveness. The face paradigm could eventually provide opportunities for early identification of families at risk (Swain et al., 2007). It would be of considerable interest to investigate the brain responses to infants of other species to see whether a similar effect is present.

Changes in Facial Expression as a Tool for Social Learning

While the hedonic processing of infant faces may be the result of powerful evolutionary programming, humans are also perfectly capable of social learning. In particular, changes in facial expression are powerful social stimuli to make us change our behavior. A similar role is filled by secondary reinforcers such as monetary reward. We initially investigated learning-dependent decision-making in a visual discrimination reversal task, where subjects had to associate an arbitrary stimulus with monetary wins or losses, and then rapidly reverse these associations when the reinforcement contingencies altered (O'Doherty et al., 2001a). Probabilistic reward and punishment schedules were used such that selecting either the currently rewarded stimulus or the unrewarded stimulus can lead to a monetary gain or loss, but only consistent selection of the currently rewarded stimulus results in overall monetary gain. An fMRI study of this task in normal subjects found a dissociation of activity in the medial and lateral parts of the OFC: activity in the medial OFC correlated with how much money was won on single trials, and activity of the lateral OFC correlated with how much money was lost on single trials.

There was, however, an inherent problem in this monetary reversal-learning task, where the probabilistic nature of the task meant that the magnitude of negative reinforcers (money loss) was slightly confounded by the reversal event per se. We therefore designed a reversal-learning study without this problem, where the overall goal is to keep track of the mood of two people presented in a pair with neutral facial expressions and as much as possible to select the image of a "happy" person who will then smile, while the "angry" person will show an angry face. The task is to continue to select the "happy" person and receive smiles in return.

Just like in normal social interactions, the "happy" person will suddenly change her mood into angry, while the other person will now become the "happy" person. Now, one has to learn to select the image of the other person and unlearn to select the image of other previously happy person. Most of us find such a task quite easy, if slightly distressing, and will learn very quickly to select the other person when seeing an angry face where we expected a happy smile.

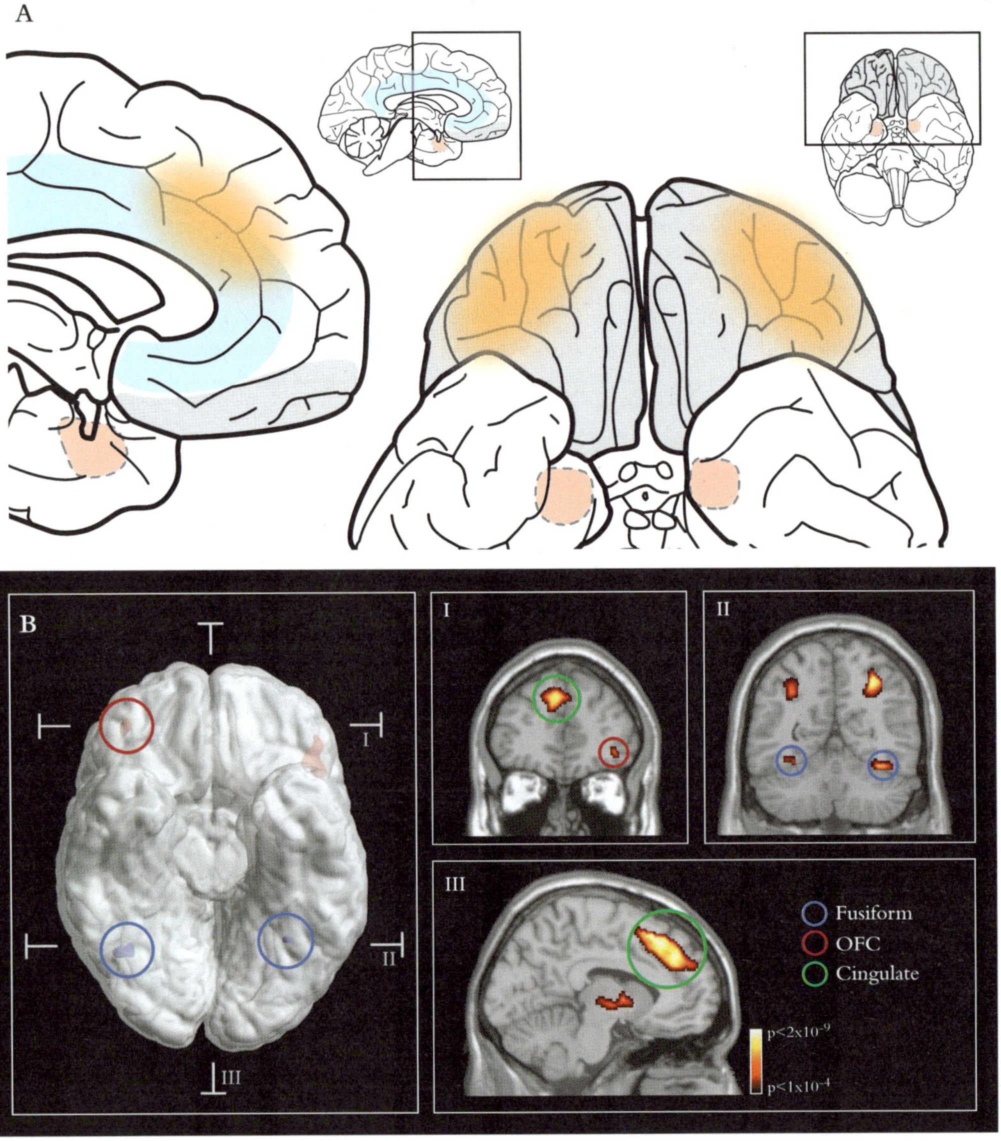

Figure 12.5 Social interactions and the case of reversal learning. (A) The lateral orbitofrontal and parts of the anterior cingulate cortices in the rostral cingulate zone are often found to be coactive in neuroimaging studies (with the regions superimposed in red). Most often this is found in tasks where the subjects have to evaluate negative stimuli which when detected may lead to a change in current behavior. (B) A recent neuroimaging study found that the lateral orbitofrontal and the anterior cingulate/paracingulate cortices are together responsible for changing behavior in an object reversal-task (Kringelbach and Rolls, 2003). This task was setup to model aspects of human social interactions (see text for full description of task). Subjects were required to keep track of the faces of two people and to select the "happy" person, who would change mood after some time, and subjects had to learn to change, reverse, their behavior to choose the other person. The most significant activity during the reversal phase was found in the lateral orbitofrontal and cingulate cortices (red and green circles), while the main effects of faces were found to elicit activity in the fusiform gyrus and intraparietal sulcus (blue circles).

We gained a rather precise understanding of which parts of the brain that are linked to flexibly changing social behavior. This task elicited significant brain activity in the frontal part of the brain and specifically in the lateral orbitofrontal and anterior cingulate cortices (Kringelbach and Rolls, 2003) (see Figure 12.5). In contrast, we found brain activity only in the posterior parts of the brain when just watching faces without having to change behavior.

Overall, these neuroimaging studies demonstrate that faces are important stimuli to help understand how the social pleasures might govern behavior. In particular, they show that the sensory and social pleasures share a similar network of interacting brain regions.

Deep Brain Pleasures

The sensory and social pleasures can be bypassed by direct stimulation of the brain (Kringelbach et al., 2007b; Green et al., Chapter 18, this book). DBS is generally thought to have started with the demonstration of the localized electrical excitability of the motor cortex by Fritsch and Hitzig (1870). It was, however, only with the invention of the Horsley–Clarke frame for stereotactic neurosurgery (Horsley and Clarke, 1908) that DBS became practical for subcortical structures and the full potential awaited the adaptation of this frame for humans by Spiegel et al. (1947).

Early usage included alleviation of movement disorders, where mostly the globus pallidus and the ventral thalamus were targeted. In parallel, human psychosurgery from Moniz and Freedman's lobotomies to Heath's electrical stimulation in schizophrenics and homosexuals drew sharp public criticism (Baumeister, 2000; Valenstein, 1973). Most of this research proceeded without the use of a stereotactic frame and the exact brain targets of these early electrical stimulation studies were never clear (Smith et al., Chapter 1, this book).

After an initial flourish, stereotactic surgery for Parkinson's disease (PD) was largely abandoned in the 1960s when the link was found to the degeneration of the dopamine cells of the substantia nigra pars compacta and L-dopa became widely used for treatment. However, L-dopa often has very serious side effects and in the 1990s lesions of the globus pallidus internus (GPi) were reintroduced for PD dyskinesia. Lesions of the subthalamic nucleus (STN) can cause hemiballismus, and instead DBS of the STN and GPi at 130–180 Hz has been shown as effective and comparably safe (Aziz et al., 1991).

The current DBS targets for pain are in the brainstem (periventricular gray [PVG] and periaqueductal gray [PAG]) and in the thalamus (Nandi et al., 2002). The targets for PD are in the STN (Bittar et al., 2005a), GPi (Bittar et al., 2005a; Krack et al., 2003), and pedunculopontine nucleus in the brainstem (Jenkinson et al., 2005; Mazzone et al., 2005; Plaha and Gill, 2005). The current target for cluster headache is in the hypothalamus (Leone et al., 2004). Some promising targets for depression have been found in the inferior thalamic peduncle (Andy and Jurko, 1987), the nucleus accumbens (Schlaepfer et al., 2007), and the subgenual cingulate cortex (Mayberg et al., 2005). Programmable stimulators are implanted subcutaneously and hundreds of patients have been restored to near normal lives (Perlmutter and Mink, 2006).

Mood changes linked to changes in reward and hedonic processing such as unipolar depression are found in up to 40% of PD patients often starting before the onset of PD symptoms (Cummings, 1992). This is perhaps not surprising given the important dual role of the basal ganglia not only in movement but also in affect. The technique of stereotactic DBS thus has wide-reaching therapeutic applications clinically and in the neurosciences generally.

Patients with chronic pain who have DBS of the PVG/PAG report experiencing much less pain (Bittar et al., 2005b,c). The PVG/PAG receives noxious input from ascending spinothalamic pathways and descending regulatory input from higher brain structures such as the orbitofrontal cortex. Electrical stimulation of the PAG induces "stimulation-produced analgesia" in animals and humans (Boivie and Meyerson, 1982; Reynolds, 1969). This effect is ascribed to a release of endogenous opioids, since the effects are reversible with the administration of the opioid antagonist naloxone (Akil et al., 1976; Hosobuchi et al., 1977), and also to the activation of descending inhibitory systems that depress spinal noxious transmission (Fields and Basbaum, 1999).

Measuring Whole Brain Activity from DBS

What is particularly exciting about DBS is that it offers the potential for causally changing brain activity and thus potentially can inform us about the fundamental mechanisms of brain function (Kringelbach et al., 2007c). This is particularly true when combined with noninvasive whole-brain neuroimaging techniques such as MEG (Kringelbach et al., 2007a).

In some select patients, this chronic pain can be significantly changed over a short period of time with the DBS stimulation. This subjective change can be measured with MEG when changing the DBS stimulation from effective to noneffective, while acquiring repeated subjective measurements on a visual scale. This can then be used in the data analysis to reveal the brain regions which mediate the change in subjective hedonic experience.

We were the first group to use MEG to make direct measurements of the whole brain elicited by DBS. When DBS was turned off, the participant reported significant increases in subjective pain. During the pain relief, we found corresponding significant changes in brain activity in a network that comprises the regions of the hedonic brain and includes the mid-anterior orbitofrontal and subgenual cingulate cortices (Figure 12.3d) (Kringelbach et al., 2007a). We found similar brain changes in a patient with depression and in a patient with intractable cluster headache (Ray et al., 2007). This is strong evidence linking the OFC to pain relief and thus pleasure.

These findings raise some pertinent questions about the nature of direct brain stimulation. While stimulation of the PVG/PAG brings about pain relief, which is clearly pleasurable, it is not obvious if this would also be the case in humans without chronic pain. It is well known that while low-frequency stimulation of the PVG/PAG can bring about pain relief in chronic pain patients, high-frequency stimulation has the opposite effect and actually makes the pain worse (Green et al., Chapter 18, this book). Anecdotally, some DBS patients with chronic pain relief report that the pain is still there but that the stimulation makes them care less about the pain (Aziz, personal communication).

Interestingly, a PET study investigating analgesia and placebo, which found that the opioid-rich brain structures—lateral orbitofrontal and anterior cingulate cortices—are coactive in placebo responders, suggesting that the pain relief effect of the placebo might be related to these two brain areas being coactive (Petrovic and Ingvar, 2002; Petrovic et al., 2002) (see also Petrovic, Chapter 17, this book).

It is also potentially of interest to note that lesions to an output structure of the OFC, the ventral pallidum (Öngür and Price, 2000), can lead to anhedonia (Miller et al., 2006). Similar evidence of anhedonia linked to lesions of some parts of the pallidum was also found in a large case series of 117 patients undergoing pallidotomies for movement disorders (Aziz, personal communication). This is of particular importance since lesions of the posterior ventral pallidum in rats abolishes and replaces 'liking' reactions to sweetness with bitter-type 'disliking' instead (e.g., gapes) (Cromwell and Berridge, 1993).

Similar, DBS of the nucleus accumbens, which is another output structure of the OFC, can alleviate anhedonia in patients with treatment-resistant depression (Schlaepfer et al., 2007). These results are not surprising given the previous animal literature on lesions and brain stimulation effects (see Smith et al., Chapter 1, this book). It opens up a number of interesting avenues of research, and it would, for example, be of considerable interest to investigate the effects of DBS of the mid-anterior OFC as well as of the ventral pallidum.

But does DBS actually produce pleasure? One could plausibly argue that DBS may help to modulate malignant oscillatory activity in the brain based on what is known about the neural mechanisms of DBS (Kringelbach et al., 2007b). The evidence would suggest that while DBS may help to restore the brain's normal equilibrium in pathological states, DBS might have little effect on long-term pleasure in the normal brain. It might of course be possible for DBS to perturb the brain's equilibrium in the normal state but such perturbations are likely to be short-lived, similar to those induced by various drugs. For now, DBS remains a very useful technique for alleviating the acute symptoms of anhedonia such that people can again come to appreciate the everyday sensory and social pleasures.

Conclusions

The scientific study of pleasure and hedonic processing in humans remains in its infancy. Some progress has been made in understanding the putative brain structures involved in emotion and pleasure, mostly based on animal models but also to some extent based on human neuroimaging, neuropsychological, and neurosurgical studies. Animal models using primarily rodents have convincingly shown that the hypothalamus, nucleus accumbens, ventral pallidum, and various brainstem nuclei such as the periaqueductal gray are important for hedonic processing (see Smith et al., Chapter 1, this book; Aldridge and Berridge, Chapter 3, this book). Human neuroimaging research has implicated primarily the orbitofrontal and cingulate cortices as well as the amygdala and the insular cortices. The subcortical brain regions identified with animal models provide some of the necessary input and output systems for multimodal association regions such as the OFC that are involved in representing and learning about the reinforcers that elicit emotions and conscious feelings (Kringelbach, 2005).

The recent convergence of findings from neuroimaging, neuropsychology, neurophysiology, and neurosurgery has demonstrated that the human OFC is best thought of as an important nexus for sensory integration, emotional processing, and hedonic experience (see Figures 12.6 and 12.7).

A model for the functions of the OFC could be following: The posterior parts process the sensory information for further multimodal integration. The reward value of the reinforcer is assigned in more anterior parts of the OFC from where it can be modulated by hunger and other internal states and can be used to influence subsequent behavior (in lateral parts of the anterior OFC with connections to anterior cingulate cortex), stored for monitoring, learning, and memory (in medial parts of the anterior OFC) and made available for subjective hedonic experience (in mid-anterior OFC). At all times, there is important reciprocal information flowing between the various regions of the OFC and other brain regions subserving hedonic processing including the anterior cingulate cortex, the amygdala, the nucleus accumbens, and the ventral pallidum. Lateralization does not appear to play a major role for the functions of the human OFC as shown by the largest meta-analysis of its involvement by neuroimaging studies (Kringelbach and Rolls, 2004).

This model does not posit that medial OFC only codes for the valence of positive reinforcers and vice versa for the lateral parts. Instead, the evidence from

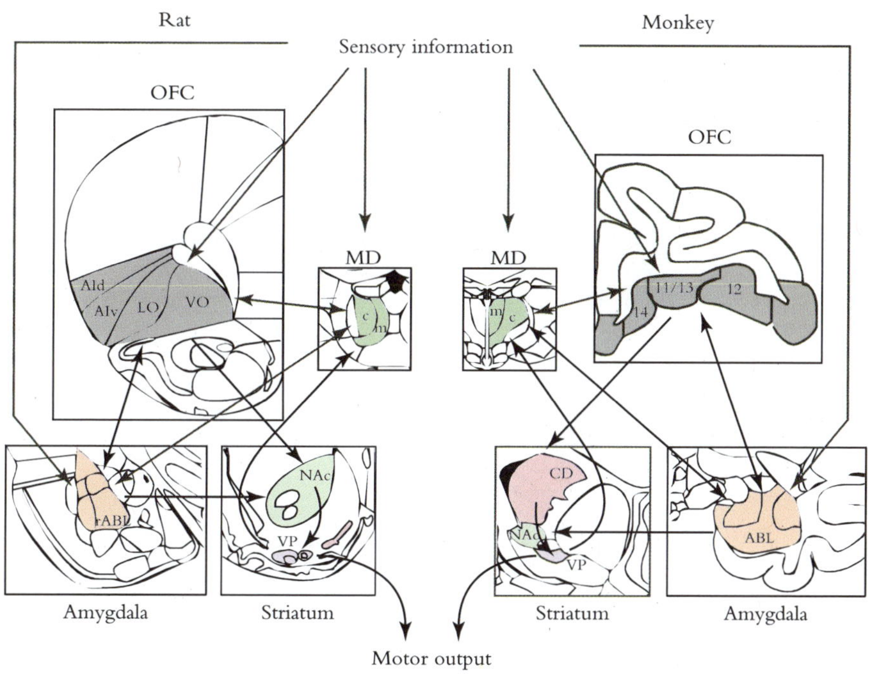

Figure 12.6 OFC comparison in rats and primates. While the prefrontal cortex has clearly expanded in primates, there still appears to be homology between the prefrontal cortex in rat (orbital and agranular insular areas) and primates (OFC, gray), as indicated by their similar patterns of connectivity with the mediodorsal thalamus (MD, green), amygdala (light red) and striatum/accumbens/pallidum system (pink, light green, and light purple). In both species, the OFC receives sensation input from sensory cortices and associative information from the amygdala and in both sends motor and limbic outputs to the striatum and nucleus accumbens. A coronal example is shown in each box. Abbreviations: AId, dorsal agranular insula; AIv, ventral agranular insula; c, central; CD, caudate; LO, lateral orbital; m, medial; NAc, nucleus accumbens core; rABL, rostral basolateral amygdala; VO, ventral orbital, including ventrolateral and ventromedial orbital regions; VP, ventral pallidum. (Modified, with permission, from Schoenbaum et al. 2006).

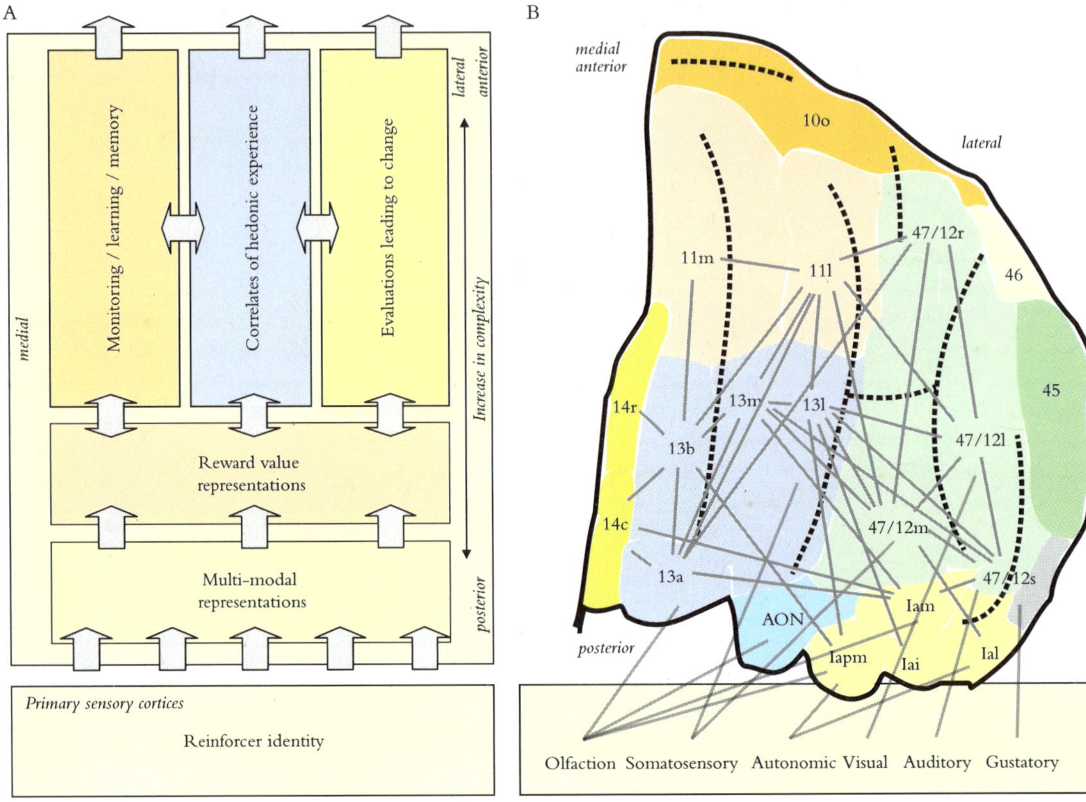

Figure 12.7 Model of the functions of the OFC. The proposed model shows the interactions between sensory and hedonic systems in the OFC using as an example one hemisphere of the OFC (Kringelbach, 2004b). Information is flowing from bottom to top on the figure. Sensory information arrives from the periphery to the primary sensory cortices, where the stimulus identity is decoded into stable cortical representations. This information is then conveyed for further multimodal integration in brain structures in the posterior parts of the OFC. The reward value of the reinforcer is assigned in more anterior parts of the OFC from where it can then be used to influence subsequent behavior (in lateral parts of the anterior OFC with connections to anterior cingulate cortex), stored for learning/memory (in medial parts of the anterior OFC), and made available for subjective hedonic experience (in mid-anterior OFC). The reward value and the subjective hedonic experience can be modulated by hunger and other internal states. In addition, there is important reciprocal information flowing between the various regions of the OFC and other brain regions involved in hedonic processing.

neuroimaging would seem to suggest that the valence of pleasures can be represented differently in different subparts of the orbitofrontal cortex. The activity (as indexed by the BOLD signal) in the medial OFC would appear to correlate with the valence of reinforcers such that positive reinforcers elicit a higher BOLD signal than negative reinforcers, which is consistent with a monitoring role for the medial OFC. The inverse appears to be true for the lateral parts of the OFC but with the important caveat that only the lateral parts are mostly concerned with those negative reinforcers that can bring about a change in behavior. Finally, the mid-anterior region of the OFC would appear to integrate the valence with state-dependent mechanisms such as selective satiation and orgasms, and is thus a candidate region for taking part in the mediation of subjective hedonic experience.

The proposed link to subjective hedonic processing places the OFC as an important gateway to subjective conscious experience. One possible way to conceptualize the role of the orbitofrontal and anterior cingulate cortices would be as part of a global workspace for access to consciousness with the specific role of evaluating the affective valence of stimuli (Dehaene et al., 1998). In this context, it is interesting that the medial parts of the OFC are part of a proposed network for the baseline activity of the human brain at rest (Gusnard and Raichle, 2001), as this would place the OFC as a key node in the network subserving consciousness. This could potentially explain why all experiences have a hedonic and emotional tone.

There are many interesting and important issues in pleasure research, which are not yet fully understood. We have still to understand the exact interactions and oscillations of the network of brain regions subserving hedonic processing. In particular, it is presently unclear which brain regions are necessary and sufficient for pleasure. It is also clear that although conscious appraisal of pleasure is usually what we mean by referring to pleasure, many emotional stimuli can be processed on a nonconscious level as demonstrated by subliminal priming (Naccache et al., 2005; Winkielman et al., 2005).

The most difficult question facing pleasure research remains the nature of the subjective experience of pleasure, and while some progress has been made, it is important not to overinterpret mere correlations from neuroimaging studies with the elusive qualities of subjective experience. It is still not clear how pleasure and happiness are linked (Schooler and Mauss, Chapter 14, this book), and in particular we have still not made substantial progress toward understanding the functional neuroanatomy of happiness.

Happiness cannot be reduced to pleasure alone, and pleasure is but a fleeting moment in the state, which is happiness. But the attainment of happiness must surely include the ready capacity for pleasure-elicited reactions. Some have suggested that "true" happiness or bliss might be a state of 'liking' without 'wanting'—which with the current available neuroscientific evidence is becoming a testable hypothesis.

In summary, pleasure and emotions are evolutionarily important for animals (including humans) in evaluating and preparing for appropriate actions. The evolution of conscious pleasure and emotion in humans could be adaptive, because they allow us to consciously appraise our emotions and actions, and subsequently to learn to manipulate these appropriately. Pleasure and emotion may be some of evolution's most productive breakthroughs, constantly reminding us that we are still animals at heart, but endowed with the possibility of enjoying our limited time on this planet and with the enhanced control of our subjective experience that comes with it.

Acknowledgments

I thank Kent C. Berridge, Predrag Petrovic, and J. Wayne Aldridge for helpful comments on earlier versions of this chapter. The research was supported by the Wellcome Trust, the Charles Wolfson Trust, the MRC, and the TrygFonden Charitable Foundation. I would like to thank Caroline Andrews, Tipu Aziz, Annie Cattrell, Phil Cowen, Ivan De Araujo, Alex Green, Peter Hobden, Julia Hornak, John O'Doherty, Edmund Rolls, Alan Stein, Birgit Völlm, and James Wilson for their collaboration on some of the work reviewed here.

References

Adolphs, R. (2003) Cognitive neuroscience of human social behavior. *Nat. Rev. Neurosci.* 4, 165–178.

Adolphs, R., Tranel, D., Damasio, H. and Damasio, A. (1994) Impaired recognition of emotion in facial expressions following bilateral damage to the human amygdala. *Nature* 372, 669–672.

Aharon, I., Etcoff, N., Ariely, D., Chabris, C. F., O'Connor, E. and Breiter, H. C. (2001) Beautiful faces have variable reward value: fMRI and behavioral evidence. *Neuron* 32, 537–551.

Akil, H., Mayer, D. J. and Liebeskind, J. C. (1976) Antagonism of stimulation-produced analgesia by naloxone, a narcotic antagonist. *Science* 191, 961–962.

Aldridge, J. W. and Berridge (this book) Neural coding of pleasure: "Rose-tinted glasses" of the ventral pallidum. In: *Pleasures of the Brain*. Eds. M. L. Kringelbach, K. C. Berridge. Oxford University Press: Oxford.

Anderson, A. K., Christoff, K., Stappen, I., Panitz, D., Ghahremani, D. G., Glover, G., Gabrieli, J. D. and Sobel, N. (2003) Dissociated neural representations of intensity and valence in human olfaction. *Nat. Neurosci.* 6, 196–202.

Andy, O. J. and Jurko, F. (1987) Thalamic stimulation effects on reactive depression. *Appl. Neurophysiol.* 50, 324–329.

Aziz, T. Z., Peggs, D., Sambrook, M. A. and Crossman, A. R. (1991) Lesion of the subthalamic nucleus for the alleviation of 1-methyl-4-phenyl-1,2,3,6-tetrahydropyridine (MPTP)-induced parkinsonism in the primate. *Mov. Disord.* 6, 288–292.

Baumeister, A. A. (2000) The Tulane Electrical Brain Stimulation Program a historical case study in medical ethics. *J. Hist. Neurosci.* 9, 262–278.

Baxter, M. G., Parker, A., Lindner, C. C., Izquierdo, A. D. and Murray, E. A. (2000) Control of response selection by reinforcer value requires interaction of amygdala and orbital prefrontal cortex. *J. Neurosci.* 20, 4311–4319.

Baylis, L. L. and Gaffan, D. (1991) Amygdalectomy and ventromedial prefrontal ablation produce similar deficits in food choice and in simple object discrimination learning for an unseen reward. *Exp. Brain Res.* 86, 617–622.

Berridge, K. C. (1996) Food reward: Brain substrates of wanting and liking. *Neurosci. Biobehav. Rev.* 20, 1–25.

Berridge, K. C. and Kringelbach, M. L. (2008) Affective neuroscience of pleasure: Reward in humans and animals. *Psychopharmacology* 199, 457–480.

Berridge, K. C. and Robinson, T. E. (1998) What is the role of dopamine in reward: Hedonic impact, reward learning, or incentive salience? *Brain Res. Rev.* 28, 309–369.

Bindra, D. (1978) How adaptive behavior is produced: a perceptual-motivational alternative to response-reinforcement. *Behav. Brain Sci.* 1, 41–91.

Bittar, R. G., Otero, S., Carter, H. and Aziz, T. Z. (2005c) Deep brain stimulation for phantom limb pain. *J. Clin. Neurosci.* 12, 399–404.

Bittar, R. G., Burn, S. C., Bain, P. G., Owen, S. L., Joint, C., Shlugman, D. and Aziz, T. Z. (2005a) Deep brain stimulation for movement disorders and pain. *J. Clin. Neurosci.* 12, 457–463.

Bittar, R. G., Kar-Purkayastha, I., Owen, S. L., Bear, R. E., Green, A., Wang, S. and Aziz, T. Z. (2005b) Deep brain stimulation for pain relief: a meta-analysis. *J. Clin. Neurosci.* 12, 515–519.

Blair, R. J. (2003) Facial expressions, their communicatory functions and neuro-cognitive substrates. *Philos. Trans. R. Soc. Lond. B. Biol. Sci.* 358, 561–572.

Blood, A. J. and Zatorre, R. J. (2001) Intensely pleasurable responses to music correlate with activity in brain regions implicated in reward and emotion. *Proc. Natl Acad. Sci. USA.* 98, 11818–11823.

Blood, A. J., Zatorre, R. J., Bermudez, P. and Evans, A. C. (1999) Emotional responses to pleasant and unpleasant music correlate with activity in paralimbic brain regions. *Nat. Neurosci.* 2, 382–387.

Bodamer, J. (1947) Die Prosop-Agnosie. *Arch. Psychiatr Nerve.* 179, 6–53.

Boivic, J. and Meyerson, B. A. (1982) A correlative anatomical and clinical study of pain suppression by deep brain stimulation. *Pain* 13, 113–126.

Bornstein, W. S. (1940a) Cortical representation of taste in man and monkey. I. Functional and anatomical relations of taste, olfaction, and somatic sensibility. *Yale J. Biol. Med.* 12, 719–736.

Bornstein, W. S. (1940b) Cortical representation of taste in man and monkey. II. The localization of the cortical taste area in man, a method of measuring impairment of taste in man. *Yale J. Biol. Med.* 13, 133–156.

Bowlby, J. (1957) An ethological approach to research in child development. *Br. J. Med. Psychot.* 30, 230–240.

Bowlby, J. (1969) *Attachment and Loss, Vol 1: Attachment.* Hogarth Press: London.

Breiter, H. C., Gollub, R. L., Weisskoff, R. M., Kennedy, D. N., Makris, N., Berke, J. D., Goodman, J. M., Kantor, H. L., Gastfriend, D. R., Riorden, J. P., Mathew, R. T., Rosen, B. R. and Hyman, S. E. (1997) Acute effects of cocaine on human brain activity and emotion. *Neuron* 19, 591–611.

Bruce, C., Desimone, R. and Gross, C. G. (1981) Visual properties of neurons in a polysensory area in superior temporal sulcus of the macaque. *J. Neurophysiol.* 46, 369–384.

Butter, C. M., Mishkin, M. and Rosvold, H. E. (1963) Conditioning and extinction of a food-rewarded response after selective ablations of frontal cortex in rhesus monkeys. *Exp. Neurol.* 7, 65–75.

Cabanac, M. (1971) Physiological role of pleasure. *Science* 173, 1103–1107.

Cabanac, M. (this book) The dialectics of pleasure. In: *Pleasures of the Brain.* Eds. M. L. Kringelbach, K. C. Berridge. Oxford University Press: Oxford.

Cerf-Ducastel, B. and Murphy, C. (2001) fMRI activation in response to odorants orally delivered in aqueous solutions. *Chem. Senses* 26, 625–637.

Chalmers, D. (1995) Facing up to the problem of consciousness. *J. Conscious. Stud.* 2, 200–219.

Cohn, J. F. and Tronick, E. Z. (1983) Three-month-old infants' reaction to simulated maternal depression. *Child Dev.* 54, 185–193.

Cooper, P. J. and Murray, L. (1998) Postnatal depression. *BMJ* 316, 1884–1886.

Critchley, H. D., Mathias, C. J. and Dolan, R. J. (2002) Fear conditioning in humans: the influence of awareness and autonomic arousal on functional neuroanatomy. *Neuron* 33, 653–663.

Cromwell, H. C. and Berridge, K. C. (1993) Where does damage lead to enhanced food aversion: the ventral pallidum/substantia innominata or lateral hypothalamus? *Brain Res.* 624, 1–10.

Cummings, J. L. (1992) Depression and Parkinson's disease: a review. *Am. J. Psychiatry* 149, 443–454.

Darwin, C. (1872) *The Expression of the Emotions in Man and Animals.* University of Chicago Press: Chicago, IL.

De Araujo, I. E. T., Kringelbach, M. L., Rolls, E. T. and Hobden, P. (2003a) The representation of umami taste in the human brain. *J. Neurophysiol.* 90, 313–319.

De Araujo, I. E. T., Kringelbach, M. L., Rolls, E. T. and McGlone, F. (2003b) Human cortical responses to water in the mouth, and the effects of thirst. *J. Neurophysiol.* 90, 1865–1876.

De Araujo, I. E. T., Rolls, E. T., Kringelbach, M. L., McGlone, F. and Phillips, N. (2003c) Taste-olfactory convergence, and the representation of the pleasantness of flavour, in the human brain. *Eur. J. Neurosci.* 18, 2059–2068.

Dehaene, S., Kerszberg, M. and Changeux, J. P. (1998) A neuronal model of a global workspace in effortful cognitive tasks. *Proc. Natl. Acad. Sci. USA* 95, 14529–14534.

Desimone, R. and Gross, C. G. (1979) Visual areas in the temporal cortex of the macaque. *Brain Res.* 178, 363–380.

Ekman, P. and Friesen, W.-V. (1971) Constants across cultures in the face and emotion. *J. Pers. Soc. Psychol.* 17(2), 124–129.

Fields, H. L. and Basbaum, A. (1999) Central nervous system mechanisms of pain modulation. In: *Textbook of Pain.* pp. 309–329. Eds. P. D. Wall, R. Melzack. Churchill Livingstone: Edinburgh.

Frey, S., Kostopoulos, P. and Petrides, M. (2000) Orbitofrontal involvement in the processing of unpleasant auditory information. *Eur. J. Neurosci.* 12, 3709–3712.

Frijda, N. (this book) On the nature and function of pleasure. In: *Pleasures of the Brain.* Eds. M. L. Kringelbach, K. C. Berridge. Oxford University Press: Oxford.

Frith, C. D. (2006) The value of brain imaging in the study of development and its disorders. *J. Child. Psychol. Psychiatry* 47, 979–982.

Fritsch, G. and Hitzig, E. (1870) Über die elektrische Erregbarkeit des Grosshirns. *Arch. Anat. Physiol.* 37, 300–332.

Geake, J. G. and Hansen, P. C. (2005) Neural correlates of intelligence as revealed by fMRI of fluid analogies. *Neuroimage* 26, 555–564.

Georgiadis, J. R. and Kortekaas, R. (this book) The sweetest taboo: Functional neurobiology of human sexuality in relation to pleasure. In: *Pleasures of the Brain.* Eds. M. L. Kringelbach, K. C. Berridge. Oxford University Press: Oxford.

Georgiadis, J. R., Kortekaas, R., Kuipers, R., Nieuwenburg, A., Pruim, J., Reinders, A. A. and Holstege, G. (2006) Regional cerebral blood flow changes associated with clitorally induced orgasm in healthy women. *Eur. J. Neurosci.* 24, 3305–3316.

Gottfried, J. A. (this book) Olfaction and its pleasures: Human neuroimaging perspectives. In: *Pleasures of the Brain*. Eds. M. L. Kringelbach, K. C. Berridge. Oxford University Press: Oxford.

Gottfried, J. A. and Dolan, R. J. (2003) The nose smells what the eye sees: Crossmodal visual facilitation of human olfactory perception. *Neuron* 39, 375–386.

Gottfried, J. A., O'Doherty, J. and Dolan, R. J. (2003) Encoding predictive reward value in human amygdala and orbitofrontal cortex. *Science* 301, 1104–1107.

Gottfried, J. A., Deichmann, R., Winston, J. S. and Dolan, R. J. (2002) Functional heterogeneity in human olfactory cortex: an event-related functional magnetic resonance imaging study. *J. Neurosci.* 22, 10819–10828.

Green, A. L., Pereira, E. A. and Aziz, T. Z. (this book) Deep brain stimulation and pleasure. In: *Pleasures of the Brain*. Eds. M. L. Kringelbach, K. C. Berridge. Oxford University Press: Oxford.

Grill, H. J. and Norgren, R. (1978) The taste reactivity test. II. Mimetic responses to gustatory stimuli in chronic thalamic and chronic decerebrate rats. *Brain Res.* 143, 281–297.

Gusnard, D. A. and Raichle, M. E. (2001) Searching for a baseline: Functional imaging and the resting human brain. *Nat. Rev. Neurosci.* 2, 685–694.

Hasselmo, M. E., Rolls, E. T., Baylis, G. C. and Nalwa, V. (1989) Object-centred encoding by face-selective neurons in the cortex in the superior temporal sulcus of the the monkey. Experimental *Brain Res.* 75, 417–429.

Haxby, J. V., Horwitz, B., Ungerleider, L. G., Maisog, J. M., Pietrini, P. and Grady, C. L. (1994) The functional organization of human extrastriate cortex: a PET–rCBF study of selective attention to faces and locations. *J. Neurosci.* 14, 6336–6353.

Heath, R. G. (1954) Psychiatry. *Annu. Rev. Med.* 5, 223–236.

Holstege, G., Georgiadis, J. R., Paans, A. M., Meiners, L. C., van der Graaf, F. H. and Reinders, A. A. (2003) Brain activation during human male ejaculation. *J. Neurosci.* 23, 9185–9193.

Hornak, J., Bramham, J., Rolls, E. T., Morris, R. G., O'Doherty, J., Bullock, P. R. and Polkey, C. E. (2003) Changes in emotion after circumscribed surgical lesions of the orbitofrontal and cingulate cortices. *Brain* 126, 1671–1712.

Horsley, V. and Clarke, R. H. (1908) The Structure and functions of the cerebellum examined by a new method. *Brain* 31, 45–124.

Hosobuchi, Y., Adams, J. E. and Linchitz, R. (1977) Pain relief by electrical stimulation of the central gray matter in humans and its reversal by naloxone. *Science* 197, 183–186.

Hull, C. L. (1951) *Essentials of Behavior*. Yale University Press: New Haven, CT.

James, W. (1890) *The Principles of Psychology*. Henry Holt: New York.

Jenkinson, N., Nandi, D., Aziz, T. Z. and Stein, J. F. (2005) Pedunculopontine nucleus: a new target for deep brain stimulation for akinesia. *Neuroreport* 16, 1875–1876.

Kinomura, S., Kawashima, R., Yamada, K., Ono, S., Itoh, M., Yoshioka, S., Yamaguchi, T., Matsui, H., Miyazawa, H., Itoh, H. and et al. (1994) Functional anatomy of taste perception in the human brain studied with positron emission tomography. *Brain Res.* 659, 263–266.

Komisaruk, B. R., Whipple, B. and Beyer, C. (this book) Sexual pleasure. In: *Pleasures of the Brain*. Eds. M. L. Kringelbach, K. C. Berridge. Oxford University Press: Oxford.

Krack, P., Batir, A., Van Blercom, N., Chabardes, S., Fraix, V., Ardouin, C., Koudsie, A., Limousin, P. D., Benazzouz, A., LeBas, J. F., Benabid, A. L. and Pollak, P. (2003) Five-year follow-up of bilateral stimulation of the subthalamic nucleus in advanced Parkinson's disease. *N. Engl. J. Med.* 349, 1925–1934.

Kringelbach, M. L. (2004a) Emotion. In: *The Oxford Companion to the Mind*, 2nd ed. pp. 287–290. Ed. R. L. Gregory. Oxford University Press: Oxford.

Kringelbach, M. L. (2004b) Food for thought: Hedonic experience beyond homeostasis in the human brain. *Neuroscience* 126, 807–819.

Kringelbach, M. L. (2004c) *Hjernerum. Den følelsesfulde hjerne*. People's Press: Copenhagen.

Kringelbach, M. L. (2005) The orbitofrontal cortex: Linking reward to hedonic experience. *Nat. Rev. Neurosci.* 6, 691–702.

Kringelbach, M. L. (2006) Cortical systems involved in appetite and food consumption. In: *Appetite and Body Weight: Integrative Systems and the Development of Anti-Obesity Drugs*. pp. 5–26. Eds. S. J. Cooper, T. C. Kirkham. Elsevier: London.

Kringelbach, M. L. (2009) *The Pleasure Center. Trust your animal instincts*. Oxford University Press: Oxford.

Kringelbach, M. L. and Rolls, E. T. (2003) Neural correlates of rapid context-dependent reversal learning in a simple model of human social interaction. *Neuroimage* 20, 1371–1383.

Kringelbach, M. L. and Rolls, E. T. (2004) The functional neuroanatomy of the human orbitofrontal cortex: Evidence from neuroimaging and neuropsychology. *Prog. Neurobiol.* 72, 341–372.

Kringelbach, M. L., de Araujo, I. E. and Rolls, E. T. (2004) Taste-related activity in the human dorsolateral prefrontal cortex. *Neuroimage* 21, 781–788.

Kringelbach, M. L., Owen, S. L. F. and Aziz, T. Z. (2007c) Deep brain stimulation. *Future Neurol.* 2, 633–646.

Kringelbach, M. L., Jenkinson, N., Owen, S. L. F. and Aziz, T. Z. (2007b) Translational principles of deep brain stimulation. *Nat. Rev. Neurosci.* 8, 623–635.

Kringelbach, M. L., O'Doherty, J., Rolls, E. T. and Andrews, C. (2003) Activation of the human orbitofrontal cortex to a liquid food stimulus is correlated with its subjective pleasantness. *Cereb. Cortex* 13, 1064–1071.

Kringelbach, M. L., Jenkinson, N., Green, A. L., Owen, S. L. F., Hansen, P. C., Cornelissen, P. L., Holliday, I. E., Stein, J. and Aziz, T. Z. (2007a) Deep brain stimulation for chronic pain investigated with magnetoencephalography. *Neuroreport* 18, 223–228.

Kringelbach, M. L., Lehtonen, A., Squire, S., Harvey, A. G., Craske, M. G., Holliday, I. E., Green, A. L., Aziz, T. Z., Hansen, P. C., Cornelissen, P. L. and Stein, A. (2008) A specific and rapid neural signature for parental instinct. *PLoS ONE* 3, e1664. doi:1610.1371/journal.pone.0001664.

Layard, R. (2005) *Happiness: Lessons from a New Science*. Penguin Press: London.

Leone, M., Franzini, A., Broggi, G., May, A. and Bussone, G. (2004) Long-term follow-up of bilateral hypothalamic stimulation for intractable cluster headache. *Brain* 127, 2259–2264.

Lorenz, K. (1943) Die angeborenen Formen Möglicher Erfahrung. [Innate forms of potential experience]. *Z. Tierpsychol.* 5, 235–519.

Lorenz, K. (1971) *Studies in Animal and Human Behavior,* Vol. II. Methuen: London.

Mayberg, H. S., Lozano, A. M., Voon, V., McNeely, H. E., Seminowicz, D., Hamani, C., Schwalb, J. M. and Kennedy, S. H. (2005) Deep brain stimulation for treatment-resistant depression. *Neuron* 45, 651–660.

Mazzone, P., Lozano, A. M., Stanzione, P., Galati, S., Scanati, E., Peppe, A. and Stefani, A. (2005) Implantation of human pedunculopontine nucleus: a safe and clinically relevant target in Parkinson's disease. *Neuroreport* 16, 1877–1881.

Miller, J. M., Vorel, S. R., Tranguch, A. J., Kenny, E. T., Mazzoni, P., van Gorp, W. G. and Kleber, H. D. (2006) Anhedonia after a selective bilateral lesion of the globus pallidus. *Am. J. Psychiatry* 163, 786–788.

Murray, L., Fiori-Cowley, A., Hooper, R. and Cooper, P. (1996) The impact of postnatal depression and associated adversity on early mother-infant interactions and later infant outcome. *Child. Dev.* 67, 2512–2526.

Naccache, L., Gaillard, R., Adam, C., Hasboun, D., Clemenceau, S., Baulac, M., Dehaene, S. and Cohen, L. (2005) A direct intracranial record of emotions evoked by subliminal words. *Proc. Natl. Acad. of Sci. USA* 102, 7713–7717.

Nandi, D., Liu, X., Joint, C., Stein, J. and Aziz, T. (2002) Thalamic field potentials during deep brain stimulation of periventricular gray in chronic pain. *Pain* 97, 47–51.

Neveu Kringelbach, H. (2005) Encircling the Dance: Social Mobility Through the Transformation of Performance in Urban Senegal. D. Phil. Thesis. University of Oxford: Oxford.

O'Doherty, J., Kringelbach, M. L., Rolls, E. T., Hornak, J. and Andrews, C. (2001a) Abstract reward and punishment representations in the human orbitofrontal cortex. *Nat. Neurosci.*4, 95–102.

O'Doherty, J., Rolls, E. T., Francis, S., Bowtell, R. and McGlone, F. (2001b) Representation of pleasant and aversive taste in the human brain. *J. Neurophysiol.* 85, 1315–1321.

O'Doherty, J., Rolls, E. T., Francis, S., Bowtell, R., McGlone, F., Kobal, G., Renner, B. and Ahne, G. (2000) Sensory-specific satiety-related olfactory activation of the human orbitofrontal cortex. *Neuroreport* 11, 893–897.

O'Hara, M. W. and Swain, A. M. (1996) Rates and risk of postpartum depression—a meta-analysis. *Int. Rev. Psychiatry* 8, 37–54.

Öngür, D. and Price, J. L. (2000) The organization of networks within the orbital and medial prefrontal cortex of rats, monkeys and humans. *Cereb.Cortex* 10, 206–219.

Perlmutter, J. S. and Mink, J. W. (2006) Deep brain stimulation. *Annu. Rev. Neurosci.* 29, 229–257.

Perrett, D. I., Rolls, E. T. and Caan, W. (1982) Visual neurons responsive to faces in the monkey temporal cortex. *Exp. Brain Res.* 47, 329–342.

Petrovic, P. and Ingvar, M. (2002) Imaging cognitive modulation of pain processing. *Pain* 95, 1–5.

Petrovic, P., Kalso, E., Petersson, K. M. and Ingvar, M. (2002) Placebo and opioid analgesia—imaging a shared neuronal network. *Science* 295, 1737–1740.

Pierce, J. and Halpern, B. P. (1996) Orthonasal and retronasal odorant identification based upon vapor phase input from common substances. *Chem. Senses* 21, 529–543.

Plaha, P. and Gill, S. G. (2005) Bilateral deep brain stimulation of the pedunculopontine nucleus for idiopathic Parkinson's disease. *Neuroreport* 16, 1883–1887.

Pritchard, T. C., Edwards, E. M., Smith, C. A., Hilgert, K. G., Gavlick, A. M., Maryniak, T. D., Schwartz, G. J. and Scott, T. R. (2005) Gustatory neural responses in the medial orbitofrontal cortex of the old world monkey. *J. Neurosci.* 25, 6047–6056.

Ranote, S., Elliott, R., Abel, K. M., Mitchell, R., Deakin, J. F. and Appleby, L. (2004) The neural basis of maternal responsiveness to infants: an fMRI study. *Neuroreport* 15, 1825–1829.

Ray, N. J., Kringelbach, M. L., Jenkinson, N., Owen, S. L. F., Davies , P., Wang, S., De Pennington, N., Hansen, P. C., Stein, J. and Aziz, T. Z. (2007) Using magnetoencephalography to investigate deep brain stimulation for cluster headache. *BIIJ.* 3, e25.

Reck, C., Hunt, A., Fuchs, T., Weiss, R., Noon, A., Moehler, E., Downing, G., Tronick, E. Z. and Mundt, C. (2004) Interactive regulation of affect in postpartum depressed mothers and their infants: an overview. *Psychopathology* 37, 272–280.

Reynolds, D. V. (1969) Surgery in the rat during electrical analgesia induced by focal brain stimulation. *Science* 164, 444–445.

Rolls, E. T. (1999) *The Brain and Emotion.* Oxford University Press: Oxford.

Rolls, E. T., Kringelbach, M. L. and de Araujo, I. E. T. (2003a) Different representations of pleasant and unpleasant odors in the human brain. *Eur. J. Neurosci.*18, 695–703.

Rolls, E. T., O'Doherty, J., Kringelbach, M. L., Francis, S., Bowtell, R. and McGlone, F. (2003b) Representations of pleasant and painful touch in the human orbitofrontal and cingulate cortices. *Cereb. Cortex* 13, 308–317.

Royet, J. P., Hudry, J., Zald, D. H., Godinot, D., Gregoire, M. C., Lavenne, F., Costes, N. and Holley, A. (2001) Functional neuroanatomy of different olfactory judgments. *Neuroimage* 13, 506–519.

Rozin, P. (2001) Food Preference. In: *International Encyclopedia of the Social & Behavioral Sciences.* Eds. N. J. Smelser, P. B. Baltes. Elsevier: Amsterdam.

Ryle, G. (1954) Pleasure. *Proc. Aristote. Soc.* 28, 135–146.

Saper, C. B., Chou, T. C. and Elmquist, J. K. (2002) The need to feed: Homeostatic and hedonic control of eating. *Neuron* 36, 199–211.

Schlaepfer, T. E., Cohen, M. X., Frick, C., Kosel, M., Brodesser, D., Axmacher, N., Joe, A. Y., Kreft, M., Lenartz, D. and Sturm, V. (2007) Deep brain stimulation to reward circuitry alleviates anhedonia in refractory major depression. *Neuropsychopharmacology* 33, 368–377.

Schoenbaum, G., Roesch, M. R. and Stalnaker, T. A. (2006) Orbitofrontal cortex, decision-making and drug addiction. *Trends Neurosci.* 29, 116–124.

Schooler, J. W. and Mauss, I. B. (this book) To be happy and to know it: the experience and meta-awareness of pleasure. In: *Pleasures of the Brain.* Eds. M. L. Kringelbach, K. C. Berridge. Oxford University Press: Oxford, UK.

Scott, T. R., Yaxley, S., Sienkiewicz, Z. J. and Rolls, E. T. (1986) Gustatory responses in the frontal opercular cortex of the alert cynomolgus monkey. *J. Neurophysiol.* 56, 876–890.

Sergent, J. and Villemure, J. G. (1989) Prosopagnosia in a right hemispherectomized patient. *Brain* 112 (Pt 4), 975–995.

Skov, M. (this book) The pleasures of art. In: *Pleasures of the Brain.* Eds. M. L. Kringelbach, K. C. Berridge. Oxford University Press: Oxford.

Small, D. M., Gerber, J. C., Mak, Y. E. and Hummel, T. (2005) Differential neural responses evoked by orthonasal versus retronasal odorant perception in humans. *Neuron* 47, 593–605.

Small, D. M., Jones-Gotman, M., Zatorre, R. J., Petrides, M. and Evans, A. C. (1997) Flavor processing: More than the sum of its parts. *Neuroreport* 8, 3913–3917.

Small, D. M., Zatorre, R. J., Dagher, A., Evans, A. C. and Jones-Gotman, M. (2001) Changes in brain activity related to eating chocolate: From pleasure to aversion. *Brain* 124, 1720–1733.

Small, D. M., Gregory, M. D., Mak, Y. E., Gitelman, D., Mesulam, M. M. and Parrish, T. (2003) Dissociation of neural representation of intensity and affective valuation in human gustation. *Neuron* 39, 701–711.

Small, D. M., Zald, D. H., Jones-Gotman, M., Zatorre, R. J., Pardo, J. V., Frey, S. and Petrides, M. (1999) Human cortical gustatory areas: a review of functional neuroimaging data. *Neuroreport* 10, 7–14.

Smith, K. S., Mahler, S. V., Pecina, S. and Berridge, K. C. (this book) Hedonic hotspots: Generating sensory pleasure in the brain. In: *Pleasures of the Brain.* Eds. M. Kringelbach, K. C. Berridge. Oxford University Press: Oxford.

Spiegel, E. A., Wycis, H. T., Marks, M. and Lee, A. J. (1947) Stereotaxic apparatus for operations on the human brain. *Science* 106, 349–350.

Stein, A., Gath, D. H., Bucher, J., Bond, A., Day, A. and Cooper, P. J. (1991) The relationship between post-natal depression and mother-child interaction. *Br. J. Psychiatry* 158, 46–52.

Steiner, J. E., Glaser, D., Hawilo, M. E. and Berridge, K. C. (2001) Comparative expression of hedonic impact: Affective reactions to taste by human infants and other primates. *Neurosci. Biobehav. Rev.* 25, 53–74.

Swain, J. E., Lorberbaum, J. P., Kose, S. and Strathearn, L. (2007) Brain basis of early parent-infant interactions: Psychology, physiology, and in vivo functional neuroimaging studies. *J. Child Psychol. Psychiatry* 48, 262–287.

Tronick, E. D., Als, H. and Brazelton, T. B. (1977) Mutuality in mother-infant interaction. *J. Commun.* 27, 74–79.

Valenstein, E. S. (1973) *Brain Control: a Critical Examination of Brain Stimulation and Psychosurgery.* Wiley-Interscience: London.

Veldhuizen, M. G., Rudenga, K. J. and Small, D. M. (this book) The pleasure of taste, flavor and food. In: *Pleasures of the Brain.* Eds. M. L. Kringelbach, K. C. Berridge. Oxford University Press: Oxford, U.K.

Völlm, B. A., de Araujo, I. E. T., Cowen, P. J., Rolls, E. T., Kringelbach, M. L., Smith, K. A., Jezzard, P., Heal, R. J. and Matthews, P. M. (2004) Methamphetamine activates reward circuitry in drug naïve human subjects. *Neuropsychopharmacology* 29, 1715–1722.

Vuust, P. and Kringelbach, M. L. (this book) The pleasure of music. In: *Pleasures of the Brain.* Eds. M. L. Kringelbach, K. C. Berridge. Oxford University Press: Oxford.

Watson, K. K., Shepherd, S. V. and Platt, M. L. (this book) Neuroethology of pleasure. *In: Pleasures of the Brain.* Eds. M. L. Kringelbach, K. C. Berridge. Oxford University Press: Oxford.

Winkielman, P., Berridge, K. C. and Wilbarger, J. L. (2005) Unconscious affective reactions to masked happy versus angry faces influence consumption behavior and judgments of value. *Pers. Soc. Psychol. Bull.* 31, 121–135.

Zald, D. H. and Pardo, J. V. (1997) Emotion, olfaction, and the human amygdala: Amygdala activation during aversive olfactory stimulation. *Proc. Natl. Acad. Sci. USA* 94, 4119–4124.

13

The Neurobiology of Desire: Dopamine and the Regulation of Mood and Motivational States in Humans

MARCO LEYTON

In most contemporary theories of emotion, the processes of desire, happiness, and pleasure are separable yet affect each other (Lazarus, 1991; Rolls, 2005) (see also Bindra, 1969, 1978; LeDoux, 1989, 2000; Toates, 1986). Desire reflects the focused interest in a goal object and the drive to obtain it. Happiness is an affective state linked to the appraisal that progress is being made. Pleasure, in comparison, is a positive response to obtaining the goal.

Mesocorticolimbic dopamine (DA) transmission has been proposed to influence each of the above processes. Recent work, however, suggests that DA strongly influences sustained interest and approach, weakly influences positive emotions, and affects pleasure only tenuously, if at all. This chapter reviews this material, focusing on studies in humans.

Background

Fifty years ago Arvid Carlsson and colleagues (1957) reported that administration of the immediate DA precursor, 3,4-dihydroxyphenylalanine (L-DOPA), restored motor activity in immobile reserpinized rabbits, and did so by replenishing DA rather than norepinephrine (NE). By the 1970s, it was evident that DA also contributed to a more specific type of movement, approach toward rewards (Janssen et al., 1968; Phillips and Fibiger, 1973; Ungerstedt, 1971; Wilson and Schuster, 1972; Yokel and Wise, 1975). The subsequent observation that lowered DA transmission could disrupt goal-directed behaviors in the absence of motor deficits led to the proposal that the amine could affect reward processing directly (Mogenson et al., 1979; Yokel and Wise, 1975; Zarevics and Setler, 1979).[1]

A number of mechanisms by which DA could affect reward-related behaviors have been suggested. One of the first proposals was that DA mediates the pleasure commonly associated with reward (Wise, 1982). This bold hypothesis caught the imagination of many, both animal researchers and those who studied humans. The proposal was largely based on two main observations. First, interactions with pleasurable stimuli (e.g., food, rewarding drugs, etc.) increased limbic DA release (Di Chiara and Imperato, 1988; Hernandez and Hoebel, 1988). Second, DA receptor antagonists and lesions disrupted instrumental responding for rewards, and did so with distinctive features (de Wit and Wise, 1977; Wise et al., 1978; Yokel and Wise, 1975). Low doses of DA antagonists increased bar pressing for psychostimulant drugs, seeming to do so as if the pleasure had been "diluted," while high doses of DA antagonists elicited an extinction-like response, initially increasing and then decreasing bar pressing, as if the animal was discovering that the "fun" was now gone completely (Yokel and Wise, 1975).

Subsequent studies provided additional support for the proposition that DA transmission was important for the ability of reward objects to sustain interest, but this did not appear to be due to an effect on pleasure. For example, by the 1990s, it was clear that there were circumstances when DA cell firing and release were elicited by previously learned cues that predicted reward but not by actual receipt of the reward even though the reward was, presumably, still pleasurable (Gratton and Wise, 1994; Kiyatkin and Gratton, 1994; Schultz and Romo, 1990). Particularly influential was evidence that DA antagonists and lesions had no effect on an orofacial index of pleasure responses to taste stimuli (Berridge et al., 1989; Berridge, 2007b).

Based on the above and additional observations, a number of alternate hypotheses of DA's role have been suggested. The best studied and supported of these alternate hypotheses are that DA influences the expectation of reward (Redgrave et al., 1999; Schultz, 1998), the ability of reward-paired cues to elicit and sustain interest (Beninger, 1983; Berridge, 2007a; Stewart, 1992), and the learning of associations between rewards and their predictive cues (Beninger, 1983; Everitt et al., 1999; Schultz et al., 1997; Wise, 2004). An integrated model suggested here is that phasic bursts in DA cell firing shift attention toward events that might be worth getting; enduring increases in DA transmission, in comparison, sustain interest in the reward-related stimuli, and, as a consequence facilitate learned associations and goal-directed behavior (Beninger, 1983; Berridge, 2007a; Grace et al., 2007; Redgrave et al., 1999; Robbins et al., 1989; Salamone, 1994; Salamone et al., 2007; Schultz, 2007; Stewart, 1992). These DA-dependent changes in the incentive salience of reward-related cues might intersect with non-DA mediated pleasure (e.g., opioid, gamma-amino butyric acid [GABA], endocannabinoid) (Kringelbach, 2005; Peciña et al., 2006; Smith and Berridge, 2007; Zhang and Kelley, 2000). The result, it is proposed, is a coordinated attentional-cognitive-appetitive process for the generation and expression of positive mood and motivational states. With the development of new methods for measuring and manipulating DA release, it has become possible to test critical components of this model in humans. This chapter reviews the evidence.

What Causes Dopamine Release in Humans?

In the model proposed above, mesolimbic DA transmission sustains interest in rewards and reward-paired stimuli. This focused interest directs behavior toward the goal object. The evaluation that progress is being made toward this goal—conscious or otherwise—augments desire and elicits positive emotions. If the goal is obtained and enjoyed, pleasure ensues, an effect that then modulates interest, desire, and happiness. In this way, DA is directly related to sustained interest only, but influences—and is influenced by—the other states.

If the proposed model is correct, one would predict that a wide range of motivationally relevant stimuli should increase limbic DA transmission. Recent developments in functional neuroimaging methods have made it possible to test this prediction in humans.[2]

Extracellular DA levels in the human limbic striatum are increased in response to eating palatable food (Small et al 2003), exposure to food- and drug-related cues (Boileau et al., 2007; Volkow et al., 2002, 2006; Wong et al., 2006), winning money while performing a card selection task (Zald et al., 2004), and the ingestion of drugs of abuse. This effect of abused drugs has been seen across pharmacological classes (Figure 13.1), including alcohol (Boileau et al., 2003; Yoder et al., 2007) (though see Yoder et al., 2005), tobacco (Barrett et al., 2004; Brody et al., 2004, 2006; Scott et al., 2007a), amphetamine (Abi-Dargham et al., 2003; Boileau et al., 2006, 2007; Breier et al., 1997; Casey et al., 2007; Drevets et al., 2001; Gravel et al., 2007; Laruelle et al., 1995; Leyton et al., 2002, 2003; Martinez et al., 2003, 2007; Munro et al., 2006; Oswald et al., 2005, 2007; Riccardi et al., 2006; Volkow et al., 1994, 1997, 1999, 2001), and cocaine (Cox et al., 2009; Schlaepfer et al., 1997).

In the majority of neuroimaging studies, individual differences in the magnitude of the DA response correlated with positive affective states and pleasure (Abi-Dargham et al., 2003; Barrett et al., 2004; Boileau et al., 2006, 2007; Drevets et al., 2001; Laruelle et al., 1995; Martinez et al., 2003; Munro et al., 2006; Oswald et al., 2005, 2007; Volkow et al., 1999). These associations were presented as support for the pleasure hypothesis of DA, leading to some renewed interest in this view (Abi-Dargham et al., 2003; Drevets et al., 2001; Laruelle et al., 1995; Volkow et al., 1999). In comparison, in a smaller number of the neuroimaging studies, craving responses were also measured, and in some (Evans et al., 2006; Leyton et al., 2002; Volkow et al., 2002, 2006; Wong et al., 2006), though not in all studies (Boileau et al., 2006, 2007; Cox et al., 2009; Oswald et al., 2005; Small et al., 2003), there was a stronger association with desire for the reward than pleasure (Figure 13.2).

In animal studies, DA release is not specific to events that yield pleasure. That is, the release of DA is also increased by aversive or stressful stimuli (Ikemoto and Panksepp, 1999; Salamone, 1994). One interpretation is that stress-induced DA release facilitates coping, and does so via increased motivational drive to get

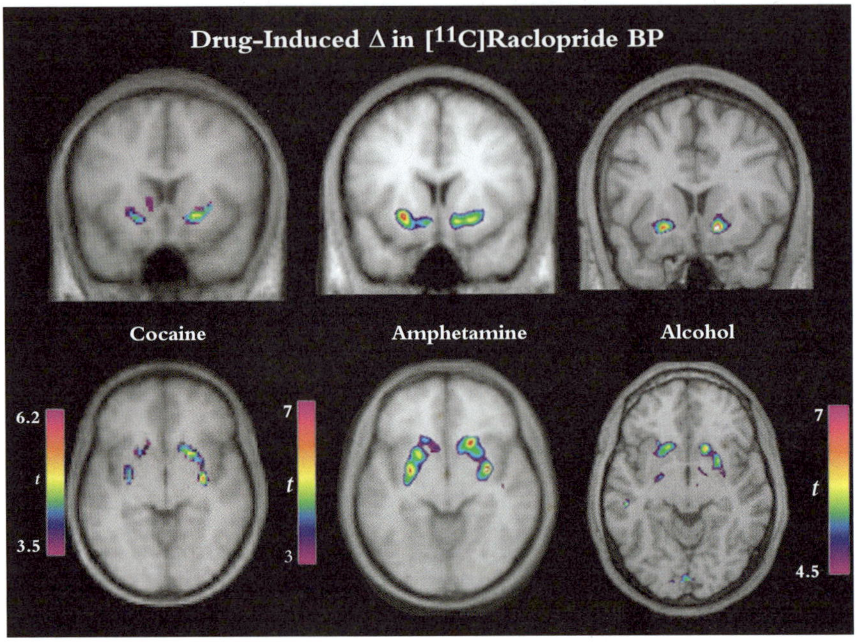

Figure 13.1 The figure depicts the ability of cocaine (left, 1.0mg/kg, taken intra-nasally), D-amphetamine (middle, 0.3 mg/kg, taken orally), and alcohol (right, 1 ml/kg, taken orally) to increase extracellular levels of dopamine in human striatum, as measured by positron emission tomography (PET) and changes in the binding of [^{11}C]raclopride. The colored images are t-maps, and they depict those regions where there were significant changes in tracer binding on the drug vs. placebo test session. The t-maps are superimposed on anatomical magnetic resonance images (MRI) from participants in the studies. Right side is on right. [Data are from Cox et al. (2009), Leyton et al. (2002, 2003) and Boileau et al. (2003).]

somewhere better (Ikemoto and Panksepp, 1999; Jensen et al., 2007; Menon et al., 2007). For example, stress-induced DA release (Abercrombie et al., 1989; Anstrom and Woodward, 2005; Wu et al., 1999; Young et al., 1998) elicits goal-directed behavior (Antelman and Szechtman, 1975; Kalivas and Stewart, 1991; Katz et al., 1980; Leyton and Stewart, 1990, 1996; Redgrave et al., 1999; Shaham and Stewart, 1996). If and when the animal gives up—for example, if the stressor is prolonged and uncontrollable, or the animal exhibits behavioral despair—then DA release decreases (Cabib and Puglisi-Allegra, 1994; Puglisi-Allegra et al., 1991; Rossetti et al., 1993). Of note, DA receptor antagonists diminish motivation to avoid aversive events and the stimuli that predict them (Blackburn et al., 1992; Killcross et al., 1997; Salamone, 1994) just as they decrease approach toward rewards and conditioned reward stimuli (Berridge and Robinson, 2003; Ikemoto and Panksepp, 1999; Phillips et al., 1991; Stewart et al., 1984).

In humans also, striatal DA release can be produced by events and activities that are not necessarily pleasurable. This includes participating in a challenging video game (Koepp et al., 1998), performing math tasks (Volkow et al., 2004), exposure to a psychosocial stressor (Booij et al., 2007; Pruessner et al., 2004), and the experimental induction of pain (Scott et al., 2006; Wood et al., 2007). In these studies, higher DA release predicted better performance during the video game (Koepp et al., 1998), greater interest in the math (Volkow et al., 2004), greater stress-induced cortisol release (Pruessner et al., 2004), greater pain (Scott et al., 2006; Wood et al., 2007), greater expectation of a placebo-induced alleviation of pain (Scott et al., 2007b), and a smaller stress-induced mood-lowering response (Booij et al., 2007).

What Are the Effects of Increasing Dopamine Transmission in Humans?

If the proposed model about DA in interest and focused attention is correct, changes to DA should alter each of

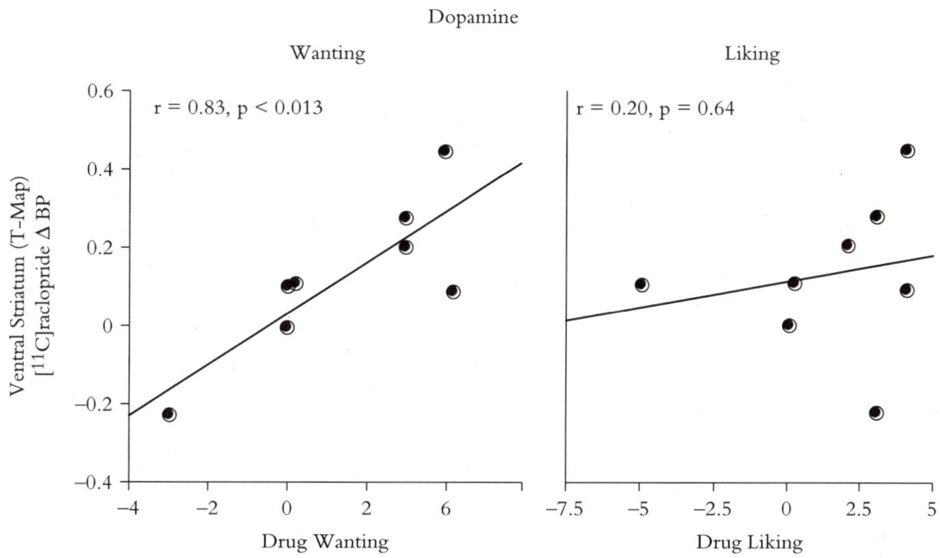

Figure 13.2 The figure depicts correlations between d-amphetamine-induced increases in extracellular dopamine levels within the ventral limbic striatum and drug-induced drug "wanting" (left) and drug-induced drug "liking" (right). Only the former association was statistically significant, suggesting that dopamine transmission in the limbic striatum is more closely related to the desire for reward than pleasure, per se. [Data are from Leyton et al. (2002).]

the following behaviors, but with successively diminishing efficacy: (i) the ability of reward stimuli to preferentially elicit and sustain interest (implicit, "unconscious" desire), (ii) conscious desire or craving, and (iii) positive emotional states. By contrast, pleasure should not be affected at all.

One approach to testing these predictions is to measure the effects of increasing DA transmission in humans. For example, the administration of psychostimulant drugs increases drug craving (Breiter et al., 1997; Jaffe et al., 1989; Leyton et al., 2005), induces positive mood states and pleasure (Leyton et al., 2005; Resnick et al., 1977), and enhances selective attention (Servan-Schreiber et al., 1998), preferential responding to reward-paired cues (de Wit et al., 2002), and delay-discounting performance (de Wit et al., 2002), a change that may reflect an improved ability to sustain interest in distal rewards.

These varied effects of stimulant drugs noted, there are strikingly few studies testing the effects of drugs that increase DA neurotransmission selectively. Contrary to what would be predicted by the anhedonia hypothesis, administration of the direct DA agonist, apomorphine, does not cause euphoria (e.g., Hollander et al., 1990; Wiesbeck et al., 1995). Nor, typically, does pergolide, though it does increase the self-rated confidence of having previously seen stimuli that had been presented at the threshold of conscious awareness (Lou, 2007). Moreover, when stimulants are given in a moderately low dose range that increases extracellular DA levels in striatal and extrastriatal limbic regions by up to 500%, the subjective effects are remarkably mild (Gravel et al., 2007; Leyton et al., 2002). Indeed, in our neuroimaging studies, most subjects who receive such doses are unable to articulate whether they have received placebo or amphetamine (0.3 mg/kg, p.o.) while laying quietly in the positron emission tomography (PET) scanner (unpublished observation). Together, these studies raise the possibility that many of the reported subjective effects of stimulant drugs are produced by actions on non-DA systems.[3]

What are the Effects of Decreasing Dopamine Transmission in Humans?

An alternative approach to investigating the role of DA is to examine the effects of selectively decreasing DA transmission. Although the literature is small, the most consistently reported effects of decreased DA transmission are on selective attention (Ahveninen

Table 13.1 Low Dopamine Related Changes to Selective Attention.

Authors	Subjects	Dopamine Challenge	Effects on Selective Attention	Other Effects/Comment
Clark (1986)	12 healthy subjects	Droperidol (0, 15μg/kg, iv) DA D2 antagonist	Decreased selective detection of target words	Slowed reaction time, and induced subjective indifference to stimuli. Methylphenidate reversed all droperidol effects. Droperidol also has antiadrenergic effects, more so than haloperidol.
Shelley (1987)	10 healthy subjects	Droperidol (0, 15μg/kg, iv)	Decreased ERP PN response to attended auditory stimuli	Reduced self-report alertness, and attention, increased depressed rating, and slowed reaction times plus decreased CPT hit rates.
Magliozzi (1989)	21 healthy subjects	Haloperidol (4, 10 mg, po)	Decreased attention on SDST	Effects dose dependent and specific to SDST suggesting effects not due to sedation.
Ahveninen (2000)	11 healthy subjects	Haloperidol (0, 2mg, po)	Decreased 40-Hz response to the attended auditory stimuli	Drug also binds to sigma adrenergic, GABAergic and serotonergic sites.
Kähkönen (2001)	12 healthy subjects	Haloperidol (0, 2mg, po)	Decreased ERP PN response to attended auditory stimuli	
Franken (2004)	17 detoxified heroin addicts	Haloperidol (0, 2mg, po)	Decreased ability of heroin cues to delay reaction times in Stroop task	Did not affect self-reported craving.
McLean (2004)	19 healthy subjects	APTD	Biased responding toward negative cues on affective Go/No-Go task	
Saeedi (2006)	59 healthy subjects	Haloperidol (0, 1, 3, 5mg po)	Decreased sustained attention on CPT	Reduced self-reported contentment.
Leyton (2007)	14 healthy subjects	APTD	Decreased preferential responding to reward paired cues on Go/No-Go	
Munafò (2007)	20 nicotine-dependent smokers	APTD	Decreased ability of smoking cues to delay reaction times in Stroop task.	Effect seen in women only.

Abbreviations: ERP, event-related potential. PN, processing negativity. CPT, continuous performance task. SDST, symbol-digit substitution test. APTD, acute phenylalanine/tyrosine depletion.

et al., 2000; Clark et al., 1986; Kahkonen et al., 2001; Shelley et al., 1997) and the ability to sustain it (Magliozzi et al., 1989; Saeedi et al., 2006) (see also Table 13.1). Lowered DA transmission also perturbs the ability to switch cognitive sets (Berger et al., 1989; Vitiello et al., 1997), decreases the ability to adjust betting behavior when reward parameters change in a gambling task (McLean et al., 2004; Roiser et al., 2005; Scarná et al., 2005), biases responding toward negatively valenced stimuli on an affective Go/No-Go task (McLean et al., 2004), and decreases the propensity to preferentially respond to reward-paired cues (Leyton 2007) (Figure 13.3). In the latter study, only responses to reward-paired cues were disturbed; responses to cues that were not paired with reward were unaffected, indicating that the perturbation was not a nonspecific effect on motor performance or vigilance. These effects on focused attention might come closest to mesocorticolimbic DA's proximal effects: affecting the ability of appetitive stimuli to elicit and sustain interest.

A second commonly reported effect of decreased DA transmission is a perturbed sense of well-being along with a detached "psychic indifference." This response was first observed in psychotic patients who were being treated with neuroleptics (Lewander, 1994; Singh, 1976; Singh and Smith, 1973). Though difficult to separate from negative symptoms of the psychiatric disorder, the syndrome develops within hours of initiating treatment and correlates with the medication's

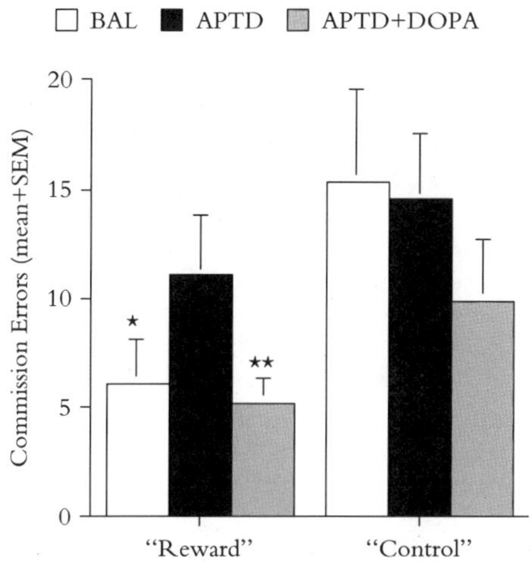

Figure 13.3 The figure depicts commission errors (CE) on a Go/No-Go task. Compared to the test session with the nutritionally balanced amino acid mixture (BAL), acute phenylalanine/tyrosine depletion (APTD) selectively increased CE when subjects were responding for cues that that had been paired with reward (win 10 cents). Performance was restored by L-DOPA. These effects of APTD and L-DOPA were not seen in the control condition. ★, $p \leq 0.005$, compared to the BAL test day; ††, $p \leq 0.001$, compared to the APTD + L-DOPA test day. [Data are from Leyton et al. (2007).]

occupancy of DA D2 receptors (Bressan et al., 2002; de Haan et al., 2000).

The initial observations of "'psychic indifference" were based on descriptions of patients receiving typical neuroleptics, drugs that also induce nonspecific sedative effects attributable to their antihistamine and antiadrenergic properties (Lewander, 1994). Subsequent studies with atypical antipsychotic drugs and other more selective DA D2 antagonists (e.g., haloperidol) observed less nonspecific sedation and more of the indifference syndrome characterized by apathy and psychomotor slowing (Lewander, 1994). Placebo-controlled studies in healthy individuals demonstrate the same response. For example, in a relatively large sample ($n = 59$), administration of the DA D2 antagonist, haloperidol (3, 5 mg, p.o.), elicited a mild, dose-dependent mood-lowering response associated with reduced contentment (Saeedi et al., 2006) (see also Clark et al., 1986; Shelley et al., 1997). As seen in patients, individual differences in neuroleptic-induced D2 receptor blockade correlate with a sense of well-being; the higher the D2 blockade, the worse the subjective effects (Mizrahi et al., 2007).

Recent work with the acute phenylalanine/tyrosine depletion (APTD) method, a method believed to preferentially decrease DA synthesis and release (Leyton et al., 2004; McTavish et al., 1999b; Montgomery et al., 2003), also suggests that changes in DA function can weakly affect subjective states. In healthy individuals, APTD has been reported to elicit a sense of bored apathy (Leyton et al., 2000b; McLean et al., 2004), an indistinct sense of unease (Harmer et al., 2001), and a potentiated stress-induced decrease in happiness (Leyton et al., 2000b) (Figure 13.4). One interpretation is that DA's influence on the motivational salience of incentive stimuli facilitates the appraisal that progress is being made toward desired goals; in turn, this could increase tendencies to cortically mediated positive affective states. This association of DA with affective states does not appear to be strong, though, and in the absence of a psychological challenge, APTD does not produce frank changes in mood (Harrison et al., 2002; Leyton et al., 2000b; Lythe et al., 2005; McTavish et al., 2004, 2005; Roiser et al., 2005).

What are the Effects of Decreasing Drug-Induced Increases in Dopamine Transmission in Humans?

A third approach to investigating the role of DA is to examine the effects of diminishing the ability of abused drugs to increase DA transmission (Table 13.2). This work began 40 years ago with a series of inpatient trials (Jönsson, 1972; Jönsson et al., 1969, 1971), and the initial results provided some of the original impetus for proposing that DA mediated pleasure (Wise, 1982). In brief, the studies suggested that euphorigenic effects of D,L-amphetamine (200 mg, i.v.) were decreased by both the acute and chronic (1 week) administration of the neuroleptics, pimozide and chlorpromazine, as well as α-methyl-para-tyrosine (αMPT), a competitive inhibitor of the rate-limiting enzyme in catecholamine synthesis, tyrosine hydroxylase. The neuroleptic effects, however, were not dose dependent, nor were they replicated when the same group tested thioridazine (Jönsson, 1972). Moreover, attempts by subsequent investigators to reproduce these effects have been predominantly negative (Brauer and de Wit, 1996; 1997; Enggasser and de Wit, 2001; Evans et al., 2001; Gawin, 1986; Sherer, 1988; Stine et al., 1997). More selective antagonism of the DA D1/D5 receptor with ecopipam has been reported to diminish

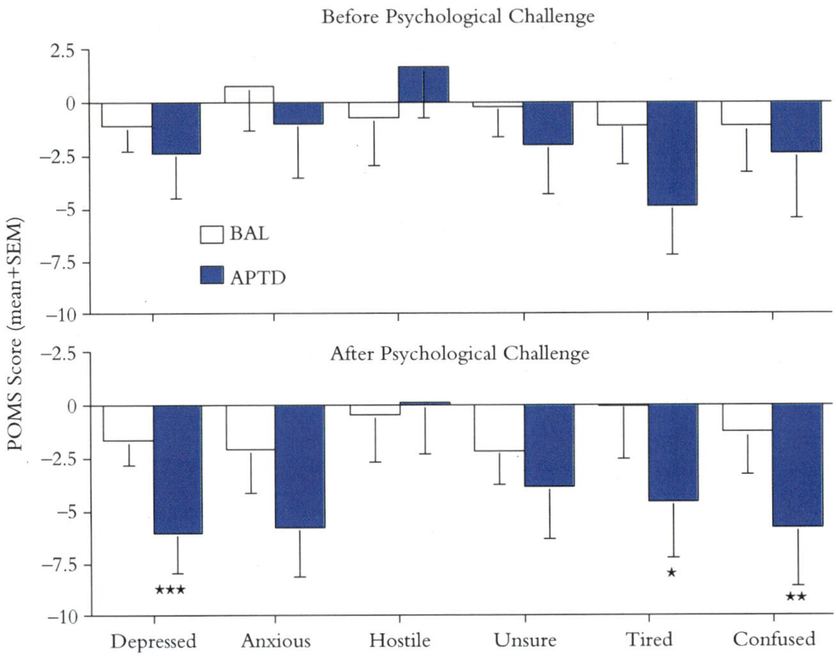

Figure 13.4 The figure depicts changes in self-reported mood states as measured with the bipolar Profile of Mood States (POMS). As illustrated, under relatively neutral conditions, changes in mood did not occur following decreases in dopamine synthesis, as produced by acute phenylalanine/tyrosine depletion (APTD). In comparison, APTD did lead to a mood-lowering response when subjects were exposed to a psychosocial stressor. [Data are from Leyton et al. (2000b).]

cocaine-induced euphoria (Romach et al., 1999) but this too has not been replicated (Haney et al., 2001; Nann-Vernotica et al., 2001).

A series of studies conducted over the past decade with the APTD method for decreasing DA transmission (Leyton et al., 2004; McTavish et al., 1999a; Montgomery et al., 2003) has yielded particularly consistent results. APTD diminishes amphetamine-induced "mind racing" (McTavish et al., 2001, 2004) and cocaine-induced "confidence" (Leyton et al., 2005), but leaves intact the pleasurable and mood altering effects of a wide range of abused substances, including alcohol (Barrett et al., 2008; Leyton et al., 2000a), tobacco (Casey et al., 2006; Munafò et al., 2007), amphetamine (Leyton, 2007; McTavish et al., 2001, 2004), and cocaine (Leyton et al., 2005) (see Figures 13.5 and 13.6). The latter of these studies is illustrative. In this experiment, nondependent regular cocaine users ingested cocaine hydrochloride in their usual fashion (3.0 mg/kg, taken intranasally). When the participants returned to the same test environment approximately 1 week later, presentation of the cocaine paraphernalia (a mirror, bag of cocaine powder, razor blade, and straw) led to significant increases in cocaine craving. Individual differences in this cue-induced craving response predicted lifetime uses of stimulant drugs, consistent with the hypothesis that, with repeated drug use, drug-paired cues become increasingly able to elicit appetitive states. Ingestion of the cocaine led to further, dose-dependent (0.6, 1.5, 3.0 mg/kg) increases in craving as well as euphoria, excitement, and self-confidence. Reduction of DA function with APTD, however, did not alter any of the drug's pleasurable or mood-elevating effects (Leyton et al., 2005). Nor, as noted above, has APTD been seen to decrease these effects of other abused substances (Table 13.3).

Some studies suggest that a more consistent effect of diminished DA function is to decrease the ability of drug-paired cues to elicit attentional biases, and this has been seen in both heroin (Franken et al., 2004) and nicotine addicts (Munafo et al., 2007). Lowered DA function can also diminish drug craving, an effect that has been seen most consistently for cocaine (Romach et al., 1999; Leyton et al., 2005) and cocaine cue-induced craving (Berger et al., 1996; Leyton et al., 2005). In comparison, selective decreases in DA transmission

Table 13.2 Low Dopamine Related Changes in Subjective Craving

Authors	Subjects	Dopamine Challenge	Effects on Subjective Craving	Other Effects / Comment
Berger (1996)	20 cocaine-dependent subjects	Haloperidol (0, 4 mg, po)	Decreased cue-induced craving	Decreased cue-induced anxiety and high
Romach (1999)	11 cocaine-dependent subjects	Ecopipam (0, 10, 25, 100 mg, po). DA D1/D5 antagonist	Decreased cocaine-induced craving	Decreased euphoria and stimulant effects
Haney (2001)	10 detoxified cocaine-dependent subjects.	Ecopipam (0, 100 mg po, 8 days)	Decreased cocaine-induced craving	Increased euphoria and self-administration
Nann-Vernotica (2001)	10 detoxified cocaine-dependent subjects.	Ecopipam (0, 10, 25, 100 mg p.o., 7 days)	No effect on cocaine-induced craving	
Franken (2004)	17 detoxified heroin addicts	Haloperidol (0, 2mg, p.o.)	No effect	Decreased effect of heroin cues on Stroop task
Leyton (2005)	8 cocaine users	APTD	Decreased cue- and cocaine-induced craving	No effect on cue- or cocaine-induced euphoria or other mood-elevating effects
Casey (2006)	14 nicotine-dependent smokers	APTD	No effect	
Munafò (2007)	20 nicotine-dependent smokers	APTD	Increased abstinence-related cigarette craving	Effect seen in men only
Barrett (2008)	16 social drinkers	APTD	No effect	Decreased progressive ratio breakpoint for alcohol beverages

APTD, acute phenylalanine/tyrosine depletion.

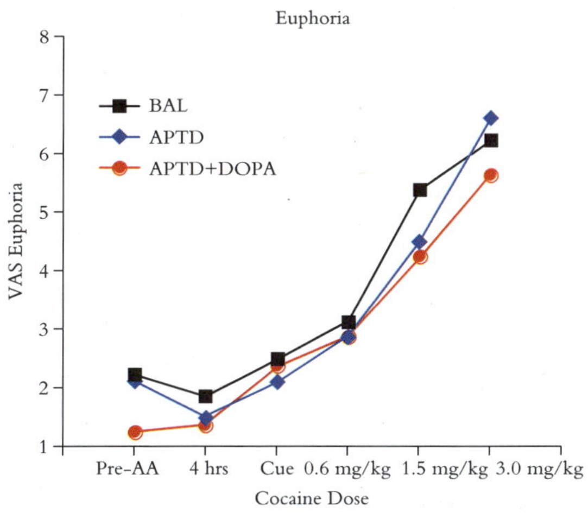

Figure 13.5 The figure depicts self-reported euphoria on a visual analog scale (VAS, 1 – 10) before and after participants have ingested three intranasal doses of cocaine hydrochloride (0.6, 1.5, 3.0 mg/kg). Compared to scores on the control test session (BAL), euphoria was not altered by either acute phenylalanine/tyrosine depletion (APTD) or APTD + L-DOPA. [Data are from Leyton et al. (2005).]

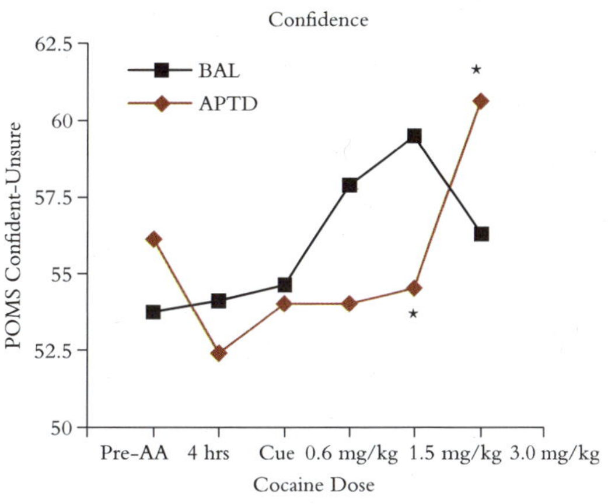

Figure 13.6 The figure depicts self-reported confidence scores on the bipolar profile of mood states (POMS) before and after participants have ingested three intranasal doses of cocaine hydrochloride (0.6, 1.5, 3.0 mg/kg). Compared to scores on the control test session (BAL), confidence scores were decreased (cocaine dose curve shifted to the right) following acute phenylalanine/tyrosine depletion (APTD). [Data are from Leyton et al. (2005).]

did not diminish self-reported craving scores in social drinkers (Barrett et al., 2008), nicotine-dependent smokers (Casey et al., 2006; Munafo et al., 2007), or heroin-dependent opiate addicts (Franken et al., 2004). These observations are consistent with the present proposal that DA transmission is more closely related to unconscious attentional biases and the ability of reward-paired cues to sustain interest than conscious craving (Figure 13.7).

Acute decreases in DA transmission can also diminish the self-administration of alcohol, both in free-choice (Enggasser and de Wit, 2001; Leyton et al., 2000a; Modell et al., 1993) and progressive ratio breakpoint paradigms (Barrett et al., 2008). Since, in rodents, it was the ability of decreased DA transmission to alter self-administration behavior in the absence of motor deficits that first led to reward-based hypotheses of DA function, the present observations in humans provide important confirmatory findings. Studies of the effects of these manipulations on the ad lib self-administration of cigarettes (Brauer et al., 2001; Caskey et al., 1999, 2002, 2007; Dawe et al., 1995) and cocaine (Leyton et al., 2005), though have been less consistent, and increases (Caskey et al., 1999, 2002; Dawe et al., 1995), decreases (Brauer et al., 2001), and no change (Casey et al., 2006; Leyton et al., 2005) have all been reported. More studies will be needed to determine whether these variable results reflect the different drugs, different populations, or different mechanisms that would regulate the self-administration of freely available substances in healthy nondependent occasional users versus substance abusers and the long-term substance dependent.

In summary, studies in humans suggest that DA neurotransmission affects the ability of appetitive stimuli to sustain focused interest and the production of motivational drive and optimistic positive affect. Of particular importance is the finding that decreased DA transmission can diminish reward 'wanting' in the absence of changes to pleasure; DA-related changes in motivational states, it seems, are not due to changes in pleasure. Despite the once prevalent notion that low DA causes anhedonia, the amine appears to have little effect on such states, altering, instead, focused

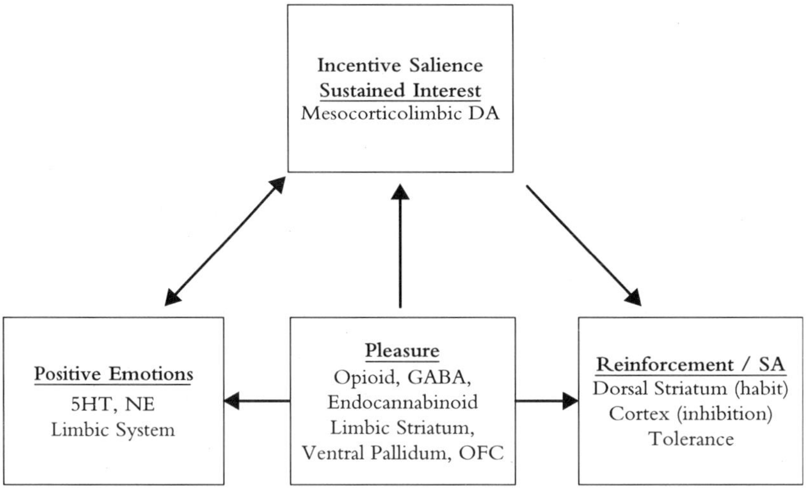

Figure 13.7 The figure depicts a speculative model of links between hypothesized neurobiological mechanisms related to incentive salience, positive mood, pleasure, and the habit-like self-administration of familiar rewards. The various processes can influence each other. For example, when obtaining a goal produces pleasure this can reinforce the appraisal that current strategies are effective. This positive appraisal, in turn, can reinstate happiness. However, pleasure is not required for desire or happiness. For example, people will choose rewards that do not produce positive subjective effects, such as familiar food (Hetherington et al., 2002), electrical stimulation of the brain (Heath, 1963), and extremely low doses of abused drugs (Lamb et al., 1991; Fischman and Foltin, 1992; Hart et al., 2001). Similarly, we can feel happy even if we do not currently have a pleasurable stimulus. Rolls (2005) makes the same point when he notes that positive emotions are not inextricably linked to current pleasure, and instead "may persist after…[the receipt of a reward] (p. 426)." DA, dopamine. 5HT, serotonin. NE, norepinephrine. OFC, orbitofrontal cortex. SA, self-administration.

attention and appetitive motivation in the absence of changes in pleasure. Together, these findings indicate that the mechanisms that regulate desire and positive affect are dissociable from those that produce pleasure even though they may influence each other under normal conditions.

Dopamine and Clinical Depression

As discussed above, the inability of DA manipulations to consistently affect drug-induced pleasure does not preclude it from having a role in affective states (Dunlop and Nemeroff, 2007; Fibiger, 1995; Willner, 1995). Happiness, it has been proposed, follows from the appraisal that "[I am] making reasonable progress toward the realization of a goal (p. 122 Lazarus, 1991)." DA neurotransmission might influence these appraisals and the linked positive affective states by sustaining goal-directed behavior and signaling—perhaps unconsciously—that progress is being made. Low DA function in comparison might increase susceptibility to the converse appraisal: that is, that progress is not being made.

The first evidence that low DA might increase susceptibility to aspects of clinical depression dates back to the 1950s and 1960s. At this time, the monoamine vesicle depletor, reserpine, was being used for the treatment of hypertension. Case reports during this period noted that, with high doses, bouts of "pseudodepression" could develop in individuals who did not have a preexisting vulnerability to mood disorders (Freis, 1954; Goodwin and Bunney, 1971). These "pseudodepressions" were characterized by sedation, decreased energy, and, occasionally, decreased motivation to interact with the environment. Although the response could have been due to the loss of any of the monoamines, the "pseudodepressions" could be reversed by L-DOPA (Degkwitz et al., 1960).

Ten years later, a series of inpatient studies in participants with bipolar mood disorders indicated that manic symptoms were diminished by administration of the catecholamine synthesis inhibitor, αMPT, and reinstated by L-DOPA (Bunney et al., 1971; Goodwin et al., 1970; Murphy et al., 1971). Particularly pronounced were αMPT-induced decreases in rate of speech and attention seeking in the manic patients and an L-DOPA-enhanced sense of hope, assertiveness, and speech in depressed patients with psychomotor retardation. These beneficial effects of αMPT in manic patients have been reproduced with a second method of decreasing DA synthesis, APTD, suggesting further that there is a significant DA component to mania (McTavish et al., 2001; Scarná et al., 2003). Whether APTD primarily decreased the hyperactivity, mood elevation, or increased goal-directed behavior was, unfortunately, not indicated.

The first formal proposition that low DA states contribute to aspects of depression susceptibility was by Randrup and colleagues (1975). As the authors noted, at least some depressed patients have low cerebrospinal fluid (CSF) concentrations of the DA metabolite, homovanillic acid (HVA), particularly those exhibiting psychomotor retardation (Banki, 1977; Banki et al., 1981; Korf and van Praag, 1971; van Praag et al., 1973). Subsequent studies confirmed these associations, particularly when HVA levels were measured with probenecid blockade of transport out of CSF (Banki, 1977; Berger et al., 1980; Bowers, 1972; Goodwin et al., 1973; Korf and van Praag, 1971; Sjostrom, 1973; Sjostrom and Roos, 1972; van Praag et al., 1973). DA's role in depression though seems insufficient to induce overtly sad, depressed mood. Unlike αMPT (Berman et al., 1999, 2002; Bremner et al., 2003), which decreases both DA and NE, more selective decreases in DA transmission with APTD do not reinstate depressive states in remitted patients with a history of major depression (McTavish et al., 2004, 2005; Roiser et al., 2005).[4]

Functional neuroimaging studies have also found evidence of disturbed DA function in depressed patients; again, though, the association is not with frank sadness. Instead, as in the CSF studies, individual differences in DA function correlate with psychomotor retardation (Meyer et al., 2001, 2006) (see also Brunswick et al., 2003; Newberg et al., 2007). Overall, then, the literatures on the effects of low DA function in humans offers little support for the anhedonia hypothesis; instead, the work suggests that DA affects the ability of environmental stimuli to sustain interest, but the ability to experience pleasure remains intact.

Dopamine-Related Neuropsychiatric Disorders: Parkinson's Disease

The two best-studied neuropsychiatric disturbances with endogenous DA disturbances are Parkinson's disease (Barbeau et al., 1961, 1962; Birkmayer and Hornykiewicz, 1961; Ehringer and Hornykiewicz, 1960) and psychosis (Carlsson and Lindqvist, 1963; Laruelle and Abi-Dargham, 1999; Seeman et al., 1975, 1976; Van Rossum, 1967). Patients with Parkinson's

Table 13.3 Low Dopamine Related Changes in Drug-Induced Euphoria (Double-Blind Studies Only)

Authors	Subjects	Dopamine Challenge	Effects on Drug-Induced Euphoria	Other Effects / Comment
Jönsson (1969)	5 amphetamine-abusing subjects	αMPT (0, 3 g, p.o.)	αMPT decreased euphorigenic effects of amphetamine	
Jönsson (1971)	29 amphetamine-abusing subjects	αMPT (0, 2, 4 g, p.o., 1, 7 days)	αMPT decreased euphorigenic effects of amphetamine	More effective after 1 day (70% reduction) than 7 (30% reduction)
Jönsson (1972)	22 amphetamine-abusing subjects	Chlorpromazine (150 mg/day, 1–13 days), Pimozide (5, 10, 20 mg), and Thioridamine (25, 50 mg)	Chlorpromazine and pimozide decreased amphetamine-induced euphoria. No effect of thioridamine	Pimozide's effects were not dose-dependent
Sherer (1988)	5 cocaine-dependent subjects	Haloperidol (0, 8 mg, i.m.)	No effect on cocaine "rush". Slightly weakened "high"	
Brauer (1996)	10 healthy subjects	Pimozide (0, 1, 2 mg, p.o.)	No effect on amphetamine's positive subjective effects	
Brauer (1997)	12 healthy subjects	Pimozide (0, 12 mg, p.o.)	No effect on amphetamine's positive subjective effects	
Stine (1997)	10 cocaine-dependent subjects	αMPT (1 g, p.o., tid)	No effect on cocaine's positive subjective effects	
Romach (1999)	11 cocaine-dependent subjects	Ecopipam (0, 10, 25, 100 mg, p.o.)	Decreased cocaine-induced euphoria and stimulant effects	
McTavish (1999b)	7 healthy subjects	APTD	No effect on amphetamine's positive subjective effects	Decreased "mind-racing"
Leyton (2000a)	12 social drinkers	APTD	No effect on alcohol's positive subjective effects	
McTavish (2001)	16 healthy subjects	APTD	No effect on methamphetamine's positive subjective effects	Decreased "mind racing"
Evans (2001)	18 detoxified cocaine-dependent subjects.	Flupenthixol (0, 2.5, 5.0 mg p.o.; 10, 20 mg, i.m.)	No effect on cocaine's positive subjective effects	
Haney (2001)	10 detoxified cocaine-dependent subjects.	Ecopipam (0, 100 mg p.o., 8 days)	Increased cocaine's positive subjective effects	
Nann-Vernotica (2001)	10 detoxified cocaine-dependent subjects.	Ecopipam (0, 10, 25, 100 mg p.o., 7 days)	No effect on cocaine's positive subjective effects	
Enggasser (2001)	17 social drinkers	Haloperidol (0, 3 mg, p.o.)	Alcohol-induced stimulant and euphoric effects were decreased in a subgroup that experienced stimulant effects of alcohol (n=8). No effect in the other 9 subjects	
Leyton (2005)	8 cocaine users	APTD	No effect on cocaine's positive induced confidence	Decreased cocaine-subjective effects
Casey (2006)	14 abstinent nicotine-dependent smokers.	APTD	No effect on cigarette's positive subjective effects	
Leyton (2007)	14 healthy subjects	APTD	No effect on amphetamine's positive subjective effects	
Barrett (2008)	16 social drinkers	APTD	No effect on alcohol's positive subjective effects	Decreased progressive ratio breakpoint for alcohol beverages

disease, as well as other disorders related to striatonigral degeneration, exhibit a wider range of symptoms than is commonly recognized (Aarsland et al., 2005; Hornykiewicz and Kish, 1984; Pillon et al., 1995; Weintraub et al., 2005). In addition to the progressively diminished ability to initiate movement, these patients experience disturbed mood, attentional capacities, and apathy. These additional symptoms correlate closely with the motor disturbances, and are responsive to treatment with L-DOPA and other DAergic medications.

Patients with Parkinson's disease also have low rates of cigarette smoking, caffeine use (Di Monte, 2003; Hernan et al., 2003; Maher et al., 2002; Ragonese et al., 2003), and possibly alcohol abuse (Ragonese et al., 2003) (though see Garwood et al., 2006; Hernan et al., 2004). It has been proposed that this might reflect diminished pleasure from the drugs. Slightly lower "good drug effect" responses to methylphenidate have been described in patients compared to healthy controls (Cantello et al., 1989; Persico et al., 1998). However, these group differences are small, and in the first study were only seen in Parkinson's patients with comorbid depression. Moreover, patients with Parkinson's disease do not have altered pleasantness or aversiveness sensitivity to positive (e.g., sweet) or negative (e.g., quinine) taste stimuli (Sienkiewicz-Jarosz et al., 2005). In comparison, patients with Parkinson's disease do exhibit deficits in selective attention (Vieregge et al., 1994) and when tested with or without L-DOPA medication exhibit greater sensitivity to positive versus negative stimuli during Go/No-Go tasks (Frank et al., 2004). On self-report measures, the patients score themselves higher on measures of apathy (Isella et al., 2003) but not on the Snaith–Hamilton Pleasure Scale (Pluck and Brown, 2002) suggesting that they experience a loss of interest and control of focused attention rather than a loss of pleasure.

Whereas low DA function in patients with Parkinson's disease is at least correlated with deficits in motivation, drive, and positive mood, treatment of the disease with DAergic agents can lead to hypomanic-like increases in goal-directed behavior and associated states. Typical features of this syndrome, sometimes referred to as "DA dysregulation syndrome" (DDS) (Giovannoni et al., 2000), include elevated mood, compulsive seeking of DAergic medicines, behavioral stereotypies ("punding"), hypersexuality, and pathological gambling. The syndrome is more common than previously suspected, occurring in an estimated 3% to 4% of patients (Pezzella et al., 2005). Pathological gambling alone is more common, developing in 8% of those taking DA agonists (e.g., ropinirole, pramipexole, pergolide) (Grosset et al., 2006), strikingly higher than its occurrence in the U.S. adult general population (0.42%) (Petry et al., 2005). Even more common is the development of psychotic symptoms, particularly visual hallucinations, and this can occur in up to 30% of patients (Cummings, 1991).

These adverse responses to DA agonist medications are thought to reflect the development of DA receptor supersensitivity in the degenerating nigrostriatal pathway as well as hyperactivation of the still intact mesolimbic pathway. Of interest for this discussion, Parkinson's patients with DDS, versus those without, exhibit greater L-DOPA-induced DA release in the ventral limbic striatum (Evans et al., 2006). Individual differences in the magnitude of this DA response correlated with drug-induced drug 'wanting' but not 'liking.'

Dopamine-Related Neuropsychiatric Disorders: Psychosis

Three compelling lines of evidence indicate that psychotic states are related to increased DA function. First, drugs that increase DA transmission can induce (Angrist et al., 1974; Bell, 1973; Connell, 1958; Ellinwood, 1967; Griffith et al., 1972; Kalayasiri et al., 2006; Post, 1975) and aggravate psychotic states (Laruelle et al., 1996; Lieberman et al., 1987). Second, DA D2 receptor antagonists are clinically effective (Agid et al., 2007; Angrist et al., 1974; Delay et al., 1952; Lehmann and Hanrahan, 1954; Seeman et al., 1975, 1976; Van Rossum, 1967;). And third, functional neuroimaging studies indicate that psychotic patients display increased striatal DA synthesis (Hietala et al., 1995, 1999; McGowan et al., 2004; Meyer-Lindenberg et al., 2002; Reith et al., 1994) (though for a negative result, see Dao-Castellana et al., 1997) and stimulated DA release (D-amphetamine, 0.2 to 0.3 mg/kg, i.v.) (Abi-Dargham et al., 1998; Breiter et al., 1997; Laruelle et al., 1996), as compared to matched controls.

The accumulating evidence that psychotic patients have increased DA function provided a clear problem for the DA pleasure hypothesis: patients with psychotic symptoms are not euphoric. One resolution may be that elevated DA transmission heightens the attribution of importance given to delusional thoughts and hallucinations (Cutmore and Beninger, 1990; Kapur, 2003). This proposal is based on a number of observations, including disrupted DA-related latent inhibition (Baruch et al., 1988; Gray et al., 1992) (though see Swerdlow et al., 2005), increased susceptibility to forming abnormal associations (Jensen et al., 2008), deficits in selective attention (Oades, 1982), elevated vulnerability to substance abuse despite experiencing greater adverse effects (Batel, 2000), and the clinical observation that the initial efficacy of neuroleptics appears to reflect a decrease in the evaluated salience of psychotic experiences as opposed to a decrease in their occurrence (Kapur, 2003).

Conclusions

Studies in the 1970s and 1980s in both human and nonhuman animals led to the hypothesis that motivation to obtain rewards was directly related to DA-mediated changes in pleasure. Despite the continued and not infrequent citation of this hypothesis, the material reviewed in the present chapter indicates that the view is no longer tenable. Instead, there appears to be at least three psychobiological processes related to the regulation of positive mood and motivational states in humans: (i) the sustained conscious or unconscious interest in a stimulus and the sense of being drawn toward it; (ii) emotion states that are linked to this sense of making progress toward a desired goal; and (iii) 'liking', the unadulterated pleasure often experienced upon obtaining the desired reward. Studies conducted in humans suggest that limbic DA transmission is most closely related to the first of these processes, enhancing the salience of reward-related cues, biasing attentional processes and sustaining focused interest. These DAergic changes have a number of consequences, influencing susceptibility to craving states and altering both the direction and persistence of goal-oriented behavior. Since mood states are related to appraisals of the environment, the sense that good things are about to happen can affect mood, but DA's role in this process seems to be at the beginning (unconscious appraisal) rather than the end (euphoria).

Acknowledgments

The author thanks Drs. Paul B. Clarke and Kent C. Berridge for feedback on earlier versions of the chapter, Fonds de la recherche en santé du Québec (FRSQ) for a salary award, and McGill University for the William Dawson Scholar addiction research chair.

Notes

1. This research was occurring against a backdrop of change within the field of motivation. Specifically, the field was switching from physiological need based drive reduction theories of goal-directed behaviors to reward-based approach theories (Bindra and Stewart, 1966; Glickman and Schiff, 1967). Whether the experience of pleasure was required for the hypothesized reward mechanisms to function, and whether dopamine was a key neurobiological component came to be central questions.
2. During the past 20 years, two main tools have been developed to measure activity in living human brain. In functional magnetic resonance imaging (fMRI) studies, brain regional activity is assessed by measuring changes in regional cerebral blood flow (rCBF). Like a muscle, the more active an area is, the more blood it needs at a given time. MRIs literally contain a large magnet, and fMRI signals respond to changes in blood flow based on the paramagnetic (attracted) and diamagnetic (repelled) properties of deoxygenated versus oxygenated hemoglobin, respectively. The temporal (100 ms) and anatomical (mm) resolution are both quite good, but the method lacks neurochemical specificity. Some evidence though suggests that signal changes in the striatum are strongly influenced by dopamine. Positron emission tomography (PET) can also be used to measure brain activity, but it is based on different principles. Subjects are administered a radioactively labeled substance that can cross the blood–brain barrier. The decaying tracer emits positrons that typically travel 0.2 to 7 mm at which point they collide with an electron. The collision destroys both, producing gamma rays that travel at nearly the speed of light in diametrically opposite directions. These simultaneous collisions are detected by coincidence detectors that encircle the subject's head, and they produce a light signal. The light signal is then converted into an electric signal that is sent to a computer which calculates how much signal is coming from where and when. If the labeled substance is water, then a measure of rCBF can be obtained. If the labeled substance is a tracer such as [^{11}C]raclopride (a D2 receptor antagonist), then the availability of D2 receptors can be determined. When extracellular dopamine levels increase, the availability of dopamine receptors for [^{11}C]raclopride goes down. Although the temporal (20 to 30 min) and anatomical resolution (5 to 10 cubic mm) is modest, the method is well validated (Laruelle, 2000).
3. Recent studies with functional magnetic resonance imaging (fMRI), which allows investigators to evaluate activations throughout the brain, suggest that conscious desire might arise when the subcortically regulated incentive salience is combined with cognitive processing of the possibility of pleasure by the

dorsolateral and limbic prefrontal cortex (Breiter et al., 1997; Wilson et al., 2004; Kringelbach, 2005; McBride et al., 2006; Franklin et al., 2007).

4. The inability of APTD to reinstate depressive symptoms in individuals with a history of mood disorders also stands in contrast to the ability of acute depletions of the serotonin amino acid precursor to produce this effect (Delgado et al., 1990; Leyton et al., 1997; Booij et al., 2002; Young and Leyton, 2002).

References

Aarsland, D., Alves, G. and Larsen, J. P. (2005) Disorders of motivation, sexual conduct, and sleep in Parkinson's disease. *Adv. Neurol.* 96, 56–64.

Abercrombie, E. D., Keefe, K. A., DiFrischia, D. S. and Zigmond, M. J. (1989) Differential effect of stress on in vivo dopamine release in striatum, nucleus accumbens, and medial frontal cortex. *J. Neurochem.* 52, 1655–1658.

Abi-Dargham, A., Kegeles, L. S., Martinez, D., Innis, R. B. and Laruelle, M. (2003) Dopamine mediation of positive reinforcing effects of amphetamine in stimulant naive healthy volunteers: Results from a large cohort. *Eur. Neuropsychopharmacol.* 13, 459–468.

Abi-Dargham, A., Gil, R., Krystal, J., Baldwin, R. M., Seibyl, J. P., Bowers, M., van Dyck, C. H., Charney, D. S., Innis, R. B. and Laruelle, M. (1998) Increased striatal dopamine transmission in schizophrenia: Confirmation in a second cohort. *Am. J. Psychiatry* 155, 761–767.

Agid, O., Mamo, D., Ginovart, N., Vitcu, I., Wilson, A. A., Zipursky, R. B. and Kapur, S. (2007) Striatal vs extrastriatal dopamine D2 receptors in antipsychotic response—a double-blind PET study in schizophrenia. *Neuropsychopharmacology* 32, 1209–1215.

Ahveninen, J., Kahkonen, S., Tiitinen, H., Pekkonen, E., Huttunen, J., Kaakkola, S., Ilmoniemi, R. J. and Jaaskelainen, I. P. (2000) Suppression of transient 40-Hz auditory response by haloperidol suggests modulation of human selective attention by dopamine D2 receptors. *Neurosci. Lett.* 292, 29–32.

Angrist, B., Sathananthan, G., Wilk, S. and Gershon, S. (1974) Amphetamine psychosis: Behavioral and biochemical aspects. *J. Psychiatr. Res.* 11, 13–23.

Anstrom, K. K. and Woodward, D. J. (2005) Restraint increases dopaminergic burst firing in awake rats. *Neuropsychopharmacology* 30, 1832–1840.

Antelman, S. M. and Szechtman H. (1975) Tail pinch induces eating in sated rats which appears to depend on nigrostriatal dopamine. *Science* 189, 731–733.

Banki, C. M. (1977) Correlation between cerebrospinal fluid amine metabolites and psychomotor activity in affective disorders. *J. Neurochem.* 28, 255–257.

Banki, C. M., Molnar, G. and Vojnik, M. (1981) Cerebrospinal fluid amine metabolites, tryptophan and clinical parameters in depression. Part 2. Psychopathological symptoms. *J. Affect. Disord.* 3, 91–99.

Barbeau, A., Murphy, G. F. and Sourkes, T. L. (1961) Excretion of dopamine in diseases of basal ganglia. *Science* 133, 1706–1707.

Barbeau, A., Sourkes, T. and Murphy, G. (1962) Les catécholamines dans la maladie de Parkinson. In: *Monoamines et systeme nerveau central*, pp. 247–262. Ed. D. Ajuriaguerra. Masson: Paris.

Barrett, S. P., Boileau, I., Okker, J., Pihl, R. O. and Dagher, A. (2004) The hedonic response to cigarette smoking is proportional to dopamine release in the human striatum as measured by positron emission tomography and [^{11}C] raclopride. *Synapse* 54, 65–71.

Barrett, S. P., Pihl, R.O., Benkelfat, C., Brunelle, C., Young, S. N. and Leyton, M. (2008) The role of dopamine in alcohol self-administration in humans: Individual differences. *Eur. Neuropsychopharmacol.,* 18, 439–447.

Baruch, I., Hemsley, D. R. and Gray, J. A. (1988) Differential performance of acute and chronic schizophrenics in a latent inhibition task. *J. Nerv. Ment. Dis.* 176, 598–606.

Batel, P. (2000) Addiction and schizophrenia. *Eur. Psychiatry.* 15, 115–122.

Bell, D. S. (1973) The experimental reproduction of amphetamine psychosis. *Arch. Gen. Psychiatry* 29, 35–40.

Beninger, R. J. (1983) The role of dopamine in locomotor activity and learning. *Brain Res.* 287, 173–196.

Berger, H. J., van Hoof, J. J., van Spaendonck, K. P., Horstink, M. W., van den Bercken, J. H., Jaspers, R. and Cools, A. R. (1989) Haloperidol and cognitive shifting. *Neuropsychologia* 27, 629–639.

Berger, P. A., Faull, K. F., Kilkowski, J., Anderson, P. J., Kraemer, H., Davis, K. L. and Barchas, J. D. (1980) CSF monoamine metabolites in depression and schizophrenia. *Am. J. Psychiatry* 137, 174–180.

Berger, S. P., Hall, S., Mickalian, J. D., Reid, M. S., Crawford, C. A., Delucchi, K., Carr, K. and Hall, S. (1996) Haloperidol antagonism of cue-elicited cocaine craving. *Lancet* 347, 504–508.

Berman, R. M., Narasimhan, M., Miller, H. L., Anand, A., Cappiello, A., Oren, D. A., Heninger, G. R. and Charney, D. S. (1999) Transient depressive relapse induced by catecholamine depletion: Potential phenotypic vulnerability marker? *Arch. Gen. Psychiatry* 56, 395–403.

Berman, R. M., Sanacora, G., Anand, A., Roach, L. M., Fasula, M. K., Finkelstein, C. O., Wachen, R. M., Oren, D. A., Heninger, G. R. and Charney, D. S. (2002) Monoamine depletion in unmedicated depressed subjects. *Biol. Psychiatry.* 51, 469–473.

Berridge, K.C. (2007a) The debate over dopamine's role in reward: the case for incentive salience. *Psychopharmacology* 191, 391–431.

Berridge, K. C. (2007b) The debate over dopamine's role in reward: the case for incentive salience. *Psychopharmacology (Berl)* 191, 391–431.

Berridge, K. C. and Robinson, T. E. (2003) Parsing reward. *Trends Neurosci.* 26, 507–513.

Berridge, K. C., Venier, I. L. and Robinson, T. E. (1989) Taste reactivity analysis of 6-hydroxydopamine-induced aphagia: Implications for arousal and anhedonia hypotheses of dopamine function. *Behav. Neurosci.* 103, 36–45.

Bindra, D. (1969) A unified interpretation of emotion and motivation. *Ann. N Y Acad. Sci.* 159, 1071–1083.

Bindra, D. (1978) How adaptive behavior is produced: a perceptual-motivation alternative to response reinforcement. *Behav Brain Sci* 1, 41–91.

Bindra, D. and Stewart, J. (1966) *Motivation: Selected Readings.* Penguin Books: Baltimore, MD.

Birkmayer, W. and Hornykiewicz, O. (1961) [The L-3,4-dioxyphenylalanine (DOPA)-effect in Parkinson-akinesia.]. *Wien Klin Wochenschr* 73, 787–788.

Blackburn, J. R., Pfaus, J. G. and Phillips, A. G. (1992) Dopamine functions in appetitive and defensive behaviours. *Prog. Neurobiol.* 39, 247–279.

Boileau, I., Dagher, A., Leyton, M., Gunn, R. N., Baker, G. B., Diksic, M. and Benkelfat, C. (2006) Modeling sensitization to stimulants in humans: an [11C]raclopride/positron emission tomography study in healthy men. *Arch. Gen. Psychiatry* 63, 1386–1395.

Boileau, I., Dagher, A., Leyton, M., Welfeld, K., Booij, L., Diksic, M. and Benkelfat, C. (2007) Conditioned dopamine release in humans: a positron emission tomography [^{11}C]raclopride study with amphetamine. *J. Neurosci.* 27, 3998–4003.

Boileau, I., Assaad, J. M., Pihl, R. O., Benkelfat, C., Leyton, M., Diksic, M., Tremblay, R. E. and Dagher, A. (2003) Alcohol promotes dopamine release in the human nucleus accumbens. *Synapse* 49, 226–231.

Booij, L., Welfeld, K., Dagher, A., Leyton, M., Boileau, I., Sibon, I., Pruessner, J., Casey, K., Diksic, M. and Benkelfat, C. (2007) Cross-sensitization between stimulants and stress in humans: Behavioral and neurochemical correlates. *Biological Psychiatry* San Diego, CA, 17–19 May.

Booij, L., Van der Does, W., Benkelfat, C., Bremner, J. D., Cowen, P. J., Fava, M., Gillin, C., Leyton, M., Moore, P., Smith, K. A. and Van der Kloot, W. A. (2002) Predictors of mood response to acute tryptophan depletion. A reanalysis. *Neuropsychopharmacology* 27, 852–861.

Bowers, M. B., Jr. (1972) Cerebrospinal fluid 5-hydroxyindoleacetic acid (5HIAA) and homovanillic acid (HVA) following probenecid in unipolar depressives treated with amitriptyline. *Psychopharmacologia* 23, 26–33.

Brauer, L. H. and de Wit, H. (1996) Subjective responses to D-amphetamine alone and after pimozide pretreatment in normal, healthy volunteers. *Biol. Psychiatry* 39, 26–32.

Brauer, L. H. and de Wit, H. (1997) High dose pimozide does not block amphetamine-induced euphoria in normal volunteers. *Pharmacol. Biochem. and Behav.* 56, 265–272.

Brauer, L. H., Cramblett, M. J., Paxton, D. A. and Rose, J. E. (2001) Haloperidol reduces smoking of both nicotine-containing and denicotinized cigarettes. *Psychopharmacology* 159, 31–37.

Breier, A., Su, T. P., Saunders, R., Carson, R. E., Kolachana, B. S., de Bartolomeis, A., Weinberger, D. R., Weisenfeld, N., Malhotra, A. K., Eckelman, W. C. and Pickar, D. (1997) Schizophrenia is associated with elevated amphetamine-induced synaptic dopamine concentrations: Evidence from a novel positron emission tomography method. *Proc Natl Acad Sci USA* 94, 2569–2574.

Breiter, H. C., Gollub, R. L., Weisskoff, R. M., Kennedy, D. N., Makris, N., Berke, J. D., Goodman, J. M., Kantor, H. L., Gastfriend, D. R., Riorden, J. P., Mathew, R. T., Rosen, B. R. and Hyman, S. E. (1997) Acute effects of cocaine on human brain activity and emotion. *Neuron* 19, 591–611.

Bremner, J. D., Vythilingam, M., Ng, C. K., Vermetten, E., Nazeer, A., Oren, D. A., Berman, R. M. and Charney, D. S. (2003) Regional brain metabolic correlates of alpha-methylparatyrosine-induced depressive symptoms: Implications for the neural circuitry of depression. *JAMA* 289, 3125–3134.

Bressan, R. A., Costa, D. C., Jones, H. M., Ell, P. J. and Pilowsky, L. S. (2002) Typical antipsychotic drugs–D(2) receptor occupancy and depressive symptoms in schizophrenia. *Schizophr. Res.* 56, 31–36.

Brody, A. L., Olmstead, R. E., London, E. D., Farahi, J., Meyer, J. H., Grossman, P., Lee, G. S., Huang, J., Hahn, E. L. and Mandelkern, M. A. (2004) Smoking-induced ventral striatum dopamine release. *Am. J. Psychiatry* 161, 1211–1218.

Brody, A. L., Mandelkern, M. A., Olmstead, R. E., Scheibal, D., Hahn, E., Shiraga, S., Zamora-Paja, E., Farahi, J., Saxena, S., London, E. D. and McCracken, J. T. (2006) Gene variants of brain dopamine pathways and smoking-induced dopamine release in the ventral caudate/nucleus accumbens. *Arch. Gen. Psychiatry* 63, 808–816.

Brunswick, D. J., Amsterdam, J. D., Mozley, P. D. and Newberg, A. (2003) Greater availability of brain dopamine transporters in major depression shown by [^{99m}Tc] TRODAT-1 SPECT imaging. *Am. J. Psychiatry* 160, 1836–1841.

Bunney, W. E., Jr., Brodie, H. K., Murphy, D. L. and Goodwin, F. K. (1971) Studies of alpha-methyl-para-tyrosine, L-dopa, and L-tryptophan in depression and mania. *Am. J. Psychiatry* 127, 872–881.

Cabib, S. and Puglisi-Allegra, S. (1994) Opposite responses of mesolimbic dopamine system to controllable and uncontrollable aversive experiences. *J. Neurosci.* 14, 3333–3340.

Cantello, R., Aguggia, M., Gilli, M., Delsedime, M., Chiardo Cutin, I., Riccio, A. and Mutani, R. (1989) Major depression in Parkinson's disease and the mood response to intravenous methylphenidate: Possible role of the "hedonic" dopamine synapse. *J. Neurol. Neurosurg. Psychiatry* 52, 724–731.

Carlsson, A. and Lindqvist, M. (1963) Effect of chlorpromazine or haloperidol on formation of 3methoxytyramine and normetanephrine in mouse brain. *Acta Pharmacol. Toxicol. (Copenh.)* 20, 140–144.

Carlsson, A., Lindqvist, M. and Magnusson, T. (1957) 3,4-Dihydroxyphenylalanine and 5-hydroxytryptophan as reserpine antagonists. *Nature* 180, 1200.

Casey, K. F., Benkelfat, C., Young, S. N. and Leyton, M. (2006) Lack of effect of dopamine precursor depletion in nicotine-dependent smokers. *Eur. Neuropsychopharmacol.* 16, 512–520.

Casey, K.F., Benkelfat, C., Dagher, A., Baker, G. and Leyton, M. (2007) Stimulant drug exposure and family environment predict the striatal dopamine response to D-amphetamine: a PET [11C]raclopride study. *Canadian College of Neuropsychopharmacology* Banff, Canada, 15–18 June.

Caskey, N. H., Jarvik, M. E. and Wirshing, W. C. (1999) The effects of dopaminergic D2 stimulation and blockade on smoking behavior. *Exp.. Clin. Psychopharmacol.* 7, 72–78.

Caskey, N. H., Jarvik, M. E., Wirshing, W. C., Madsen, D. C., Iwamoto-Schaap, P. N., Eisenberger, N. I., Huerta, L., Terrace, S. M. and Olmstead, R. E. (2002) Modulating tobacco smoking rates by dopaminergic stimulation and blockade. *Nicotine Tob. Res.* 4, 259–266.

Clark, C. R., Geffen, G. M. and Geffen, L. B. (1986) Role of monoamine pathways in the control of attention: Effects of droperidol and methylphenidate in normal adult humans. *Psychopharmacology (Berl.)* 90, 28–34.

Connell, P. (1958) *Amphetamine Psychosis, Maudsley Monograph No. 5.* Chapman and Hall: London.

Cox, S. M. L., Benkelfat, C., Dagher, A., Delaney, J. S., Durand, F., McKenzie, S. A., Kolivakis, T., Casey, K. F. and Leyton, M. (2009) Striatal dopamine responses to intranasal cocaine self-administration in humans. *Biol. Psychiatry,* In press.

Cummings, J. L. (1991) Behavioral complications of drug treatment of Parkinson's disease. *J. Am. Geriatr. Soc.* 39, 708–716.

Cutmore, T. R. and Beninger, R. J. (1990) Do neuroleptics impair learning in schizophrenic patients? *Schizophr. Res.* 3, 173–186.

Dao-Castellana, M. H., Paillere-Martinot, M. L., Hantraye, P., Attar-Levy, D., Remy, P., Crouzel, C., Artiges, E., Feline, A., Syrota, A. and Martinot, J. L. (1997) Presynaptic dopaminergic function in the striatum of schizophrenic patients. *Schizophr. Res.* 23, 167–174.

Dawe, S., Gerada, C., Russell, M. A. and Gray, J. A. (1995) Nicotine intake in smokers increases following a single dose of haloperidol. *Psychopharmacology (Berl)* 117, 110–115.

de Haan, L., Lavalaye, J., Linszen, D., Dingemans, P. M. and Booij, J. (2000) Subjective experience and striatal dopamine D(2) receptor occupancy in patients with schizophrenia stabilized by olanzapine or risperidone. *Am. J. Psychiatry* 157, 1019–1020.

de Wit, H., Enggasser, J. L. and Richards, J. B. (2002) Acute administration of D-amphetamine decreases impulsivity in healthy volunteers. *Neuropsychopharmacology* 27, 813–825.

de Wit, H. and Wise, R. A. (1977) Blockade of cocaine reinforcement in rats with the dopamine receptor blocker pimozide, but not with the noradrenergic blockers phentolamine or phenoxybenzamine. *Can. J. Psychol.* 31, 195–203.

Degkwitz, R., Frowein, R., Kulenkampff, C. and Mohs, U. (1960) [On the effects of L-dopa in man and their modification by reserpine, chlorpromazine, iproniazid and vitamin B6.]. *Klin. Wochenschr.* 38, 120–123.

Delay, J., Deniker, P. and Harl, J. M. (1952) [Therapeutic use in psychiatry of phenothiazine of central elective action (4560 RP).]. *Ann. Med. Psychol. (Paris)* 110, 112–117.

Delgado, P. L., Charney, D. S., Price, L. H., Aghajanian, G. K., Landis, H. and Heninger, G. R. (1990) Serotonin function and the mechanism of antidepressant action. Reversal of antidepressant-induced remission by rapid depletion of plasma tryptophan. *Arch. Gen. Psychiatry* 47, 411–418.

Di Chiara, G. and Imperato, A. (1988) Drugs abused by humans preferentially increase synaptic dopamine concentrations in the mesolimbic system of freely moving rats. *Proc. Natl. Acad. Sc.i USA* 85, 5274–5278.

Di Monte, D. A. (2003) The environment and Parkinson's disease: Is the nigrostriatal system preferentially targeted by neurotoxins? *Lancet Neurol.* 2, 531–538.

Drevets, W. C., Gautier, C., Price, J. C., Kupfer, D. J., Kinahan, P. E., Grace, A. A., Price, J. L. and Mathis, C. A. (2001) Amphetamine-induced dopamine release in human ventral striatum correlates with euphoria. *Biol. Psychiatry* 49, 81–96.

Dunlop, B. W. and Nemeroff, C. B. (2007) The role of dopamine in the pathophysiology of depression. *Arch. Gen. Psychiatry* 64, 327–337.

Ehringer, H. and Hornykiewicz, O. (1960) [Distribution of noradrenaline and dopamine (3-hydroxytyramine) in the human brain and their behavior in diseases of the extrapyramidal system.]. *Klin. Wochenschr.* 38, 1236–1239.

Ellinwood, E. (1967) Amphetamine psychosis: I. Description of the individuals and process. *J. Nerv. Ment. Dis.* 144, 273–283.

Enggasser, J. L. and de Wit, H. (2001) Haloperidol reduces stimulant and reinforcing effects of ethanol in social drinkers. *Alcohol. Clin. Exp. Res.* 25, 1448–1456.

Evans, A. H., Pavese, N., Lawrence, A. D., Tai, Y. F., Appel, S., Doder, M., Brooks, D. J., Lees, A. J. and Piccini, P. (2006) Compulsive drug use linked to sensitized ventral striatal dopamine transmission. *Ann. Neurol.* 59, 852–858.

Evans, S. M., Walsh, S. L., Levin, F. R., Foltin, R. W., Fischman, M. W. and Bigelow, G. E. (2001) Effect of flupenthixol on subjective and cardiovascular responses to intravenous cocaine in humans. *Drug Alcohol Depend.* 64, 271–283.

Everitt, B. J., Parkinson, J. A., Olmstead, M. C., Arroyo, M., Robledo, P. and Robbins, T. W. (1999) Associative processes in addiction and reward. The role of amygdala-ventral striatal subsystems. *Ann. NY Acad. Sci.* 877, 412–438.

Fibiger, H. (1995) Neurobiology of depression: Focus on dopamine. In: *Depression and Mania: From Neurobiology to Treatment*, pp. 1–17. Eds. W. Fratta, L. Pani, and G. Serra. Raven Press: New York.

Fischman, M. W. and Foltin, R. W. (1992) Self-administration of cocaine by humans: a laboratory perspective. In: *Cocaine: Scientific and Social Dimensions) Vol. 166*, pp. 165–173; discussion 173–180. Eds. J. Bock and J. Whelan. Wiley: New York.

Frank, M. J., Seeberger, L. C. and O'Reilly R, C. (2004) By carrot or by stick: Cognitive reinforcement learning in parkinsonism. *Science* 306, 1940–1943.

Franken, I. H., Hendriks, V. M., Stam, C. J. and Van den Brink, W. (2004) A role for dopamine in the processing of drug cues in heroin dependent patients. *Eur. Neuropsychopharmacol.* 14, 503–518.

Franklin, T. R., Wang, Z., Wang, J., Sciortino, N., Harper, D., Li, Y., Ehrman, R., Kampman, K., O'Brien, C. P., Detre, J. A. and Childress, A. R. (2007) Limbic activation to cigarette smoking cues independent of nicotine withdrawal: a perfusion fMRI study. *Neuropsychopharmacology* 32, 2301–2309.

Freis, E. D. (1954) Mental depression in hypertensive patients treated for long periods with large doses of reserpine. *N. Engl. J. Med.* 251, 1006–1008.

Garwood, E. R., Bekele, W., McCulloch, C. E. and Christine, C. W. (2006) Amphetamine exposure is elevated in Parkinson's disease. *Neurotoxicology* 27, 1003–1006.

Gawin, F. H. (1986) Neuroleptic reduction of cocaine-induced paranoia but not euphoria? *Psychopharmacology (Berl.)* 90, 142–3.

Giovannoni, G., O'Sullivan, J. D., Turner, K., Manson, A. J. and Lees, A. J. (2000) Hedonistic homeostatic dysregulation in patients with Parkinson's disease on dopamine replacement therapies. *J. Neurol. Neurosurg. Psychiatry* 68, 423–428.

Glickman, S. and Schiff, B. (1967) A biological theory of reinforcement. *Psych. Rev.* 74, 81–109.

Goodwin, F. K. and Bunney, W. E., Jr. (1971) Depressions following reserpine: a reevaluation. *Semin Psychiatry* 3, 435–448.

Goodwin, F. K., Murphy, D. L., Brodie, H. K. and Bunney, W. E., Jr. (1970) L-DOPA, catecholamines, and behavior: a clinical and biochemical study in depressed patients. *Biol. Psychiatry.* 2, 341–366.

Goodwin, F. K., Post, R. M., Dunner, D. L. and Gordon, E. K. (1973) Cerebrospinal fluid amine metabolites in affective illness: the probenecid technique. *Am J Psychiatry* 130, 73–79.

Grace, A. A., Floresco, S. B., Goto, Y. and Lodge, D. J. (2007) Regulation of firing of dopaminergic neurons and control of goal-directed behaviors. *Trends Neurosci.* 30, 220–227.

Gratton, A. and Wise, R. A. (1994) Drug- and behavior-associated changes in dopamine-related electrochemical signals during intravenous cocaine self-administration in rats. *J. Neurosci.* 14, 4130–4146.

Gravel, P., Dagher, A., Leyton, M., Gunn, R., Sibon, I., Mzengeza, S., Durand, F., Soucy, J.-P., Milot, S., Baker, G. and Benkelfat, C. (2007) Positron emission tomography (PET) imaging of extra-striatal dopamine release. *Biological Psychiatry* San Diego, CA, 17–19 May.

Gray, N., Hemsley, D. and Gray, J. (1992) Abolition of latent inhibition in acute, but not chronic, schizophrenia. *Neurol. Psychiatry Brain Res.* 1, 83–89.

Griffith, J. D., Cavanaugh, J., Held, J. and Oates, J. A. (1972) Dextroamphetamine. Evaluation of psychomimetic properties in man. *Arch. Gen. Psychiatry* 26, 97–100.

Grosset, K. A., Macphee, G., Pal, G., Stewart, D., Watt, A., Davie, J. and Grosset, D. G. (2006) Problematic gambling on dopamine agonists: Not such a rarity. *Mov. Disord.* 21, 2206–2208.

Haney, M., Ward, A. S., Foltin, R. W. and Fischman, M. W. (2001) Effects of ecopipam, a selective dopamine D1 antagonist, on smoked cocaine self-administration by humans. *Psychopharmacology (Berl)* 155, 330–337.

Harmer, C. J., McTavish, S. F., Clark, L., Goodwin, G. M. and Cowen, P. J. (2001) Tyrosine depletion attenuates dopamine function in healthy volunteers. *Psychopharmacology (Berl)* 154, 105–111.

Harrison, B. J., Olver, J. S., Norman, T. R. and Nathan, P. J. (2002) Effects of serotonin and catecholamine depletion on interleukin-6 activation and mood in human volunteers. *Hum. Psychopharmacol.* 17, 293–297.

Hart, C. L., Ward, A. S., Haney, M., Foltin, R. W. and Fischman, M. W. (2001) Methamphetamine self-administration by humans. *Psychopharmacology* 157, 75–81.

Heath, R. G. (1963) Electrical self-stimulation of the brain in man. *Am. J. Psychiatry* 120, 571–577.

Hernan, M. A., Logroscino, G. and Rodriguez, L. A. (2004) A prospective study of alcoholism and the risk of Parkinson's disease. *J. Neurol.* 251 Suppl 7, vII14– vII17.

Hernan, M. A., Chen, H., Schwarzschild, M. A. and Ascherio, A. (2003) Alcohol consumption and the incidence of Parkinson's disease. *Ann. Neurol.* 54, 170–175.

Hernandez, L. and Hoebel, B. G. (1988) Food reward and cocaine increase extracellular dopamine in the nucleus accumbens as measured by microdialysis. *Life Sci.* 42, 1705–1712.

Hetherington, M. M., Pirie, L. M. and Nabb, S. (2002) Stimulus satiation: Effects of repeated exposure to foods on pleasantness and intake. *Appetite* 38, 19–28.

Hietala, J., Syvalahti, E., Vilkman, H., Vuorio, K., Rakkolainen, V., Bergman, J., Haaparanta, M., Solin, O., Kuoppamaki, M., Eronen, E., Ruotsalainen, U. and Salokangas, R. K. (1999) Depressive symptoms and presynaptic dopamine function in neuroleptic-naive schizophrenia. *Schizophr. Res.* 35, 41–50.

Hietala, J., Syvalahti, E., Vuorio, K., Rakkolainen, V., Bergman, J., Haaparanta, M., Solin, O., Kuoppamaki, M., Kirvela, O., Ruotsalainen, U. and et al. (1995) Presynaptic dopamine function in striatum of neuroleptic-naive schizophrenic patients. *Lancet* 346, 1130–1131.

Hollander, E., Nunes, E., DeCaria, C. M., Quitkin, F. M., Cooper, T., Wager, S. and Klein, D. F. (1990) Dopaminergic sensitivity and cocaine abuse: Response to apomorphine. *Psychiatry Res.* 33, 161–169.

Hornykiewicz, O. and Kish, S. J. (1984) Neurochemical basis of dementia in Parkinson's disease. *Can. J. Neurol. Sci.* 11, 185–190.

Ikemoto, S. and Panksepp, J. (1999) The role of nucleus accumbens dopamine in motivated behavior: a unifying interpretation with special reference to reward-seeking. *Brain Res. Brain Res. Rev.* 31, 6–41.

Isella, V., Iurlaro, S., Piolti, R., Ferrarese, C., Frattola, L., Appollonio, I., Melzi, P. and Grimaldi, M. (2003) Physical anhedonia in Parkinson's disease. *J. Neurol. Neurosurg. Psychiatry* 74, 1308–1311.

Jaffe, J. H., Cascella, N. G., Kumor, K. M. and Sherer, M. A. (1989) Cocaine-induced cocaine craving. *Psychopharmacology (Berl.)* 97, 59–64.

Janssen, P. A., Niemegeers, C. J., Schellekens, K. H., Dresse, A., Lenaerts, F. M., Pinchard, A., Schaper, W. K., van Nueten, J. M. and Verbruggen, F. J. (1968) Pimozide, a chemically novel, highly potent and orally long-acting neuroleptic drug. I. The comparative pharmacology of pimozide, haloperidol, and chlorpromazine. *Arzneimittelforschung* 18, 261–279.

Jensen, J., Smith, A. J., Willeit, M., Crawley, A. P., Mikulis, D. J., Vitcu, I. and Kapur, S. (2007) Separate brain regions code for salience vs. valence during reward prediction in humans. *Hum. Brain Mapp.* 28, 294–302.

Jensen, J., Willeit, M., Zipursky, R. B., Savina, I., Smith, A. J., Menon, M., Crawley, A. P. and Kapur, S. (2008) The formation of abnormal associations in schizophrenia: Neural and behavioral evidence. *Neuropsychopharmacology* 33, 473–479.

Jönsson, L.E. (1972) Pharmacological blockade of amphetamine effects in amphetamine dependent subjects. *Eur. J. Clin. Pharmacol.* 4, 206–211.

Jönsson, L. E., Anggard, E. and Gunne, L. M. (1971) Blockade of intravenous amphetamine euphoria in man. *Clin. Pharmacol. Ther.* 12, 889–896.

Jönsson, L.E., Gunne, L. and Anggard, E. (1969) Effects of alpha-methyltyrosine in amphetamine-dependent subjects. *Pharmacol. Clin.* 2, 27–29.

Kahkonen, S., Ahveninen, J., Jaaskelainen, I. P., Kaakkola, S., Naatanen, R., Huttunen, J. and Pekkonen, E. (2001) Effects of haloperidol on selective attention: a combined whole-head MEG and high-resolution EEG study. *Neuropsychopharmacology* 25, 498–504.

Kalayasiri, R., Kranzler, H. R., Weiss, R., Brady, K., Gueorguieva, R., Panhuysen, C., Yang, B. Z., Farrer, L., Gelernter, J. and Malison, R. T. (2006) Risk factors for cocaine-induced paranoia in cocaine-dependent sibling pairs. *Drug Alcohol Depend.* 84, 77–84.

Kalivas, P. W. and Stewart, J. (1991) Dopamine transmission in the initiation and expression of drug- and stress-induced sensitization of motor activity. *Brain Res. Brain Res. Rev.* 16, 223–244.

Kapur, S. (2003) Psychosis as a state of aberrant salience: a framework linking biology, phenomenology, and pharmacology in schizophrenia. *Am. J. Psychiatry* 160, 13–23.

Katz, R. J., Roth, K. A. and Schmaltz, K. (1980) Tail pinch facilitation of self-stimulation in the rat – dependence upon dopamine and independence of opiates. *Pharmacol. Biochem. Behav.* 12, 389–391.

Killcross, A. S., Everitt, B. J. and Robins, T. W. (1997) Symmetrical effects of amphetamine and alpha-flupenthixol on conditioned punishment and conditioned reinforcement: Contrasts with midazolam. *Psychopharmacology (Berl.)* 129, 141–152.

Kiyatkin, E. A. and Gratton, A. (1994) Electrochemical monitoring of extracellular dopamine in nucleus accumbens of rats lever-pressing for food. *Brain Res.* 652, 225–234.

Koepp, M. J., Gunn, R. N., Lawrence, A. D., Cunningham, V. J., Dagher, A., Jones, T., Brooks, D. J., Bench, C. J. and Grasby, P. M. (1998) Evidence for striatal dopamine release during a video game. *Nature* 393, 266–268.

Korf, J. and van Praag, H. M. (1971) Retarded depression and the dopamine metabolism. *Psychopharmacologia* 19, 199–203.

Kringelbach, M. L. (2005) The human orbitofrontal cortex: Linking reward to hedonic experience. *Nat. Rev. Neurosci.* 6, 691–702.

Lamb, R. J., Preston, K. L., Schindler, C. W., Meisch, R. A., Davis, F., Katz, J. L., Henningfield, J. E. and Goldberg, S. R. (1991) The reinforcing and subjective effects of morphine in post-addicts: a dose–response study. *J. Pharmacol. Exp. Ther.* 259, 1165–1173.

Laruelle, M. (2000) Imaging synaptic neurotransmission with in vivo binding competition techniques: a critical review. *J. Cereb. Blood Flow Metab.* 20, 423–451.

Laruelle, M. and Abi-Dargham, A. (1999) Dopamine as the wind of the psychotic fire: New evidence from brain imaging studies. *J. Psychopharmacol.* 13, 358–71.

Laruelle, M., Abi-Dargham, A., van Dyck, C. H., Gil, R., D'Souza, C. D., Erdos, J., McCance, E., Rosenblatt, W., Fingado, C., Zoghbi, S. S., Baldwin, R. M., Seibyl, J. P., Krystal, J. H., Charney, D. S. and Innis, R. B. (1996) Single photon emission computerized tomography imaging of amphetamine-induced dopamine release in drug-free schizophrenic subjects. *Proc. Natl Acad. Sci. USA* 93, 9235–40.

Laruelle, M., Abi-Dargham, A., van Dyck, C. H., Rosenblatt, W., Zea-Ponce, Y., Zoghbi, S. S., Baldwin, R. M., Charney, D. S., Hoffer, P. B., Kung, H. F. and et al. (1995) SPECT imaging of striatal dopamine release after amphetamine challenge. *J. Nucl. Med.* 36, 1182–1190.

Lazarus, R. (1991) *Emotion and Adaptation*. Oxford University Press: New York.

LeDoux, J. E. (1989) Cognitive-emotional interactions in the brain. *Cogn. Emot.* 3, 267–289.

LeDoux, J. E. (2000) Emotion circuits in the brain. *Annu. Rev. Neurosci.* 23, 155–184.

Lehmann, H. E. and Hanrahan, G. E. (1954) Chlorpromazine; new inhibiting agent for psychomotor excitement and manic states. *AMA Arch. Neurol. Psychiatry* 71, 227–237.

Lewander, T. (1994) Neuroleptics and the neuroleptic-induced deficit syndrome. *Acta Psychiatr. Scand. Suppl.* 380, 8–13.

Leyton, M. (2007) Conditioned and sensitized responses to stimulant drugs in humans. *Prog. Neuropsychopharmacol. Biol. Psychiatry* 31, 1601–1613.

Leyton, M. and Stewart, J. (1990) Preexposure to foot-shock sensitizes the locomotor response to subsequent systemic morphine and intra-nucleus accumbens amphetamine. *Pharmacol. Biochem. Behav.* 37, 303–310.

Leyton, M. and Stewart, J. (1996) Acute and repeated activation of male sexual behavior by tail pinch: Opioid and dopaminergic mechanisms. *Physiol. Behav.* 60, 77–85.

Leyton, M., Young, S. N. and Benkelfat, C. (1997) Relapse of depression after rapid depletion of tryptophan. *Lancet* 349, 1840–1841.

Leyton, M., Benkelfat, C., Boileau, I. and Dagher, A. (2003) Amphetamine-induced striatal dopamine release in humans: Drug wanting or euphoria? . *American College of Neuropsychopharmacology Abstract,* San Juan, Puerto Rico, 6–11 December.

Leyton, M., Boileau, I., Benkelfat, C., Diksic, M., Baker, G. and Dagher, A. (2002) Amphetamine-induced increases in extracellular dopamine, drug wanting, and novelty seeking: a PET/[11C]raclopride study in healthy men. *Neuropsychopharmacology* 27, 1027–1035.

Leyton, M., Casey, K. F., Delaney, J. S., Kolivakis, T. and Benkelfat, C. (2005) Cocaine craving, euphoria, and self-administration: a preliminary study of the effect of catecholamine precursor depletion. *Behav. Neurosci.* 119, 1619–1627.

Leyton, M., Young, S. N., Blier, P., Baker, G. B., Pihl, R. O. and Benkelfat, C. (2000a) Acute tyrosine depletion and alcohol ingestion in healthy women. *Alcohol. Clin. Exp. Res.* 24, 459–464.

Leyton, M., Young, S. N., Pihl, R. O., Etezadi, S., Lauze, C., Blier, P., Baker, G. B. and Benkelfat, C. (2000b) Effects on mood of acute phenylalanine/tyrosine depletion in healthy women. *Neuropsychopharmacology* 22, 52–63.

Leyton, M., Dagher, A., Boileau, I., Casey, K., Baker, G. B., Diksic, M., Gunn, R., Young, S. N. and Benkelfat, C. (2004) Decreasing amphetamine-induced dopamine release by acute phenylalanine/tyrosine depletion: a PET/[11C]raclopride study in healthy men. *Neuropsychopharmacology* 29, 427–432.

Lieberman, J. A., Kane, J. M. and Alvir, J. (1987) Provocative tests with psychostimulant drugs in schizophrenia. *Psychopharmacology (Berl.)* 91, 415–433.

Lou, H. (2007) Dopamine and attention deficit/hyperactivity disorder. Dopamine: 50 Years, Göteborg, Sweden, 30 May to 2 June 2007, p. 40.

Lythe, K. E., Anderson, I. M., Deakin, J. F., Elliott, R. and Strickland, P. L. (2005) Lack of behavioral effects after acute tyrosine depletion in healthy volunteers. *J. Psychopharmacol.* 19, 5–11.

Magliozzi, J. R., Mungas, D., Laubly, J. N. and Blunden, D. (1989) Effect of haloperidol on a symbol digit substitution task in normal adult males. *Neuropsychopharmacology* 2, 29–37.

Maher, N. E., Golbe, L. I., Lazzarini, A. M., Mark, M. H., Currie, L. J., Wooten, G. F., Saint-Hilaire, M., Wilk, J. B., Volcjak, J., Maher, J. E., Feldman, R. G., Guttman, M., Lew, M., Waters, C. H., Schuman, S., Suchowersky, O., Lafontaine, A. L., Labelle, N., Vieregge, P., Pramstaller, P. P., Klein, C., Hubble, J., Reider, C., Growdon, J., Watts, R., Montgomery, E., Baker, K., Singer, C., Stacy, M. and Myers, R. H. (2002) Epidemiologic study of 203 sibling pairs with Parkinson's disease: the GenePD study. *Neurology* 58, 79–84.

Martinez, D., Narendran, R., Foltin, R. W., Slifstein, M., Hwang, D. R., Broft, A., Huang, Y., Cooper, T. B., Fischman, M. W., Kleber, H. D. and Laruelle, M. (2007) Amphetamine-induced dopamine release: Markedly blunted in cocaine dependence and predictive of the choice to self-administer cocaine. *Am. J. Psychiatry* 164, 622–629.

Martinez, D., Slifstein, M., Broft, A., Mawlawi, O., Hwang, D. R., Huang, Y., Cooper, T., Kegeles, L., Zarahn, E., Abi-Dargham, A., Haber, S. N. and Laruelle, M. (2003) Imaging human mesolimbic dopamine transmission with positron emission tomography. Part II: Amphetamine-induced dopamine release in the functional subdivisions of the striatum. *J. Cereb. Blood Flow Metab.* 23, 285–300.

McBride, D., Barrett, S. P., Kelly, J. T., Aw, A. and Dagher, A. (2006) Effects of expectancy and abstinence on the neural response to smoking cues in cigarette smokers: an fMRI study. *Neuropsychopharmacology* 31, 2728–2738.

McGowan, S., Lawrence, A. D., Sales, T., Quested, D. and Grasby, P. (2004) Presynaptic dopaminergic dysfunction in schizophrenia: a positron emission tomographic [18F] fluorodopa study. *Arch. Gen. Psychiatry* 61, 134–42.

McLean, A., Rubinsztein, J. S., Robbins, T. W. and Sahakian, B. J. (2004) The effects of tyrosine depletion in normal healthy volunteers: Implications for unipolar depression. *Psychopharmacology (Berl.)* 171, 286–297.

McTavish, S. F., Cowen, P. J. and Sharp, T. (1999a) Effect of a tyrosine-free amino acid mixture on regional brain catecholamine synthesis and release. *Psychopharmacology (Berl.)* 141, 182–188.

McTavish, S. F., Mannie, Z. N. and Cowen, P. J. (2004) Tyrosine depletion does not cause depressive relapse in antidepressant-treated patients. *Psychopharmacology (Berl.)* 175, 124–126.

McTavish, S. F., Mannie, Z. N., Harmer, C. J. and Cowen, P. J. (2005) Lack of effect of tyrosine depletion on mood in recovered depressed women. *Neuropsychopharmacology* 30, 786–791.

McTavish, S. F., McPherson, M. H., Sharp, T. and Cowen, P. J. (1999b) Attenuation of some subjective effects of amphetamine following tyrosine depletion. *J. Psychopharmacol.* 13, 144–147.

McTavish, S. F., McPherson, M. H., Harmer, C. J., Clark, L., Sharp, T., Goodwin, G. M. and Cowen, P. J. (2001) Antidopaminergic effects of dietary tyrosine depletion in healthy subjects and patients with manic illness. *Br. J. Psychiatry* 179, 356–360.

Menon, M., Jensen, J., Vitcu, I., Graff-Guerrero, A., Crawley, A., Smith, M. A. and Kapur, S. (2007) Temporal difference modeling of the blood-oxygen level dependent response during aversive conditioning in humans: Effects of dopaminergic modulation. *Biol. Psychiatry* 62, 765–772.

Meyer-Lindenberg, A., Miletich, R. S., Kohn, P. D., Esposito, G., Carson, R. E., Quarantelli, M., Weinberger, D. R. and Berman, K. F. (2002) Reduced prefrontal activity predicts exaggerated striatal dopaminergic function in schizophrenia. *Nat. Neurosci.* 5, 267–271.

Meyer, J. H., McNeely, H. E., Sagrati, S., Boovariwala, A., Martin, K., Verhoeff, N. P., Wilson, A. A. and Houle, S. (2006) Elevated putamen D(2) receptor binding potential in major depression with motor retardation: an [11C] raclopride positron emission tomography study. *Am. J. Psychiatry* 163, 1594–1602.

Meyer, J. H., Kruger, S., Wilson, A. A., Christensen, B. K., Goulding, V. S., Schaffer, A., Minifie, C., Houle, S., Hussey, D. and Kennedy, S. H. (2001) Lower dopamine transporter binding potential in striatum during depression. *Neuroreport* 12, 4121–4125.

Mizrahi, R., Rusjan, P., Agid, O., Graff, A., Mamo, D. C., Zipursky, R. B. and Kapur, S. (2007) Adverse subjective experience with antipsychotics and its relationship to striatal and extrastriatal D2 receptors: a PET study in schizophrenia. *Am. J. Psychiatry* 164, 630–637.

Modell, J. G., Mountz, J. M., Glaser, F. B. and Lee, J. Y. (1993) Effect of haloperidol on measures of craving and impaired control in alcoholic subjects. *Alcohol. Clin. Exp. Res.* 17, 234–240.

Mogenson, G. J., Takigawa, M., Robertson, A. and Wu, M. (1979) Self-stimulation of the nucleus accumbens and ventral tegmental area of Tsai attenuated by microinjections of spiroperidol into the nucleus accumbens. *Brain Res.* 171, 247–259.

Montgomery, A. J., McTavish, S. F., Cowen, P. J. and Grasby, P. M. (2003) Reduction of brain dopamine concentration with dietary tyrosine plus phenylalanine depletion: an [^{11}C]raclopride PET study. *Am. J. Psychiatry* 160, 1887–1889.

Munafo, M. R., Mannie, Z. N., Cowen, P. J., Harmer, C. J. and McTavish, S. B. (2007) Effects of acute tyrosine depletion on subjective craving and selective processing of smoking-related cues in abstinent cigarette smokers. *J. Psychopharmacol.* 21, 805–14.

Munro, C. A., McCaul, M. E., Wong, D. F., Oswald, L. M., Zhou, Y., Brasic, J., Kuwabara, H., Kumar, A., Alexander, M., Ye, W. and Wand, G. S. (2006) Sex differences in striatal dopamine release in healthy adults. *Biol. Psychiatry* 59, 966–974.

Murphy, D. L., Brodie, H. K., Goodwin, F. K. and Bunney, W. E., Jr. (1971) Regular induction of hypomania by L-dopa in "bipolar" manic-depressive patients. *Nature* 229, 135–136.

Nann-Vernotica, E., Donny, E. C., Bigelow, G. E. and Walsh, S. L. (2001) Repeated administration of the D1/5 antagonist ecopipam fails to attenuate the subjective effects of cocaine. *Psychopharmacology (Berl.)* 155, 338–347.

Newberg, A., Amsterdam, J. and Shults, J. (2007) Dopamine transporter density may be associated with the depressed affect in healthy subjects. *Nucl. Med. Commun.* 28, 3–6.

Oades, R. (1982) *Attention and Schizophrenia: Neurobiological Bases*. Pitman Advanced Publishing Program: London.

Oswald, L. M., Wong, D. F., McCaul, M., Zhou, Y., Kuwabara, H., Choi, L., Brasic, J. and Wand, G. S. (2005) Relationships among ventral striatal dopamine release, cortisol secretion, and subjective responses to amphetamine. *Neuropsychopharmacology* 30, 821–832.

Oswald, L. M., Wong, D. F., Zhou, Y., Kumar, A., Brasic, J., Alexander, M., Ye, W., Kuwabara, H., Hilton, J. and Wand, G. S. (2007) Impulsivity and chronic stress are associated with amphetamine-induced striatal dopamine release. *Neuroimage* 36, 153–166.

Peciña, S., Smith, K. S. and Berridge, K. C. (2006) Hedonic hot spots in the brain. *Neuroscientist* 12, 500–511.

Persico, A. M., Reich, S., Henningfield, J. E., Kuhar, M. J. and Uhl, G. R. (1998) Parkinsonian patients report blunted subjective effects of methylphenidate. *Exp. Clin. Psychopharmacol.* 6, 54–63.

Petry, N. M., Stinson, F. S. and Grant, B. F. (2005) Comorbidity of DSM-IV pathological gambling and other psychiatric disorders: Results from the National Epidemiologic Survey on Alcohol and Related Conditions. *J. Clin. Psychiatry* 66, 564–574.

Pezzella, F. R., Colosimo, C., Vanacore, N., Di Rezze, S., Chianese, M., Fabbrini, G. and Meco, G. (2005) Prevalence and clinical features of hedonistic homeostatic dysregulation in Parkinson's disease. *Mov. Disord* .20, 77–81.

Phillips, A., Pfaus, J. and Blaha, C. (1991) Dopamine and motivated behavior: Insights provided by in vivo analyses. . In: *The Mesolimbic Dopamine System: From Motivation to Action* pp. 199–224. Eds. P. Willner and J. Scheel-Krüger. John Wiley & Sons Ltd.: Chichester, UK.

Phillips, A. G. and Fibiger, H. C. (1973) Dopaminergic and noradrenergic substrates of positive reinforcement: Differential effects of D- and L-amphetamine. *Science* 179, 575–577.

Pillon, B., Gouider-Khouja, N., Deweer, B., Vidailhet, M., Malapani, C., Dubois, B. and Agid, Y. (1995) Neuropsychological pattern of striatonigral degeneration: Comparison with Parkinson's disease and progressive supranuclear palsy. *J. Neurol. Neurosurg. Psychiatry* 58, 174–179.

Pluck, G. C. and Brown, R. G. (2002) Apathy in Parkinson's disease. *J. Neurol. Neurosurg. Psychiatry* 73, 636–642.

Post, R. M. (1975) Cocaine psychoses: a continuum model. *Am. J. Psychiatry* 132, 225–231.

Pruessner, J. C., Champagne, F., Meaney, M. J. and Dagher, A. (2004) Dopamine release in response to a psychological stress in humans and its relationship to early life maternal care: a positron emission tomography study using [11C]raclopride. *J. Neurosci.* 24, 2825–2831.

Puglisi-Allegra, S., Imperato, A., Angelucci, L. and Cabib, S. (1991) Acute stress induces time-dependent responses in dopamine mesolimbic system. *Brain Res.* 554, 217–22.

Ragonese, P., Salemi, G., Morgante, L., Aridon, P., Epifanio, A., Buffa, D., Scoppa, F. and Savettieri, G. (2003) A case-control study on cigarette, alcohol, and coffee consumption preceding Parkinson's disease. *Neuroepidemiology* 22, 297–304.

Randrup, A., Munkvad, J., Fog, R., Gerlach, J., Molander, L., Kjellberg, B. and Scheel-Kruger, J. (1975) Mania, depression and brain dopamine. In: *Current Developments in Psychopharmacology*. Vol. 2, pp. 206–248. Eds. W. Essman and L. Valzelli. Spectrum: New York.

Redgrave, P., Prescott, T. J. and Gurney, K. (1999) Is the short-latency dopamine response too short to signal reward error? *Trends Neurosci.* 22, 146–151.

Reith, J., Benkelfat, C., Sherwin, A., Yasuhara, Y., Kuwabara, H., Andermann, F., Bachneff, S., Cumming, P., Diksic, M., Dyve, S. E., Etienne, P., Evans, A. C., Lal, S., Shevell, M., Savard, G., Wong, D. F., Chouinard, G. and Gjedde, A. (1994) Elevated dopa decarboxylase activity in living brain of patients with psychosis. *Proc. Natl. Acad. Sci. USA* 91, 11651–11654.

Resnick, R. B., Kestenbaum, R. S. and Schwartz, L. K. (1977) Acute systemic effects of cocaine in man: a controlled study by intranasal and intravenous routes. *Science* 195, 696–698.

Riccardi, P., Li, R., Ansari, M. S., Zald, D., Park, S., Dawant, B., Anderson, S., Doop, M., Woodward, N., Schoenberg, E., Schmidt, D., Baldwin, R. and Kessler, R. (2006) Amphetamine-induced displacement of [18F] fallypride in striatum and extrastriatal regions in humans. *Neuropsychopharmacology* 31, 1016–1026.

Robbins, T. W., Cador, M., Taylor, J. R. and Everitt, B. J. (1989) Limbic–striatal interactions in reward-related processes. *Neurosci. Biobehav. Rev.* 13, 155–162.

Roiser, J. P., McLean, A., Ogilvie, A. D., Blackwell, A. D., Bamber, D. J., Goodyer, I., Jones, P. B. and Sahakian, B. J. (2005) The subjective and cognitive effects of acute phenylalanine and tyrosine depletion in patients recovered from depression. *Neuropsychopharmacology* 30, 775–785.

Rolls, E. (2005) *Emotions Explained*. Oxford University Press: Oxford.

Romach, M. K., Glue, P., Kampman, K., Kaplan, H. L., Somer, G. R., Poole, S., Clarke, L., Coffin, V., Cornish, J., O'Brien, C. P. and Sellers, E. M. (1999) Attenuation of the euphoric effects of cocaine by the dopamine D1/D5 antagonist ecopipam (SCH 39166). *Arch. Gen. Psychiatry* 56, 1101–1106.

Rossetti, Z. L., Lai, M., Hmaidan, Y. and Gessa, G. L. (1993) Depletion of mesolimbic dopamine during behavioral despair: Partial reversal by chronic imipramine. *Eur. J. Pharmacol* 242, 313–315.

Saeedi, H., Remington, G. and Christensen, B. K. (2006) Impact of haloperidol, a dopamine D2 antagonist, on cognition and mood. *Schizophr. Res.* 85, 222–231.

Salamone, J. D. (1994) The involvement of nucleus accumbens dopamine in appetitive and aversive motivation. *Behav. Brain Res.* 61, 117–133.

Salamone, J. D., Correa, M., Farrar, A. and Mingote, S. M. (2007) Effort-related functions of nucleus accumbens dopamine and associated forebrain circuits. *Psychopharmacology (Berl.)* 191, 461–482.

Scarná, A., McTavish, S. F., Cowen, P. J., Goodwin, G. M. and Rogers, R. D. (2005) The effects of a branched chain amino acid mixture supplemented with tryptophan on biochemical indices of neurotransmitter function and decision-making. *Psychopharmacology (Berl.)* 179, 761–768.

Scarná, A., Gijsman, H. J., McTavish, S. F., Harmer, C. J., Cowen, P. J. and Goodwin, G. M. (2003) Effects of a branched-chain amino acid drink in mania. *Br. J. Psychiatry* 182, 210–213.

Schlaepfer, T.E., Pearlson, G.D., Wong, D.F., Marenco, S., and Dannals, R.F. (1997) PET study of competition between intravenous cocaine and [11C]raclopride at dopamine receptors in human subjects. *Am. J. Psychiatry.* 154, 1209–1213.

Schlaepfer, T. E., Cohen, M. X., Frick, C., Kosel, M., Brodesser, D., Axmacher, N., Joe, A. Y., Kręft, M., Lenartz, D. and Sturm, V. (2007) Deep brain stimulation to reward circuitry alleviates anhedonia in refractory major depression. *Neuropsychopharmacology* 33, 368–377.

Schultz, W. (1998) Predictive reward signal of dopamine neurons. *J. Neurophysiol.* 80, 1–27.

Schultz, W. (2007) Behavioral dopamine signals. *Trends Neurosci.* 30, 203–210.

Schultz, W. and Romo, R. (1990) Dopamine neurons of the monkey midbrain: Contingencies of responses to stimuli eliciting immediate behavioral reactions. *J. Neurophysiol.* 63, 607–624.

Schultz, W., Dayan, P. and Montague, P. R. (1997) A neural substrate of prediction and reward. *Science* 275, 1593–1599.

Scott, D. J., Heitzeg, M. M., Koeppe, R. A., Stohler, C. S. and Zubieta, J.-K. (2006) Variations in the human pain stress experience mediated by ventral and dorsal basal ganglia dopamine activity. *J. Neurosci.* 26, 10789–10795.

Scott, D. J., Domino, E. F., Heitzeg, M. M., Koeppe, R. A., Ni, L., Guthrie, S. and Zubieta, J. K. (2007a) Smoking modulation of mu-opioid and dopamine D2 receptor-mediated neurotransmission in humans. *Neuropsychopharmacology* 32, 450–457.

Scott, D. J., Stohler, C. S., Egnatuk, C. M., Wang, H., Koeppe, R. A. and Zubieta, J. K. (2007b) Individual differences in reward responding explain placebo-induced expectations and effects. *Neuron* 55, 325–336.

Seeman, P., Chau-Wong, M., Tedesco, J. and Wong, K. (1975) Brain receptors for antipsychotic drugs and dopamine: Direct binding assays. *Proc. Natl. Acad. Sci. USA* 72, 4376–4380.

Seeman, P., Lee, T., Chau-Wong, M. and Wong, K. (1976) Antipsychotic drug doses and neuroleptic/dopamine receptors. *Nature* 261, 717–719.

Servan-Schreiber, D., Carter, C. S., Bruno, R. M. and Cohen, J. D. (1998) Dopamine and the mechanisms of cognition: Part II. D-amphetamine effects in human subjects performing a selective attention task. *Biol. Psychiatry* 43, 723–729.

Shaham, Y. and Stewart, J. (1996) Effects of opioid and dopamine receptor antagonists on relapse induced by stress and re-exposure to heroin in rats. *Psychopharmacology (Berl.)* 125, 385–391.

Shelley, A. M., Catts, S. V., Ward, P. B., Andrews, S., Mitchell, P., Michie, P. and McConaghy, N. (1997) The effect of decreased catecholamine transmission on ERP indices of selective attention. *Neuropsychopharmacology* 16, 202–210.

Sherer, M. A. (1988) Intravenous cocaine: Psychiatric effects, biological mechanisms. *Biol. Psychiatry* 24, 865–885.

Sienkiewicz-Jarosz, H., Scinska, A., Kuran, W., Ryglewicz, D., Rogowski, A., Wrobel, E., Korkosz, A., Kukwa, A., Kostowski, W. and Bienkowski, P. (2005) Taste responses in patients with Parkinson's disease. *J. Neurol. Neurosurg. Psychiatry* 76, 40–46.

Singh, M. M. (1976) Dysphoric response to neuroleptic treatment in schizophrenia and its prognostic significance. *Dis. Nerv. Syst.* 37, 191–196.

Singh, M. M. and Smith, J. M. (1973) Kinetics and dynamics of response to haloperidol in acute schizophrenia: a longitudinal study of the therapeutic process. *Comp. Psychiatry* 14, 393–414.

Sjostrom, R. (1973) 5-Hydroxyindole acetic acid and homovanillic acid in cerebrospinal fluid in manic-depressive psychosis and the effect of probenecid treatment. *Eur. J. Clin. Pharmacol.* 6, 75–80.

Sjostrom, R. and Roos, B. E. (1972) 5-Hydroxyindolacetic acid and homovanillic acid in cerebrospinal fluid in manic-depressive psychosis. *Eur. J. Clin. Pharmacol.* 4, 170–176.

Small, D. M., Jones-Gotman, M. and Dagher, A. (2003) Feeding-induced dopamine release in dorsal striatum correlates with meal pleasantness ratings in healthy human volunteers. *Neuroimage* 19, 1709–1715.

Smith, K. S. and Berridge, K. C. (2007) Opioid limbic circuit for reward: Interaction between hedonic hotspots of nucleus accumbens and ventral pallidum. *J. Neurosci.* 27, 1594–1605.

Stewart, J. (1992) Neurobiology of conditioning to drugs of abuse. In: *The Neurobiology of Drug and Alcohol Addiction,* pp. 335–346. Eds. P. Kalivas and H. Samson. New York Academy of Sciences: New York.

Stewart, J., de Wit, H. and Eikelboom, R. (1984) Role of unconditioned and conditioned drug effects in the self-administration of opiates and stimulants. *Psychol. Rev.* 91, 251–268.

Stine, S. M., Krystal, J. H., Petrakis, I. L., Jatlow, P. I., Heninger, G. R., Kosten, T. R. and Charney, D. S. (1997) Effect of alpha-methyl-para-tyrosine on response to cocaine challenge. *Biol. Psychiatry* 42, 181–190.

Swerdlow, N. R., Stephany, N., Wasserman, L. C., Talledo, J., Sharp, R., Minassian, A. and Auerbach, P. P. (2005) Intact visual latent inhibition in schizophrenia patients in a within-subject paradigm. *Schizophr. Res.* 72, 169–183.

Toates, F. (1986) *Motivational Systems.* Cambridge University Press,: Cambridge.

Ungerstedt, U. (1971) Adipsia and aphagia after 6-hydroxydopamine induced degeneration of the nigro-striatal dopamine system. *Acta Physiol Scand. Suppl.* 367, 69–93.

van Praag, H. M., Korf, J. and Schut, D. (1973) Cerebral monoamines and depression. An investigation with the Probenecid technique. *Arch. Gen. Psychiatry* 28, 827–831.

Van Rossum, J. (1967) The significance of dopamine-receptor blockade for the action of neuroleptic drugs. In: *Neuropsychopharmacology: Proceedings of the 5th Collegium Internationale Neuropsychopharmacologicum*, pp. 321–329. Eds. H. Brill, J. Cole, P. Deniker, H. Hippius and P. Bradley. Excerpta Medica: Amsterdam.

Vieregge, P., Verleger, R., Wascher, E., Stuven, F. and Kompf, D. (1994) Auditory selective attention is impaired in Parkinson's disease—event-related evidence from EEG potentials. *Brain Res. Cogn. Brain Res.* 2, 117–29.

Vitiello, B., Martin, A., Hill, J., Mack, C., Molchan, S., Martinez, R., Murphy, D. L. and Sunderland, T. (1997) Cognitive and behavioral effects of cholinergic, dopaminergic, and serotonergic blockade in humans. *Neuropsychopharmacology* 16, 15–24.

Volkow, N. D., Wang, G. J., Fowler, J. S., Logan, J., Gatley, S. J., Wong, C., Hitzemann, R. and Pappas, N. R. (1999) Reinforcing effects of psychostimulants in humans are associated with increases in brain dopamine and occupancy of D(2) receptors. *J. Pharmacol. Exp. Ther.* 291, 409–415.

Volkow, N. D., Wang, G. J., Fowler, J. S., Logan, J., Gatley, S. J., Hitzemann, R., Chen, A. D., Dewey, S. L. and Pappas, N. (1997) Decreased striatal dopaminergic responsiveness in detoxified cocaine-dependent subjects. *Nature* 386, 830–833.

Volkow, N. D., Wang, G., Fowler, J. S., Logan, J., Gerasimov, M., Maynard, L., Ding, Y., Gatley, S. J., Gifford, A. and Franceschi, D. (2001) Therapeutic doses of oral methylphenidate significantly increase extracellular dopamine in the human brain. *J. Neurosci.* 21, RC121.

Volkow, N. D., Wang, G. J., Telang, F., Fowler, J. S., Logan, J., Childress, A. R., Jayne, M., Ma, Y. and Wong, C. (2006) Cocaine cues and dopamine in dorsal striatum: Mechanism of craving in cocaine addiction. *J. Neurosci.* 26, 6583–6588.

Volkow, N. D., Wang, G. J., Fowler, J. S., Logan, J., Jayne, M., Franceschi, D., Wong, C., Gatley, S. J., Gifford, A. N., Ding, Y. S. and Pappas, N. (2002) "Nonhedonic" food motivation in humans involves dopamine in the dorsal striatum and methylphenidate amplifies this effect. *Synapse* 44, 175–180.

Volkow, N. D., Wang, G. J., Fowler, J. S., Telang, F., Maynard, L., Logan, J., Gatley, S. J., Pappas, N., Wong, C., Vaska, P., Zhu, W. and Swanson, J. M. (2004) Evidence that methylphenidate enhances the saliency of a mathematical task by increasing dopamine in the human brain. *Am. J. Psychiatry* 161, 1173–1180.

Volkow, N. D., Wang, G. J., Fowler, J. S., Logan, J., Schlyer, D., Hitzemann, R., Lieberman, J., Angrist, B., Pappas, N., MacGregor, R. and et al. (1994) Imaging endogenous dopamine competition with [^{11}C]raclopride in the human brain. *Synapse* 16, 255–262.

Weintraub, D., Newberg, A. B., Cary, M. S., Siderowf, A. D., Moberg, P. J., Kleiner-Fisman, G., Duda, J. E., Stern, M. B., Mozley, D. and Katz, I. R. (2005) Striatal dopamine transporter imaging correlates with anxiety and depression symptoms in Parkinson's disease. *J. Nucl. Med.* 46, 227–232.

Wiesbeck, G. A., Mauerer, C., Thome, J., Jakob, F. and Boening, J. (1995) Neuroendocrine support for a relationship between "novelty seeking" and dopaminergic function in alcohol-dependent men. *Psychoneuroendocrinology.* 20, 755–761.

Willner, P. (1995) Dopaminergic mechanisms in depression and mania. In: *Psychopharmacology: The Fourth Generation of Progress*, pp. 921–931. Eds. F. E. Bloom and D. J. Kupfer. Raven Press: New York.

Wilson, M. C. and Schuster, C. R. (1972) The effects of chlorpromazine on psychomotor stimulant self-administration in the rhesus monkey. *Psychopharmacologia* 26, 115–126.

Wilson, S. J., Sayette, M. A. and Fiez, J. A. (2004) Prefrontal responses to drug cues: a neurocognitive analysis. *Nat. Neurosci.* 7, 211–214.

Wise, R. A. (1982) Neuroleptics and operant behavior: the anhedonia hypothesis. *Behav. Brain Sci.* 5, 39–87.

Wise, R. A. (2004) Dopamine, learning and motivation. *Nat. Rev. Neurosci.* 5, 483–494.

Wise, R. A., Spindler, J., deWit, H. and Gerberg, G. J. (1978) Neuroleptic-induced "anhedonia" in rats: Pimozide blocks reward quality of food. *Science* 201, 262–264.

Wong, D. F., Kuwabara, H., Schretlen, D. J., Bonson, K. R., Zhou, Y., Nandi, A., Brasic, J. R., Kimes, A. S., Maris, M. A., Kumar, A., Contoreggi, C., Links, J., Ernst, M., Rousset, O., Zukin, S., Grace, A. A., Lee, J. S., Rohde, C., Jasinski, D. R., Gjedde, A. and London, E. D. (2006) Increased occupancy of dopamine receptors in human striatum during cue-elicited cocaine craving. *Neuropsychopharmacology* 31, 2716–2727.

Wood, P. B., Schweinhardt, P., Jaeger, E., Dagher, A., Hakyemez, H., Rabiner, E. A., Bushnell, M. C. and Chizh, B. A. (2007) Fibromyalgia patients show an abnormal dopamine response to pain. *Eur. J. Neurosci.* 25, 3576–3582.

Wu, Y. L., Yoshida, M., Emoto, H. and Tanaka, M. (1999) Psychological stress selectively increases extracellular dopamine in the 'shell', but not in the 'core' of the rat nucleus accumbens: a novel dual-needle probe simultaneous microdialysis study. *Neurosci. Lett.* 275, 69–72.

Yoder, K. K., Constantinescu, C. C., Kareken, D. A., Normandin, M. D., Cheng, T. E., O'Connor, S. J. and Morris, E. D. (2007) Heterogeneous effects of alcohol on dopamine release in the striatum: a PET study. *Alcohol. Clin. Exp. Res.* 31, 965–973.

Yoder, K. K., Kareken, D. A., Seyoum, R. A., O'Connor S, J., Wang, C., Zheng, Q. H., Mock, B. and Morris, E. D. (2005) Dopamine D(2) receptor availability is associated with subjective responses to alcohol. *Alcohol Clin. Exp. Res.* 29, 965–970.

Yokel, R. A. and Wise, R. A. (1975) Increased lever pressing for amphetamine after pimozide in rats: Implications for a dopamine theory of reward. *Science* 187, 547–549.

Young, A. M., Ahier, R. G., Upton, R. L., Joseph, M. H. and Gray, J. A. (1998) Increased extracellular dopamine in the nucleus accumbens of the rat during associative learning of neutral stimuli. *Neuroscience* 83, 1175–1183.

Young, S. N. and Leyton, M. (2002) The role of serotonin in human mood and social interaction. insight from altered tryptophan levels. *Pharmacol. Biochem. Behav.* 71, 857–865.

Zald, D. H., Boileau, I., El-Dearedy, W., Gunn, R., McGlone, F., Dichter, G. S. and Dagher, A. (2004) Dopamine transmission in the human striatum during monetary reward tasks. *J. Neurosci.* 24, 4105–4112.

Zarevics, P. and Setler, P. E. (1979) Simultaneous rate-independent and rate-dependent assessment of intracranial self-stimulation: Evidence for the direct involvement of dopamine in brain reinforcement mechanisms. *Brain Res.* 169, 499–512.

Zhang, M. and Kelley, A. E. (2000) Enhanced intake of high-fat food following striatal mu-opioid stimulation: Microinjection mapping and fos expression. *Neuroscience* 99, 267–277.

14

To Be Happy and to Know It: The Experience and Meta-Awareness of Pleasure

JONATHAN W. SCHOOLER AND IRIS B. MAUSS

The refrain of an old favorite children's song goes "if you're happy and you know it, clap your hands." Implicit in this popular lyric is the curious observation that at least in principle one might be happy but not know it. Although embedded in the folk wisdom of popular culture, the possibility that people might not necessarily know whether or not they are happy is often overlooked in scientific discussions of happiness and pleasure. While researchers who study subjective well-being acknowledge that there are limitations to self-report measures, they generally take individuals' assessments of their happiness at face value. As Myers, one of the foremost purveyors of this research observes: "By definition, the final judge of someone's subjective well-being is whoever lives inside that person's skin. 'If you feel happy' noted Jonathan Freedman (1978) 'you are happy—that's all we mean by the term'" (Myers, 2000).

There are, of course, a number of good reasons why we might want to trust individuals' ability to decipher their experience of pleasure. First, who could possibly be a better arbiter of the hedonic quality of subjective experience than the person who is having that experience? Moreover, surely nothing could be more necessary for survival than an ability to accurately evaluate which experiences are reinforcing and which are not. Finally, and perhaps most importantly as illustrated in the above quote, there is a certain definitional self-evidence to our ability to assess the pleasure that we derive from experiences. The dictionary defines pleasure as "a feeling of happiness, delight, or satisfaction." Thus, if there is no feeling, an experience simply cannot be pleasurable, at least not as the term is commonly understood. What then could it mean for an individual to experience pleasure if they were not aware of it?

Although the notion of unconscious pleasure seems to undermine the very meaning of the term, we argue that it is still possible that individuals could experience pleasure without being aware of it. We approach this problem by distinguishing between experiential consciousness (i.e., the contents of ongoing experience) and meta-awareness (i.e., one's explicit awareness of the contents of consciousness) (Schooler, 2001, 2002; Schooler et al., 2003; Schooler and Schreiber, 2004). Central to this distinction is the claim that we can have experiences (experiential consciousness) without being contemporaneously aware of the nature of those experiences (meta-awareness). Recent neuroscientific evidence lends some support to this notion: the brain may register valenced responses to events (e.g., subliminally presented stimuli) for which the hedonic reaction is not consciously experienced (e.g., Winkielman and Berridge, 2004).

The dissociation of experiential and meta-awareness is illustrated by the case of mind-wandering during reading. All readers are familiar with the experience of suddenly realizing that despite the best of intentions, one's mind has wandered, and one has no idea what one has been reading. What is so striking about this experience is that although one consciously experiences the contents of the mind-wandering episode, one fails to notice that one's mind has wandered. Otherwise, one would have either stopped reading or stopped daydreaming. The fact that both activities continue demonstrates the absence of awareness that one is daydreaming even though that is precisely what is occupying one's minds at the time. In short, the common everyday experience of mind-wandering during reading illustrates that we can have an experience without being explicitly aware (i.e., meta-conscious) of the fact that we are having that experience.

Recent laboratory studies demonstrated the ubiquity of mind-wandering during reading, and by extension the ease with which individuals can be unaware of the contents of their own experience (Schooler et al., 2004). Participants read passages and were asked to press a button every time they caught their mind-wandering ("zoning out"). On average, people caught themselves zoning out five times during a 45-min period. In addition, participants were intermittently probed and asked whether at that particular moment they had been zoning out. Despite the fact that a central component of this task was to actively monitor mind-wanderings, on more than 11% of the probe trials, participants were still caught zoning out. Moreover, the frequency of these unaware flights of thought was a strong predictor of ultimate comprehension. This finding suggests that the individuals who were zoning out without awareness during the sampling procedure similarly failed to notice other zoning-out episodes that were never caught at all. Thus, these individuals were ultimately unprepared to answer questions about text that was read when their mind was elsewhere.

If individuals can have conscious, lucid, and perhaps even quite pleasurable mind-wandering experiences during reading without meta-awareness of what they are thinking about, then it seems quite plausible that many other experiences, including pleasurable ones, may also occur in the absence of explicit appraisal. If so, then the notion that individuals might often lack explicit awareness of their states of pleasure shifts from a logical impossibility to a phenomenon that may occur all the time. Indeed, when we consider the available evidence, it seems that many of our most pleasurable experiences occur with little meta-awareness of the fact that we are experiencing pleasure.

Dissociations Between Experience And Meta-Awareness of Pleasure

Two phenomena are particularly well suited to illustrate dissociations between experience and meta-awareness of pleasure, namely the experience of flow and the influence of forced meta-awareness on judgments.

Experience of Flow

One of the most effective ways of assessing the occurrence of pleasure in everyday life is the experience-sampling technique in which participants are equipped with a pocket computer that intermittently probes them regarding what they are doing and how much they are enjoying it (Csikszentmihalyi and LeFevre, 1989). Using this methodology with over 1000 participants, Csikszentmihalyi and LeFevre (1989) found that many of most pleasurable moments occur when individuals are in what Csikszentmihalyi terms a state of "flow." The flow state occurs when one is deeply absorbed in a task that is both highly challenging yet also accomplishable. What is so striking about research on the flow states is the fact that it indicates that individuals' most positive experiences occur when they are not thinking about themselves, but are rather deeply absorbed in the activity itself. Indeed the flow state is so absorbing that individuals do not have the attentional resources to explicitly notice that they are happy at the time. As Csikszentmihalyi (1999) puts it:

"Strictly speaking, during the experience [of flow] people are not necessarily happy because they are too involved in the task to have the luxury to reflect on their subjective states. Being happy would be a distraction, an interruption of flow. But afterwards, when the experience is over, people report having been in as positive a state as it is possible to feel" (p. 825).

Thus, the conclusion of one of the most extensive investigations of individuals' actual experiences of happiness suggests that people experience the greatest pleasure when they are not reflecting on the fact that they are happy. Importantly, however, as Csikszentmihalyi notes, as soon as individuals in a flow state direct their attention to their hedonic state, they readily report that they were experiencing pleasure. In other words, the flow state illustrates a "temporal dissociation of meta-awareness" (Schooler, 2002), in which an individual goes for a period of time without taking explicit stock of what they are experiencing. However, as soon as the experiential state is explicitly considered, the experience of pleasure is readily acknowledged.

The observation of temporal dissociations between having an experience and explicitly noticing that experience raises the possibility of another type of dissociation between experience and meta-awareness, termed as "translation dissociation" (Schooler,

2002) in which in the process of re-representing the quality of an experience to oneself, one distorts or omit critical elements of the experience, thereby misconstruing it. Although clearly more controversial than temporal dissociations, a variety of findings suggest that individuals may sometimes misrepresent the quality of their own subjective experience to themselves.

Impact of Reflection on the Assessment of Pleasure

If the process of re-representing an experience to oneself could in principle lead to errors in characterizing the experience, then it follows that encouraging extensive elaboration of an experience might be particularly apt to introduce such distortions. In fact, a number of studies suggest that reflection can interfere with people's ability to assess their experience. For example, in a study by Wilson and Schooler (1991), participants sampled five different strawberry jams. In the reflection condition, participants were then asked to reflect on their evaluation, listing the reasons why they felt the way they did about each jam. All participants were then asked to rate the five jams. The correlations between participants' jam ratings and expert judges' ratings (provided by Consumer Reports) were then assessed. Wilson and Schooler found that whereas control subjects provided ratings that were closely aligned with that of the experts (r=.51), the judgments of the participants who analyzed their reasons were completely unrelated (r=.16) to those of the experts. Within the current context, the findings of Wilson and Schooler can be interpreted as suggesting that reflection caused participants to "lose touch with their feelings," providing ratings that did not correspond to the actual pleasure that they, and others unbiased by reflection, derived from the jams.

One possible concern with Wilson and Schooler's findings is that it used experts' opinions as its normative basis for assessing the quality of participants' hedonic judgments. Failing to agree with an expert does not necessarily mean that one's opinions are unreflective of the pleasure that one derives from an experience. In other words, participants in the self-reflection condition might simply have had different hedonic experiences, which were equally well captured by their self-reports. A follow-up study by Wilson et al. (1993) however argues against this interpretation. In this study, participants examined various different art posters. Participants in the reflection condition analyzed why they felt the way they did about the posters and then rated them. Participants in the control condition simply rated the posters without reflection. Participants were then given the opportunity to select a poster and take it home. Two weeks later, participants were contacted and asked various questions to assess their postchoice satisfaction, including how much they now liked the poster and whether they had hung it up. Wilson et al. found that participants who had selected posters in the reflection condition were less satisfied with their choices and less likely to have hung them up than participants who had simply gone with their gut. The fact that participants who engaged in reflection were ultimately less satisfied with their selections suggests that reflection did not change the pleasure they experienced. Rather these findings suggest that reflection actually undermined people's ability to decipher the pleasure that they had actually experienced and which they re-experienced after the impact of self-reflection had worn off.

The above findings provide just a sampling of the numerous studies that indicate that self-reflection may impair people's ability to decipher the hedonic value of experience. Other studies have found similar effects of self-reflection on people's ability to judge the pleasure they derive from courses (Wilson and Schooler, 1991), beverages (Wilson and et al., 1984), and even relationships (Wilson et al., 2000). Moreover, additional studies have found that when self-reflection is minimized by forcing individuals to make very quick hedonic judgments, assessments become realigned with actual experience. For example, Wilson and Lindsey (as reported in Wilson et al., 2000) had participants evaluate the quality of their relationship with a significant other (romantic partner). Some participants engaged in self-reflection, analyzing their reasons for their evaluations, whereas other simply gave an overall rating. As in prior studies, they found that self-reflection reduced people's ability to adequately gauge the quality of their relationship, as revealed by the fact that those who analyzed their reasons were less able to predict the quality of their relationship at a later date, relative to the control subjects who did not engage in self-reflection. Importantly, however, Wilson and Lindsey included an additional condition in which, following self-reflection, participants made very quick (2s) evaluations. In this condition, the correlations between participants' ratings of their relationship, and their later-reported ratings, were as high as it was for participants who did not engage in self-reflection at all. Apparently, when self-reflection is discouraged, individuals are able to get a more direct "read-off" of their actual subjective state.

Assessing Accuracy of Meta-Awareness

The suggestion that people can be inaccurate in characterizing their hedonic experience raises the important question of whether there might exist some independent method for assessing individuals' hedonic state, and, by extension, for assessing the degree to which individuals' meta-awareness of their hedonic states is "accurate." In principle, one way of gauging the accuracy of individuals' meta-awareness of their own affective state is to assess the extent to which the self-reported hedonic experience correlates, or coheres, with behavioral or physiological measures of affect (Schooler and Schreiber, 2004). Such an approach is premised on the notion that behavioral measures such as facial behavior, facial electromyography (EMG), or autonomic physiological responses such as heart rate or skin conductance, can provide an accurate gauge of underlying hedonic response. If such measures could be shown to reflect actual hedonic experience, then it could be assumed that the greater the coherence between self-report and other covert measures of hedonic state the greater the accuracy of meta-awareness.

There is, of course, a fundamental logical challenge to validating the use of behavioral and physiological measures as a yardstick for assessing individuals' meta-awareness of their underlying experience (Gilbert, 2006). The only way to demonstrate that such measures tap actual hedonic experience is to show that they systematically covary with self-reports, or with situations that reliably differ in the type of self-reports that they invoke. But if self-reports are themselves suspect, then how can we ever establish the validity of an alternative measure? If the claim were that self-reports rarely if ever adequately capture the hedonic quality of an experience, then this concern would clearly be inescapable. However, the argument is not that self-reports have never any bearing on underlying experience. On the contrary, there are clearly situations in which it is self-evident that people's capacity to self-report their hedonic state is quite reasonable. Who, for example, would question that when someone cries out in pain after hitting their finger with a hammer that they are indeed suffering, or when a child squeals in glee after receiving a long-begged-for gift that she is experiencing genuine pleasure? The more modest claim that we are making is that, under some specified circumstances (as, for example, when individuals engage in extensive reflective analysis), the correspondence between self-reports and underlying experience can be somewhat discrepant. Accordingly, if self-reports are generally in line with underlying hedonic experience, then measures that are found to consistently covary with self-reported hedonic state may be assumed to serve as a reasonable proxy for underlying hedonic experience. Once such independent proxies of hedonic experience are identified, we will be able to examine situational and individual fluctuations in the accuracy of meta-awareness by assessing the conditions under which self-reports show greater versus less coherence with other measures. This would open interesting avenues for further research because we could then examine the correlates and the potential functions of accurate meta-awareness.

Challenges in Finding Coherence between Self-Report and Covert Measures

Unfortunately, much of the past research on coherence between self-reports and other potential measures of hedonic response have observed only weak correlations between self-report and physiological measures (Hodgson and Rachman, 1974; Mandler et al., 1961; Stemmler, 1992; Weinstein et al.,1968). Studies that have assessed experiential, behavioral, and physiological measures in the context of various affective states have similarly found relatively modest correlations (Bradley and Lang, 2000; Hubert and de Jong-Meyer, 1991; Lang, 1988; Rachman, 1978) (for a review, see Barrett, 2006). In general, links between self-reported hedonic experience and facial behavior have been strongest (Ekman et al.,1980, 1990; Rosenberg and Ekman, 1994), but again, findings are inconsistent across studies (Adelmann and Zajonc, 1989; Blumberg and Izard, 1991; Bonanno and Keltner, 2004; Reisenzein, 2000; Ruch, 1995) (for a review, see Fridlund et al., 1987). Even with sensitive EMG measures of facial behavior, correlations between self-reports of hedonic experience and facial behavior are only low to moderate (Brown and Schwarz, 1980; Cacioppo et al.,1988; Lang et al.,1993). Still more challenging for the use of covert affective measures for appraising the accuracy of meta-awareness is the fact that some studies have found no (Edelmann and Baker, 2002; Fernandez-Dols et al., 1997; Fridlund, 1991; Jakobs et al., 2001; Mauss et al., 2004) or even negative associations between self-reports of hedonic experience and other measures (Buck, 1977; Lacey, 1967; Lang, 1988).

Thus, at first blush, it seems that indirect measures of hedonic response offer little promise for providing a yardstick by which to assess the accuracy of meta-awareness of hedonic experience. However, this conclusion—at least in its general form—might be premature. After all, one systematic review of coherence studies by Ruch (1995) suggests a range of possible findings. Across 25 studies, correlations between funniness ratings and facial expressions of amusement ranged from −.30 to nearly 1.0. Also, a nonnegligible number of studies have reported substantial correlations between self-reported hedonic experience and other measures (Casey, 1993; Chovil, 1991; Gross et al., 2000; Lazarus et al., 1966). This range of findings suggests that perhaps methodological features of prior studies substantively influenced their outcomes. Indeed, some prior studies feature methods that may have made it difficult to detect associations between self-reported experience and other measures. Four of these methodological factors appear particularly relevant.

Factors That May Reduce Coherence between Self-Reports and Other Measures

The first factor that could have contributed to the variability in coherence estimates found in prior research is the intensity of hedonic state induced. The likely target state has to be sufficiently intense in order to find coherence among responses (Davidson, 1992; Rosenberg and Ekman, 1994; Tassinary and Cacioppo, 1992). Thus, some of the low estimates of coherence may have been due to the fact that only weak hedonic states were induced.

The second factor that influences coherence estimates is which measures are assessed and how well they are matched to the hedonic state under investigation. For example, some studies investigating pleasure have found surprisingly low correlations between self-reported feelings of pleasure and laughter (e.g., Bonanno and Keltner, 2004). However, laughter may reflect amusement or relief from a negative emotion rather than pleasure, and thus, might not be an appropriate index of pleasure. This example illustrates that it is important to carefully select one's response measures.

The third important methodological factor is whether coherence has been assessed at the *between-individual* or the *within-individual* level. In the between-individual approach, an individual who reports greater hedonic experience than other individuals would also be expected to exhibit greater behavioral and physiological responses than other individuals. The alternative approach is to investigate within-individual correlations among responses across time. In this approach, one would expect greater behavioral and physiological responding in time periods when an individual self-reports greater hedonic experience relative to time periods when the same individual self-reports less hedonic experience. As several researchers have noted, the within-participant design is often more sensitive to detecting coherence than the between-participants design because it minimizes sources of between-individual variance (Lazarus et al., 1963; Pennebaker, 1982; Reisenzein, 2000; Rosenberg and Ekman, 1994; Ruch, 1995). In addition, between-individual analyses might be conceptually irrelevant to the question of how tightly responses are associated (Buck, 1980; Cacioppo et al., 1992; Lacey, 1967; Stemmler, 1992). Within-individual as compared to between-individual associations more closely denote accuracy of meta-awareness in the sense that self-reported hedonic experience should be associated with other measures *within individuals* and *across time*.

The fourth factor that affects indices of coherence consists of the timing of measures and their temporal resolution. When measuring self-reported hedonic experience, researchers have often relied on *retrospective* and *aggregated* ratings because rather than assessing emotional experience *online* and *moment-by-moment* (Gottman and Levenson, 1985; Rosenberg and Ekman, 1994). However, assessing experience ratings *after* a hedonic event might lead to measurement error due to processes such as memory biases or defensive mechanisms (Barrett, 1997; Kahneman, 2000; Rosenberg and Ekman, 1994). Thus, low associations between self-reported experience and other measures might be the result of suboptimal measures of self-reported experience. Additionally, prior studies have sometimes neglected to take into account varying lags among measures of emotional responding. This also might artificially decrease indices of coherence because it might lead one to miss responses outside the window under investigation, especially if the responses involved are short-lived (e.g., Kettunen et al., 2000).

Finding Greater Coherence between Self-Reports and Other Measures

Together, these methodological factors might have resulted in the inconsistent and relatively low coherence findings in prior studies. A recent study addressed these methodological considerations in four ways

(Mauss et al., 2005). First, it assessed a positive hedonic state (amusement) induced at relatively high intensity levels using a well-validated film. Amusement is a positive hedonic state especially conducive to detecting coherence because it appears to recruit behavioral as well as physiological responses (Ruch, 1995). Second, the study sampled several important responses systems, including self-reported experience, behavior, and autonomic physiological responses (cardiovascular responding and skin conductance). Third, the study employed a within-individual design by assessing responses to a film continuously across time. Fourth, issues of resolution and timing were addressed by assessing self-reported amusement experience *moment-by-moment* using a variant of the rating dial method introduced and validated by Levenson and Gottman (1983) (see also Gottman and Levenson, 1985). This method minimized measurement error in self-reported amusement experience. In addition, it ensured that measures of self-reported experience, of behavior, and of physiological responses were matched with respect to temporal resolution. Lastly, it enabled a time-series approach that took into account varying lags among measures.

While the rating dial method thus provides a number of advantages when assessing self-reported emotion experience, it raises an important concern. Before we turn to the main results, this concern needs to be addressed. As noted above, instructing participants to report on their hedonic states may alter those hedonic states themselves under certain conditions. Might providing continuous reports of experienced amusement thus distort the very phenomenon under observation? In order to address this question, the study assessed two groups of participants. One group provided continuous reports of amusement ("Adjust the dial so as to indicate how much amusement you feel at each moment.") as well as "traditional" retrospective ratings of amusement after the film clip ("What was the greatest amount of amusement you felt during the film clip"?). The other group only provided retrospective ratings of amusement after the film clip. By comparing retrospective ratings, facial behavior, and physiological responses between the two groups, it could be assessed whether providing continuous ratings distorted the experience of amusement. Results revealed that the two groups did not differ significantly with respect to retrospective amusement experience, facial behavior, or autonomic physiological responding (Mauss et al., 2005), suggesting that providing ratings with the dial did not alter participants' actual hedonic state.

Another recent study argued that perhaps these outcome measures were not sensitive enough to detect group differences. This study thus used a similar design to ascertain whether providing continuous ratings of one's hedonic experience alters brain activation associated with emotional responding (Hutcherson et al., 2005). Findings suggested that providing continuous ratings of hedonic experience did not significantly alter activation of brain areas associated with amusement experience (e.g., temporal cortex, insula). It may be that after some practice continuous ratings using the dial do not require participants' attention. Together, these studies suggest that continuous ratings using a rating dial provide a viable method for assessing meta-awareness.

So how closely then does this continuous measure of meta-awareness track other, more indirect measures of amusement? Results from the study described above indicated average disattenuated cross correlations of .89 between self-reported and facially expressed amusement, of .25 between self-reported amusement and cardiovascular activation, and of .57 between self-reported amusement and skin conductance level (Mauss et al., 2005). In other words, when assessed across time and when taking into account lags between measures, meta-awareness shared moderate to high amounts of variance with other measures. These results suggest that, when using appropriate methods, facial behavior and some measures of autonomic physiological responding (most notably skin conductance level) converge with an index of hedonic experience. In other words, when adequate methods are used, these indirect measures of affect may indeed provide an alternative window on individuals' hedonic experience.

Importantly, beyond these *average* indices of coherence, this study suggests that even under ideal conditions *individuals vary considerably* with respect to how closely their meta-awareness tracks other measures of hedonic experience. For example, disattenuated cross-correlations between self-reported amusement and facial amusement behavior ranged from 0.21 to 1.32, and disattenuated cross-correlations between self-reported amusement and skin conductance level (SCL) ranged from -0.22 to 0.96 across individuals. What are we to make of variations in coherence between self-reports and other measures? Could it be, as intimated above, that those individuals who show greater coherence are more meta-aware of their underlying experience, and that the accuracy of individuals' meta-awareness has functional implications?

Do Variations in Coherence between Self-Reports and Physiological Measures of Pleasures Reflect Differences in Accuracy of Meta-Awareness?

A recent study by Sze et al. (2007) suggests that variations in coherence between self-report and indirect measures may indeed reflect variations in individuals' meta-awareness of their hedonic state. Specifically, these researchers found that Vipassana (body-awareness) meditators as compared to advanced dancers and demographically matched controls exhibited greater coherence between self-reported hedonic states and heart rate during emotionally evocative film clips. In Vipassana meditation, practitioners are trained to increase awareness of physical sensations in the body. These results suggest that teaching individuals to attend to their internal state increases the accuracy of their meta-awareness and thus the coherence between indirect measures and self-reports.

Is Accurate Meta-Awareness Adaptive?

Theoretically, it seems reasonable that greater accuracy of meta-awareness of hedonic states (i.e., greater coherence of self-reported with indirect measures) would be associated with greater socioemotional functioning. Indeed, a variety of lines of research support such a relationship. For example, the emotion regulation literature suggests that in order to effectively regulate one's emotions, one must be able to both promptly notice and correctly identify one's emotional experiences (Barrett et al., 2001; Gross and Thompson, 2007). Similarly, from a communication perspective, individuals who possess accurate meta-awareness might communicate their emotional states better to others, which might in turn produce positive and avoid negative social outcomes (Ciarrochi et al., 2002; Mayer et al., 2004; Roter and Ewart, 1992). Research also suggests that avoiding meta-experience of hedonic states (as is the case in repression or experiential avoidance) is generally associated with negative outcomes for well-being, social outcomes, and health (Gratz et al., 2006; Kashdan et al., 2006; Marx and Sloan, 2005). In contrast, acceptance, reappraisal, and some automatic forms of emotion regulation—emotion regulation strategies that bring in line conscious and meta-aware experience of emotions—appear to be generally associated with positive outcomes (Gross, 1998; Gross and John, 2003; Hayes et al., 2006; Mauss et al., 2007). Collectively, these studies suggest that individuals who are more "in touch" with (i.e., more meta-aware of) their feelings may experience socioemotional benefits. Accordingly, if coherence between self-report and covert indices of emotions taps the accuracy of individuals' meta-awareness of their emotions, then we would expect a relationship between coherence measures and socioemotional functioning.

Although very little research has directly explored this issue, a recent study offers preliminary evidence that coherence between self-reports and indirect measures of positive emotions may indeed be associated with greater socioemotional functioning. In 150 participants, we assessed coherence between emotional behavior and self-reported hedonic state during an amusing film clip, using a within-participants approach (Mauss et al., in preparation), individuals differed widely in accuracy of their meta-awareness. Participants' well-being (as indexed by depressive symptoms) was assessed 2 years later to examine whether individual differences in accuracy of meta-awareness would predict well-being. Indeed, greater coherence between self-reports and indirect measures was associated with greater well-being. In addition, in line with the idea that individuals who possess accurate meta-awareness might communicate their emotions more effectively, the association between coherence and well-being was mediated by social support. The conclusion that accurate meta-awareness might be adaptive is consistent with the studies described above, which suggested that when dissociations between consciousness and meta-awareness are induced by forcing individuals to extensively reflect on their experiences, they make less apt choices and judgments (Wilson et al.,1984; Wilson et al.,2000; Wilson and Schooler, 1991).

In sum, although more research in this area is clearly needed, the extant literature on coherence suggests that: (1) when adequate methodological considerations are taken into account, indirect measures of hedonic states reasonably cohere with self-reported measures; (2) individuals vary widely in the degree to which their self-reports correspond to their indirect measures of emotion; (3) higher levels of coherence appear to reflect greater emotional meta-awareness; and (4) more accurate meta-awareness might be generally adaptive. Together these findings suggest that coherence measures may provide a useful tool for assessing fluctuations in the accuracy of individuals' meta-awareness of their hedonic state.

Some Implications of Dissociations Between Experience and Meta-Awareness of Pleasure

The claim that there are fluctuations (both across situations and individuals) in accuracy of meta-awareness offers a potentially fresh perspective on variety of domains of hedonic experience. We briefly consider two such domains: (1) failures to pursue flow and (2) failures in affective forecasting.

If Flow Feels So Good, Why Don't People Pursue It More Often?

One puzzling finding in research on flow is that although individuals generally experience maximum pleasure when they are engaged in flow experiences, their leisure time preferences do not reflect this fact, as individuals tend to devote their leisure time to passive activities, such as watching television, that do not promote flow. The riddle that Csikszentmihalyi ponders is why, if flow states are so positive, do people not seek them out more reliably (Csikszentmihalyi and LeFevre, 1989). Within the present context, the answer to this question seems relatively straightforward. People fail to seek out flow experiences because they lack meta-awareness about the fact that such experiences are the most positive. The absence of reflection during flow, though it may enhance individuals' experience of the moment, may also undermine their ability to remember what a wonderful time they are having. As a consequence, individuals may tend to seek out experiences that they have come to believe will make them happy (perhaps through cultural immersion) rather than in engaging in the behaviors that actually have made them happy.

Failures in Affective Forecasting

People's frequent failure to pursue flow despite the pleasure that they derive from such experiences illustrates one of the many situations in which individuals inadequately anticipate the hedonic quality of future experiences. A large body of work reveals numerous situations in which people show a remarkable lack of insight regarding the pleasures and displeasures that will be gleaned from future events. In general, people tend to overestimate both joys and sorrows. With respect to joys, people overestimate the happiness they will gain from increased earnings (Kahneman et al., 2006), a favorable dormitory room (Dunn et al., 2003), or how much they will enjoy a drink if they have just exercised (Van Boven and Loewenstein, 2003). With respect to sorrows, people overestimate how upset they will feel following their team losing a football game, receiving negative feedback about their performance on a test, and failing to receive tenure (Wilson et al., 2000).

The distinction between experience and meta-awareness may help to illuminate one of the most puzzling aspects of affective forecasting errors, namely, why it is that people do not learn? For example, Wilson et al. (2000) found that individuals reliably overestimate how long they will remain upset following the loss of a home team. They interpreted this finding as suggesting a process of "focalism" whereby people fail to take into account the larger context in which this particular negative event occurred, and thus, overweigh its impact on their lives. Although Wilson et al. (2000) provide compelling evidence that an excessive focus on the impact of a single event contributes to many affective forecasting errors, this account fails to explain one important thing. If (as seems certain) everyone who cares about their home team has experienced big game losses, why do they fail to learn how quickly other events distract them from the pain of the loss? From the present perspective, one reasonable explanation is that individuals' frequent lack of meta-awareness of their hedonic states prevents them from noticing how quickly they move on, and thus, from factoring the richness of their lives into their predictions.

The distinction between experience and meta-awareness also raises potential concerns about how to interpret affective forecasting findings. Importantly, affective forecasting errors are revealed by discrepancies between what individuals predict they will feel prior to an event, and what they report experiencing after the event. However, if the veracity of individuals' self-reports of their hedonic responses can vary, then discrepancies between predicted and experienced affect may not only stem from errors in the affective forecast, but may also result from errors in reporting the hedonic experience itself. For example, one potential method for overcoming the hardship of a negative experience may be to downplay how upsetting it is. Accordingly, people's seemingly exaggerated forecasts of the magnitude and duration of negative response to learning that they did particularly badly on a test, might be at least partially due to participants not wanting admit to themselves the displeasure they are actually experiencing. If underreporting of experienced affect contributes to affective forecasting discrepancies, then the inclusion of behavioral and physiological measures of hedonic experience (such as those described earlier) might reveal covert evidence of hedonic responses that are more in tune with people's predictions then their self-reports.

Although the distinction between experience and meta-awareness raises the possibility that self-reports may exaggerate affective forecasting errors, it also suggests that in some cases self-reports might actually underestimate the magnitude of such errors. Specifically, one hypothesized source of dissociations between experience and meta-awareness are faulty theories about how people think they ought to be feeling (Schooler and Schreiber, 2004). If this is the case, then it seems quite possible that people would consult the very same theories they use to generate their predictions about how they will feel in the future, when they actually come to make appraisals of their current state. If one has a theory which predicts he or she should be feeling bad in a particular situation (i.e., "I feel unhappy when people tell me I have done poorly on a test"), then this theory may color the appraisal of that experience, leading them to report being unhappy longer then they really are. Once again, the evidence reviewed earlier that indirect measures can be used as a metric for assessing the accuracy of meta-awareness suggests that we may now be poised to assess the situations in which self-reports exaggerate, underestimate, and accurately characterize affective forecasting errors.

Final Thought: The Relative Merit of the Experience Versus Meta-Awareness of Pleasure

The suggestion that people may experience pleasure without realizing that they are doing so raises the fundamental issue of the relative merit of having an experience of pleasure versus knowing that you are having it. Consider two situations: you can have an experience that you would rate a "9" if only you stopped to consider it, or one that is an "8" but that you are actually able to stop and savor as it occurs. Does the fact that you can attend to a pleasure as it as it happens somehow give it greater value, even if it is of lesser sheer hedonic quality? Or is the memory of an intense pleasure, even if it was not acknowledged as such at the time, ultimately of greater importance? Furthermore, if you did not actually attend to the quality of the pleasure at the time, how confident can you be that it really was as good as it is remembered? As you recall the thrill of going down that roller coaster, you may remember it as intense pleasure, but perhaps this is just a reframing of the sheer fear that you actually experienced as you plummeted down the ramp. And if it is the meta-awareness that is remembered, should we live our lives to maximize the actual fleeting pleasure of experiences, or the more enduring, if flawed, retrospective appraisal of it? Although resolving the relative merit of maximizing the experience versus meta-awareness of pleasure is clearly a difficult task, recognizing that there may be sizable differences between the two is certainly an important first step.

References

Adelmann, P. K. and Zajonc, R. B. (1989) Facial efference and the experience of emotion. *Annu. Rev. Psychol.* 40, 249–280.

Barrett, L. F. (1997) The relationships among momentary emotion experiences, personality descriptions, and retrospective ratings of emotion. *Pers Soc. Psychol.Bull.* 23, 1100–1110.

Barrett, L. F. (2006) Are emotions natural kinds? *Perspect. Psychol.Sci.*1, 28–58.

Barrett, L. F., Gross, J., Christensen, T. C. and Benvenuto, M. (2001) Knowing what you're feeling and knowing what to do about it: Mapping the relation between emotion differentiation and emotion regulation. *Cogn. Emot.*15, 713–724.

Blumberg, S. H. and Izard, C. E. (1991) Patterns of emotion experiences as predictors of facial expressions of emotion. *Merrill Palmer* Q. 37, 183–197.

Bonanno, G. A. and Keltner, D. (2004) Brief report: the coherence of emotion systems: Comparing "online" measures of appraisal and facial expressions, and self-report. *Cogn.Emot.* 18, 431–444.

Bradley, M. M. and Lang, P. J. (2000) Measuring emotion: Behavior, feeling, and physiology. In: *Cognitive Neuroscience of Emotion*, pp. 242–276. Eds. R. D. Lane and L. Nadel. Oxford University Press: New York, NY.

Brown, A. and Schwarz, G. E. (1980) Relationship between facial electromyography and subjective experience during affective imagery. *Biol.Psychol.* 11, 49–62.

Buck, R. (1977) Nonverbal communication of affect in preschool children: Relationships with personality and skin conductance. *J. Pers.Soc.Psychol.* 35, 225–236.

Buck, R. (1980) Nonverbal behavior and the theory of emotion: the facial feedback hypothesis. *J.Pers.Soc.Psychol.* 38, 811–824.

Cacioppo, J. T., Martzke, J. S., Petty, R. E. and Tassinary, L. G. (1988) Specific forms of facial EMG response index emotions during an interview: From Darwin to the continuous flow hypothesis of affect-laden information processing. *J.Pers.Soc. Psychol.*54, 592–604.

Cacioppo, J. T., Uchino, B. N., Crites, S. L., Snydersmith, M. A., Smith, G., Berntson, G. G. and Lang, P. J. (1992) Relationship between facial expressiveness and sympathetic activation in emotion: a critical review, with emphasis on modeling underlying mechanisms and individual differences. *J.Pers.Soc.Psychol.* 62, 110–128.

Casey, R. J. (1993) Children's emotional experience: Relations among expression, self-report, and understanding. *Dev.Psychol.* 29, 119–129.

Chovil, N. (1991) Social determinants of facial displays. *J.Nonverbal Behav.*15, 141–154.

Ciarrochi, J., Deane, F. P., Wilson, C. J. and Rickwood, D. (2002) Adolescents who need help the most are the least likely to seek it: the relationship between low emotional competence and low intention to seek help. *Br.J.Guid. Counc.*30, 173–188.

Csikszentmihalyi, M. (1999) If we are so rich, why aren't we happy? *Am.Psychol.* 54, 821–827.

Csikszentmihalyi, M. and LeFevre, J. (1989) Optimal experience in work and leisure. *J.Pers.Soc.Psychol.*56, 815–822.

Davidson, R. J. (1992) Prolegomenon to the structure of emotion: Gleanings from neuropsychology. *Cogn. Emot.*6, 245–268.

Dunn, E. W., Wilson, T. D. and Gilbert, D. T. (2003) Location, location, location: the misprediction of satisfaction in housing lotteries. *Pers.Soc.Psychol.Bull.*29, 1421–1432.

Edelmann, R. J. and Baker, S. R. (2002) Self-reported and actual physiological responses in social phobia. *Br.J.Clin. Psychol.* 41, 1–14.

Ekman, P., Davidson, R. J. and Friesen, W. V. (1990) The Duchenne smile: Emotional expression and brain physiology: II. *J. Pers.Soc.Psychol.*58, 342–353.

Ekman, P., Freisen, W. V. and Ancoli, S. (1980) Facial signs of emotional experience. *J.Pers.Soc.Psychol.*39, 1125–1134.

Fernandez-Dols, J.-M., Sanchez, F., Carrera, P. and Ruiz-Belda, M.-A. (1997) Are spontaneous expressions and emotions linked? an experimental test of coherence. *J.Nonverbal Behav.*21, 163–177.

Fridlund, A. J. (1991) Sociality of solitary smiling: Potentiation by an implicit audience. *J.Pers.Soc.Psychol.*60, 229–240.

Fridlund, A. J., Ekman, P. and Oster, H. (1987) Facial expressions of emotion. In: *Nonverbal Behavior and Communication* (2nd ed.), pp. 143–223. Eds. A. W. Siegman, S. Feldstein. Lawrence Erlbaum Associates: Hillsdale, NJ.

Gilbert, D. T. (2006) *Stumbling on Happiness.* Knopf: New York.

Gottman, J. M. and Levenson, R. W. (1985) A valid procedure for obtaining self-report of affect in marital interaction. *J.Consult. Clin.Psychol.*53, 151–160.

Gratz, K. L., Rosenthal, M. Z., Tull, M. T., Lejuez, C. W. and Gunderson, J. G. (2006) An experimental investigation of emotion dysregulation in borderline personality disorder. *J.Abnorm.Psychol.*115, 850–855.

Gross, J. J. (1998) Antecedent- and response-focused emotion regulation: Divergent consequences for experience, expression, and physiology. *J.Pers.Soc.Psychol.*74, 224–237.

Gross, J. J. and John, O. P. (2003) Individual differences in two emotion regulation processes: Implications for affect, relationships, and well-being. *J.Pers.Soc.Psychol.* 85, 348–362.

Gross, J. J., John, O. P. and Richards, J. M. (2000) The dissociation of emotion expression from emotion experience: a personality perspective. *Pers. Soc.Psychol.Bull.*26, 712–726.

Gross, J. J. and Thompson, R. A. (2007) Emotion regulation: Conceptual foundations. In: *Handbook of Emotion Regulation*, pp. 3–24. Ed. J. J. Gross. Guilford Press: New York.

Hayes, S. C., Luoma, J. B., Bond, F. W., Masuda, A. and Lillis, J. (2006) Acceptance and commitment therapy: Model, processes and outcomes. *Behav.Res.Ther.*44, 1–25.

Hodgson, R. and Rachman, S. (1974) II. Desynchrony in measures of fear. *Behav. Res. Ther.* 12, 319–326.

Hubert, W. and deJong-Meyer, R. (1991) Psychophysiological response patterns to positive and negative film stimuli. *Biol. Psychol.* 31, 73–93.

Hutcherson, C. A., Goldin, P. R., Ochsner, K. N., Gabrieli, J. D., Barrett, L. F. and Gross, J. J. (2005) Attention and emotion: Does rating emotion alter neural responses to amusing and sad films? *NeuroImage* 27, 656–668.

Jakobs, E., Manstead, A. S. R. and Fischer, A. H. (2001) Social context effects on facial activity in a negative emotional setting. *Emotion* 1, 51–69.

Kahneman, D. (2000) Experienced utility and objective happiness: a moment-based approach. In: *Choices, Values, and Frames*, pp. 673–692. Eds. D. Kahneman and A. Tversky. Cambridge University Press: New York.

Kahneman, D., Krueger, A. B., Schkade, D., Schwarz, N. and Stone, A. A. (2006) Would you be happier if you were richer? A focusing illusion. *Science* 312, 1908–1910.

Kashdan, T. B., Barrios, V. and Forsyth, J. P. (2006) Experiential avoidance as a generalized psychological vulnerability: Comparisons with coping and emotion regulation strategies. *Behav.Res.Ther.*44, 1301–1320.

Kettunen, J., Ravaja, N. and Keltikangas-Jarvinen, L. (2000) Smoothing facilitates the detection of coupled responses in psychophysiological time series. *Int.J. Psychophysiol.*14, 1–10.

Lacey, J. I. (1967) Somatic response patterning and stress: Some revisions of activation theory. In: *Psychological Stress: Issues in Research,* pp. 14–42. Eds. M. H. Appley and R. Trumbull. Appleton-Century-Crofts: New York.

Lang, P. J. (1988) What are the data of emotion? In: *Cognitive Perspectives on Emotion and Motivation*, pp. 173–191. Eds. V. Hamilton, G. H. Bower, and N. H. Frijda. Kluwer Academic/Plenum Publishers: New York.

Lang, P. J., Greenwald, M. K., Bradley, M. M. and Hamm, A. O. (1993) Looking at pictures: Affective, facial, visceral, and behavioral reactions. *Psychophysiology* 30, 261–273.

Lazarus, R. S., Opton, E. and Tomita, M. (1966) A cross-cultural study of stress-reaction patterns in Japan. *J.Pers. Soc.Psychol.* 4, 622–633.

Lazarus, R. S., Speisman, J. C. and Mordkoff, A. M. (1963) The relationship between autonomic indicators of psychological stress: Heart rate and skin conductance. *Psychosom.Med.* 25, 19–30.

Levenson, R. W. and Gottman, J. M. (1983) Marital interaction: Physiological linkage and affective exchange. *J.Pers. Soc.Psychol.* 45, 587–597.

Mandler, G., Mandler, J. M., Kremen, I. and Sholiton, R. (1961) The response to threat: Relations among verbal and physiological indices. *Psychol.Monogr.*75(No. 513).

Marx, B. P. and Sloan, D. M. (2005) Peritraumatic dissociation and experiential avoidance as predictors of post-traumatic stress symptomatology. *Behav.Res.Ther.*43, 569–583.

Mauss, I. B., Cook, C. L. and Gross, J. J. (2007) Automatic emotion regulation during anger provocation. *J.Exp.Soc. Psychol.*43, 698–711.

Mauss, I. B., Levenson, R. W., McCarter, L., Wilhelm, F. H. and Gross, J. J. (2005) The tie that binds? Coherence among emotion experience, behavior, and physiology. *Emotion* 5, 175–190.

Mauss, I. B., Shallcross, A. J., Caston, A. T., Ferrer, E., Wilhelm, F. H., John, O. P., and Gross, J. J. (in preparation) Socioemotional correlates of emotional response system coherence.

Mauss, I. B., Wilhelm, F. H. and Gross, J. J. (2004) Is there less to social anxiety than meets the eye? Emotion experience, expression, and bodily responding. *Cogn. Emot.*18, 631–662.

Mayer, J. D., Salovey, P. and Caruso, D. R. (2004) Emotional intelligence: Theory, findings, and implications. *Psychol. Inq.*15, 197–215.

Myers, D. G. (2000) *The American Paradox: Spiritual Hunger in An Age of Plenty.* Yale University Press: New Haven, NJ.

Pennebaker, J. W. (1982) *The Psychology of Physical Symptoms.* Springer: New York.

Rachman, S. (1978) Human fears: a three systems analysis. *Scand.J.Behav.Ther.* 7, 237–245.

Reisenzein, R. (2000) Exploring the strength of association between the components of emotion syndromes: the case of surprise. *Cogn.Emot.*14, 1–38.

Rosenberg, E. L. and Ekman, P. (1994) Coherence between expressive and experiential systems in emotion. *Cogn. Emot.*8, 201–229.

Roter, D. L. and Ewart, C. K. (1992) Emotional inhibition in essential hypertension: Obstacle to communication during medical visits? *Health Psychol.*11, 163–169.

Ruch, W. (1995) Will the real relationship between facial expression and affective experience please stand up: the case of exhilaration. *Cogn.Emot.* 9, 33–58.

Schooler, J. W. (2001) Discovering memories in the light of meta-awareness. *Journal of Aggression, Maltreatment and Trauma* 4, 105–136.

Schooler, J. W. (2002) Re-representing consciousness: Dissociations between experience and meta-consciousness. *Trends Cogn. Sci.* 6, 339–344.

Schooler, J. W., Ariely, D. and Loewenstein, G. (2003) The pursuit and monitoring of happiness can be self-defeating. In: *Psychology and Economics,* pp. 41–70. Eds. J. Carrillo and I. Brocas. Oxford University Press: Oxford.

Schooler, J. W., Reichle, E. D. and Halpern, D. V. (2004) Zoning out while reading: Evidence for dissociations between experience and meta-consciousness. In: *Thinking and Seeing: Visual Metacognition in Adults and Children,* pp. 203–226. Ed. D. T. Levin. MIT Press: Cambridge, MA.

Schooler, J. W. and Schreiber, C. A. (2004) Experience, meta-consciousness, and the paradox of introspection. *Journal of Consciousness Studies* 11, 17–39.

Stemmler, G. (1992) *Differential Psychophysiology: Persons in Situations.* Springer-Verlag: Berlin.

Sze, J. A., Yuan, J. W., Gyurak, A. and Levenson, R. W. (2007) Poster presented at the annual meeting of the Society for Personality and Social Psychology, Memphis, TN.

Tassinary, L. G. and Cacioppo, J. T. (1992) Unobservable facial actions and emotion. *Psychol.Sci.*3, 28–33.

Van Boven, L. and Loewenstein, G. (2003) Social projection of transient drive states. *Pers.Soc.Psychol.Bull.*29, 1159–1168.

Weinstein, J., Averill, J. R., Opton, E. M., Jr. and Lazarus, R. S. (1968) Defensive style and discrepancy between self-report and physiological indexes of stress. *J.Pers.Soc. Psychol.*10, 406–413.

Wilson, T. D., Dunn, D. S., Bybee, J. A., Hyman, D. B., and Rotondo, J. A. et al. (1984) Effects of analyzing reasons on attitude-behavior consistency. *J. Pers. Soc. Psychol.* 47, 5–16.

Wilson, T. D., Lindsey, S. and Schooler, T. Y. (2000) A model of dual attitudes. *Psychol.Rev.*107, 101–126.

Wilson, T. D., Lisle, D. J., Schooler, J. W., Hodges, S. D. and et al. (1993) Introspecting about reasons can reduce postchoice satisfaction. *Pers. Soc. Psychol. Bull.*19, 331–339.

Wilson, T. D. and Schooler, J. W. (1991) Thinking too much: Introspection can reduce the quality of preferences and decisions. *J. Pers. Soc. Psychol.* 60, 181–192.

Winkielman, P. and Berridge, K. C. (2004) Unconscious emotion. *Curr.Dir.Psychol..Sci.*13, 120–123.

15

The Pleasure of Music

PETER VUUST AND MORTEN L. KRINGELBACH

As shown in this book, many pleasures are shared between species, but there are clearly species-specific routes to this pleasure. When asked about which of the pleasures of life they would miss the most, most people consistently rate music as one of the most important, yet the ability to derive pleasure from music listening and performance would appear to be a trait that is unique to humans. While other animals are clearly capable of hearing the sounds that make up music, they appear unable to take pleasure in music (McDermott and Hauser, 2007). There is clearly a difference between the brain regions participating in decoding the sounds that make up music perception (Griffiths, 2001; Hauser and McDermott, 2003) and subsequent pleasure and emotional processing evoked by music (Blood and Zatorre, 2001; Blood et al., 1999; Green et al., 2008; Griffiths et al., 2004).

Studies in other animals including nonhuman primates have consistently failed in showing any sort of pleasure or displeasure related to music-like activity or perception (Bates and Horvath, 1971; Hauser and McDermott, 2003; Steele, 2006)—although there is recent evidence that music can act in conjunction with other cues as a significant aversive noisy stressor even in rats (Reynolds and Berridge, 2008). While studies in some of our closest cousins have shown that some of the basic abilities underlying music perception, such as octave recognition, may be in place (Hauser and McDermott, 2003), the monkeys are unaffected by dissonance and consonance (McDermott and Hauser, 2004) and they do not appear to take pleasure in music overall (McDermott and Hauser, 2007).

Music is an integral part of life's highly pleasurable activities such as having good dinners with friends, dancing, going to the movies, and hanging out with friends. In fact, the very thought of not having music present at these activities may significantly devalue the activity to the extent that most of the pleasure may no longer be present.

Music can evoke a range of different emotions, which are both similar and different to those found in other activities. These include everyday emotions such as happiness, sadness, surprise, and nostalgia, as well as emotions that are unique to music such as for instance the sensation of swing.

From an evolutionary perspective, it is hard to imagine that music would have survived as a human cognitive ability, if music did not confer an adaptive advantage on our emotional state and our well-being. In this sense, the benefit from a better understanding of human emotional processing are clear to most areas of music research trying to understand what music is and how it works.

It has, however, proven nontrivial to study how music, consisting of organized sequences of sounds, is translated to the wealth of emotions that we can experience and from which we can derive pleasure. Until the recent advent of modern neuroimaging techniques, the lack of any good animal models for the study of music has made it difficult to investigate the neural foundation of musical emotions and pleasure, and even today remarkably little is known about how music evokes emotions and pleasures.

In this chapter, we are reviewing our current knowledge of the neural mechanisms of how sounds and music are translated to the subjective experience of emotion and pleasure in both performers and listeners.

First, we present some recent theoretical frameworks for understanding the relation between music and human emotions. One important model holds that anticipation/prediction is the basic mechanism that drives music perception, which in turn maps nicely onto predictive coding theories of how the brain might work.

Second, we review the insights into the pleasure of music gained from neuroimaging research. We focus on how musical anticipation allows for the sensation of pleasure evoked by music exemplified by two very common sensations related to musical experience: the so-called "chill" or "shivers down the spine" reactions to particular pieces of music and the sensation of swing.

One underlying assumption is that the pleasure induced by music could be related to an increase of dopamine release but this link has never been proven directly. This link between dopamine and the pleasure of music is probably misleading, as suggested by the evidence from other human neuroimaging studies of pleasure (Leyton, Chapter 13, this book). Dopamine might be closer linked to prediction/anticipation mechanisms of pleasure-inducing music rather than the hedonic experience per se. If, however, dopamine release is important to understanding components of why music is pleasurable and motivating, this may also hold the key to understanding the main problem in relation to music in an evolutionary perspective: Does music have survival value and why is it unique to humans?

Translating Music into Pleasure and Emotion

Central to this chapter is the fundamental question of how it is possible for music to induce pleasure and emotions at all. The most common explanations fall into three categories: (1) Music evokes survival-related responses connected to the way sound is processed by the auditory system, such as for example how brainstem responses to loud sounds can trigger fear responses; (2) music links to some extramusical space that carries the particular emotion; and (3) music establishes, fulfills or disappoints anticipatory neural structures and mechanisms which are setup within the music itself.

At first glance, the first two kinds of explanations may seem easiest to fit with contemporary understanding of emotion. However, the third kind of explanation may in fact be more fundamental to the experience of music as such. At the very least, as we shall see later, it offers an explanation to understanding music-specific emotions such as, for example, the "sensation of swing" that is not easily explained by the other two kinds of explanations.

Let us perhaps begin with discussing a recent overview of the different ways in which the human brain might carry out the translational process from music to emotion (Juslin and Västfjäll, 2008). Those authors point out that the study of emotions evoked by music has suffered from a neglect of the underlying psychological mechanisms evoking these emotions and propose that these mechanisms could be summarized into at least six categories: (a) brainstem reflexes, (b) evaluative conditioning, (c) emotional contagion, (d) visual imagery, (e) episodic memory, and (f) musical expectancy. Some of these mechanisms may be dissociable but could also be shared, since it is clear that the first of their proposed mechanism falls in the first category of explanations we mentioned, while the next four fall in the second category and the final, musical expectancy, falls into the third category.

Brainstem Reflexes

Each of these mechanisms may be important in their own right and provide important pieces to the puzzle of why music induces emotion and pleasure. According to Juslin and Västfjäll, *brainstem reflexes* refer to "a process whereby an emotion is induced by music because one or more fundamental acoustical characteristics of the music are taken by the brainstem to signal a potentially important and urgent event." These reactions are hard-wired and automatic and may explain why sound in itself can have an emotional effect; the rough sound from the brakes of an approaching car is a sound that in itself evokes arousal and maybe fear (although perhaps in the case of car sounds depending on learning). The level of arousal is regulated by the brainstem and the arousal-regulating properties of music has been suggested by Berlyne (1971) to be a main factor in our preference for certain music at certain times (DeNora, 2000; Thayer, 1996). Note that while music processing may be reliant on these brainstem reflexes, they are not sufficient for experiencing music as music.

Evaluative Conditioning

Evaluative conditioning is a form of classical conditioning in which a formerly neutral stimulus (in this case the music) is turned into a conditioned stimulus by repeated pairing with an emotionally charged stimulus, the unconditioned stimulus. This leads to an emotional response to the conditioned stimulus, even in the absence of the unconditioned stimulus, a process that is often nonconscious (Field and Moore, 2005; Hammerl and Fulcher, 2005).

For example, we will often pair the emotion evoked by a movie with its musical theme, which may in part explain famous music artists' willingness to write title songs for movies. Animal research has shown that the key brain structures involved in evaluative conditioning are likely to be subcortical brain regions such as the amygdala and the cerebellum (Balleine and Killcross, 2006; Dickinson and Balleine, Chapter 4, this book; Johnsrude et al., 2000; Sacchetti et al., 2005).

Emotional Contagion

Another proposed mechanism, emotional contagion, refers (in relation to music) to the process in which the listener internally mimics the emotional expression of the music, thereby reproducing the emotion felt by the performer. Emotional contagion has been observed in relation to facial expressions, where expressed emotions such as, for example, anger and fear activate facial muscles in the observer (Lundqvist, 1995).

Accordingly, music listening has been shown to evoke a premotor response (Koelsch et al., 2006), which has been hypothesized to reflect an auditory mirror-neuron system at work (Iacoboni et al., 2005; Rizzolatti and Craighero, 2004), such that the emotions of the listener correlates with the expressed emotions in the music, be it in vocal or instrumental music. Admittedly, there is not much evidence to support the existence of this mechanism and the knowledge about the associated brain systems is very sparse, apart from the possible involvement of premotor and mirror systems. However, it can help to explain various phenomena related to music psychology such as, for example, the strong identification between musical fans and their musical heroes.

Visual Imagery

Visual imagery is a very commonly reported reaction to music (Tan and Kelly, 2004). Even though the images that may occur in the mind of the listener when exposed to a specific musical piece are very individual and also probably not reproducible between exposures, visual imagery is a known effect that is also used by composers deliberately from time to time such as the bombs going off in Tchaikovsky's *1812 ouverture* or the strange organ of The Beatles song *To the benefit of Mr. Kite*, designed to conjure up images of a traveling circus show.

Even though listeners, when prompted, often associate more or less similar visual images to certain musical pieces (Osborne, 1989), composers often use "real sounds" such as bird song (e.g., in the song *Blackbird* by The Beatles) to evoke specific images in reaction to certain musical pieces. Visual imagery is thought to induce emotions resembling the emotions attached to the images itself (Kolers, 1983), and has been used extensively in relation music therapy (Bonny, 1997; Toomey, 1996) and for medical use (e.g., http://www.musicure.com).

Episodic Memory

Another commonly proposed mechanism, episodic memory is related to the episodic emotional memory of hearing the music for the first time, and has also been called the "Honey, they are playing our tune" phenomenon (Davies, 2001). It is experienced by most people more or less frequently. The specific emotions that can be induced by episodic memory are the emotion associated with the memory itself and are therefore often connected to a sort of nostalgia (O'Neill and Sloboda, 2001).

Episodic memories from childhood and adolescence are more vivid than from later stages in life; music interest peaks in adolescence and people generally prefer music from this part of their lives (Schulkind et al., 1999). Given the strong ties between music and emotion, it is therefore likely that this process involves the memory-encoding brain structures such as the amygdala and the hippocampus, and may be part of the reason why singing in certain instances can help elderly people suffering from dementia to recall words that are not accessible to them in other ways.

Musical Expectancy

The above mechanisms all account for some aspects of how music may evoke emotion and pleasure in the brain. But the perhaps most interesting mechanism relates to musical expectancy, which is the process whereby an emotion is induced in a listener because a specific feature of the music violates, delays, or confirms the listener's expectations about the continuation of the music.

As a special case of this, creation of expectations in music is linked to the motion over time between opposites such as for example in harmony, the shift in the authentic cadence from dominant to the expected tonic (Sadie, 2001), and in melody, the motion from approach notes to target notes (Friedman et al., 2001). In the very restricted sense that Juslin and Västfjäll (2008) use this terminology, musical expectation is almost entirely related to the notion of musical syntax (Lerdahl, 1971).

This is exemplified in harmony by the expectation for chords appearing in different places in the tonal cadence. The authentic cadence is composed of tonic (T), dominant (D), and subdominant (S) chords in the following order: T S D T (Bharucha and Krumhansl, 1983). The dominant is perceived as tension-creating, demanding resolution to the stable position of the tonic, whereas the subdominant reflects an intermediate position between these two oppositions. Chords incorporating notes outside the prevailing harmonic context usually demand resolution to more stable harmonies of the system (Bharucha and Krumhansl, 1983). Chords breaking these harmonic expectations may be perceived as colorful, interesting, or simply erroneous, and give rise to different emotions such as, for example, surprise or sometimes even pleasure (Berent and Perfetti, 1993; Meyer, 1956).

The neural foundation of violations of harmonic expectancy is one of the more well-understood areas of brain processing of music. Stefan Koelsch and colleagues have in a number of studies investigated the neural processing of inappropriate chords inserted in the authentic cadence (Koelsch and Friederici, 2003; Koelsch et al., 2000). Using electroencephalopathy (EEG)/magnetoencephalography (MEG), they have found that a brain response termed the *early right anterior negativity* is elicited when a harmonically incongruous chord is inserted within or at the end of a musical sequence and localized the source of this response to the inferior frontal cortex; more specifically to a component of Broca's region, or Brodmann area 44, and its right hemispheric homologue; an area often associated with processing of syntax in language. However, the impact on the emotional brain of these unexpected chords still needs to be investigated.

If we consider the vast amount of neuroscientific research in music that has been published in recent years, it is certainly true that the study of music emotions seems to be pointing in different directions. Consider, for instance, the very different activation patterns reported in studies of major and minor mode music supposedly evoking very simple emotions (happy/sad) (Green et al., 2008; Khalfa et al., 2005; Pallesen et al., 2005). Even though these results may be due to many different factors contributing to the emotional state of the subjects under different experimental conditions, we agree with the Justlin and Västfjäll that one of the reasons for these contradictory results may be found in the lack of a coherent theoretical framework.

Linking Music and Psychology

The Juslin and Västfjäll framework gives a nice overview of the different possible mechanisms to translate between music and brain suggested by the literature. Still, it remains unclear, how these different mechanisms relate to each other. The problem with the framework is that the categories are not ordered hierarchically, are not mutually exclusive, and only the category *musical expectancy* directly links musical and psychological mechanisms as such. This limits the scope of the proposed framework somewhat, especially if its purpose is to act as a guideline for researchers trying to understand the modularity of brain structures involved in the processing of emotions in music.

As we shall see shortly, the balance of the data on music in the brain suggests that musical expectancy is the fundamental mechanism, which underlies other translating mechanisms, and it could reorganize the six categories of the Juslin and Västfjäll framework in a hierarchical fashion.

Expectancy in Music

It is hard to imagine that musical emotions are evoked without some sort of musical meaning assigned to what is heard, unless we think of emotions such as, for example, fear evoked by the mere advent of a sudden loud, scary sound—in which case it is questionable whether one would define this as music. Most music theoreticians consider musical anticipation as one of the principal means by which music conveys musical meaning and emotion. According to this point of view, understanding music (Cooper and Meyer, 1960; Lerdahl, 1971; Lerdahl and Jackendoff, 1999; Meyer, 1956; Monelle, 1992) is related to the anticipatory interplay between local auditory events and a deeper structural layer partly inherent in the music as such and partly provided by mental structures in the listeners induced by the music (Palmer and Krumhansl, 1990; Vuust et al., 2006a).

In short, the musical experience is dependent on the structures of the actual music as well as on the expectations of the brain that interprets it. These expectations are dependent upon long-term learning of musical structures (which could be called *culture-dependent statistical learning*), familiarity with a particular piece of music, short-term memory for the immediate musical history while listening to a musical piece as well as on deliberate listening strategies (Huron, 2006; Vuust et al., 2006b). Brain structures underlying musical expectation are thus shaped by culture as well as personal listening history and musical training (Vuust et al., 2005). Moreover, as soon as one hears the first sound of a musical piece, anticipatory structures such as meter, tonality, and memory for particular musical pieces seem to be in place already and unavoidable (see, e.g., Brochard et al., 2003). Thus, it is difficult to imagine any of the above proposed mechanisms acting without the involvement of musical expectation.

According to Juslin and Västfjäll's definition, musical expectation develops slowly over time during listening experience and is not fully developed until the age of 5–11 years. This is correct if musical expectation is restricted to anticipation of complex musical structures such as the hierarchy of harmony dependent on long-term learning (Leino et al., 2007). However, expectation of more simple repetitive sound patterns, which is a part of all music, such as pitch deviants in successive pitch trains has been detected even before birth, as indicated by the mismatch negativity (MMN) measured by EEG/MEG (Huotilainen et al., 2005). Moreover, in an elegant study, Winkler et al. (1996) showed that the auditory predictive model is updated for each new acoustic event in the sound environment, indicating that the anticipatory structures of music are in constant flux during the listening experience.

Furthermore, even though the authors claim that the degree of volitional influence on musical anticipation is low, we recently conducted a study in which musicians were asked to maintain either the main meter or a counter meter while listening to Sting's *The Lazarus Heart* (Vuust et al., 2006b). In this experiment (which we will return to), the subjects volitionally imposed two very different anticipatory frameworks onto the music by tapping different but related rhythms. Another example of volitional control of the anticipatory framework in music would be to deliberately listen to a melody from the perspective of two different tonalities.

Thus, Juslin and Västfjäll use a very narrow definition of musical expectancy encompassing only predictive structures that develops over time. We would like to broaden up the definition of musical expectation to include any kind of auditory/musical patterning with potential to create predictive musical structures that can be fulfilled or broken. Hence, music should be seen as constantly evolving wickerwork of expectancies created in different layers of the musical structure (Bharucha and Stoeckig, 1986; Meyer, 1956; Monelle, 1992; Sloboda, 1985).

These expectation structures of tension and relief depend critically on the timing structure of music and can develop in music on a timescale that is much smaller than what is required by harmony. The predictive structures that underlie the anticipation of timing in music are provided by the meter, based on a fundamental opposition between strong and weak beats. A 3/4 meter represents a strong beat followed by two weak beats, whereas the accents in a 4/4 meter are "strong–weak–intermediate–weak." The alternating structure of a meter is replicated on the global level of the musical form (Cooper and Meyer, 1960; Vuust, 2000), but in principle also at smaller levels of subdivisions of the pulse. Meter therefore provides the listener with a temporal, hierarchical expectancy structure, underlying the perception of music, in which each musical time-point encompasses a conjoint prediction of timing and strength (Large and Kolen, 1994). This is in accordance with behavioural studies that indicate a human predisposition for temporal regularity (Drake and Bertrand, 2001; Drake et al., 2000). When the expectancy structure of meter is violated, however, this may be followed by a strong perceptual response depending on the degree of violation (Jones and Boltz, 1989; Vuust, 2000) to laughter (Huron, 2004). Importantly, the hierarchical structure of the meter underlies all other expectancy structures in music, for example, rhythm, harmony, melody, intensity, in that it influences perception of any musical event. Hence, anticipatory structures such as the meter (but also, e.g., tonality) provide the listener with a framework for interpreting and remembering music.

But how can musical expectations be translated into emotions other than those related to surprise?

Expectancy and Emotion

The relationship between musical expectancy and emotion was originally explored by Meyer (1956) and has recently been elaborated in a very convincing way by Huron in his book *Sweet Anticipation* (Huron,

2006). Huron proposes a general model of the anticipatory process leading to an event. His so-called ITPRA model (mnemonic for five expectation-related responses: Imagination response; Tension response; Prediction response; Reaction response; Appraisal response) describes how different possible phases of the anticipatory process can trigger survival-related responses that give rise to different sets of emotions, and in particular how music can exploit all of these responses.

The *imagination* response (I) describes the phase occurring some time before the event takes place, where people imagine the emotion related to the future event. The ability to predict emotions linked to events in the future is important to humans when deciding which of many different paths to take. In music, the imaginary response is often related to anticipation to changes in the musical form, but can also be imagination of salient parts of pieces, such as, for example, upcoming "chills."

The *tension* response (T) describes the increase in tension just before the event, resulting in an increase in arousal in order to prepare for the event. In music theory, tension and relief is the basis of the theory of harmony as well as of the meter (see above), hence the tension response is directly in accordance with classical music theory.

After the event has happened, there is an obligatory response related to the *precision* of the prediction (P). If the event is correctly predicted, this response is a positive reinforcement of the correct prediction. This helps the learning of correct predictions of the future. Notably the prediction response is positive even if the outcome of the event should be considered as bad.

The event, however, is also followed by an equally fast response termed the *reaction* response (R) as well as a response that takes longer to unfold, the *appraisal* response (A), both related to the assessment of the pleasantness or the unpleasantness of the outcome.

The *reaction* response is fast and always assumes worst-case scenario. It is negative to unpleasant events and the result is increased arousal. The survival value of the reaction response, of course, is to make the nervous system alert to events that constitute possible dangers that may have to be avoided. Implications of the reaction response are often surprise, increased awareness, or even fear. The reaction response often proceeds nonconsciously and is not always available for conscious introspection.

The *appraisal* response is an assessment of the long-term implication of the event. The appraisal response is the evaluation of the outcome seen in the larger context and need not be congruent with the reaction response. Events that are considered dangerous by the reaction response, such as the sight of a tiger, may on evaluation become delightful when one realizes that it is in a cage. The appraisal response is almost always a conscious appraisal process, which may come to reinterpret the responses.

Huron suggests that the web of predictions arising specifically in music is based on different memory systems: semantic memory, episodic memory, and short-term memory. Semantic memory, or schemes, are statistically learned through cultural exposure. Even though listening history is highly individual, prediction schemes have been shown to be rather consistent within a given culture. Schemes in music provide very strong prediction for music even when we do not know the particular piece of music beforehand. Schematic predictions in different aspects or layers of the music can work together or against each other. Harmonic expectation can point to a specific endpoint whereas rhythmic or melodic expectations can point to another.

Episodic memory, or in Huron's terminology, *veridical expectation*, is based on the memory for a particular piece of music. Knowing a certain piece of music well means having formed very precise expectations for harmony, melody, and rhythmic structures. These predictions may, and often will, go against schematic predictions, which gives rise to more complex patterns of emotions. This possible contradiction between veridical and schematic expectations may explain why it is possible to listen to certain pieces over and over again. Even though we may be very familiar with a certain piece of music, we still experience ITPRA responses based on our knowledge about musical structure based on long-term learned musical schemes, which are very strongly encoded in the brain.

Short-term memory is a third memory component responsible for formation of predictions in music. Within a certain piece, phrases will form expectations for the phrases to follow. Music is highly repetitive. Empirical research has shown that 93% of all phrases occurring in music are repeated more that once during musical performances (Huron, 2006). Therefore, when a certain motive is repeated, we expect this pattern to continue. This takes place both at the local level of small repetitive patterns, as well as at the level of the musical form.

The predictions arising from the various ITPRA responses can be weaved into each other in a prediction network, which may give rise to emotions. At the level of schematic predictions, this may create a complex set of predictive patterns. Positive emotions are usually associated with resolving tension, as from the dominant chord to the tonic.

On the other hand, negative emotions can arise when predictions fail, such as in the so-called deceptive cadence where the dominant chord is resolved to the subtonic, the tension builds, resulting in an even stronger prediction response at the later arrival of the tonic. This effect can be enhanced rhythmically by slowing down the beats while approaching the resolution of the cadence and delaying the onset of the final chord (ritardando) so that harmonic and rhythmic anticipation works together.

In addition to such schematic predictions, it is possible to add predictions based on veridical and short-term memory, which altogether creates a potentially very complex system of predictions.

These complex expectations can, according to Huron, facilitate the generation of a great variety of emotions. The prediction response is one of the most important pathways from expectations to emotions. Correct predictions are rewarded by the brain for survival-related reasons, evoking positive emotions that are attributed to the music itself. These positive emotions are mediated through the reward system, which plays a central role in relation to the experience of pleasure (Berridge and Kringelbach, 2008; Kringelbach, Chapter 12, this book; Smith et al., Chapter 1, this book). We will return to the available evidence showing the critical involvement of the reward systems in relation to musical pleasure, but first we will present a brief overview of how the brain might handle predictions.

Predictive Coding of Music

If music expectation/anticipation is viewed as the fundamental mechanism for musical experience, this can be made to map nicely onto recent theories of predictive coding in the brain. One such promising model of brain function was proposed by Friston (2005) where predictive coding is the central principle of brain function. It provides an account of how the brain identifies and categorizes the causes of its sensory inputs (Friston, 2002; Shepard, 2001; Tononi and Edelman, 1998). The model posits a hierarchical organization whereby lower level brain regions estimate predictions of their expected input based on contextual information through feedback connections from higher level regions. A comparison between prediction and actual input produces an error term that, if sufficiently large, will try to force an update of the model. This generates a recursive process, which aims at minimizing the difference between input and prediction. As the representational capacity of any neuronal assembly in this model is dynamic and context-sensitive, this, among other issues, addresses the problem of top-down control (Frith and Dolan, 1997; Roepstorff and Frith, 2004).

Recently, we have argued that, given the anticipatory nature of music, violations of musical anticipation in different aspects of the music may be good substrates for testing the predictive coding hypothesis (Vuust et al., 2008). One such example is our recent MEG experiment with simple rhythm sequences of increasing rhythmic incongruence and measured brain responses (event-related potentials), which were used to test the hypothesis that preattentive neural responses to increasing rhythmical incongruity could be identified and would be congruent with an error term and subsequent evaluation.

Moreover, if predictive coding schemes of general brain organization are correct, then this would mean that skilled musicians should have acquired a more detailed expectancy structure (through learning) than nonmusicians. This might influence both neuronal markers of the prediction error and the neuronal markers of evaluation.

We therefore compared rhythmically unskilled nonmusicians with expert jazz musicians. Jazz musicians use challenging rhythmic material in their musical performance and are therefore ideal candidates for identifying putative competence-dependent differences in the processing of metric violations.

Rhythmic incongruities elicited the magnetic counterpart of the mismatch negativity (MMNm), an event-related field, peaking around at 110–130 ms from change onset, an index of preattentive detection of change in some repetitive aspect of auditory stimulation (Naatanen, 1992), accompanied by a later component, the P3am, peaking around 80 ms after the MMNm in expert jazz musicians and some of the rhythmically unskilled subjects, as well as responses to more subtle rhythmic incongruities in most of the expert musicians.

The MMN and the P3a are thought to reflect two survival-related stages of an attention catching process. The MMN is a brain response, occurring locally in the auditory cortices, to change in the auditory environment, whereas the P3a is associated with the evaluation of that change for subsequent behavioral action and believed to indicate activity in a network which contains frontal, temporal, and parietal sources (Friedman et al., 2001).

In our study, the MMNms were localized to the auditory cortices, whereas the P3am showed greater variance in localization between individual subjects. MMNms of expert musicians were stronger in the left hemisphere than in the right hemisphere in contrast to P3ams showing a slight nonsignificant right lateralization. Thus the observed MMNm and P3am could be interpreted as an error term generated in the locally followed by its subsequent evaluation in a broader network including generators in the auditory cortex as well as higher level neuronal sources.

This is in keeping with expectations based on predictive coding schemes and suggests that there is congruence between perceptual experience of musical expectancy and the way that these are processed by the brain. Furthermore, we found enhanced and earlier processing of rhythmic deviants in expert musicians compared to rhythmically unskilled nonmusicians both at the level of the MMNm and the P3am, as well as a left lateralization of the MMNm in experts compared to nonmusicians to both subtle and strong metric violation, consistent with earlier suggestions of music being left lateralized in musicians (Altenmüller, 2001; Bever and Chiarello, 1974; Ohnishi et al., 2001).

This indicates that the difference in lateralization and strength of the MMNm in neural response between experts and rhythmically unskilled nonmusicians reflects a stronger metrical predictive structure in the experts, which again affects the evaluation of the error as reflected by the stronger P3am. This again fits with the predictive coding model according to which the size of the error term is dependent on the nature of the prediction.

Thus, anticipatory structures in music seem to be translated directly by the human brain, which is geared especially to this kind of processing. In the above experiment, we only observed brain response to these anticipatory musical structures in cortical brain areas. This was, however, to be expected due to the limitations of the applied method (MEG without magnetometers and using dipole modeling), which is not particularly sensitive toward detecting activity in subcortical brain structures.

In the following, we review some of the very sparse existing evidence of the involvement of the reward system in relation to music processing and try to explain the role of prediction in these experiments.

The Pleasure of Music, and in Particular Chills

One of the difficulties in studying emotional responses to music is that these are clearly individual and not stable in a listener during listening or even across several instances of listening to a musical piece. While neuroimaging lends itself to what is perhaps best termed neophrenology, it is not particularly meaningful to measure the average brain activity over seconds and even minutes in participants listen to music evoking pleasant feelings, although such a pilot study has been carried out in nonmusicians using positron emission tomography (PET; Brown et al., 2004). Such studies ignore the important temporal dynamics of music, which one functional magnetic resonance imaging (fMRI) study tried to redress by contrasting the effects on brain activity of pleasant and unpleasant music (Koelsch et al., 2006) but other neuroimaging modalities with much better time resolution such as MEG would seem better suited to untangle the hedonic valence associated with music.

One way to address some of these problems is to concentrate on the more stable emotional reactions to music, the so-called "chills" or "shivers down the spine," which are particularly salient (Blood and Zatorre, 2001; Goldstein, 1980; Panksepp, 1995; Sloboda, 1991). Chills denote the sensation of shivers running up and down the spine, goose pimples, and hair standing up on your arms that can accompany especially delightful musical listening experiences.

Psychologically, chills are related to the survival mechanisms of what has been called the four Fs of life ("fighting," "fleeing," "feeding," and "reproduction") and in particular to surprise. In most species, these fundamental responses involve both subcortical and cortical mechanisms and are fast and mostly automatic. As an example, the characteristics of the fight response are immediate, increased arousal, aggressive display, which is seen in the hair standing on ends (usually more efficient for a cat than for a human) and in case the slower appraisal response determines a real danger can help the urge to fight as well. In the case of music, the subsequent appraisal process always determines that the surprising event does not imply any real danger and this leaves the delightful feeling of shivers down the spine.

Surprise always indicates a biological failure to predict future events, and thus the chill effect is directly linked to musical expectation. According to the Huron (2006), the delight from the "chills" stems from a contrast between a "fast track" response (the reaction/prediction response) mediated by subcortical structures in the brain, which is substituted by a "slower track" response (the appraisal response) mediated through cortical structures (LeDoux, 1989). The fast response to the surprise is quick and has a negative valence. The slower appraisal responses follow quickly thereafter and are, in the case of music, often has a neutral or positively valence, resulting in an overall positive feeling of pleasure. The chill response is therefore a much desired quality of music.

The chill response can be measured using psychophysical and behavioral measures. Grewe et al. (2005, 2007) used a combination of skin conductance measures, button presses, and subjective reports of goose pimples to determine chill responses. This group, headed by Eckart Altenmüller, developed the EMuJoy software to allow for a participant's continuous self-report of feelings in a two-dimensional emotional space while listening to music in order to combine methods of continuous measurement of physiology and motor responses (Nagel et al., 2007). They measured 38 participants and found that "chills" as a paradigm for strong emotional responses to music is dependent on familiarity with a musical style and on personality factors, such as "reward dependence" or "sensation seeking."

Chills were also found to be related to changes in loudness; however, no distinct acoustical pattern could be identified that induces chills in a reflex-like way and suggested that chills are bodily reactions based on subjective feelings. Interestingly, they found that even though a distinct chill-triggering acoustical pattern could not be found, important musical factors seem to be harmonic sequences, the entrance of a voice, and the beginning of a new part, which is a violation of expectancies.

Since musical chills are fast survival-related responses, pointing to the involvement of phylogenetically ancient brain structures, and reported to be linked to the pleasure of music, it has been obvious to test the hypothesis that musical chills is associated with the reward system in the brain. However, the subjective nature of the chill response has made it a difficult problem to investigate.

In an elegantly designed study, Anne Blood and Robert Zatorre (2001) investigated chill responses in 10 music students from the Department of Classical Music at McGill university while scanning them with PET, measuring heart rate, skin conductance, and respiration. On the basis of their self-reports, each of the participants chose a classical piece of music that elicited strong emotions and chill experiences. Each of these self-chosen pieces was then used as control situation for another participant. Participants' reported chills that correlated with changes in the psychophysical measures during listening to their own pieces compared to the control pieces. Regression analysis assessing the relationship between increasing intensity ratings related to chills and PET measurements of regional changes in blood flow–identified changes in brain structures that are thought to be involved in reward, motivation, emotion, arousal, and pleasure. These included the structures discussed elsewhere in this book such as the ventral striatum (nucleus accumbens), midbrain, amygdala, and the orbitofrontal cortex.

This result indicates that listening to music can in certain instances induce intense pleasure in the reward systems of the brain and suggests that music has an ability to tap directly into these survival-related brain mechanisms. The authors proposed that even though music is not obviously necessary for human survival, it may have psychological benefits.

The Blood and Zatorre experiment was a groundbreaking step forward in understanding the neural foundation of musical pleasure in linking "chills" to brain structures involved in reward. However, very few experiments have added new knowledge to this very important aspect of musical behavior.

And even though the authors make a convincing argument of their findings of brain structures related to reward, the poor time resolution of PET is still only indirect evidence that the brain does, in fact, reward music listening.

Pleasurable Music on the Brain

On balance, these pilot neuroimaging studies of the emotions evoked by music show activity in the reward regions of the brain including the orbitofrontal cortex, anterior cingulate cortex, the nucleus accumbens, the insula, and the amygdala. As shown in various chapters of this book, these brain regions appear to code for the pleasure of many different stimuli.

Similar to many other sensory systems, there are functionally and anatomically separable neural systems mediating music perception and emotion. The perception of music involves superior temporal regions including the auditory cortices as well as the inferior frontal regions, while the emotional processing engages the reward systems. But none of the neuroimaging studies can provide information about the causality of any of these brain regions in the emotional processing of the pleasure of music.

Some light on this question has, however, been shed by a recent case report of a 52-year-old man suffering a stroke, which only affected his music appreciation and *not* his sound perception (Griffiths et al., 2004).

The stroke affected mostly his left insula although the potential additional damage to fiber pathways by his stroke was not assessed. At the very least, this finding would seem to indicate that regions of the left insular cortex are involved in normal musical emotional processing of music.

In addition to these questions regarding necessary and sufficient brain systems for experiencing the pleasure of music, there are many different neurotransmitters linked in nontrivial ways to reward, as shown elsewhere in this book. Dopamine is linked mostly to 'wanting' (or expectancy) rather than 'liking' (Abler et al., 2006; Fiorillo et al., 2003; Leyton, Chapter 13, this book; Shizgal and Arvanitogiannis, 2003; Zald et al., 2004). Dopamine is increasingly thought to be a key player in relation to reinforcement, learning, and in reward-seeking behavior, but not to pleasure per se (Leyton, Chapter 13, this book).

A future step toward a better understanding of the brain mechanisms and neurotransmitters involved in emotional processing of music would be to measure dopamine release in the brain directly, as already demonstrated for other types of stimuli (Leyton, Chapter 13, this book).

Another obvious question that remains to be investigated in the context of the Blood and Zatorre experiment is whether the reward system is involved in *musical performance* rather than *listening*. Until now, research has mainly been concentrating on emotions involved in musical listening but is performing music different from listening to music?

In other words, why do musicians play? One possible explanation may be the euphoria that many musicians report to experience occasionally when they play and which is an important motivational factor possibly linked both to the music and to social factors (Berliner, 1994; Monson, 1997). In a preliminary questionnaire investigation, 111 out of 129 Danish conservatory students enrolled in programs designed to make them professional musicians reported to be "feeling high" when playing music. It seems to be a plausible hypothesis that the reward system and dopamine is involved also when musicians play. This interesting question, however, remains to be tested experimentally.

The Pleasure of Swing

The "chill" effect is only one of many responses that are associated with the pleasure of music. Another distinct emotional response to music is the sensation of "swing," which is often intensely pleasurable.

In the following, we shall only speak about swing in the context of jazz, rock, and related styles of music because swing is the ontological center in these styles of music (Pressing et al., 1996; Vuust, 2000), even though classical music can certainly also swing. The sensation of "swing" or "groove" is something that most people intuitively know what is, but which is surprisingly difficult to define, despite many attempts (Berliner, 1994; Pressing et al., 1996).

Usually the sensation of swing is associated with music with a regular pulse (Grewe et al., 2007; Temperley and Sleator, 1999) and in western culture in most instances also a regular meter such as 3/4 or 4/4. Swing is what makes you move your feet or even feel like dancing. In many contemporary styles of music, the feeling of swing is so important that it subordinates other aspects of the music such as the form, the melodic and harmonic content.

For example, this is found in some of the dance music played in discothèques, where it matters less when the music begins or ends than how it *feels*. If circumstances, music and the mood are right (which, of course, is highly individualistic as in the case of "chills"), one can end up in an almost euphoric state.

Actually, when the music really swings, you never want it to stop. The claim that we are making here is that the sensation of swing depends on the relationship between the meter and the actual rhythm, or rather the tension created between the predictive timing model and the actual rhythm.

Rhythms provide the sensation of a meter. Even the sound of a metronome can make the brain of the listeners go tick-tock as measured by electroencephalography (Brochard et al., 2003). This effect does not, however, make the metronome *swing*. Swing requires that the rhythm is organic and interesting to the brain.

In the early 1980s, it became fashionable to use drum machines in studios instead of real drummers. However, it soon became obvious that this did not make the music swing. As a consequence, drum machines were manufactured with an extra knob, a so-called humanizer that makes the exact timing of the drum beats randomly more or less unpredictable. Today, drum machines are rarely used, and almost never without a human touch or in combination with a real, live drummer.

One of the things known to musicians—albeit not scientifically investigated—is that in order for a rhythm to swing, drummers tend to play certain beats slightly out of sync with the underlying metric subdivisions. When slowing down recordings of the jazz drummer

Tony Williams, you will notice that his eighth note feel is somewhere in between triplets and 16th notes. Also rock drummers tend to delay their snare drum beat some milliseconds compared to the prediction of the meter emerging from the high hat. This creates a constant feel of tension that is intensified or decreased during musical performances in symbiosis with the changing part of the melody or solos and this organic developing tension between the predictive model of the meter, provided by the brain, and the actual music is what creates the sensation of swing. At the same time, it reinforces the feeling of the meter, which is what makes us tap our foot. In rock music, the changes in tension between the music and the meter are often relatively small.

In jazz, the tension changes between meter and rhythms can be very dramatic. One main way in which modern jazz musicians create tension between rhythm and meter is through the use of polyrhythmic (/metric) structures to challenge the listener's sense of the meter. This creates tension between a foreground counter meter and the background main meter felt by the listener (Pressing, 2002). Experiencing this counter meter provided by the music forces the listener either to shift the metric feeling to the counter meter, or to reinforce the sense of the original meter, occasionally by tapping with the foot.

Thus polyrhythms can be seen as an extreme form of swing. In order to investigate the neural foundation of swing, we conducted an experiment in which we asked 20 professional jazz/rock musicians to tap the main meter, while listening to an excerpt from Sting's *The Lazarus Heart* that contains a challenging epoch of polyrhythm (Vuust et al., 2006b). This is an ecologically valid example of a bistable percept created by polyrhythm, and it is particularly well-suited to an experimental block design in that it is played on top of a drum machine and thus contains epochs of equal length main meter (M) and countermeter (C). We recorded the whole brain blood oxygen level–dependent (BOLD) responses using fMRI and contrasted the effects of keeping the main pulse in the context of the counter meter (tapM/C) and in the context of the main meter (tapM/M).

Our results showed that a part of the inferior frontal gyrus in the frontal lobe (Brodmann area 47, BA47) was active bilaterally as well as right BA40 in the parietal lobe, when musicians kept the rhythm during the intense polyrhythmic tension. When the musicians on the other hand produced a polyrhythm on top of a regular meter provided by music without polyrhythm, BA47 was active in the left hemisphere and there was also activity in the anterior cingulate cortex. Together, these two tapping experiments show that the activation of BA47 is linked to polyrhythmic tension as such independently of whether it is created by a tapping task or provoked by the musical stimulus in itself.

As mentioned, BA47 is part of the inferior frontal gyrus typically defined to comprise of BA44/45, BA46, and BA47, which in the left hemisphere traditionally are thought of as language areas. In relation to language, it has recently been proposed that cognition of syntax, phonology, and semantics are interdependent and that the inferior frontal gyrus should be seen as a unification center (Hagoort, 2005), with a gradient toward areas BA44/45 for syntactic processing and a gradient toward BA47 for semantic processing. According to this theory, the inferior frontal gyrus is not language specific but acts as a single unification space, integrating the semantic consequences of a broader range of cognitive domains than previously thought, for example, gestures (Ozyurek et al., 2007; Willems and Hagoort, 2007). The results from our music study extend these domains to the music-semantically relevant cognitive music processing of polyrhythmic tension.

Unexpectedly, apart from the finding of activation of the anterior cingulate cortex, we did not find activity in the subcortical system during this task which subjects reported to be associated with strong, yet somewhat individual emotions such as, for example, fear, off balance, and tension. This could be due to a number of factors including the individual variability in the evoked emotions, the very conservative statistical thresholding of our data analysis, and the imaging parameters, which may have caused dropout in key regions such as the orbitofrontal cortex. However, the activity in the BA47 did extend into the anterior insula, which is thought to be a mediator between cortical structures and the subcortical systems (Berridge and Kringelbach, 2008). The question of further cortical and subcortical involvement awaits further investigation.

Keeping the rhythm, while listening to music, is fundamental to the experience of musical tension between different layers of the musical structure and crucial for music to appear as a coherent, meaningful expression. This finding is an experimental specification of the claim made, but not proven, by Levitin and Menon (2003). They also found activity in BA47 when contrasting musical excerpts with nonmusical excerpts. One obvious feature of scrambled music is the absence of meter. Hence, their contrast between non-scrambled and scrambled musical excerpts included, but could not be reduced to, a contrast between some unspecified level of metric tension and the complete absence of tension, resembling the contrast between the two epochs of the Sting stimulus.

In their original paper, Menon and Levitin (2005) did not demonstrate activity in subcortical structures. However, in a reanalysis of the data, using functional and effective connectivity analyses, they showed that listening to music with a regular meter modulated activity in a network of subcortical mesolimbic structures involved in reward processing including the nucleus accumbens and the ventral tegmental area, as well as the hypothalamus and the cortical insula, which are thought to be involved in regulating autonomic and physiological responses to rewarding and emotional stimuli (Kringelbach, Chapter 12, this book; Leknes and Tracey, Chapter 19, this book). The correlation between the nucleus accumbens and the ventral tegmental area might indicate a possible subcortical association between dopamine release and responses in the nucleus accumbens in relation to the metrically well-defined stimulus used in the study and thus to expectations and violations of those expectations.

Thus, it is possible that the sensation of meter, tapping the foot hence the feeling of swing, in addition to the neocortical activation of BA47, activates the subcortical reward system and results in the release of dopamine. In contrast to the "chill" response, keeping the rhythm also involves the motor system including cerebellum, premotor areas, and basal ganglia. Dopamine is crucially involved in motor action at the level of the basal ganglia or striatum and midbrain substantia nigra and ensures the ability to execute smooth, controlled movements. The interaction between the dopamine released by music rewards and motion is still far from understood (Leyton, Chapter 13, this book). However, it is clear that the pleasure of swing compared to, for example, "chills" is special and often found with music.

Conclusions

In summary, despite the current paucity of available experimental evidence, we have tried to review the actual and putative musical and neural mechanisms that allow music to be translated into emotion and pleasure. We have proposed that anticipation/prediction could act as some of the fundamental mechanisms underlying musical structuring and that this taps into the way that the brain works on different levels with a capacity to evoke pleasure in humans.

If we consider music from the viewpoint of music theory, it works by way of predictive structures in all possible layers of its structure. These range from simple acoustical patterns to melodic, harmonic, rhythmic hierarchical anticipatory patterns of greater complexity being established, confirmed, delayed, or violated. These anticipatory structures are stored in different kinds of memory systems: schemes (predictions of how music normally develops) are related to semantic memory, veridical anticipation (predictions of music that we have heard before) is stored in long-term memory and memory for musical events that has occurred earlier while listening to a particular piece of music is stored in short-term memory.

The human brain decoding all this information is rather good at processing such predictive information, and it could be that predictive coding is one of the fundamental ways in which the brain integrates information between different brain regions. In relation to music, the brain appears to be constantly scanning the auditory input for predictive patterns and responds strongly to deviations.

Musical anticipation stimulates the brain in two basic ways underlying our perception of the emotional content. First, anticipatory structures such as tonality and meter is the basis of memorization and learning of musical material in that they provide the background for musical surface structure such as melody, chord changes, and rhythms. For instance, it is impossible to learn and remember a complex rhythm if you do not know the meter.

Second, the predictive patterns act directly on the emotional brain by way of different survival-related responses to anticipation, in particular the prediction response rewarding correct predictions in order to reinforce correct predictions of the future. Brainstem reflexes, evaluative conditioning, emotional contagion, visual imagery, and episodic memory in relation to music are all dependent on the basic anticipatory structuring of music, described above, allowing for interpretation, memory, and learning of music.

Hence, we would propose that the emotion-evoking mechanisms described by Juslin and Västfjäll (2008) act on top of the general principle of musical anticipation and may help to identify how music can influence the reward systems of the brain.

All the different emotions evoked by music—both positive or negative—are potentially pleasurable. Investigations of the neural underpinning of musical pleasure are, however, still in their infancy. While briefly reviewing the neuroimaging correlates of listening to music, we have focused on two distinct responses to music that are widely associated with pleasure and relatively stable: the so-called "chills" or "shivers down the spine," and the sensation of swing.

The ultimate hedonic evaluation of both of these responses to music would appear to be mediated through the reward system and is as such related to the underlying proposed principles of musical expectancy. Music-induced emotions are unlikely to be different from other emotions evoked by other types of biological stimuli. The pleasure related to music listening and performance is therefore likely ultimately to be mediated through the same pleasure and reward systems as described elsewhere in this book.

Thus, the hedonic potential of music is linked to the ability of music to help fulfill the Darwinian imperatives of survival and procreation by creating anticipation, fulfilment, or violation. This pleasure is subsequently attributed to the music itself.

This pleasure is very important to most people although some scientists see music as an artifact, a by-product of the evolution of the human brain that "could vanish from our species and the rest of our lifestyle would be virtually unchanged" (Pinker, 1997). A more refined version of this viewpoint is that music is a form of nonadaptive pleasure-seeking, merely exploiting existing brain mechanisms, perhaps to be likened to a drug with no side effects. Others, however, consider music parallel to speech as a language for emotions having great importance for social cohesion and interaction (Huron, 2001).

We would argue that while music may be an accidental by-product of our species-specific acoustic abilities and as such may be a higher pleasure, which could be unique to humans, it is a vital pleasure that we would be foolish not to enjoy as a perfect counterpart to many of life's other sensory, sexual, and social pleasures.

References

Abler, B., Walter, H., Erk, S., Kammerer, H. and Spitzer, M. (2006) Prediction error as a linear function of reward probability is coded in human nucleus accumbens. *Neuroimage* 31, 790–795.

Altenmüller, E. (2001) The neurobiology of music perception: Cortical processing of time- and pitch-structures. The Günne Conference, 2001.

Balleine, B. W. and Killcross, S. (2006) Parallel incentive processing: an integrated view of amygdala function. *Trends Neurosci.* 29, 272–279.

Bates, F. C. and Horvath, T. (1971) Discrimination learning with rhythmic and nonrhythmic background music. *Percept. Motor Skills* 33, 1123–1126.

Berent, I. and Perfetti, C. A. (1993) An on-line method in studying music parsing. *Cognition* 46, 203–222.

Berliner, P. F. (1994) *Thinking in Jazz: the Infinite Art of Improvisation.* The University of Chicago Press, Chicago, IL.

Berlyne, D. E. (1971) *Aesthetics and Psychobiology.* Appleton-Century-Croft: New York.

Berridge, K. C. and Kringelbach, M. L. (2008) Affective neuroscience of pleasure: Reward in humans and animals. *Psychopharmacology*, 199, 457–480

Bever, T. G. and Chiarello, R. J. (1974) Cerebral dominance in musicians and nonmusicians. *Science* 185, 537–539.

Bharucha, J. and Krumhansl, C. L. (1983) The representation of harmonic structure in music: Hierarchies of stability as a function of context. *Cognition* 13, 63–102.

Bharucha, J. J. and Stoeckig, K. (1986) Reaction time and musical expectancy: Priming of chords. *J. Exp. Psychol. Hum. Percept. Perform.* 12, 403–410.

Blood, A. J. and Zatorre, R. J. (2001) Intensely pleasurable responses to music correlate with activity in brain regions implicated in reward and emotion. *Proc. Natl Acad. Sci. USA* 98, 11818–11823.

Blood, A. J., Zatorre, R. J., Bermudez, P. and Evans, A. C. (1999) Emotional responses to pleasant and unpleasant music correlate with activity in paralimbic brain regions. *Nat. Neurosci.* 2, 382–387.

Bonny, H. L. (1997) The state of the art of music therapy. *Arts Psychother.* 24, 1–73.

Brochard, R., Abecasis, D., Potter, D., Ragot, R. and Drake, C. (2003) The "ticktock" of our internal clock: Direct brain evidence of subjective accents in isochronous sequences. *Psychol.Sci* 14, 362–366.

Brown, S., Martinez, M. J. and Parsons, L. M. (2004) Passive music listening spontaneously engages limbic and paralimbic systems. *Neuroreport.* 15, 2033–2037.

Cooper, G. W. and Meyer, L. B. (1960) *The Rhythmic Structure of Music.* The University of Chicago Press: Chicago, IL.

Davies, S. (2001) Philosophical perspectives on music's expressiveness. In: *Music and Emotion: Theory and Research* Eds. P. N. Juslin and J. A. Sloboda, pp. 23–44.. Oxford University Press: New York.

DeNora, T. (2000) *Music in Everyday Life.* Cambridge University Press: Cambridge.

Dickinson, A. and Balleine, B. (this book) Hedonics: the cognitive–motivational interface. In: *Pleasures of the Brain.* Eds. M. L. Kringelbach and K. C. Berridge. Oxford University Press: Oxford.

Drake, C. and Bertrand, D. (2001) The quest for universals in temporal processing in music. *Ann NY Acad.Sci* 930, 17–27.

Drake, C., Jones, M. R. and Baruch, C. (2000) The development of rhythmic attending in auditory sequences: Attunement, referent period, focal attending. *Cognition* 77, 251–288.

Field, A. P. and Moore, A. C. (2005) Dissociating the effects of attention and contingency awareness on evaluative conditioning effects in the visual paradigm. *Cognition & Emotion* 19, 217–243.

Fiorillo, C. D., Tobler, P. N. and Schultz, W. (2003) Discrete coding of reward probability and uncertainty by dopamine neurons. *Science* 299, 1898–1902.

Friedman, D., Cycowicz, Y. M. and Gaeta, H. (2001) The novelty P3: an event-related brain potential (ERP) sign of the brain's evaluation of novelty. *Neurosci. Biobehav. Rev.* 25, 355–373.

Friston, K. (2002) Beyond phrenology: What can neuroimaging tell us about distributed circuitry? *Annu. Rev. Neurosci.* 25, 221–250.

Friston, K. (2005) A theory of cortical responses. *Philos. Trans. R. Soc. Lond B Biol. Sci.* 360, 815–836.

Frith, C. and Dolan, R. J. (1997) Brain mechanisms associated with top-down processes in perception. *Philos. Trans. R. Soc. Lond B Biol. Sci.* 352, 1221–1230.

Goldstein, A. (1980) Thrills in response to music and other stimuli. *Physiol. Psychol.* 8, 126–129.

Green AC, Baerentsen KB, Stodkilde-Jorgensen H, Wallentin M, Roepstorff A, Vuust P (2008) Music in minor activates limbic structures: a relationship with dissonance? *Neuroreport* 19, 711–715..

Grewe, O. L. I. V., Nagel, F. R. E. D., Kopiez, R. E. I. N. and Altenmuller, E. C. K. A. (2005) How does music arouse "chills"? Investigating strong emotions, combining psychological, physiological, and psychoacoustical methods. *Ann. NY Acad. Sci.* 1060, 446–449.

Grewe, O., Nagel, F., Kopiez, R. and Altenmuller, E. (2007) Emotions over time: Synchronicity and development of subjective, physiological, and facial affective reactions to music. *Emotion*. 7, 774–788.

Griffiths, T. D. (2001) The neural processing of complex sounds. *Ann. NY Acad. Sci.* 930, 133–142.

Griffiths, T. D., Warren, J. D., Dean, J. L. and Howard, D. (2004) "When the feeling's gone": a selective loss of musical emotion. *J. Neurol. Neurosurg. Psychiatry* 75, 344–345.

Hagoort, P. (2005) On Broca, brain, and binding: a new framework. *Trends Cogn. Sci.* 9, 416–423.

Hammerl, M. and Fulcher, E. P. (2005) Reactance in affective-evaluative learning: Outside of conscious control? *Cognition Emotion* 19, 197–216.

Hauser, M. D. and McDermott, J. (2003) The evolution of the music faculty: a comparative perspective. *Nat. Neurosci.* 6, 663–668.

Huotilainen, M., Kujala, A., Hotakainen, M., Parkkonen, L., Taulu, S., Simola, J., Nenonen, J., Karjalainen, M. and Naatanen, R. (2005) Short-term memory functions of the human fetus recorded with magnetoencephalography. *Neuroreport* 16, 81–84.

Huron, D. (2001) Is music an evolutionary adaptation? *Ann. NY Acad. Sci.* 930, 43–61.

Huron, D. (2004) Music-engendered laughter: an analysis of humor devices in PDQ Bach. Proceedings of the 8th International Conference of Music Perception and Cognition, Evanston, IL.

Huron, D. (2006) *Sweet Anticipation*. The MIT Press: Cambridge, MA.

Iacoboni, M., Molnar-Szakacs, I., Gallese, V., Buccino, G., Mazziotta, J. C. and Rizzolatti, G. (2005) Grasping the intentions of others with one's own mirror neuron system. *PLoS.Biol.* 3, e79.

Johnsrude, I. S., Owen, A. M., White, N. M., Zhao, W. V. and Bohbot, V. (2000) Impaired preference conditioning after anterior temporal lobe resection in humans. *J. Neurosci.* 20, 2649–2656.

Jones, M. R. and Boltz, M. (1989) Dynamic attending and responses to time. *Psychol. Rev.* 96, 459–491.

Juslin, P.N. and Vastfjall D. (2008) Emotional responses to music: the need to consider underlying mechanisms. *Behav. Brain Sci.* 31, 559–575.

Khalfa, S., Schon, D., Anton, J. L. and Liegeois-Chauvel, C. (2005) Brain regions involved in the recognition of happiness and sadness in music. *Neuroreport* 16, 1981–1984.

Koelsch, S. and Friederici, A. D. (2003) Toward the neural basis of processing structure in music. Comparative results of different neurophysiological investigation methods. *Ann. NY Acad. Sci.* 999, 15–28.

Koelsch, S., Gunter, T., Friederici, A. D. and Schroger, E. (2000) Brain indices of music processing: "Nonmusicians" are musical. *J. Cogn. Neurosci.* 12, 520–541.

Koelsch, S., Fritz, T., Dy, V. C., Muller, K. and Friederici, A. D. (2006) Investigating emotion with music: an fMRI study. *Hum.Brain Mapp.* 27, 239–250.

Kolers, P. A. (1983) Perception and representation. *Ann. Rev. Psychol.* 34, 129–166.

Kringelbach, M. L. (this book) The hedonic brain: a functional neuroanatomy of human pleasure. In: *Pleasures of the Brain*. Eds. M. L. Kringelbach and K. C. Berridge. Oxford University Press: Oxford.

Large, E. W. and Kolen, J. F. (1994) Resonance and the perception of musical meter. *Connect. Sci.* 6 177–208.

LeDoux, J. E. (1989) Cognitive–emotional interactions in the brain. *Cogn. Emotion* 3, 267–289.

Leino, S., Brattico, E., Tervaniemi, M. and Vuust, P. (2007) Representation of harmony rules in the human brain: Further evidence from event-related potentials. *Brain Res.* 1142, 169–177.

Leknes, S. and Tracey, I. (this book) Pleasure and pain: Masters of mankind. In: *Pleasures of the Brain*. Eds. M. L. Kringelbach and K. C. Berridge. Oxford University Press: Oxford.

Lerdahl F. and Jackendoff R. (1977) Toward a formal theory of tonal music. *J. Music Theory* 21, 111–171..

Lerdahl, F. and Jackendoff, R. (1999) *A Generative Theory of Music*. MIT Press Cambridge, MA.

Levitin, D. J. and Menon, V. (2003) Musical structure is processed in "language" areas of the brain: a possible role for Brodmann Area 47 in temporal coherence. *Neuroimage* 20, 2142–2152.

Leyton, M. (this book) The neurobiology of desire: Dopamine and the regulation of mood and motivational states in humans. In: *Pleasures of the Brain*. Eds. M. L. Kringelbach and K. C. Berridge. Oxford University Press: Oxford.

Lundqvist, L. O. D. U. (1995) Facial expressions are contagious. *J. Psychophysiol.* 9, 203–211.

McDermott, J. and Hauser, M. (2004) Are consonant intervals music to their ears? Spontaneous acoustic preferences in a nonhuman primate. *Cognition* 94, B11–B21.

McDermott, J. and Hauser, M. D. (2007) Nonhuman primates prefer slow tempos but dislike music overall. *Cognition* 104, 654–668.

Menon, V. and Levitin, D. J. (2005) The rewards of music listening: Response and physiological connectivity of the mesolimbic system. *Neuroimage* 28, 175–184.

Meyer, L. (1956) *Emotion and Meaning in Music*. University of Chicago Press.: Chicago, IL.

Monelle, R. (1992) *Linguistics and Semiotics in Music*. Harwood Academic Publishers: Chur, Schwitzerland.

Monson, I. (1997) *Saying Something: Jazz Improvisation and Interaction*, The University of Chicago Press: Chicago, IL.

Naatanen, R. (1992) Attention and brain function. In: *Attention and Brain Function*, pp. 102–211. Lawrence Erlbaum Associates: London.

Nagel, F., Kopiez, R., Grewe, O. and Altenmuller, E. (2007) EMuJoy: Software for continuous measurement of perceived emotions in music. *Behav. Res. Methods* 39, 283–290.

O'Neill, S. A. and Sloboda, J. (2001) Emotions in everyday listening to music. In: *Music and Emotion: Theory and Research*, pp. 23–44. Eds. P. N. Juslin and J. A. Sloboda. Oxford University Press: New York.

Ohnishi, T., Matsuda, H., Asada, T., Aruga, M., Hirakata, M., Nishikawa, M., Katoh, A. and Imabayashi, E. (2001) Functional anatomy of musical perception in musicians. *Cerebral Cortex* 11, 754–760.

Osborne, J. W. (1989) A phenomenological investigation of the musical representation of extra-musical ideas. *J. Phenomenol. Psychol.* 20, 151–175.

Ozyurek, A., Willems, R. M., Kita, S. and Hagoort, P. (2007) On-line integration of semantic information from speech and gesture: Insights from event-related brain potentials. *J. Cogn Neurosci.* 19, 605–616.

Pallesen, K. J., Brattico, E., Bailey, C., Korvenoja, A., Koivisto, J., Gjedde, A. and Carlson, S. (2005) Emotion processing of major, minor, and dissonant chords: a functional magnetic resonance imaging study. *Ann. NY Acad Sci.* 1060, 450–453.

Palmer, C. and Krumhansl, C. L. (1990) Mental representations for musical meter. *J. Exp. Psychol. Hum. Percept. Perform.* 16, 728–741.

Panksepp, J. (1995) The emotional sources of "chills" induced by music. *Music Perception*, 171–207.

Pinker, S. (1997) *How the Mind Works.* Allen Lane: London.

Pressing, J. (2002) Black Atlantic rhythm1: Its computational and transcultural foundations. *Music Percept.* 19 285–310.

Pressing, J., Summers, J. and Magill, J. (1996) Cognitive multiplicity in polyrhythmic pattern performance. *J. Exp. Psychol. Hum. Percept. Perform.* 22, 1127–1148.

Reynolds, S. M. and Berridge, K. C. (2008) Emotional environments retune the valence of appetitive versus fearful functions in nucleus accumbens. *Nat. Neurosci,* 11, 423–425.

Rizzolatti, G. and Craighero, L. (2004) The mirror-neuron system. *Annu. Rev. Neurosci.* 27, 169–192.

Roepstorff, A. and Frith, C. (2004) What's at the top in the top-down control of action? Script-sharing and 'top-top' control of action in cognitive experiments. *Psychol. Res.* 68, 189–198.

Sacchetti, B., Scelfo, B. and Strata, P. (2005) The cerebellum: Synaptic changes and fear conditioning. *Neuroscientist* 11, 217–227.

Sadie, S. (2001) *The New Grove Dictionary of Jazz.* Macmillian: London.

Schulkind, M. D., Hennis, L. K. and Rubin, D. C. (1999) Music, emotion, and autobiographical memory: They're playing your song. *Mem.Cognit.* 27, 948–955.

Shepard, R. N. (2001) Perceptual-cognitive universals as reflections of the world. *Behav. Brain Sci.* 24, 581–601.

Shizgal, P. and Arvanitogiannis, A. (2003) Neuroscience: Gambling on dopamine. *Science* 299, 1856–1858.

Sloboda, J. (1985) *The Musical Mind.* Oxford University Press: Oxford.

Sloboda, J. (1991) Music structure and emotional response: Some empirical findings. *Psychol. Music* 19, 110–120.

Smith, K. S., Mahler, S. V., Pecina, S. and Berridge, K. C. (this book) Hedonic hotspots: Generating sensory pleasure in the brain. In: *Pleasures of the Brain.* Eds. M. Kringelbach and K. C. Berridge. Oxford University Press: Oxford.

Steele, K. M. (2006) Unconvincing evidence that rats show a Mozart effect. *Music Percept.* 23, 455–458.

Tan, S.-L. and Kelly, M. E. (2004) Graphic representations of short musical compositions. *Psychol. Music* 32, 191–212.

Temperley, D. and Sleator, D. (1999) Modeling meter and harmony: a preference-rule approach. *Computer Music J.* 23, 10–27.

Thayer, R. E. (1996) *The Origin of Everyday Moods. Managing Energy, Tension, and Stress.* Oxford University Press: New York.

Tononi, G. and Edelman, G. M. (1998) Consciousness and the integration of information in the brain. *Adv. Neurol.* 77, 245–279.

Toomey, L. (1996) Literature review: the Bonny method of guided imagery and music. *J. Assoc. Music Imagery* 5, 75–103.

Vuust, P. (2000) *Polyrhythm and Metre in Modern Jazz—a Study of the Miles Davis' Quintet of the 1960'ies (Danish).* Royal Academy of Music: Aarhus, Denmark.

Vuust, P., Pallesen, K. J., Bailey, C., Van Zuijen, T. L., Gjedde, A., Roepstorff, A. and Ostergaard, L. (2005) To musicians, the message is in the meter pre-attentive neuronal responses to incongruent rhythm are left-lateralized in musicians. *Neuroimage.* 24, 560–564.

Vuust, P., Østergaard, L. and Roepstorff, A. (2006a) Polyrhythmic communicational devices appear as language in the brains of musicians. ICMPC9—International Conference on Music Perception and Cognition, ESCOM, Bologna, pp. 1159–1167.

Vuust, P., Roepstorff, A., Wallentin, M., Mouridsen, K. and Ostergaard, L. (2006b) It don't mean a thing...Keeping the rhythm during polyrhythmic tension, activates language areas (BA47). *Neuroimage.* 31, 832–841.

Vuust, P., Ostergaard, L., Pallesen, K. J., Bailey, C. and Roepstorff, A. (2008) Predictive coding of music—brain responses to rhythmic incongruity. *Cortex.* 45, 80–92.

Willems, R. M. and Hagoort, P. (2007) Neural evidence for the interplay between language, gesture, and action: a review. *Brain Lang.* 101, 278–289.

Winkler, I., Karmos, G. and Naatanen, R. (1996) Adaptive modeling of the unattended acoustic environment reflected in the mismatch negativity event-related potential. *Brain Res.* 742, 239–252.

Zald, D. H., Boileau, I., El Dearedy, W., Gunn, R., McGlone, F., Dichter, G. S. and Dagher, A. (2004) Dopamine transmission in the human striatum during monetary reward tasks. *J. Neurosci.* 24, 4105–4112.

16

The Pleasure of Art

MARTIN SKOV

No doubt the perceptive powers of man and the lower animals are so constituted that brilliant colours and certain forms [. . .] give pleasure and are called beautiful; but why this should be so, we know no more than why certain bodily sensations are agreeable and others disagreeable.

Charles Darwin (1871, p. 353)

Art is a vital part of human existence, perhaps more vital than we normally realize. Because we are still very much the conceptual heirs to the 18th century notion of "fine arts"—the idea that the term "art" refers to a special class of objects, distinctly separable from other kinds of objects (Kristeller, 1951, 1952)—we have a tendency to view art as something refined and extraordinary associated with exceptional occasions—for example, concerts and museum exhibitions, or church visits and other social occasions. However, this is a rather limited conception of art. Art is a much more integral part of daily life. We listen to music (or play or sing ourselves) as we work or socialize. We read novels or watch movies or simply tell each other stories as a daily way of recreation. We live in buildings that have been carefully designed by architects, in rooms that more often than not are decorated with a variety of posters, knick-knack, or other types of ornaments. If we reject the idea that "art" only amounts to a select class of objects— paintings, sculptures, architecture, music, and poetry—it immediately becomes clear that the role of art in human life is highly ubiquitous. For instance, to continue the examples already listed, we cultivate our gardens and parks, we color, cut, and embroider our clothes, and shape and design the artefacts and tools we make use of. We even go to great lengths to tinker with the appearance of our own bodies, styling our hair, or decorating parts of our body through the use of makeup, jewellery, and other "accessories."[1] Taken in this sense the notion of "art" can be said to refer to any kind of object that has been created, or modified, by humans with the intention of making them appear in a specific way—and the forms of behavior associated with making and experiencing art may well counted among the cardinal traits of the human species.

Why do humans take precious time out of their lives to engage in this peculiar form of behavior when doing so take resources away from potentially more important endeavors (such as finding something to eat)? Why are we not content with just covering our bodies in whatever suitable material that nature sees fit to bestow us with, but need to embellish, color, and otherwise manipulate it? Why must the locations we inhabit and the tools we employ look a specific way? Why is it not enough that they, their appearance notwithstanding, are suitable to do the job we invented them to do? In short, what is the root cause of the human species' curious preoccupation with modifying the physical world in "artful" ways? Clearly, this question will not have just one answer. For instance, sometimes we make art in order to signal social standings or values (as when the business manager puts on his or her suit, or when the monarch's palace dwarfs the houses of his subjects). At other times, we seek to impart information to other people, or to simply give material form to mental imaginings.[2]

However, at the heart of all these motivations for creating art, there is probably always a desire to affect the observer in some manner (where the observer could be the artist herself). In other words, whatever else may be the reason for creating art, it is probably at the same time the case that we endeavor to create clothes, architecture, industrial design, or church cantatas, as it might be, with the intention of affecting the observer's hedonic reaction to the object we manipulate: clothes can be seen to signal social standing but they are also meant to be *stylish* or *elegant*; we choose the new iPod over other available MP3 players not just because it serves a purpose but also because it is *cool*; and, of course, Bach's cantatas are clearly not only part of a religious ritual, but also *moving* or *beautiful* in and of themselves. I believe that evoking such hedonic reactions is one of the most important functions of art, if not simply the most important. Indeed, there is some evidence to suggest that art in fact ceases to hold an interest for patients that, due to a disruption of selected brain regions, no longer generate appropriate emotional responses to the perception of works of art. For instance, one patient with an infarction involving the left insula, extending anteriorly into the left frontal lobe and inferiorly into the left amygdala, reported having stopped caring for the music by Rachmaninov that previously had been a source of immense pleasure (Griffiths et al., 2004). Interestingly, this patient showed normal scores on the perception of scale, contour, interval, meter, and incidental memory, indicating that his ability to form a perceptual and cognitive representation of Rachmaninov's music remained intact. (Unfortunately, the authors does not report if the patient's loss of interest in music was general or limited to Rachmaninov.) Naturally, one case study is not conclusive evidence, and the authors of the report never tested the patient's appreciation for art in a more controlled manner, but the fact that an interest in art can fully disappear due solely to lesions in the limbic system is indicative that the capacity of art to inflict a hedonic reaction in observers is a crucial aspect of its nature.

Among philosophers interested in understanding the arts, it has long been recognized that hedonic responses play a pivotal role in the experience of art (Beardsley, 1975). Since the time of Plato and Aristotle, attempts have been made to unravel why objects exhibiting certain properties are experienced as pleasurable. In the 18th century, it was proposed by Francis Hutcheson and others that the human mind might be endowed with an esthetic sense that produces the experience of beauty as well as other esthetic feelings when triggered by specific types of objects (Hutcheson, 1973). Hence, the core mental mechanism defining this putative esthetic sense was thought to be an ability to relate processes producing esthetic feelings to processes underlying the perception of objects. However, how this mechanism actually worked, and what might be the physiological processes underlying it, were always questions beyond philosophical analysis. Lacking the proper investigative techniques to probe the functional machinery of the esthetic sense, first philosophers, and later psychologists, had to contend themselves with detailing what *kinds* of properties, or what type of *arrangements* of such properties, would suffice to "excite" the esthetic sense enough to produce feelings of esthetic pleasure. This approach yielded a number of "rules," suggesting, inter alia, that objects must be harmonic, or symmetric, or proportional, or "uniform amidst variety" (Hutcheson's proposal) to be experienced as beautiful (Tatarkiewicz, 1972). Unfortunately, it was always possible for some enterprising philosopher to come up with awkward counterexamples to these rules (Stolnitz, 1961). For example, it turns out that not all people find a particular symmetrical object beautiful, or that, embarrassingly some people actually find (some) nonsymmetrical objects beautiful. This variety in taste indicates that, although we clearly react to some, and not all, perceptual features or qualities with feelings of pleasure, there is no clear-cut relation between perceiving object properties and having an esthetic reaction. Just as it has been proposed that hedonic affects are a "gloss" on behavior and sensation (Aldridge and Berridge, Chapter 3, this book), it can be hypothesized that esthetic feelings are the result of a complex interaction between hedonic and perceptual processes. To understand this interaction in detail, it is necessary to open the black box of the philosophers' esthetic sense—that is, we have to begin investigating the neural processes underlying the esthetic appreciation of objects.

Among the many questions in need of answering, the most pressing is whether or not there actually is such a thing as an "esthetic sense." Is there a dedicated neural system computing the esthetic value of objects with neurons functioning as neural generators of esthetic feelings? As explained in other chapters of this book, we already know that neural generators of hedonic affect exist in relation to primary reinforcers (Aldridge and Berridge, Chapter 3, this book; Kringelbach, Chapter 12, this book; Smith et al., Chapter 1, this book). An important question is therefore if the experience of esthetic values is grounded in these more basic hedonic processes or invokes its own distinct set of neural processes. Is beauty simply another name for sensory pleasure? (This question is also important because it can help illuminate the evolutionary question if the appreciation of art is rooted in neurobiological systems that evolved very early to imbue food, drink, smells, and sexual partners with hedonic value, or in systems that might have evolved more recently.) Indeed, it is not presently understood what is the precise physiological and phenomenological nature of the experiential value we call beauty, or even if it is the only or the most important "esthetic feeling." In a recent study, Thomas Jacobsen and colleagues asked German college students to produce adjectives they associate with the esthetics of objects (Jacobsen et al., 2004). Although beauty was by far the most frequent word generated by these students (produced by 91.6% of the participants), all in all they reported 590 different adjectives! (The average number of adjectives generated by each participant was 9.4.) It remains to be seen if this variety of names for putative esthetic feelings—ugly, nice, pretty, elegant, repulsive, fascinating, charming, groovy, sublime, and so on—map unto a distinct set of neural processes, or if they are different ways of conceptualizing the experiential result of a more basic, and perhaps singular, neural system for computing subjective preferences. Finally, it is essential to understand why, as noted above, the human population exhibits such a distinctive variation in taste; in other words: what makes the "esthetic sense," if it exists, react differently to the same perceptual input in different people?

Thanks to the invention of noninvasive neuroimaging techniques, it has become possible to begin experimental examination of some of these questions (Skov and Vartanian, 2009). Using positron emission tomography (PET), functional magnetic resonance imaging (fMRI), and magnetoencephalopathy (MEG), it is today possible to image normal subjects as they experience works of art and other objects that elicit esthetic reactions. By correlating reports of subjective preference for art objects with neural activity, a growing number of researchers have begun mapping the brain areas involved in forming such preferences. In this chapter, I review a number of these imaging papers and propose a preliminary model of the brain systems involved in assessing the esthetic value of objects.[3] I refer to the functional job of this system as "esthetic preference formation" (EPF) so as to avoid the presupposition of specific esthetic feelings—such as beauty, "attractiveness," or "cuteness"—the existence of which we cannot be sure of. (Such special esthetic feelings may turn out to have a distinctive physiological and psychological nature, but we need more data to settle this matter conclusively.) The notion "esthetic preference" is taken to simply signify that the brain assess the value of the objects it perceive and represent—that is, how much I prefer an object, either on its own or compared to other objects. What is more precisely meant by "value," for instance if some instances of high preference entail a specific emotional content, is for future experiments to determine. Together, I believe the selected set of studies considered here support the idea that esthetic values are the result of the interaction of several brain systems, including perceptual, cognitive, and reward processes—hence the proposal that esthetic preferences are *formed*. The papers I discuss either use "fine arts" objects as stimuli (music and visual art), or objects usually considered within the purview of esthetics (human faces, human-made objects). This, as already noted, may turn out to be an artificial distinction. It is clear that the human brain also forms preferences for smells (Fulbright et al., 1998), taste (Kringelbach et al., 2003), touch (Rolls et al., 2003), social attachments and behaviors (Fehr and Camerer, 2007), money, and conceptual ideas, including political beliefs (Zysset et al., 2002). It remains to be determined how these types of sensory preference formation differ from, or overlaps with, the assessments of beauty or esthetic excellence involved in EPF.

Brain Systems Mediating Esthetic Preference Formation

In general, neuroimaging studies looking for brain correlates of esthetic appraisal expose subjects to stimuli varying in esthetic value and compare brain activity related to different levels of preference, either by identifying brain areas where neural activity correlates to ratings on a single valence dimension from highly negative to highly positive, or by directly contrasting brain activation correlating with bivalent positive and negative preferences (alternatively contrasting positive or negative activations with some baseline). As an example of the first approach, Vartanian and Goel (2004) showed subjects representational and abstract paintings in different formats and asked them to rate these pictures, while in the scanner, on a scale from 0 to 4. A parametric analysis revealed that the right caudate nucleus decreased in response to the subjects' decreasing preference for the paintings, while activity in the left cingulate sulcus, bilateral occipital gyri, right fusiform gyrus, and bilateral cerebellum increased in response to preference for paintings. (Importantly, due to signal drop-out, in this study the researchers were unable to assess activity in the orbitofrontal cortex (OFC).) As an example of the other approach, Kawabata and Zeki (2004) instructed subjects to rate four different categories of paintings (abstract, still life, landscape, and portrait) as beautiful, neutral, or ugly. In their analysis, they then compared beauty reports with ugly reports and found that beautiful > ugly produced activity in OFC, whereas the opposite contrast, ugly > beautiful, produced activity in bilateral motor cortex. The authors also contrasted beautiful with neutral, which yielded activity in the anterior cingulate cortex (ACC) and the left parietal cortex, and ugly with neutral which produced no activity.[4] Together, these two studies indicate that in assessing their esthetic preference for paintings, subjects activate both brain structures known to be implicated in reward processing (basal ganglia, OFC, and the cingulate cortex) and structures that underlie perceptual processing (occipital cortex and the fusiform gyrus).

Studies using music have so far mostly attempted to compare pleasurable music with some kind of unpleasant contrast. To my knowledge, no study has yet imaged subjects as they engage in active judgment of music. Therefore, parametric modulation has not been an analytic possibility. Thus, in the first study to investigate the esthetic reaction to music, a PET study by Blood and colleagues (1999), the subjects listened passively to versions of a musical sequence varying in degree of dissonance. In a post hoc behavioral rating task, sequences with the highest degree of dissonance were judged more unpleasant than harmonic sequences. Analysis of the PET data demonstrated that activity in right parahippocampal gyrus and precuneus regions correlated with increasing dissonance (and with post hoc ratings of displeasure) whereas activity in OFC, the subcallosal cingulate and frontal polar cortex correlated with decreasing dissonance (= post hoc pleasure).

In another PET study, in an attempt to directly induce pleasure, Blood and Zatorre (2001) compared regional cerebral blood flow (rCBF) changes when subjects were presented with pieces of their own, preselected favorite music and the favorite music of other participants participating in the experiment. Subjects reported (again post hoc) that listening to their own favorite music elicited a highly pleasurable experience ("chills") compared to the favorites of the other subjects; these reports correlated with changes in heart rate, respiration, and electromyogram. Stimuli associated with chills produced increasing rCBF in the insula, ventral striatum, ACC, and OFC and decreasing rCBF in amygdala, precuneus, hippocampus, and medial prefrontal cortex.

In contrast, Brown et al. (2004), also using PET, investigated emotional reactions to unfamiliar music deemed pleasurable in a post hoc questionnaire. Passive music listening was compared to a rest condition, producing increases in rCBF in primary auditory cortex, auditory association cortex, superior temporal sulcus, superior temporal pole, anterior insula, the subcallosal cingulate and ACC, hippocampus, and part of nucleus accumbens (NAcc). Finally, in two fMRI studies, subjects were presented with excerpts of classical music, deemed pleasant (post hoc), and manipulated counterparts (deemed unpleasant post hoc).

In Menon and Levitin (2005), passive listening to the original pieces of music compared to scrambled music produced activation in NAcc, the ventral tegmental area (VTA), hypothalamus, inferior frontal cortex (IFC), OFC, and ACC. Interestingly, in addition to contrasting these two conditions, Menon and Livitin also performed an effective connectivity analysis, looking for causal effects of these structures on each other. They were especially interested in the modulatory effects of the VTA–NAcc pathway, so their model looked for brain areas exhibiting VTA-mediated interaction with the NAcc. Analysis demonstrated VTA-dependent NAcc interactions with the hypothalamus, insula, OFC, IFC, and middle and superior temporal gyri.

In Koelsch et al. (2006), excerpts of classical music alternated with alternative versions produced by changing the pitch of the original versions. In between listening to these pieces of music, the subjects were asked to rate how (un)pleasant they felt on a 5-point scale. However, although these ratings were used to demonstrate that the original excerpts were experienced as significantly more pleasant than the manipulated counterparts, they authors did not model this part of the time series as an event in the fMRI analysis. Instead, they simply contrasted the original with the manipulated excerpts, yielding activity in the hippocampus, parahippocampus, and temporal pole for the manipulated > original contrast, and activity in Herschl's gyrus, IFG, and anterior superior insula for the original > manipulated contrast. Since each excerpt was presented twice, in separate blocks, Koelsch and colleagues also analyzed the contrast between original and manipulated excerpts with data taken only from the second presentation. Here, the original > manipulated contrast produced higher activation in the Rolandic operculum, IFG, and ventral striatum, while the manipulated > original contrast yielded activity in amygdala. This difference between these two contrasts could be taken to suggest that the participants anticipate how likable the music will be in the second run based on the earlier exposure and pleasantness rating, resulting in higher activation of the ventral striatum and amygdala, respectively.

Together, these five music studies demonstrate that listening to passages of music that are experienced as pleasurable produces activity linked with the processing of reward and emotion, especially in OFC, ACC, the ventral striatum (NAcc), the insula, and amygdala. Of these brain areas, the first three may be involved in computing various reward properties of the stimuli, whereas the insula activity could reflect the subjective feeling of pleasure induced by listening to nice music, or alternatively the experience of visceral chills (in line with theories positing the anterior insula as a crucial hub for the re-representation of interoceptive states; cf. Craig, 2009). It is interesting that, in Menon and Levitin's experiment, insula activity did not show up in the ordinary main effect, contrasting music with scrambled counterparts, but was found in the effective connectivity analysis as one of the structures modulated by the VTA–NAcc pathway. As a hypothesis, this finding could possibly reflect a difference in function, where the dopaminergic pathway (VTA >> NAcc/basal ganglia/amygdala >> OFC/ACC) first computes the reward properties of what is heard, and then provokes insula activity if a positive hedonic experience results from this processing. Obviously, it would be interesting to address this question more directly in a new study. Furthermore, a recent neuropsychological study by Nathalie Gosselin and colleagues (2006) provides evidence that the parahippocampal activity found in Blood et al. (1999) and Koelsch et al. (2006) to correlate with conditions of unpleasant music appears to be specifically involved in the computation of the emotional reaction to dissonance. Gosselin and colleagues presented music varying in dissonance to patients with excisions of the medial temporal lobe (including the parahippocampus) and to healthy controls. Patients with substantial resections of parahippocampal cortex rated the dissonant music as slightly pleasant whereas the control group judged it as wholly unpleasant. In contrast, there was no difference in how patients and controls rated consonant—that is, pleasant—music. Moreover, the larger the removals of the parahippocampal cortex, the more unresponsive to dissonance were the patients. Finally, the authors also tested the music on one subject with selective damage to the amygdala who rated the dissonant music as just as unpleasant as the normal subjects did. This suggests that the parahippocampus could play a rather specific role in the processing of the emotional reaction to dissonance, but may otherwise not contribute to EPF.

In Vartanian and Goel's experiment, using paintings as stimuli increasing ratings of preference produced enhanced activity in both brain areas associated with emotion and brain areas associated with visual perception (occipital gyrus and fusiform gyrus). As mentioned, this result suggests that experiencing paintings as pleasing—at least engaging in situations where paintings are actively judged as pleasing—recruits perceptual processes in addition to reward and/or emotional processes. It is worth noticing that the music studies just reviewed does not quite exhibit this same kind of dual activation of reward processes and perceptual processes. Although Brown et al. (2004) and Koelsch et al. (2006) also found elevated activity in structures involved in auditory perception—Heschl's gyrus, superior temporal sulcus, and IFG—these areas show up as a result of a comparison of passive listening with a rest condition in Brown et al. (2004) and as the result of a contrast of original music with altered versions in Koelsch et al. (2006).

Therefore, it is uncertain if this activity is related to conditions of assessing the esthetic value of the stimuli—as is clearly the case in Vartanian and Goel (2004). Regrettably, since none of the imaging studies using music as a stimulus have asked their subjects to make an active preference judgment while in the scanner, we have yet to see if there is a difference between passive listening tasks and active rating tasks, although this, of course, on the face of it seems likely.[5]

It could be speculated that situations where you are asked to determine if a series of stimuli match different levels of preference (as when an experimenter instructs you to report how a number of paintings rate on a scale from 1 to 5) require additional perceptual processing in order to decide just how likable or unlikable the individual stimulus is.

Although the seven studies reviewed here differ in methodology as well as in the kind of stimuli, tasks, and ways of analyzing the data employed, together they appear to indicate that esthetic preferences for paintings and music—in the form of subjective reports of pleasantness, chills, beauty, and likability—are primarily associated with the activation of brain structures assumed to subserve appetitive behavior, motivational mechanisms, and reward processing, that is, structures located in the striatum, amygdala, OFC, and cingulate cortex. Neurons situated in this pathway have consistently been associated with the processing of primary reinforcers such as food, drink, and odors

(Rolls, 2006; Verhagen, 2007) or sexual cues (Becker et al., 2001; Rolls, 2005), and is believed to be crucial for guided, adaptive behavior (Swanson, 2000). Thus, one possible interpretation of the available data is that works of art acquire their esthetic value by functioning as secondary reinforcers, tapping into more primitive affective states (the so-called common currency hypothesis).

Supportive of this idea is the fact that imaging studies trying to identify the neural correlates of facial attractiveness—faces being both a natural stimulus, highly important for human mating behavior (Gangestad and Scheyd, 2005; Rhodes, 2006), and a prominent artistic motif—find activity in many of the same structures, including OFC, ACC, amygdala, NAcc, and the basal ganglia (Aharon et al., 2001; Nakamura et al., 1998; O'Doherty et al., 2003; Winston et al., 2007).

Naturally, we must be mindful of the fact that activation blobs unearthed by neuroimaging techniques, often several cubic millimetres in size, might very well hide more subtle functional specializations. But the overlap of activations in imaging studies reporting neural correlates of esthetic preferences for different types of stimuli is at least suggestive that we are talking about a common system based on primary reward processes. Indeed, to consider non-"fine arts" objects as well, a study comparing pictures of (subjectively determined) highly attractive cars with pictures of less attractive cars found higher activity in the ventral striatum, OFC, and ACC (Erk et al., 2002)—in addition to activity in occipital cortex and the fusiform gyrus, as in Vartanian and Goel (2004).

And similarly, Paulus and Frank (2003), comparing preference judgments of pictures of soft drinks with visual discrimination trials, found higher activity in the ventromedial cortex (centred on OFC), ACC, and anterior insula during the preference judgments. Although assessing the esthetic value of an object clearly recruit other neural systems than reward processes, it is remarkable how consistently OFC and the ACC, and to some degree also striatal structures, are activated across different preference formation conditions and stimulus classes (see Table 16.1).

It will be an important task for future studies to tease out more precisely what kind of functional mechanisms EPF employs and correlate these functions with physiological processes. The different methods, tasks, and analyses utilized in the experiments discussed above indicate, though, that the computation of an objects' esthetic value is rather context sensitive. For example, not only is the brain able to appraise different stimulus categories, but the studies summarized in Table 16.1 are evidence that it is also sensitive to a number of different stimulus dimensions, including real art versus simplified or manipulated objects, excerpts of work of art versus whole art objects, and well-known pieces versus novel or unknown pieces.

There is already considerable behavioral evidence that level of complexity, mere exposure, and a host of other perceptual factors influence esthetic preferences; it stands to reason that such differences in stimulus properties and anticipatory situations affect the neural processes taking place in structures under consideration here in distinct ways. Furthermore, while some of the studies here present several hundred events (Kawabata and Zeki, 2004; Vartanian and Goel, 2004), others present considerably fewer. It is highly likely that it is not the same task having to compare a large number of stimuli in order to reach an esthetic judgment—"is this painting really *very* good if the last one was merely good?"—as having to just listen to two instrumental songs (as in Brown et al., 2004). This factor is also related to the question just raised of whether or not the subject making the evaluation of esthetic value is engaged in an active process of judgment, and if so, what kind of response he or she is required to make. Does having to decide if an object is beautiful engage different neural processes than simply having to assess if an object is likable or pleasurable? Is it easier to differentiate between a simple dichotomy of choice (for instance, like vs. dislike) than between multiple choices (continuous scales, such as 1–5, for example)? All these types of judgment (and passive experience) are used in one or another of the studies discussed above. It remains to be seen if they rely on different neural networks or not.

When we talk about EPF, then we presumably talk about a number of different neural processes depending upon the presence or absence of different factors: for instance, the physical character of the object being evaluated; how familiar the object is; if the object is novel or not (as is often the case with works of art); whether or not it is assessed in isolation or in comparison with other objects; the situational context (am I evaluating the esthetic value of a display of cars in a show room, intending to buy one of them, or am I idly leafing through a car catalogue in the comfort of my armchair?); the character of esthetic judgment being elicited, and so on. Moreover, if it turns out to be true that reward processing is an important part of the EPF system, we know from countless experiments that things such as associative learning, predicting or anticipating gains and losses (Knutson

Table 16.1 Reported Activations in Neuroimaging Studies Reviewed in the Present Chapter

Stimuli	Reference	Frontal Cortical	Subcortical-Limbic	Posterior Cortical
Faces	Aharon et al. (2001)	OFC	Amygdala, VTA, NAcc	
	O'Doherty et al. (2003)	OFC, postCC, medPFC, IFG	Insula, paracing.	
	Nakamura et al. (1998)	OFC, medPFC	Insula, NC, thalamus	Inf. temp. gyrus
	Winston et al. (2007)	OFC, medPFC	Insula, amygdala, paracing	STS
Music	Blood and Zatorre (1999)	OFC	Paracing., subcalCing.	Precuneus
	Blood et al. (2001)	OFC, ACC, medPFC	NAcc, thal., insula, amygdala	Hippocampus, precuneus
	Koelsch et al. (2005)	IFG	Insula	Heschl's gyrus, hippocampus, parahippo., temp. pole
	Brown et al. (2004)	ACC	Insula, subcalCing	Hippocampus, inferior parietal lobule
	Menon and Levitin (2005)		VTA, NC	
Paintings	Kawabata and Zeki (2004)	OFC, ACC		Parietal cortex
	Vartanian & Goel 2005	Cing. sulcus		Occipital cortex, fusiform gyrus
Geometrical figures	Jacobsen et al. (2006)	OFC, ACC, postCC, IFG		Precuneus, temp. pole
Cars	Erk et al. (2002)	OFC, ACC	NAcc	Occipital cortex
Soft drinks	Paulus and Frank (2003)	OFC, ACC	Insula	Posterior parietal

These studies report neural correlates of subjects' subjective preferences for various types of stimuli.

Abbreviations: OFC, orbitofrontal cortex; ACC, anterior cingulate cortex; postCC, posterior cingulate cortex; medPFC, medial prefrontal cortex; IFG, inferior frontal gyrus; DLPFC, dorsolateral prefrontal cortex; VTA, ventral tegmental area; NAcc, nucleus accumbens; paracing., paracingulate cortex; NC, nucleus caudatus; subcalCing., subcallosal cingulate; thal., thalamus; inf.temp.gyrus, inferior temporal gyrus; STS, superior temporal sulcus; parahippo., parahippocampus; temp.pole, temporal pole; cing. sulcus, cingulate sulcus.

and Greer, 2008), and the signaling of reward errors (Schultz et al., 2000), the attribution of incentive salience to perceived objects (Berridge and Robinson, 1998), tracking (Kringelbach, 2005) and comparing (Padoa-Schioppa, 2007) reward outcomes, and decision-making (Wallis, 2007) all may play functionally different roles. The next step for the research on how esthetic preferences are formed is to systematically manipulate these factors and establish the precise involvement of neural structures in different contextual situations.

Recently, my colleagues and I have tried to ascertain what happens to the EPF system under some of these circumstances. In one experiment, for instance, we decided to test if judgments of beauty are strictly determined by the object's valence (Skov et al., 2009). If esthetic preferences are rooted in basic pleasure mediated by appetitive, emotional systems, one hypothesis would be that positive esthetic values are driven by more basic positive affective responses to the object being appraised (and negative esthetic values, conversely, by basic negative affective responses); we simply like objects that are pleasurable and dislike objects that are unpleasant. However, at the same time, it is well known from the history of art that artists often gestalts unpleasant subjects so that they appear as beautiful. Think of the teaspoon and saucer covered in fur by the Surrealist Meret Oppenheim, or the grotesque human figures in Max Beckmann's paintings. How can a depiction of a hanged man getting his arm broken (cf. Beckmann's *The Night*) be beautiful? One reason might be that different properties of Beckmann's painting give rise to different affective responses: the *content* of his painting might be unpleasant, but the lines, the color, or other formal properties, are perceived as beautiful. Such cases raise the possibility that the EPF system reacts differently to pictures that are in some way unpleasant, if it is to reach the conclusion that they are also beautiful, than to pictures that are predominantly pleasant.

To find out if this is indeed true, we sampled a number of pleasant, neutral, and unpleasant photographs from the *The International Affective Picture System* (IAPS). Pictures from this stimulus set have been used to elicit positive and negative emotional responses in a number of imaging studies (Lane et al., 1997). Without informing our subjects that the photos they were going to see in the scanner varied systematically in valence, we asked them to rate 300 IAPS pictures as beautiful, neutral, or ugly in an event-related fMRI design. In this way, by looking at the interaction between valence and esthetic ratings, we could contrast the neural activation correlating with assessments of unpleasant photographs as beautiful and assessments of pleasant photographs as beautiful. As predicted, this comparison did reveal a difference in activity between the two conditions. Beautiful unpleasant pictures contrasted with beautiful pleasant pictures produced higher activation in the fusiform gyrus, the inferior and superior temporal sulcus, OFC, cingulate cortex, and the caudate nucleus.

Interestingly, the main effect of beauty, that is, the contrast between all beautiful responses (regardless of valence) and all ugly responses produced higher activity in fusiform gyrus, OFC, the cingulate, and putamen, but *not* in the temporal structures associated with the first contrast. This result suggests that valence is just one factor influencing the formation of esthetic preferences, and can possibly be interpreted to mean that the reward system can be modulated by this dimension of the stimulus through differential activation of perceptual processes. Possibly, this ability to experience an object with unpleasant properties as beautiful is a crucial aspect of art. At least, art is one of the comparatively few classes of objects where we willingly subject ourselves to highly unpleasant emotions (sadness, fear, nausea, shock, etc.).

In another study, the idea was to investigate the effect of expertise on EPF (Kirk et al., 2009a). It is well-known that art-related expertise modulates the esthetic evaluation of art objects (Hekkert and van Wieringen, 1996). Hence, expertise—innate or acquired—could be one of the factors explaining variance in taste: experts' brains handle stimulus information differently than the brains of nonexperts. To test if there is such a neural difference, we recruited a group of architects and a group of nonarchitects. In an event-related fMRI design, the two groups were showed pictures of buildings and pictures of faces and asked to rate their preference of each on a scale from 1 to 5. Since faces have been demonstrated to elicit highly similar judgments of attractiveness across population groups and cultures (Rhodes, 2006), we expected the buildings, but not the faces, to affect both esthetic ratings and neural activity differently in the two groups.

As it turned out, we did not find a difference in rating behavior between the two groups. However, a parametric analysis where the subject-specific behavioral responses were used to model the interaction between the two groups and the two stimulus conditions yielded higher activity in medial OFC and ACC for the experts than for the nonexperts. We further modeled second-order polynomial expansions of the subject-specific esthetic judgments. This type of analysis is interesting because it is able to capture voxels where the blood oxygen–level dependent (BOLD) signal does not respond linearly across the valence space. More specifically, the second-order model captures nonlinear responses where structures increase activity in response to highly or lowly regarded stimuli but not to neutral stimuli. Thus, if a structure responds both to positive and negative valence, it will "disappear" in a normal contrast analysis, but turn up in the nonlinear model. (There is some evidence that this, for instance, in some situations is the case with the amygdala; see Winston et al., 2007). We did not find group-specific activations as result of an interaction between groups and stimuli, but a conjunction analysis yielded significant activity in left NAcc. In other words, whereas medial OFC and ACC were modulated by expertise—that is, displayed higher activity in the architects compared to the nonarchitects when the subjects rated buildings but not faces—left NAcc increased its response to pleasant and unpleasant ratings, but not to neutral events, for *both* buildings and faces and in both groups of subjects. This clearly suggests that medial OFC, ACC, and NAcc perform different functions as part of the EPF system.

Finally, we also wanted to test if contextual information can influence the formation of esthetic preferences (Kirk et al., 2009b). Since Duchamp, there has been an ongoing discussion about the role played by institutions in art appreciation. Does a urinal become a (beautiful?) work of art by being displayed in a museum? Mounting evidence suggest that, in fact, various contextual information such as brand information (being told what type of cola you are drinking; McClure et al., 2004), semantic labeling (de Araujo et al., 2005), frames (Deppe et al., 2005) and the price of a product (Plassmann et al., 2008) modulates EPF, even when the physical properties of the stimulus being appraised are kept unchanged. We therefore surmised that manipulating the assumed provenance of a work of art would be sufficient to modulate subjects' appreciation of it. To this end, we recruited art-naïve subjects whom we could assume not to be familiar with the set of abstract paintings we asked them to rate.

Before going into the scanner, we informed the subjects that they were to rate a series of paintings where half had been acquired from a very prestigious gallery in Denmark, and the other half produced by ourselves (that is, the experimenters). None of this was true; all the paintings were found in art databases on the Internet. In the scanner, each painting was accompanied by either the label "gallery" or the label "computer" to indicate provenance. Critically, across subjects, all paintings were alternatively labeled as gallery paintings and computer paintings. Hence, the physical input was held constant, only the context under which the paintings were assessed changed.

The behavioral ratings of the subjects demonstrated that paintings assumed to originate from the prestigious gallery were rated as more likeable than the paintings thought to be produced by the experimenters. Furthermore, a parametric regression analysis demonstrated that this contextual modulation of esthetic rating correlated with activity in the medial OFC. A zero-order analysis, comparing gallery paintings with computer paintings irrespective of esthetic rating, showed that the context shift in itself correlated with activations of the temporal pole and entorhinal cortex. Together, these two results suggest that stimulus information is integrated with memory-related information when a stimulus is experienced in different contexts, and that the medial OFC presumably is critical for integrating these difference sources of information with reward processes.

Experimental designs such as the three briefly introduced above go some way toward demonstrating that esthetic preferences in all probability are constructed as a result of the integration of different neural processes, sensitive to a range of factors including sensory identity, valence, differences in personality, and situational context. Thus, the first of our studies shows that the behavioral task of making a specific type of judgment (is this photograph beautiful?) gives rise to the activation of different neural networks depending on the valence of the stimulus. It also reveals that the same behavioral end result, pronouncing a painting beautiful, may be the result of distinct neural processes, again supporting the idea that esthetic preferences for objects are formed. The second study demonstrates that level of expertise modulates responses in OFC and ACC to the elements belonging to the class of stimulus the expert is an expert on. And the third study suggests that contextual information, in some instances, is integrated with perceptual and reward processes. I therefore put forward the hypothesis that EPF variably combines different functional processes, prominently among these the processing of sensory identity, conceptual knowledge, memory, reward, integration of information from these mechanisms, and behavioral control systems (see Figure 16.1), depending upon the task at hand.

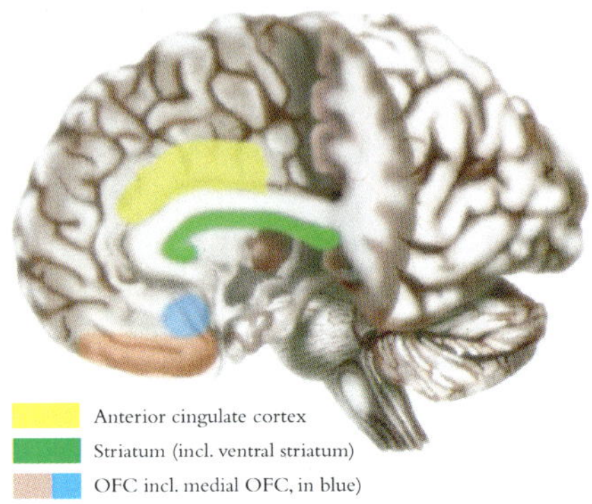

Figure 16.1 Location of core anatomical structures involved in esthetic preference formation. Possible functional roles of brain structures involved in esthetic preference formation include behavioral control (decision-making and motor control) centred on prefrontal structures (ACC; yellow); tracking reward outcomes based on multiple factors and translating these outcomes into subjective hedonic experience (OFC; red and blue); and the processing of reward (striatum, nucleus accumbens, amygdala; green). Also important is structures in the posterior parts of the cortex computing percepetual properties and relating these to the brain's memory and conceptual systems (colors not shown).

Each of the neural structures underlying these functions in principle plays a role in EPF. However, one of the major questions confronting us is how exactly they interact under different circumstances. For example, is the activity we see in Vartanian and Goel (2004) and Skov et al. (2009) in structures belonging to the visual system a *result* of processing going on in reward structures or the antecedent of reward processing? Naturally, to view a painting, you need the contribution of visual perception to get emotion and reward processing going in the first place. But it is possible that the ensuing esthetic value attributed to the perceived painting feeds back into the visual system. It is also a possibility that extra perceptual processing is needed to clarify conceptual problems of different sorts ("is this unpleasant representation of a man being hanged really beautiful?"). It would be interesting to see connectivity analyses try to address these questions, and, in fact, we are working on this at the moment.

Another big question is what role learning plays in EPF. Psychometric tests indicate that familiarity with a work of art influence our preference for it (Valentine, 1962). The difference in activity seen between the first and second exposures to the stimulus set in Koelsch et al. (2006) indicates that, when subjects assess the esthetic value of an object, parts of the reward system (amygdala and NAcc) are sensitive to familiarity. Similarly, the nonlinear response in our expertise study to high and low value, but not neutral value, of both buildings and faces in NAcc could possibly be interpreted as a function of reward prediction, that is, the anticipation of reward. Neurons in ventral striatum and amygdala have repeatedly been implicated in associative learning so if this turns out to be true it would not be surprising. Yet, systematic manipulation of exposure, repetition, familiarity, and so on in neuroimaging studies of EPF is still a thing of the future.

Our gallery study indicates that the EPF system is subject to top-down modulation. Contextual information impacts the appraisal of abstract paintings in naïve subjects. Presumably this modulatory influence relies on the integration of sensory and reward information with conceptual, and therefore possibly also memory-related, information. The OFC–ACC network appears to be important for this integration to take place. Animal studies suggest that OFC is a nexus for control of appetitive behavior, for instance, in guiding foraging and ingestion (Rolls, 2006; Wallis, 2007), and the ACC is thought to constitute an interface for drive, motor control, and cognition (Paus, 2001). EPF may tap into this system. For example, our experience of a work of art can clearly be influenced by being party to knowledge about the artist or the art work ("this painting is the artists' attempt to convey his remembrance of his mother; it was painted when they hadn't talked to each other for twenty years"), or by the nature of the behavioral task the esthetic value seeks to inform (my listening to a casual song on the radio or listening to a sample of CDs in a record shop with the objective of determining which CD to purchase). However, the precise computational mechanisms employed by OFC and ACC in integrating perceptual information, reward properties, cognition, and behavioral task await illumination.

One possible mechanism subserved by OFC activation could be the integration of multiple sources of information, including comparisons of different reward properties, which is then related to the hedonic experience of the outcome of this integration (Kringelbach, 2004, 2005). The studies reviewed here demonstrate that OFC activity reflect the modulation of EPF by contextual information and expertise—presumably two different neural properties. The contrast in our IAPS study between unpleasant beautiful photos and pleasant beautiful photos also produced enhanced activity in OFC, suggesting that OFC also integrates different levels of valence.

Moreover, recent studies on the modulation of face perception by the sexual preference of the observer by Ishai and colleagues (Ishai, 2007; Kranz and Ishai, 2006) have revealed that the OFC is also sensitive to an interaction between the gender of the face and sexual orientation when subjects are asked to rate the attractiveness of faces. Thus, not only cognition but also stimulus properties and personality traits appear to intersect with reward processing in the OFC. Since works of art in general are rather complex objects, consisting of a diffuse arrangement of material properties and novel representations, it seems likely that OFC, and perhaps especially the medial part of OFC, constitutes an absolutely necessary node of the EPF system. It could be suggested that, in contrast to the assessment of the reward value of primary reinforcers, EPF is much more reliant on a complex integration of a variety of sources of information (Figure 16.2), with a possible higher premium on mediating conflicting factors ("I like the color in this painting, but I have never been partial to his use of broad brushstrokes").

An important question not addressed in the studies I have reviewed here is to what extent the EPF system is influenced by the physiological state of the neural structures involved. As explained by Berridge and colleagues in their chapters in this book, by inducing a temporary state of sodium deficiency rats can be made to "like" salt. Ingestion of salt in this physiological condition produces hedonic 'liking' reactions and changed patterns of neural firing in the ventral pallidum. Where the firing rates of these neurons to the taste of salt is low in normal homeostatic physiological states, in physiological states of sodium deficiency, firing rates rise to the equal of tastes of sugar (Aldridge and Berridge, Chapter 3, this book). The reason for this changed behavior is the fact that the assessment of the reward value of primary reinforcers takes place within the framework of motivational behavior—that is, computing the need for, for example, drink or food based on the homeostatic state of the organism. The reward system takes this physiological state into account, and neurons in various parts of brain change

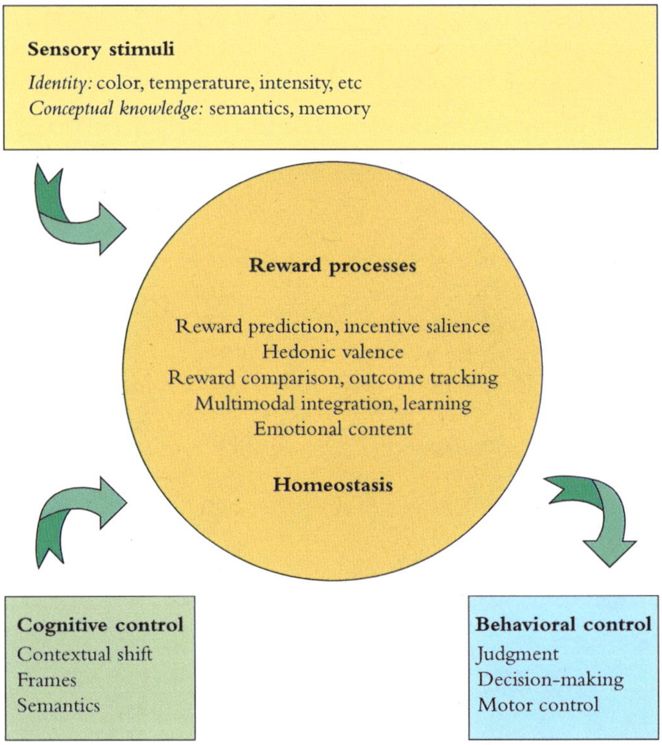

Figure 16.2 Parsing esthetic preference formation. This model relates factors known to influence esthetic preference formation to each other.

their firing pattern accordingly. However, as has been widely accepted since Immanuel Kant's popularization of the idea of *disinterestedness* in his book *Kritik der Urteilskraft*, it is not clear that works of art hold any practical interest for us. On the other hand, based on my own experiences, I would venture the guess that some forms of homeostatic states, for instance mood, do play a role in the appreciation of art. If I am depressed art may fail to interest me at all—although being down could also provoke me to seek out a quick "pick me up" in the form of a funny comedy. Being drunk and happy, and in the company of thousands of other people, at a concert surely makes music sound differently. Conversely, being melancholic or unhappy, often have the effect that I prefer some forms of music over others ("sad" music). Indeed, since art is very much a vehicle for the "artificial" inducement of emotion, as argued in this paper, it is more likely than not that the reward value of a work of art is influenced by the physiological and homeostatic state in which it is experienced. It would be fairly easy to test this hypothesis through the use of neuropharmalogical challenges.

Conclusion

Neuroimaging research on how esthetic values for works of art and other objects arise is still in its infancy. Many important questions abound; yet we are beginning to understand what neural mechanisms and structures are involved in computing esthetic preferences (Figures 16.1 and 16.2; see also the discussions in Nadal et al., 2008; Vartanian, 2009; Vartanian and Nadal, 2007; Zaidel, 2005).

The studies available today indicate that neural structures associated with the processing of reward form a core part of the EPF system. At the same time, perceptual, memory, and cognitive processes contribute some functions, at least in some situations. It would be of great interest to have neuropsychological evidence informing the nature of this diverse system. Unfortunately, there is an absolute dearth of patient studies investigating the relation between lesions and EPF in a systematic manner. The few papers I have been able to find (e.g., Adolphs and Tranel, 1999; Geroldi et al., 2000; Gosselin et al., 2006; Griffiths et al., 2004; Halpern et al., 2008; Peretz et al., 1998) are either inconclusive case studies or studies with a very limited perspective (in, for instance, only testing one stimulus modality). However, together they do tend to support the notion that structures such as striatum, amygdala, insula, and OFC are necessary for computing the affective reaction to work of art.

On the question of the relation of EPF to the processing of primary reinforcers, I take the view that, at present, it is most parsimonious to assume that EPF is rooted in more primitive reward processes. It might be an unique feature of the human brain, though, that it is able to relate primitive reward processing to a great number of different factors, including novel perceptual properties, contextual information, conflicting valence, and personality traits. Anatomical and functional properties of the OFC (Kringelbach, 2005) might be crucial to this ability. The existence of art could be seen as evidence that the reward system is more flexible and widely applicable in humans than in other animals.

Notes

1. We may even describe certain types of behavior as "artful." Think of the way we intentionally shape and ritualize body movements in interpersonal interactions (shaking hands, kissing on the cheek, waving the hand as a farewell, etc.). Moreover, we have a natural tendency to categorize people's behavior esthetically: Lisa always walks very gracefully; Tom is a klutz; Johnny has no manners. Indeed, the concept of "good manners" would probably be meaningless without this esthetic aspect of human behavior.
2. There is now also a rapidly growing literature arguing that the human propensity for art is en evolutionary adaptation (see, for different theoretical approaches, Dissanayake, 1979, 1982; Grammer et al., 2003; Miller, 2001; Thornhill, 2003). Manipulating material forms and behavior so that they elicit positive hedonic reactions can be highly rewarding in many situations of life. Improving the attractiveness of your body may secure more potential mates (Andersson and Simmons, 2006; Gangestad and Sheyd, 2005), as may the use of an erotic tone of speech in conversation (Ethofer et al., 2007). The facial appearance of politicians is known to affect voting (Little et al., 2007), just as marketing and appealing designs will sell more products, or sell them at a higher price.
3. Since I am primarily concerned with detailing a model of which brain structures that underlie EPF, I focus on PET and fMRI papers here, excluding from my discussion a number of interesting EEG and MEG studies (e.g., Cela-Conde et al., 2004). I should also note that two new papers of interest appeared after the draft for this chapter was written (Calvo-Merino et al., 2008; Di Dio, Macaluso and Rizzolatti, 2007). Therefore, I am not reviewing the results reported in these papers either.
4. It should be noted that these results were based on a fixed effect analysis where the motor cortex activity in the ugly > beautiful contrast is only driven by a subset of the subjects.
5. EEG data from Thomas Jacobsen's laboratory, not discussed here, also strongly indicate that there is a difference between passively and actively engaging works of art.

References

Adolphs, R. and Tranel, D. (1999) Preferences for visual stimuli following amygdala damage. *J. Cogn. Neurosci.* 11, 610–616.

Aharon, I., Etcoff, N., Ariely, D., Chabris, C. F., O'Connor, E. and Breiter, H. C. (2001) Beautiful faces have variable reward value: fMRI and behavioral evidence. *Neuron* 32, 537–551.

Aldridge, J. W. and Berridge, K. C. (Chapter 3, this book) Neural coding of pleasure: "Rose-tinted glasses" of the ventral pallidum. In: *Pleasures of the Brain.* Eds. M. L. Kringelbach and K. C. Berridge. Oxford University Press: Oxford.

Andersson, M. and Simmons, L. W. (2006) Sexual selection and mate choice. *Trends Ecol. Evol.* 21, 296–302.

Beardsley, M. C. (1975) *Aesthetics from Classical Greece to the Present: A Short History.* The University of Alabama Press: Tuscaloosa, AZ.

Becker, J. B., Rudick, C. N. and Jenkins, W. J. (2001) The role of dopamine in the nucleus accumbens and striatum during sexual behavior in the female rat. *J. Neurosci.* 21, 3236–3241.

Berridge, K. C. and Robinson, T. E. (1998) What is the role of dopamine in reward: Hedonic impact, reward learning, or incentive salience? *Brain Res. Brain Res. Rev.* 28, 309–369.

Blood, A. J. and Zatorre, R. J. (2001) Intensely pleasurable responses to music correlate with activity in brain regions implicated in reward and emotion. *Proc. Natl Acad. Sci. USA* 98, 11818–11823.

Blood, A. J., Zatorre, R. J., Bermudez, P. and Evans, A. C. (1999) Emotional responses to pleasant and unpleasant music correlate with activity in paralimbic brain regions. *Nat. Neurosci.* 2, 382–387.

Brown, S., Martinez, M. J. and Parsons, L. M. (2004) Passive music listening spontaneously engages limbic and paralimbic systems. *Neuroreport* 15, 2033–2037.

Calvo-Merino, B., Jola, C., Glaser, D. E. and Haggard, P. (2008) Towards a sensorimotor aesthetics of performing art. *Conscious.Cogn.*17, 911–922.

Cela-Conde, C. J., Marty, G., Maestu, F., Ortiz, T., Munar, E., Fernandez, A., Roca, M., Rossello, J. and Quesney, F. (2004) Activation of the prefrontal cortex in the human visual aesthetic perception. *Proc. Natl Acad. Sci. USA* 101, 6321–6325.

Craig, A.D. (2009) How do you feel – now? The anterior insula and human awareness. *Nat. Rev. Neurosci.* 10, 59–70.

Darwin, C. (1871) *The Descent of Man, and Selection in Relation to Sex*. John Murray: London.

de Araujo, I. E., Rolls, E. T., Velazco, M. I., Margot, C. and Cayeux, I. (2005) Cognitive modulation of olfactory processing. *Neuron* 46, 671–679.

Deppe, M., Schwindt, W., Kramer, J., Kugel, H., Plassmann, H., Kenning, P. and Ringelstein, E. B. (2005) Evidence for a neural correlate of a framing effect: Bias-specific activity in the ventromedial prefrontal cortex during credibility judgments. *Brain Res. Bull.* 67, 413–421.

Di Dio, C., Macaluso, E. and Rizzolatti, G. (2007) The golden beauty: Brain response to classical and renaissance sculptures. *PLoS ONE*. 2, e1201.

Dissanayake, E. (1979) An ethological view of ritual and art in human evolutionary history. *Leonardo* 12, 27–31.

Dissanayake, E. (1982) Aesthetic experience and human evolution. *J. Aesthetics Art Criticism* 41, 145–155.

Erk, S., Spitzer, M., Wunderlich, A. P., Galley, L. and Walter, H. (2002) Cultural objects modulate reward circuitry. *Neuroreport* 13, 2499–2503.

Ethofer, T., Wiethoff, S., Anders, S., Kreifelts, B., Grodd, W. and Wildgruber, D. (2007) The voices of seduction: Cross-gender effects in processing of erotic prosody. *Soc. Cog. Affective Neurosci.* 2, 334–337.

Fehr, E. and Camerer, C. F. (2007) Social neuroeconomics: the neural circuitry of social preferences. *Trends Cogn. Sci.* 11, 419–427.

Fulbright, R. K., Skudlarski, P., Lacadie, C. M., Warrenburg, S., Bowers, A. A., Gore, J. C. and Wexler, B. E. (1998) Functional MR imaging of regional brain responses to pleasant and unpleasant odors. *AJNR Am. J. Neuroradiol.* 19, 1721–1726.

Gangestad, S. W. and Scheyd, G. J. (2005) The evolution of human physical attractiveness. *Annu. Rev. Anthropol.* 34, 523–548.

Geroldi, C., Metitieri, T., Binetti, G., Zanetti, O., Trabucchi, M. and Frisoni, G. B. (2000) Pop music and frontotemporal dementia. *Neurology* 55, 1935–1936.

Gosselin, N., Samson, S., Adolphs, R., Noulhiane, M., Roy, M., Hasboun, D., Baulac, M. and Peretz, I. (2006) Emotional responses to unpleasant music correlates with damage to the parahippocampal cortex. *Brain* 129, 2585–2592.

Grammer, K., Fink, B., Møller, A. P. and Thornhill, R. (2003) Darwinian aesthetics: Sexual selection and the biology of beauty. *Biol. Rev. Camb. Philos. Soc.* 78, 385–407.

Griffiths, T. D., Warren, J. D., Dean, J. L. and Howard, D. (2004) "When the feeling's gone": a selective loss of musical emotion. *J. Neurol. Neurosurg. Psychiatry* 75, 344–345.

Halpern, A. R., Ly, J., Elkin-Frankston, S. and O'Connor, M. G. (2008) "I know what I like": Stability of aesthetic preference in Alzheimer's patients. *Brain Cogn.* 66, 65–72.

Hekkert, P. and van Wieringen, P. C. W. (1996) Beauty in the eye of expert and nonexpert beholders: a study in the appraisal of art. *Am. J. Psychol.* 109, 389–407.

Hutcheson, F. (1973) *An Inquiry Concerning Beauty, Order, Harmony, Design*. Springer: New York.

Ishai, A. (2007) Sex, beauty and the orbitofrontal cortex. *Int. J. Psychophysiol.* 63, 181–185.

Jacobsen, T., Buchta, K., Kohler, M. and Schroger, E. (2004) The primacy of beauty in judging the aesthetics of objects. *Psychol. Rep.* 94, 1253–1260.

Kawabata, H. and Zeki, S. (2004) Neural correlates of beauty. *J. Neurophysiol.* 91, 1699–1705.

Kirk, U., Skov, M., Christensen, M. S. and Nygaard, N. (2009a) Brain correlates of aesthetic expertise. A parametric fMRI study. *Brain and Cognition* 69, 309–315.

Kirk, U., Skov, M., Hulme, O., Christensen, M. S. and Zeki, S. (2009b) Modualtion of aesthetic value by semantic context: an fMRI study. *Neuroimage* 44, 1125–1132.

Knutson, B. and Greer, S.M. (2008) Anticipatory affect: Neural correlates and consequences for choice. *Phil. Trans. R. Soc. B* 363, 3771-3786.

Koelsch, S., Fritz, T., Dy, V. C., Muller, K. and Friederici, A. D. (2006) Investigating emotion with music: an fMRI study. *Hum. Brain Mapp.* 27, 239–250.

Kranz, F. and Ishai, A. (2006) Face perception is modulated by sexual preference. *Curr. Biol.* 16, 63–68.

Kringelbach, M. L. (2004) Food for thought: Hedonic experience beyond homeostasis in the human brain. *Neuroscience* 126, 807–819.

Kringelbach, M. L. (2005) The human orbitofrontal cortex: Linking reward to hedonic experience. *Nat. Rev. Neurosci.* 6, 691–702.

Kringelbach, M. L., O'Doherty, J., Rolls, E. T. and Andrews, C. (2003) Activation of the human orbitofrontal cortex to a liquid food stimulus is correlated with its subjective pleasantness. *Cereb. Cortex* 13, 1064–1071.

Kringelbach, M. L. (Chapter 12, this book) The hedonic brain: a functional neuroanatomy of human pleasure. In: *Pleasures of the Brain*. Eds. M. L. Kringelbach and K. C. Berridge. Oxford University Press: Oxford.

Kristeller, P. O. (1951) The modern system of the arts: a study in the history of aesthetics I. *J. History Ideas* 12, 496–527.

Kristeller, P. O. (1952) The modern system of the arts: a study in the history of aesthetics II. *J. History Ideas* 13 17–46.

Lane, R. D., Fink, G. R., Chau, P. M. and Dolan, R. J. (1997) Neural activation during selective attention to subjective emotional responses. *Neuroreport* 8, 3969–3972.

McClure, S. M., Li, J., Tomlin, D., Cypert, K. S., Montague, L. M. and Montague, P. R. (2004) Neural correlates of behavioral preference for culturally familiar drinks. *Neuron* 44, 379–387.

Menon, V. and Levitin, D. J. (2005) The rewards of music listening: Response and physiological connectivity of the mesolimbic system. *Neuroimage*. 28, 175–184.

Nadal, M., Munar, E., Capó, M. A., Rosselló, J. and Cela-Conde, C. J. (2008) Towards a framework for the study of the neural correlates of aesthetic preference. *Spatial Vision* 21, 379–396.

Nakamura, K., Kawashima, R., Nagumo, S., Ito, K., Sugiura, M., Kato, T., Nakamura, A., Hatano, K., Kubota, K., Fukuda, H. and Kojima, S. (1998) Neuroanatomical correlates of the assessment of facial attractiveness. *Neuroreport* 9, 753–757.

O'Doherty, J., Winston, J., Critchley, H., Perrett, D., Burt, D. M. and Dolan, R. J. (2003) Beauty in a smile: the role of medial orbitofrontal cortex in facial attractiveness. *Neuropsychologia* 41, 147–155.

Padoa-Schioppa, C. (2007) Orbitofrontal cortex and the compuatation of economic value. *Ann. N.Y. Acad. Sci.* 1121, 232–253.

Paulus, M. P. and Frank, L. R. (2003) Ventromedial prefrontal cortex activation is critical for preference judgments. *Neuroreport* 14, 1311–1315.

Paus, T. (2001) Primate anterior cingulate cortex: Where motor control, drive and cognition interface. *Nat. Rev. Neurosci.* 2, 417–424.

Peretz, I., Gagnon, L. and Bouchard, B. (1998) Music and emotion: Perceptual determinants, immediacy, and isolation after brain damage. *Cognition* 68, 111–141.

Plassmann, H., O'Doherty, J., Shiv, B. and Rangel, A. (2008) Marketing actions can modulate neural representations of experienced pleasantness. *Proc. Natl Acad. Sci. USA* 105, 1050–1054.

Rhodes, G. (2006) The evolutionary psychology of facial beauty. *Annu. Rev. Psychol.* 57, 199–226.

Rolls, E. T. (2005) *Emotion Explained.* Oxford University Press: Oxford.

Rolls, E. T. (2006) Brain mechanisms underlying flavour and appetite. *Philos. Trans. R. Soc. Lond B Biol. Sci.* 361, 1123–1136.

Rolls, E. T., O'Doherty, J., Kringelbach, M. L., Francis, S., Bowtell, R. and McGlone, F. (2003) Representations of pleasant and painful touch in the human orbitofrontal and cingulate cortices. *Cereb. Cortex* 13, 308–317.

Schultz, W., Tremblay, L. and Hollerman, J. R. (2000) Reward processing in primate orbitofrontal cortex and basal ganglia. *Cereb. Cortex* 10, 272–284.

Skov, M. (2009) Neuroaesthetic problems: a framework for neuroaesthetic work. In: *Neuroaesthetics* (pp. 9–26). Eds. M. Skov and O. Vartanian. Baywood Publishers: Amityville, NY.

Skov, M., Christensen, M. S., Rowe, J. B. and Paulson, O. B. (2009) When beauty is not pleasant: Aesthetic judgment of affective pictures. Under review.

Skov, M. and Vartanian, O. (2009) *Neuroaesthetics.* Baywood Publishers: Amityville, NY.

Smith, K. S., Mahler, S. V., Pecina, S. and Berridge, K. C. (Chapter 1, this book) Hedonic hotspots: Generating sensory pleasure in the brain. In: *Pleasures of the Brain.* Eds. M. Kringelbach and K. C. Berridge. Oxford University Press: Oxford.

Stolnitz, J. (1961) "Beauty": Some stages in the history of an idea. *J. History Ideas* 22, 185–204.

Swanson, L. W. (2000) Cerebral hemisphere regulation of motivated behavior. *Brain Res.* 886, 113–164.

Tatarkiewicz, W. (1972) The great theory of beauty and its decline. *J. Aesthetics Art Criticism* 31, 165–180.

Thornhill, R. (2003) Darwinian aesthetics informs traditional aesthetics. In: *Evolutionary Aesthetics* (pp. 9–35). Eds. E.Voland and K. Grammer Springer: Berlin.

Valentine, C. (1962) *The Experimental Psychology of Beauty.* Methuen & Co Ltd.: London.

Vartanian, O. and Goel, V. (2004) Neuroanatomical correlates of aesthetic preference for paintings. *Neuroreport* 15, 893–897.

Vartanian, O. and Nadal, M.,. (2007) A biological approach to a model of aesthetic experience. In: *Aesthetics and Innovation* (pp. 429-444). Eds. L. Dorfman, C. Martindale and V. Petrov. Cambridge Scholars Press: Cambridge.

Vartanian, O. (2009) Conscious experience of pleasure in art. In: *Neuroaesthetics* (pp. 261–273). Eds. M. Skov and O. Vartanian. Baywood Publishers: Amityville, NY.

Verhagen, J. V. (2007) The neurocognitive bases of human multimodal food perception: Consciousness. *Brain Res. Rev.* 53, 271–286.

Wallis, J. D. (2007) Orbitofrontal cortex and its contribution to decision-making. *Annu. Rev. Neurosci.* 30, 31–56.

Winston, J. S., O'Doherty, J., Kilner, J. M., Perrett, D. I. and Dolan, R. J. (2007) Brain systems for assessing facial attractiveness. *Neuropsychologia* 45, 195–206.

Zaidel, D. W. (2005) *The Neuropsychology of Art: Neurological, Cognitive, and Evolutionary Perspectives.* Psychology Press: Hove, UK.

Zysset, S., Huber, O., Ferstl, E. and von Cramon, D. Y. (2002) The anterior frontomedian cortex and evaluative judgment: an fMRI study. *Neuroimage* 15, 983–991.

PART III

CLINICAL APPLICATIONS

17

Placebo Analgesia and the Brain

PREDRAG PETROVIC

Although the placebo phenomenon has fascinated humans for centuries, we are just beginning to understand some of its components. Well-structured behavioral studies, combined with functional imaging methods and ideas from cognitive theories, have recently greatly improved our understanding of the underlying mechanisms. The placebo process has been studied more in relation to pain than any other sensation or experience. However, we are postulating that this is a general phenomenon where our belief system can affect any experience in the human mind. In the present chapter, I will outline the research on how placebo treatment modulates pain and relate these findings to possible underlying processes in the brain. I will also generalize these ideas to other dimensions outside the field of pain and relate them to pleasure processing.

Behavioral Studies of the Placebo Effect

Placebo treatment suggests that a nonspecific treatment without any physiological effects on the biological system still can have a measurable treatment effect. This phenomenon has fascinated people for centuries probably since there has been a close association to magic and healing. And needless to say, the placebo effect has been used and has had a great impact on humankind since thousand of years through rituals and healing herbs. But is it all just in our mind? Does the patient just believe that the placebo treatment has an effect or does it have effect? Is there a difference? While early research suggested that placebo treatment really worked, much of the used methods have been heavily criticized for not using well-controlled control groups and not taking account of natural history, that is, the possibility that everyone actually will feel better given some time (Wall, 1999). Other criticisms were that we actually only reported a regression to the mean, that it all could be explained with signal detection theory, and that patients only reported what the doctors wanted them to report (Allan and Siegel, 2002; Klosterhalfen and Enck, 2006).

The real scientific breakthrough in placebo research came when it was discovered that the placebo effect on the pain experience depended on the endogenous opioid system (Levine et al., 1978), a finding that has been replicated in several experimental studies (Amanzio and Benedetti, 1999; Benedetti et al., 1999a,b; Pollo et al., 2003). These studies showed that the placebo-induced reduction of the pain experience could be blocked when the subjects were pretreated with an opioid-receptor blocker, that is, naloxone. Not only were the studies well-controlled for the confounds described above, but for the first time, they linked neuroscience to the placebo effect. This discovery has boosted the studies on how placebo can change our experience of pain. The second major breakthrough was when it was discovered that it was possible to manipulate what people expected from a treatment (Montgomery and Kirsch, 1997; Price et al., 1999;

Voudouris et al., 1989; 1990) and that this expectation directly correlated with the magnitude of the placebo effect (Price et al., 1999). Thus, by now, two key features had been discovered in the placebo effect, that is, a specific neuromodulatory system was involved and a cognitive process could directly affect the degree of the placebo effect. Later behavioral studies on the placebo effect have tended to include larger groups and use better study design. These studies have brought the placebo effect even closer to neuroscience since they have shown that not only is experience changed by placebo treatment but more objective measurements also show a change, such as heart rate (Pollo et al., 2003) and breathing (Benedetti et al., 1999a). As for the experience also, these effects can be reversed by blocking the opioid system using naloxone. It has also been shown that there is a somatotopic aspect of the placebo effect; thus if we expect that a treatment should affect just the left leg, it will specifically change how we perceive pain in the left leg but not in any other limb (Benedetti et al., 1999b). Another phenomenon that has recently been studied is the negative placebo effect, that is, the nocebo effect. If people expect an aversive effect induced by a treatment, they will also perceive it as more aversive (Benedetti et al., 1997; Benedetti et al., 2006; Johansen et al., 2003). As an example, a painful stimulation will be perceived as more unpleasant and intense if it is expected that a treatment will increase the pain experience. The most fascinating point about the nocebo effect is that it seems to be controlled by another neuromodulatory system than the placebo effect, that is, the cholecystokinin (CCK) system (Benedetti et al., 1997; Benedetti et al., 2006). These systems are in several ways opposite. CCK activation will induce anxiety and increase pain perception, while the opioid system will induce well-being and decrease pain sensation (Hebb et al., 2005). Thus, there are two opposing neuromodulatory systems that both work in close relation to our expectations.

Recently, an important step has been taken in placebo research, that is, unravelling the underlying neuronal correlates involved in the placebo analgesia response using positron emission tomography (PET) or functional magnetic resonance imaging (fMRI) (Bingel et al., 2004; Kong et al., 2006; Lieberman et al., 2004; Petrovic et al., 2002; Wager et al., 2004; Zubieta et al., 2005). Using these functional imaging tools, we can study networks underlying specific components of the placebo response. We can describe which regions are activated in both opioid and placebo response, and which regions are specifically involved in the placebo response. We can also study specific cognitive subcomponents of the placebo response. Since we know that the opioid system is involved in the placebo response, we will start to discuss the opioid system in the human brain. But first it has to be mentioned that only few studies have focused on whether placebo actually modulates pain processing (Lieberman et al., 2004; Wager et al., 2004; Price et al 2007). Such findings are important since they strongly suggest that the placebo effect is not just a bias in the report but a real suppression of pain processing. These studies have observed a decrease in pain-related activation in thalamus, anterior cingulate cortex (ACC), insula, somatosensory cortex, that is, in large parts of the network that processes pain. However, while one study assessed chronic pain (Lieberman et al., 2004), and another study assessed induced pain in patients (Price et al 2007), only one study assessed the effects in experimental pain and found only effect in pain related activation after the noxious stimulation itself (Wager et al., 2004). Thus, more studies focusing on how placebo interacts with pain processing are needed.

Opioid System in the Brain

Opioid receptors in the central nervous system (CNS) are found in the entire neuroaxis, for example, in the cortex, subcortical structures (such as the thalamus, amygdala, and the nucleus accumbens), brainstem, and spinal cord (Atweh and Kuhar, 1983; Fields, 2004; Fields and Basbaum, 1999; Pfeiffer et al., 1982; Sadzot et al., 1991; Wamsley et al., 1982). Although these receptors are widespread throughout the CNS, their localization is highly regional. It has been proposed that networks containing opioid receptors may exert analgesic effects through several different mechanisms (Jensen, 1997) including modulation of spinal noxious input (in the dorsal horn and in the ascending pathways), direct control of cortical and brainstem structures that are involved in pain processing, or regulation of the ascending forebrain systems. At present, the modulation of the spinal cord has been best described (Fields, 2004; Fields and Basbaum, 1999). The brainstem opioid system consists of a network of regions, including the periaqueductal gray (PAG), the parabrachial nucleus, and the rostral ventromedial medulla, which are critically involved in descending opioid-dependent analgesia (Fields, 2004; Fields and Basbaum, 1999). The opioid receptor network is less characterized in the cortex, especially in the more developed human brain. However, autoradiographic studies of post-mortem human and primate brains and PET studies of opioid receptors in the human subjects have started to reveal a cortical opioid system. The autoradiographic studies indicate high concentrations of opioid receptors not only in the brainstem, for example, PAG, intralaminar, and medial thalamic nuclei, but also in the cingulate cortex and prefrontal cortex (Pfeiffer et al., 1982; Sadzot et al., 1991; Wamsley et al., 1982). These and other animal studies have suggested that the ACC has one of the highest levels of opioid receptor bindings in the cortex (Vogt et al., 1993).

PET studies using radioactive opioid [^{11}C]-diprenorphine, which indicates the mu-, delta-, and kappa-opioid receptor availability, have confirmed previous animal and human autoradiography findings (Jones et al., 1991; Willoch et al., 2004; Willoch et al., 1999). Although no quantitative analysis of the entire brain has been presented, the opioid receptor images from these PET studies suggest a high opioid receptor concentration in the insula and the frontal cortex. Raw data indicate that the binding potential is highest in the rostral parts of the anterior cingulate cortex (rACC) (Jones et al., 1991; Willoch et al., 2004; Willoch et al., 1999). The receptor imaging studies have also indicated high opioid receptor binding potential in basal ganglia and thalamus.

A similar network has showed increased neuronal activity (using PET or fMRI, and measuring the regional blood flow as an index of the underlying neuronal activity) during treatment with opioid-receptor agonists, such as remifentanil and fentany (Adler et al., 1997; Casey et al., 2000; Firestone et al., 1996; Petrovic et al., 2002; Wagner et al., 2001). In three of the studies, it was shown that remifentanil, a mu-opioid compound with a short half-life, increases the activity in several regions known to be involved in pain processing and containing a high concentrations of opioid receptors (e.g., Figures 17.1 and 17.2) (Petrovic et al., 2002; Wagner et al., 2001). Opioid-dependent increases were observed in the caudal and rostral ACC, midanterior insula, stretching into the orbitofrontal/temporopolar cortex, and in the brainstem. Common for all previously presented functional imaging studies was an observed activity increase in the rACC in response to opioid treatment. Thus, it may be suggested that a relationship exists between the opioid system and the rACC. In summary, autoradiographic studies, opioid-receptor imaging studies, and functional imaging studies all indicate that a specific opioid-rich network exists in the cortex that includes the ACC and the anterior insula.

Conditions that Activate the Endogenous Opioid System

The complexity of the endogenous opioid network indicates that it is an important regulatory system for

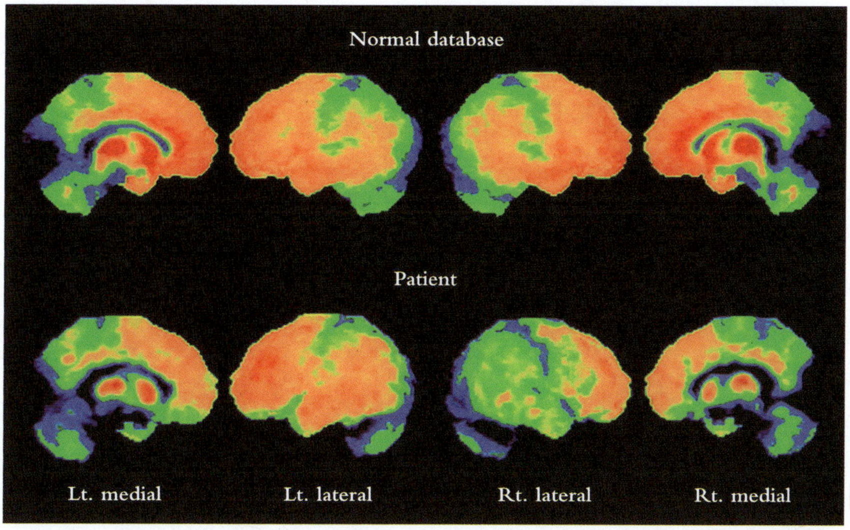

Figure 17.1 PET studies using radioactive opioid [^{11}C]-diprenorphine indicates biding potential of opioid receptors. A large binding potential (red) indicates a large availability of free receptors and therefore may indicate a high concentration of receptor. Data from Willoch et al. (1999) indicates that rostral anterior cingulate cortex and the insula (not shown in the image) have a high binding potential of opioid receptors.

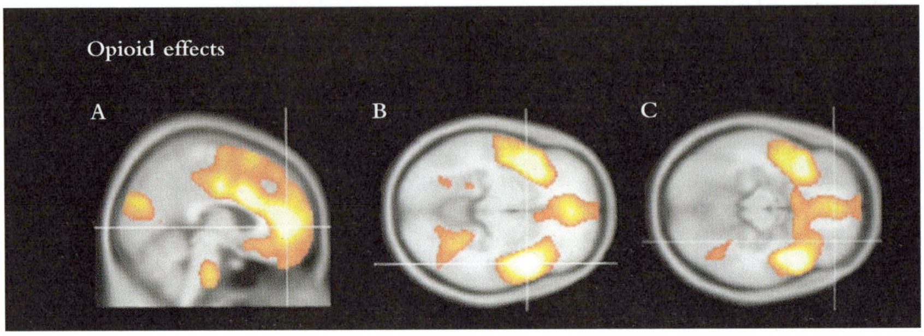

Figure 17.2 Remifentanil, a mu-selective opioid, increases the activation in cingulate cortex (including the rostral ACC (a) and insula bilaterally (b) but show no major activation in lateral orbitofrontal cortex (c), overlapping with the areas showing a large opioid receptor binding potential. This suggests that areas with large opioid concentrations in the cortex also show a strong activation when subjects are treated with opioids. Data adapted from Petrovic et al. (2002).

the organism. However, it is less elaborated in which natural situations these systems are active and their role for increasing the chances of the survival of the organism. In animal studies, contexts that induce fear and stress have been described as important for the activation of the endogenous opioid system, that is, fear- or stress-related analgesia (Fanselow, 1994). This state has been induced either using noxious shocks, conditioning a noxious shock with an neutral stimulation (i.e., conditioned stress induced analgesia), or setting up a context in which fear is thought to be induced in the animal, for example, putting an animal next to its predator (see Fanselow, 1994, for details). These triggers will activate an opioid system mediated via the amygdala, which then activates the brainstem opioid system via the PAG (Fanselow, 1994). It has been hypothesized that several contexts may be important for activating endogenous opioid systems for the induction of analgesia in humans, although a great deal of variability exists in these studies (Fields, 2004; Price, 1999a).

The placebo response is probably the best-described experimental situation in which the endogenous opioid system in humans is naturally activated and, thus, research on the placebo response is of general importance also for understanding the endogenous opioid system. Apart from the analgesic function, the opioid subsystems may be specifically involved in different motivational and mood regulations. Emotional responses in rats have been suggested to be dependent on the opioid system (Fanselow, 1994; Filliol et al., 2000). Species specific 'liking' behavior, a component of the pleasure response, is regulated by opioid receptors in the nucleus accumbens (Berridge, 1996). Even more intriguing, it has been suggested that the mu-opioid system may be involved in emotional processing and regulation in humans (Kennedy et al., 2006; Liberzon et al., 2002; Zubieta et al., 2003b). Thus, its role is not just to regulate pain perception but to modulate the general state of the organism from inducing an experience of pleasure and its associated behavior to alleviating aversive experiences such as negative emotional states and pain. This role implies that the opioid effect can be understood only in relation to the external context and the internal state (e.g., motivation and emotion).

Placebo Analgesia and the Opioid Network

Given the behavioral experiments implicating opioid mechanisms in placebo analgesia and a very specific opioid system, the next obvious step would be to understand whether the same systems are activated in the placebo response. We therefore performed a PET study in which an increased neuronal activity was observed during placebo analgesia in the same region of rostral ACC as maximally activated when the same subjects were treated with opioids (Petrovic et al., 2002) (Figure 17.3) and which has been implicated in opioid response in several other studies (Adler et al., 1997; Firestone et al., 1996; Leppa et al., 2006; Wagner et al., 2001). Placebo-dependent activation of the rACC has been replicated in several studies (Bingel et al., 2004; Kong et al., 2006; Wager et al., 2004) (Figure 17.4).

In the research by Wager et al. (2004), two placebo studies were performed while the underlying neuronal

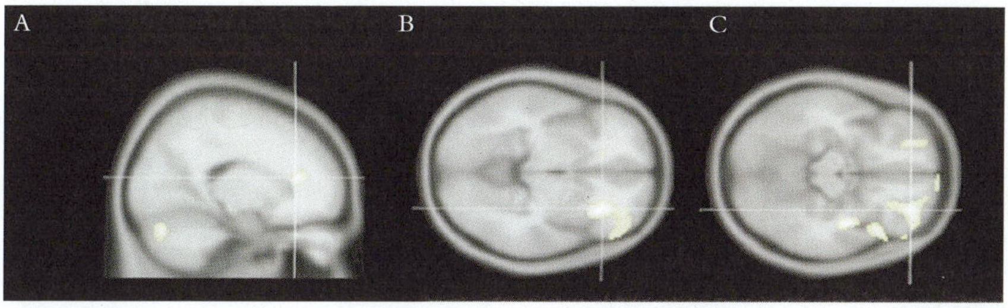

Figure 17.3 Placebo analgesia showed increased activity in the rostral anterior cingulate cortex (A) and anterior insula (B), which overlaps with opioid-rich areas, but also in the lateral orbitofrontal cortex (C), which is not as highly associated with the opioid system. Data adapted from Petrovic et al. (2002).

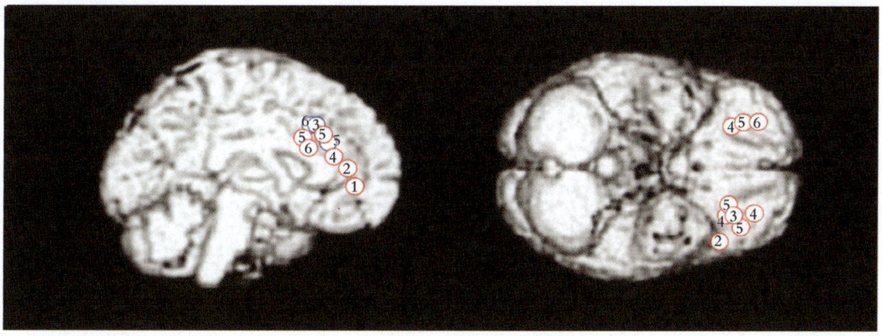

Figure 17.4 Several PET and fMRI studies have shown increased activations (red circles) in the rACC associated with the placebo response (although also decreased activations have been indicated in somewhat more caudal parts of the ACC in average—blue circles) (A). Most of these studies also show a robust activation of the lateral orbitofrontal cortex (red circles) (B) Approximate localization of the placebo dependent activations overlayed on a SPM-template (http://www.fil.ion.ucl.ac.uk/spm/). (Bingel et al., 2004 (1); Kong et al., 2006 (2); Lieberman et al., 2004 (3); Petrovic et al., 2002 (4); Wager et al., 2004 (5—study 1) and (6—study 2).

activity was studied using fMRI in both the anticipation and the induction phase of a painful stimulus. Both studies showed placebo-dependent activations of the rACC in the anticipation phase. However, these activations were not observed during the pain stimulation itself. Instead, decreased activations in the ACC were observed in the placebo pain phase. The authors suggest that when there is a preparatory phase, the placebo-dependent modulatory effect may be induced before the actual painful stimulation, leading to a decreased pain processing during the stimulation. This is a fascinating suggestion, indicating that we are able to activate the endogenous system in a preperceptive manner. The decreased activation observed somewhat later would then indicate an attenuated processing of pain unpleasantness processed in the ACC.

In the study by Lieberman et al. (2004), no increased placebo-dependent activation was observed in the rACC; instead, a negative correlation was observed between the placebo degree and the activity in mid-caudal ACC (a more posterior region of the ACC). However, in this study, the placebo effect was induced during a 2-week period and acute modulatory effects were not studied.

The two last described articles (Lieberman et al., 2004; Wager et al., 2004) point to some of the difficulties in the study of higher cognitive modulations of pain, that is, similar regions in the ACC are involved in processing perception of pain unpleasantness (possibly inducing relative decrease of ACC activity during placebo treatment) as well as in pain regulation (possibly inducing relative increases of ACC activity during placebo treatment). Although the pain regulatory activations seem to be somewhat more rostral (Petrovic and Ingvar, 2002), the two areas of activity may partly overlap and the pain-related decreases may attenuate

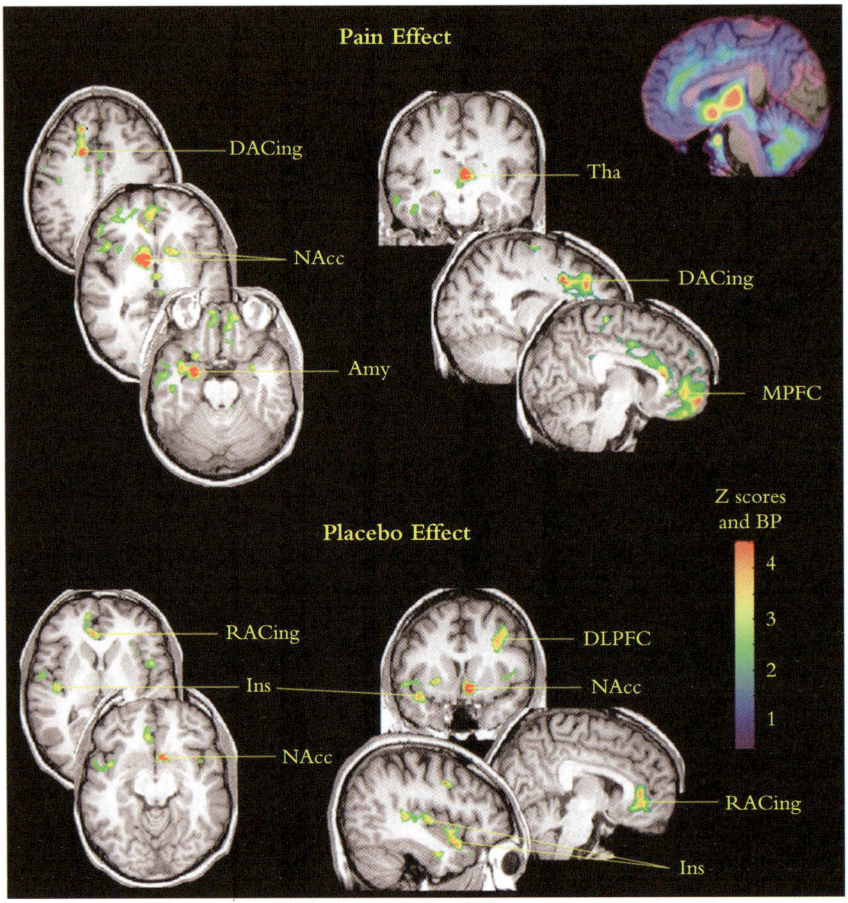

Figure 17.5 Functional receptor imaging study by Zubieta et al. (2005) shows placebo-associated decrease in opioid-binding potential (suggesting an increase in endogenous opioid activation) in rACC (RACing) and insula (Ins).

placebo-induced increase in activity. Nevertheless, five of the six presented placebo experiments (Bingel et al., 2004; Kong et al., 2006; Petrovic et al., 2002; Wager et al., 2004—study 1 and 2) indicate the activation of the rACC in placebo effect (Figure 17.4).

Overlapping activations in placebo and opioid analgesia could at best indicate in an involvement of regions containing a high opioid receptor concentration in the placebo response. However, no causal relationship between the placebo response and the opioid system in the rACC can be inferred. A more direct approach was taken by Zubieta et al. (2005) who used a dynamic version of opioid receptor PET imaging described previously (Figure 17.5). They measured the occupancy of opioid receptors in the brain by an injected radioactive opioid tracer (carfentanil) and observed that the occupancy decreased in rACC and insula in a placebo condition. This is in line with an increased endogenous occupancy of the opioid receptors in those regions. In other words, the results indicate that the placebo condition actually activated the opioid system in rACC and the insula. Although this is an important and promising study, it has a caveat in that also more intense nociceptive stimulation was induced in the placebo condition—a condition that in itself may activate the cerebral opioid system (Zubieta et al., 2001). A recent study from the same lab showed a similar placebo related opioid activation in rACC even when this bias was controlled for (Wager et al., 2007).

Placebo Responders and the Opioid System

Several studies have indicated that there is a large variability between individual subjects in the placebo

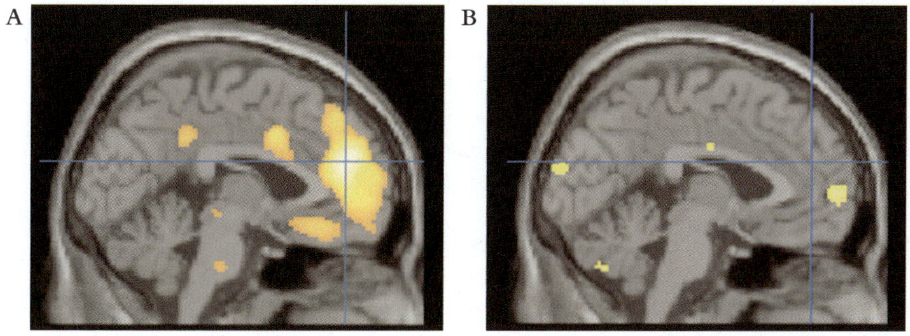

Figure 17.6 Placebo responders (A) showed a larger increase in the anterior cingulate cortex (especially the rostral part) when they were treated with the opioid Remifentanil than placebo nonresponders (B). This suggests that placebo responders may have a more effective opioid system and possibly higher concentrations of opioid receptors than nonresponders although there probably are also other factors that will determine the magnitude of the placebo response. Data adapted from Petrovic et al. (2002).

response. The reason for this is probably multifactorial. One factor that may crucially affect the placebo analgesic response is the underlying opioid system, that is, it would not be possible to induce opioid-dependent placebo analgesia in a subject who is lacking an endogenous opioid system. Data indicate that the opioid system possibly vary in different subjects (Zubieta et al., 2001). Although this finding may have different causes, genetic factors are probably highly involved. In line with this suggestion, one study has indicated that the *COMT* gene is directly involved in regulating the opioid system and the functionality of the opioid system in human subjects (Zubieta et al., 2003a). In our placebo analgesia study (Petrovic et al., 2002), we crudely divided the subjects into placebo responders and placebo nonresponders. We then compared the activations elicited by the opioids (i.e., remifentanil) and observed that the placebo responders activated the rACC whereas the nonresponders did not (Figure 17.6). This analysis was powerful because we specifically studied the opioid response (and not placebo response) in placebo responders versus nonresponders. The results are in line with a more effective opioid system in the placebo responders, possibly due to a larger concentration of opioid receptors in the rACC. However, this subject sample was very small and must be repeated in a larger sample.

Interaction Between rACC and Brainstem Opioid Systems

It has previously been proposed that the opioid system in the ACC has access to the powerful opioid system in the brainstem (Fields, 2004; Vogt et al., 1993). We suggested that this is one of the mechanisms by which the higher cognitive systems in the ACC may influence nociceptive input to the brain and thereby pain perception (Petrovic et al., 2002). We performed a regression analysis and observed that both in placebo analgesia and opioid analgesia, the rACC activity correlated with the brainstem, a finding that was not observed for the pain condition without treatment (Figure 17.7). Although no causality can be shown, the data indicate that a relation between rACC and the brainstem exists. A similar regression has been replicated in the placebo analgesia study by Bingel et al. (2004). Moreover, a similar functional connection has been shown in pain distraction (Valet et al., 2004) and prolonged tonic pain (Petrovic et al., 2004), suggesting that other pain regulatory conditions may involve ACC-mediated opioid-dependent control of the brainstem opioid system. Wager and coworkers did not supply such an analysis with the rACC but showed that also the orbitofrontal cortex has a similar functional connectivity with the brainstem (Wager et al., 2004). This supports the conclusion that both regions work in the same modulatory network.

Placebo, Cognition, and the Opioid System

The placebo effect must be accessible via higher cognitive processes because it is clearly affected by beliefs, attitudes, and conscious expectations (Price, 1999b; Wall, 1999). Nonspecific stimuli in the context, which are coded in higher cognitive networks, may both induce and modulate placebo analgesia. An impressing and well-studied higher cognitive process

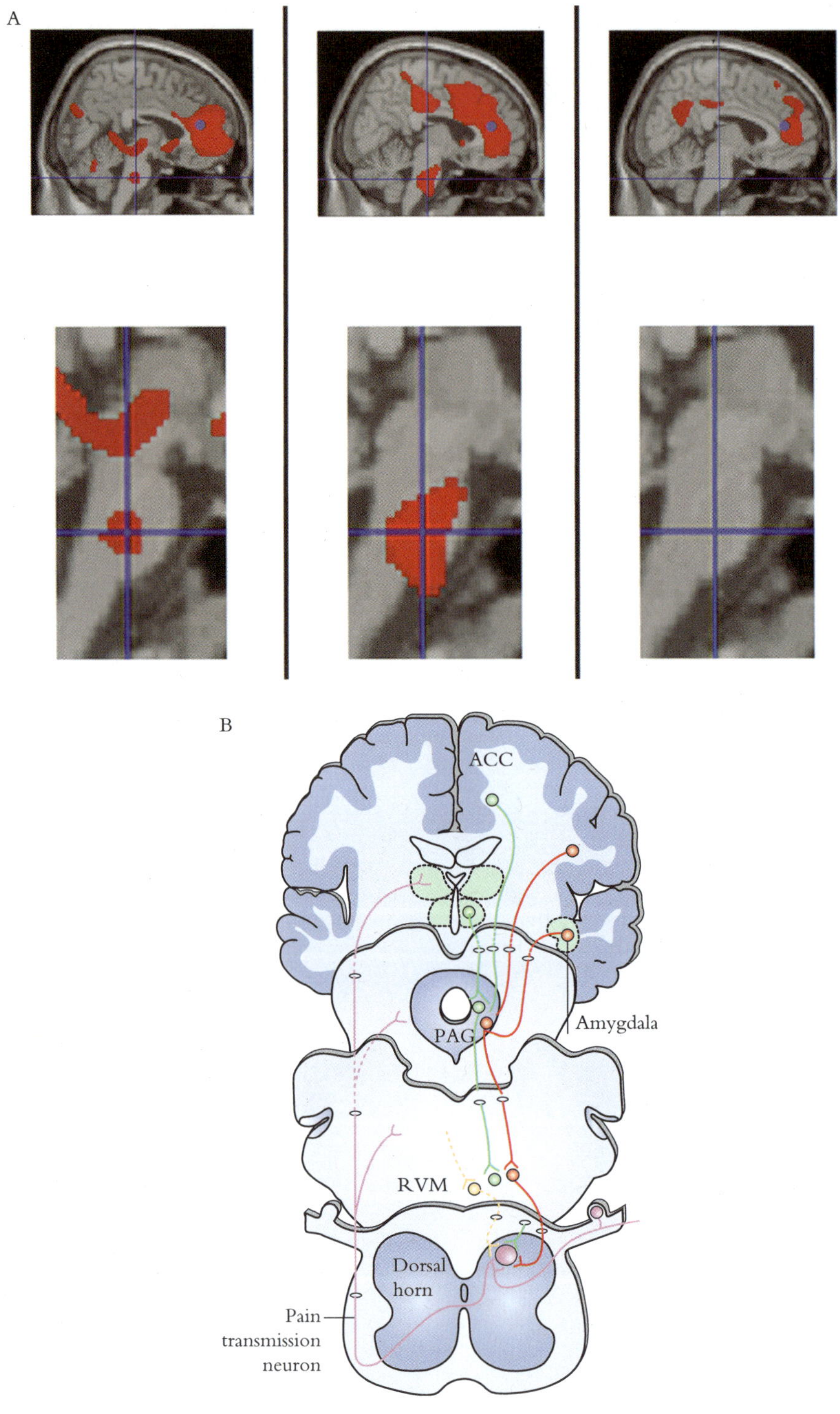

Figure 17.7 We showed that the rACC (blue dot) interacts with opioid rich regions the brainstem (including the PAG and the pons) after treatment with opioids (first column), in the placebo response (second column) but not in a control pain condition (third column) (A). These regions are rich in opioid receptors and the connection between them is in line with known descending opioid tracts (depicted in green) from the ACC to the PAG and the RVM (B). [A is adapted from Petrovic et al. (2002) and B is from Fields (2005).]

that influences placebo response is expectations of a treatment outcome (Amanzio and Benedetti, 1999; Montgomery and Kirsch, 1997; Price et al., 1999; Voudouris et al., 1989; 1990). Thus, the belief that a treatment is effective directly correlates with the degree of placebo analgesia (Price et al., 1999). Therefore, the placebo effect must also contain a cognitive process that converts the expectation to a changed pain processing induced by top-down modulation.

Apart from containing a large concentration of opioid receptors, the ACC is involved in higher cognitive attentional tasks. Modern theories of attention suggest that the ACC is involved in conflict monitoring or conflict resolution (Bush et al., 2000; Carter et al., 1999; Paus, 2001). A conflict implies that two different processes compete for the attentional space in the brain and therefore must be controlled (Stuss et al., 1995). The meta-analysis by Bush et al. (2000) showed that the ACC may be divided into a cognitive (midcaudal ACC) and an emotional region (rACC) (Figure 17.8). Moreover, the authors suggested that the rACC also is involved attentional tasks but in the emotional–motivational domain. Because pain may be viewed as a homeostatic emotion (Craig, 2003) and a therefore a conflicting emotional process competing for the attentional space, it may be suggested that the rACC is involved in attentional processing on the pain experience (Petrovic and Ingvar, 2002). If pain is a process that is controlled by cognitive mechanisms, the question is how this may happen. We have suggested that cognitive processes may directly activate opioid systems in the ACC (Petrovic and Ingvar, 2002; Petrovic et al., 2002). Using this view, the rACC is involved in the interaction between attention and the opioid system in placebo analgesia.

The ACC contains dense concentration of other receptor systems (Paus, 2001), including the CCK system known to be involved in the nocebo response (Benedetti et al., 1997, 2006). This suggests that the ACC may use various specific neuromodulatory systems to regulate pain processing in order to accomplish attentional goals. Thus, rACC may be viewed as the region in which higher cognitive attentional processes have access to a specific neuromodulatory system.

Orbitofrontal Cortex and the Placebo Effect

As stated previously, expectation of a treatment effect is an important component underlying the placebo response (Amanzio and Benedetti, 1999; Montgomery and Kirsch, 1997; Price et al., 1999; Voudouris et al., 1989, 1990) and correlates with the magnitude of analgesic effect (Price et al., 1999). Expectation is also a key aspect of the imaging studies on placebo analgesia, although these studies did not directly focus on this effect. However, the effect of expectations on pain has been directly studied in imaging experiments, which have shown that expectations of a more intense pain

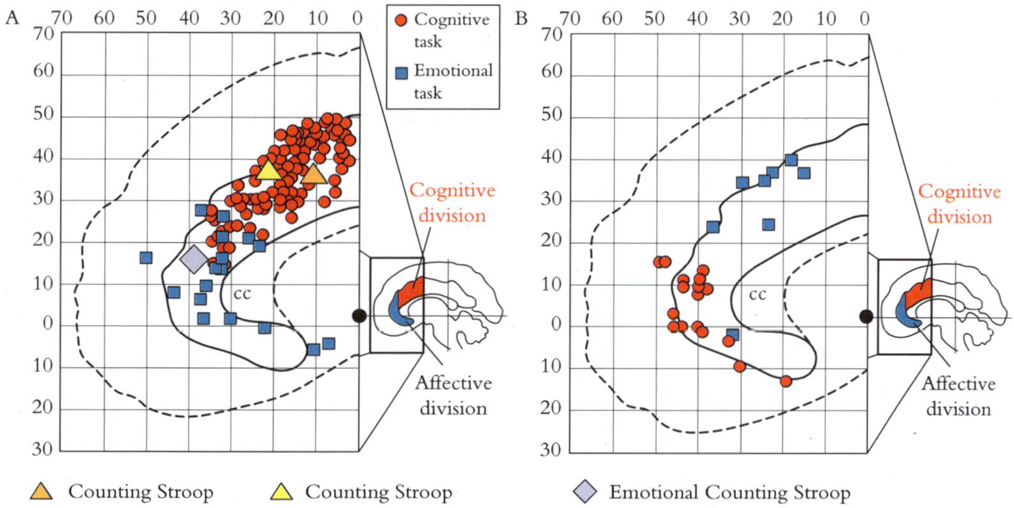

Figure 17.8 A meta-analysis by Bush et al. (2000) showed that cognitive tasks in the emotional motivational (in blue) domain activated the rACC (A) and suppressed mid-caudal ACC (B), while nonemotional tasks had the opposite effect (red).

may directly increase pain perception and pain processing in regions such as ACC and insula (Koyama et al., 2005), while expectation of a lowered pain may decrease both the perception of pain and suppress pain-induced activation in ACC (Keltner et al., 2006). It has been postulated that expectations of emotional and motivational outcomes are processed and monitored in the orbitofrontal cortex and ventrolateral prefrontal cortex (vlPFC). The Obfc may also modulate the motivational value of external stimuli based on their coupling to primary reinforcers (Gottfried et al., 2003; Kringelbach and Rolls, 2003; Morris and Dolan, 2004; O'Doherty et al., 2002; Remijnse et al., 2005; Tremblay and Schultz, 1999) to perform a goal-directed behavior (Schoenbaum and Setlow, 2001). Few functional imaging studies on the opioid effect have described extensive or any opioid-related activations of the Obfc, and thus it seems less likely to be involved directly in the opioid response. As an example, we observed placebo-dependent increases in the Obfc that were clearly more extensive than the opioid-dependent increases (Figures 17.2 and 17.3) (Petrovic et al., 2002). The opioid-induced Obfc activation was present only on the border between the temporopolar/insular activations. In fact, the activations observed in the Obfc may more represent a smoothing effect of the data.

The Obfc and vlPFC have shown reliable placebo-related activation in PET and fMRI studies. Several of the presented functional imaging experiments indicate a placebo-dependent activation of this region (Figure 17.4) with a preference for the right Obfc (Kong et al., 2006; Lieberman et al., 2004; Petrovic et al., 2002; Wager et al., 2004). Because the Obfc is involved in attributing external stimuli affective-motivational value depending on the internal states and external contexts, it is tempting to suggest that this region is involved in inducing the expectation of treatment and how the expectation should affect the pain experience. This process may precede the opioid-dependent modulation of pain processing. Such a bias signal may then be used by the ACC, which is in a position to interact with the specific neuromodulatory systems, including the endogenous opioid system. The placebo-induced activity in the Obfc also stretched into the temporopolar region/anterior insula, which is interesting because the insula has a high concentration of opioid receptors, as indicated by both receptor imaging and functional imaging studies (as mentioned previously) and also has been suggested to be involved in a meta-representation of the state of the body associated with emotional awareness (Craig, 2003).

An Integrated Network

In this chapter, I have discussed ACC, anterior insula, amygdala, and a brainstem network including the PAG, parabrachial nucleus, and the rostral ventromedial cortex. These regions are all tightly connected with each other and contain a large concentration of opioid receptors. Imaging data suggest that that they work together in an integrated fashion in order to accomplish an opioid-dependant modulation of pain processing during placebo analgesia. However, this is only a part of the mechanism involved in the placebo response as cognitive processes preceding the analgesic induction must be as important. Although the ACC is a region where attentional processes and the opioid system may interact, there seems to be other regions that are even more specifically involved in these higher order cognitive processes. One such network may be the lateral orbitofrontal cortex and the adjacent ventrolateral prefrontal cortex, which are important for motivational and goal-directed processes and where expectation is a major driving factor. Since we know that expectations are highly relevant for the placebo response, we suggest that these prefrontal regions may process those aspects preceding the placebo response. It is well known that these regions have dense anatomical connections to the ACC, allowing a driving signal to reach the ACC. In line with these ideas, we have observed a functional connection between the ventrolateral PFC and the ACC during placebo, but not during opioid treatment (Petrovic et al., 2007, unpublished data). Also, other regions such as the basal ganglia, which are involved in processing anticipatory reward signals in tight connection with prefrontal regions, may be important in the cognitive processes preceding the opioid-dependent placebo response. Thus, we are suggesting how a whole network may interact in the processing of all stages from expectation to endogenous analgesia underlying the placebo response (Figure 17.9).

Generalization of the Placebo Concept

Given that general cognitive processes such as expectations are major factors contributing to the placebo response, it may be suggested that the placebo effect represents a modulation of how any external input is processed and experienced. In other words, the placebo effect should also apply to other dimensions than pain. We directly tested this hypothesis in the domain of emotion processing using similar expectation modulations

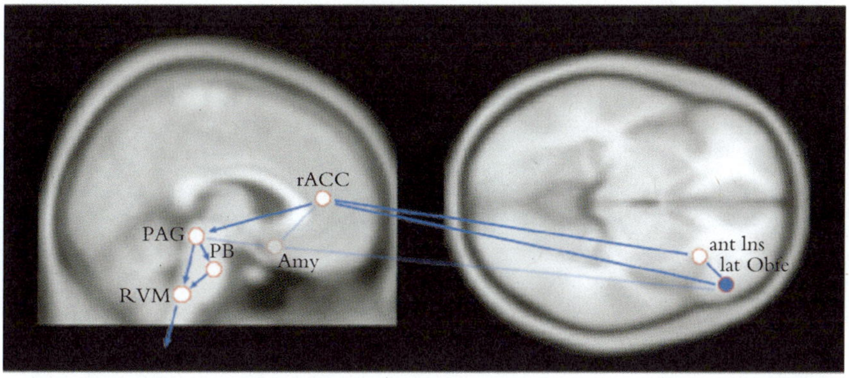

Figure 17.9 A network involved in the placebo response consists of opioid rich areas in the rACC, anterior insula, the brainstem and networks containing less opioid receptors such as the orbitofrontal cortex. [Adapted from Petrovic et al. (2005).]

as have been used previously in experiments of placebo analgesia (Petrovic et al., 2005). Subjects had to come for two testing days, and we modulated expectations on the first day and performed the placebo experiment on the second day. The subjects watched and rated how unpleasant pictures were experienced before and after a small dose of anxiolytic treatment, which was followed by an antagonist treatment that abolished the previously induced anxiolytic effect. As expected, the pictures were rated as less unpleasant after the anxiolytic treatment, but more unpleasant after the treatment with the antagonist. In this way, we were able to induce a firm expectation that we can pharmacologically manipulate how unpleasant pictures are experienced, that is, a manipulation used in placebo analgesia studies. On the second day, the subjects returned and had to watch similar pictures while the brain response was simultaneously recorded. They were also told that they would be treated with the same anxiolytic drug and antagonist. However, the subjects were only treated with saline intravenously. Nevertheless, they experienced the unpleasant pictures to be less unpleasant during the placebo treatment and showed a suppressed activity in emotional processing of these pictures. Moreover, both the rACC and the Obfc were activated in the placebo context. These brain effects correlated both with the magnitude of the behavioral placebo response and the magnitude of treatment expectation.

Thus, a similar process seems to be involved in emotional placebo as in placebo analgesia. This is not a surprising finding in the light of the high placebo response in antidepressive treatment (Walsh et al., 2002), which also seems to suppress depression-related activity in the brain (Mayberg et al., 1991). Another disorder in which the placebo effect may be even more robust and has been well studied is Parkinson's disease (PD) (de la Fuente-Fernandez et al., 2004). Expectation processes are highly associated with PD since these patients have a failure of the dopamine system, which is involved both in generating a prediction error and anticipation of an outcome in order to learn new associations between predictors and rewards (Schultz, 2004). It has been shown using both functional imaging of the dopamine system (de la Fuente-Fernandez et al., 2001) and electrophysiology (Benedetti et al., 2004; Pollo et al., 2002) that dopamine areas in the striatum and in the subthalamic nucleus response respond with a heightened activity to placebo treatment, and that these objectively measured processes correlate with the functional improvement. A receptor imaging study has shown an involvement of the dopamine system in expectations of caffeine (Kaasinen et al., 2004), suggesting that the dopamine system is normally involved in expectation processes in healthy humans, and that these effects are not specific to patient populations. These studies suggest that other specific neuromodulatory systems than the opioid system are involved also in the placebo analgesia, possibly at an earlier stage such as in the processing expectation responses, and mirrors the interaction between the opioid and the dopamine system in reward processing (Berridge, 1996).

Placebo and Pleasure

Placebo derives from the Latin word for "I will please" but does it have any relation to pleasure processes? The placebo effect has often been studied in the context of reducing an aversive effect, but not inducing pleasure. It has been debated whether emotional processes are a continuum from extremely negative (aversive) to highly positive (pleasure) experiences (Larsen, 2001; Rolls, 1995; Schimmack, 2001). If there is an emotional continuum, it may be suggested that the placebo effect is biasing the emotional state toward pleasure. In neurobiological research, there are several suggestions that this may be the case. For example, the opioid system is a key actor in both pleasure (Berridge, 1996) and the placebo effect (Price, 1999b; Wall, 1999). As have been described in detail in other chapters of this book, 'liking behavior' in relation to a reward is dependent on the opioid system in contrast to 'wanting behavior,' which is more dependant of the dopamine system. While 'liking behavior' is mediated by the opioid receptors in nucleus accumbens (Berridge, 1996), large areas in the cortex such as the ACC and the anterior insula also have abundant concentrations of opioid receptors (Willoch et al., 1999; Jones et al., 1991). Although it has not been shown how these cortical opioid networks are involved in processes underlying pleasure, they are activated in reward processing and it may well be that they underlie the experience of pleasure. In fact, a recent study has suggested that both the experience of pleasure and reward-related activity in ACC is suppressed by treatment with an opioid blocker such as Naloxone during a gambling task (Petrovic et al., 2008). Moreover, it has been suggested that the opioid activity in ACC is suppressed in negative fear states such as sadness (Kennedy et al., 2006; Liberzon et al., 2002; Zubieta et al., 2003b). Thus, these studies indicate that there is a continuum at least in the response of the opioid system from negative to positive emotional states in the ACC. Similar areas in the ACC are activated in the placebo response (Bingel et al., 2004; Kong et al., 2006; Petrovic et al., 2002; Wager et al., 2004) and this activation seems to be opioid-dependent (Kennedy et al., 2006; Liberzon et al., 2002; Zubieta et al., 2003b). This suggests that the placebo effect is interacting with processes that also are involved in processing pleasure. One of the core ideas presented in this chapter is that the placebo effect is the consequence of expectation-like processes, suggesting a general underlying mechanism. We have previously shown a similar effect in suppressing the emotional experience and processing of unpleasant pictures (Petrovic et al., 2005). But it should be equally possible to boost the experience of pleasure by modulating expectations. This crucial experiment still needs to be performed.

Conclusion

Recent imaging studies have started to disclose a functional–anatomical relationship between the opioid and placebo processes in the human brain. An emerging view suggests that a complex network in the brain underlies the placebo response (Figure 17.9), consisting of both opioid-rich regions in ACC and insula (involved in subprocesses of the placebo effect such as attentional modulation and emotional awareness) and also regions not activated by opioids such as the orbitofrontal cortex (involved in the processing of expectation). This network induces top-down control of a powerful opioid network in the brainstem. Of course, other regions also may be involved in the placebo process, for example, the dorsolateral prefrontal cortex is probably involved in short-term memory, holding the nonemotional context supporting the placebo response online. Another region possibly belonging to this network is the amygdala because it is involved in a similar opioid-dependent control of the brainstem in the fear response. In the near future, displacement studies of the opioid receptor system will further show whether the opioid network may be directly involved in the placebo response. Moreover, a combination between genetic and imaging studies may indicate whether different genotypes affecting the opioid system also affects the opioid phenotype and thereby the placebo response.

References

Adler, L. J., Gyulai, F. E., Diehl, D. J., Mintun, M. A., Winter, P. M. and Firestone, L. L. (1997) Regional brain activity changes associated with fentanyl analgesia elucidated by positron emission tomography [published erratum appears in *Anesth. Analg.* 1997 May;84(5):949]. *Anesthesia Analgesia* 84, 120–126.

Allan, L. G. and Siegel, S. (2002) A signal detection theory analysis of the placebo effect. *Eval. Health Prof.* 25, 410–420.

Amanzio, M. and Benedetti, F. (1999) Neuropharmacological dissection of placebo analgesia: Expectation-activated opioid systems versus conditioning-activated specific subsystems. *J. Neurosci.* 19, 484–494.

Atweh, S. F. and Kuhar, M. J. (1983) Distribution and physiological significance of opioid receptors in the brain [review]. *Br. Med. Bull.* 39, 47–52.

Benedetti, F., Amanzio, M., Casadio, C., Oliaro, A. and Maggi, G. (1997) Blockade of nocebo hyperalgesia by the cholecystokinin antagonist proglumide. *Pain* 71, 135–140.

Benedetti, F., Amanzio, M., Baldi, S., Casadio, C. and Maggi, G. (1999a) Inducing placebo respiratory depressant responses in humans via opioid receptors. *Eur. J. Neurosci.* 11, 625–631.

Benedetti, F., Arduino, C. and Amanzio, M. (1999b) Somatotopic activation of opioid systems by target-directed expectations of analgesia. *J. Neurosci.* 19, 3639–3648.

Benedetti, F., Colloca, L., Torre, E., Lanotte, M., Melcarne, A., Pesare, M., Bergamasco, B. and Lopiano, L. (2004) Placebo-responsive Parkinson patients show decreased activity in single neurons of subthalamic nucleus. *Nat. Neurosci.* 7, 587–588.

Benedetti, F., Amanzio, M., Vighetti, S. and Asteggiano, G. (2006) The biochemical and neuroendocrine bases of the hyperalgesic nocebo effect. *J. Neurosci.* 26, 12014–12122.

Berridge, K. C. (1996) Food reward: Brain substrates of wanting and liking. *Neurosci. Biobehav. Rev.* 20, 1–25.

Bingel, U., Lorenz, J., Gläscher, J., Weiller, C. and Büchel, C. (2004) Cognitive control of pain: a common system for pain expectation and placebo-analgesia—a single trial fMRI study. *Neuroimage* 22, 202.

Bush, G., Luu, P. and Posner, M. I. (2000) Cognitive and emotional influences in anterior cingulate cortex. *Trends Cogn. Sci.* 4, 215–222.

Carter, C. S., Botvinick, M. M. and Cohen, J. D. (1999) The contribution of the anterior cingulate cortex to executive processes in cognition [review]. *Rev. Neurosci.* 10, 49–57.

Casey, K. L., Svensson, P., Morrow, T. J., Raz, J., Jone, C. and Minoshima, S. (2000) Selective opiate modulation of nociceptive processing in the human brain. *J. Neurophysiol.* 84, 525–533.

Craig, A. D. (2003) A new view of pain as a homeostatic emotion. *Trends Neurosci.* 26, 303–307.

de la Fuente-Fernandez, R., Ruth, T. J., Sossi, V., Schulzer, M., Calne, D. B. and Stoessl, A. J. (2001) Expectation and dopamine release: Mechanism of the placebo effect in Parkinson's disease. *Science* 293, 1164–1166.

de la Fuente-Fernandez, R., Schulzer, M. and Stoessl, A. J. (2004) Placebo mechanisms and reward circuitry: Clues from Parkinson's disease. *Biol. Psychiatry* 56, 67–71.

Fanselow, M. S. (1994) Neural organization of the defensive behavior system responsible for fear. *Psychonom. Bull. Rev.* 1, 429–438.

Fields, H. (2004) State-dependent opioid control of pain. *Nat. Rev. Neurosci.* 5, 565–575.

Fields, H. and Basbaum, A. (1999) Central nervous system mechanisms of pain modulation. In: *Textbook of Pain*, pp. 309–329. Eds. P. Wall and R. Melzack. Churchill Livingstone: London.

Filliol, D., Ghozland, S., Chluba, J., Martin, M., Matthes, H. W., Simonin, F., Befort, K., Gaveriaux-Ruff, C., Dierich, A., LeMeur, M., Valverde, O., Maldonado, R. and Kieffer, B. L. (2000) Mice deficient for delta- and mu-opioid receptors exhibit opposing alterations of emotional responses. *Nat. Genet.* 25, 195–200.

Firestone, L. L., Gyulai, F., Mintun, M., Adler, L. J., Urso, K. and Winter, P. M. (1996) Human brain activity response to fentanyl imaged by positron emission tomography. *Anesth. Analgesia* 82, 1247–1251.

Gottfried, J. A., O'Doherty, J. and Dolan, R. J. (2003) Encoding predictive reward value in human amygdala and orbitofrontal cortex. *Science* 301, 1104–1107.

Hebb, A. L., Poulin, J. F., Roach, S. P., Zacharko, R. M. and Drolet, G. (2005) Cholecystokinin and endogenous opioid peptides: Interactive influence on pain, cognition, and emotion. *Prog. Neuropsychopharmacol. Biol. Psychiatry* 29, 1225–1238.

Jensen, T. S. (1997) Opioids in the brain: Supraspinal mechanisms in pain control. [Review] [61 refs]. *Act. Anaesthesiolog. Scand.* 41, 123–132.

Johansen, O., Brox, J. and Flaten, M. A. (2003) Placebo and Nocebo responses, cortisol, and circulating beta-endorphin. *Psychosom. Med.* 65, 786–790.

Jones, A. K., Qi, L. Y., Fujirawa, T., Luthra, S. K., Ashburner, J., Bloomfield, P., Cunningham, V. J., Itoh, M., Fukuda, H. and Jones, T. (1991) In vivo distribution of opioid receptors in man in relation to the cortical projections of the medial and lateral pain systems measured with positron emission tomography. *Neurosci. Lett.* 126, 25–28.

Kaasinen, V., Aalto, S., Nagren, K. and Rinne, J. O. (2004) Expectation of caffeine induces dopaminergic responses in humans. *Eur. J. Neurosci.* 19, 2352–2356.

Keltner, J.R., Furst, A,. Fan, C., Redfern, R., Inglis, B., and Fields. H.L. (2006) Isolating the modulatory effect of expectation on pain transmission: a functional magnetic resonance imaging study. *J. Neurosci.* 26(16), 4437–4443.

Kennedy, S. E., Koeppe, R. A., Young, E. A. and Zubieta, J. K. (2006) Dysregulation of endogenous opioid emotion regulation circuitry in major depression in women. *Arch. Gen. Psychiatry* 63, 1199–208.

Klosterhalfen, S. and Enck, P. (2006) Psychobiology of the placebo response. *Auton. Neurosci.* 125, 94–99.

Koyama, T., McHaffie, J. G., Laurienti, P. J., and Coghill, R. C. (2005) The subjective experience of pain: Where expectations become reality. *Proc Natl Acad Sci U S A* 102(36), 12950–129505. Epub 2005 Sep 6.

Kong, J., Gollub, R. L., Rosman, I. S., Webb, J. M., Vangel, M. G., Kirsch, I. and Kaptchuk, T. J. (2006) Brain activity associated with expectancy-enhanced placebo analgesia as measured by functional magnetic resonance imaging. *J. Neurosci.* 26, 381–388.

Kringelbach, M. L. and Rolls, E. T. (2003) Neural correlates of rapid reversal learning in a simple model of human social interaction. *Neuroimage* 20, 1371–1383.

Larsen, J. T., McGraw, A. P., Cacioppo, J. T. (2001) Can people feel happy and sad at the same time? *J. Pers. Soc. Psychol.* 81, 684–696.

Leppa, M., Korvenoja, A., Carlson, S., Timonen, P., Martinkauppi, S., Ahonen, J., Rosenberg, P. H., Aronen, H. J. and Kalso, E. (2006) Acute opioid effects on human brain as revealed by functional magnetic resonance imaging. *Neuroimage* 31, 661–669.

Levine, J. D., Gordon, N. C. and Fields, H. L. (1978) The mechanism of placebo analgesia. *Lancet* 2, 654–657.

Liberzon, I., Zubieta, J. K., Fig, L. M., Phan, K. L., Koeppe, R. A. and Taylor, S. F. (2002) mu-Opioid receptors and limbic responses to aversive emotional stimuli. *Proc. Natl Acad. Sci. U S A* 99, 7084–7089.

Lieberman, M. D., Jarcho, J. M., Berman, S., Naliboff, B. D., Suyenobu, B. Y., Mandelkern, M. and Mayer, E. A. (2004) The neural correlates of placebo effects: a disruption account. *Neuroimage* 22, 447–455.

Mayberg, H. S., Parikh, R. M., Morris, P. L. and Robinson, R. G. (1991) Spontaneous remission of post-stroke depression and temporal changes in cortical S2-serotonin receptors. *J. Neuropsychiatry Clin. Neurosci.* 3, 80–3.

Montgomery, G. H. and Kirsch, I. (1997) Classical conditioning and the placebo effect. *Pain* 72, 107–113.

Morris, J. S. and Dolan, R. J. (2004) Dissociable amygdala and orbitofrontal responses during reversal fear conditioning. *Neuroimage* 22, 372–380.

O'Doherty, J. P., Deichmann, R., Critchley, H. D. and Dolan, R. J. (2002) Neural responses during anticipation of a primary taste reward. *Neuron* 33, 815–826.

Paus, T. (2001) Primate anterior cingulate cortex: Where motor control, drive and cognition interface. *Nat. Rev. Neurosci.* 2, 417–424.

Petrovic, P. and Ingvar, M. (2002) Topical review: Imaging cognitive modulation of pain processing. *Pain* 95, 1–5.

Petrovic, P., Kalso, E., Petersson, K. M. and Ingvar, M. (2002) Placebo and opioid analgesia—imagine a shared neuronal network. *Science* 295, 1737–1740.

Petrovic, P., Petersson, K. M., Hansson, P. and Ingvar, M. (2004) Brainstem involvement in the initial response to pain. *Neuroimage* 22(2), 995–1005.

Petrovic, P., Dietrich, T., Fransson, P., Andersson, J., Carlsson, K. and Ingvar, M. (2005) Placebo in emotional processing – induced expectations of anxiety relief activate a generalized modulatory network. *Neuron* 46, 957–969.

Petrovic, P., Pleger, B., Seymour, B., Klöppel, S., De Martino, B., Critchley, H., and Dolan, R. J. (2008) Blocking central opiate function modulates hedonic impact and anterior cingulate response to rewards and losses. *J Neurosci* 28(42), 10509–10516.

Pfeiffer, A., Pasi, A., Mehraein, P. and Herz, A. (1982) Opiate receptor binding sites in human brain. *Brain Res.* 248, 87–96.

Pollo, A., Torre, E., Lopiano, L., Rizzone, M., Lanotte, M., Cavanna, A., Bergamasco, B. and Benedetti, F. (2002) Expectation modulates the response to subthalamic nucleus stimulation in Parkinsonian patients. *Neuroreport* 13, 1383–1386.

Pollo, A., Vighetti, S., Rainero, I. and Benedetti, F. (2003) Placebo analgesia and the heart. *Pain* 102, 125–133.

Price, D. (1999a) General analgesia of pain modulation. In: *Psychological Mechanisms of Pain and Analgesia*, pp. 137–153. Ed. D. Price. IASP Press: Seattle, WA.

Price, D. (1999b) Placebo analgesia. In: *Psychological Mechanisms of Pain and Analgesia*, pp. 155–181. Ed. D. Price. IASP Press: Seattle, WA.

Price, D. D., Milling, L. S., Kirsch, I., Duf, f. A., Montgomery, G. H. and Nicholls, S. S. (1999) An analysis of factors that contribute to the magnitude of placebo analgesia in an experimental paradigm. *Pain* 83, 147–156

Price, D. D, Craggs J., Verne G. N., Perlstein W. M., Robinson M. E (2007). Placebo analgesia is accompanied by large reductions in pain-related brain activity in irritable bowel syndrome patients. *Pain.* 127(1–2), 63–72.

Remijnse, P. L., Nielen, M. M., Uylings, H. B. and Veltman, D. J. (2005) Neural correlates of a reversal learning task with an affectively neutral baseline: an event-related fMRI study. *Neuroimage* 26, 609–618.

Rolls, E. (1995) A theory of emotion and consciousness, and its application to understandning the neural basis of emotion. In: *The Cognitive Neurosciences: A Bradford Book*,. Editor in chief, M.S. Gazzaniga. MIT Press. Cambridge, MA, London.

Sadzot, B., Price, J. C., Mayberg, H. S., Douglass, K. H., Dannals, R. F., Lever, J. R., Ravert, H. T., Wilson, A. A., Wagner, H. N., Jr. and Feldman, M. A. (1991) Quantification of human opiate receptor concentration and affinity using high and low specific activity and 2 over black square]; [1 and 2 over black square]1C]diprenorphine and positron emission tomography. [erratum appears in *J. Cereb. Blood Flow Metab.* 1992 Sep;12(5):885]. *J. Cereb. Blood Flow Metabolism* 11, 204–219.

Schimmack, U. (2001) Pleasure, displeasure, and mixed feelings: Are semantic opposites mutually exclusive? *Cognit. Emotion* 15, 81–97.

Schoenbaum, G. and Setlow, B. (2001) Integrating orbitofrontal cortex into prefrontal theory: Common processing themes across species and subdivisions. *Learning Memory* 8, 134–147.

Schultz, W. (2004) Neural coding of basic reward terms of animal learning theory, game theory, microeconomics and behavioural ecology. *Curr. Opin. Neurobiol.* 14, 139–147.

Stuss, D. T., Shallice, T., Alexander, M. P. and Picton, T. W. (1995) A multidisciplinary approach to anterior attentional functions. [Review] [87 refs]. *Ann. NY Acad. Sci.* 769, 191–211.

Tremblay, L. and Schultz, W. (1999) Relative reward preference in primate orbitofrontal cortex [see comments]. *Nature* 398, 704–708.

Valet, M., Sprenger, T., Boecker, H., Willoch, F., Rummeny, E., Conrad, B., Erhard, P. and Tolle, T. R. (2004) Distraction modulates connectivity of the cingulo-frontal cortex and the midbrain during pain—an fMRI analysis. *Pain* 109, 399–408.

Vogt, B. A., Sikes, R. W. and Vogt, L. J. (1993) Anterior cingulate cortex and the medial pain system. In: *Neurobiology of Cingulate Cortex and Limbic Thalamus: A Comprehensive Handbook.*, pp. 313–344. Eds. B. A. Vogt and M. Gabriel. Birkhäuser: Boston, MA.

Voudouris, N. J., Peck, C. L. and Coleman, G. (1989) Conditioned response models of placebo phenomena: Further support. *Pain* 38, 109–116.

Voudouris, N. J., Peck, C. L. and Coleman, G. (1990) The role of conditioning and verbal expectancy in the placebo response. *Pain* 43, 121–128.

Wager, T. D., Rilling, J. K., Smith, E. E., Sokolik, A., Casey, K. L., Davidson, R. J., Kosslyn, S. M., Rose, R. M. and Cohen, J. D. (2004) Placebo-induced changes in FMRI in the anticipation and experience of pain. *Science* 303, 1162–1167.

Wager, T. D., Scott, D. J. and Zubieta, J. K. (2007) Placebo effects on human mu-opioid activity during pain. *Proc. Natl. Acad. Sci. U S A* 104(26), 11056–11061.

Wagner, K. J., Willoch, F., Kochs, E. F., Siessmeier, T. and Tölle, T. R. (2001) Dose-dependent regional cerebral blood flow changes during remifentanil infusion in humans. A positron emission tomography study. Anesthesiology 94, 732–739.

Wall, P. (1999) The placebo and the placebo response. In: *Textbook of Pain*, pp. 1419–1430. Eds. P. Wall and R. Melzack. Churchill Livingstone: London.

Walsh, B. T., Seidman, S. N., Sysko, R. and Gould, M. (2002) Placebo response in studies of major depression: Variable, substantial, and growing. *Jama* 287, 1840–1847.

Wamsley, J. K., Zarbin, M. A., Young, W. S., 3rd and Kuhar, M. J. (1982) Distribution of opiate receptors in the monkey brain: an autoradiographic study. *Neuroscience* 7, 595–613.

Willoch, F., Schindler, F., Wester, H. J., Empl, M., Straube, A., Schwaiger, M., Conrad, B. and Tolle, T. R. (2004) Central poststroke pain and reduced opioid receptor binding within pain processing circuitries: a [^{11}C] diprenorphine PET study. *Pain* 108, 213–220.

Willoch, F., Tolle, T. R., Wester, H. J., Munz, F., Petzold, A., Schwaiger, M., Conrad, B. and Bartenstein, P. (1999) Central pain after pontine infarction is associated with changes in opioid receptor binding: a PET study with 11C-diprenorphine. *AJNR: Am. J. Neuroradiol.* 20, 686–690.

Zubieta, J.-K., Smith, Y. R., Bueller, J. A., Xu, Y., Kilbourn, M. R., Jewett, D. M., Meyer, C. R., Koeppe, R. A. and Stohler, C. S. (2001) Regional Mu opioid receptor regulation of sensory and affective dimensions of pain. *Science* 293, 311–315.

Zubieta, J. K., Heitzeg, M. M., Smith, Y. R., Bueller, J. A., Xu, K., Xu, Y., Koeppe, R. A., Stohler, C. S. and Goldman, D. (2003a) COMT val158met genotype affects mu-opioid neurotransmitter responses to a pain stressor. *Science* 299, 1240–1243.

Zubieta, J. K., Ketter, T. A., Bueller, J. A., Xu, Y., Kilbourn, M. R., Young, E. A. and Koeppe, R. A. (2003b) Regulation of human affective responses by anterior cingulate and limbic mu-opioid neurotransmission. *Arch. Gen. Psychiatry* 60, 1145–1153.

Zubieta, J. K., Bueller, J. A., Jackson, L. R., Scott, D. J., Xu, Y., Koeppe, R. A., Nichols, T. E. and Stohler, C. S. (2005) Placebo effects mediated by endogenous opioid activity on mu-opioid receptors. *J. Neurosci.* 25, 7754–7762.

18

Deep Brain Stimulation and Pleasure

ALEXANDER L. GREEN, ERLICK A. C. PEREIRA, AND TIPU Z. AZIZ

I dreamt of you last night as operating on Hughlings Jackson. The great principle you said in cerebral surgery is to create a commotion by which the association paths are restored. You took off the scalp—like a p. m. incision—made a big hole over the cerebellum and put in a Christ-Church whipped-cream wooden instrument and rotated it rapidly. Then put back the bone & sewed him up. H.J. seemed very comfortable after the operation and bought three oranges from a small Neapolitan who strolled into the Queen-Square amphitheatre! I have been studying my dreams lately and have come to the conclusion that just one third of my time is spent in an asylum—or should be!

Osler to Cushing (April 1911 in Cushing, 1925)

Ethics and Orgasms

A century on from Sir William Osler's conviction that neurosurgery on the otherwise well patient constitutes insanity, electrical stimulation of the human nervous system to elicit pleasure has begun to emerge as a distinct possibility, concomitant with the concept of neuroenhancement—the science of enhancing an undiseased nervous system—and the establishment of neuroethics as an academic discipline (Illes and Raffin, 2002; Moreno, 2003).

Media and the arts have long played upon themes of neuroenhancement to manufacture bionic humans from Yul Brynner's gunslinger in *Westworld*, *Robocop*, and *Universal Soldier* to Arnold Schwarzenneger's *Terminator.* Considering pleasure, there has been a cinematic infatuation with the notion of the "Orgasmatron," a device that can deliver orgasms first depicted in the fictional future society of Woody Allen's 1973 film *Sleeper* and later in films such as *Flesh Gordon* (1974) and the 1993 films *Coneheads* and *Demolition Man*.

Life recently imitated art with the use of an electrical stimulator implanted over the theca containing the sacral spinal nerves being marketed in the United States not only as a treatment for the putative disorder of female anorgasmia, but tentatively suggested as a means to enhance orgasms (Meloy, 2007). In contrast, deep brain stimulation (DBS) is a treatment involving brain surgery and therefore not without risk, but one which has been extensively applied to movement disorders, psychiatric disorders, and chronic pain refractory to medicines and a neurosurgical therapy whose relationship to pleasure we explore here.

Despite the speculative insights gained from DBS for Tourette's syndrome into alterations of libido and the control of erections (Temel et al., 2004), we do not propose DBS as a cure for anorgasmia, whether male or female. However, its relief of debilitating chronic pain and restoration of quality of life in movement disorders give pleasure to patient and clinician alike. Furthermore DBS uniquely enables the invasive electrophysiological study of conscious humans in various states of arousal and thus holds much promise to shed light on profound philosophical questions, in particular those of subjective qualia—what David Chalmers described as the "hard problem of consciousness" and Thomas Nagel the sense of "what it is like to be" (Chalmers, 1998; Nagel, 1986). Finally, as Morten Kringelbach highlights elsewhere in this book, the therapeutic intervention of DBS enables causal inference to be made from the "neuromodulation" of a programmable and reversible deep brain–targeted intervention from which we may learn much about the neural mechanisms of subjective emotions like pleasure.

Indications

DBS is neurosurgery that enables deep brain structures to be stimulated electrically by a pacemaker implanted under the skin. It has been popularized for Parkinson's disease (PD) that is unresponsive to drugs or showing side effects from them. Its efficacy has been demonstrated robustly by clinical trials with multiple novel brain targets having been discovered recently. Many other indications for DBS now exist such as tremor and dystonia in movement disorders; psychiatric disorders like obsessive-compulsive disorder (OCD), depression and Tourette's syndrome; cluster headache, epilepsy, and chronic pain including amputation, stroke, trigeminal neuralgia, and multiple sclerosis (Pereira et al., 2007b). Novel indications of orthostatic hypotension and hypertension also show experimental promise that relates DBS to autonomic control and fight, flight, fear, and pleasure (see later). Table 18.1 estimates numbers of patients worldwide treated by DBS for each indication.

Parkinson's Disease

PD is a slowly progressive, neurodegenerative disease characterized by tremor, rigidity, bradykinesia, and postural instability. It is common in middle or late life with prevalence rising to 1% in people older than 60 years. Established basal ganglia brain structures currently targeted for PD DBS include the globus pallidus interna (GPi), ventralis intermedius nucleus of the thalamus (VIm), and subthalamic nucleus (STN); over 30,000 patients having been implanted to date (Pereira and Aziz, 2006b).

GPi has traditionally been targeted mainly for dyskinesia symptoms, STN for levodopa refractory patients, and VIm for tremor. Despite its smaller size, the STN recently gained dominance over GPi as the surgical target of choice for PD due to reports of favorable motor outcomes (Kleiner-Fisman et al., 2006). A 156-patient randomized, controlled, multicenter trial of STN DBS versus medical treatment alone showed a 25% benefit in motor function and 22% improvement in quality-of-life outcomes at 6 months after surgery (Deuschl et al., 2006b). Sustained benefit with STN DBS has also been described after 5 years of follow-up (Krack et al., 2003; Schupbach et al., 2005). GPi and STN have been compared at 4-year follow-up (Rodriguez-Oroz et al., 2005), however long-term, back to back, randomized, blinded, controlled trials of the two surgical targets are yet to be completed (Okun and Foote, 2005).

The pedunculopontine nucleus (PPN) has been discovered in the last decade as a deep brain target, stimulation of which reduces gait abnormalities and postural instability (Jenkinson et al., 2004). Like the STN, its clinical utility has been realized by nonhuman primate research (Nandi et al., 2002a; Pereira and Aziz, 2006a). Initial results favor its use in PD patients blighted most by akinesia and postural instability, in PD-plus syndromes of multiple system atrophy and progressive supranuclear palsy, and in those with symptoms not ameliorated by STN stimulation alone (Stefani et al., 2007).

Tremor

Tremor is the involuntary, rhythmic oscillation of a body part. Essential tremor prevalence varies greatly throughout the world and can be up to 2%. DBS can alleviate contralateral limb tremor in essential tremor, Holmes' tremor, cerebellar tremor, tremulous multiple sclerosis, and tremor after head injury (Deuschl and Bain, 2002). For trunk, head, and voice tremors, bilateral DBS is considered (Taha et al., 1999). Brain targets considered in patients refractory to medication are the VIm and the zona incerta (ZI).

Sustained and consistent motor improvements with VIm DBS have been shown 6 years after surgery in 19 patients with essential tremor (Sydow et al., 2003). Quality-of-life improvements have also been demonstrated in 40 patients 1 year after surgery (Fields et al., 2003). In multiple sclerosis, patient selection is paramount (Alusi et al., 2001). Distal limb tremor responds best to VIm DBS and proximal limb tremor to ZI DBS (Nandi et al., 2002b). Postoperative benefits in motor function for 88% of patients and in daily functioning for 76% have been shown in a systematic review of 75 multiple sclerosis patients (Wishart et al., 2003). Brain targets in DBS for head injury depend upon the prevailing movement disorder with excellent results described in the small numbers of cases reported (Krauss and Jankovic, 2002).

Dystonia

Dystonias are disorders of involuntary sustained muscle contractions that can affect certain body regions or be generalized. They may begin in childhood or young adulthood, often progressing from focal limb involvement to a severe generalized form, or manifest in later adulthood when they are usually focal or segmental and frequently craniocervical (spasmodic torticollis). Prevalence of early onset dystonias is up to 50 per million with a greater, up to 0.01%, prevalence of the late-onset type. DBS is considered for children refractory to medical therapy, usually by anticholinergic, dopaminergic, or benzodiazepine treatments, and adults refractory to botulinum toxin injections.

Stimulation of the posteroventral GPi is performed for primary dystonias (Krauss et al., 2004). GPi DBS is particularly effective in childhood dystonias (Parr et al., 2007), and in those patients carrying a mutation in the *DYT1* gene (Coubes et al., 2004). Secondary dystonias are less responsive (Eltahawy et al., 2004). Moderate benefits have also been observed with VIm but not STN DBS (Detante et al., 2004). Motor improvements are often not fully realized until weeks to months later (Yianni et al., 2003). Sustained motor and quality-of-life improvements without cognitive impairment have been shown 3 months after surgery in a prospective, multicenter trial of 40 patients (Kupsch et al., 2006), and 3 years after surgery in 58% of patients in a trial of 22 patients (Vidailhet et al., 2007).

Epilepsy

Epilepsy is a debilitating neurological condition affecting 50 per 100,000 people with higher prevalence in children and the elderly. Symptomatic epilepsy is estimated to reduce life expectancy by up to two decades. Sudden death in medically refractory epilepsy is 0.5% and highest in adolescents and young adults. Neurosurgical treatment is considered after poor seizure control despite trial of at least three antiepileptic medications.

DBS of the anterior thalamic nuclei has been undertaken by several groups. In one study, five of six patients had improvements in their seizures over an average follow-up period of 5 years (Andrade et al., 2006). In another study, four of five patients showed significant reductions in frequency and severity of seizures after 6–36 months without adverse complications (Kerrigan et al., 2004). A third study showed significantly reduced seizures in all four patients over an average 44-month follow-up period (Lim et al., 2007). Putative targets of stimulation may depend upon seizure localization and also include the STN, caudate, hippocampus, cerebellum, hypothalamus, and medial intralaminar thalamic nuclei (Theodore and Fisher, 2004), but further controlled trials underway with long-term follow-up are required before DBS in epilepsy can be considered appropriate for patients as an alternative to resective surgery (Oommen et al., 2005).

Depression

Depression is extremely common. Lifetime prevalence for major depressive disorder has been estimated at 16%, half of all patients having reduced function and role impairment. Patients with major depressive disorder are twice as likely to die as those who are not depressed. One trial of DBS in drug refractory depression targeted the subgenual cingulate cortex bilaterally, with four out of the six patients showing improvement (Mayberg et al., 2005). Another targeted the anterior limb of the internal capsule as for OCD in five patients (Greenberg et al., 2005), with three patients showing a greater than 50% symptom improvement. Both studies were uncontrolled and had less than 1 year of follow-up. While DBS for severe depression appears promising, further studies are required to confirm efficacious targets and successful outcomes.

Obsessive-Compulsive Disorder

OCD can manifest at any age, but first onset is usually in a person's third decade. Prevalence is 0.8% in adults and lower in children. About 10% of patients are refractory to pharmacotherapy, usually selective serotonin reuptake inhibitors, and frequently become housebound.

The anatomical target for DBS derives from the success of the lesional procedure of anterior capsulotomy that improves symptoms in approximately half of the patients treated (Jenike, 1998). Long-term outcomes for bilateral DBS of the anterior limb of the internal capsule and adjacent ventral striatum have been reported by two groups. In one study, blinded assessment of four patients followed up for at least 21 months after surgery revealed significant improvements in three patients (Nuttin et al., 2003). In another study, of 10 patients evaluated 3 years after surgery, seven showed a one-third or greater percentage reduction in symptoms and six had an improvement in activities of daily living (Greenberg et al., 2006).

Tourette's Syndrome

Tourette's syndrome has 0.1%–1% prevalence, usually affecting children and adolescents. It is more common in people with autistic spectrum disorders and is characterized by motor and vocal tics. Simple tics typically involve one muscle group and complex tics may mimic a purposeful movement such as an obscene gesture. Simple vocal tics are sounds or noises like grunting and complex vocalizations include echolalia and coprolalia, the latter affecting 10% of patients. For most sufferers, symptoms decline in adulthood, but DBS may be considered for those with debilitating tics refractory to drugs such as neuroleptics and anticonvulsants.

Brain regions targeted for DBS have included the medial intralaminar thalamic (centromedian and parafascicular) nuclei (three patients) (Visser-Vandewalle et al., 2003) and case reports of stimulation of the anterior limb of the internal capsule (Flaherty et al., 2005) and GPi (Ackermans et al., 2006; Diederich et al., 2005; Houeto et al., 2005). With this initial experience and experience of ablative surgery for Tourette's syndrome (Temel and Visser-Vandewalle, 2004), criteria for DBS suitability have been proposed (Mink et al., 2006). Alongside clarifying the quality and duration of failed medical and behavioral treatments, the criteria suggest a minimum age of 25 years. A lower age limit for the procedure lies juxtaposed against the benefits seen from early treatment of medically refractory childhood dystonias, yet may be appropriate while numbers of patients treated remain small and long-term outcomes yet to be determined.

Cluster Headache

Cluster headache is characterized by severe unilateral periorbital pain with concomitant autonomic sequelae of vasodilatation and periorbital edema. Prevalence is less than 1% with men more commonly affected.

DBS can be performed for cluster headache refractory to medical treatments, targeting the ipsilateral posterior hypothalamus. After mean follow-up of almost 2 years, 13 out of 16 patients were reported symptom-free or almost headache-free in the largest study to date (Leone et al., 2006). Another group has reported that three of six patients considerably improved (Schoenen et al., 2005). The procedure appears extremely successful (Leone, 2006), but risks of hemorrhage remain a pervading concern in targeting such a deep brain structure bilaterally.

Chronic Pain

Chronic pain presents a considerable burden to society, occurring in cancer, stroke, trauma, and failed surgery. It may affect as many as one in five people. DBS has been undertaken for almost four decades. Targets have included the internal capsule and medial intralaminar thalamic nuclei, but most current treatments target the ventral posterolateral and ventral posteromedial thalamic nuclei (VPL/VPM) and periaqueductal and periventricular gray matter (PAG/PVG).

DBS for pain had reported cases of 1300 recipients (Levy, 2003; Pereira et al., 2007c). Chronic pain etiologies with good outcomes in contemporary series are stroke (Owen et al., 2006a), amputation (Bittar et al., 2005b), anasthesia dolorosa (Green et al., 2003, 2006a), and plexopathies with success also seen in multiple sclerosis (Hamani et al., 2006) and malignancy (Owen et al., 2006b).

Equipment and Procedure

At present, only one commercial manufacturer (Medtronic Inc., Minneapolis, MN) produces widely available deep brain electrodes suitable for DBS. Two models are currently available—the 3387 model and the 3389 model. Both are quadripolar electrodes, having four contacts. The 3387 electrode contains four 1.5 mm cylindrical electrodes separated by 1.5 mm insulation. In contrast, the 3389 model reduces insulation between contacts to 0.5 mm, enabling enhanced spatial and stimulation resolution. The 3389 overall contact length is reduced from 10.5 to 7.5 mm. It therefore yields greater flexibility theoretically for stimulation field manipulation within smaller target areas like the STN. However, the 3387 enables greater current spread, a facility that is particularly useful in compensating for inaccuracies in deep brain targeting, thus it remains preferred by many surgeons.

Several stimulation parameters can be altered in DBS. Using Medtronic Kinetra equipment, stimulation can be monopolar or bipolar over any combination of the four contacts of each electrode and multiple contacts can be specified as anodes or cathodes. Voltage can be altered from 0 to 9.5 V, frequency from 3 to 250 Hz, and width of the square wave pulse from 60 to 450 μs. Stimulation may also be either continuous or cycled between on and off states.

The DBS electrode is secured to skull and connected to a lead that is tunneled to the chest or abdomen where a pulse generator (pacemaker) is implanted under the skin (see Figure 18.1). Recent developments in devices include commercially available transcutaneously rechargeable pulse generators and the

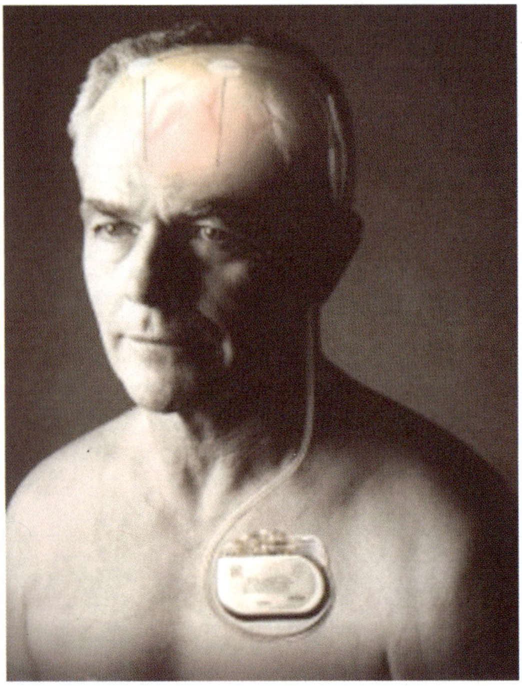

Figure 18.1 Schematic diagram of a deep brain stimulator and implantable pulse generator in situ (courtesy of Medtronic™).

prospective entry of other device manufacturers to the expanding clinical domain of DBS from related fields like spinal cord stimulation.

The technique of DBS involves "stereotaxically" implanting a fine electrode into an area of the brain to be stimulated. The procedure is often performed using just local anesthetic and light sedation to enable the patient to report their subjective experiences and hence therapeutic efficacy on the operating table. A "stereotactic" frame is fitted to the patient's head prior to surgery (see Figure 18.2a).

A brain scan, either magnetic resonance (MR) or computed tomography (CT) depending on the frame, is then performed. Various software packages are then used to create a "target" and trajectory based on the frame's coordinates—if CT is used (see Figure 18.2b), this can be "fused" to a preoperative MR (Figure 18.2c). The electrode is then inserted via a small scalp incision and a small drill hole (termed a "craniostomy") using the calculated targets. Detailed device and procedural issues are described elsewhere (Bittar et al., 2005a; Rezai et al., 2006).

How Does it Work?

Much remains to be understood about the mechanisms of DBS (Kringelbach et al., 2007b). Pioneering research by Benabid on the use of high-frequency stimulation in the 120–180 Hz frequency range for PD showed that DBS has a similar effect to neurosurgical thermocoagulative ablation of the same area (the VIm) (Benabid et al., 1987, 1991). Similarly, its use in dystonia resembled that of successful pallidotomy (Laitinen et al., 1992). Thus, it was initially thought that high-frequency stimulation somehow inhibits the area being stimulated and therefore emulates a surgical lesion, albeit a reversible one. However, the use of a variety of experimental modalities including neural modeling, neuronal recording, microdialysis, gene expression studies, and functional imaging during and after DBS have demonstrated considerably more complexity to the mechanisms than mere inhibition (Lozano and Eltahawy, 2004; McIntyre et al., 2007).

A number of hypotheses to explain DBS have been formulated: depolarization blockade (alterations in voltage-gated currents that block neural output); synaptic inhibition (acting at the axon terminals in the region of the electrode); synaptic depression (due to depletion of transmitter vesicles during increased activity secondary to stimulation); and modulation of pathological network activity. Within these studies, there are many conflicting results such as the demonstration that high-frequency DBS of the STN in rats increases striatal dopamine release (Meissner et al., 2002), but the negative findings of no increase in striatal dopamine in humans using positron emission tomography (PET) studies (Hilker et al., 2003). DBS almost certainly works via a combination of the aforementioned mechanisms. It has also been suggested that it works differentially on somatic and axonal elements (Walter and Vitek, 2004), thus providing a "functional decoupling" of somatic and axonal activity. This would help to explain how stimulation may have one type of effect on the area immediately surrounding the electrode and a different effect on more distant structures. Indeed, functional imaging studies show both increases and decreases in activity of distant brain regions during DBS (Ceballos-Baumann et al., 2001).

Although a detailed discussion of the mechanisms of DBS is beyond the scope of this chapter, it is important to mention that most of the studies investigating this topic have concentrated on high-frequency stimulation in PD. Some conditions, such as chronic pain are traditionally treated using much lower frequencies (5–50 Hz) (Bittar et al., 2005a; Nandi et al., 2003).

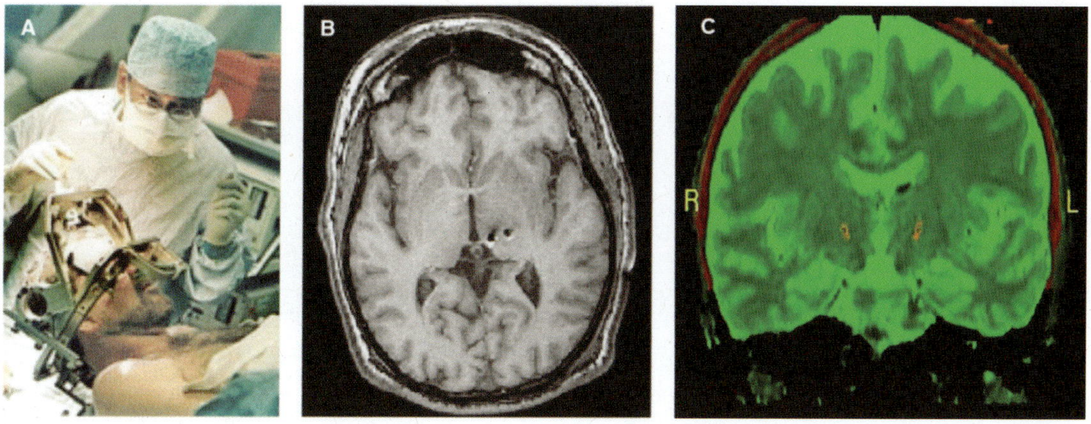

Figure 18.2 (A) Intraoperative awake deep brain stimulation. (B) Axial MRI of deep brain stimulators for pain in situ. (C) Fused coronal CT and MRI showing subthalamic nuclei targeted in deep brain stimulation for Parkinson's disease. (After Pereira et al., 2007.)

However, whole brain functional neuroimaging studies that have been performed using low-frequency DBS in chronic pain still show differential increases or decreases in activation of distant brain areas, consistent with the hypothesis that DBS works not only by affecting the area stimulated, but also related brain regions including reward processing areas involved in pleasure-seeking behavior such as the insula, amygdala, anterior cingulate, and orbitofrontal cortex (Kringelbach et al., 2007a; Pereira et al., 2007a). DBS for movement disorders is postulated to act upon the motor aspects of parallel, segregated corticostriatal loops (Alexander and Crutcher, 1990), and DBS for disorders of affect disorders is likely to act upon their emotional equivalent (Figure 18.3a).

Hedonic Electrical Stimulation

In the first half of the 20th century, a number of researchers looked at the "eliciting" effects of stimulation of the brain in animals. These included eliciting laughter-like vocalizations and fear in monkeys (Brown, 1915; Delgado et al., 1956a), and behavioral effects of stimulation of various areas such as the frontal lobes (Rosvold and Delgado, 1956) and limbic system (Maclean and Delgado, 1953). In 1954, Olds and Milner at the McGill University in Montreal investigated the "reinforcing" effects of stimulation in the rat brain (Olds and Milner, 1954). Their experiments were designed to assess the reward aspects of stimulation by allowing a rat to "self-stimulate." A rat that persistently depressed a lever to deliver electrical stimulation to an area of its brain was presumably experiencing "reward." thereby leading to the desire to keep pressing it. The reaction to stimulation of the septal area was particularly strong. These experiments were extended so that the rat had to run across a hot plate to get to the lever. This meant that the reward from pressing the lever had to be greater than the level of pain experienced to do it. Again, the septal area gave the strongest reaction.

Whether the rats in Olds and Milner's experiments were experiencing "pleasure" or not cannot be proven as a rat cannot report its emotions, if indeed a rat is capable of experiencing pleasure! Therefore, it was not until humans had electrodes implanted in their brains that stimulation sufficient to cause "pleasure" could be unequivocally demonstrated by patient self-reporting of pleasurable sensations. In this regard, Robert Heath, at Tulane University in New Orleans, conducted pioneering studies. His initial experiments (Heath, 1954), alongside those of Sem-Jacobsen (1959), Delgado and others (Delgado et al., 1956b; Mahl et al., 1964), described self-reporting of pleasurable subjective experiences with human brain stimulation.

In 1963, Heath, together with his coworkers, Bishop and Elder, emulated Olds and Milner's earlier rodent experiments by describing a case of human intracranial self-stimulation (ICSS) (Bishop et al., 1963). The patient was able to stimulate his own brain using a switch incorporated into the stimulation circuit of his implanted DBS. In their study, in which a 35-year-old male chronic catatonic schizophrenic was implanted with multiple depth electrodes, several brain areas were associated with reward and positive reinforcement.

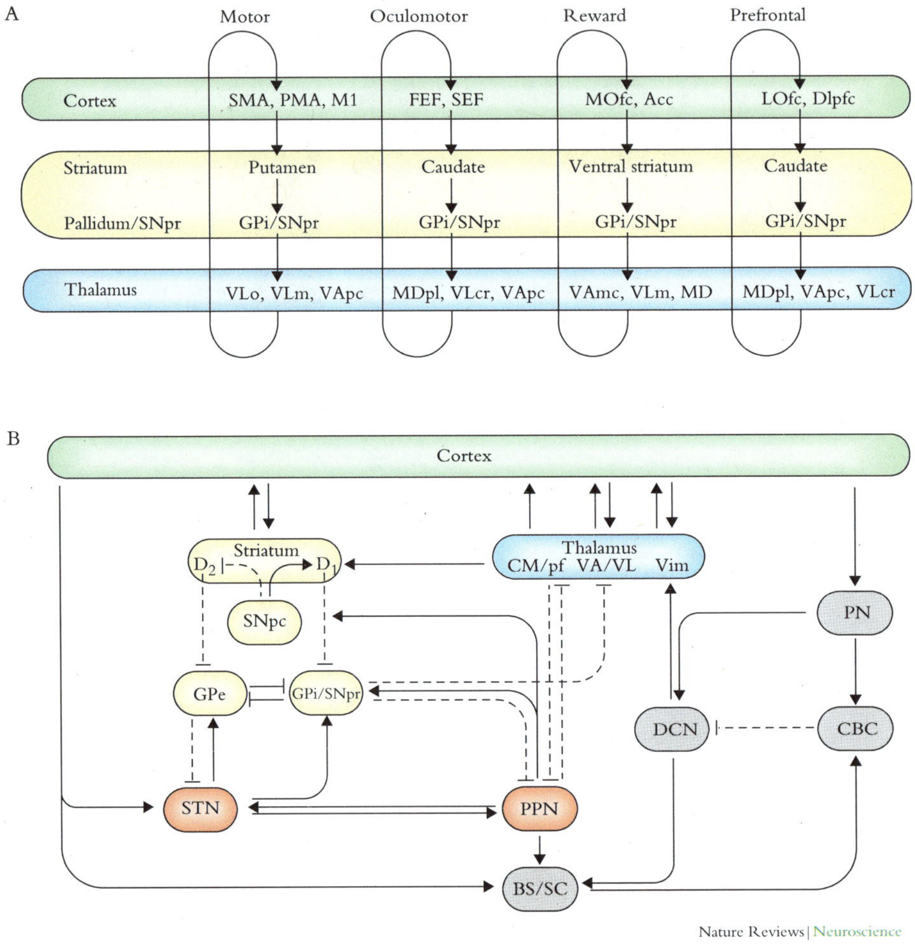

Figure 18.3 (A) Segregated, parallel thalamocortical motor, oculomotor, affective, and cognitive neural circuitryders. (B) An anatomical model of basal ganglia function and DBS (after Kringelbach et al., 2007). Abbreviations Acc, anterior cingulate cortex; BS/SC, brain stem/spinal cord; CBC, cerebellar cortex; CM, centromedian nucleus of the thalamus; DCN, deep cerebellar nuclei; Dlpfc, dorsolateral prefrontal cortex; FEF, frontal eye fields; LOfc, lateral orbitofrontal cortex; GABA, gamma-aminobutyric acid; GPi, internal globus pallidus; GPe, external globus pallidus; M1, primary motor cortex; MD, mediodorsal nucleus of the thalamus; MDpl, mediodorsal nucleus of the thalamus, pars lateralis; MOfc, medial orbitofrontal cortex; Pf, parafascicular nucleus of the thalamus; PMA, premotor area; PN, pontine nucleus; PPN, pedunculopontine nucleus; SEF, supplementary eye field; SMA, supplementary motor area; SNpc, substantia nigra pars compacta; SNpr, substantia nigra pars reticulata; STN, subthalamic nucleus; VA, ventral anterior nucleus of the thalamus; VAmc, ventral anterior nucleus of the thalamus, pars magnocellularis; VApc, ventral anterior nucleus of the thalamus, pars parvocellularis; Vim, ventral intermediate nucleus of the thalamus; VL, ventrolateral nucleus of the thalamus; VLcr, ventrolateral nucleus of the thalamus, pars caudalis, rostral division; VLm, ventrolateral nucleus of the thalamus, pars medialis; VLo, ventrolateral nucleus of the thalamus, pars oralis.

These regions included the caudate nucleus, amygdala, mid-hypothalamus, posterior hypothalamus, and the boundary of the hypothalamus–tegmentum, although the area that gave the greatest effect was the septal area, as in Olds and Milner's work. They also found that the effect was only apparent at certain current intensities, above which stimulation related to "aversive" behavior.

Subsequent more detailed observations reported a variety of reactions to septal stimulation where it was described as a "good" or "pleasurable" feeling and that it was "as if he were building up to a sexual orgasm" (Heath, 1963). Thus, although implantable stimulators had been shown to produce pleasurable sensations, their clinical efficacy in the treatment of the socially prevalent psychiatric burdens upon asylums of the time—schizophrenia and depression—at this stage remained to be seen. Furthermore, as Smith et al. detail in Chapter 1, pleasurable sensations with self-stimulation were far from consistent or ubiquitous across Heath's patients. While, by and large, patients continued to seek to press the buttons to switch on the stimulator as often as possible, the reported sensations were frequently ambivalent. While Bechterewa reported a woman who achieved orgasm with ventrolateral thalamic stimulation and became addicted to it, accosting the scientists administering the stimulation, Sem-Jacobsen reported a female patient who kept smiling and sometimes laughed with stimulation and thus it was assumed enjoyed the experience, until one day she told him that she was "fed up" and "did not enjoy these stimulations at all." On enquiry, it transpired that the stimulation was eliciting a rhythmic contraction in the patient's pelvic muscles, tickling her and causing undesired mirth. Such findings and others led Sem-Jacobsen to remark that "in man, curiosity is probably the most dominant factor in initiating self-stimulation" (Valenstein, 1973).

We cannot conclude from these human experiments half a century ago whether such so-called pleasurable sensations were 'wanting' (as Smith et al. [Chapter 1, this book] suggest) or genuine 'liking'. Our current mechanistic knowledge suggests that the predominantly septal stimulation of Heath and others is likely to have influenced mesolimbic dopaminergic systems in the adjacent ventral striatum involved in appetitive behavior, that is, 'wanting'. However, that some patients derived a subjective sensation of "pleasure" is irrefutable. Clearly much remains to be delineated about both targets and mechanisms in man (and woman—the literature suggests that they are likely to differ subtly between genders) and in how and why they differ between individuals and in certain disease conditions like chronic pain (see below).

Ablative Psychosurgery

Perhaps the experiments by Heath, Delgado, and other pioneers would have reached larger clinical trials if not for the social and political contextualization of ablative psychosurgery during that epoch. Neurosurgery for psychiatric disorders was first established by the Portuguese neurologist Antònio Egas Moniz in the 1930s (Moniz, 1936), against a background of the Adolf Meyer's conceptualization of psychiatry as a scientific specialty, with heated debates dividing Freudian psychotherapists from Kraepelinian advocates of somatic treatments like electroconvulsive shock treatment, cold water baths, and insulin-induced comas. Moniz applied prefrontal leucotomy to depressive disorders. The procedure was popularized and adapted to an outpatient technique by the American neurologist Walter Freeman in the 1940s (see Figure 18.4). Its clinical and scientific popularity was greatest during the late 1940s when Moniz won a Nobel prize largely for work in psychosurgery (see Figure 18.5). That psychosurgery did not cure, but merely altered, was clear even to Moniz and his ardent disciple Freeman who believed reintroduction of the individual to society rather than restoration of emotional faculties to be the most pragmatic marker of efficacy, quipping that "apparently it doesn't take much frontal lobe to hold down a job" (Freeman and Watts, 1950). Within just two decades, more than 10,000 patients in the United Kingdom and 60,000 in the United States had undergone psychosurgery (Tooth and Newton, 1961; Valenstein, 1973). The outcomes were interpreted as true medical benefits, although the intervention was considered a "last resort" in which recipients "paid a price" (El-Hai, 2005; Pressman, 1998).

Yet unique historical circumstances including overburdened asylums and lack of adequate pharmacological treatments placed many patients within this category of last resort, and framed mental health such that the operation's apparent benefits to society were amplified and its costs to the individual downplayed. Although both Heath's experiments and MacLean's conceptualization of the limbic system were driven by research fueled by the tremendous clinical interest in psychosurgery (Maclean, 1955), it is a combination of the backlash against psychosurgery and concerns about its potential use for social control together with the emerging role of pharmacotherapy for psychiatric disorders that led to the eventual demise of psychosurgery at around the time of Heath's experiments and obviated the widespread clinical uptake of DBS for psychiatric disorders half a century ago.

From Ablation to DBS

PD, the dominant clinical indication for DBS, is particularly pertinent in the context of pleasure as it involves alterations in the mesolimbic and

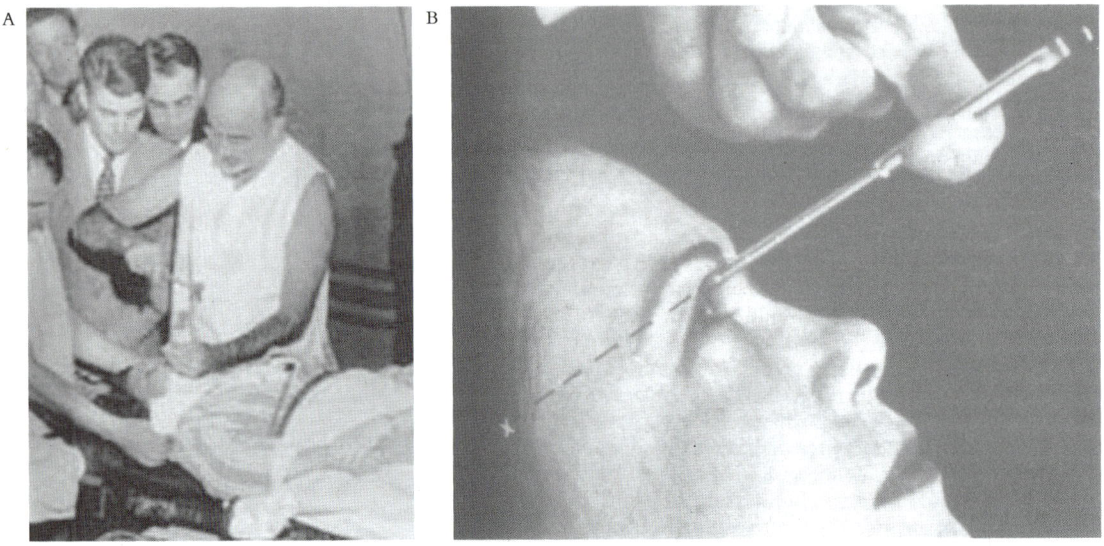

Figure 18.4. Freeman and Watts' transorbital leucotomy, the "ice pick procedure," demonstrated by Freeman, not a neurosurgeon himself, in nonsterile conditions. (After Freeman and Watts, 1950.)

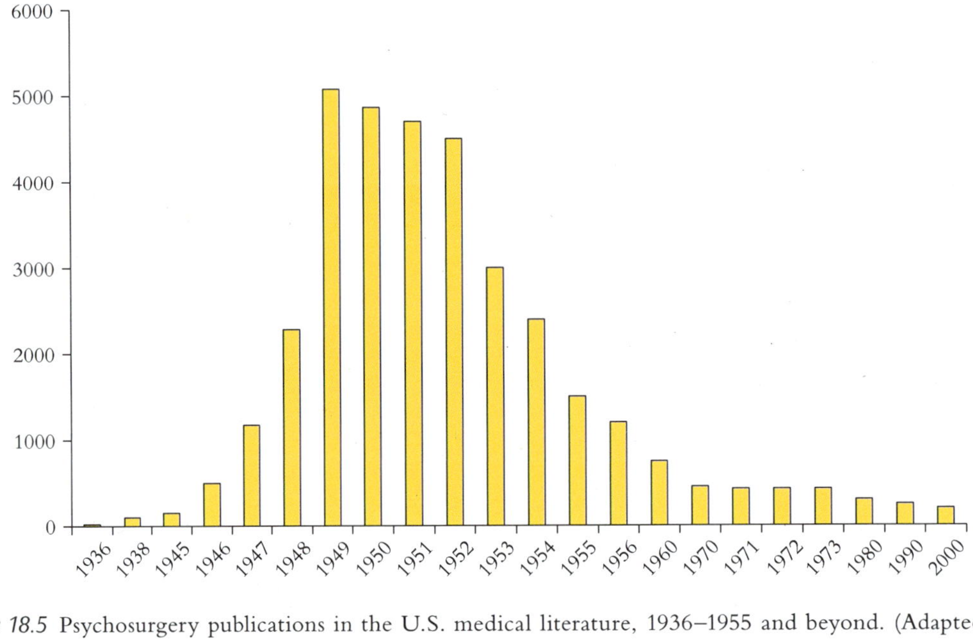

Figure 18.5 Psychosurgery publications in the U.S. medical literature, 1936–1955 and beyond. (Adapted from Valenstein, 1973.)

mesocortical dopaminergic neural circuitry also implicated in reward behavior. Levodopa was established 40 years ago as an effective treatment for PD (Cotzias et al., 1967). Dopamine acting at both D1 and D2 receptors has been shown to be important in reward-dependent actions such as the reaction time of saccadic eye movements in primates (Nakamura and Hikosaka, 2006), alongside other dopamine receptor subtypes being related to reward-related behavior and individual differences (Dalley et al., 2007).

Two decades ago, it was demonstrated that dopaminergic projections are important in the reward-related behavior of medial forebrain self-stimulation in animals (Gallistel, 1986). In the maladaptive situation of addiction, dysfunction in dopaminergic pathways involving brain regions, such as the ventral striatum, has been related to craving, impulsivity, and attentional deficits (Everitt et al., 2001; Robbins, 2003).

In PD, there is degeneration of the dopamine-producing cells in the substantia nigra. This lack of dopamine leads to changes in activity of the areas in the basal ganglia that the substantia nigra projects to, leading to the symptoms of bradykinesia (slowness of movement), tremor, and rigidity. However, in addition to the "motor" symptoms, PD sufferers also exhibit reductions in the activations of dopaminergic pathways that are related to motivational modulation of task performance. Thus, although they display increasing task times with increasing rewards, their overall performance is reduced (Gerent et al., 2004).

As summarized above, numerous deep brain targets are targeted for stimulation in the treatment of PD. Given that high-frequency stimulation is used in STN DBS, one parsimonious "anatomical" hypothesis is that the effects of stimulation decrease activity in the STN that has become overactive secondary to reduced activity in the inhibitory pathways leading from the GPe, which in turn is related to reduced dopaminergic activity in the SNr/GPi (see Figure 18.3b). The reality is almost certainly more complex, and in fact there is increasing evidence that abnormal oscillatory activity in the basal ganglia in PD may be modified by DBS, leading to the proposal of an elegant "kinetic" model (Brown, 2003). However, in relation to DBS and reward, the fundamental question is whether STN stimulation leads to an increase in the motivational modulation of task performance and if so, how?

These fundamental questions are yet to be fully elucidated but there is multimodal evidence for changes in dopaminergic pathways in response to STN stimulation. For example, PET studies have revealed attenuation of fluctuating striatal synaptic dopamine levels with DBS (Nimura et al., 2005). Such a finding may partly explain the role of STN DBS in reducing pathological gambling that can occasionally result from dopaminergic drug therapy in PD (Bandini et al., 2007). Moreover, considering clinical outcomes, many studies have demonstrated improvements in mood, mood stability, improved quality of life, and even impairment of fear recognition with stimulation (Berney et al., 2007; Biseul et al., 2005; Castelli et al., 2007; Lyons and Pahwa, 2005). Although some of these improvements may be related to the improvement of motor symptoms (Troster et al., 2003), it is likely that neuromodulation of the limbic components of STN and the affective corticostriatal loop (see Figure 18.3a) is partly responsible.

Pleasure, Pain, and Melancholia

It could be said that the extreme melancholia that is experienced in depression is diametrically opposed to the experience of pleasure, if indeed there is a continuum. The same could be said of pain, the pain–pleasure principle being discussed elsewhere in this book, although perhaps pain is more specific and less mood-related than depression.

Electrical treatment to improve the symptoms of depression has been used in the form of electroconvulsive therapy since 1938. Prior to then, convulsions were induced in depressed patients by rendering them hypoglycaemic, the idea of Manfred Sakel at the Lichterfelder sanitarium near Berlin. His theory was that this somehow reduced the energy expenditure of the nerve cell thus "reinforcing" it later. In 1938, Ugo Cerletti, professor of neuropsychiatry in Rome, made the conceptual leap that the convulsions could be induced by passing an electrical current through the patient's head. Today, there is still much controversy as to the mechanism of action of electroconvulsive therapy (ECT) and its benefits and risks (UK ECT Review Group, 2003). However, ECT remains an important part of the armamentarium available to psychiatrists in their treatment of drug-resistant depression.

As discussed above, DBS has been used to treat depression. Electrodes have been targeted by Mayberg and Lozano at the subgenual cingulate area (Brodmann area Cg25; Figure 18.6). The rationale behind choosing area Cg25 comes from previous research, highlighting this area as being important in treatment-resistant depression. Neuroimaging studies in humans revealed altered activity in Cg25 in both "acute sadness" and antidepressant treatment (Mayberg et al., 1999; Seminowicz et al., 2004). Reduction of activity in Cg25 has also been found after successful treatment with cognitive behavioral therapy (Goldapple et al., 2004), ECT (Nobler et al., 2001), ablative surgery (Malizia, 1997), and transcranial magnetic stimulation (Mottaghy et al., 2002). Anatomical pathways have been demonstrated between area Cg25 and many other brain regions implicated in the genesis of depression. For example, Cg25 shows reciprocal connectivity to the orbitofrontal and medial prefrontal cortices, areas implicated in the linking of reward to hedonic experience (Kringelbach, 2005). It

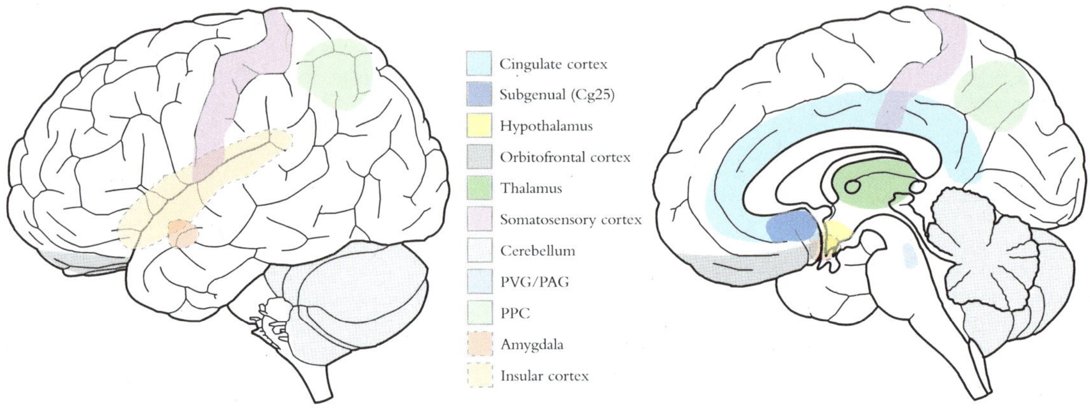

Figure 18.6 Brain structures in the pain neuromatrix and Cg 25, a brain target in DBS for depression.

also connects to areas implicated in the disturbances of circadian rhythm associated with depression, such as the brainstem, hypothalamus, and insular cortices (Freedman et al., 2000).

Cingulotomy has been a relatively successful treatment of refractory psychiatric disorders such as depression for half a century (Cosgrove and Rauch, 2003). This alongside the aforementioned neuroimaging and anatomical studies provided a robust rationale for DBS for depression. In addition, the supposition that high-frequency stimulation inhibits activity in aberrantly "overactive" brain circuitry added extra impetus to the argument. Mayberg and coworkers, in their first report of Cg25 stimulation in depressive patients, describe interesting intraoperative changes in patients' mood and well-being that occur over a timescale of several minutes of stimulation (Mayberg et al., 2005), including "sudden calmness or lightness," "disappearance of the void," "sense of heightened awareness," "connectedness," and "sudden brightening of the room" together with a quickening of psychomotor function. The immediate changes described are uncannily reminiscent of Heath's original experiments with septal stimulation half a century earlier in which patients would describe self-stimulation with phrases like "I like the sexy button that goes all the way down!" The preliminary long-term results of DBS for depression also appear promising with only one-third not improving in a group of patients refractory to other treatments.

Depression is a complex pathological condition that almost certainly has a complex underlying neurochemical and neurophysiological basis, and for that reason, cingulate cortex stimulation no doubt acts at many more subtle levels than simply to "provide pleasure." However, when successful, the end result is a transformation from severe melancholia to an increased awareness and a lifting of mood perhaps equivalent to induction of a form of pleasure.

DBS for pain arose from rodent stimulation experiments, suggesting PVG/PAG regions as DBS targets (Reynolds, 1969)—these findings were translated to humans in 1977 (Hosobuchi et al., 1977; Richardson and Akil, 1977). Evidence supporting VPL/VPM thalamic nuclei and adjacent structures as putative targets for DBS came from ablative surgery (Mark and Ervin, 1965), leading facial pain to be treated by thalamic DBS (Hosobuchi et al., 1973). Pioneering researchers also targeted the internal capsule and more medial thalamic nuclei including the centromedian–parafascicular complex (Adams et al., 1974; Thoden et al., 1979).

Historically, clinical approaches to DBS have sought to categorize patients first by cause of pain and second by dichotomizing the pain into such categories as nociceptive or deafferentation, "epicritic" or "protopathic," and peripheral or central. Such distinctions are largely unhelpful to patient selection as functional neuroimaging and electrophysiological evidence suggests that chronic pain arises concomitant with centrally mediated changes related to neuronal plasticity (Coderre et al., 1993). Thus, chronic pain refractory to medical treatment is largely central and thus neuropathic. While the mechanisms are complex and poorly understood, chronic pain amenable to DBS can thus, like melancholia, be conceived as the polar opposite of pleasure and its amelioration by neuromodulation a recalibration of a patient's pain–pleasure continuum.

While the analgesic mechanisms of DBS are unknown, altered rhythmic activity in VPL/VPM and PVG/PAG neurons is likely to play an important role in pain pathophysiology. Analgesic DBS may therefore augment pathologically diminished low-frequency synchronous oscillations in the thalamic and reticular components of a reticulo-thalamo-corticofugal pain neuromatrix (see Figure 18.6). A positive correlation has been shown between analgesic efficacy at either DBS site and the amplitude of slow frequency (<1 Hz) VPL/VPM local field potentials using deep brain recording (Nandi et al., 2002c, 2003). Patients in pain also have characteristically enhanced low-frequency (8–15 Hz) power spectra of both PVG/PAG and VPL/VPM local field potentials (Pereira et al., 2007d). Further research is required to elucidate if such neuronal signatures could aid patient selection or enable "smart" demand-driven stimulation, in particular if combined with technical advances in noninvasive electrophysiological techniques to characterize functional neuronal connectivity. However, it is not implausible to conceive that such "smart" stimulation could theoretically deliver pleasurable sensations or a feeling of well-being on demand, whether stimulating PVG/PAG, Cg25 or other components of the human emotional neural circuitry.

The "Four F's"—Fight, Flight, Fear, and Procreation

The promising but still experimental indication of DBS for disorders of blood pressure includes malignant hypertension and medically refractory orthostatic hypotension as putative indications. Suggestive findings from PAG/PVG DBS in 16 patients with chronic pain augur well for autonomic applications of DBS (Green et al., 2005, 2006b Pereira et al., 2007c). The PVG/PAG is a structure optimally sited anatomically to integrate interoceptive function, both from adjacent mesencephalic cardiovascular centers and more distal pain processing areas. Its autonomic effects have been well studied in animals (Behbehani, 1995), and changes noted with DBS (Young and Rinaldi, 1997). In mammals, the region is instrumental to "defense" reactions (Bittencourt et al., 2004), integrating descending responses from forebrain to cardiovascular effector organs to assist survival by modulating active sympathetic "flight" or passive "withdrawal" responses (Carrive, 1993; Hunsperger, 1956).

Whether the coping is active or passive seems to determine whether the sympathetic response is pressor or hypotensive in animals (Bandler et al., 2000). Electrical stimulation of the PAG in animals elicits such defence reactions, dorsal regions being associated with active coping and hypertensive effects and ventral regions with passive coping and hypotensive effects (Bandler et al., 1991). Thus, it is likely that PAG stimulation affects not only pain modulation pathways, but also cardiovascular autonomic pathways. PAG DBS has been related to hypertensive and chronotropic cardiovascular effects (Bendok et al., 2003; Young and Rinaldi, 1997). Current research confirms cardiovascular effects including blood pressure changes (Green et al., 2005, 2006b,c).

Notably, a positive correlation has been shown between degree of analgesia in patients receiving PVG/PAG DBS and magnitude of blood pressure reduction (Green et al., 2006c). While the mechanisms again remain to be fully elucidated, it is clear that subjective sensations of analgesia relate to autonomic responses and by corollary that the neuromodulatory alteration of sympathetic drive may not only alter fear responses but also have the potential to augment pleasurable sensations and their somatic sequelae, leading us full circle back to notions of the fabled orgasmatron.

Hedonic DBS

> *The Japanese had already forgotten more neurosurgery than the Chinese had ever known. The black clinics of Chiba were cutting edge, whole bodies of technique supplanted monthly, and still they couldn't repair the damage he'd suffered in that Memphis hotel. A year here and he still dreamed of cyberspace, hope fading nightly.*
>
> Gibson (1986)

William Gibson's iconic science fiction work *Neuromancer* paints a stark view of neuroenhancement and its deleterious consequences. Indeed, as neurosurgeons, we are appropriately limited in any quest toward hedonic DBS not by technology but by ethical constraints and the Hippocratic doctrine *primum non nocere*.

Rare psychiatric complications of DBS for PD begin to illustrate such ethical dilemmas, such as the predicament of a patient who became manic, and mentally incompetent, when stimulated, but remained debilitated by motor side effects when unstimulated (Leentjens et al., 2004). As the authors report:

> *Ultimately, there seemed to be only two alternatives: to admit the patient to a nursing home because of serious*

invalidity , but mentally in good condition, or to admit the patient to a chronic psychiatric ward because of a manic state, but with acceptable motor capacity... When not being stimulated, the patient was considered competent to decide about his own treatment; in this condition the patient chose for the second option. In accordance with his own wishes he was therefore legally committed to a chronic ward in the regional psychiatric hospital.

Ethically, psychosurgical procedures exist in a continuum of psychiatric interventions to which medical paternalism seems justified where it comports with beneficence and nonmaleficence. Treatment decisions should be made by physicians with "best intentions" made secondary to their "substituted judgement" of the incompetent patient's perceived wishes. History has demonstrated proxy consent to be potentially open to sociopolitical abuse (Mark and Ervin, 1970), therefore appropriate patient selection must be ensured by multidisciplinary teams of experts. However, in considering hedonic DBS, the individual is invariably competent yet requesting a neurosurgical procedure not without risk—where there is no obvious impairment or disability requiring restoration of function.

It should be emphasized that DBS is brain surgery and therefore brings small but significant risks including stroke (1–3%), seizures (<1%), death (0.1%), skin erosion, lead breakage, and the need for implantable pulse generator revision surgery every 1 to 10 years depending upon indication, and infection (3%)—a small proportion of cases requiring complete removal of the DBS system (Joint et al., 2002; Lyons et al., 2004; Pereira et al., 2007c; Yianni et al., 2005). Patients should also be counseled for the possibility that they may derive no benefit from DBS or not tolerate it well, again necessitating its removal. The likelihood of this varies between indications, but it should be emphasized when treating less established indications. Indication and target specific complications can arise (Bittar et al., 2005a; Deuschl et al., 2006a), for example dysarthria with bilateral VIm DBS (Limousin et al., 1999), altered libido with medial thalamic stimulation (Visser-Vandewalle et al., 2003), and anxiety with PAG/PVG DBS (Levy, 2003).

At present, surgeon experience and clinical evidence point toward whom to offer DBS and when, which targets to select, and tentatively toward prognostication for many emerging and some enduring clinical indications. For all DBS targets, the relative contributions of local interactions and wider functional neuroanatomical circuitry are yet to be fully elucidated. Yet clinical results are frequently spectacular and life-transforming.

Half a century on from Heath's first experiments, we know considerably more about electrical stimulation of the brain in humans, but much remains to be learnt. Intriguingly, Heath commented that "the most intense pleasurable responses occurred in patients stimulated while they were suffering intense pain, whether emotional and reflected by despair, anguish, intense fear or rage, or physical, such as that caused by metastatic carcinoma... Patients who felt well at the time of stimulation, on the other hand, experienced only slight pleasure" (Valenstein, 1973). Our considerable clinical experience of DBS for chronic pain does not concur with Heath; our patients rarely report pleasure with stimulation, but we stimulate different brain targets now from those he stimulated then (Pereira et al., 2007c). Nevertheless, like Heath, we suspect that the greatest insights from DBS into pleasure are likely to come from intensive studies of DBS for chronic pain, perhaps with novel brain targets for DBS like Cg25 in those patients refractory to thalamic and PAG/PVG DBS yet not necessarily able to translate to people not in pain. DBS for pleasure may only be attainable for those at those who lie at the extremes of the pain–pleasure continuum.

As brain surgery, DBS is often regarded as a last resort treatment. However, as evidence of efficacy and our mechanistic knowledge improve and the number of indications increases, interest in the technique will continue to yield insights and answers. DBS gives many insights into the neural mechanisms of pleasurable experience and pleasure-seeking behavior. However, hedonic DBS, DBS for pleasure alone, cannot be clinically justified. What differentiates the safe neurosurgeon from the dexterous psychopath is not technical ability, but awareness of when intervention is both timely and appropriate. In other words, a good neurosurgeon should know when *not* to operate. Lars Leksell, the father of stereotactic surgery, put it most succinctly with his pithy aphorism, "a fool with a tool is still a fool." With current technologies, only a fool would use a deep brain stimulator as a tool to make an orgamsatron. Osler's conviction has stood the test of time this past century, but we cannot predict if it will hold true over the next.

References

Ackermans, L., Temel, Y., Cath, D., van der Linden, C., Bruggeman, R., Kleijer, M., Nederveen, P., Schruers,

K., Colle, H., Tijssen, M. A. and Visser-Vandewalle, V. (2006) Deep brain stimulation in Tourette's syndrome: Two targets? *Mov. Disord.* 21, 709–713.

Adams, J. E., Hosobuchi, Y. and Fields, H. L. (1974) Stimulation of internal capsule for relief of chronic pain. *J. Neurosurg.* 41, 740–744.

Alexander, G. E. and Crutcher, M. D. (1990) Functional architecture of basal ganglia circuits: Neural substrates of parallel processing. *Trends Neurosci.* 13, 266–271.

Alusi, S. H., Aziz, T. Z., Glickman, S., Jahanshahi, M., Stein, J. F. and Bain, P. G. (2001) Stereotactic lesional surgery for the treatment of tremor in multiple sclerosis: a prospective case-controlled study. *Brain* 124, 1576–1589.

Andrade, D. M., Zumsteg, D., Hamani, C., Hodaie, M., Sarkissian, S., Lozano, A. M. and Wennberg, R. A. (2006) Long-term follow-up of patients with thalamic deep brain stimulation for epilepsy. *Neurology* 66, 1571–1573.

Bandini, F., Primavera, A., Pizzorno, M. and Cocito, L. (2007) Using STN DBS and medication reduction as a strategy to treat pathological gambling in Parkinson's disease. *Parkinsonism Relat. Disord.* 13, 369–371.

Bandler, R., Carrive, P. and Zhang, S. P. (1991) Integration of somatic and autonomic reactions within the midbrain periaqueductal grey: Viscerotopic, somatotopic and functional organization. *Prog. Brain Res.* 87, 269–305.

Bandler, R., Keay, K. A., Floyd, N. and Price, J. (2000) Central circuits mediating patterned autonomic activity during active vs. passive emotional coping. *Brain Res. Bull.* 53, 95–104.

Behbehani, M. M. (1995) Functional characteristics of the midbrain periaqueductal gray. *Prog. Neurobiol.* 46, 575–605.

Benabid, A. L., Pollak, P., Louveau, A., Henry, S. and de Rougement, J. (1987) Combined (thalamotomy and stimulation) stereotactic surgery of the VIM thalamic nucleus for bilateral Parkinson disease. *Appl. Neurophysiol.* 50, 344–346.

Benabid, A. L., Pollak, P., Gervason, C., Hoffmann, D., Gao, D. M., Hommel, M., Perret, J. E. and de Rougemont, J. (1991) Long-term suppression of tremor by chronic stimulation of the ventral intermediate thalamic nucleus. *Lancet* 337, 403–406.

Bendok, B. R., Levy, R. M. and Onibukon, A. (2003) Deep brain stimulation for the treatment of intractable pain. In: *Textbook of Neurological Surgery: Principles and Practice*, Vol. 3, pp. 2673–2681. Eds. H. H. Batjer and C. M. Loftus. Lippincott Williams & Wilkins: Philadelphia, PA.

Berney, A., Panisset, M., Sadikot, A. F., Ptito, A., Dagher, A., Fraraccio, M., Savard, G., Pell, M. and Benkelfat, C. (2007) Mood stability during acute stimulator challenge in Parkinson's disease patients under long-term treatment with subthalamic deep brain stimulation. *Mov. Disord.* 22, 1093–1096.

Biseul, I., Sauleau, P., Haegelen, C., Trebon, P., Drapier, D., Raoul, S., Drapier, S., Lallement, F., Rivier, I., Lajat, Y. and Verin, M. (2005) Fear recognition is impaired by subthalamic nucleus stimulation in Parkinson's disease. *Neuropsychologia* 43, 1054–1059.

Bishop, M. P., Elder, S. T. and Heath, R. G. (1963) Intracranial self-stimulation in man. *Science* 140, 394–396.

Bittar, R. G., Burn, S. C., Bain, P. G., Owen, S. L., Joint, C., Shlugman, D. and Aziz, T. Z. (2005a) Deep brain stimulation for movement disorders and pain. *J. Clin. Neurosci.* 12, 457–463.

Bittar, R. G., Otero, S., Carter, H. and Aziz, T. Z. (2005b) Deep brain stimulation for phantom limb pain. *J. Clin. Neurosci.* 12, 399–404.

Bittencourt, A. S., Carobrez, A. P., Zamprogno, L. P., Tufik, S. and Schenberg, L. C. (2004) Organization of single components of defensive behaviors within distinct columns of periaqueductal gray matter of the rat: Role of *N*-methyl-D-aspartic acid glutamate receptors. *Neuroscience* 125, 71–89.

Brown, T. G. (1915) Note on the physiology of the basal ganglia and mid-brain of the anthropoid ape, especially in reference to the act of laughter. *J. Physiol.* 49, 195–207.

Brown, P. (2003) Oscillatory nature of human basal ganglia activity: Relationship to the pathophysiology of Parkinson's disease. *Mov. Disord.* 18, 357–363.

Carrive, P. (1993) The periaqueductal gray and defensive behavior: Functional representation and neuronal organization. *Behav. Brain Res.* 58, 27–47.

Castelli, L., Lanotte, M., Zibetti, M., Caglio, M., Rizzi, L., Ducati, A., Bergamasco, B. and Lopiano, L. (2007) Apathy and verbal fluency in STN-stimulated PD patients : an observational follow-up study. *J. Neurol.* 54(9), 1238–1243.

Ceballos-Baumann, A. O., Boecker, H., Fogel, W., Alesch, F., Bartenstein, P., Conrad, B., Diederich, N., von Falkenhayn, I., Moringlane, J. R., Schwaiger, M. and Tronnier, V. M. (2001) Thalamic stimulation for essential tremor activates motor and deactivates vestibular cortex. *Neurology* 56, 1347–1354.

Chalmers, D. J. (1998) The problems of consciousness. *Adv. Neurol.* 77, 7–16; discussion 16–18.

Coderre, T. J., Katz, J., Vaccarino, A. L. and Melzack, R. (1993) Contribution of central neuroplasticity to pathological pain: Review of clinical and experimental evidence. *Pain* 52, 259–285.

Cosgrove, G. R. and Rauch, S. L. (2003) Stereotactic cingulotomy. *Neurosurg. Clin. N. Am.* 14, 225–235.

Cotzias, G. C., Van, W. M. and Schiffer, L. M. (1967) Aromatic amino acids and modification of parkinsonism. *N. Engl. J. Med.* 276, 374–379.

Coubes, P., Cif, L., El Fertit, H., Hemm, S., Vayssiere, N., Serrat, S., Picot, M. C., Tuffery, S., Claustres, M., Echenne, B. and Frerebeau, P. (2004) Electrical stimulation of the globus pallidus internus in patients with primary generalized dystonia: Long-term results. *J. Neurosurg.* 101, 189–94.

Cushing, H. (1925) *The Life of Sir William Osler.* Clarendon Press: Oxford.

Dalley, J. W., Fryer, T. D., Brichard, L., Robinson, E. S., Theobald, D. E., Laane, K., Pena, Y., Murphy, E. R., Shah, Y., Probst, K., Abakumova, I., Aigbirhio, F. I., Richards, H. K., Hong, Y., Baron, J. C., Everitt, B. J. and Robbins, T. W. (2007) Nucleus accumbens D2/3 receptors predict trait impulsivity and cocaine reinforcement. *Science* 315, 1267–1270.

Delgado, J. M., Rosvold, H. E. and Looney, E. (1956a) Evoking conditioned fear by electrical stimulation of subcortical structures in the monkey brain. *J. Comp. Physiol. Psychol.* 49, 373–380.

Delgado, J. R., Hamlin, H., Higgins, J. W. and Mahl, G. F. (1956b) Behavioral changes during intracerebral electrical stimulation. *AMA Arch. Neurol Psychiatry* 76, 399–419.

Detante, O., Vercueil, L., Krack, P., Chabardes, S., Benabid, A. L. and Pollak, P. (2004) Off-period dystonia in Parkinson's disease but not generalized dystonia is improved by high-frequency stimulation of the subthalamic nucleus. *Adv. Neurol.* 94, 309–314.

Deuschl, G. and Bain, P. (2002) Deep brain stimulation for tremor [correction of trauma]: Patient selection and evaluation. *Mov. Disord.* 17, S102–S111.

Deuschl, G., Herzog, J., Kleiner-Fisman, G., Kubu, C., Lozano, A. M., Lyons, K. E., Rodriguez-Oroz, M. C., Tamma, F., Troster, A. I., Vitek, J. L., Volkmann, J. and Voon, V. (2006a) Deep brain stimulation: Postoperative issues. *Mov. Disord.* 21, S219–S237.

Deuschl, G., Schade-Brittinger, C., Krack, P., Volkmann, J., Schafer, H., Botzel, K., Daniels, C., Deutschlander, A., Dillmann, U., Eisner, W., Gruber, D., Hamel, W., Herzog, J., Hilker, R., Klebe, S., Kloss, M., Koy, J., Krause, M., Kupsch, A., Lorenz, D., Lorenzl, S., Mehdorn, H. M., Moringlane, J. R., Oertel, W., Pinsker, M. O., Reichmann, H., Reuss, A., Schneider, G. H., Schnitzler, A., Steude, U., Sturm, V., Timmermann, L., Tronnier, V., Trottenberg, T., Wojtecki, L., Wolf, E., Poewe, W. and Voges, J. (2006b) A randomized trial of deep-brain stimulation for Parkinson's disease. *N. Engl. J. Med.* 355, 896–908.

Diederich, N. J., Kalteis, K., Stamenkovic, M., Pieri, V. and Alesch, F. (2005) Efficient internal pallidal stimulation in Gilles de la Tourette syndrome: a case report. *Mov. Disord.* 20, 1496–1499.

El-Hai, J. (2005) *The Lobotomist: a Maverick Medical Genius and His Tragic Quest to Rid the World of Mental Illness.* Wiley: Hoboken, NJ.

Eltahawy, H. A., Saint-Cyr, J., Giladi, N., Lang, A. E. and Lozano, A. M. (2004) Primary dystonia is more responsive than secondary dystonia to pallidal interventions: Outcome after pallidotomy or pallidal deep brain stimulation. *Neurosurgery* 54, 613–619; discussion 619–621.

Everitt, B. J., Dickinson, A. and Robbins, T. W. (2001) The neuropsychological basis of addictive behaviour. *Brain Res. Brain Res. Rev.* 36, 129–138.

Fields, J. A., Troster, A. I., Woods, S. P., Higginson, C. I., Wilkinson, S. B., Lyons, K. E., Koller, W. C. and Pahwa, R. (2003) Neuropsychological and quality of life outcomes 12 months after unilateral thalamic stimulation for essential tremor. *J. Neurol. Neurosurg. Psychiatry* 74, 305–311.

Flaherty, A. W., Williams, Z. M., Amirnovin, R., Kasper, E., Rauch, S. L., Cosgrove, G. R. and Eskandar, E. N. (2005) Deep brain stimulation of the anterior internal capsule for the treatment of Tourette syndrome: Technical case report. *Neurosurgery* 57, E403; discussion E403.

Freedman, L. J., Insel, T. R. and Smith, Y. (2000) Subcortical projections of area 25 (subgenual cortex) of the macaque monkey. *J. Comp. Neurol.* 421, 172–188.

Freeman, W. and Watts, J. W. (1950) *Psychosurgery in the Treatment of Mental Disorders and Intractable Pain.* C. C. Thomas: Springfield, IL.

Gallistel, C. R. (1986) The role of the dopaminergic projections in MFB self-stimulation. *Behav. Brain Res.* 22, 97–105.

Gibson, W. (1986) *Neuromancer.* Grafton: London.

Goerendt, I. K., Lawrence, A. D. and Brooks, D. J. (2004) Reward processing in health and Parkinson's disease: Neural organization and reorganization. *Cereb. Cortex* 14, 73–80.

Goldapple, K., Segal, Z., Garson, C., Lau, M., Bieling, P., Kennedy, S. and Mayberg, H. (2004) Modulation of cortical-limbic pathways in major depression: Treatment-specific effects of cognitive behavior therapy. *Arch. Gen. Psychiatry* 61, 34–41.

Green, A. L., Nandi, D., Armstrong, G., Carter, H. and Aziz, T. (2003) Post-herpetic trigeminal neuralgia treated with deep brain stimulation. *J. Clin. Neurosci.* 10, 512–514.

Green, A. L., Wang, S., Owen, S. L., Xie, K., Liu, X., Paterson, D. J., Stein, J. F., Bain, P. G. and Aziz, T. Z. (2005) Deep brain stimulation can regulate arterial blood pressure in awake humans. *Neuroreport* 16, 1741–1745.

Green, A. L., Owen, S. L., Davies, P., Moir, L. and Aziz, T. Z. (2006a) Deep brain stimulation for neuropathic cephalalgia. *Cephalalgia* 26, 561–567.

Green, A. L., Wang, S., Owen, S. L., Paterson, D. J., Stein, J. F. and Aziz, T. Z. (2006b) Controlling the heart via the brain: a potential new therapy for orthostatic hypotension. *Neurosurgery* 58, 1176–1183; discussion 1176–1183.

Green, A. L., Wang, S., Owen, S. L., Xie, K., Bittar, R. G., Stein, J. F., Paterson, D. J. and Aziz, T. Z. (2006c) Stimulating the human midbrain to reveal the link between pain and blood pressure. *Pain* 124, 349–359.

Greenberg, B. D., Friehs, G. M., Carpenter, L. L., Tyrka, A., Malone, D., Rezai, A. R., Shapira, N., Fotte, K., Okun, M. S., Goodman, W. K., Rasmussen, S. A. and Price, L. H. (2005) Deep brain stimulation: Clinical findings in intractable depression and OCD. *Neuropsychopharmacology* 29, S32.

Greenberg, B. D., Malone, D. A., Friehs, G. M., Rezai, A. R., Kubu, C. S., Malloy, P. F., Salloway, S. P., Okun, M. S., Goodman, W. K. and Rasmussen, S. A. (2006) Three-year outcomes in deep brain stimulation for highly resistant obsessive-compulsive disorder. *Neuropsychopharmacology* 31, 2384–2393.

Hamani, C., Schwalb, J. M., Rezai, A. R., Dostrovsky, J. O., Davis, K. D. and Lozano, A. M. (2006) Deep brain stimulation for chronic neuropathic pain: Long-term outcome and the incidence of insertional effect. *Pain* 125, 188–196.

Heath, R. G. (1954) *Studies in Schizophrenia.* Harvard University Press: Cambridge, MA.

Heath, R. G. (1963) Electrical self-stimulation of the brain in man. *Am. J. Psychiatry* 120, 571–577.

Hilker, R., Voges, J., Ghaemi, M., Lehrke, R., Rudolf, J., Koulousakis, A., Herholz, K., Wienhard, K., Sturm, V. and Heiss, W. D. (2003) Deep brain stimulation of the subthalamic nucleus does not increase the striatal dopamine concentration in parkinsonian humans. *Mov. Disord.* 18, 41–48.

Hosobuchi, Y., Adams, J. E. and Rutkin, B. (1973) Chronic thalamic stimulation for the control of facial anesthesia dolorosa. *Arch. Neurol.* 29, 158–161.

Hosobuchi, Y., Adams, J. E. and Linchitz, R. (1977) Pain relief by electrical stimulation of the central gray matter in humans and its reversal by naloxone. *Science* 197, 183–186.

Houeto, J. L., Karachi, C., Mallet, L., Pillon, B., Yelnik, J., Mesnage, V., Welter, M. L., Navarro, S., Pelissolo, A., Damier, P., Pidoux, B., Dormont, D., Cornu, P. and Agid, Y. (2005) Tourette's syndrome and deep brain stimulation. *J. Neurol. Neurosurg. Psychiatry* 76, 992–995.

Hunsperger, R. W. (1956) [Affective reaction from electric stimulation of brain stem in cats.]. *Helv. Physiol. Pharmacol. Acta* 14, 70–92.

Illes, J. and Raffin, T. A. (2002) Neuroethics: an emerging new discipline in the study of brain and cognition. *Brain Cogn.* 50, 341–344.

Jenike, M. A. (1998) Neurosurgical treatment of obsessive-compulsive disorder. *Br. J. Psychiatry Suppl.*, 79–90.

Jenkinson, N., Nandi, D., Miall, R. C., Stein, J. F. and Aziz, T. Z. (2004) Pedunculopontine nucleus stimulation improves akinesia in a Parkinsonian monkey. *Neuroreport* 15, 2621–2624.

Joint, C., Nandi, D., Parkin, S., Gregory, R. and Aziz, T. (2002) Hardware-related problems of deep brain stimulation. *Mov. Disord.* 17, S175–S180.

Kerrigan, J. F., Litt, B., Fisher, R. S., Cranstoun, S., French, J. A., Blum, D. E., Dichter, M., Shetter, A., Baltuch, G., Jaggi, J., Krone, S., Brodie, M., Rise, M. and Graves, N. (2004) Electrical stimulation of the anterior nucleus of the thalamus for the treatment of intractable epilepsy. *Epilepsia* 45, 346–354.

Kleiner-Fisman, G., Herzog, J., Fisman, D. N., Tamma, F., Lyons, K. E., Pahwa, R., Lang, A. E. and Deuschl, G. (2006) Subthalamic nucleus deep brain stimulation: Summary and meta-analysis of outcomes. *Mov. Disord.* 21 (Suppl 14), S290–304.

Krack, P., Batir, A., Van Blercom, N., Chabardes, S., Fraix, V., Ardouin, C., Koudsie, A., Limousin, P. D., Benazzouz, A., LeBas, J. F., Benabid, A. L. and Pollak, P. (2003) Five-year follow-up of bilateral stimulation of the subthalamic nucleus in advanced Parkinson's disease. *N. Engl. J. Med.* 349, 1925–1934.

Krauss, J. K. and Jankovic, J. (2002) Head injury and post-traumatic movement disorders. *Neurosurgery* 50, 927–939; discussion 939–940.

Krauss, J. K., Yianni, J., Loher, T. J. and Aziz, T. Z. (2004) Deep brain stimulation for dystonia. *J. Clin. Neurophysiol.* 21, 18–30.

Kringelbach, M. L. (2005) The human orbitofrontal cortex: Linking reward to hedonic experience. *Nat. Rev. Neurosci.* 6, 691–702.

Kringelbach, M. L., Jenkinson, N., Green, A. L., Owen, S. L., Hansen, P. C., Cornelissen, P. L., Holliday, I. E., Stein, J. and Aziz, T. Z. (2007a) Deep brain stimulation for chronic pain investigated with magnetoencephalography. *Neuroreport* 18, 223–228.

Kringelbach, M. L., Jenkinson, N., Owen, S. L. and Aziz, T. Z. (2007b) Translational principles of deep brain stimulation. *Nat. Rev. Neurosci.* 8, 623–635.

Kupsch, A., Benecke, R., Muller, J., Trottenberg, T., Schneider, G. H., Poewe, W., Eisner, W., Wolters, A., Muller, J. U., Deuschl, G., Pinsker, M. O., Skogseid, I. M., Roeste, G. K., Vollmer-Haase, J., Brentrup, A., Krause, M., Tronnier, V., Schnitzler, A., Voges, J., Nikkhah, G., Vesper, J., Naumann, M. and Volkmann, J. (2006) Pallidal deep-brain stimulation in primary generalized or segmental dystonia. *N. Engl. J. Med.* 355, 1978–1990.

Laitinen, L. V., Bergenheim, A. T. and Hariz, M. I. (1992) Leksell's posteroventral pallidotomy in the treatment of Parkinson's disease. *J. Neurosurg.* 76, 53–61.

Leentjens, A. F., Visser-Vandewalle, V., Temel, Y. and Verhey, F. R. (2004) [Manipulation of mental competence: an ethical problem in case of electrical stimulation of the subthalamic nucleus for severe Parkinson's disease]. *Ned. Tijdschr Geneeskd.* 148, 1394–1398.

Leone, M. (2006) Deep brain stimulation in headache. *Lancet Neurol.* 5, 873–877.

Leone, M., Franzini, A., Broggi, G. and Bussone, G. (2006) Hypothalamic stimulation for intractable cluster headache: Long-term experience. *Neurology* 67, 150–152.

Levy, R. M. (2003) Deep brain stimulation for the treatment of intractable pain. *Neurosurg. Clin. N. Am.* 14, 389–399, vi.

Lim, S. N., Lee, S. T., Tsai, Y. T., Chen, I. A., Tu, P. H., Chen, J. L., Chang, H. W., Su, Y. C. and Wu, T. (2007) Electrical stimulation of the anterior nucleus of the thalamus for intractable epilepsy: a long-term follow-up study. *Epilepsia* 48, 342–347.

Limousin, P., Speelman, J. D., Gielen, F. and Janssens, M. (1999) Multicentre European study of thalamic stimulation in parkinsonian and essential tremor. *J. Neurol. Neurosurg. Psychiatry* 66, 289–296.

Lozano, A. M. and Eltahawy, H. (2004) How does DBS work? *Suppl. Clin. Neurophysiol.* 57, 733–736.

Lyons, K. E., Wilkinson, S. B., Overman, J. and Pahwa, R. (2004) Surgical and hardware complications of subthalamic stimulation: a series of 160 procedures. *Neurology* 63, 612–616.

Lyons, K. E. and Pahwa, R. (2005) Long-term benefits in quality of life provided by bilateral subthalamic stimulation in patients with Parkinson disease. *J. Neurosurg.* 103, 252–255.

Maclean, P. D. (1955) The limbic system ("visceral brain") and emotional behavior. *AMA Arch. Neurol. Psychiatry* 73, 130–134.

Maclean, P. D. and Delgado, J. M. (1953) Electrical and chemical stimulation of frontotemporal portion of limbic system in the waking animal. *Electroencephalogr. Clin. Neurophysiol.* 5, 91–100.

Mahl, G. F., Rothenberg, A., Delgado, J. M. and Hamlin, H. (1964) Psychological responses in the human to intracerebral electrical stimulation. *Psychosom. Med.* 26, 337–368.

Malizia, A. L. (1997) The frontal lobes and neurosurgery for psychiatric disorders. *J. Psychopharmacol.* 11, 179–187.

Mark, V. H. and Ervin, F. R. (1965) Role of thalamotomy in treatment of chronic severe pain. *Postgrad. Med.* 37, 563–571.

Mark, V. H. and Ervin, F. R. (1970) *Violence and the Brain.* Harper & Row: New York.

Mayberg, H. S., Liotti, M., Brannan, S. K., McGinnis, S., Mahurin, R. K., Jerabek, P. A., Silva, J. A., Tekell, J. L., Martin, C. C., Lancaster, J. L. and Fox, P. T. (1999) Reciprocal limbic-cortical function and negative mood: Converging PET findings in depression and normal sadness. *Am. J. Psychiatry* 156, 675–682.

Mayberg, H. S., Lozano, A. M., Voon, V., McNeely, H. E., Seminowicz, D., Hamani, C., Schwalb, J. M. and Kennedy, S. H. (2005) Deep brain stimulation for treatment-resistant depression. *Neuron* 45, 651–660.

McIntyre, C. C., Miocinovic, S. and Butson, C. R. (2007) Computational analysis of deep brain stimulation. *Expert Rev. Med. Devices* 4, 615–622.

Meissner, W., Harnack, D., Paul, G., Reum, T., Sohr, R., Morgenstern, R. and Kupsch, A. (2002) Deep brain stimulation of subthalamic neurons increases striatal dopamine metabolism and induces contralateral circling in freely moving 6-hydroxydopamine-lesioned rats. *Neurosci. Lett.* 328, 105–108.

Meloy, S. (2007) Neurally augmented sexual function. *Acta Neurochir. Suppl.* 97, 359–363.

Mink, J. W., Walkup, J., Frey, K. A., Como, P., Cath, D., Delong, M. R., Erenberg, G., Jankovic, J., Juncos, J., Leckman, J. F., Swerdlow, N., Visser-Vandewalle, V. and Vitek, J. L. (2006) Patient selection and assessment recommendations for deep brain stimulation in Tourette syndrome. *Mov. Disord.* 21, 1831–1838.

Moniz, E. (1936) *Tentatives Opératoires dans le Traitement de Certaines Psychoses.* Masson: Paris.

Moreno, J. D. (2003) Neuroethics: an agenda for neuroscience and society. *Nat. Rev. Neurosci.* 4, 149–153.

Mottaghy, F. M., Keller, C. E., Gangitano, M., Ly, J., Thall, M., Parker, J. A. and Pascual-Leone, A. (2002) Correlation of cerebral blood flow and treatment effects of repetitive transcranial magnetic stimulation in depressed patients. *Psychiatry Res.* 115, 1–14.

Nagel, T. (1986) *The View from Nowhere*. Oxford University Press: New York.

Nakamura, K. and Hikosaka, O. (2006) Role of dopamine in the primate caudate nucleus in reward modulation of saccades. *J. Neurosci.* 26, 5360–5369.

Nandi, D., Aziz, T. Z., Giladi, N., Winter, J. and Stein, J. F. (2002a) Reversal of akinesia in experimental parkinsonism by GABA antagonist microinjections in the pedunculopontine nucleus. *Brain* 125, 2418–2430.

Nandi, D., Chir, M., Liu, X., Bain, P., Parkin, S., Joint, C., Winter, J., Stein, J., Scott, R., Gregory, R. and Aziz, T. (2002b) Electrophysiological confirmation of the zona incerta as a target for surgical treatment of disabling involuntary arm movements in multiple sclerosis: Use of local field potentials. *J. Clin. Neurosci.* 9, 64–68.

Nandi, D., Liu, X., Joint, C., Stein, J. and Aziz, T. (2002c) Thalamic field potentials during deep brain stimulation of periventricular gray in chronic pain. *Pain* 97, 47–51.

Nandi, D., Aziz, T., Carter, H. and Stein, J. (2003) Thalamic field potentials in chronic central pain treated by periventricular gray stimulation—a series of eight cases. *Pain* 101, 97–107.

Nimura, T., Yamaguchi, K., Ando, T., Shibuya, S., Oikawa, T., Nakagawa, A., Shirane, R., Itoh, M. and Tominaga, T. (2005) Attenuation of fluctuating striatal synaptic dopamine levels in patients with Parkinson disease in response to subthalamic nucleus stimulation: a positron emission tomography study. *J. Neurosurg.* 103, 968–973.

Nobler, M. S., Oquendo, M. A., Kegeles, L. S., Malone, K. M., Campbell, C. C., Sackeim, H. A. and Mann, J. J. (2001) Decreased regional brain metabolism after ect. *Am. J. Psychiatry* 158, 305–308.

Nuttin, B. J., Gabriels, L. A., Cosyns, P. R., Meyerson, B. A., Andreewitch, S., Sunaert, S. G., Maes, A. F., Dupont, P. J., Gybels, J. M., Gielen, F. and Demeulemeester, H. G. (2003) Long-term electrical capsular stimulation in patients with obsessive-compulsive disorder. *Neurosurgery* 52, 1263–1272; discussion 1272–1274.

Okun, M. S. and Foote, K. D. (2005) Subthalamic nucleus vs globus pallidus interna deep brain stimulation, the rematch. *Arch. Neurol.* 62, 533–536.

Olds, J. and Milner, P. (1954) Positive reinforcement produced by electrical stimulation of septal area and other regions of rat brain. *J. Comp. Physiol. Psychol.* 47, 419–427.

Oommen, J., Morrell, M. and Fisher, R. S. (2005) Experimental electrical stimulation therapy for epilepsy. *Curr. Treat. Options Neurol.* 7, 261–271.

Owen, S. L., Green, A. L., Stein, J. F. and Aziz, T. Z. (2006a) Deep brain stimulation for the alleviation of post-stroke neuropathic pain. *Pain* 120, 202–206.

Owen, S. L. F., Green, A. L., Nandi, D., Bittar, R. G., Wang, S. and Aziz, T. Z. (2006b) Deep brain stimulation for neuropathic pain. *Neuromodulation* 9, 100–106.

Parr, J. R., Green, A. L., Joint, C., Andrew, M., Gregory, R. P., Scott, R. B., McShane, M. A. and Aziz, T. Z. (2007) Deep brain stimulation in childhood: an effective treatment for early onset generalised idiopathic dystonia. *Arch. Dis. Child.* 92(8), 708–711.

Pereira, E. A. and Aziz, T. Z. (2006a) Parkinson's disease and primate research: Past, present, and future. *Postgrad. Med. J.* 82, 293–299.

Pereira, E. A. and Aziz, T. Z. (2006b) Surgical insights into Parkinson's disease. *J. R. Soc. Med.* 99, 238–244.

Pereira, E. A., Green, A. L., Bradley, K. M., Soper, N., Moir, L., Stein, J. F. and Aziz, T. Z. (2007a) Regional cerebral perfusion differences between periventricular grey, thalamic and dual target deep brain stimulation for chronic neuropathic pain. *Stereotact. Funct. Neurosurg.* 85, 175–183.

Pereira, E. A., Green, A. L., Nandi, D. and Aziz, T. Z. (2007b) Deep brain stimulation: Indications and evidence. *Expert Rev. Med. Devices* 4, 591–603.

Pereira, E. A. C., Green, A. L. and Aziz, T. Z. (2007c) Deep brain stimulation for blood pressure control. In: *Textbook of Neuromodulation*. Eds. E. Krames, P. H. Peckham and A. R. Rezai. Oxford University Press: New York.

Pereira, E. A. C., Green, A. L., Wang, S., Ray, N., Kringelbach, M., Owen, S. L., Jenkinson, N., Nandi, D., Stein, J. F. and Aziz, T. Z. (2007d) Stimulating the brain to relive pain: From homunculi to consciousness and deep brain recording to magnetoencephalography. *Neurosurgery* 61, 221.

Pressman, J. D. (1998) *Last Resort: Psychosurgery and the Limits of Medicine*. Cambridge University Press: Cambridge.

Reynolds, D. V. (1969) Surgery in the rat during electrical analgesia induced by focal brain stimulation. *Science* 164, 444–445.

Rezai, A. R., Kopell, B. H., Gross, R. E., Vitek, J. L., Sharan, A. D., Limousin, P. and Benabid, A. L. (2006) Deep brain stimulation for Parkinson's disease: Surgical issues. *Mov. Disord.* 21, S197–S218.

Richardson, D. E. and Akil, H. (1977) Long term results of periventricular gray self-stimulation. *Neurosurgery* 1, 199–202.

Robbins, T. W. (2003) Dopamine and cognition. *Curr. Opin. Neurol.* 16, S1–S2.

Rodriguez-Oroz, M. C., Obeso, J. A., Lang, A. E., Houeto, J. L., Pollak, P., Rehncrona, S., Kulisevsky, J., Albanese, A., Volkmann, J., Hariz, M. I., Quinn, N. P., Speelman, J. D., Guridi, J., Zamarbide, I., Gironell, A., Molet, J., Pascual-Sedano, B., Pidoux, B., Bonnet, A. M., Agid, Y., Xie, J., Benabid, A. L., Lozano, A. M., Saint-Cyr, J., Romito, L., Contarino, M. F., Scerrati, M., Fraix, V. and Van Blercom, N. (2005) Bilateral deep brain stimulation in Parkinson's disease: a multicentre study with 4 years follow-up. *Brain* 128, 2240–2249.

Rosvold, H. E. and Delgado, J. M. (1956) The effect on delayed-alternation test performance of stimulating or destroying electrically structures within the frontal lobes of the monkey's brain. *J. Comp. Physiol. Psychol.* 49, 365–372.

Schoenen, J., Di Clemente, L., Vandenheede, M., Fumal, A., De Pasqua, V., Mouchamps, M., Remacle, J. M. and de Noordhout, A. M. (2005) Hypothalamic stimulation in chronic cluster headache: a pilot study of efficacy and mode of action. *Brain* 128, 940–947.

Schupbach, W. M., Chastan, N., Welter, M. L., Houeto, J. L., Mesnage, V., Bonnet, A. M., Czernecki, V., Maltete, D., Hartmann, A., Mallet, L., Pidoux, B., Dormont, D., Navarro, S., Cornu, P., Mallet, A. and Agid, Y. (2005) Stimulation of the subthalamic nucleus in Parkinson's disease: a 5-year follow up. *J. Neurol. Neurosurg. Psychiatry* 76, 1640–1644.

Sem-Jacobsen, C. W. (1959) Depth-electrographic observations in psychotic patients: a system related to emotion and behavior. *Acta Psychiatr. Scand. Suppl.* 34, 412–416.

Seminowicz, D. A., Mayberg, H. S., McIntosh, A. R., Goldapple, K., Kennedy, S., Segal, Z. and Rafi-Tari, S. (2004) Limbic-frontal circuitry in major depression: a path modeling metanalysis. *Neuroimage* 22, 409–418.

Stefani, A., Lozano, A. M., Peppe, A., Stanzione, P., Galati, S., Tropepi, D., Pierantozzi, M., Brusa, L., Scarnati, E. and Mazzone, P. (2007) Bilateral deep brain stimulation of the pedunculopontine and subthalamic nuclei in severe Parkinson's disease. *Brain* 130, 1596–1607.

Sydow, O., Thobois, S., Alesch, F. and Speelman, J. D. (2003) Multicentre European study of thalamic stimulation in essential tremor: a six year follow up. *J. Neurol. Neurosurg. Psychiatry* 74, 1387–1391.

Taha, J. M., Janszen, M. A. and Favre, J. (1999) Thalamic deep brain stimulation for the treatment of head, voice, and bilateral limb tremor. *J. Neurosurg.* 91, 68–72.

Temel, Y. and Visser-Vandewalle, V. (2004) Surgery in Tourette syndrome. *Mov. Disord.* 19, 3–14.

Temel, Y., van Lankveld, J. J., Boon, P., Spincemaille, G. H., van der Linden, C. and Visser-Vandewalle, V. (2004) Deep brain stimulation of the thalamus can influence penile erection. *Int. J. Impot. Res.* 16, 91–94.

Theodore, W. H. and Fisher, R. S. (2004) Brain stimulation for epilepsy. *Lancet Neurol.* 3, 111–118.

Thoden, U., Doerr, M., Dieckmann, G. and Krainick, J. U. (1979) Medial thalamic permanent electrodes for pain control in man: an electrophysiological and clinical study. *Electroencephalogr. Clin. Neurophysiol.* 47, 582–591.

Tooth, J. and Newton, M. (1961) *Leucotomy in England and Wales 1942–1954: Report on Public Health and Medical Subjects No. 104.* Her Majesty's Stationery Office: London.

Troster, A. I., Fields, J. A., Wilkinson, S., Pahwa, R., Koller, W. C. and Lyons, K. E. (2003) Effect of motor improvement on quality of life following subthalamic stimulation is mediated by changes in depressive symptomatology. *Stereotact. Funct. Neurosurg.* 80, 43–47.

UK ECT Review Group (2003) Efficacy and safety of electroconvulsive therapy in depressive disorders: a systematic review and meta-analysis. *Lancet* 361, 799–808.

Valenstein, E. S. (1973) *Brain Control: a Critical Examination of Brain Stimulation and Psychosurgery.* Wiley-Interscience: London.

Vidailhet, M., Vercueil, L., Houeto, J. L., Krystkowiak, P., Lagrange, C., Yelnik, J., Bardinet, E., Benabid, A. L., Navarro, S., Dormont, D., Grand, S., Blond, S., Ardouin, C., Pillon, B., Dujardin, K., Hahn-Barma, V., Agid, Y., Destee, A. and Pollak, P. (2007) Bilateral, pallidal, deep-brain stimulation in primary generalised dystonia: a prospective 3-year follow-up study. *Lancet Neurol.* 6, 223–229.

Visser-Vandewalle, V., Temel, Y., Boon, P., Vreeling, F., Colle, H., Hoogland, G., Groenewegen, H. J. and van der Linden, C. (2003) Chronic bilateral thalamic stimulation: a new therapeutic approach in intractable Tourette syndrome. Report of three cases. *J. Neurosurg.* 99, 1094–1100.

Walter, B. L. and Vitek, J. L. (2004) Surgical treatment for Parkinson's disease. *Lancet Neurol.* 3, 719–728.

Wishart, H. A., Roberts, D. W., Roth, R. M., McDonald, B. C., Coffey, D. J., Mamourian, A. C., Hartley, C., Flashman, L. A., Fadul, C. E. and Saykin, A. J. (2003) Chronic deep brain stimulation for the treatment of tremor in multiple sclerosis: Review and case reports. *J. Neurol. Neurosurg. Psychiatry* 74, 1392–137.

Yianni, J., Bain, P. G., Gregory, R. P., Nandi, D., Joint, C., Scott, R. B., Stein, J. F. and Aziz, T. Z. (2003) Post-operative progress of dystonia patients following globus pallidus internus deep brain stimulation. *Eur. J. Neurol.* 10, 239–247.

Yianni, J., Green, A. L., McIntosh, E., Bittar, R. G., Joint, C., Scott, R., Gregory, R., Bain, P. G. and Aziz, T. Z. (2005) The costs and benefits of deep brain stimulation surgery for patients with dystonia: an initial exploration. *Neuromodulation* 8, 155–161.

Young, R. F. and Rinaldi, P. C. (1997) *Brain Stimulation.* Springer-Verlag: New York.

19

Pain and Pleasure: Masters of Mankind

SIRI LEKNES AND IRENE TRACEY

Nature has placed mankind under the governance of two sovereign masters, pain and pleasure. It is for them alone to point out what we ought to do, as well as to determine what we shall do.

Jeremy Bentham, 1789

For Bentham, the study of pleasures and pains was the study of hedonic feelings. All good feelings were pleasures, and "pain" could describe all that humankind sought to avoid (Bentham, 1789). Everyday expressions ("what a pain!", "my pleasure" etc.) indicate that this type of categorization of positive and negative hedonic feelings is still in use today. In modern-day science, however, the terms reward and punishment have largely replaced Bentham's pleasures and pains. For instance, a recent PubMed search of "reward and brain" yielded over 20 times more entries than "pleasure and brain." Reward and punishment are defined as something an animal will work to achieve or avoid, thus effectively circumventing the hedonic aspect inherent in pleasures and pains. This has allowed for a flourishing behavioral neuroscience literature on positive and negative reinforcement. The probability that a previously rewarded (or punished) response is emitted is considered an objective measure of the reinforcement value of that reward. In contrast, the hedonic value of a reward or punishment is by definition subjective.

The hedonic quality of pleasures and pains is the subject matter of this chapter. Exploring the relationship between the brain and subjective hedonic feelings (qualia) is necessary to understand "what it is like" to be a sentient being (Nagel, 1974). Here, we shall consider a simple continuum of hedonic feelings spanning from the extremely unpleasant through to the extremely pleasant (Figure 19.1). In general, it holds true that punishments feel unpleasant and generate negative affect (NA), whereas rewards cause pleasure and positive affect (PA). This relationship is not always straightforward, however. Importantly, even primary rewards or punishments are not always rewarding or aversive, although some affective responses to them may be hardwired. For instance, sweet tastes and tissue damage are associated with innate reflexive reactions such as smiling and withdrawal; this holds true across a range of species (Berridge, 2003). Nonetheless, having already devoured six big chocolate bars, the seventh would not feel like much of a reward; satiation has made its taste aversive (Small et al., 2001). Conversely, scratching an itchy bit of skin until it looks red and irritated can feel extremely pleasant despite causing tissue damage (Craig, 2003). This scenario can also illustrate the complex relationship between hedonic feelings and positive affect and negative affect. For eczema sufferers, scratching often exacerbates the skin condition (Carroll et al., 2005). Thus the pleasure of scratching may be diminished by fear and guilt, while resisting the "irresistible" itch can cause positive affect (Leknes et al., 2007). Here, we will consider reward, punishment, and positive affect/negative affect only in as much as these give rise to or affect hedonic feelings.

Pain and Hedonic Feelings

"Pleasure and pain were the earliest forms of emotion to evolve" (Jaak Panksepp, as cited in Phillips, 2003).

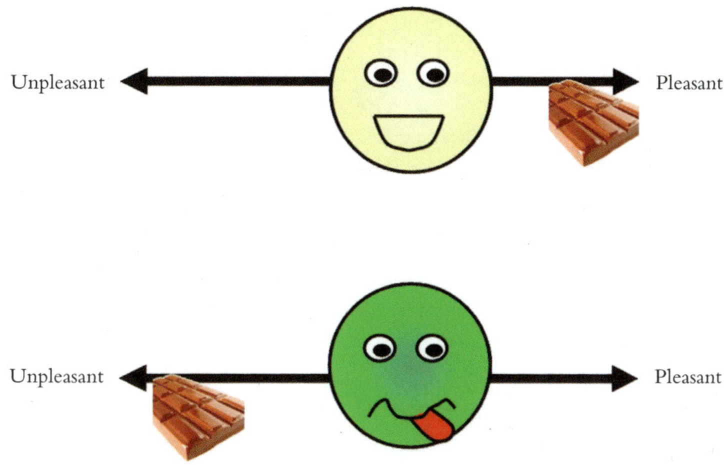

Figure 19.1 The inner state determines the pleasantness of a stimulus. While chocolate and other sweet foods are pleasurable under normal circumstances, the opposite may be true for someone with nausea.

Unlike pleasure (but like reward), pain is the subject of a vast field of neuroscientific and medical research. The International Association for the Study of Pain (IASP) defines pain as "an unpleasant sensory and emotional experience associated with actual or potential tissue damage, or described in terms of such damage." Note that unlike for Bentham, pain no longer describes negative hedonic feelings such as unhappiness, irritation, and itch. Instead, pain research has mainly concerned itself with pain related to nociception. Pain is reliably induced by stimuli that activate nociceptive receptors in the skin, muscle, gut, and others. Although sensory pains vary qualitatively (consider a sharp pin-prick pain vs. a dull muscle ache), these feelings are similar enough to be classified as pain both in hedonic and physiological terms. Chronic pain syndromes such as poststroke and phantom pain are examples of painful conditions where the pain is not caused by stimulation of nociceptors in the periphery. Nevertheless, the subjective hedonic feeling of these pain syndromes mimics sensory pains, and central pain syndromes are thus encompassed by the IASP definition of pain. Most of the pain studies described in this chapter have used nociceptive stimulation to induce pain. Numerous attempts have been made to find objective measures for pain. Animal pain research largely relies on measures of avoidance behaviors such as tail flick latency, although some quantifications of suffering behavior have also been reported (Dickinson and Dearing, 1979; Szechtman et al., 1981). Although many human pain studies use subjective pain ratings to indicate the level of pain, others choose to assess nociceptive signalling, for example, by measuring electrical activity from peripheral neurons (Raja et al., 1999). Some still argue for objective measures of pain, such as quantification of reflexes (Gerdelat-Mas et al., 2007). With the advent of functional brain imaging, many hope that a technique that will provide the elusive objective measurement of pain has been found. A number of brain regions light up in neuroimaging studies of pain. Some, notably the insula, thalamus, and dorsal anterior cingulate cortex (ACC), are reported with great consistency (Tracey, 2005). Rainville and colleagues (1997) used hypnotic suggestion to show that activity in the ACC varies with the affective component of pain processing, leaving the thalamus and the insula as the main candidate regions for an objective marker of nociceptive input. Direct electrical stimulation of insular cortex in epilepsy patients causes intense feelings of pain (Ostrowsky et al., 2002). Interestingly, however, both the insula and the thalamus have recently been shown to activate during hypnotic suggestion of pain in the absence of nociceptive stimulation (Raij et al., 2005).

The role of subjective interpretation of pain as the determinant of the hedonic pain experience is becoming increasingly recognized within the pain field, especially in the study of the factors that increase pain unpleasantness (Fairhurst et al., 2007; Gracely et al., 2004; Wiech et al., 2006). However, little research focuses on the role of pleasure or positive emotion for pain (but see Strand et al., 2006, 2007). Medical treatment for pain is concerned with reducing negative emotion (analgesia) more than increasing positive emotion. Although opiate and other analgesics are frequently abused and are known to induce euphoria (Franklin,

1998), few studies have systematically assessed positive affect related to pain or pain relief. We believe that the field of pain research may benefit from looking to Bentham's wider definition of pain as well as his focus on subjective hedonic feelings. For instance, comparing pain with unpleasant sensations such as itch and nausea, and also with pleasant sensations and emotions, could elucidate common emotional components of sensory hedonic feelings. Similarly, studying the interactions between pain and other hedonic emotions may further our understanding of both pains and pleasures. To our knowledge, neuroimaging studies of pain have not identified a single brain region that has not also been implicated in aspects of reward processing. The insula encodes taste and food cravings (Pelchat et al., 2004; Small and Apkarian, 2006); the ACC represents reward size (Koyama et al., 2001; Rogers et al., 2004); and the amygdala is involved in anticipation of pleasant taste (O'Doherty et al., 2002) and in the experience of intense pleasure when listening to music (Blood and Zatorre, 2001). The thalamus is involved in drug cravings and dysregulation of reward motivation (Volkow and Fowler, 2000). In addition, opioids and dopamine, which are perhaps the two most well-defined neurotransmitter systems involved in modulation of pain (Fields, 2004; Scott et al., 2006; Wood, 2006; Zubieta et al., 2005), are also crucial for positive hedonic processing (Robinson and Berridge, 2001; Schultz, 2004). Much remains to be learned about the function of these neurotransmitter systems in mediating pleasure–pain interactions.

In terms of evolutionary psychology, both seeking pleasures and avoiding pains are important for survival and may compete for preference within the brain (Fields, 2006). In the face of a large food reward, which can only be obtained at the cost of a small amount of pain, for instance, it would be beneficial if the pleasurable food reduced pain unpleasantness. Cabanac (2002) argues that the brain must contain a common currency that allows motivations for pleasures and pains to be weighed against each other. This chapter will summarize the research on interactions between pleasure and pain and other factors influencing the hedonic quality of pains and pleasures. The most important of these is homeostasis.

Homeostasis and Opponent Process Theory

The state of the body and the mind determines the pleasantness or unpleasantness we experience when we perceive a stimulus. The seventh chocolate bar eaten in a row is aversive because the body is already more than sated on cocoa, sugar, and fat. For Bentham (1789), the key to pleasures and pains is subjective utility. If overeating and skin damage are not useful to you, they should not be pleasant. These ideas are conceptualized in homeostatic theory.

All organisms strive to maintain optimal internal equilibrium. The notion of homeostasis was first introduced in relation to automatic regulatory processes such as thermoregulation (Cannon, 1929). Later findings have highlighted the relationship between homeostasis and emotion. Michel Cabanac showed that the pleasantness of a stimulus increases the more effective that stimulus is in restoring bodily homeostasis (Cabanac, 1979). When someone's core temperature is too low, stimuli that would normally feel too hot become pleasant (Cabanac, 1971). In other words, homeostatic utility determines the hedonic value of a stimulus. This effect is well-documented for primary rewards such as food and drink, which taste better when relieving a hunger or thirst state (de Araujo et al., 2003; Kringelbach et al., 2003; Small et al., 2001). Similarly, pain unpleasantness increases with greater perceived threat (Price et al., 1987).

In a certain sense, hedonic feelings exist to encourage the constant optimization of our internal homeostatic balance. Unpleasant sensations such as pain and itch have probably evolved as homeostatic alarm signals, notifying us of imbalances in the mechanical, thermal, or chemical status of the tissues of the body (Stante et al., 2005). Unfortunately, when itch and pain become chronic, these sensations retain the interruptive quality of alarm signals, constantly pulling attention toward the unpleasantness of the condition and disrupting other thoughts and activities (Eccleston and Crombez, 1999). In contrast, positive hedonic feelings more often signal that a goal has been reached, and pleasure does not seem to have the same interruptive effect on attention.

According to the opponent process theory, the homeostatic system strives to neutralize any deviation from the optimal balance of the organism, whether pleasurable or aversive, both externally or internally generated: "There are certain systems in the brain, the business of which is to suppress or reduce all excursions from hedonic neutrality" (Solomon and Corbit, 1974, p. 143).

In this model, an unpleasant stimulus or emotion would trigger not only a negative affective reaction, but also a process of opposite valence, which has a slower onset and offset (Figure 19.2A,B). If the

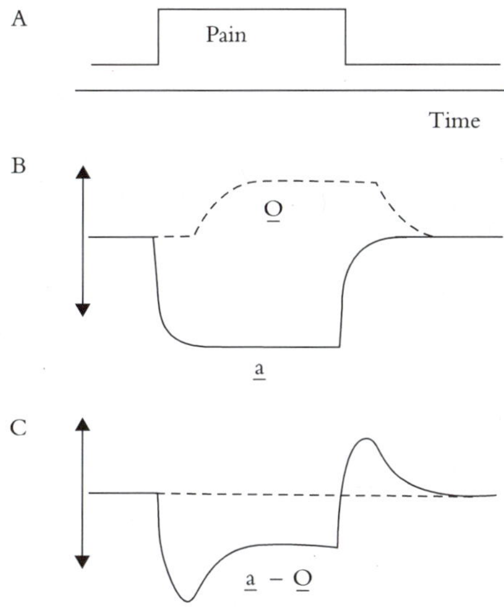

Figure 19.2 The opponent process theory. A outlines the event. The arrows in B and C signal hedonic valence. a is the primary process, reflecting negative valence. o is the opponent process. Panel C shows the net result of the two opposing processes. The late peak reflects pleasant relief. (Adapted from Solomon and Corbit, 1974.)

unpleasant sensation is suddenly terminated, the activity of the positive affective process causes us to perceive positive emotion (Figure 19.2C). The opposite pattern is proposed for positive hedonic experiences, which, according to Solomon and Corbit, are followed by a dysphoric "low" when abruptly terminated. Although the opponent process model is almost certainly too simplistic, it provides a putative mechanism for explaining such phenomena as anticlimaxes and the euphoria of risk-taking. Similarly, it accounts for the pleasure of relieving an itch (for most people) as well as the pleasure of not doing so (for someone anxious about the consequences of scratching). The idea that pleasure can be caused by relief from something unpleasant is not a new one: in his *Discourse on the Nature of Pleasures and Pain,* Verri holds that all pleasures are the results of relief from pain, and that pleasure is limited by the quantity of the pain it removes (Guidi, 1994 cites Verri, P., 1781). The neuroscience of pleasurable relief is discussed further in the next section.

Solomon and Corbit (1974) made no distinction between homeostatic control of externally and internally generated feelings. Inspired perhaps by the popularity of classic and operant conditioning research in the 1960s and 1970s, the authors emphasized the role of homeostatic mechanisms in learning and expectation. In brief, learning that A precedes B ultimately leads to the affective reaction to B shifting forward in time and becoming associated with A. The opponent process initially associated with B also shifts forward in time. For instance, an eczema sufferer may learn that scratching during the day increases itchiness in the evening, and so feels guilt and fear when he scratches. This shift of affect toward the "cue" (A) means that if B fails to occur after A, B's opponent process will take place instead. For our eczema sufferer, this effect is apparent as the pleasant relief he feels when for once the itch *has not* increased after a day of scratching. Within neuroscience, the reward learning literature has provided evidence for opponent hedonic reactions to counterfactual outcomes. When an expected monetary or food reward is omitted, the resulting negative affective reaction has been called "frustrative non-reward" (Siegrist et al., 2005; Tucker et al., 2005).

Anticipated Emotions, Frustration, and Relief

In fact, since animals will work to avoid it, frustrative nonreward matches the definition of a punishment. Similarly, the effort we are willing to exert to relieve something unpleasant tells us that relief is a type of reward. As to the hedonic aspect of relief and frustration/disappointment, it is clear that these belong to opposite ends of the hedonic continuum. In our laboratory, we have found evidence for a positive hedonic reaction associated with relief from pain (Leknes et al., 2008). In brief, the results from our laboratory confirm the predictions stipulated in the opponent process theory as follows: (1) The sudden termination of a painful sensation elicits positive affect, as measured by subjective ratings of relief pleasantness. (2) The relief associated with the offset of pain increases with the intensity of the pain sensation. (3) The pleasantness of relief from pain increases with the efficacy and speed of return to homeostatic balance, as evidenced by the higher positive hedonic ratings when the skin is gently cooled after burning heat pain. In addition, findings from a study by Donald Price and colleagues (1980) are consistent with the notion that when pain is signaled by a cue, both the negative pain-related affective reaction and its opponent process are shifted forward in time to be elicited by the cue. In fact, this effect is so strong that subjects even reported positive affect when their skin was heated to a painful level—mild pain had become a relief relative to the expected severe pain stimulus (Price et al., 1980).

A recent study has investigated the hedonic value of waiting for pain, which they call dread (although despite the obvious hedonic connotations of this term, the authors present their findings largely in terms of neuroeconomical utility rather than hedonic displeasure). Berns and colleagues (2006) report compelling evidence for unpleasantness conferred onto the pain cue: several subjects found waiting for the painful electric shocks so aversive that they opted for increasing pain intensity just to shorten the wait. Little research has focused on the pleasure of waiting for something good, despite theoretical suggestions that the anticipation may even be the best part. In a comment to Berns et al., Loewenstein (2006, p. 305) enthuses over the idea that we "derive pleasure and pain directly from information, rather than from any material benefits that the information procures." Similarly, Rozin emphasizes the role of anticipatory pleasure: "In terms of real life, most pleasure may come from memory or anticipation, as opposed to online experience" (Rozin, 1999, p. 113). A recent suggestion that the ability to enjoy anticipatory pleasure depends on personality traits may spur more research in this area (Gard et al., 2006).

The above findings on anticipated emotions and counterfactual outcomes may also inform the interpretation of data from extinction learning paradigms. The process of "unlearning" or extinguishing the association between a cue and an outcome is generally more time-consuming and less successful when the cue is only partially predictive of the outcome. During partially reinforced reward learning, frustration from reward omission can become "counterconditioned" by the occasional reward (Tucker et al., 2005). Because of this, the animal trained on a variable reward schedule more easily tolerates later frustrations: it does not give up, "hoping" for a pleasant food reward despite repeated disappointment. An intriguing thought is that extinction learning in general, far from being "unlearning" or associating a conditioned stimulus with no outcome, instead involves the forming of new associations with the opposite valence. Extinction of fear conditioning would thus entail appetitive learning, in which the cue previously associated with fear becomes predictive of pleasant relief. Presumably then, as cue-related fear dissolved, relief would decrease until eventually the cue would elicit little hedonic affect. In support of this prediction, an important region associated with extinction of fear associations in human subjects is the ventromedial prefrontal cortex (Phelps et al., 2004), a region consistently implicated in the encoding of positive hedonic feelings (Knutson et al., 2003; O'Doherty et al., 2003). The ventromedial prefrontal cortex has a very high opiate receptor density (Baumgartner et al., 2006), and activation in this region has been shown to decrease in response to unexpected reward failure (Knutson et al., 2001a,b; Ramnani et al., 2004).

Another prediction of the opponent process theory supported by neuroscience research is based on what Solomon and Corbit call Pavlovian "backward conditioning." Here, the conditioned stimulus follows the termination of the unconditioned stimulus closely in time and becomes associated with the relief or dysphoria following a pain or a pleasure. An elegant study of fruit flies illustrates this concept. Tanimoto and colleagues (2004) varied the interval between a neutral odor and an electric shock, and showed that if the odor precedes the shock, the fruit flies will avoid this odor. When the odor was presented immediately after the shock, however, it became a signal of safety from pain. The flies would later approach this odor as they would approach the smell of food. Although we may never know what pleasure feels like for a fruit fly, the results from this study nevertheless illustrate the similarity between primary rewards and obtaining relief from primary punishers.

Dopamine and Opioid Involvement in Pleasure and Pain

The title of a recent paper sums up current thinking on the role of these two neurotransmitters: "Opioids for Hedonic Experience and Dopamine To Get Ready for It" (Barbano and Cador, 2007). The role of dopamine in reward processing is well established, and for a long time, dopamine was dubbed "the pleasure molecule." This hypothesis is no longer supported, however (Salamone et al., 1997). Instead, it appears that opioids underpin hedonic 'liking', whereas dopamine's role is primarily in motivation or 'wanting' (Berridge, 2007). The opioid-driven pleasure circuit overlaps considerably with the dopamine system, to the point where some cells take part in both circuits. But its role and chemistry are quite different (Berridge, 2003). While dopamine neurons signal salient events even when these are unrelated to primary rewards (Blatter and Schultz, 2006), endogenous opioids have been shown to encode relative taste preference (Taha et al., 2006), and microinjection of opioids into the nucleus accumbens (NAc), the ventral pallidum, the ventral tegmental area (VTA), and the periacqueductal gray (PAG) increase pleasure-related facial expressions in rodents (Pecina and Berridge, 2000; Smith and Berridge, 2007). Perhaps reflecting the fact that things pleasurable often induce 'wanting', opioids may increase dopamine release in the NAc through the inhibition of GABAergic neurons at the VTA (GABA disinhibition).

Not all opioids are involved in positive hedonic processing, however. While opiate agonists that bind preferentially to μ-opiate receptors cause a feeling of elation, kappa selective opiates generate negative affect in humans (Burgdorf and Panksepp, 2006). Of the endogenous μ-specific opiates, endomorphin-2 appears to have stronger rewarding effects than endomorphin-1 (Huang et al., 2004; Wilson et al., 2000; Zangen et al., 2002). All known μ-opioid subtypes have potent analgesic effects, however. Microinjection of μ-opioids directly into the NAc has been shown to induce antinociception, and microinjection of naloxone into the NAc attenuates the antinociceptive effect of systemically administered morphine (Dill and Costa, 1977). Dopamine activity can also cause pain suppression (Shimizu et al., 2004; Wood, 2006).

The close overlap between the rewarding and analgesic effect of opioids forms the basis of the affective analgesia hypothesis, which in its weakest form holds that pleasure ("reinforcement") can drive the neural substrate of analgesia (Franklin, 1998). A stronger form of the hypothesis—that pleasure and analgesia are identical—has been rejected. For one thing, there are multiple mechanisms of analgesia, including the classic descending inhibitory control system driven by aversive events (Fields, 2004). In fact, the biological significance of endogenous pain control is generally seen in the context of behavioral conflicts where the injured individual must disengage from pain in order to fight or escape (Melzack and Casey, 1968). Both pain- and pleasure-induced analgesia appear to be mediated via the mesolimbic reward system as well as brainstem modulatory nuclei and can be blocked by opioid and/or dopamine antagonists (Forsberg et al., 1987; Gear et al., 1999; Reboucas et al., 2005). According to the motivation-decision model of pain, both types of analgesia act via an all-or-none decision circuit, exerting bidirectional control over pain (Fields, 2006). The circuit consists of ON- and OFF-cell populations in midbrain and medullary pain-modulatory nuclei (the PAG and the rostroventral medulla [RVM]). The cells have a reciprocal activity pattern where OFF-cell silence *permits* a pain response and ON-cell activity facilitates it. Conversely, OFF-cell activity blocks responses to noxious stimuli (Fields, 2006). Taken together, the opioid and dopaminergic mechanisms underlying pleasure- and pain-related analgesia lend support to the idea of a "common currency," which helps the brain make decisions to optimize survival.

Self-Harming, Chilli Peppers, and Masochism

Often, however, the decisions we make seem paradoxical. If it is really better to avoid pain, why do so many people engage in painful and/or potentially harmful activities? Boxers, marathon runners, and soldiers are obvious examples, but even the pleasure of scratching a mosquito bite and eating a spicy meal may be a direct consequence of the tissue damage these activities bring about. The desire for relief from an unpleasant homeostatic state may be key to understanding at least some of these activities, as has been suggested in the case of skydiving (Seymour et al., 2005) and self-cutting (Korner et al., 2007). Subjective reports from borderline personality patients who self-harm imply that the physical pain provides relief from the mental pain they are experiencing (Korner et al., 2007). As is described in more detail below, skydivers experience more anhedonia and derive less pleasure from everyday rewards than individuals who do not take part in extreme sports (Franken et al., 2006). It is certainly possible that thrilling activities like skydiving provide "relief" from an otherwise flat affective state. Interestingly, both the above paradoxical behaviors have been related to changes in dopamine and opioid neurotransmitter systems. Treatment with the predominantly μ-opioid antagonist naltrexone reduces self-harming (Symons et al., 2004), suggesting that opioid release caused by the physical pain is key to maintaining this behavior. Evidence from positron emission tomography (PET) studies of the role of the μ-opioid in emotion regulation suggests a possible mechanism by which self-cutting may relieve mental pain. When people are feeling sad, μ-opioid neurotransmission is reduced in several brain regions, including the rostral ACC (Zubieta et al., 2003). Physical pain, on the other hand, increases μ-opioid activation in this and other regions, especially when subjects believe their pain is being reduced (Zubieta et al., 2005). Both sadness and anhedonia are also associated with disruptions of dopaminergic signaling in the brain (Tremblay et al., 2005).

Another interesting case of paradoxical and potentially self-injurious behavior is the frequent human consumption of chilli peppers. Chillies "burn" in the mouth and on the skin because they contain the irritant capsaicin, and preparations containing this substance are used in pain research to cause burning pain and hyperalgesia (Dirks et al., 2003; Zambreanu et al., 2005). Some capsaicin creams are also used in the treatment of persistent pain, as the initial burning sensation is followed by a period of antinociception (Dray, 1992). Since capsaicin is neurotoxic to certain sensory neurons in the skin (Chard et al., 1995; Hail, 2003), it makes sense from a homeostatic point of view that applying it to the skin or eating it in a hot curry should feel unpleasant. In fact, plants such as chilli peppers and garlic appear to use thermoTRP-activating capsaicin and allicin to deter mammalian predators from consuming the plant (Dhaka et al., 2006). So why do so many people enjoy foods containing these irritants? One interesting explanation is suggested in Harold McGee's book on *the Science and Lore of the Kitchen* (McGee, 2004). Since many "spicy" compounds induce a temporary inflammation in the mouth, they may enhance pleasure by making eating more *sensual* and intense; the mouth and tongue are tender and more sensitive to touch and temperature.

"I have just run the hot water tap and put my hands underneath it, with the water as hot as I could bear for as long as I could bear. The water was probably hotter than most people could stand, certainly beyond the temperature to cause pain. That was why I did it" (Launer, 2004, p. 383). Dermatologist and eczema sufferer John Launer explains that when the itch becomes intolerable, most people with eczema know that the only thing that will "crack" the itch is to subject themselves to pain. While applying cool water to the skin will relieve the itch only briefly, pain—even when applied to an unaffected body part—will provide longer-lasting itch relief (Mochizuki et al., 2003; Ward et al., 1996). This is probably the reason that scratching a mosquito bite until the skin looks red and flared can often feel pleasurable; the tissue damage caused by nails biting into the skin "cracks" the attention-grabbing and unpleasant itch sensation. The unpleasant homeostatic imbalance caused by itch is thus an example of a state for which pain becomes beneficial (Craig, 2003).

For many athletes, whether ballet dancers, boxers, or football players, entry into the athletic community is marked by willingness to endure pain (Downey, 2007). The role of pain in athletic activities is not well understood and is unlikely to be explained by a single factor. It is possible that the pain endured during physical activity, like the painful burn of capsaicin cream or the exquisite hurt of self-cutting, produces relief from another, ongoing pain. It may also be that this type of pain enhances other, pleasurable sensory experiences, like a hot curry increasing sensitivity in the mouth. What is known is that during training, many athletes learn to distinguish between "normal" and other pain signals—some should be endured and others paid attention to (Downey, 2007). Janal and colleagues (1994) investigated the pain sensitivity of runners and nonrunners in their paper "Are Runners Stoical?" A significant difference between the two groups was found only for cold pain, a sensation the athletes were accustomed to from running in cold weather. Changes in athletes' pain sensitivity during competitions has been attributed to stress-induced endogenous opioid analgesia (Sternberg et al., 1998). It is likely that the endogenous opioid system is involved in antinociception during all kinds of athletic pain, and endorphins are generally thought to be responsible for the feeling of well-being, which often follows vigorous exercise (Morgan, 1985).

As a positive hedonic feeling, post-training well-being may in turn make athletes remember their pain as less unpleasant. Daniel Kahneman and colleagues (1993) have found that adding a better end to a painful procedure changes subjects' memories of the pain they experienced. A few minutes of less intense pain added to the end of a medical procedure led subjects to remember less overall pain, and although this procedure involved experiencing pain for a longer time, it was the subjects' preferred option. Finally, while on the topic of paradoxical behaviors, let us not forget that enduring some discomfort is an efficient way of increasing the pleasure of returning to homeostasis. Who has not tried fasting for a few extra hours before a feast, or stayed in the sun until almost unbearably hot before jumping into a refreshing pool?

Whether similar homeostatic mechanisms are involved in sexual masochism—perhaps the most paradoxical of all human behaviours—is not known. For all the activities outlined above, although there is evidence that the hedonic feelings associated with the pain become less intense, there is little proof that the pain is experienced as pleasurable *in itself.* While the overall experience adds up to be pleasurable, the painful sensation itself is seldom seen as the direct cause of pleasure. This appears to hold true even in extreme pain-seeking subcultures of a nonsexual nature. Such pain-seeking behavior appears to be similar to the type of thrill-seeking seen in extreme sports, and stress- and fear-induced analgesia undoubtedly diminishes the pain intensity of these activities. In contrast, although little study has been done in this field, we know of at least one subculture where even extremely intense pains may cause pleasure, and where the main mechanism for analgesia may be pleasant sexual arousal. Contrary to popular belief, however, pain may not be the main goal for sexual masochists. Instead, it has been argued that the main purpose of pain in sadomasochistic (SM) interactions is to denote power (Cross and Matheson, 2006). In support of this notion, many important symbols and activities in SM relations are more to do with handing over personal autonomy (i.e., bondage) than with pain per se (Moser and Kleinplatz, 2006). The peculiar mixture of pain, power games, and sex central to the SM subculture is distasteful to many, and this dichotomy was exploited in an elegant neuroimaging study on disgust and sexual arousal (Stark et al., 2005). Two groups of subjects viewed disgust-inducing, erotic, neutral, and SM-related images. The researchers report a striking resemblance between brain activation patterns in healthy subjects viewing erotic images compared with SM subjects looking at images of SM interactions (see Figure 19.3). When the non-SM group viewed the SM-related images, however, their brain activation was more similar to that of either group feeling disgust.

Pain, Reward, Mood, and Placebo

Many theories have been put forth to explain why sadomasochists associate pain with sexual arousal and pleasure, but little empirical evidence exists to support these (Cross and Matheson, 2006). What is clear is that for most sexual masochists, pain is only pleasurable within an SM context. The context in which pain is experienced has been shown to strongly influence pain perception both in studies of pain in both experimental and clinical populations (Benedetti et al., 2005; Moseley and Arntz, 2007; Price et al., 1987). As has been suggested above, when pain occurs within a context of available rewards, the pain modulatory system may induce antinociception (Fields, 2006). In this section, we present evidence for the analgesic effects of both primary and secondary rewards, and discuss a possible interaction with positive mood.

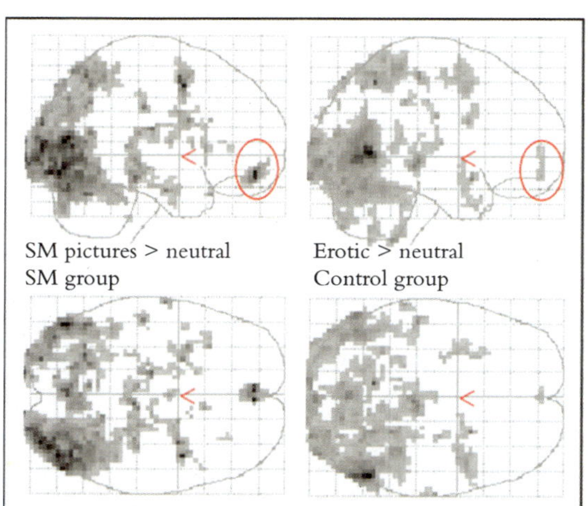

Figure 19.3 "Glass brains" from SPM analysis of the SM and control groups viewing SM-related and erotic images, respectively. The circles indicate the ventromedial prefrontal cortex. Other common regions were identified using region of interest analysis, notably the amygdala. (Image adapted from Stark et al, 2005.)

Although it cannot fully explain pain-seeking in sexual masochists, endogenous opioid release appears to play a role in sexual behavior in humans and in animals (Coolen et al., 2004), and may be the main mechanism of antinociception in sexual contexts. Endogenous opioids underpin positive hedonic feelings in sexual contexts. Treatment with the μ-opioid antagonist naloxone reduced subjective arousal and reported pleasure in human male orgasms (Murphy et al., 1990). Naloxone blocks mating-induced conditioned place preference in both female and male rats (Paredes and Martinez, 2001) and may even bring about conditioned place aversion (Agmo and Berenfeld, 1990). The link between sexual behavior and endogenous opioids may be phylogenetically ancient; as naloxone was also found to reduce appetitive sexual responses in quails (Holloway et al., 2004). Furthermore, in the somewhat provocatively titled article "Vaginal Stimulation-Produced Analgesia in Rats and Women," Komisaruk and Whipple (1986) review evidence for the antinociceptive effects of sexual behavior in females. Szechtman and colleagues (1981) gave copulating male rats electric shocks and reported a pattern of antinociception consistent with increasing levels of pleasure and endogenous opioid release before and during mating. For a brief period immediately after ejaculation, the rats were more sensitive to pain, consistent with the notion a opioid-opponent "dysphoric" low following the positive hedonic feelings of mating (Solomon and Corbit, 1974). In addition, sexual behavior–induced antinociception in male rats is naloxone-reversible (Forsberg et al., 1987). A final point of interest from the literature on pain in sexual contexts relates to the roughly 20% of laboratory rats, which fail to initiate mating within a certain time window when placed next to female rat. Painful procedures such as the tail pinch have proven efficient for speeding up mating in these "noncopulators" (King and Alexander, 2000). Since endogenous opioids are released during pain (Zubieta et al., 2001), this arousing effect of pain may also be mediated by opioids.

Endogenous opioids are also involved in both the hedonic 'liking' and analgesic effect of eating. As mentioned, microinjection of opioids into the brain's reward network increases 'liking' behavior in rodents (Berridge and Robinson, 2003). In humans, naloxone reduces the hedonic value of sweet high-fat snacks (Drewnowski et al., 1992; Fantino et al., 1986) and decreases consumption of palatable foods in binge-eaters (Drewnowski et al., 1995). Sweet foods and drinks increase pain tolerance, especially in neonates (Blass and Hoffmeyer, 1991), women (Mercer and Holder, 1997), and people with low blood pressure (Lewkowski et al., 2003). In rats, sweet substance–induced analgesia is naltrexone-reversible (Reboucas et al., 2005). Furthermore, Pearce and Dickinson (1975) showed that when hungry rats learn that a painful electric shock predicts a food reward, this Pavlovian pairing will attenuate the defensive reactions elicited by the shock. The authors contend that this *counterconditioning* reduced shock aversiveness, "because after such conditioning the shock activates both the positive and negative systems, thus allowing the positive system to attenuate the activity in the negative one" (Pearce and Dickinson, 1975, p. 177). They also showed that the antinociceptive effect of the counterconditioning is dependent on the balance between the hedonic value of the food reward and the shock, and is ineffective if the shock is too severe. Note that these rats were kept at 80% of their normal bodyweight, such that the subjective utility and pleasantness of the food reward was likely to "outweigh" a substantial level of pain.

Several studies of pain and emotion in humans have used positive and negative affective pictures to compare pain sensitivity in pleasant and unpleasant contexts. The findings point to a linear pain modulation effect where subjective pain ratings are reduced in the positive image context and enhanced in the negative context compared to when they view neutral images. This pain modulatory effect is echoed in the amplitude of brain event-related potentials (Kenntner-Mabiala and Pauli, 2005) and appears to extend into clinical populations (Rhudy et al., 2006). Villemure and colleagues (2003) found a similar effect using pleasant and unpleasant odors to modulate pain perception. This study went one step further and also showed that the odors modulated pain through their effects on mood. It is not unlikely that the pain-modulatory effects of viewing pleasant images, partaking in sexual behaviors, or eating tasty foods also rely on the influence of mood. In addition, pleasant music was recently shown to decrease experimental thermal pain ratings (Roy et al., in press). The hedonic value of the music had no effect on perception of innocuous warm stimulation, however.

Perhaps the best-studied effect of context on pain perception is the placebo analgesic effect. Typically, subjects in placebo studies are led to believe that an inactive substance (the placebo treatment) has potent analgesic effects. This *meaning response* (Moerman and Jonas, 2002) has been remarkably effective across a range of studies using different methodologies in various clinical and nonclinical populations (Benedetti et al., 2005; Bingel et al., 2006; de la Fuente-Fernandez and Stoessl, 2002; Levine et al., 1978; Mayberg et al., 2002; Petrovic et al., 2002, 2005; Wager et al., 2004; Zubieta et al., 2005). Placebo treatment rarely causes complete pain relief, but even a weak effect makes it the better of two hedonic outcomes. According to psychologist Barbara Mellers, such comparison effects are very powerful-enough to make a loss that is the better of two losses more pleasurable than a gain that is the worse of two gains (Mellers, 2001). Neuroimaging data supports the notion that expectation of placebo is equivalent to expectation of a positive hedonic outcome. In 2001, de la Fuente-Fernandez and colleagues reported dopamine release in the nigrostriatal system during placebo treatment of patients with Parkinson's disease. The authors argued that the dopamine release underpinned expectation of reward—in this case, the anticipation of therapeutic benefit. Similarly, placebo treatment of anxiety is thought to rely on reward expectation (Petrovic et al., 2005). In support of this idea, a neuroimaging study of pain and cooling relief reported that anticipation of pain relief was processed in the same way as reward expectation in the brain (Seymour et al., 2005).

Pleasure-Seeking, Evolution, and Morality

The close relationship between homeostatic processes and hedonic feelings point to an evolutionary benefit for pleasure-seeking, pain-avoiding individuals. This contrasts sharply with the pleasure-conservative views expressed in many scientific papers, religious scriptures, etc. In an influential paper in *Science*, Koob and Le Moal (1997) describe pleasure as a "limited resource" (p. 56) and argue for the benefits of a "hedonic Calvinistic" approach where the use of the reward system is restricted. These researchers are experts on drug addiction, and their view of pleasure is likely influenced by the detrimental consequences of "pleasure-seeking" for drug addicts. It is important to keep in mind, however, that the pleasures sought by nonaddicts are significantly more diverse than those of a dedicated drug fiend (Kelley and Berridge, 2002). One important difference between drugs of abuse and natural rewards relates to satiety. Although it is well established that addictive drugs exert their positive hedonic effects via the brain's naturally evolved "reward system" (Volkow and Fowler, 2000), the sensory-specific satiety effects triggered by natural rewards such as food appear to be missing. Sensory-specific satiety is the reason the pleasure of eating chocolate will diminish as we work our way through bar after bar, while at the same time increasing the tastiness of other foods (Kringelbach et al., 2003). This mechanism prevents one-tracked pleasure-seeking and ensures consumption of diverse rewards, edible and otherwise. Through this diversifying effect, sensory-specific satiety appears to foster important health benefits: for instance, a varied diet enhances the cancer-fighting ability of antioxidant metabolites (Halvorsen et al., 2006).

In contrast, addictive drugs such as cocaine fail to trigger a decrease in positive affect with repeated exposure. Instead, frequent drug use appears to modify the brain's reward system, increasing affinity for the addictive drug and decreasing the pleasure associated with food, social rewards, and other normally enjoyable activities (Volkow et al., 2002). Research on the increasing obesity epidemic in the Western world points to startling similarities with drug addiction, suggesting that high-energy foods such as sugar and fat may sometimes override the brain's natural satiety system in the same way as a recreational drug (Volkow and Wise, 2005). Another class of nonpharmacological rewards, which may exert drug-like effects on the brain's reward system, is extreme sports activity. Here, the main link is anhedonia: "the inability to find enjoyment in food, sex, physical recreational pastimes, as well as socializing, humor and achievement" (Marbach and Lund, 1981, p 75). The idea is simple: because of the similarity between drug-induced euphoria and the intense hedonic feelings experienced by skydivers, their enjoyment of everyday pleasures is lower than that of, for example, rowers, who do not experience such "natural high" on a regular basis (Franken et al., 2006). However, in terms of their adverse effects, "addiction" to extreme sport thrills and drug addiction are not comparable.

It appears, then, that so long as we avoid addictive drugs and seek a variety of rewards, we are not at risk of exhausting our "limited resources" of pleasure. The "Calvinistic" focus on moderation or even abstinence of pleasures has deep roots within Western culture, however. A prevailing belief is that a dichotomy exists between virtue and morality on the one hand, and self-interest or pleasure on the other (Wringe, 1999). In the pertinently named essay "Shame, Pleasure and the Divided Soul," Moss (2005) explains that for Plato, feelings of shame can separate a person's judgments about what is pleasant from his judgments about what is good: "appeals to a person's feelings of shame and admiration may be able to succeed, when rational arguments have failed, in bringing him to see that a harmful pleasure is to be avoided, or that a beneficial pain is to be pursued" (Moss, 2005, p 3).

Unlike some religious fanatics, however, Plato was not opposed to pleasures in general. His "harmful pleasures" and "beneficial pains" highlight the potential conflict between immediate and delayed gratification. Many believe that what sets us apart from animals is our ability to override our instinct for immediate rewards in favor of some greater good. Neuroimaging data supports the idea that the human capacity for appreciating delayed or abstract rewards is processed separately from the phylogenetically older immediate gratification system (McClure et al., 2004). In animals, both delays and effort discount the hedonic value of a reward, possibly due to increased uncertainty of reward receipt (Rudebeck et al., 2006). For humans, an important part of childhood development is learning to resist the lure of immediate pleasures in order to obtain greater and often more abstract rewards. The ability to delay or restrict pleasures is often considered a virtue. In the words of writer Paul Bischke (2003), "the motive of virtue is to enact and embody what is good, right, and fitting." The use of the term "fitting" is important: the definition of a virtue changes between cultures and within cultures over time. For instance, "temperance" to all pleasures is advocated in the Bible, but whether this refers to complete abstinence from pleasures or merely refraining from overindulgence is still subject to debate. Interestingly, even the reputedly frugal Calvin did not advocate absolute abstinence: "it is permissible to use wine not only for necessity, but also to make us merry" (as cited in Bischke, 2003). All in all, the strong historical association between virtue on the one hand, and shame, guilt, and pleasure on the other, may help explain the apparent preference for formulating scientific research questions in terms of reward rather than pleasure.

Meaningful Suffering?

While temperance is considered a virtue with respect to pleasures, stoicism in the face of pains is also highly regarded in many cultures (Downey, 2007; Janal et al., 1994; Tudor, 2001). For instance, Harper (2006) reports that within the British Armed Forces, a prevalent view is that physical activity will only be beneficial if it is painful. The everyday expression "no pain, no gain" implies that this type of belief is not restricted to a military context. Self-flagellation is another example of paradoxical pain-seeking human behaviors, motivated perhaps by the abstract pleasure of feeling closer to God. According to Parker (1997), God may bring new self-awareness, love for the divine, or the reconciliation of enemies through suffering: "these gifts are immeasurably wonderful and more than compensate for any suffering that one might endure" (Parker, 1997, p. 207). The belief that suffering has real, though perhaps incomprehensible, meaning is comforting and is generally preferred over the belief that suffering is meaningless. In a paper published in the journal *Pastoral Psychology*, the therapists Driscoll and Edwards (1983) discuss various Christian takes on meaningful suffering. They claim that a popular misconception among Christians is that there is automatic merit to suffering, and that by suffering one shows oneself to be a better Christian. Suffering is part of life, to be endured to enter heaven (Davidhizar and Giger, 2004). In contrast, one of the earliest concepts portrayed in the Old Testament is that of a God who blesses the righteous and afflicts the wicked, so that suffering was the consequence of a violation of God's will (Driscoll and Edwards, 1983). Tudor, writing in the context of prisoners' accepting their punishment as meaningful suffering, divides suffering into different categories, including the following: (1) the cost to achieve a goal; (2) a necessary condition for virtue, such as courage, fortitude, stoicism; and (3) a necessary condition for understanding other people and the world through empathy. Interestingly, a study of congenital insensitivity to pain showed that a normal personal experience of pain is not required for perceiving and feeling empathy for others' pain (Danziger et al., 2006). C.S. Lewis formulated a moderate Christian interpretation of suffering, which was also based on empathy: through our actions, we cause both suffering and joy for ourselves and for others, and our actions therefore have personal and ethical importance, which otherwise they would not have. It is easy to understand that God would allow our actions to matter, and there is no need in this view to suggest that God in any way wants us to suffer (C.S. Lewis, as cited in Driscoll and Edwards, 1983).

As discussed above in relation to positive hedonic feelings, the context in which we experience pain determines what meaning we assign to it (Moseley and Arntz, 2007). As we have seen, the diminished suffering experienced during placebo analgesia directly relates to the more positive context of the treatment, perhaps via the same antinociceptive mechanism, which mediates pain relief from viewing pleasant images, smelling delicious scents, or listening to lovely music. The meaning of suffering is especially important in the case of persistent, chronic pain, which often constitutes a threat to the patients' identity and sense of self (McCracken et al., 2004). Like excessive pleasure, persistent suffering is associated with shame: "I am not the kind of woman who complains of everything" (Werner et al., 2004). Research from Rita Charon's laboratory has shown the importance of narrative and sense of self in the treatment of chronic pain. Patients in their clinic who were allowed time to relate a coherent story of their pain and suffering to their doctor showed more improvement from treatment than did the patients who were allocated a regular appointment with the pain clinician (Charon, 2006).

Concluding Remarks

In this chapter, we have used theories of homeostasis as a framework for understanding the *how* and *why* of hedonic feelings. This framework allows us to explain a range of paradoxical human behaviors, notably the many situations in which people willingly subject themselves to pain. It is important to remember that purely "mental" (emotional) changes are considered deviations from homeostasis in the same way as changes in physiological body state. Another important consideration is that homeostatic processing is often subconscious: we are not aware of most homeostatic imbalances, since they are easily remedied through subconscious processes. A pertinent example is posture changes, which most people make frequently with little awareness (Gallagher and Cole, 1995). Thus the pleasures we experience and can introspect on may in fact inform us of underlying needs that we would otherwise not be aware of.

Much is known about the physiology of nociceptive signalling from the periphery, but it is clear that to understand how the subjective, hedonic feeling of pain arises in the brain, how nociception-related signalling is *interpreted* into pain, we must extend our focus to include a number of other factors. Both negative and positive hedonic feelings influence the interpretation of nociceptive signals in the brain. More research is warranted on how expectation, learning, and memory for pain influence the hedonic pain experience. For instance, expectation has a value in itself (Berns et al., 2006; Gard et al., 2006; Mellers, 2001), *and* changes interpretation of nociceptive signals (Fairhurst et al., 2007; Koyama et al., 2005). Expectation and context interact in a powerful way: the negative hedonic feelings are higher when the pain is worse than expected, or even if it is simply the worse of two or more alternative outcomes. Conversely, placebo analgesia treatment represents the best of two outcomes and is characterized by decreased unpleasantness. The same mechanisms are at work for positive hedonic feelings: people are frustrated when they get less than they expect (Abler et al., 2005), less than other people, or simply less than they could have received (e.g., Prisoner's Dilemma; Singer et al., 2004), and this diminishes their pleasure (Mellers, 2001; Ursu and Carter, 2005). The context can also be defined by motivation for abstract goals, such as reducing another's (or others') suffering, which can also decrease the hedonic component of pain by changing its meaning—it hurts but "no pain, no gain," remember?

Much work remains before we will fully understand the interactions between negative and positive hedonic feelings. As reviewed here, however, the indications that pleasures can reduce pain-related suffering are already numerous. As we have seen, the opioid and dopamine systems play an important role for both pleasures and pains. These neurotransmitters are likely to underpin pain and pleasure interactions, such as decreases in pain in the context of pleasures (food, music, sex, etc). More research on these effects could lead to novel treatments of pain and suffering. In the chronic pain field, it is becoming increasingly clear that the effect of persistent pain on patients' sense of self is key to their suffering (Jensen, 2007). To understand the human self, we must know *what it is like* to be a sentient being: we must study hedonic feelings, pleasures as well as pains. Taken together, the available evidence suggests that seeking a variety of pleasures is beneficial not only in itself, but also because pleasures may reduce pain and suffering.

Acknowledgments

The authors would like to thank Lorimer Moseley, John Keltner, and the editors of this book for their insights and helpful comments on earlier versions of this manuscript.

References

Abler, B., Walter, H. and Erk, S. (2005) Neural correlates of frustration. *Neuroreport* 16, 669–672.

Agmo, A. and Berenfeld, R. (1990) Reinforcing properties of ejaculation in the male rat: Role of opioids and dopamine. *Behav. Neurosci.* 104, 177–182.

Barbano, M. and Cador, M. (2007) Opioids for hedonic experience and dopamine to get ready for it. *Psychopharmacology* 191, 497–506.

Baumgartner, U., Buchholz, H. G., Bellosevich, A., Magerl, W., Siessmeier, T., Rolke, R., Hohnemann, S., Piel, M., Rosch, F., Wester, H. J., Henriksen, G., Stoeter, P., Bartenstein, P., Treede, R. D. and Schreckenberger, M. (2006) High opiate receptor binding potential in the human lateral pain system. *Neuroimage* 30, 692–699.

Benedetti, F., Mayberg, H. S., Wager, T. D., Stohler, C. S. and Zubieta, J.-K. (2005) Neurobiological mechanisms of the placebo effect. *J. Neurosci.* 25, 10390–10402.

Bentham, J. (1789) *An Introduction to the Principles of Morals and Legislation*. Library of Economics and Liberty. Retrieved September 16, 2005 from http://www.econlib.org/library/Bentham/bnthPML1.html.

Berns, G. S., Chappelow, J., Cekic, M., Zink, C. F., Pagnoni, G. and Martin-Skurski, M. E. (2006) Neurobiological substrates of dread. *Science* 312, 754–758.

Berridge, K. C. (2003) Pleasures of the brain. *Brain Cogn.* 52, 106–128.

Berridge, K. (2007) The debate over dopamine's role in reward: the case for incentive salience. *Psychopharmacology* 191, 391–431.

Berridge, K. C. and Robinson, T. E. (2003) Parsing reward. *Trends Neurosci.* 26, 507–513.

Bingel, U., Lorenz, J., Schoell, E., Weiller, C. and Buchel, C. (2006) Mechanisms of placebo analgesia: rACC recruitment of a subcortical antinociceptive network. *Pain* 120, 8–15.

Bischke, P. M. (2003) Pleasure drugs and classical virtues: Temperance and abstinence in U.S. religious thought. *Int. J. Drug Policy* 14, 273–278.

Blass, E. M. and Hoffmeyer, L. B. (1991) Sucrose as an analgesic for newborn infants. *Pediatrics* 87, 215–218.

Blatter, K. and Schultz, W. (2006) Rewarding properties of visual stimuli. *Exp. Brain Res.* 168, 541–546.

Blood, A. J. and Zatorre, R. J. (2001) Intensely pleasurable responses to music correlate with activity in brain regions implicated in reward and emotion. *Proc Natl Acad Sci USA* 98, 11818–11823.

Burgdorf, J. and Panksepp, J. (2006) The neurobiology of positive emotions. *Neurosci. Biobehav. Rev.* 30, 173–187.

Cabanac, M. (1971) Physiological role of pleasure. *Science* 173, 1103–1107.

Cabanac, M. (1979) Sensory pleasure. *Q. Rev. Biol.* 54, 1–29.

Cabanac, M. (2002) What is emotion? *Behav. Processes* 60, 69–83.

Cannon, W. B. (1929) Organization for physiological homeostasis. *Physiol. Rev.* 9, 399–431.

Carroll, C. L., Balkrishnan, R., Feldman, S. R., Fleischer, A. B. and Manuel, J. C. (2005) The burden of atopic dermatitis: Impact on the patient, family, and society. *Pediatr. Dermatol.* 22, 192–199.

Chard, P. S., Bleakman, D., Savidge, J. R. and Miller, R. J. (1995) Capsaicin-induced neurotoxicity in cultured dorsal root ganglion neurons: Involvement of calcium-activated proteases. *Neuroscience* 65, 1099–1108.

Charon, R. (2006) Suffering, storytelling, and community: an approach to pain treatment from Columbia's Program in Narrative Medicine for Pain. In: *Proceedings of the 11th World Congress on Pain*. Eds. H. Flor, E. Kalso and J. O. Dostrovsky. IASP Press: Seattle, WA.

Coolen, L. M., Fitzgerald, M. E., Yu, L. and Lehman, M. N. (2004) Activation of [mu] opioid receptors in the medial preoptic area following copulation in male rats. *Neuroscience* 124, 11–21.

Craig, A. D. (2003) A new view of pain as a homeostatic emotion. *Trends Neurosci.* 26, 303–307.

Cross, P. and Matheson, K. (2006) Understanding sadomasochism: an empirical examination of four perspectives *J. Homosex* 50, 133–166.

Danziger, N., Prkachin, K. M. and Willer, J. C. (2006) Is pain the price of empathy? The perception of others' pain in patients with congenital insensitivity to pain. *Brain* 129, 2494–2507.

Davidhizar, R. and Giger, J. N. (2004) A review of the literature on care of clients in pain who are culturally diverse. *International Nursing Review* 51, 47–55.

de Araujo, I. E., Kringelbach, M. L., Rolls, E. T. and McGlone, F. (2003) Human cortical responses to water in the mouth, and the effects of thirst. *J. Neurophysiol.* 90, 1865–1876.

de la Fuente-Fernandez, R. and Stoessl, A. J. (2002) The placebo effect in Parkinson's disease. *Trends Neurosci.* 25, 302–306.

de la Fuente-Fernandez, R., Ruth, T. J., Sossi, V., Schulzer, M., Calne, D. B. and Stoessl, A. J. (2001) expectation and dopamine release: Mechanism of the placebo effect in Parkinson's disease. *Science* 293, 1164–1166.

Dhaka, A., Viswanath, V. and Patapoutian, A. (2006) TRP ion channels and temperature sensation. *Annu. Rev. Neurosci.* 29, 135–161.

Dickinson, A. and Dearing, M. F. (1979) Appetitive-aversive interactions and inhibitory processes. In: *Mechanisms of Learning and Motivation*, pp. 203–231. Eds. A. Dickinson and R. A. Boakes. Erlbaum: Hillsdale, NJ.

Dill, R. E. and Costa, E. (1977) Behavioural dissociation of the enkephalinergic systems of nucleus accumbens and nucleus caudatus. *Neuropharmacology* 16, 323–326.

Dirks, J., Petersen, K. and Dahl, J. (2003) The heat/capsaicin sensitization model: a methodologic study. *J. Pain* 4, 122–128.

Downey, G. (2007) Producing pain: Techniques and technologies in no-holds-barred fighting. *Soc. Studies Sci.* 37, 201–226.

Dray, A. (1992) Mechanism of action of capsaicin-like molecules on sensory neurons. *Life Sci.* 51, 1759–1765.

Drewnowski, A., Krahn, D. D., Demitrack, M. A., Nairn, K. and Gosnell, B. A. (1992) Taste responses and preferences for sweet high-fat foods: Evidence for opioid involvement. *Physiol. Behav.* 51, 371–379.

Drewnowski, A., Krahn, D. D., Demitrack, M. A., Nairn, K. and Gosnell, B. A. (1995) Naloxone, an opiate blocker, reduces the consumption of sweet high-fat foods in obese and lean female binge eaters. *Am. J. Clin. Nutr.* 61, 1206–1212.

Driscoll, R. and Edwards, L. (1983) The misconception of Christian suffering. *Pastoral Psychol.* 32, 34–48.

Eccleston, C. and Crombez, G. (1999) Pain demands attention: a cognitive–affective model of the interruptive function of pain. *Psychol. Bull.* 125, 356–366.

Fairhurst, M., Wiech, K., Dunckley, P. and Tracey, I. (2007) Anticipatory brainstem activity predicts neural processing of pain in humans. *Pain* 128, 101–110.

Fantino, M., Hosotte, J. and Apfelbaum, M. (1986) An opioid antagonist, naltrexone, reduces preference for sucrose in humans. *Am. J. Physiol.* 251, R91–R96.

Fields, H. L. (2004) State-dependent opioid control of pain. *Nat. Rev. Neurosci.* 5, 565–575.

Fields, H. L. (2006) A motivation-decision model of pain: the role of opioids. In: *Proceedings of the 11th World Congress on Pain*. Eds. H. Flor, E. Kalso and J. O. Dostrovsky. IASP Press: Seattle, WA.

Forsberg, G., Wiesenfeld-Hallin, Z., Eneroth, P. and Sodersten, P. (1987) Sexual behavior induces naloxone-reversible hypoalgesia in male rats. *Neurosci. Lett.* 81, 151–154.

Franken, I. H. A., Zijlstra, C. and Muris, P. (2006) Are non-pharmacological induced rewards related to anhedonia? A study among skydivers. *Prog. Neuro-Psychopharmacol. Biol. Psychiatry* 30, 297–300.

Franklin, K. B. J. (1998) Analgesia and abuse potential: an accidental association or a common substrate? *Pharmacol. Biochem. Behav.* 59, 993–1002.

Gallagher, S. and Cole, J. (1995) Body schema and body image in a deafferented subject. *J. Mind Behav.* 16, 369–390.

Gard, D. E., Gard, M. G., Kring, A. M. and John, O. P. (2006) Anticipatory and consummatory components of the experience of pleasure: a scale development study. *J. Res. Personality* 40, 1086–1102.

Gear, R. W., Aley, K. O. and Levine, J. D. (1999) Pain-induced analgesia mediated by mesolimbic reward circuits. *J. Neurosci.* 19, 7175–7181.

Gerdelat-Mas, A., Simonetta-Moreau, M., Thalamas, C., Ory-Magne, F., Slaoui, T., Rascol, O. and Brefel-Courbon, C. (2007) Levodopa raises objective pain threshold in Parkinson's disease: a RIII reflex study. *J. Neurol. Neurosurg. Psychiatry*, 78(10), 1140–1142 (jnnp.2007.120212).

Gracely, R. H., Geisser, M. E., Giesecke, T., Grant, M. A. B., Petzke, F., Williams, D. A. and Clauw, D. J. (2004) Pain catastrophizing and neural responses to pain among persons with fibromyalgia. *Brain* 127, 835–843.

Guidi, M. E. L. (1994) Pain and human action: Locke to Bentham. In: *Fourth Conference of the International Society for Utilitarian Studies*. 27–29 August, Tokyo.

Hail, N., Jr. (2003) Mechanisms of vanilloid-induced apoptosis. *Apoptosis* 8, 251–62.

Halvorsen, B. L., Carlsen, M. H., Phillips, K. M., Bohn, S. K., Holte, K., Jacobs, D. R., Jr. and Blomhoff, R. (2006) Content of redox-active compounds (i.e., antioxidants) in foods consumed in the United States. *Am. J. Clin. Nutr.* 84, 95–135.

Harper, P. (2006) No pain, no gain: Pain behaviour in the armed forces. *Br. J. Nurs.* 15, 548–551.

Holloway, K. S., Cornil, C. A. and Balthazart, J. (2004) Effects of central administration of naloxone during the extinction of appetitive sexual responses. *Behav. Brain Res.* 153, 567–72.

Huang, E. Y., Chen, C. M. and Tao, P. L. (2004) Supraspinal anti-allodynic and rewarding effects of endomorphins in rats. *Peptides* 25, 577–583.

Janal, M. N., Glusman, M., Kuhl, J. P. and Clark, W. C. (1994) Are runners stoical? An examination of pain sensitivity in habitual runners and normally active controls. *Pain* 58, 109–116.

Jensen, M. P. (2007) Motivational aspects of pain. In: *Encyclopedia of Pain*, pp. 13–13. Eds. F. R. Schmidt and W.D. Willis, W.D. Springer: Berlin.

Kahneman, D., Fredrickson, B. L., Schreiber, C. A. and Redelmeier, D. A. (1993) When more pain is preferred to less: Adding a better end. *Psychol. Sci.* 4, 401–405.

Kelley, A. E. and Berridge, K. C. (2002) The neuroscience of natural rewards: Relevance to addictive drugs. *J. Neurosci.* 22, 3306–3311.

Kenntner-Mabiala, R. and Pauli, P. (2005) Affective modulation of brain potentials to painful and nonpainful stimuli. *Psychophysiology* 42, 559–567.

King, B. E. and Alexander, G. M. (2000) Pain sensitivity and individual differences in self-reported sexual behavior. *J. Comp. Psychol.* 114, 193–199.

Knutson, B., Adams, C. M., Fong, G. W. and Hommer, D. (2001a) Anticipation of increasing monetary reward selectively recruits nucleus accumbens. *J. Neurosci.* 21, 159RC.

Knutson, B., Fong, G., Adams, C., Varner, J. and Hommer, D. (2001b) Dissociation of reward anticipation and outcome with event-related fMRI. *Neuroreport* 12, 3683–3687.

Knutson, B., Fong, G. W., Bennett, S. M., Adams, C. M. and Hommer, D. (2003) A region of mesial prefrontal cortex tracks monetarily rewarding outcomes: Characterization with rapid event-related fMRI. *NeuroImage* 18, 263–272.

Komisaruk, B. R. and Whipple, B. (1986) Vaginal stimulation-produced analgesia in rats and women. *Ann. NY Acad. Sci.* 467, 30–39.

Koob, G. F. and Moal, M. L. (1997) Drug abuse: Hedonic homeostatic dysregulation. *Science* 278, 52–58.

Korner, A., Gerull, F., Stevenson, J. and Meares, R. (2007) Harm avoidance, self-harm, psychic pain, and the borderline personality: Life in a "haunted house". *Comprehens. Psychiatry* 48, 303–308.

Koyama, T., Kato, K., Tanaka, Y. Z. and Mikami, A. (2001) Anterior cingulate activity during pain-avoidance and reward tasks in monkeys. *Neurosci. Res.* 39, 421–430.

Koyama, T., McHaffie, J. G., Laurienti, P. J. and Coghill, R. C. (2005) The subjective experience of pain: Where expectations become reality. *Proc. Natl Acad. Sci.* 102, 12950–12955.

Kringelbach, M. L., O'Doherty, J., Rolls, E. T. and Andrews, C. (2003) Activation of the human orbitofrontal cortex to a liquid food stimulus is correlated with its subjective pleasantness. *Cereb. Cortex* 13, 1064–1071.

Launer, J. (2004) The itch. *QJM* 97, 383–384.

Leknes, S. G., Bantick, S., Willis, C. M., Wilkinson, J. D., Wise, R. G. and Tracey, I. (2007) Itch and motivation to scratch: an investigation of the central and peripheral correlates of allergen- and histamine-induced itch in humans. *J. Neurophysiol.* 97, 415–422.

Leknes, S. G., Brooks, J. C. W., Wiech, K. and Tracey I (2008) Pain relief as an opponent process: a psychophysical investigation. *Eur J Neurosci* 28, 794–801.

Levine, J. D., Gordon, N. C. and Fields, H. L. (1978) The mechanism of placebo analgesia. *Lancet* 2, 654–657.

Lewkowski, M. D., Ditto, B., Roussos, M. and Young, S. N. (2003) Sweet taste and blood pressure-related analgesia. *Pain* 106, 181–186.

Loewenstein, G. (2006) Social science: the pleasures and pains of information. *Science* 312, 704–706.

Marbach, J. J. and Lund, P. (1981) Depression, anhedonia and anxiety in temporomandibular joint and other facial pain syndromes. *Pain* 11, 73–84.

Mayberg, H. S., Silva, J. A., Brannan, S. K., Tekell, J. L., Mahurin, R. K., McGinnis, S. and Jerabek, P. A. (2002) The functional neuroanatomy of the placebo effect. *Am. J. Psychiatry* 159, 728–737.

McClure, S. M., Laibson, D. I., Loewenstein, G. and Cohen, J. D. (2004) Separate neural systems value immediate and delayed monetary rewards. *Science* 306, 503–507.

McCracken, L. M., Carson, J. W., Eccleston, C. and Keefe, F. J. (2004) Acceptance and change in the context of chronic pain. *Pain* 109, 4–7.

McGee, H. (2004) *On Food and Cooking: The Science and Lore of the Kitchen*. Scribner: New York.

Mellers, B. A. (2001) Anticipated Emotions as Guides to Choice. *J. Current Dir. Psychol. Sci.* 10, 210–214.

Melzack, R. and Casey, K. L. (1968) Sensory, motivational, and central control determinants of pain. In: *The Skin Senses*, pp. 423–439. Ed. D. R. Kenshalo. Thomas: Springfield, IL.

Mercer, M. E. and Holder, M. D. (1997) Antinociceptive effects of palatable sweet ingesta on human responsivity to pressure pain. *Physiol. Behav.* 61, 311–318.

Mochizuki, H., Tashiro, M., Kano, M., Sakurada, Y., Itoh, M. and Yanai, K. (2003) Imaging of central itch modulation in the human brain using positron emission tomography. *Pain* 105, 339–346.

Moerman, D. E. and Jonas, W. B. (2002) Deconstructing the placebo effect and finding the meaning response. *Ann. Intern. Med.* 136, 471–476.

Morgan, W. P. (1985) Affective beneficence of vigorous physical activity. *Med. Sci. Sports Exerc.* 17, 94–100.

Moseley, G.L. and Arntz, A. (2007) The context of a noxious stimulus affects the pain it evokes. *Pain* 133, 64–71..

Moser, C. and Kleinplatz, P. J. (2006) Introduction: the state of our knowledge on SM. *J. Homosex* 50, 1–15.

Moss, J. (2005) Shame, pleasure and the divided soul. In: *Oxford Studies in Ancient Philosophy,* Vol. 29. Ed. D. Sedley. Oxford University Press: New York.

Murphy, M. R., Checkley, S. A., Seckl, J. R. and Lightman, S. L. (1990) Naloxone inhibits oxytocin release at orgasm in man. *J. Clin. Endocrinol. Metab.* 71, 1056–1058.

Nagel, T. (1974) What is it like to be a bat? *Philos. Rev,* 83, 435–50.

O'Doherty, J. P., Deichmann, R., Critchley, H. D. and Dolan, R. J. (2002) Neural responses during anticipation of a primary taste reward. *Neuron* 33, 815–826.

O'Doherty, J., Winston, J., Critchley, H., Perrett, D., Burt, D. M. and Dolan, R. J. (2003) Beauty in a smile: the role of medial orbitofrontal cortex in facial attractiveness. *Neuropsychologia* 41, 147–155.

Ostrowsky, K., Magnin, M., Ryvlin, P., Isnard, J., Guenot, M. and Mauguiere, F. (2002) Representation of pain and somatic sensation in the human insula: a study of responses to direct electrical cortical stimulation. *Cereb. Cortex* 12, 376–385.

Paredes, R. G. and Martinez, I. (2001) Naloxone blocks place preference conditioning after paced mating in female rats. *Behav. Neurosci.* 115, 1363–1367.

Parker, P. P. (1997) Suffering, prayer, and miracles. *J. Religion Health* 36, 205–220.

Pearce, J. M. and Dickinson, A. (1975) Pavlovian counterconditioning: Changing the suppressive properties of shock by association with food. *J. Exp. Psychol. Anim. Behav. Process.* 1, 170–177.

Pecina, S. and Berridge, K. C. (2000) Opioid site in nucleus accumbens shell mediates eating and hedonic 'liking' for food: Map based on microinjection Fos plumes. *Brain Res.* 863, 71–86.

Pelchat, M. L., Johnson, A., Chan, R., Valdez, J. and Ragland, J. D. (2004) Images of desire: Food-craving activation during fMRI. *NeuroImage* 23, 1486–1493.

Petrovic, P., Kalso, E., Petersson, K. M. and Ingvar, M. (2002) Placebo and opioid analgesia—imaging a shared neuronal network. *Science* 295, 1737–1740.

Petrovic, P., Dietrich, T., Fransson, P., Andersson, J., Carlsson, K. and Ingvar, M. (2005) Placebo in emotional processing – induced expectations of anxiety relief activate a generalized modulatory network. *Neuron* 46, 957–969.

Phelps, E. A., Delgado, M. R., Nearing, K. I. and LeDoux, J. E. (2004) Extinction learning in humans: Role of the amygdala and vmPFC. *Neuron* 43, 897–905.

Phillips, H. (2003) The pleasure seekers. *New Scientist* 2416, 36–40.

Price, D. D., Barrell, J. J. and Gracely, R. H. (1980) A psychophysical analysis of experiential factors that selectively influence the affective dimension of pain. *Pain* 8, 137–149.

Price, D. D., Harkins, S. W. and Baker, C. (1987) Sensory-affective relationships among different types of clinical and experimental pain. *Pain* 28, 297–307.

Raij, T. T., Numminen, J., Narvanen, S., Hiltunen, J. and Hari, R. (2005) Brain correlates of subjective reality of physically and psychologically induced pain. *Pro Natl Acad Sci USA* 102, 2147–2151.

Rainville, P., Duncan, G. H., Price, D. D., Carrier, B. and Bushnell, M. C. (1997) Pain affect encoded in human anterior cingulate but not somatosensory cortex. *Science* 277, 968–971.

Raja, S., Meyer, R., Ringkamp, M. and Campbell, J. (1999) Peripheral neural mechanisms of nociception. In: *Textbook of Pain*, pp. 11–58. Eds. P. Wall and R. Melzack. Harcourt Publishers Ltd.: London.

Ramnani, N., Elliott, R., Athwal, B. S. and Passingham, R. E. (2004) Prediction error for free monetary reward in the human prefrontal cortex. *NeuroImage* 23, 777–786.

Reboucas, E. C., Segato, E. N., Kishi, R., Freitas, R. L., Savoldi, M., Morato, S. and Coimbra, N. C. (2005) Effect of the blockade of mu1-opioid and 5HT2A-serotonergic/alpha1-noradrenergic receptors on sweet-substance-induced analgesia. *Psychopharmacology (Berl)* 179, 349–355.

Rhudy, J. L., Dubbert, P. M., Parker, J. D., Burke, R. S. and Williams, A. E. (2006) Affective modulation of pain in substance-dependent veterans. *Pain Med.* 7, 483–500.

Robinson, T. E. and Berridge, K. C. (2001) Incentive-sensitization and addiction. *Addiction* 96, 103–114.

Rogers, R. D., Ramnani, N., Mackay, C., Wilson, J. L., Jezzard, P., Carter, C. S. and Smith, S. M. (2004) Distinct portions of anterior cingulate cortex and medial prefrontal cortex are activated by reward processing in separable phases of decision-making cognition. *Biological Psychiatry* 55, 594–602.

Roy, M., Peretz, I. and Rainville, P. Emotional valence contributes to music-induced analgesia. *Pain*, in press.

Rozin, P. (1999) Preadaptation and the puzzles and properties of pleasure. In: *Well-being: The Foundations of Hedonic Psychology*, pp. 109–133. Eds. D. Kahneman, E. Diener and N. Schwartz. Russell Sage Foundation: New York.

Rudebeck, P. H., Walton, M. E., Smyth, A. N., Bannerman, D. M. and Rushworth, M. F. (2006) Separate neural pathways process different decision costs. *Nat. Neurosci.* 9, 1161–1168.

Salamone, J. D., Cousins, M. S. and Snyder, B. J. (1997) Behavioral functions of nucleus accumbens dopamine: Empirical and conceptual problems with the anhedonia hypothesis. *Neurosci. Biobehav. Rev.* 21, 341–359.

Schultz, W. (2004) Neural coding of basic reward terms of animal learning theory, game theory, microeconomics and behavioural ecology. *Curr. Opin. Neurobiol.* 14, 139–147.

Scott, D. J., Heitzeg, M. M., Koeppe, R. A., Stohler, C. S. and Zubieta, J.-K. (2006) Variations in the human pain stress experience mediated by ventral and dorsal basal ganglia dopamine activity. *J. Neurosci.* 26, 10789–10795.

Seymour, B., O'Doherty, J. P., Koltzenburg, M., Wiech, K., Frackowiak, R., Friston, K. and Dolan, R. (2005) Opponent appetitive-aversive neural processes underlie predictive learning of pain relief. *Nat. Neurosci.* 8, 1234–1240.

Shimizu, T., Iwata, S., Morioka, H., Masuyama, T., Fukuda, T. and Nomoto, M. (2004) Antinociceptive mechanism of L-DOPA. *Pain* 110, 246–249.

Siegrist, J., Menrath, I., Stocker, T., Klein, M., Kellermann, T., Shah, N. J., Zilles, K. and Schneider, F. (2005) Differential brain activation according to chronic social reward frustration. *Neuroreport* 16, 1899–1903.

Singer, T., Kiebel, S. J., Winston, J. S., Dolan, R. J. and Frith, C. D. (2004) Brain responses to the acquired moral status of faces. *Neuron* 41, 653–662.

Small, D. M. and Apkarian, A. V. (2006) Increased taste intensity perception exhibited by patients with chronic back pain. *Pain* 120, 124–130.

Small, D. M., Zatorre, R. J., Dagher, A., Evans, A. C. and Jones-Gotman, M. (2001) Changes in brain activity related to eating chocolate: From pleasure to aversion. *Brain* 124, 1720–1733.

Smith, K. S. and Berridge, K. C. (2007) Opioid limbic circuit for reward: Interaction between hedonic hotspots of nucleus accumbens and ventral pallidum. *J. Neurosci.* 27, 1594–1605.

Solomon, R. L. and Corbit, J. D. (1974) An opponent-process theory of motivation: I. Temporal Dynamics of affect. *Psychol. Rev.* 81, 119–145.

Stante, M., Hanna, D. and Lotti, T. (2005) Itch, pain, and metaesthetic sensation. *Dermatol. Ther.* 18, 308–313.

Stark, R., Schienle, A., Girod, C., Walter, B., Kirsch, P., Blecker, C., Ott, U., Schafer, A., Sammer, G., Zimmermann, M. and Vaitl, D. (2005) Erotic and disgust-inducing pictures—differences in the hemodynamic responses of the brain. *Biol. Psychol.* 70, 19–29.

Sternberg, W. F., Bailin, D., Grant, M. and Gracely, R. H. (1998) Competition alters the perception of noxious stimuli in male and female athletes. *Pain* 76, 231–238.

Strand, E. B., Kerns, R. D., Christie, A., Haavik-Nilsen, K., Klokkerud, M. and Finset, A. (2007) Higher levels of pain readiness to change and more positive affect reduce pain reports—a weekly assessment study on arthritis patients. *Pain* 127, 204–213.

Strand, E. B., Zautra, A. J., Thoresen, M., Odegard, S., Uhlig, T. and Finset, A. (2006) Positive affect as a factor of resilience in the pain–negative affect relationship in patients with rheumatoid arthritis. *J. Psychosom. Res.* 60, 477–484.

Symons, F., Thompson, A. and Rodriguez, M. (2004) Self-injurious behavior and the efficacy of naltrexone treatment: a quantitative synthesis. *Ment. Retard. Dev. Disabil. Res. Rev.* 10, 193–200.

Szechtman, H., Hershkowitz, M. and Simantov, R. (1981) Sexual behavior decreases pain sensitivity and stimulated endogenous opioids in male rats. *Eur. J. Pharmacol.* 70, 279–285.

Taha, S. A., Norsted, E., Lee, L. S., Lang, P. D., Lee, B. S., Woolley, J. D. and Fields, H. L. (2006) Endogenous opioids encode relative taste preference. *Eur. J. Neurosci.* 24, 1220–1226.

Tanimoto, H., Heisenberg, M. and Gerber, B. (2004) Experimental psychology: Event timing turns punishment to reward. *Nature* 430, 983.

Tracey, I. (2005) Nociceptive processing in the human brain. *Curr. Opin. Neurobiol.* 15, 478–487.

Tremblay, L. K., Naranjo, C. A., Graham, S. J., Herrmann, N., Mayberg, H. S., Hevenor, S. and Busto, U. E. (2005) Functional neuroanatomical substrates of altered reward processing in major depressive disorder revealed by a dopaminergic probe. *Arch. Gen. Psychiatry* 62, 1228–1236.

Tucker, D., Luu, P. and Derryberry, D. (2005) Love hurts: the evolution of empathic concern through the encephalization of nociceptive capacity. *Dev. Psychopathol.* 17, 699–713.

Tudor, S. (2001) Accepting one's punishment as meaningful suffering. *Law Philos.* 20, 581–604.

Ursu, S. and Carter, C. S. (2005) Outcome representations, counterfactual comparisons and the human orbitofrontal cortex: Implications for neuroimaging studies of decision-making. *Cogn. Brain Res.* 23, 51–60.

Villemure, C., Slotnick, B. M. and Bushnell, M. C. (2003) Effects of odors on pain perception: Deciphering the roles of emotion and attention. *Pain* 106, 101–108.

Volkow, N. D. and Fowler, J. S. (2000) Addiction, a disease of compulsion and drive: Involvement of the orbitofrontal cortex. *Cereb. Cortex* 10, 318–325.

Volkow, N. D. and Wise, R. A. (2005) How can drug addiction help us understand obesity? *Nat. Neurosci.* 8, 555–560.

Volkow, N. D., Fowler, J. S., Wang, G.-J. and Goldstein, R. Z. (2002) Role of dopamine, the frontal cortex and memory circuits in drug addiction: Insight from imaging studies. *Neurobiol. Learning Memory* 78, 610–624.

Wager, T. D., Rilling, J. K., Smith, E. E., Sokolik, A., Casey, K. L., Davidson, R. J., Kosslyn, S. M., Rose, R. M. and Cohen, J. D. (2004) Placebo-induced changes in fMRI in the anticipation and experience of pain. *Science* 303, 1162–1167.

Ward, L., Wright, E. and McMahon, S. B. (1996) A comparison of the effects of noxious and innocuous counterstimuli on experimentally induced itch and pain. *Pain* 64, 129–138.

Werner, A., Isaksen, L. W. and Malterud, K. (2004) 'I am not the kind of woman who complains of everything': Illness stories on self and shame in women with chronic pain. *Soc. Sci. Med.* 59, 1035–1045.

Wiech, K., Kalisch, R., Weiskopf, N., Pleger, B., Stephan, K. E. and Dolan, R. J. (2006) Anterolateral prefrontal cortex mediates the analgesic effect of expected and perceived control over pain. *J. Neurosci.* 26, 11501–11509.

Wilson, A. M., Soignier, R. D., Zadina, J. E., Kastin, A. J., Nores, W. L., Olson, R. D. and Olson, G. A. (2000) Dissociation of analgesic and rewarding effects of endomorphin-1 in rats. *Peptides* 21, 1871–1874.

Wood, P. B. (2006) Mesolimbic dopaminergic mechanisms and pain control. *Pain* 120, 230–234.

Wringe, C. (1999) Being good and living well: Three attempts to resolve an ambiguity. *J. Philos. Educ.* 33, 287–293.

Zambreanu, L., Wise, R. G., Brooks, J. C. W., Iannetti, G. D. and Tracey, I. (2005) A role for the brainstem in central sensitisation in humans. Evidence from functional magnetic resonance imaging. *Pain* 114, 397–407.

Zangen, A., Ikemoto, S., Zadina, J. E. and Wise, R. A. (2002) Rewarding and psychomotor stimulant effects of endomorphin-1: Anteroposterior differences within the ventral tegmental area and lack of effect in nucleus accumbens. *J Neurosci* 22, 7225–7233.

Zubieta, J.-K., Smith, Y. R., Bueller, J. A., Xu, Y., Kilbourn, M. R., Jewett, D. M., Meyer, C. R., Koeppe, R. A. and Stohler, C. S. (2001) Regional mu opioid receptor regulation of sensory and affective dimensions of pain. *Science* 293, 311–315.

Zubieta, J.-K., Ketter, T. A., Bueller, J. A., Xu, Y., Kilbourn, M. R., Young, E. A. and Koeppe, R. A. (2003) Regulation of human affective responses by anterior cingulate and limbic {micro}-opioid neurotransmission. *Arch Gen Psychiatry* 60, 1145–1153.

Zubieta, J.-K., Bueller, J. A., Jackson, L. R., Scott, D. J., Xu, Y., Koeppe, R. A., Nichols, T. E. and Stohler, C. S. (2005) Placebo effects mediated by endogenous opioid activity on {micro}-opioid receptors. *J. Neurosci.* 25, 7754–7762.

Index